11th Joint Conference on Chemistry 2016

Held in Conjunction with the 4th Regional Biomaterials Scientific Meeting

IOP Conference Series: Materials Science and Engineering
Volume 172

Purwokerto, Indonesia
15 - 16 September 2016

ISBN: 978-1-5108-3675-4
ISSN: 1757-8981

Printed from e-media with permission by:

Curran Associates, Inc.
57 Morehouse Lane
Red Hook, NY 12571

Some format issues inherent in the e-media version may also appear in this print version.

Printed by Curran Associates, Inc. (2017)

Institute of Physics
Dirac House, Temple Back
Bristol BS1 6BE UK

Phone: 44 1 17 929 7481
Fax: 44 1 17 920 0979

techtracking@iop.org

Additional copies of this publication are available from:

Curran Associates, Inc.
57 Morehouse Lane
Red Hook, NY 12571 USA
Phone: 845-758-0400
Fax: 845-758-2633
Email: curran@proceedings.com
Web: www.proceedings.com

Table of contents

Volume 172

11th Joint Conference on Chemistry in Conjunction with the 4th Regional Biomaterials Scientific Meeting

15–16 September 2016, Purwokerto, Indonesia

Accepted papers received: 25 January 2017
Published online: 17 February 2017

Preface

Papers

IOP Conference Series: Materials Science and Engineering

JCC 2016 | INTERNATIONAL CONFERENCE

11th Joint Conference on Chemistry

in conjunction with

The 4th Regional Biomaterial Scientific Meeting

"Material Chemistry Development for Future Medicine, Industry, Environmental and Biomaterial Application"

September, 15 - 16, 2016
Purwokerto - Indonesia

Organized by:

IOP Publishing

IOP Conf. Series: Materials Science and Engineering **172** (2017) 011001 doi:10.1088/1757-899X/172/1/011001

Welcome Message From Chairman of Conference

On behalf of the Committee, I am very pleased that the 11th Joint Chemistry Conference in Conjunction with The 4th Regional Biomaterials Scientific Meeting has attracted many scientist from Indonesia, Malaysia, Thailand, Bangladesh, Egypt, Japan as well as other countries.

This international conference is attended by more than 150 participants covering wide variety subject grouped as theoretical chemistry and educational, material synthesis and modification, bioscience and analysis, and also Industrial and environmental chemistry. The given oral and poster presentation would showing outputs for future need as indicated in the conference theme of "Material Chemistry Development for Future Medicine, Industry, Environmental and Biomaterial Application"

The success of the Conference would not have been attained without strong supports from contributing scientists and our partner institutes including Diponegoro University, Semarang State University, Sebelas Maret University, Satya Wacana Christian University and Indonesian Biomaterial Society. I would like to thank all of them for helping to make a very successful conference.

Chairman,

Amin Fatoni, Ph.D

IOP Conf. Series: Materials Science and Engineering **172** (2017) 011001 doi:10.1088/1757-899X/172/1/011001

Committee

Chairman : Amin Fatoni, Ph.D (*Universitas Jenderal Soedirman, Indonesia*)
Secretary : Anung Riapanitra, M.Sc. (*Universitas Jenderal Soedirman, Indonesia*)
Dr. Hartiwi Diastuti (*Universitas Jenderal Soedirman, Indonesia*)
Treasurer : Senny W, M.Si. (*Universitas Jenderal Soedirman, Indonesia*)
Dian Riana N, M.Si. (*Universitas Jenderal Soedirman, Indonesia*)
Dian Windy Dwiasi, M.Sc. (*Universitas Jenderal Soedirman, Indonesia*)
Moch. Chasani, M.Si. (*Universitas Jenderal Soedirman, Indonesia*)
Kapti Riyani, M.Si. (*Universitas Jenderal Soedirman, Indonesia*)
Scientific : Uyi Sulaiman, Ph.D. (*Universitas Jenderal Soedirman, Indonesia*)
Dadan Hermawan, Ph.D. (*Universitas Jenderal Soedirman, Indonesia*)
Dr. Suwandri (*Universitas Jenderal Soedirman, Indonesia*)
Puji Lestari, M.Si. (*Universitas Jenderal Soedirman, Indonesia*)
Mardiyah K, M.Sc. (*Universitas Jenderal Soedirman, Indonesia*)
Zusfahair, M.Si. (*Universitas Jenderal Soedirman, Indonesia*)
Program : Ari Asnani, Ph.D. (*Universitas Jenderal Soedirman, Indonesia*)
Dr. Ponco Iswanto (*Universitas Jenderal Soedirman, Indonesia*)
Roy Andreas, Ph.D. (*Universitas Jenderal Soedirman, Indonesia*)
Suyata, M.Si. (*Universitas Jenderal Soedirman, Indonesia*)
Tien Setyaningtyas, M.Si. (*Universitas Jenderal Soedirman, Indonesia*)
Eva Vaulina, M.Si. (*Universitas Jenderal Soedirman, Indonesia*)
Purwati, M.Si. (*Universitas Jenderal Soedirman, Indonesia*)
Publication : Dwi Kartika, M.Sc. (*Universitas Jenderal Soedirman, Indonesia*)
Ely Setiawan, M.Si. (*Universitas Jenderal Soedirman, Indonesia*)
Irmanto, M.Si. (*Universitas Jenderal Soedirman, Indonesia*)

IOP Conf. Series: Materials Science and Engineering **172** (2017) 011001 doi:10.1088/1757-899X/172/1/011001

International Scientific Board

IOP Conf. Series: Materials Science and Engineering **172** (2017) 011002 doi:10.1088/1757-899X/172/1/011002

Peer review statement

All papers published in this volume of *IOP Conference Series: Materials Science and Engineering* have been peer reviewed through processes administered by the proceedings Editors. Reviews were conducted by expert referees to the professional and scientific standards expected of a proceedings journal published by IOP Publishing.

Interlayer free – nickel doped silica membranes for desalination

A Darmawan[1,*], L Karlina[1], Y Astuti[1], Sriatun[1], D K Wang[2], J Motuzas[2], J C D da Costa[2]

[1]Diponegoro University, Department of Chemistry, Semarang 50275, Indonesia.
[2]The University of Queensland, FIM2Lab-Functional Interfacial Materials and Membranes Laboratory, School of Chemical Engineering, Brisbane Qld 4072, Australia.

*corresponding author e-mail: adi_darmawan@undip.ac.id
Tel.:+62-8222-121-9817

Abstract. This work shows for the first time the potential of nickel oxide silica membranes for desalination applications. Nickel oxide silica xerogels were synthesised via a sol–gel method including TEOS, nickel nitrate with and without addition of hydrogen peroxide. The effects of nickel addition (5% - 50 mol %) on the structure–property relationship of the silica materials were systematically studied. The membrane performance were tested as a function of feed salt concentration (0.3–3.5 wt% NaCl) and temperature (27–60 °C). The membranes which were prepared using equal sol–gel conditions to the xerogel samples showed the raised feed temperatures resulted in increased water fluxes, whilst increasing the salt concentration resulted in decreased water fluxes. The membranes with addition of hydrogen peroxide exhibited better performances than their H_2O_2 absence counterpart. The salt rejection was in excess of 90% and the maximum flux observed was 7.3 kg m^{-2} h^{-1} at 60°C for a 0.3 wt% NaCl feed concentration.

Introduction

The world is facing a global water crisis due to population growth and climate change [1]. With increasingly limited water resources, the use of infinite number of sea water becomes one choice in the provision of clean and fresh water, hence the desalination technology is one of solution[2]. Currently, approximately 43.5% of the desalination systems in the world are still using thermal technology. Membrane technology becomes an alternative instead of thermal desalination [3] because it uses less energy and produce greater water fluxes. In commercial applications, the polymeric membrane are still dominating the market despite of the polymeric membranes have some limitations, especially in their thermal and chemical stability, hence some researchers tried to find a new alternative using inorganic membranes [4-10]

There are two types of inorganic membranes that have been tested for molecular sieve application, zeolite membranes and silica membranes. In general, silica membranes are developed primarily for gas separation [11, 12]. However, some studies have shown that silica membranes coated on alumina support could reject salt ranging from 92% to 99% [8, 10, 13-15]. However, one of the main obstacles in using silica membranes for water purification is due to their hydro instability. To overcome this

problem, some researches have been done i.e. by adding the ligand and surfactant [10, 13], by carbonization a non-ligand template into a silica matrix [10, 16-18], and doping silica membrane with metal both as single metal such as cobalt [19], niobium [20], iron [21], nickel [22, 23] or binary metal such as cobalt-iron [24, 25], palladium-cobalt [26] and lanthanum-cobalt [27, 28]. Amongst these works, cobalt oxide silica membranes have been extensively examined for enhancing the hydrothermal stability, mainly for high temperature wet gas separations. Cobalt oxide silica membranes were also investigated for desalination [15, 29]. A common features of silica based membranes is an asymmetric configuration with a combination of γ-/α-alumina interlayers which their primary roles are to reduce the surface irregularities and roughness of the substrate which may cause stresses leading to film defects, cracks and generally poor membrane performance [30]. However, adding intermediate layers to the membrane fabrication process, not only increases costs but also the processing time and manual handling requirements. Recently, Elma and co-worker [29, 31] and Liu et al [32] reported the preparation of interlayer-free membranes which improved transport to high water fluxes.
In this research, it will be studied the synthesis of nickel oxide impregnated silica membranes for desalination applications as well as optimization of membrane synthesis variables against expected characters. Also, it will be assessed the performance of the resulting membrane for desalination of sea water.

Experimental Section

1. Preparation of nickel doped silica sol and xerogel

The nickel oxide silica sol was prepared similar to metal oxide silica sol methods synthesis according to the method of Darmawan et al [21] and Lin et al [15], the major difference being the use of nickel nitrate instead of iron nitrates or cobalt nitrates. Briefly, nickel oxide sols were synthesized by the hydrolysis and condensation of tetraethyl orthosilicate (TEOS, Aldrich) in absolute ethanol (EtOH) with and without 30% aqueous H_2O_2 with nickel nitrate nonahydrate ($Ni(NO_3)_2.6H_2O$). An initial molar ratio of 255 EtOH: 4 TEOS: x $Ni(NO_3)_2.6H_2O$: yH_2O_2: z H_2O was mixed and vigorously stirred for 3 hours in an ice-cooled bath to avoid partial hydrolysis as suggested by de Vos and Verweij [33]. The x molar concentration of $Ni(NO_3)_2.6H_2O$ was varied 0.2; 0.4, 1, 1.4 and 2 to give Ni/Si molar ratio of 5%, 10%, 25%, 35% and 50%. The y molar of H_2O_2 was varied of 0 and 9 for absence and presence of hydrogen peroxide. The z molar ratio of H_2O was varied to maintain the water mol ratio of 40 where the water of crystallization was taken into account. The aim of this preparation procedure was to recognize the effect of nickel content and hydrogen peroxide to character of xerogel produced. Other variables such as stirring time, drying temperature (60 °C), calcination ramp rate (2 °C/min) and a final calcination temperature (500°C) were kept constant. The sol was then further prepared for characterisation.
An aliquot of each sol sample was dried in a temperature-controlled drying oven at 60 °C under normal atmospheric conditions to form a xerogel. The xerogel samples were crushed finely and calcined at temperature 500°C with ramp rates of 2°C min^{-1} in ambient pressure air and held for 2 h at the preferred temperature.

2. Membrane Synthesis

Thin membrane films were coated on macroporous alumina substrates (α-Al_2O_3 tubular support ($\phi \approx 97$ mm), Ceramic Oxide Fabricates) via a dip-coating process with a dwell time of 2 min and a dipping and withdrawal rate of 10 cm min^{-1}. Each layer was calcined separately after dip coating at 500 °C in a temperature controlled furnace, with a hold time of 2.5 h and a ramp rate of 2 °C min^{-1}. The dip-coating and calcination procedure were repeated five times to reduce the probability of thin film defects and to obtain high quality membranes with high salt rejection

3. Characterization of nickel oxide impregnated silica membrane

Nickel doped silica membrane xerogels in powder form were characterised by using a Shimadzu IRAffinity-1 Fourier-transform infrared (FTIR) spectrometer with a Pike MIRacle attenuated total reflectance accessory (ATR-FTIR) at wavelength range 400–4000 cm^{-1}. Deconvolution of the FTIR spectra was performed using the Fityk computer program with Gaussian peak fitting as the preferred curve. The same number of peaks was used in all the spectral peak fitting. The half width at half maximum (HWHM) was fixed for each peak, while the peak position was allowed to change slightly to realize qualified fitting. The peak area for silanol at 950 cm^{-1} was normalized against the peak area of the siloxane bridges at 1080 cm^{-1}. The gravimetric analyses of the nickel doped silica membrane xerogels were performed on a Shimadzu thermogravimetric analyser TGA-50 using air flow rate of 60 mL min^{-1} and heating rate 2°C min^{-1}. Nitrogen adsorption analysis was performed at 77 K and 1 bar using Micromeritic TriStar 3020 instrument of samples degassed under vacuum for >6 h at 150°C. The specific surface area was determined from the Brunauer, Emmett and Teller (BET) method and total pore volume was taken from the last point of the isotherm. Adsorption average pore width (4V/A by BET) methods was taken to determine the average pore sizes. The porosity (ϵ) can be calculated from this volume of adsorbed nitrogen, according to [34]

$$\epsilon(\%) = \frac{V_p}{V_p + (\frac{1}{\rho})} \times 100\% \quad (1)$$

in which V_p is the pore volume (ml/g), and ρ is the density of the solid phase (g/ml). A skeletal density of 2.2 g/cm^3 was used for amorphous calcined silica [34]. The pore volume is calculated from the adsorbed gas volume [ml(STP)/g, at 1 atm and 0°C] assuming ideal gas behaviour and taking the density of liquid nitrogen, ρ = 0.808 g/ml, which yields

$$V_p = V_{ads} \times 1.547 \times 10^{-3} \quad (2)$$

with V_p in ml liquid N_2/g and V_{ads} in ml (STP)/g.

4. Membrane performance

Nickel oxide impregnated silica membrane performance for the desalination of sea water is tested by measuring the produced water flux and rejected salt concentration. The salt concentration and temperature of the feed water are kept constant. The membrane was assembled into a filter set-up for desalination experiments. The membrane was assembled into a classical pervaporation set-up for desalination experiments as shown in Figure 1 according to the method of Elma [31]. The membrane was immersed in a tank containing saline feed solution. The membrane was blocked at one end and connected to a cold trap (immersed in liquid nitrogen) and vacuum pump operating at 1.5 kPa at the other end. The feed solutions were prepared with sodium chloride (NaCl) dissolved in deionised water with concentrations ranging from 0.3 and 3.5 wt% and temperature was varied from room temperature, 40 °C and 60 °C. The feed solution was under constant recirculation to prevent concentration polarisation on the contact membrane side.

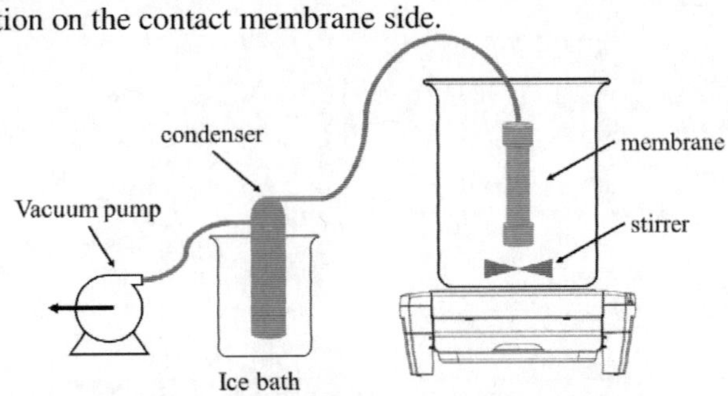

Figure 1. A membrane desalination set up

The water flux, F (kg $m^{-2}h^{-1}$), was determined based on the equation $F = \frac{m}{A\Delta t}$, where m is the mass of permeate (kg) retained in the container, A is the surface-active area (m^2) of the membrane and Δt is the time measurement (h). The salt rejection, R (%), was calculated as $R = \frac{C_f - C_p}{C_f} \times 100\%$, where C_f and C_p are the feed and permeate concentrations of salt (wt%). The salt concentrations were correlated to conductivities of the retentate and permeate solutions determined by a conductivity meter.

IOP Conf. Series: Materials Science and Engineering **172** (2017) 012001 doi:10.1088/1757-899X/172/1/012001

Results and Discussion

1. Nitrogen Adsorption

The physical and textural properties of the nickel doped silica xerogels were studied by nitrogen sorption which provides important qualitative information regarding the microstructure of the resulting molecular sieving membranes. The adsorption-desorption isotherms are shown in Figure 2. The BET surface areas, total pore volumes and average pore diameters are summarised in Figure 3. In porous inorganic membrane, the pore volume generally provides an indication of the membrane flux while the pore size is related with the separation of molecules or purity. In general, pore sizes with diameters greater than 50 nm are classified as macroporous; between 2 and 50 nm are mesoporous and smaller than 2 nm are microporous.

For all samples with addition of hydrogen peroxide, the isotherms of nickel doped silica xerogels resulted in microporous materials with a type I isotherm as indicated by a very strong initial adsorption at very low partial pressures ($P/P_o < 0.2$) followed by saturation and are thus characteristic of microporous materials, which produced the low surface area (~330-390 $m^2\ g^{-1}$), total pore volume (0.17-0.21 $cm^3\ g^{-1}$) and the average pore radius remained similar at around 2.1 nm (±0.05 nm). This consistency is similar with the result of iron doped [21] and cobalt doped silica membranes [15, 32, 35]. Although pore volumes and surface areas varied depending upon the Ni/Si ratio.

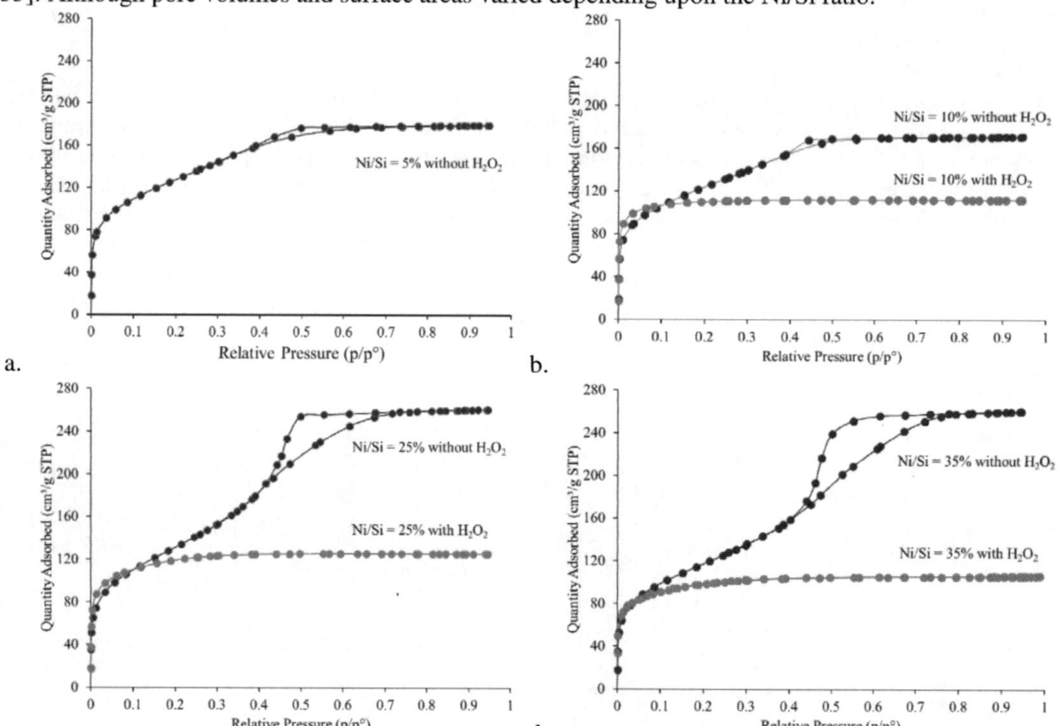

IOP Conf. Series: Materials Science and Engineering **172** (2017) 012001 doi:10.1088/1757-899X/172/1/012001

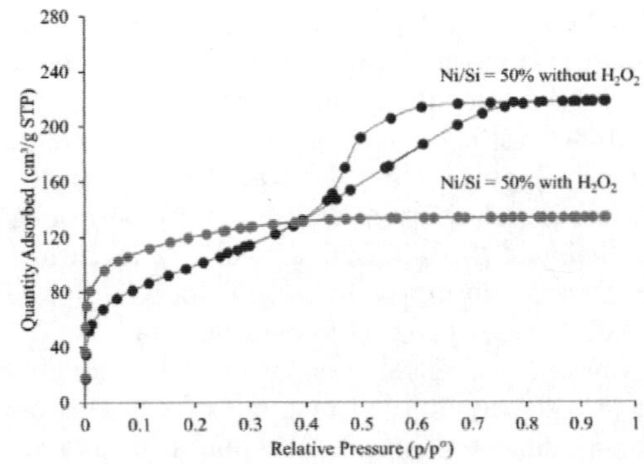

e.

Figure 2. Nitrogen adsorption–desorption isotherms the nickel doped silica xerogels (a) Ni/Si=5% (b) Ni/Si=10% (c) Ni/Si=25% (d) Ni/Si=35% (e) Ni/Si=50%

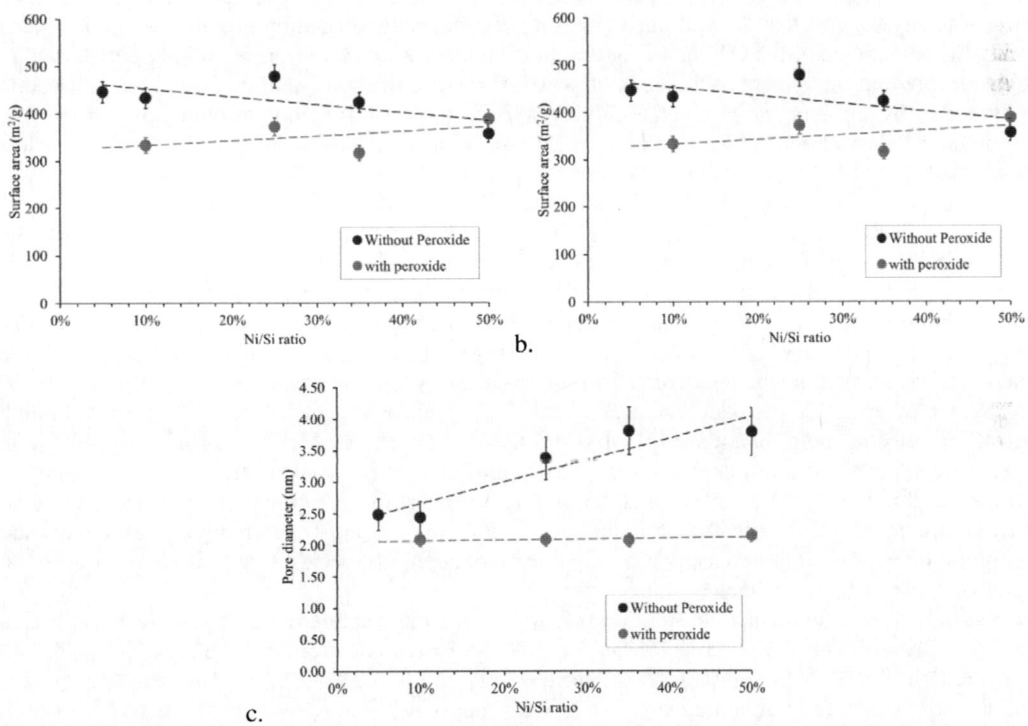

Figure 3. Surface properties of bulk nickel doped silica xerogels (a) surface area (b) pore volume (c) pore diameter

It can be seen that the pore structures tend to shift to mesoporous region as Ni/Si molar ratio increases. This result differs from other studies showing that the addition of metal oxides lowers surface area and pore volume [21] as the metal oxide generally have a low surface area as compared to polymeric silica although similar observations are typically observed in cobalt-silica materials [36]. This result is suggesting that nickel oxide opposes the densification of the silica matrix by stiffening the silica aggregates against the collapse during drying. However, the amorphous silica is still dominantly responsible for the textural properties of the nickel doped oxide silica matrix. It appears to be expected that the amorphous silica structure is enveloped the nickel oxide particles.

Interestingly, in the absence of hydrogen peroxide, a significant structural re-arrangement is observed where the nickel doped silica xerogels tended to form mesoporous materials with a higher adsorption

saturation capacity above 0.4 p/p° and a hysteresis with a smooth inflection indicating a type IV isotherm. The surface area, pore volume and pore diameter increases of 30% up to twice. The absence of hydrogen peroxide in the nickel doped silica samples contributed to mesoporous formation (average pore radius > 2 nm), lead to higher surface area (>400 $m^2 \cdot g^{-1}$) and total pore volume (up to 0.4 $cm^3 \cdot g^{-1}$). It is interesting to observe in Figure 2 that the hysteresis loops due to escalation in multilayer adsorption are increasing by the increase of the nickel which confirming the broadening of the pore sizes. Moreover, when nickel/silica ratio reached 25 wt%, it is clearly detected that the isotherm profiles are truly type IV isotherms and mesoporous as evidenced by their large extent of hysteresis near p/p°=0.5 indicating the occurrence of capillary condensation.
The position of inflection is an increasing function of the pore diameter [37]. Since by increase of nickel content, the position of inflection shifted to a higher value of p/p°, the pore sizes should be increase. Significantly large pore diameters reached the values close to 3.8 nm. The change of porosity should correspond to the decrease in selectivity of the membrane. This result clearly indicating that the structural modifications in the template xerogel are directly attributed to the hydrogen peroxide molecules which has a far greater impact on the resultant structure than originally thought. The results above strongly suggest that the hydrogen peroxide facilitates the formation of polymeric silica species during the early stage of the sol-gel process and the condensation occurs preferentially between silanol groups located on monomers or the ends of polymers as occurring on acid-catalysed sol-gel reaction. Conversely, by absence of H_2O_2, the preferred reaction sites for condensation are on the most deprotonated silanol bonds of the Q^3 Si atoms [38] and condensation reaction via the monomer–cluster growth pathway [19] which tends to produce mesoporous silica materials.
The change of textural properties of these silica based matrices can also be rationalized by the change of pH of the sol as H_2O_2 depresses the pH in 2 digit below their H_2O_2 absence counterpart. Such changes in the sol pH have been well known had a dramatic effect on the microstructure of the silica matrices as reported elsewhere [29, 38, 39]. When the sol is acidified by adding the H_2O_2, hydrolysis reaction is promoted which long chains of siloxane bonds, forming a weakly branched polymer. This type of bond can densify relatively easily upon drying and calcination forming a mesh like structure where the connected interchain apertures form pores of 2-4Å [40-42].
By absence of hydrogen peroxide, the nickel doped silica sol pHs were between (4-5) which is higher than the isoelectric point boundary (pH 1–3) of the silica species [43]. At this instance, the silanol species are expected to be all deprotonated participating in the polycondensation reaction generating a significant amounts of silica monomers constantly forming and reforming, in a process known as Ostwald ripening, allowing the structure to settle to its most thermodynamically favoured configuration with a large concentration of highly condensed species (e.g., siloxane bridges) and thereby creating a highly branched silica network.
Interestingly, it was found that the solution pH shifted to a more acidic region by the increasing nickel content in the solution. This pH shift was found to be systematic from pH 4.62 (Ni/Si = 5%), 4.25 (Ni/Si = 10%), 4.23 (Ni/Si = 25%), 4.14 (Ni/Si = 35%) to pH 4.05 (Ni/Si = 50%). But conversely, it lead the pore structure to be more mesoporous. This fact is in contrary to what had been believed that the increase of acidity causes the silica matrix structure to more microporous [29, 38, 39]. This fact strengthen the idea which has been described above that the nickel oxide inhibits densification of the silica matrix.
Although pH decreased (acidity increased), the increasing of mesoporosity can be justified that although the pHs of nickel doped silica solution remained quite close to, but they were still above the isoelectric point, at the which there is no electrostatic particle- particle repulsion [29] hence the effect of pH was not more dominant than the effect of dispersion of metal ions in solution.
Compared with the elements in the same transition group (Group VIII), indeed, the nickel is less acidic than iron and cobalt. The acidity sequence is Fe > Co > Ni. Iron is more acidic as the solution containing 3+ ion tends to have pH's in the range from 1 to 3 whereas the solutions containing 2+ ions have higher pH's which typically around 4-5. The ions that have smaller ionic radius will be less acidic and as nickel ionic radius is smaller than cobalt, its acidity is lower than cobalt. This explains why the

iron doped silica was very acidic (pH<2) and had a microporous structure as reported by Darmawan [21] whilst the cobalt doped silica with the addition of peroxide had a pH of 3 and produced microporous structure as reported by Lin [15]. While at higher pH, cobalt doped silica produced mesoporous structures [29]. However, it seems there is no definite relationship between pH ranges with microporosity or mesoporosity silica matrix resulted. Elma and co-workers [39] reported the silica xerogel without the addition of transition metals produced a solution with a pH = 7 and the resulting structure was mesoporous. It appears that textural properties of silica matrix was the resultant effect of metal, hydrogen peroxide and the pH.

As can be seen in Figure 4, the porosity increases with increasing nickel content as calculated from the volume of adsorbed nitrogen, as expressed in Equation (1). Hence the silica matrix structure with the higher nickel content will be more polymerized and more branched thus more porous. The porosities for sample with absence of H_2O_2 are always higher than their presence of H_2O_2 counterpart. In the presence of H_2O_2, the structures would be expected to pack very well and form structures with very low porosities.

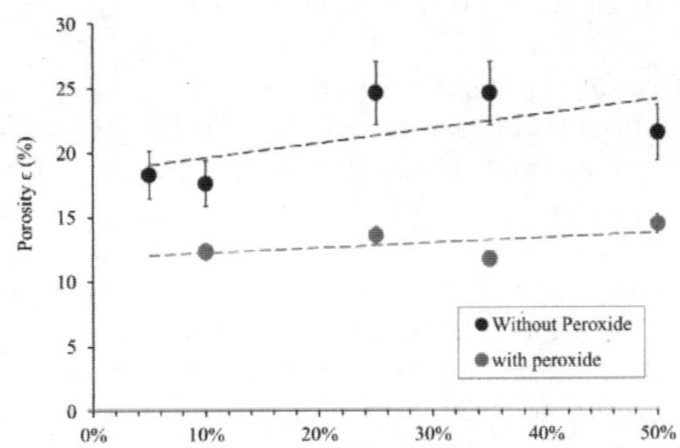

Figure 4. Effect of Ni/Si molar ratio on porosity

2. Infrared Spectroscopy analysis

Figures 5(a) and 3(b) show the IR spectra of xerogel samples between wave number 650 and 1450 cm^{-1}. It is observed that regardless of the Ni/Si molar ratio variation, the presence or absence of H_2O_2, the IR spectra of the xerogel samples remained almost persistent.

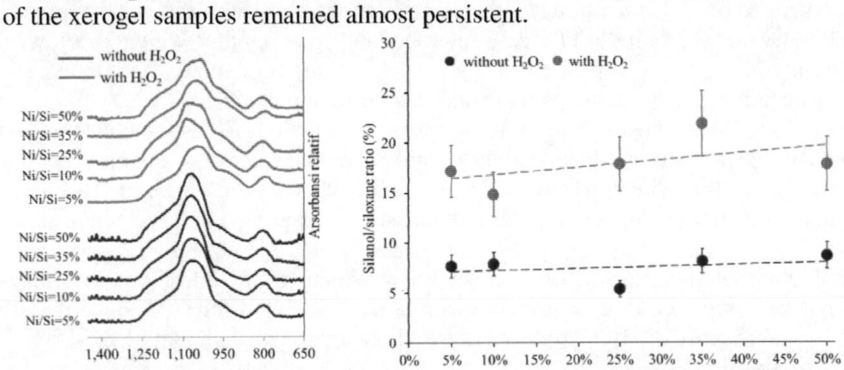

a. b.

Figure 5. (a-left) FTIR spectra of samples as a function of Ni/Si molar ratio where (i) 5%, (ii) 10%, (iii) 25% (iv) 35% (v) 50% (vi) 10% with H_2O_2 (vii) 25% with H_2O_2 (viii) 50% with H_2O_2, (b-right) Silanol and siloxane ratio as of samples as a function of Ni/Si molar ratio

These FTIR results suggest that all nickel doped silica xerogels had similar surface chemical. These results suggest that no significant variation of silica functional groups was reached by introduction of nickel. The formation of the silanol and siloxane groups was still governed by the silica hydrolysis and condensation reactions. It has been widely reported that the dissolution of metal species in the sol–gel process produces metal nanoparticles in the porous structure of silica [24, 44].

The bands observed are conventionally found in silica materials which include the bands at 800 cm^{-1} and 1070 cm^{-1} with a shoulder near 1200 cm^{-1} corresponding to different modes of siloxane stretching bonds (Si-O-Si): symmetric and asymmetric, respectively. The shoulder in the region of 960 cm^{-1} is assigned to the stretching vibration of Si–OH [45, 46]. At the absence of hydrogen peroxide, the peak of Si–O band (1070 cm^{-1}) position has been observed to be similar in the very narrow range (1068-1070 cm^{-1}), however in the presence of hydrogen peroxide there was a significant shift at the asymmetric Si-O band from 1086 cm^{-1} (Ni/Si=10% + H_2O_2), 1070 cm^{-1} (Ni/Si=10% + H_2O_2) to 1032 cm^{-1} (Ni/Si=10% + H_2O_2), indicating that Si–O–M (M is the metal atom) hetero-linkages were probably formed [47]. If the number of Si–O–M linkages increases, the number of Si atoms in Si-O-Si linkages decrease and consequently in the silanol groups at the surface will increase. Hence the intensity ratio of the 960 cm^{-1} band to 1070 cm^{-1} band could be used to measure relative heterogeneity of nickel and silica.

To understand further the effect of water in the formation of functional groups of the iron oxide silica matrices, the FTIR spectra in Figure 5a were deconvoluted using a Gaussian method and the result is showed in Figure 5b. The deconvolution allows the calculation of a silanol (~960 cm^{-1}) to siloxane (~1070 cm^{-1}) ratio. Regardless the presence and absence of hydrogen peroxide, the trend is consistent in the sense that the silanol/siloxane ratio is constant by increasing the nickel/silica ratio with concern to the error bar. This result suggests that the nickel oxide is not chemically bound to the silica, and remains as particles embedded into the silica matrix. The existence of nickel does not give any effect to either accelerate or retard the condensation reaction of silanol into siloxane.

Remarkably, the nickel doped silica xerogels with the presence of H_2O_2 show the silanol/siloxane ratio which is always higher than their same Ni/Si content counterpart. These results clearly indicate that the lower silanol concentrations, or likewise the higher siloxane bridge concentrations were achieved with the xerogels prepared with addition of hydrogen peroxide. Hence, it could be strongly proposed that the presence of H_2O_2 actually inhibits the condensation reaction and stabilize the presence of silanol from the hydrolysis reaction. These barriers lead to insufficient energy to drive the reaction to full condensation. As silanols are pendulous functional groups which not chemically inter-linked to other silica structures like siloxane bridges, therefore silanols are recognized as unreserved structures. Hence, silanols have the proclivity to freely interpenetrate one another which is known as weakly branched systems [48]. It was explained that weak branching combined with limited condensation during film formation promotes dense packing and low pore volume [49].

By absence of H_2O_2, the synthesis conditions are more favourable towards the condensation reactions which is proved by the lower silanol/siloxane ratio. This then lead to interconnect the siloxane structures with one another which is known as strongly branched systems. The increase in the siloxane bridges is parallel connected with any increase of surface area and pore volume which accord with the fractal theory.

An additional point of importance here is that the production of high siloxane bridges in silica membranes has been reported to oppose hydro-instability [50, 51]. Therefore, since the nickel doped silica xerogels with absence of H_2O_2 produces more siloxane groups, it should be more hydro-stable.

3. Thermal Analysis

Typical thermograms and weight change derivatives obtained for nickel oxide silica xerogels are shown in Figures 6 (a)(b) and Figure 7 respectively.

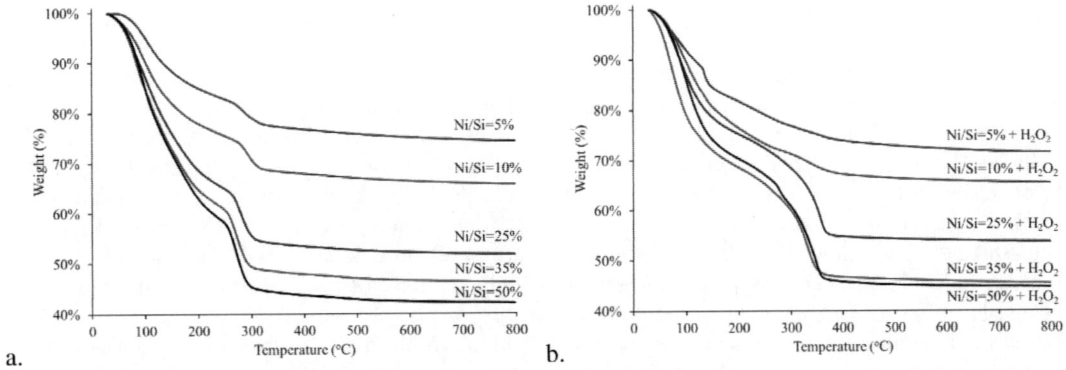

Figure 6. Mass loss curves of samples as a function of Ni/Si molar ratio (a-left) without addition of H_2O_2 and (b-right) with addition of H_2O_2

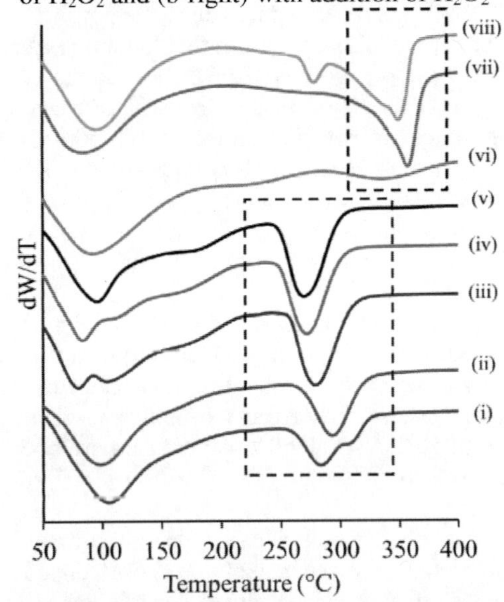

Figure 7. Differential of the TGA for samples prepared with varying Ni/Si molar ratios where (i) 5%, (ii) 10%, (iii) 25% (iv) 35% (v) 50% (vi) 10% with H_2O_2 (vii) 25% with H_2O_2 (viii) 50% with H_2O_2

The weight reduction increased gradually by increasing the amount of nickel. These results strongly suggest that the addition of iron is affecting the sol-gel processes. All the samples displayed a two stage decomposition behaviour i.e. two major weight loss regions, (i) in the temperature range of 35-245°C (ii) in the range of 245°–800°C are seen. From 80-120°C, the weight loss is generally caused by the loss of free (not chemically bound) H_2O and ethanol molecules. The sharp decrease in weight is related with the increase of Ni/Si molar ratio, suggesting that an increasing nickel content results in increased H_2O within the uncalcined xerogel matrix. As nickel in $Ni(NO_3)_2$ is surrounded by 6 H_2O molecules consequently increasing the nickel content in the sol-gel will equally increase the amount of water in the sample. The decline in the mass observed at 150-350°C corresponds to the completion of dehydration and the dehydroxylation due to the condensation of the OH groups in the silica particles [52, 53], and the oxidation of the residual carbons (from ethanol and reaction by-product) which may be trapped inside the silica network. This trend remains until the next weight loss peaks between 150-350°C, whereby total decreasing of weight correspond with increasing nickel content. This may be considered as nitrate decomposition, related to NO and NO_2 deformation which may take place around 245-350 °C.

Comparing the Figure 6 a and b, it can be found that generally with the same amount of nickel content, the quantity of weight loss is almost similar. This result indicates that the contribution of weight loss is dominated by the nickel nitrate decomposition. Conversely, the weight loss contours are different. In the nickel doped silica with absence of H_2O_2, there is a constant curve decline up to 250 °C, while by the presence of H_2O_2, the curves decrease sharply to 150 °C and from there on become slighter. In contrast, with absence of H_2O_2, the curves decrease sharply at a temperature of 250-325 °C. This may occur since the addition of H_2O_2 lead the silica structure to be more bulky to facilitate the deformation process run faster at lower temperatures. Meanwhile, with presence of H_2O_2, the silica matrix structure is denser and thus require a higher temperature to deform.

The trend of mass loss as a function of the nickel to silica ratio and presence of H_2O_2 was further analysed as per Figure 6. It is evident that the major losses occurred up to 150°C, which is mainly attributed to physisorbed water and ethanol retained in the silica matrix. As mentioned above that nickel ion is surrounded by 6 H_2O molecules which further aiding in mass loss. The samples with higher nickel content contain larger amounts of fluid in their structure (per unit weight of the sample). Therefore, this structure has to be more porous than the structures of samples with lower nickel content. This conclusion is in accordance with results from the nitrogen adsorption measurements in which samples with increasing nickel content show an increase in porosity as shown in Figure 4.

For nickel doped silica in the absence of H_2O_2, the bottom highlighted box inside the Figure 7 shows that by increasing of nickel content, the deformation temperatures decrease which marked by the peak shift to the lower temperature. This result is in line with the FTIR results which show there is no bond between nickel oxide and silica hence the increase of nickel content in silica matrix should reduce the deformation temperature.

Interestingly, the top highlighted box inside the Figure 7 shows the deformation temperatures of nickel doped silica xerogels in the absence of H_2O_2 are always higher than their H_2O_2 presence counterpart. This could be understood because the absence of H_2O_2 nickel doped silica xerogels produce denser structures as described from the results of surface morphology which is marked by smaller pore diameters. Whilst the presence of H_2O_2 nickel silica generates more open structure which causes the nitrate group to decompose more easily.

4. Desalination measurements

Membrane performance is commonly evaluated based on the permeating water flux and the salt rejection. Figure 8 and 9 shows the performance of the nickel doped silica as a function of salt concentration and temperature. In general, all the nickel doped silica membranes produced moderate water fluxes up to 7.5 kg m^{-2} h^{-1} with good salt rejections reaching values up above 90% in all testing condition. This series of membranes show that the desalination performance does not follow gradual sequence as function of nickel content. However, the membranes with presence of H_2O_2 always demonstrated higher performance in salt rejection but lower in water flux compared to their H_2O_2 absence silica membrane counterpart in all testing condition. For all membranes, greater permeating water flux was observed at temperatures by increasing of temperature. However, there is a decreasing in salt rejection as the temperature raised. The loss of performance due to increasing of temperature is comparable to the results reported for cobalt silica membrane [15, 29]. Raising the temperature of the salt solution permitted for an increasing of the vapour pressure hence increasing the driving force for the water permeation.

The flux dropped when the membrane were tested in higher NaCl concentration, indicating some effect of NaCl within the larger pores of the membrane. The increased driving force of NaCl across the membrane leads to this flux increase. The separation performance of the Ni/Si=10% and 25% membrane with presence of H_2O_2 showed the highest rejection was possible at 0.3 wt% of around 99.9%. This decreased slightly when NaCl concentration at 3.5% to around 97.5%. The ability to desalinate to potable standards of below 0.05 wt% was achievable by all membranes at 0.3 wt% feed. At 3.5 wt% feed simulating seawater, for Ni/Si=10% membrane with presence of H_2O_2, the NaCl

IOP Conf. Series: Materials Science and Engineering **172** (2017) 012001 doi:10.1088/1757-899X/172/1/012001

concentration in the permeate reduced to 0.04 wt% which is below to the 0.05 wt% drinking water standard. It is likely that Ni/Si=10% silica membrane with presence of H_2O_2 could transport seawater to potable standards in a single pass. When compared to the performance of the cobalt oxide silica membranes [15], the water fluxes produced by the nickel membranes are much higher (300%) although the salt rejections are slightly lower. This result suggests that the absence of γ-alumina interlayer gave a significant increase of water fluxes.

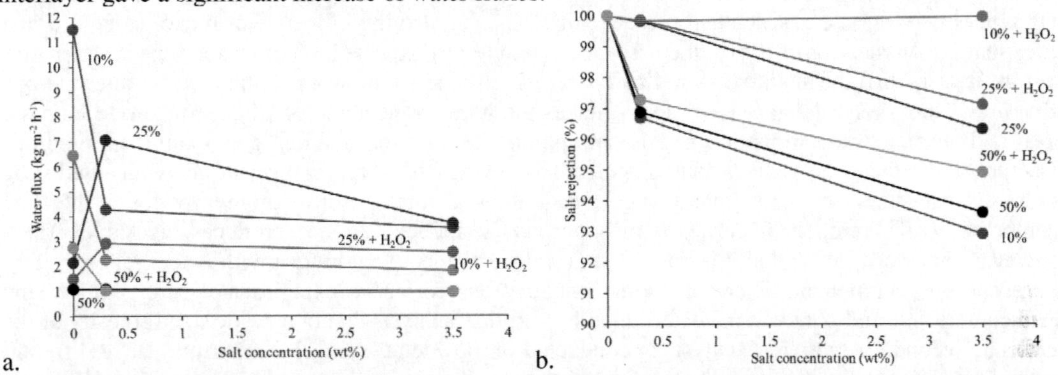

Figure 8. Membrane desalination performance as function of salt concentration toward (a) water flux (b) salt rejection at room temperature

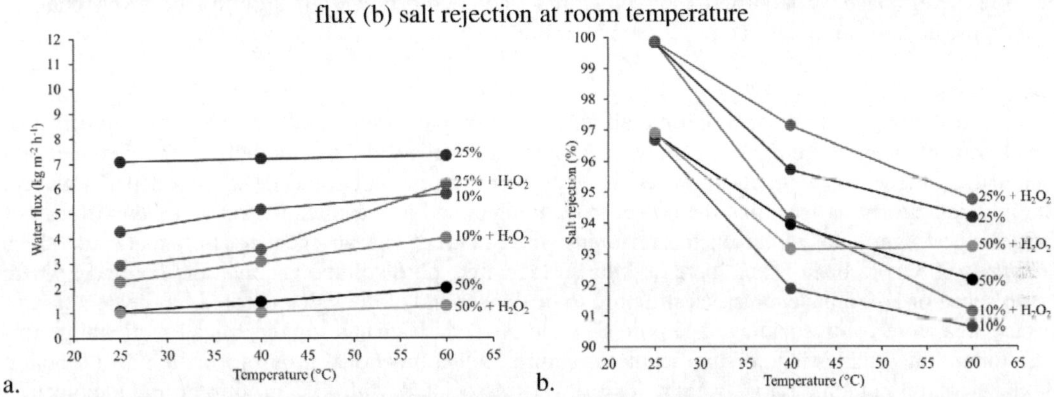

Figure 9. Membrane desalination performance as function of temperature toward (a) water flux (b) salt rejection at 0.3% salt concentration

It seems obviously there is a close relationship between the membrane performance and the surface area as recognised by N_2 sorption as shown in Table 1. For example, in a given fixed salt concentration 0.3 wt% and temperature 25°C, the water flux values are (4.3, 2.3); (7.1, 2.9) and (1.1, 1.03) kg m^{-2} h^{-1} whereas the surface area value are (433, 333); (478, 372) and (356, 388) m^2/g for the pair of (10%, 10%+H_2O_2); (25%, 25% + H_2O_2) and (50%, 50% + H_2O_2) membranes, respectively. This behaviour also likely directly related to the pore volume and average pore size of the nickel doped silica membranes. It has been well known that the pore volume controls the water flux while the pore diameter controls the sieving. The salt rejection is generally controlled by the pore size where large pores cause pore wetting due to the hydrophilic properties of silica and consequently low rejection whilst small pores deliver high salt rejection.

Table 1 obviously show that the nickel doped silica membranes resulted in average pore sizes measured between 2.08 and 3.8 nm. Actually, these average pore sizes are too big to conduct a molecular sieving to separate water molecules from solution but usually the thin films have smaller pore sizes than their bulk counterparts. Although the original chemistry that govern silica polymeric growth and gelation are principally the same for films as bulk gels, however there are other factors that influence structural evolution in films [43, 54] such as due to non-equivalent gelation and drying

conditions. If all nickel doped silica pores were all mesoporous (pore diameter >2 nm), the membranes should not be able to reject any hydrated salt ions. As the desalination results in this work show that there is a salt rejection, it can be said that the nickel doped silica silica membranes were dominated by a vast majority of pore sizes less than the size of hydrated salt ions (Na^+ : d_k = 7.2 Å and Cl^- : d_k = 6.6 Å) and above water molecule (d_k = 2.6 Å).
The nitrogen sorption isotherms results (Figure 2) strongly show that there were much structure differences between the absence and presence of H_2O_2 and also the influence of nickel content to the silica matrix. As mentioned above, the presence of H_2O_2 was expected to consolidate the silica matrix uniformly, establish the silanol stability and increasing the silica network dominated by microporous structures. Oppositely, the absence of H_2O_2 makes the pore arrangement of silica structure to be more open and accelerate the formation of siloxane bond. The fact that the xerogels exhibit mesoporous structure when the membrane produces molecular sieving with high salt rejection is an interesting finding. It is often mentioned that xerogels displaying a microporous region produce a selective membrane while xerogels featuring a mesoporous component do not produce this same effect. However, this work shows that it is possible to produce a selective thin film with a xerogel exhibiting a mesoporous component. There are some explanations for these experimental results. First, the average pore size and pore volume determined by sorption analysis do not provide the tortuosity of the material. Second, the sorption analysis is conducted on powdered xerogels which may have different structures to the final membrane. Third, as the membrane thin layer having a thickness of 150-200 nm, the pores overlapped or stratified each other that led to the downsizing of the filtration channel because the underneath pores is covered by upper pores.

Conclusions

The nickel doped silica xerogels and membranes were prepared using sol–gel method. The effect of nickel content ranging from 5% to 50 wt% into silica matrix was investigated for the physical and chemical properties of xerogels followed by membrane morphology and desalination performance. The findings clearly indicate that the nickel oxide content and in the silica matrix and the existence of H_2O_2 have a significant effect on the resultant silica matrix. The absence of H_2O_2 made the silica structure to be more open hence increased the surface area, pore volume and average pore size where the presence of H_2O_2 made the silica structure to be denser and close which marked by the decrease of surface area, pore volume and average pore size. No observable changes in the silica functional groups were found, thus suggesting no significant variation of silica functional group formation as a function of nickel oxide content. Large weight losses were associated with OH groups or H_2O molecules derived from the decomposition of iron nitrate nonahydrate in addition to solvents (ethanol) and organics (R groups from TEOS). Furthermore, nitrate decomposition under heat treatment led to NO and NO_2 formation, further contributing to weight losses. In other words, increasing the Ni/Si molar ratio likewise increases the weight losses. By increasing of nickel content in silica matrix, the surface morphology parameters such as surface area, pore volume and average pore size increased. This result is counterintuitive as the increase in nickel oxide content should decrease the total surface area since metal oxide particles have low surface area than polymeric silica structures, these results strongly suggest that the nickel oxides played a beneficial role in opposing the silica structure collapse.
The nickel doped silica membranes showed good salt rejection (>90%) for all testing temperatures and salt concentration. Water fluxes and salt rejection varied depending on the concentration of nickel oxide in the silica film, feed salt concentration and temperature. Nickel doped silica membrane with absence of H2O2 possessed larger pore sizes which yielded higher fluxes but lower salt rejection compared to their H2O2 presence counterpart.

References

[1] C.J. Vörösmarty, P. Green, J. Salisbury, R.B. Lammers, Global Water Resources: Vulnerability from Climate Change and Population Growth, Science, 289 (2000) 284-288.

[2] K.V. Reddy, N. Ghaffour, Overview of the cost of desalinated water and costing methodologies, Desalination, 205 (2007) 340-353.
[3] N. Voutchkov, Desalination – water for the next generation, Filtration & Separation, 42 (2005) 14-25.
[4] L. Cot, A. Ayral, J. Durand, C. Guizard, N. Hovnanian, A. Julbe, A. Larbot, Inorganic membranes and solid state sciences, Solid State Sciences, 2 (2000) 313-334.
[5] T. Tsuru, Inorganic Porous Membranes for Liquid Phase Separation, Separation & Purification Reviews, 30 (2001) 191-220.
[6] J. Wang, T. Tsuru, Cobalt-doped silica membranes for pervaporation dehydration of ethanol/water solutions, Journal of Membrane Science, 369 (2011) 13-19.
[7] M.M. Teoh, T.S. Chung, Membrane distillation with hydrophobic macrovoid-free PVDF-PTFE hollow fiber membranes, Separation and Purification Technology, 66 (2009) 229-236.
[8] G.-T. Lim, H.-G. Jeong, I.-S. Hwang, D.-H. Kim, N. Park, J. Cho, Fabrication of a silica ceramic membrane using the aerosol flame deposition method for pretreatment focusing on particle control during desalination, Desalination, 238 (2009) 53-59.
[9] N. Misdan, W.J. Lau, A.F. Ismail, Seawater Reverse Osmosis (SWRO) desalination by thin-film composite membrane—Current development, challenges and future prospects, Desalination, 287 (2012) 228-237.
[10] M.C. Duke, S. Mee, J.C.D. da Costa, Performance of porous inorganic membranes in non-osmotic desalination, Water Research, 41 (2007) 3998-4004.
[11] G.Q. Lu, J.C. Diniz da Costa, M. Duke, S. Giessler, R. Socolow, R.H. Williams, T. Kreutz, Inorganic membranes for hydrogen production and purification: A critical review and perspective, Journal of Colloid and Interface Science, 314 (2007) 589-603.
[12] J. Yang, J. Chen, Hydrophobic modification and silver doping of silica membranes for H2/CO2 separation, Journal of CO2 Utilization, (2013).
[13] S. Wijaya, M.C. Duke, J.C. Diniz da Costa, Carbonised template silica membranes for desalination, Desalination, 236 (2009) 291-298.
[14] Z. Xie, M. Hoang, T. Duong, D. Ng, B. Dao, S. Gray, Sol–gel derived poly(vinyl alcohol)/maleic acid/silica hybrid membrane for desalination by pervaporation, Journal of Membrane Science, 383 (2011) 96-103.
[15] C.X.C. Lin, L.P. Ding, S. Smart, J.C. Diniz da Costa, Cobalt oxide silica membranes for desalination, Journal of Colloid and Interface Science, 368 (2012) 70-76.
[16] M.C. Duke, J.C.D. da Costa, D.D. Do, P.G. Gray, G.Q. Lu, Hydrothermally Robust Molecular Sieve Silica for Wet Gas Separation, Advanced Functional Materials, 16 (2006) 1215-1220.
[17] M.C. Duke, J.C.D. da Costa, G.Q. Lu, C. Malde, P.G. Grayyy, Hydrothermal Stability Analysis of Carbonised Template Molecular Sieve Silica Membranes, in: C. Fell (Ed.) 2004 ECI Conference on Separations Technology VI, The Berkeley Electronic Press, Kingsher Resort, Fraser Island, Queensland, Australia, 2006.
[18] M.C. Duke, J.C. Diniz da Costa, G.Q. Lu, M. Petch, P. Gray, Carbonised template molecular sieve silica membranes in fuel processing systems: permeation, hydrostability and regeneration, Journal of Membrane Science, 241 (2004) 325-333.
[19] D. Uhlmann, S. Liu, B.P. Ladewig, J.C. Diniz da Costa, Cobalt-doped silica membranes for gas separation, Journal of Membrane Science, 326 (2009) 316-321.
[20] V. Boffa, J.E. ten Elshof, R. Garcia, D.H.A. Blank, Microporous niobia-silica membranes: Influence of sol composition and structure on gas transport properties, Microporous and Mesoporous Materials, 118 (2009) 202-209.
[21] A. Darmawan, S. Smart, A. Julbe, J.C. Diniz da Costa, Iron Oxide Silica Derived from Sol-Gel Synthesis, Materials, 4 (2011) 448-456.
[22] Y.H. Ikuhara, H. Mori, T. Saito, Y. Iwamoto, High-Temperature Hydrogen Adsorption Properties of Precursor-Derived Nickel Nanoparticle-Dispersed Amorphous Silica, Journal of the American Ceramic Society, 90 (2007) 546-552.

[23] M. Kanezashi, T. Fujita, M. Asaeda, Nickel- Doped Silica Membranes for Separation of Helium from Organic Gas Mixtures, Separation Science and Technology, 40 (2005) 225-238.

[24] A. Darmawan, J. Motuzas, S. Smart, A. Julbe, J.C. Diniz da Costa, Binary iron cobalt oxide silica membrane for gas separation, Journal of Membrane Science, 474 (2015) 32-38.

[25] A. Darmawan, J. Motuzas, S. Smart, A. Julbe, J.C. Diniz da Costa, Temperature dependent transition point of purity versus flux for gas separation in Fe/Co-silica membranes, Separation and Purification Technology, 151 (2015) 284-291.

[26] B. Ballinger, J. Motuzas, S. Smart, J.C. Diniz da Costa, Palladium cobalt binary doping of molecular sieving silica membranes, Journal of Membrane Science, 451 (2014) 185-191.

[27] B. Ballinger, J. Motuzas, C.R. Miller, S. Smart, J.C. Diniz da Costa, Nanoscale assembly of lanthanum silica with dense and porous interfacial structures, Scientific Reports, 5 (2015) 8210.

[28] B. Ballinger, J. Motuzas, S. Smart, J.C. Diniz da Costa, Gas permeation redox effect on binary lanthanum cobalt silica membranes with enhanced silicate formation, Journal of Membrane Science, 489 (2015) 220-226.

[29] M. Elma, D.K. Wang, C. Yacou, J. Motuzas, J.C. Diniz da Costa, High performance interlayer-free mesoporous cobalt oxide silica membranes for desalination applications, Desalination, 365 (2015) 308-315.

[30] A.J. Burggraaf, Chapter 8 Fundamentals of membrane top-layer synthesis and processing, in: A.J. Burggraaf, L. Cot (Eds.) Membrane Science and Technology, Elsevier, 1996, pp. 259-329.

[31] M. Elma, D.K. Wang, C. Yacou, J.C. Diniz da Costa, Interlayer-free P123 carbonised template silica membranes for desalination with reduced salt concentration polarisation, Journal of Membrane Science, 475 (2015) 376-383.

[32] L. Liu, D.K. Wang, D.L. Martens, S. Smart, J.C. Diniz da Costa, Interlayer-free microporous cobalt oxide silica membranes via silica seeding sol–gel technique, Journal of Membrane Science, 492 (2015) 1-8.

[33] R.M. de Vos, H. Verweij, Improved performance of silica membranes for gas separation, Journal of Membrane Science, 143 (1998) 37-51.

[34] R.S.A. De Lange, Microporous Sol–Gel Derived Ceramic Membranes for Gas Separation, in, University of Twente, The Netherlands, 1993.

[35] C. Yacou, S. Smart, J.C. Diniz da Costa, Long term performance cobalt oxide silica membrane module for high temperature H2 separation, Energy & Environmental Science, 5 (2012) 5820-5832.

[36] D.L. Martens, D.K. Wang, J. Motuzas, S. Smart, J.C.D. da Costa, Modulation of microporous/mesoporous structures in self-templated cobalt-silica, Scientific Reports, 5 (2015) 7970.

[37] K.-K. Kang, H.-K. Rhee, Synthesis and characterization of novel mesoporous silica with large wormhole-like pores: Use of TBOS as silicon source, Microporous and Mesoporous Materials, 84 (2005) 34-40.

[38] C.J. Brinker, Hydrolysis and condensation of silicates: Effects on structure, Journal of Non-Crystalline Solids, 100 (1988) 31-50.

[39] M. Elma, C. Yacou, J. Diniz da Costa, D. Wang, Performance and Long Term Stability of Mesoporous Silica Membranes for Desalination, Membranes, 3 (2013) 136.

[40] W.J. Elferink, B.N. Nair, R.M. de Vos, K. Keizer, H. Verweij, Sol-Gel Synthesis and Characterization of Microporous Silica Membranes: II. Tailor-Making Porosity, Journal of Colloid and Interface Science, 180 (1996) 127-134.

[41] B.N. Nair, J.W. Elferink, K. Keizer, H. Verweij, Preparation and Structure of Microporous Silica Membranes, J Sol-Gel Sci Technol, 8 (1997) 471-475.

[42] R. Leboda, E. Mendyk, V.A. Tertykh, Effect of medium pH on hydrothermal treatment of silica gels (xerogels) in an autoclave, Materials Chemistry and Physics, 43 (1996) 53-58.

[43] C.J. Brinker, G.W. Scherer, Sol-Gel Science: The Physics and Chemistry of Sol-Gel Processing, Academic Press, Inc, San Diego, 1990.

[44] D. Uhlmann, S. Smart, J.C. Diniz da Costa, H2S stability and separation performance of cobalt oxide silica membranes, Journal of Membrane Science, 380 (2011) 48-54.
[45] A. Bertoluzza, C. Fagnano, M. Antonietta Morelli, V. Gottardi, M. Guglielmi, Raman and infrared spectra on silica gel evolving toward glass, Journal of Non-Crystalline Solids, 48 (1982) 117-128.
[46] A. Duran, C. Serna, V. Fornes, J.M. Fernandez Navarro, Structural considerations about SiO2 glasses prepared by sol-gel, Journal of Non-Crystalline Solids, 82 (1986) 69-77.
[47] S. Araki, Y. Kiyohara, S. Imasaka, S. Tanaka, Y. Miyake, Preparation and pervaporation properties of silica-zirconia membranes, Desalination, 266 (2011) 46-50.
[48] B.B. Mandelbrot, Fractals: form, chance, and dimension, WH Freeman San Francisco, 1977.
[49] C.J. Brinker, A.J. Hurd, K.J. Ward, Fundamentals of Sol-Gel Thin Film Formation, in: J.D. Mackenzie, D.R. Ulrich (Eds.) Ultrastructure Processing of Advanced Ceramics, Wiley, New York, 1988, pp. 223-240.
[50] R. Igi, T. Yoshioka, Y.H. Ikuhara, Y. Iwamoto, T. Tsuru, Characterization of Co-Doped Silica for Improved Hydrothermal Stability and Application to Hydrogen Separation Membranes at High Temperatures, Journal of the American Ceramic Society, 91 (2008) 2975-2981.
[51] T. Tsuru, R. Igi, M. Kanezashi, T. Yoshioka, S. Fujisaki, Y. Iwamoto, Permeation properties of hydrogen and water vapor through porous silica membranes at high temperatures, AIChE Journal, 57 (2011) 618-629.
[52] J.M. Kim, S.M. Chang, S.M. Kong, K.-S. Kim, J. Kim, W.-S. Kim, Control of hydroxyl group content in silica particle synthesized by the sol-precipitation process, Ceramics International, 35 (2009) 1015-1019.
[53] J.C. Diniz da Costa, G.Q. Lu, V. Rudolph, Y.S. Lin, Novel molecular sieve silica (MSS) membranes: characterisation and permeation of single-step and two-step sol–gel membranes, Journal of Membrane Science, 198 (2002) 9-21.
[54] C.J. Brinker, G.C. Frye, A.J. Hurd, C.S. Ashley, Fundamentals of sol-gel dip coating, Thin Solid Films, 201 (1991) 97-108.

Acknowledgments

A. Darmawan, Y. Astuti and Sriatun gratefully acknowledge financial support from Diponegoro University via the Research for International Scientific Publications Award (Number: 315-04/UN7.5.1/PG/2015). Linda Karlina specially thanks for FIM2lab at the University of Queensland for their support for all experimental work.

Distributions and pollution assessment of heavy metals Pb, Cd and Cr in the water system of Kendari Bay, Indonesia

A Armid[1], R Shinjo[2], R Ruslan[3] and Fahmiati[1]

[1] Department of Chemistry, Faculty of Science, Halu Oleo University, Anduonohu Kendari 93232, Indonesia
[2] Department of Physics and Earth Sciences, Faculty of Science, University of the Ryukyus, 1 Senbaru, Nishihara, Okinawa 903-0213, Japan
[3] Department of Mathematics, Faculty of Science, Halu Oleo University, Anduonohu Kendari 93232, Indonesia

alrum.armid@gmail.com

Abstract. The concentrations of heavy metals Pb, Cd and Cr in the coastal waters of Kendari Bay were analyzed to assess their pollution status. Water samples from 32 sampling points were analyzed for dissolved heavy metals concentrations by using inductively coupled plasma mass spectrometry (ICP-MS). The RSD(%) of each metal was accounted to analyze the diversity of the heavy metals among 32 sampling points. The results demonstrate that the dissolved heavy metal Pb had the highest concentrations (0.009 to 0.549 μg/L, average = 0.210 μg/L) followed by Cr (0.085 to 0.386 μg/L, average = 0.149 μg/L), and Cd (0.001 to 0.015 μg/L, average = 0.008 μg/L). Based on the the RSD values (Pb = 87.8%, Cd = 45.2% and Cr = 41.3%), it is suggested that the antropogenic activities controls the high diversity of concentrations for heavy metal Pb relative to those of Cd and Cr. Comparing the data with the mean oceanic concentrations, only the concentrations of Pb exceed the mean oceanic level (210 folds). Therefore, the water system of Kendari Bay is severely polluted with heavy metal Pb. More management and treatment should be introduced to protect the marine environment in the study area, especially from Pb pollution.

1. Introduction

The increase of human populations and industrial activities have considerably contributed to the weakening in water quality, including accumulation of heavy metals in the coastal waters [1-3]. Heavy metals are among the most common environmental pollution and their occurrence in water system indicate the presence of natural or anthropogenic sources, such as geologic weathering, mining practices, industrial activities, urban development, as well as deposition from the atmosphere [4-7]. The behaviors of heavy metals within aquatic systems has been an issue of increasing concern in environmental studies over the past few years because of the metals' ecotoxicity features, persistence, and bioaccumulation as well as bio-magnification in marine systems [8-10].

Kendari Bay is a fringing bay located in Kendari City, in the Province of South East Sulawesi, Indonesia. The bay has attracted increased human activity over the years, and industries such as tourism, transportation, fishing and shipping thrive along the coastal area in the city. The Wanggu River system flow, which is connected directly to the western part of the bay, may become a source for the input of heavy metals to the bay. To date, research into the heavy metal distribution in this location has been scarce. Accordingly, it is quite important to understand the distribution and pollution status of heavy metals in the water environments of Kendari Bay in order to manage the quality of water in estuarine and coastal areas. The aim of the current study is to investigate the distribution of the heavy metals Pb, Cd and Cr in coastal waters of Kendari Bay in order to consider their sources and pollution status.

2. Materials and methods

2.1. Study location

Kendari Bay is a fringing bay located at 3°58′ S and 122°34′ E and has an area of 10.84 km^2 (figure 1). The Wanggu River, the largest river in Kendari, has a mouth at the western part of the bay. There are three ports located at the eastern part of the bay, i.e. Ferry, Perikanan Samudera and Nusantara Ports, which are directly connected to the Sunda Sea. The ports are mainly used for commercial purposes such as ferry service (Ferry Port) and fish landing center (Perikanan Samudera Port). Loading and unloading of industrial containers from big cargo ships also undergoes at Nusantara Port. Coral reefs are found to survive at the Bokori Isle, which is located at the eastern part of the bay. Bokori Isle is used as a tourist destination which is managed by local government.

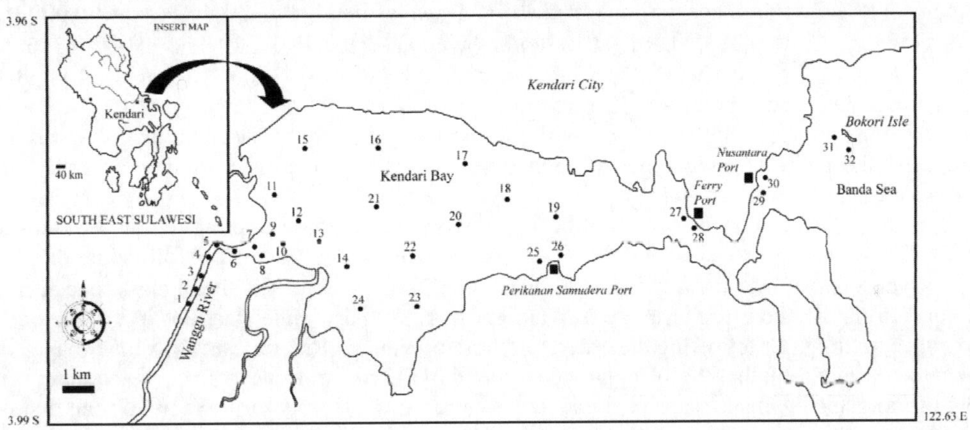

Figure 1. Map showing Kendari Bay, South East Sulawesi, Indonesia. Filled circles indicate the sampling points

2.2. Sampling

All polyethylene vessels used to collect and store water samples were pre-washed by 10% nitric acid, and rinsed with Milli-Q water before they were oven-dried at 40°C. Water samples were collected during October 2014 from 32 sampling points ranging from the lower reaches of the Wanggu River to the offshore region of Kendari Bay (figure 1). The points are clustered into 5 sampling regions, including Wanggu River (1, 2, 3, 4, 5, 6, 7, 8, 9 and 10), estuary (11, 12, 13, 14, 15, 23 and 24), inner bay (16, 17, 18, 19, 20, 21 and 22), ports (25, 26, 27, 28, 29 and 30) and offshore (31 and 32). All samples were filtered through a Millipore HA filter (0.45 μm pore size), added with concentrated nitric acid (Kanto Chemical Co., Inc., Tokyo, Japan) to meet the final solution of 1% nitric acid, and then preserved at 4°C until analysis.

2.3. Analysis

Samples were measured directly without pre-concentration by inductively coupled plasma mass spectrometry (ICP-MS) (Thermo Scientific™ XSERIES 2 ICP-MS, Germany). Each measurement of heavy metals was run in triplicate and the concentrations of metals were quantified by calibration methods using SPEX XSTC-331 as a multi-elemental standard solution. The detection limits for Pb,

Cd and Cr were 0.001, 0.004 and 0.019 μg/L, respectively. The accuracy of the measurement was assessed using NASS-5, the seawater reference material for trace metals from National Research Council (NRC), and yielded 97, 105 and 95% recoveries for Pb, Cd and Cr, respectively (table 1). The relationship between heavy metal distributions and sampling points was statistically interpreted by principal components analysis (PCA) and hierarchical clustering analysis (HCA) using the computer program MINITAB 17.1.0.0 (Minitab Inc., State College, Pennsylvania, USA).

Table 1. Measurement results of NASS-5

Metal	Certified	Measured ±SD (n = 10)	Recovery (%)
Pb (ppb)	0.008	0.0078±0.002	97
Cd (ppb)	0.023	0.0242±0.008	105
Cr (ppb)	0.110	0.1044±0.011	95

3. Results and discussion

3.1. Trends in heavy metal concentrations

The distributions of heavy metals in the coastal waters of Kendari Bay in each sampling point are tabulated in Table 2. The concentrations of dissolved heavy metals Pb, Cd and Cr are 0.009-0.549 μg/L (average = 0.210 μg/L), 0.001-0.015 μg/L (average = 0.008 μg/L) and 0.085-0.386 μg/L (average = 0.149 μg/L), respectively. The relative standard deviation (RSD, %) was used to analyze the diversity of heavy metals among 32 sampling points. The RSD of heavy metal Pb (~88%) was higher than those of Cd (~45%) and Cr (~41%); suggesting that the distribution of Pb was more controlled by the anthropogenic input at each sampling point and yielded higher diversity of Pb concentrations than other two metals.

The fluctuations of mean concentrations of heavy metals based on the sampling region are showed in figure 2. The trends of Pb concentrations associated with sampling region following the order: Wanggu River > inner bay > estuary > ports ≈ offshore. For the case of Cd, the highest concentrations were found in the offshore regions, while Wanggu River, estuary, inner bay and ports regions has similar concentrations. Cr following the order: Offshore > Wanggu River > estuary > inner bay > ports. Taking into consideration that the Wanggu River, the biggest river in Kendari City, as the main source of input of inorganic elements (such as metals) to Kendari Bay [1], it is surprisingly to note that there are significant elevations of heavy metals Cd and Cr in the offshore region beyond to the levels in Wanggu River (figure 2).

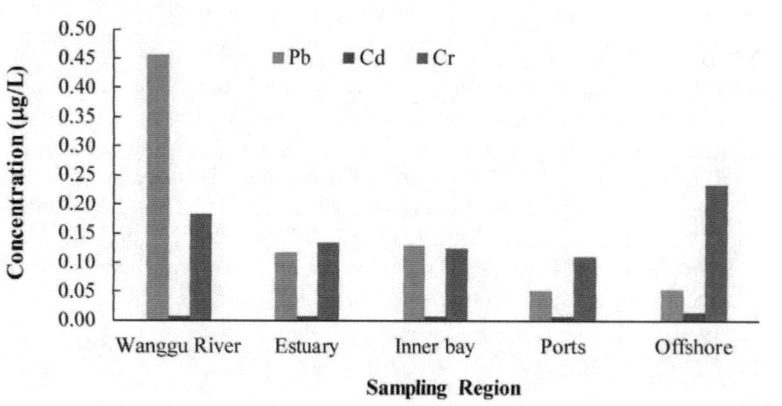

Figure 2. Mean concentrations of heavy metals in each sampling region

As the main entry of inorganic elements, the high accumulation of heavy metals (mainly Pb) in the samples derived from Wanggu River may come from anthropogenic inputs such as untreated sewage predominantly from mining activities, besides agricultural runoff and household waste. It has to be noted that there are recently 150 ore mining companies located in the districts of North Konawe, Konawe and South Konawe which they are in the passageway by the Wanggu River [11]. Accordingly, the waste generated from mining activities can be entered into Wanggu River directly or indirectly, and increase the accumulation of heavy metals in the water system of the lower reaches of the river. Additionally, the highest Pb concentration found in the Wanggu River may originated from the atmospheric loading (such as combustion and road-side dust) which is transported to the river system during runoff and rainfall events [12-14].

Table 2. Heavy metals Pb, Cd and Cr concentrations in the coastal waters of Kendari Bay

Sampling region	Sampling point	Concentration (µg/L)		
		Pb	Cd	Cr
Lower reaches of Wanggu River	1	0.501	0.007	0.157
	2	0.413	0.005	0.133
	3	0.511	0.004	0.187
	4	0.433	0.008	0.135
	5	0.473	0.007	0.386
	6	0.549	0.010	0.190
	7	0.357	0.008	0.158
	8	0.322	0.008	0.143
	9	0.48	0.003	0.135
	10	0.522	0.007	0.196
	Average	0,456	0,007	0,182
Estuary	11	0.215	0.004	0.119
	12	0.104	0.012	0.122
	13	0.021	0.001	0.119
	14	0.054	0.012	0.086
	15	0.218	0.008	0.261
	23	0.087	0.007	0.115
	24	0.117	0.012	0.116
	Average	0,117	0,008	0,134
Inner bay	16	0.186	0.009	0.132
	17	0.275	0.006	0.160
	18	0.200	0.009	0.155
	19	0.025	0.012	0.103
	20	0.054	0.004	0.123
	21	0.113	0.007	0.107
	22	0.052	0.012	0.094
	Average	0,129	0,008	0,125
Ports	25	0.069	0.003	0.136
	26	0.060	0.005	0.095
	27	0.023	0.008	0.122
	28	0.061	0.006	0.104
	29	0.009	0.005	0.085
	30	0.095	0.014	0.122
	Average	0,053	0,007	0,111
Offshore	31	0.042	0.014	0.225
	32	0.068	0.015	0.244
	Average	0,055	0,015	0,235
	Total Average	0.210	0.008	0.149
	Total RSD (%)	87.75	45.24	41.28

Figure 2 further shows that, except for Cd, there is a decrease in heavy metals concentrations significantly by 74.34% (for heavy metal Pb) and 26.37% (for Cr) in the estuarine areas in comparison to the lower reaches of the Wanggu River. Estuary region of Kendari Bay is an area populated mostly by mangroves. A decrease in the concentrations of heavy metals in the estuarine water likely to be caused by the so-called phytoremediation, a process of absorption of dissolved metals from the surrounding environment to the mangrove tissues. The ability of mangroves to perform the phytoremediation process has been widely investigated by earlier studies (e.g. [15-17]). [18] has emphasized specifically the capability of mangrove trees in the Brazilian coastal areas to uptake metals ion like Pb, Cd and Cr in their bodies. Nevertheless, the phytoremediation ability by mangrove is different between one species to another, besides it is also controlled by the concentration of heavy metal exists in the surrounding environment [19]. The very low concentration of Cd distributed in the seawater of the estuary region (0.008 mg/L in average, table 2) is likely to cause mangrove species that grow in that region cannot perform phytoremediation optimally.
Entering the inner bay, the concentrations of heavy metal Pb are increased by ~11% and the Cr concentrations decreased ~7%, while the Cd remains constant relative to the estuary levels. The middle area of Kendari Bay is highly correlated with the household wastes that come from the surrounding areas. The distribution of heavy metals in the inner bay may be affected by the tide and physicochemical recycling generated by the water mass movement in this area, increasing or decreasing the heavy metals concentrations in the aquatic system. Moreover, figure 2 shows that there are declines in heavy metals Pb and Cr concentrations from the inner bay to the ports areas of Kendari Bay. The Pb concentrations are reduced by 59%, while the Cr is reduced by 11%. These data indicate that several anthropogenic activities in the ports, such as transportation, shipping and fishery, do not enhance the accumulation of heavy metal concentrations in the water system of the region. In general, therefore, it seems that the heavy metals from water system were removed to the ports sediments. The heavy metals (including Pb, Cd and Cr) might be active in the reduction-oxidation processes whereby they dominated the soluble fraction under oxygenated conditions and precipitated straightaway when post-oxic conditions were encountered [20].

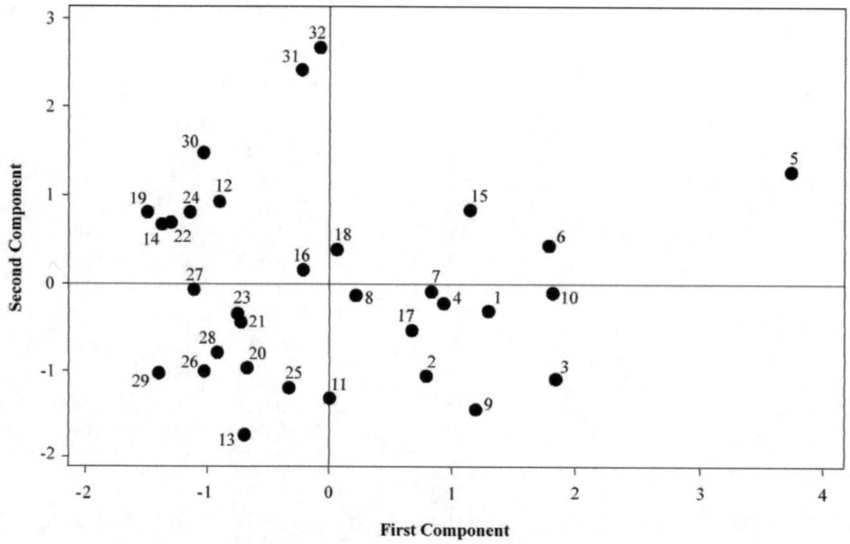

Figure 3. Plots of PC1 and PC2 scores on the distribution of Mn, Fe, Co, Ni, Cu and Zn in 25 sampling points of Kendari Bay

Finally, figure 2 remarkably shows that there are increase concentrations of heavy metals Cd and Cr in the offshores beyond their levels in the lower reaches of Wanggu River. The Cd concentrations are increased by 59%, while the Cr is increased by 11% relative to the Wanggu River samples. It is well thought-out that the offshore region which is closed to Bokori Isle is relatively not affected by the heavy metal drive from the entry point at Wanggu River as the isle is located far away from the lower reaches of the river. In addition, the open water activity and the influence of seawater current movements from the Banda Sea might wash away and diluted the heavy metals present in offshore region. An implication of this is the possibility that the anthropogenic activities around Bokori Isle, such as tourism, transportation and fishing practices, have substantial amounts of wastes containing heavy metals Cd and Cr discharged into coastal waters around Bokori Isle and led to the high accumulation of those metals in the offshores.

Principal Components Analysis (PCA) and Hierarchical Clustering Analysis (HCA) have been performed on the data comprising 32 sampling points and 3 heavy metals (Pb, Cd and Cr). Three principal components have been taken out. The first two principal components PC1 and PC2 are able to describe suitably the distribution of 32 sampling points in Kendari Bay with 85.6% of cumulative variance (figure 3). Figure 3 shows that the sampling points 1, 2, 3, 4, 6, 7, 8, 9 and 10 (Wanggu River), as well as 16, 17 and 18 (Inner bay) have similar characteristics on the distributions of Pb, Cd and Cr, indicated by the location of PC1 and PC2 adjacent to each other. Four sampling points, i.e., 5 (Wanggu River), 15 (estuary), 31 and 32 (offshores), have different characteristics on Pb, Cd and Cr distributions compared to the other 32 sampling sites, as indicated by higher scores on PC1. Similar characteristics regarding to the heavy metals distributions are also displayed among estuary (samples 12, 14 and 24), inner bay (samples 19 and 22), and port (sample 30), which are differentiated by negative scores on the PC1 and positive scores on PC2. Furthermore, the same characteristics on the heavy metals trends are found in the samples 11, 13 and 23 (estuary), 20 and 21 (inner bay), 25, 26, 27, 28 and 29 (ports), where both PC1 and PC2 scores are negative.

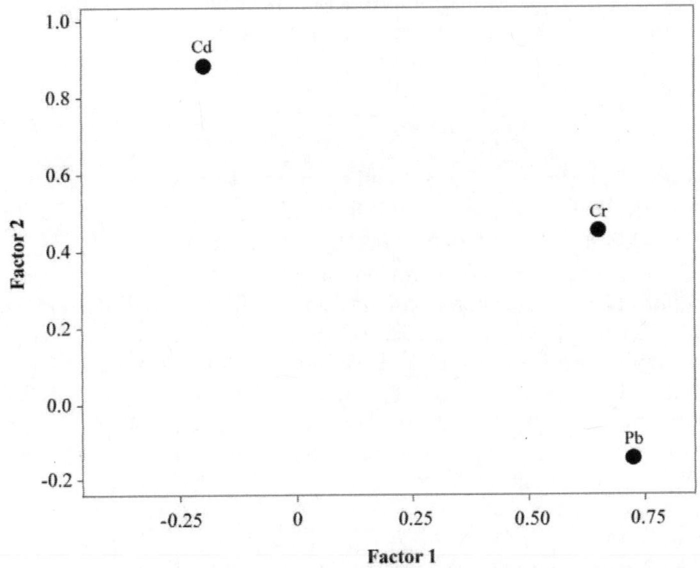

Figure 4. The loads of variables on the two principal components

Loading variables on the two principal components demonstrate that Pb and Cr have similar characteristics in 32 sampling points, specified by adjacent plots (figure 4). Such similarity is supported by a linear correlation coefficient (*r*) 0.454 ($p<0.01$). The result of HCA by using Ward's method and applied to the Euclidean distances indicates 4 clusters formed based on the homogeneity of Pb, Cd and Cr concentrations. The first cluster comprises the sampling points 1, 2, 3, 4, 6, 7, 8, 9, 10, 16, 17 and 18); the second cluster involves the sampling points 5, 15, 31 and 32; the third is 12, 14, 19, 22, 24 and 30; and the fourth cluster consists of the sampling points 11, 13, 20, 21, 23, 25, 26, 27, 28 and 29. Please note that, based on HCA, samples 31 and 32 are located on the similar cluster;

supporting the idea that the anthropogenic activities play an important role in the heavy metals distributions in this region, particularly the heavy metals Cd and Cr, as is mentioned earlier. Both PCA and HCA yield a similar conclusion on the clustering of sampling points distribution in Kendari Bay coastal waters.

3.2. Heavy metals pollution
In order to infer the pollution status, the dissolved heavy metals concentrations of the surface waters in the Kendari Bay were compared with the Seawater Quality Standard of Indonesia [21] as well as the normal composition of seawater [22], as shown in table 3.

Table 3. Heavy metals Pb, Cd and Cr concentrations in Kendari Bay in comparison with the Seawater Quality Standard of Indonesia as well as the normal composition of seawater

Reference	Point	Concentration (μg/L)		
		Pb	Cd	Cr
This study	Kendari Bay	0.210*)	0.008*)	0.149*)
[21]	Seawater quality standard of Indonesia	≤5	≤2	≤2
[22]	Mean oceanic concentration	0.001	0.07	0.33

*) = average concentration from 32 sampling points (Table 2)

Table 3 shows that the average concentrations of heavy metals Pb, Cd and Cr found in Kendari Bay coastal waters did not exceed the seawater quality standard of Indonesia [21]. However, comparing the data with the mean oceanic concentrations [22], the concentration of heavy metal Pb was considerably higher than its mean oceanic level by 210 folds; implying a serious pollution. This research has thrown up in need of further management works in the study area in order to protect marine environment from the pollution of heavy metals, especially Pb.

4. Conclusions
This study investigates the distribution of heavy metals Pb, Cd and Cr in the coastal waters around Kendari Bay. The overall concentrations of Pb, Cd and Cr were 0.009 to 0.549 μg/L (average = 0.210 μg/L), 0.001 to 0.015 μg/L (average = 0.008 μg/L) and 0.085 to 0.386 μg/L (average = 0.149 μg/L), respectively, and followed the Pb > Cr > Cd sequence. Wanggu River, the biggest river in the city, is the main entry of heavy metals (especially Pb) to the water system in Kendari Bay, in addition to the agricultural runoff, household waste and atmospheric loading. Based on the RSD values, this study shown that the antropogenic activities control the high diversity of concentrations for heavy metal Pb relative to those of Cd and Cr. One of the more significant results to emerge from this study is that the concentration of heavy metal Pb in the water system of Kendari Bay is considerably higher by 210 folds than the mean oceanic level. Accordingly, further management works is required to protect marine environment in Kendari Bay from the Pb pollution.

References
[1] Armid A, Shinjo R, Zaeni A, Sani A and Ruslan R 2014 *Mar. Pollut. Bull.* **84** 373
[2] Li S and Zhang Q 2010 *J. Hazard. Mater.* **176** 579
[3] Varol M 2013 *Chemosphere* **93** 954
[4] Mohiuddin K, Ogawa Y, Zakir H M, Otomo K and Shikazono H W 2011 *Int. J. Environ. Sci. Tech.* **8** 723
[5] Naser HA 2013 *Mar. Pollut. Bull.* **72** 6
[6] Pan J, Pan J -F and Wang M 2014 *Mar. Pollut. Bull.* **89** 427
[7] Wang J, Liu R H, Yu P, Tang A K, Xu L Q and Wang J Y 2012 *Proced. Environ. Sci.* **13** 1507
[8] Geng J, Wang Y and Luo H 2015 *Mar. Pollut. Bull.* **101** 914

[9] Wang H, Wang J, Liu R, Yu W and Shen Z 2015 *Mar. Pollut. Bull.* **93** 250
[10] Ismail A, Toriman M E, Juahir H, Zain S M, Habir N L A, Retnam A, Kamaruddin M K A, Umar R and Azid A 2016 *Mar. Pollut. Bull.* **106** 292
[11] CBSSS 2015 *Southeast Sulawesi in Figures 2015* Provincial Office of Central Bureau of Statistics Southeast Sulawesi Indonesia
[12] Schneider B, Ceburnis D, Marks R, Munthe J, Petersen G and Sofiev M 2000 *Mar. Chem.* **71** 297
[13] Wijaya A R, Ouchi A K, Tanaka K, Shinjo R and Ohde S 2012 *J. Geochem. Explor.* **118** 68
[14] Wijaya A R, Ouchi A K, Tanaka K, Cohen M D, Sirirattanachai S, Shinjo R and Ohde S 2013 *J. Geochem. Explor.* **126–127** 45
[15] MacFarlane G F, Koller C E and Blomberg S P 2007 *Chemosphere* **69** 1454
[16] Paz-Alberto A M, Celestino A B and Sigua G C 2004 *J. soils sediments* **14** 251
[17] Weis J S and Weis P 2004 *Environ. Int.* **30** 685
[18] Ramos e Silva C A, da Silva A P and de Oliveira S R 2006 *Mar. Chem.* **99** 2
[19] Pahalawattaarachchi V, Purushothaman C S and Vennila A 2009 *Indian J. Mar. Sci.* **38** 178
[20] Thomson J, Nixon S, Croudace I W, Pedersen T F, Brown L, Cook G T 2001 Earth Planet. Sci. Lett. **184** 53
[21] MNLH 2004 *Decree of the Minister of Environment of the Republic of Indonesia Number 51 of 2004 on Seawater Quality Standard of Indonesia*
[22] Broecker W S and Peng T H 1982 *Tracers in the Sea* Lamont-Doherty Geological Observatory Columbia University Palisades New York USA

Acknowledgments
The authors wish to acknowledge gratefully the financially support provided by the Ministry of Research Technology and Higher Education of Indonesia through the International Research Collaboration and Scientific Publication Project 2016 (Contract number: 043/SP2H/LT/DRPM/III/2016).

Albumen Glue, New Material for Conjunctival Graft Surgery, an Animal Experiment

A Kartiwa[1], R Miraprahesti[1], I Sovani[1], S Enus[1], A Boediono[2]
[1]Ophthalmology Department, Medical Faculty, Padjadjaran University, Bandung, Indonesia
[2]Anatomy, Physiology and Pharmacology Department, Veterinary Faculty, Bogor Agricultural University, Bogor Indonesia

E-mail : anggakartiwa@gmail.com

Abstract
Attach conjunctival graft commonly used are suture technique and fibrin glue. This study was to investigate albumen glue as an alternative to suture technique in attaching conjunctival graft in rabbits. Aim of this study was to compare the conjunctival wound healing between albumen glue and suture technique in rabbit eye as a model.There was an experimental animal study included 32 eyes (16 rabbits) in PT. Bio Farma (Persero) and Histology Laboratory, Faculty of Medicine, Padjadjaran University from March 2014 to July 2104. The study consisted of albumen glue group and suture technique group. The examination included the comparison of conjunctival graft attachment and histologic examination by microscopically was done to obtain the wound gap, then analyze by Mann-Whitney test.The results indicated that the graft attachment was significantly better-using albumen glue (grade 4) compared to suture (grade 2-3) on day-1 after surgery (p=0,000). The wound gap was smaller using albumen glue (0-0,33 µm versus 5,33-14 µm ; p=0,0005) on 10 minutes after surgery and 0 µm versus 0,33-4 µm ; p=0,0005 on day-7 after surgery.In conclusion, the graft attachment using albumen glue was better and the wound gap was smaller using albumen glue than suture technique.
Keyword : albumen glue, conjunctival wound healing, suture

1. Introduction

Conjunctival graft has been used in the field ophthalmology. In pterygium surgery, this technique was first promoted by Kenyon et al. in 1985 [1,2]. Pterygium is one of eye problem in Indonesia, as this country is located in the equatorial and has tropical climates. Morbidity of pterygium in Indonesia ranks as second (13.9%) after cataract. Total of pterygium patients visit in Cicendo Eye Hospital in Bandung during 2011 was 2158, and the number of bulbar conjunctival graft using sutures technique was 358 eyes [3]. Bulbar conjunctival graft technique may decrease pterygium grows after the operation, but may cause dissatisfaction for operators and patients due to its sutures [4]. In addition to sutures, methods used today for attaching conjunctival graft can also be using fibrin glue [5].

Bulbar conjunctival graft attachment using a suture technique is micro-surgically complex, requiring a high level of operator learning, long operation time and wound healing, and also can cause complications including retraction tissue grafts, granuloma formation, infection, and death of tissue grafts. Additionally, using a needle, cord, and suture knot will damage the tissue graft, which can cause poor adhesion tissue grafts. Another disadvantage of this technique, it will cause more tissue manipulation so the graft tissue adhesion and healing will be time-consuming [6]. Albumen is a clear liquid contained in an egg [7]. Albumen can repair burns injury due to its natural collagen form. Collagen is the main component of extracellular matrix (ECM) that supports the growth of cells and tissues. Besides the ability to support, ECM strongly affects embryonic development, cell migration flows, signaling, and supply growth factor [4,8].

The main protein consisting in egg albumen is including ovalbumin, ovotransferrin, ovomucoid, ovomucin, and lysozyme. Ovalbumin is the highest content in egg albumen, which is 54% and acts as a cell adhesion substrate that serves to support the attachment and cell adhesion [9]. Tissue transglutaminase contained on the surface of conjunctival cells, play a role in cell adhesion processes and can interact with the ECM. It also plays a role in wound healing [10]. Transglutaminase bond with the matrix protein, such as egg white can enhance cell adhesion.

2. Materials and Methods

This research is an experimental animal study to determine the degree of rabbit conjunctival graft tissue attachment to the essential tissue (sclera) and comparison slit wounds (wound gap) between the albumen glue and suture technique. The inclusion criteria New Zealand White rabbits breeding Bio Farma age 2-3 months, weight 2-2.5 kg, healthy physical (including health both eyes), duck egg albumen healthy local cultures derived from eggs are not older than 3 days. Exclusion criteria were to have two eyes with congenital anatomical abnormalities seen in general, only one eye is healthy, weight loss occurs when quarantined.

The research was conducted after the approval of the Commission and Use of Laboratory Animal Welfare (Institutional Animal Care and Use Committee) PT. Bio Farma (Persero). Place of research at PT Bio Farma Bandung and Histology Section of Faculty of Medicine, University of Padjadjaran, the period from March to July 2014. The subject of the study 16 rabbits (32 eyes) appropriate inclusion criteria and quarantined for one week more, then divided randomly into two groups: group I : albumen glue technique then suture technique; Group II: suture technique then albumen glue technique. The operation includes the manufacture of conjunctival graft, graft attachment in an open sclera same eye using albumen glue were made sterile suture technique using 8-0 vycril thread three pieces. The degree of adhesion conjunctival graft between albumen glue and suture technique assessed on day-1 after surgery by a veterinarian who has received training from an ophthalmologist. After one week, the other eye treated the same as the previous eye. Ten minutes the surgery, the ball left, and right eye of each rabbit was extracted in the anesthetized state. Histological preparations were created by 0.5-1 cm conjunctival tissue of the eye with hematoxylin eosin staining and viewed with a light microscope using a magnification of 100x by an expert doctor histology. The assessment is based on the size of the wound gap the rabbit conjunctival graft between albumen glue and suture technique on 10 minutes after and on day-7 after surgery.

Statistical test using the Mann-Whitney test with consideration of the small sample size (n = 8).

3. Results and Discussion

The result assessed by comparing the grade of bulbar conjunctiva attachment on day-1 after surgery and wound gap at ten minutes and on day-7 after surgery between albumen glue and suture technique.

3.1 Bulbar Conjunctiva Attachment Grading in Rabbit Eyes

We put the result on conjunctival graft tissue attachment to base tissue (sclera) on the criteria that consists of grade 1, 2, 3, and 4, which conducted on day-1 after surgery.

Table 1. Analysis of Bulbar Conjunctival Graft Attachment Grade in Rabbit Eye Between Albumen Glue Technique and Suture Technique on Day-1 After Surgery

Group	Number of eyes and attachment grade				Mann-Whitney Test	
	1	2	3	4	Z	p
Albumen glue technique (8 eyes)	0	0	0	8	-3,771	0,000
Suture Techniques (8 eyes)	0	1	7	0		

Description: p: the value of significance
Grade 1 of attachment : percentage of attachment ≤25%
Grade 2 of attachment : percentage of attachment> 25% - 50%
Grade 3 of attachment : percentage of attachment> 50% - 75%
Grade 4 of attachment : percentage of attachment> 75%

Table 1 shows the significant differences in the grade of attachment of the bulbar conjunctival graft on day-1 after surgery between albumen technique (grade 4) and suture technique (the value range between grade 2 and 3) with a p-value = 0.000.

3.2 Wound gap result
Histologically, we compared wound gap on the use of albumen glue technique and suture technique by assessing the microscopic picture and measured gap between the donor and recipient tissue.

Table 2. Comparative Analysis of Wound Gap between Albumen Glue Technique and Suture Technique at 10 minutes After Surgery (µm)

Rabbit	Average Albumen (n=8 eyes)	Suture (n=8 eyes)	P
1	0	11,67	0,0005*
2	0	5,33	
3	0	14,00	
4	0	9,67	
5	0	12,33	
6	0,33	12,33	
7	0	11,00	
8	-	10,67	

Description: p: the value of significance (Mann-Whitney)
Sign (-) indicates that wound gap of albumen glue cannot measured, then dropped from the analysis

Table 2 shows there is a significant difference between the average size of the wound gap in albumen glue technique (0 to 0.33 m) and suture technique (5,33-14µm) at 10 minutes after surgery with p-value = 0.0005.

Table 3. Comparative Analysis of Wound Gap between Albumen Glue Technique and Suture Techniques on day-7 After Surgery (µm)

Rabbit	Average Albumen (n=8 eyes)	Suture (n=8 eyes)	P
1	0	4,00	0,0005*
2	0	0,67	
3	0	2,67	
4	0	0,33	
5	0	-	
6	0	-	
7	0	-	
8	0	-	

Description: p: the value of significance (Mann-Whitney)
Sign (-) indicates that wound gap of albumen glue cannot measured, then dropped from the analysis

Table 3 shows there is a significant difference between the average size of the wound gap between albumen glue technique (0 m) and suture technique (0,33-4μm) on day-7 after surgery with a p-value = 0.0005.

Table 4. Analysis of Wound Gap Difference between Luka Gap Between Albumen Glue Technique and Suture Technique at 10 minutes and on day-7 After Surgery

	Wound Gap 10 minutes after surgery	**Wound gap On day-7 after surgery**	**p**	**Delta**
Albumen glue technique	0,05 ± 0,13 μm	0,00 ± 0,00 μm	0,1425	0,05 μm
Suture technique	10,88 ± 2,59 μm	1,92 ± 1,73 μm	0,003	8,96 μm
P	0,0005	0,0005		
Delta	-10,83 μm	-1,92 μm		

Description: p: the value of significance (Mann-Whitney)

Table 4 shows that wound gap in albumen glue technique on day-7 after surgery was not significantly smaller than the 10 minutes after surgery, with p-value = 0.1425. The average difference (delta) wound gap between 10 minutes and on day-7 after surgery for albumen glue technique was 0.05-0.00 = 0.05 μm. While in suture technique, wound gap on day-7 after surgery was less than on 10 minutes after surgery. The average difference (delta) between 10 minutes and on day-7 after surgery for suture technique was 10.88 to 1.92 = 8.96 μm.

3.3 Hypothesis testing

Hypothesis 1: The grade of conjunctival graft attachment in the rabbit with albumen glue technique is better than suture technique. Testing : based on Mann-Whitney test, grade of bulbar conjunctival graft attachment on day-1 after surgery in albumen glue technique is significantly better than suture technique (Table 1). Mann-Whitney test results obtained statistical value of U = 0.000 and Z = -3.771 with a chance of error or p-value = 0.000. It appears that (Z = -3.771) < (Z tabel = -1.645) or (p-value = 0.000) < (α = 0.05): significant. Z table values obtained from normal Z table, in error rate of 5% of 1-side type test. This finding shows that using albumen glue technique to transplant the bulbar conjunctival graft proven to be proficient in producing better attachment grade than suture technique on day-1 after surgery.

Conclusion: Based on the above test, the first hypothesis is accepted (tested).

Hypothesis 2: Wound gap of conjunctival graft in the rabbit with albumen glue technique is smaller than suture technique.

Testing:

a. Results of Mann-Whitney U statistical test = 0.000 and Z = -3.350 with a chance of error or p-value = 0.0005. It appears that (Z = -3.350) < (-Z table = -1.645) or (p-value = 0.001) <(α = 0,05): significant. Z table values obtained from normal Z table, in error rate of 5% of 1-side type test. This finding shows that albumen glue technique in bulbar conjunctival graft was proven to have 10-minute postoperative wound gap smaller than suture techniques.
b. Results of Mann-Whitney U statistical test = 0.000 and Z = -3.233 with a chance of error or p-value = 0.0005. It appears that (Z = -3.233) < (-Z table = -1.645) or (p-value = 0.0005) <(α = 0,05): significant. Z table values obtained from normal Z table, in error rate of 5% of 1-side type test. This finding shows that albumen glue technique in bulbar conjunctival graft was proven to have wound gap smaller than suture techniques on day-7 after surgery.

   Conclusion: based on test results above, the second hypothesis is accepted (tested).

3.4 Tissue Attachment

The attainment of the bulbar conjunctival graft is tempted by the attachment of graft tissue to the base tissue, so vascularization occurs optimally and tissue grafts can survive and be healthy. Treatment of bulbar conjunctival graft using suture technique should be carefully done because the tissue graft is very thin and easily torn by the withdrawal of sutures, and the knot will damage the conjunctival graft of donor and recipient. Thus, the more manipulation conducted in this technique, will lead to failure of tissue graft attachment and slow healing process. In contrast to the use of albumen glue as an alternative to suture technique, bulbar conjunctival will stick to base tissue adequately because ovalbumin, which is the most numerous protein in albumen acts as a substrate adhesion functioning in support cell attachment [9].

In this study, albumen was taken from the duck egg. The duck egg is known to have more albumen than chicken eggs by 60%. Based on these explanations, the researchers decided to use duck egg albumen. Duck eggs were used in this study were less than three days of age, as according to Q Huang et al. study, stated that ovalbumin would gradually turn into S-ovalbumin, which has a lower viscosity than ovalbumin and may affect adhesion process. The formation of S-ovalbumin affected by pH, temperature, duration or life-length of eggs. S-ovalbumin increased to 91.24% at the age of 12 days, with a 37°C of temperature [11].

In this study, the grade of attachment of rabbit conjunctival graft with albumen glue technique obtained to be better than suture on day-1 after surgery, p-value = 0.000. There were no similar studies reported, but in medicine research, there have been several studies reported to using ovalbumin of albumen glue tissues such as research conducted by Gabrielle et al. that used ovalbumin to reconstructed bone tissue. It was reported that ovalbumin could improve bone healing because of this protein biopolymer, which consisting of 380 amino acid, is equivalent to 10% amino acid content in human serum albumin. Gabrielle et al. study showed that ovalbumin as an adhesion tissue substrate that affects cell growth, proliferation, and differentiation of preosteoblast cells for bone tissue regeneration [12]. Kaipparettu et al. showed a growth of mammary epithelial cells in humans and mice using albumen as a three-dimensional cell culture media. It described that albumen has a function as an ECM that support growth and cell development. Sergey et al. reported that there were transglutaminase and ECM protein bond that play a role in cell adhesion. Transglutaminase is a soluble enzyme in cells and plays a role in cell adhesion. Transglutaminase activity increases when injuries occur, so we can say that the transglutaminase and albumen bond as ECM proteins increases cell adhesion and will encircle wounds that help to protect cells from further damage so that wound healing can occur [9,10].

In this study, tissue attachment on albumen glue technique is denser than suture technique, so conjunctival tissue donor was possible to adequately attached to the base tissue without tissue retraction.

In tissue graft, cell and surrounding tissue injury will occur, the surgical trauma that occurs will depend on the performed technique and will result in tissue healing. Acceleration of tissue healing is strongly induced by the quality of the tissue attachment. The phase of wound healing on the first day that stands out is the inflammatory phase. Injury due to surgical incision in the conjunctiva is the extracellular matrix, cells, and blood vessels. This will lead to the formation of excess cytokines, causing irregularities in wound healing [4,14]. Chronic wounds will form necrotic tissue or contained foreign material that cannot be phagocytosis during acute inflammation. Lymphocytes, monocytes, and macrophages will stay on the sites of inflammation. The macrophages attract fibroblasts and in a long time will produce amount of collagen, form mass slowly from the fibrous tissue called granuloma [15]. Suture technique is considered as additional trauma to the conjunctival graft due to needle cut and is also considered as trauma for the cell, ECM, and blood vessels while the cord is a foreign body and knot can cause irritation and inflammation on the tissue. It can also occur granuloma formation and prevent settlement of tissue grafts and slow healing process [4,6]. Angiogenesis, growth, and proliferation of new blood vessels is one of the special marks on the histological tissue remodeling of inflammation which caused by allergies. Clinically, there are symptoms such as itching, redness, photophobia, and watery [16]. However, this study did not assess the symptoms mentioned above.

3.5 Wound Gap

The healing process of the conjunctiva is faster than healing process of the skin. Kenyon et al. observed that conjunctival graft has been attached to the sclera and got vascularization within 3-5 days postoperatively [17]. Conjunctival epithelial is permeable epithelium because its close attachment between epithelial cells and goblet cells, so the protein that is administered topically will seep easily into the lumen of the conjunctiva capillaries [18]. In this study, 10 minutes and seven days postoperatively wound gap for albumen glue technique were obtained, and it was significantly smaller than the suture technique.

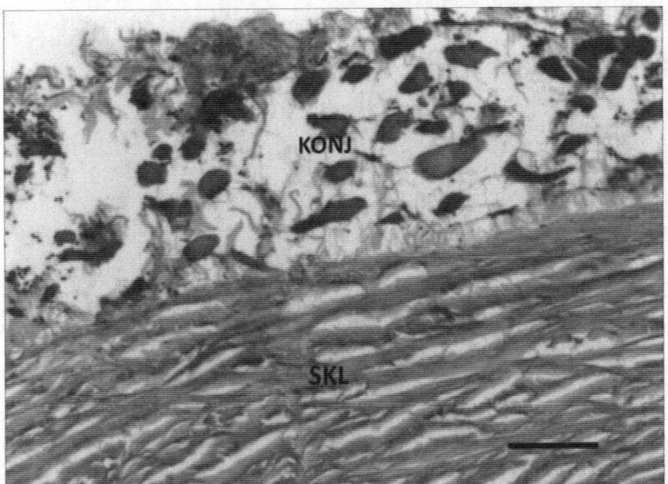

Figure 1. Histological figure of wound gap 10 minutes after surgery in albumen glue technique
Description: Konj = conjunctiva; skl = sclera; Bar = 10 μm

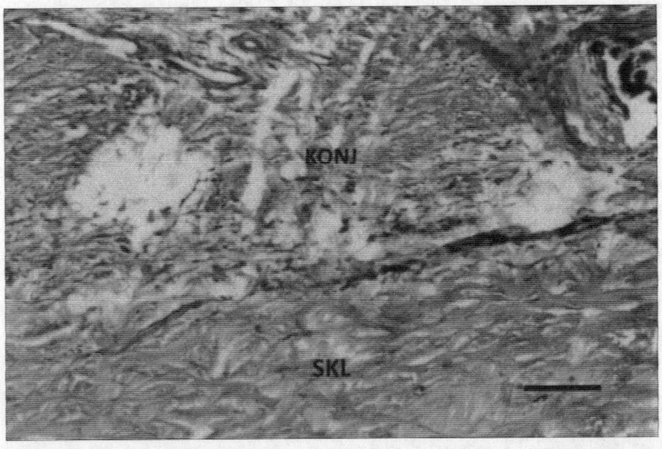

Figure 2. Histological figure of wound gap 10 minutes after surgery in suture technique

Description: Konj = conjunctiva; skl = sclera; Bar = 10 μm

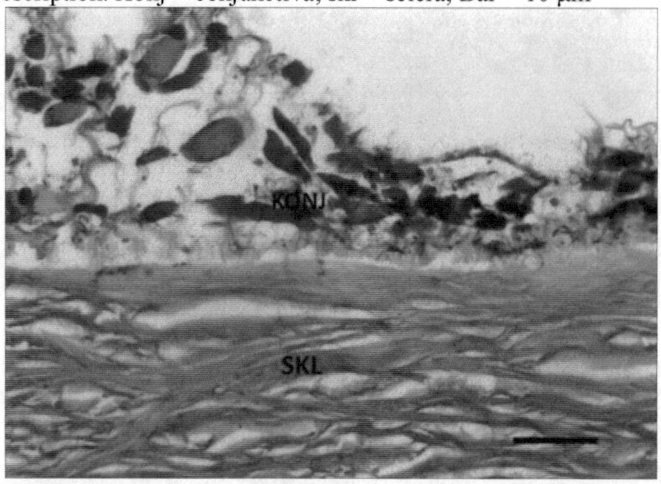

Figure 3. Histological figure of wound gap on day-7 after surgery in albumen glue technique
Description: Konj = conjunctiva; skl = sclera; Bar = 10 μm

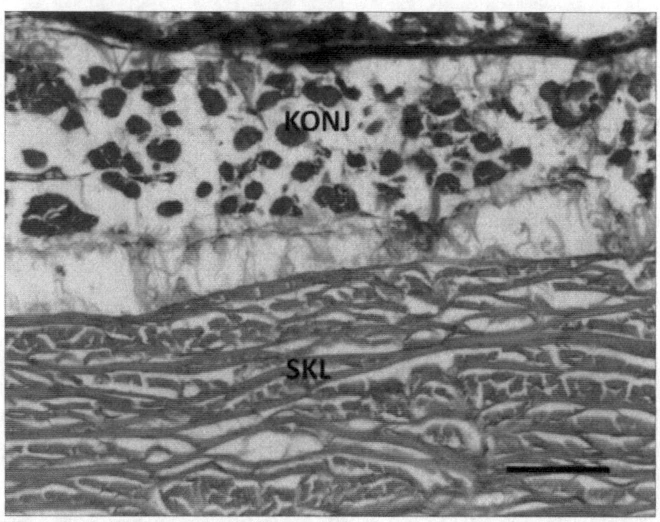

Figure 4. Histological figure of wound gap on day-7 after surgery in suture technique
Description: Kon = conjunctiva; skl = sclera; Bar = 10 μm

In this study, there were six albumen glue technique eye preparations that did not have wound gap on 10 minutes after surgery observation. Very small gap wound was observed at 10 minutes after surgery in one eye preparations of albumen glue technique, probably was caused by an irregular applicant of albumen glue or air contaminated, so the graft did not attached well. While one preparation cannot be assessed, possibly due to the thick albumen applied to the conjunctival graft had changes in protein viscosity due to temperature changes. Albumen is a gel or a viscous-elastic matrix, and it has thermolabile characteristic, so it can be transformed into a dilute when the temperature changes [18].

The wider size of the wound gap in suture technique was possibly due to less exposure of both tissues. Both tissues did not attach well maybe caused by manipulation that slower graft tissue adhesion and healing process and cause a gap between the tissues [4]. Healing process includes four phases : hemostasis, inflammation, proliferation, and remodeling. In the early stages of wound healing, vessel is having vasoconstriction about 10-15 minutes, then going vasodilation and increased capillary permeability, so plasma proteins and neutrophils go out into the wound area in 24 hours followed by lymphocytes and macrophages. Macrophages produce growth factors and new ECM, accelerate the process of angiogenesis that is critical to the further healing process. The process of epithelialization, granulation and angiogenesis is an important process in the proliferation stage. Granulation tissue formation phase is characterized by the activity of fibroblasts and new blood vessel formation. Fibroblasts are mostly active at the wound edges at 3-5 days post-traumatic and around the end of the first week. Fibroblasts are the main cells in the the wound. Granulation tissue is needed to fill wound gap during inflammatory process and continues until the base wound is healed, and it reaches maximum state on the fifth day of the wound. When there is a large wound gap, then more granulation tissue are needed, and more infiltration of inflammatory cells there and granulation tissue formation process will take longer [14]. Gilang et al. presented the histological figure that showed the greater grade of epithelialization, thinner granulation tissue, larger angiogenesis in fibrin glue technique rather than suture technique on first week of postoperative observation. This finding is related to the function of granulation tissue injury, which is known as a gap filler. In smaller wound gap of autologous fibrin glue technique, then the granulation tissue is thinner, while suture technique wound gap that larger will have more granulation tissue [16]. This condition can also be seen in wound healing between albumen glue and suture technique, which smaller wound gap showed in albumen glue technique compared to suture technique. In this study only observed the histological wound gap between albumen glue and suture techniques, not observing the thickness of granulation tissue.

In remodeling phase, there is balance change of extracellular matrix composition. Imbalances can be happened due to excessive collagen production by fibroblasts that can be happened because of mechanical pressure and leads to excessive delays of wound healing or tissue fibrosis or even keloid scarring [14]. Based on the theory, it can be said that the excessive formation of collagen by fibroblasts due to suture will delay wound healing or excessive fibrotic tissue formation, so that wound gap in suture technique is larger than albumen glue technique. This theory is in accordance with this study results in Table 4 that showed the average difference (delta) between 10 minutes after surgery wound gap and on day-7 after surgery albumen glue technique was 0.05 μm, while for suture technique was 8.96 μm. It shows that the wound gap in albumen glue technique is smaller than suture technique. In suture technique, complete epithelialization may not fully occur in the first week of observation because wound gap in suture technique that cause unification tissue process and vascularization did not successfully complete, so the graft was in a loose attachment.

4. Conclusion

There was a conjunctival graft tissue attachment grade in albumen glue, which was better than suture technique. Wound gap between the sclera graft conjunctiva and the layer beneath was smaller in albumen glue compared to suture technique.

References:

[1] Huerva V, March A and Alonso MM 2012 Pterygium surgery by means of conjunctival autograft : long term folow-up *Arq Bras Ophthalmol* **75**(4) 251-5

[2] Rashid H, Rahman F and Khan SB 2010 Results of conjunctival auto-transplant in pterygium surgery *Gomal Journal of Medical Sciences* **8**(1) 82-4

[3] Miraprahesti NR 2013 *Karakteristik Tumbuh Ulang Pterigium Pasca Eksisi di Pusat Mata Nasional, Rumah Sakit Mata Cicendo tahun 2010-2011*(Bandung: Universitas Padjadjaran)

[4] Enus S 2009 *Peran Lem Fibrin Otologus Terhadap Penempelan Tandur Konjungtiva Bulbi Melalui Ekspresi Gen Fibronektin dan Integrin pada Mata Kelinci* (Bandung: Universitas Padjadjaran)

IOP Conf. Series: Materials Science and Engineering **172** (2017) 012003 doi:10.1088/1757-899X/172/1/012003

[5] Panda A, Kumar S and Kumar A 2010 Fibrin glue in ophthalmology *Indian Journal of Ophthalmology* **58**(2) 371-9

[6] Enus S, Dalimoenthe NZ, Kartiwa A and Lessi N 2012 Aplikasi lem fibrin pada cangkok konjungtiva bulbi setelah eksisi bedah pterygium *Majalah Kedokteran Bandung* **44** 31-5

[7] Narahari D 2009 In praise of the humble egg *Poultry International* **9** 24-6

[8] Kaipparettu BA, Kuiatse I, Tak Yee Chan B, Lee AV and Osterreich S 2008 Novel egg white based 3D cell culture system *Bio Techniques* **45** 165-171.

[9] Kim JY, Park SE, Song KY and Park SC 1998 Glutathione transferase (class π) expression in pterigia *Korean J Ophthalmol* **12** 6-13

[10] Akimov SS, Krylov D, Fleischman LF and Belkin AM 2000 Tissue transglutaminase is an integrin-binding adhesion coreceptor for fibronectin *The Journal of Cell Biology* **148** 825-38

[11] Huang Q, Qiu N and Ma MH 2012 Estimation of egg freshness using s-ovalbumin as an indicator *Poultry Science* **91** 739-43

[12] Farrar G, Barone J and Morgan A 2010 Ovalbumin based porous scaffolds for bone tissue regeneration *Journal of Tissue Engineering* **61** 1-6

[13] Reddy GAK, Priyanka B, Saranya and Kumar 2012 Wound healing potential of Indian medicinal plants *International Journal of Pharmacy Review and Research* **2** 75-8

[14] Li J, Chen J and Kirsner R 2007 Pathophysiology of acute wound healing *Clinics in Dermatology* **25** 9-18

[15] El-Asrar A, Al-Mansouri S, Tabbara KF and Missotten L 2006 Immunopathogenesis of conjunctival remodelling in vernal keratoconjunctivitis *Eye* **20** 71-9

[16] Mutiara G 2007 *Perbandingan Gambaran Histologis Penyembuhan Luka Antara Teknik Lem Fibrin Otologus dan Jahitan pada Cangkok Konjungtiva Kelinci* (Bandung: Universitas Padjadjaran)

[17] Norzalehawati S, Ropilah AR, Ruszyma BHI, Norzana AG, Chua KH, Faridah HA, and J Che-Hamzah J 2013 In vitro cell expansion and growth kinetics of human bulbar, fornix, and palpebral conjunctival epithelium *Regenerative Research 2* **1** 14-22

[18] Omana DA, Liang Y, Kav NN and Wu J 2011 Proteamic analysis of egg white proteins during storage *Proteomic* **11** 144-153

Operculum bone carp (*cyprinus carprio sp.*) scaffold is a new potential xenograft material*: a preliminary study*

A Kartiwa[1], B Abbas[2], P Pandansari[2], A Prahasta[1], M Nandini[3], M Fadhlillah[4], T Subroto[4], R Panigoro[5]

[1]Ophthalmology Department, Medical Faculty, Padjadjaran University, Bandung, Indonesia
[2]National Nuclear Energy Agency of Indonesia, Jakarta, 40000, Indonesia
[3]Oral Surgical Department, Dentistry Faculty, Padjadjaran University, Bandung, Indonesia
[4]Chemistry Department, Mathematics and Science Faculty, Padjadjaran University, Bandung, Indonesia
[5]Biochemistry Department, Medical Faculty, Padjadjaran University, Bandung, Indonesia

E-mail: anggakartiwa@gmail.com

Abstract. Orbital floor fracture with extensive bone loss, would cause herniation of the orbital tissue into the maxillary sinus. Graft implantation should be done on the orbital fracture with extensive bone loss. Different types of grafts have their own characteristics and advantages. Xenograft has been widely studied for use in bone defects. This study was to investigate *cyprinus carprio sp.* opercula bone as a potential xenograft. The aim of this study was to investigate based on EDS chemical analysis using a ZAF Standardless Method of Quantitative Analysis (Oxide) and SEM examination conducted in the laboratory of Mathematics, Institute of Technology Bandung. Particularly the mass ratio of Ca and P (5.8 /3:47), the result is 1.67. This is equivalent to the stoichiometric Hydroxyapatite (HA) (Aoki H, 1991, Science and medical applications of hydroxyapatite, Tokyo: Institute for Medical and Engineering, Tokyo Medical and Dental University). C N O that there is an element of protein/amino acid collagen compound, serves as a matrix together with HA. As shown in the SEM analysis that the matrix is a porous sheet-shaped (oval) that interconnect with each other, which is good scaffold. The pore is composed of large pores >200 microns and smaller pores between the large pores with a size smaller or equal to 10 microns that can serve for the attachment of osteoblast cell. In conclusion, Opercula bone carp (*cyprinus carprio sp.*) scaffold could be a new potential xenograft material.

1. Introduction

Orbital floor fracture with extensive bone loss, would cause herniation of the orbital tissue into the maxillary sinus. Graft implantation should be done on the orbital floor fracture with extensive bone loss are aimed as a filling on fracture defects that were spacious, promoting the process of osteogenesis, providing tissue bridge on the process of bone remodeling and improve quality of life by eliminating complaints of double vision [1,2].

Selection of an ideal bone graft depends on several factors such as tissue viability, size of the defect, biomechanical and biological characteristics, and is associated with complications that will occur.

Bone grafts are expected to stimulate bone healing, through the process of osteogenesis, osteoinduction, osteoconduction, osteovascularization, osteoincorporation and osteomineralization [3,4]. Different types of grafts have their own characteristics and advantages. Autograft is preferred by most plastic and craniofacial surgeon, but the tissue sampling procedure is associated with increased morbidity such as bleeding, infection and pain in the area of donor tissue grafts. Allograft associated with the transmission of hepatitis and AIDS. Xenograft associated with the transmission of disease from animals, if not processed properly. Alloplastic such as silicone, polyethylene, hydroxyapatite, and plate metal alloy is an option but such materials are expensive [3,4]. Xenograft is associated with the transmission of disease from animals if not processed properly. There are no standard regulations on the sterility of the bone graft substitutes, but many techniques can be used for the sterilization of bone graft material [5]. Xenograft grafts have been widely studied for use in bone defects due to abundant, it can be processed at an affordable cost, and have quite physicochemical properties as a graft [6]. Now xenograft materials commonly used in bone grafts with excellent results. Various xenograft materials have different chemical properties, morphology and particle size but all are osteoconductive and manageable to fill bone defect [7].

Graft derived from goldfish has economic value and easy to be found in Indonesia. Operculum bones carp used as grafts are plat of bone that is thin and slightly curved in the medio-lateral direction. In Indonesia, Goldfish's operculum grafts used as xenograft graft material has not been studied yet to cover basic orbital bone defect. Some studies using xenograft as a bone substitute alternative materials, such as Yadao, 2004, used the shells of ostrich eggs as an alternative material to cover the basic bone defect of the orbit [8]. Costa, 2005, compared the use of the scleral graft taken from cow with silicon sheets to cover the basic orbital bone defect in rabbit [9]. Okumus 2005, once used spinal squid as xenograft material to close the bone defect [1]. Ozel, 2015, used a heterologous cortical bone as an option for the reconstruction of bone defects in orbital floor fracture [10]. This study is an initial study aimed to compare the morphology of bone operculum carp with cow bone xenografts that could potentially be used as xenografts in orbital floor bone defect.

2. Materials and Methods

Xenograft taken from the operculum carp *cyprinus carprio sp.* with a weight of 500 mg, which has been in process in NNEAI (BATAN) Research Tissue Bank through two ways: deproteinization, then lyophilized and finally sterilized by means of radiation at a dose of 25 kGy [11]. Entire stage of the production process is validated to get quality products [12].

Morphological examination is carried out using Scanning Electron Microscopy (SEM) SM-6510A / JSM-6510LA (JEOL Ltd., Tokyo, Japan) at the operational voltage 15 kV and a magnification of 100x and 300x. JEOL SEM is equipped with Energy Dispersive X-Ray Spectroscopy (EDS) for checking the composition of the sample. Xenograft taken from cow is used as a comparison to goldfish's bone operculum in this study.

Calcium level measurement is done by atomic absorption of spectrophotometer after destructing sample in 10 mL of 6.5% nitric acid and then diluted 10 times. Phosphorus content measurement is done with UV-Vis spectrophotometer at a wavelength of 400 nm after destructing sample in 10 mL of concentrated sulfuric acid and then diluted 10 times.

EDS chemical analysis using a ZAF Standardless Method of Quantitative Analysis (Oxide), and SEM examination was conducted in the laboratory of Mathematics And Science, Institute of Technology Bandung.

IOP Conf. Series: Materials Science and Engineering **172** (2017) 012004 doi:10.1088/1757-899X/172/1/012004

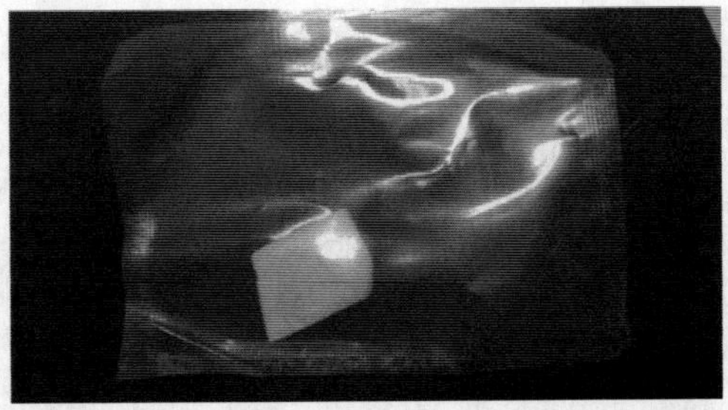

Figure 1. Goldfish's operculum bone that has been processed at BATAN

3. Results and Discussion

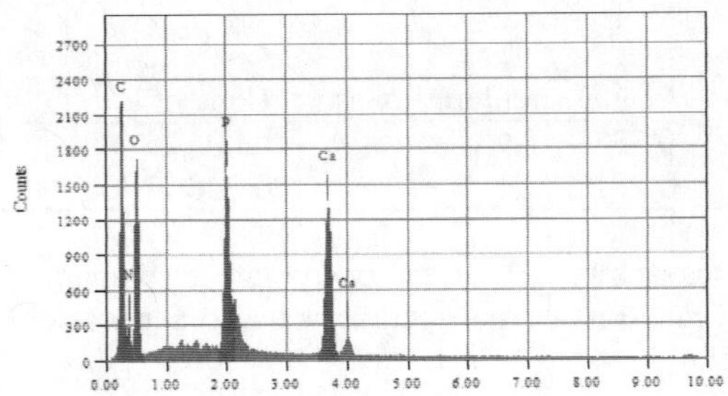

Acquisition Parameter
Instrument : 6510(LA)
Acc. Voltage : 15.0 kV
Probe Current: 1.00000 nA
PHA mode : T3
Real Time : 51.12 sec
Live Time : 50.00 sec
Dead Time : 2 %
Counting Rate: 2910 cps
Energy Range : 0 - 20 keV

ZAF Method Standardless Quantitative Analysis

Fitting Coefficient : 0.4787

Element	(keV)	Mass%	Error%	Atom%	Compound	Mass%	Cation	K
C K	0.277	27.74	0.13	33.92				26.3989
N K	0.392	34.36	0.67	36.03				47.3582
O K	0.525	28.62	0.39	26.20				9.7143
P K	2.013	3.47	0.10	1.64				5.9625
Ca K	3.690	5.81	0.16	2.13				10.5662
Total		100.00		100.00				

Graphic 1. Chemical analysis using a ZAF Standardless Method of Quantitative Analysis (Oxide)

SEM analysis shows the surface morphology of the bone with a porous structure (Figure below). The pore size depends on the nature of the source of the bone and xenograft processing stages. Three hundred folds magnification imaging showed the size of the pore diameter at both the bone is estimated at 5-25μm. EDS examination results showed womb-like composition such as C, O, N, Ca and P as in general the bone.

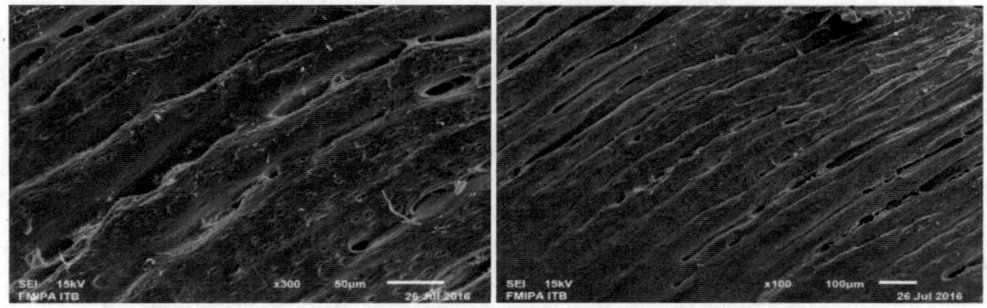

Figure 2. SEM examination with 100x and 300x magnification

Table 1. Comparison of calcium and phosphor component of fish operculum and bone of cow

No.	Composition	Fish Operculum Bones	Bone of cow
1	Calcium	0,555%	0,008%
2	Phosphor	0,171%	0,028%

The composition of calcium (Ca) and phosphorus (P) on the operculum carp appear higher than cow bone xenograft. More high content of calcium and phosphorus are also supported by the results of measurements with a spectrophotometer.

Along with the age, the incidence of orbital floor trauma, become more frequent, some of them with bone loss. In the National Eye Centre, Cicendo Eye Hospital Bandung during 2014, there was an increasing number of cases compared to the previous year. At the Singapore National Eye Centre which is a tertiary referral hospital in Asia, recorded 63 patients with a fracture of the base of the orbit with the additional complaint of the eye that has undergone therapy between November 1992 to March 2005 [2]. Clauser et al reported on 80 patients (52 cases with orbitozygomatico bone fracture and 28 cases with post-traumatic delayed enophthalmic) has done therapy between January 1998 and January 2005 [3]. He D et al, 2012, reported 64 delayed enoftalmic patients with orbitozygomatico bone fracture who had undergone terapi [12]. Ramnes et al, 2014, reported 18 enophthalmic patients with orbito-zigomatic bone fracture and enophthalmos (14%) is the most common complication [1].

Scaffold is biomaterial made in mass with a medical purpose to repair or replace tissue or organ function. The tissue is expected to form porous and can absorb (like a sponge) that can be used in tissue regeneration and therapeutic gen [13,14].

In tissue regeneration, the scaffold has several important roles including: as a pattern for the migration of cells to cover the defect of tissue; acts as a matrix for adhesion of endogenous exogenous cells; as a carrier of exogenous cells, growth factors and genes; maintain the shape of the tissue defect and to prevent the distortion of surrounding tissues; and prevent the infiltration of surrounding tissue that will prevent further healing process [13,15,18].

Damaged bone tissue can be repaired with autograft or allograft bone tissue. Tissues that can be transplanted are including allograft cortical, cancellous, cortical-cancellous, and osteoarticular. The healing process in the transplanted bone tissue includes two important aspects: the unification of the 'host-graft' junctions and incorporation of graft [14].

Goldfish or carp is a freshwater fish that has important economic value and can be found in Indonesia. In Indonesia, the carp began to be maintained around the 1920s. Carp that can be found in Indonesia are carp brought from China, Europe, Taiwan and Japan. Besides "goldfish *Punten*" and "carp

Majalaya" are selections results in Indonesia. Until now there are 10 carps which can be identified based on the morphology characteristics [19,20].

Operculum bones carp is a plate of bone that is thin and slightly curved in the medio-lateral direction. Having three different tissue zones, observed under a light microscope is reflected both in the coronal portion (operculum divided into rostral and tail section) or a cross-section (split into backs and ventral). The lateral (outside) and medial (inside) thirds of the thickness of the bones show a clear flattened organization [21,22].

Fishbone consists of the cellular and acellular parts. The difference of the two types of bone is the presence or absence of osteocytes and osteoclasts. Cellular bone has the same structure as in mammals, with the absence of osteon, so it is suspected that the acellular bone is actually has osteocytes which then die in progress. Osteoblasts grow on the part of the mineralized but never caught between osteocyte [23,24]. Operculum of goldfish and xenograft of cow's bone showed fibrous material similar to collagen and tiny particles such as hydroxyapatite. The condition is caused by composition of bone material consists of collagen, organic molecules, and calcium hydroxyapatite [5]. EDS examination results showed peak levels of calcium and phosphorus operculum fish were higher than cow bone (Figure on the results). The data was also supported by the results of the measurement of calcium levels and phosphor of each bone calculated a spectrophotometer (Table on the results). Calcium phosphate salts, in particular hydroxyl apatite, is a component of bone that supports the properties of biocompatibility and osteoconduction. Ions of calcium and phosphorus from xenograft important in supporting osteogenesis. Moreover, the pores are also important in repairing bone deficiencies and help osteoconduction due to bone surfaces become wider in order to facilitate the penetration of tissue repair and new bone formation [6].

Calcium level taken from a variety of fish bones including bones of catfish are 187 + 21 mg / g, bluegill 217 + 19 mg / g, goldfish 177 + 6 mg / g, walleye 206 + 15 mg / g, and tilapia 218 + 16 mg / g, compared to calcium levels of mouse's bone 182 + 19 mg /g.22,23,24 chemical analysis using a ZAF, mass ratio of Ca / P (5.81 / 3:47) the result is 1.67 not different according to research conducted by Mustafa et al, that had Ca / P ratio of 1.669.[25] There was also consist of C, N, and O which are the building blocks of the amino acids along with hydroxyapatite can be matrix constituent biomaterial. Hydroxyapatite is an important material found in human bones and teeth along with calcium and phosphorus. This material will be more stable if the ratio Ca/P approaching 1.67 and bioactivity would be greater [28].

As shown in the SEM analysis, the porous matrix had a shape of sheet (with the oval-shaped pores) that was interconnected to one another. These characteristics is a good material for the Scaffold. There were large pores (> 200 microns) and a small pores (the size of approximately less than or equal to 10 microns) which are among the large pore. Pores are useful for attachment osteoblast [29].

Osteocalcin (Gla bone protein) is an extracellular matrix protein synthesized by osteoblasts cell which is a marker of bone and play a role in the process of bone mineralization and calcium ion hemostasis process. Additional osteocalcin serves to form a bond with the hydroxyapatite mineral, amino acid, also a secondary structure. Hydroxyapatite bond is reinforced by calcium [29,30,31,32].

In carp, osteocalcin is an abundant component, whereof more than 35% of the total extracted protein, consisting of 45 amino acid polypeptide, which is highly homologous to the levels of osteocalcin other teleost fish types. Comparison of the levels of osteocalcin and phosphorus (P) in the bone carp, in the dorsal 2.385: 10.15 µg/mg, spine 2.197: 7.71 µg/mg, rib 1.835: 7.49 µg/mg, bone operculum 1.504: 5.81 µg/mg, and scales 0.0528: 0.302 µg/mg [30,31,32,33].

This publication is a preliminary one, and still requires histological examination, tissue responses between the host and the donor as well as further molecular biological examination, so that the bone operculum carp can be used as an alternative xenograft.

4. Conclusion

Operculum of bone carp (*cyprinus carpio sp.*) is one of potential alternative xenograft material can be used for repairing defect in the orbital floor.

References:

[1] Ramnes G et al 2014 A retrospective audit of hundred patients of orbitozygomatic fractures with brain injury *J Clin Diagn Res* **8**(7) NC04-06

[2] Almousa R et al 2010 *Blowout Fractures : Surgical Outcome in Relation to Age, Time of Intervention, and Other Preoperative Risk Factors* Craniomaxillofacial Trauma & Reconstruction **3** 131- 36

[3] Clauser L, Galie M, Paliaro F and Tieghi R 2008 Posttraumatic enophthalmos: etiology, principles of reconstruction, and correction *J Craniofac Surg* **121**(3) 948-55

[4] Costa et al 2005 Bovine sclerae graft for orbital floor fracture in rabbits *Arq. Otorrinolaringol* **9**(3) 181-9

[5] Greenspan DC 2012 *Physical and Chemical Properties of Commercially Available Mineralized Bone Allograft* (Zimmer Dental Inc)

[6] Accorsi-Mendoça T, Conz MB, Barros TC, Ágata de Sena L, Soares GA and Granjeiro JM 2008 Physicochemical characterization of two deproteinized bovine xenografts *Braz Oral Res* **22**(1) 5-10

[7] Baldini M, DeSanctis M and Ferrari M 2011 Deproteinized bovine bone in periodontal and implant surgery *Dent Mater* **27** 61-70

[8] Cantor LB, Rapuano CJ and Cioffi GA 2015 *Orbit, Eyelids, and Lacrimal System* Basic clinical science course Section 7 (American Academy of Ophthalmology)

[9] Wong TY, Tielsch JM 1999 A Population based study on the incidence of severe ocular trauma in Singapore *American Journal of Ophthalmology* **128**(3) 345-51

[10] Rey C, Combes C, Drouet C, and Glimcher MJ Bone Mineral: Update on chemical composition and structure *Osteopros Int* **20**(6) 1013-21

[11] Sfeir C et al 2005 *Bone Regeneration and Repair: Biology and Clinical Applications* (Totowa, NJ : Human Press Inc)

[12] He D et al 2012 Orbitozygomatic fractures with enophthalmos: analysis of 64 cases treated late *J oral Maxillofac Surg* **70**(3) 562-76

[13] Spector M 2008 *Basic Principles of Scaffolds in Tissue Engineering in Issue Engineering Applications in Oral and Maxillofacial Surgery and Periodontics* (China: Quintessence Publishing and Co)

[14] Nather A, Han YY 2010 *Biology of Healing of Bone Allografts in Allograft Procurement, Processing, and Transplantation: A Comprehensive Guide for Tissue Banks* (Singapore: Stallion Press)

[15] Nagase H, Visse R, Murphy G 2006 Structure and function of matrix metalloproteinase and TIMPs *Cardiovascular Research* **69** 562-73

[16] Nather A, Lee CW 2005 *Preparation of Tissue Grafts During Transplantation in Bone Tissue Engineering* (Florida: CRC Press)

[17] Lorenzo J, Horowitz M, Choi Y 2008 Osteoimmunology: interactions of the bone and immune system *Endocrine Reviews* **29**(4) 403-40

[18] Parkin JL. Stevens MD, Stringham JC 1987 Absorsable gelatin film verses silicone rubber sheeting in orbital fracture treatment *The laryngoscope* **97** (1) 1-3

[19] Pearce AI, Richards RG, Milz S, Schneider E and Pearce SG 2007 Animal models for implant biomaterial research in bone *European Cells and Materials* **13** 1-10

[20] Freyhof J, Kotellat M 2008 *Cyprinus Carprio IUCN Red List of Threatened Species International Union for Conservation of Nature*

[21] Qu Q, Zhu M, Wang W Scales and dermal skeletal histology of an early bony fish *psarolepsis romeri* and their bearing on the evolution of rhombic scales and hard tissues *PLoS ONE* **8**(4)

[22] McCormick SD 1994 *Opercular Membranes and Skin Hiochachka and Mommsen* (eds) Biochemistry and Molecular Biology of Fishes 3

[23] Weiss RE, Waatabe N 1979 Studies on biology of the fish bone *Calcif Tissue Int* **28** 43-56

[24] Horton JM, Summers AP The materials properties of acelullar bone in a teleost fish *The Journal of Experimental Biology* **212** 1413-20

[25] Zhang J, Duan R, Ye C and Konno K 2010 Isolation and characterization of collagens from scale of silver carp (*hypophthalmichyths molitrix*) *Journal of Food Biochemistry* **34** 1343-54

[26] Suzuki Y, Nakamura R, Ueda T 1972 Accumulation of strontium and calcium in freshwater fishes in Japan *J.Radiat.Res* **13** 199-207

[27] Hoang QQ 2003 Crystal structure and hydroxyapatite binding of porcine osteocalcin Thesis to the School of Graduate Studies of McMaster University for the Degree of Doctor of Philosophy

[28] Mustafa N, Ibrahim MHI, Asmawi R and Amin AM 2015 Hydroxyapatite extracted from waste fish bones and scales via calcination method *Applied Mechanics and Materials* **773-774** 287-90

[29] Venkantesan J, Kim SW 2010 Effect of Temperature on isolation and characterization of hydroxyapatite from tuna (*thunnus obessus*) Bone *Materials* **3** 4761-772

[30] Nagai T, Suzuki N 2000 Isolation of Collagen from Fish Waste Materials-Skin, Bone, and Fins *Food Chemistry* **68** 277-81

[31] Da Silva RSG, Bandeira SF, Petry FC and Pinto LA Effect of bone particle size of carp (*cyprinus carpio*) heads in gelatin extraction *Articl Vetor, Rio Grande* **21** (2) 34-43

[32] Goren et al 2004 Production of hydroxyapatite from animal bone *Key engineering materials* **264-268** 1949-52

[33] Nishimoto SK, Waite JH, Nishimoto M and Kriwacki RW 2003 Structure, activity, and distribution of fish osteocalcin *The Journal of Biological Chemistry* **278**(14) 11843-8

IOP Conf. Series: Materials Science and Engineering **172** (2017) 012005 doi:10.1088/1757-899X/172/1/012005

Photo-electro-catalytic performance of highly ordered nitrogen doped TiO_2 nanotubes array photoanode

Anthoni B Aritonang[1,2], Hedi Surahman[1], Yuni K Krisnandi[1] and Jarnuzi Gunlazuardi[1*]

[1]Departement of Chemistry, Faculty of Mathematics and Science, University of Indonesia Depok 16424, Indonesia

[2] Departement of Chemistry, Faculty of Mathematics and Science, University of Tanjungpura Pontianak78381, Indonesia

*E-mail address : jarnuzi@ui.ac.id

Abstract. Highly ordered nitrogen doped TiO_2 nanotube arrays (N-TNTAs) were synthesized via a one step anodization method at 40 V for 1 hour, in the electrolyte containing ammonium fluoride (NH_4F), water and triethylamine solution, followed calcination under N_2 atmosphere at 450°C for 3 h. The obtained samples were characterized by means of FE-SEM image showed that the N-TNTs are in a highly ordered array, having inner diameters, wall thickness, tube length of 65 nm, 30 nm and 900 nm, respectively. The X-ray diffraction (XRD) patterns of N-TNTAs and undoped TiO_2 nanotubes arrays (TNTAs) are identical consists of anatase phase, which suggests that the doping of N does not affect the crystalline structure. X-ray photoelectron spectroscopy (XPS), revealed that N atom was incorporated into the lattice of a TiO_2 nanotube array film. The infrared spectra, showed a new peak at 1240 cm^{-1} may indicate the incorporation of N into the lattice of TiO_2 through substituting O atoms, in the form of ~N-Ti-O~. A red shift of the absorption edge toward the visible region of N-TNTAs are observed by diffuse reflectance spectroscopy (DRS), which is corresponding to a band gap of 2.8 eV. The photo-electro-catalytic (PEC) degradation efficiency toward methylene blue solution under visible light illumination of the N-TNTAs electrode was 89%, in which the rate constant of N-TNTAs electrode was 8 times better compared to that the undoped TNTAs photo-electrode.

1. Introduction

Titanium dioxide (TiO_2) based electrode have been great attention to be promising in photo-electrochemical applications to degrade various organic pollutants, such as dyes, aromatic compound and pesticides, under UV light irradiation [1,2]. Using the TiO_2 photo-electrode have many advantages, chemical stability, high efficiency, non toxicity and low cost [3]. In contrast to nanostructure, highly ordered TiO_2nanotube arrays (TNTAs) not only have large surface to volume ratios and adsorptive capacity, but also have good photocatalytic activity, because of its nanotubes array structure, which enhances the electron transfer, promotes ion diffusion in the TNTAs electrode and electrolyte interface [4]. Upon the TNTAs surface irradiated with UV light, the electrons are excited from valence band to the conduction band, generating the photoelectron-hole. The positive hole can react with the water molecule produce the hydroxyl radicals. The positive hole and radical hydroxyl are powerful oxidants for degradation the organic compounds adsorbed onto the TNTAs electrode surface. However, one of the disadvantages of TiO_2 is wide band (3.0-3.2 eV) can absorb only the UV light, which limits the application of TNTAs electrode in organic pollutant degradation to utilize visible light. Another problem the application of TiO_2 photocatalyst is low efficiency of photocatalytic degradation, because high recombination rate of the photoelectron and hole during the photocatalytic process [5].

Many researchers have been attempt to extend the absorption range of TiO_2 to visible light region by non metal doping such as N, B and S into TiO_2 lattice [6-9]. Among the non metal, N doping is one most promising to increase the efficiency for solar energy utilization.

Some studies shown that N doped TiO_2 would increase its visible light photo-response. Asahi et al, [6] prepared N-doped TiO_2 with good photocatalytic activity under visible light irradiation. The N dopants incorporated into TiO_2 lattice by annealing under NH_3 or N_2 atmosphere. Substitution of O by N atoms into to TiO_2 lattice is more effective due to the comparable atom radii and electronegativity [10]. In this case, the oxygen vacancies can be formed by mixing N (2p) and O (2p) atomic orbitals which decrease the band gap energy of TiO_2 [11]. In recent years, various methods were developed to preparation of the N-doped TiO_2 nanotubes arrays (N-TNTAs) film, such as sputtering, hydrothermal, chemical vapour deposition and electrochemical oxidation. The common methods were electrochemical oxidation, which is anodizing of titanium foil in glycerol electrolyte containing NH_4Cl solution, followed calcination can be used to achieve TiO_2 nanotubes arrays [12]. The electrochemical oxidation is a simple method and convenient approach to achieve a uniform of TiO_2 nanotubes array film on titanium substrate.

Recently, the TiO_2 nanotube array has been applied as electrode in photo-electrochemical system for degradation of organic pollutants [13].The photo-electro-catalytic degradation efficiency could be further improved by applying a bias potential across the photo-electrode. The bias potential applied to photo-anode would enhance photoelectron transfer from photo-anode to counter electrode results in the improved separation of photoelectron and hole pairs [14].

In the present work, a simple method with one-step anodization was used to synthesize the N-TNTAs electrode from titanium foils using ethylene glycol electrolyte and NH_4F containing of triethylamine ($C_6H_{15}N$) followed calcination process under N_2 atmosphere at 450°C for 3 h. We have studied the effects of N doping on the properties of the TNTAs, including crystal structure and optical properties. The effects of N doping on the photo-electrochemical properties are investigated. Preliminary photocatalytic activity is examined using photo-electro-catalytic (PEC) method for degradation of methylene blue solution under visible light illumination.

2. Experiment

2.1. Synthesis of TNTAs and N-TNTAs

Titanium (Ti) foils (purity 99.6%, 0.3 mm thickness) was purchased from Baoji Jinsheng Metal Material Co. Prior to anodization, the Ti foils (1.5 cm x 3.0 cm), were first mechanically polished with sand papers 1200 grits and then degreased by sonication in acetone, ethanol, deionized water for 10 min, respectively and finally dried in nitrogen stream. The anodization was carried out in a two-electrode configuration, where Ti foil was used as an anode and stainless steel foil as a cathode. The distance between the anode and the cathode was 1.5 cm. The electrolyte consisted of NH_4F (0.3 wt%) in ethylene glycol and deionized water (2.0 wt%) and placed in a quarz semi-silinder reactor [15-16]. The anodization potential was controlled at 40 V for 1 h using a DC power supply [16]. After anodization process, the obtained sample rinsed with deionized water and dried in nitrogen stream. For preparation of N-doped TiO_2 (N-TNTAs), similar process was performed in the electrolyte consisting of triethylamine 1.5 mol.L^{-1}. The crystallization of the N-TNTAs was carried out by thermal annealing in a furnace at 450°C for 3 h with a heating rate of 5.0°C min^{-1}, held for 3 h under N_2 atmosphere [17] and cooled naturally to room temperature. This simple method comprises simultaneously N doping and fabrication of TiO_2 nanotubes.

2.2. Characterization of the TNTAs and N-TNTAs electrode

The morphology of the N-TNTAs was characterized by field emission scanning electron microscope (FESEM-FEI-Inspect F50) with accelerating voltage of 20 kV. The crystal structures of the N-TNTAs and TNTAs were analyzed by X-ray diffraction (XRD-PHILIPS PW 1710) using Cu Kα as radiation source in the range 10-80°(2θ). The elemental composition and chemical states were examined by X-ray photoelectron spectroscopy (XPS, K-alpha Thermo Sciencetific) using a monochromatic Al Kα radiation source (15 mA, 14 kV).The functional groups present in the N-TNTAs and TNTAs were determined by FTIR analysis (Shimadzu IR Prastige-21) over a wavenumber range of 400-4000 cm^{-1}.The UV-visible diffuse reflectance spectroscopy (DRS) analysis was used to calculate the band gap energy of the photocatalysts using spectrophotometer (Shimadzu 2450) with software UV Probe (DRS-8000 Shimadzu).

2.3. Photocurrent and Photo-electro-catalytic activity measurements

The photoelectrochemical properties were performed in a semi-cylindrical quartz reactor (3.0 cm x 4.0 cm) using a three-electrode with a N-TNTAs or TNTAs film (with effective area 1.5 cm x 1.5 cm) as the working electrode, an Ag/AgCl electrode as the reference, and a platinum foil as the counter electrode in 0.1 mol L^{-1} KOH as the electrolyte. The photo-anode was exposed to the visible light through a window. A 50 W Hollolite lamp (Philip, China) was used as the light source (dominant wavelength 425 nm, contain UV light a portion of ~10%), which provided an irradiation intensity of 3.7 $mWcm^{-2}$ was used as the visible light source. The light meters were used (Krisbow KW 06-288) was used to measure the intensity of the incident visible light. A scanning potentiostat (eDAQ, version 9.1) was used to perform a potentiodynamic scan form -1.0 to 1.0 V at a rate of 10 mV S^{-1} and measure the generated photocurrent. The photocurrent density was also measured at a fixed bias potential 0.5 V, with a light pulse of 1 min 40 S, using multi pulse amperometry method under visible light irradiation.

The photo-eletro-catalytic (PEC) activities of N-TNTAs or TNTAs as photo-anode, were evaluated by analyzing PEC degradation of methylene blue (MB) in aqueous solution as a model organic pollutants under following condition : vigorous stirring, 0.1 M KOH solution as electrolyte and pH 5.5.The MB solution with an initial concentration of 10 mg L^{-1} and the reaction solution was 30 mL. Prior to PEC degradation, the electrode was soaked in MB solution with a concentration of 10 mg L^{-1} in the dark for 30 min to establish the adsorption-desorption equilibrium. During the PEC reaction, the concentration of MB was determined at a time interval of every 10 min by measuring the absorbance using a UV-visible spectrophotometer (Shimadzu 2450) at wavelength 655 nm. The reaction solution (~2.0 mL) was quickly withdrawn at a given reaction intervals, and was quickly returned to the reactor after being analyzed. The Photocatalytic (PC) degradation of MB was examined also using the same of the functional electrodes without applying an external bias potential.

3. Results and discussion

3.1. Characterization of the electrode

The FESEM images of N-TNTAs shown in Fig.1. As shown in Fig.1, the morphology of the N-TNTAs are open at the top, with inner diameter and wall thickness average 65 nm, and 30 nm, respectively.

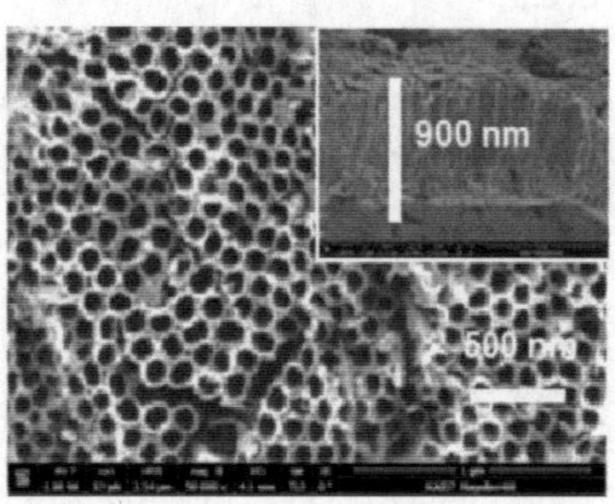

Figure 1. FESEM top view image and cross section (inset picture) of N-TNTAs

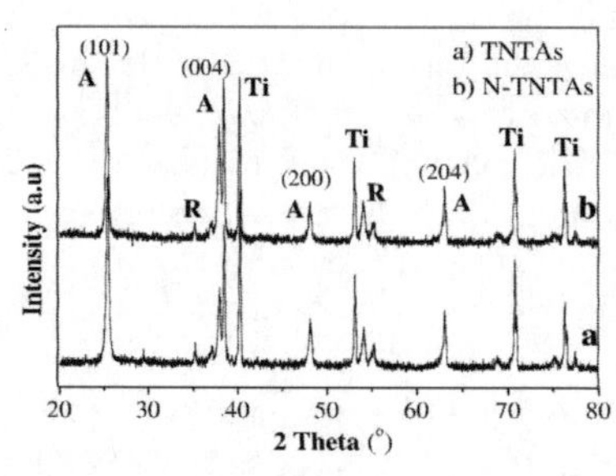

Figure 2. XRD patterns of the (a) TNTAs and (b) N-TNTAs

The cross-sectional image of the N-TNTAs sample shown the well-aligned nanotube arrays are vertical to the substrate, with height are approximated to 900 nm. There are clearly observed that the morphology of the N-TNTs on surface are homogenously have highly ordered tubes shape.

The XRD patterns of TNTAs and N-TNTAs are compared in Fig.2. The crystal structures of TNTAs and N-TNTAs after annealing at 450°C for 3 h gave peaks the diffraction of the field (101), (004), (200), (204), which correspond predominantly to anatase TiO_2 (JCPDS card No.21-1272). It can be observed that the XRD patterns of the both samples are identical, which may be implied that the N incorporated to TiO_2 lattice does not significantly affect the TNTAs crystalline structure. The Ti peaks from substrate characteristic are observed for the TNTAs and N-TNTAs, which suggests that synthesis of nanotubes TiO_2 on the surface of Ti substrate was successfully.

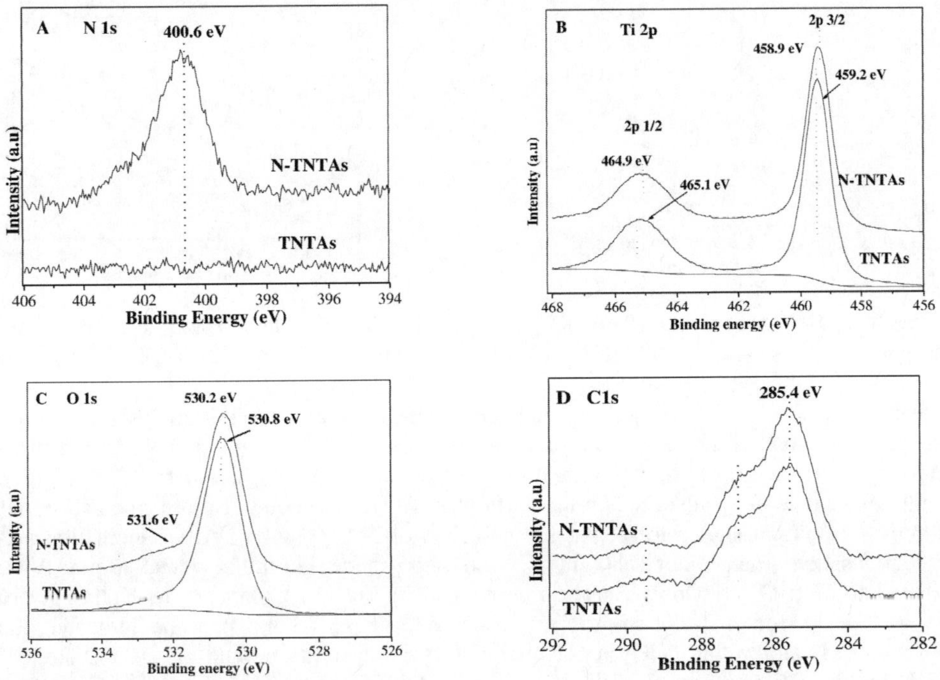

Figure 3. High-resolution XPS spectra of N 1s (A), Ti 2p (B), O 1s (C), and C 1s (D) of the TNTAs and N-TNTAs electrode

The XPS spectra of N 1s, Ti 2p, and O 1s TNTAs and N-TNTAs after annealed at 450°C under N_2 atmosphere are presented in Fig. 3. Figure 3.A. shows that N 1s peak of N-TNTAs observed at 400.6 eV, whereas no appearing distinct peak at the same binding energy for TNTAs. The N 1s peak localized around at 406 eV could be assigned to indicate the presence of N incorporated in TiO_2 lattice formed N-Ti-O linkages, which is consistent with many researchers [7,18,19]. The amount of N atomic introduced to the TiO_2 nanotubes arrays approximate 1.58 at.%. The electronegativity of N doped into TiO_2 lattice is lower than O element, leading to reducing of electron density on the N [18]. This indicates that annealing process indeed leads to substitution of N atoms for O sites in TiO_2 lattice. It shows that oxygen vacancies would be formed in TiO_2 surface and the impurity energy level was intruded between band gap of TiO_2. The Ti 2p spectra are presented in Fig.3.B. Compares the Ti $2p_{3/2}$ peak of N-TNTAs and TNTAs it was found that downshift to approximately 0.3 eV.

The red shift binding energy may be implied a higher electron density [20] and might be attributed to the presence of Ti^{3+} centre in N-TNTAs [21]. This further testifies that N atomic is incorporated into the lattice and substitutes for oxygen [18]. Figure 3.C. shows O 1s spectra of

TNTAs and N-TNTAs photocatalysts. The peak O 1s for N-TNTAs appears around 530.2 eV and 531.6 eV, indicate the presence of two states of oxygen, formed Ti-O-Ti and N-Ti-O linkages, respectively [18]. According to the N 1s region of N-TNTAs (Fig.3.A), suggests that the peak at 400.6 eV, should be attributed to the reduced nitrogen of N-Ti–O. Also significant changes upon N incorporation to TiO_2 lattice, the O1s peak at 530.8 comes from Ti-O-Ti linkages in TNTAs and a shift to 530.2 eV which is observed for N-TNTAs. These results suggest that the N atomic were successfully introduced to TiO_2 lattice. Figure 3.D. shows the C 1s spectra for TNTAs and N-TNTAs. The C atomic may be incorporated to TiO_2 lattice as residual from electrolyte during anodization process, which consistent with the published value of the C 1s peak for N-doped TiO_2 [7,22].

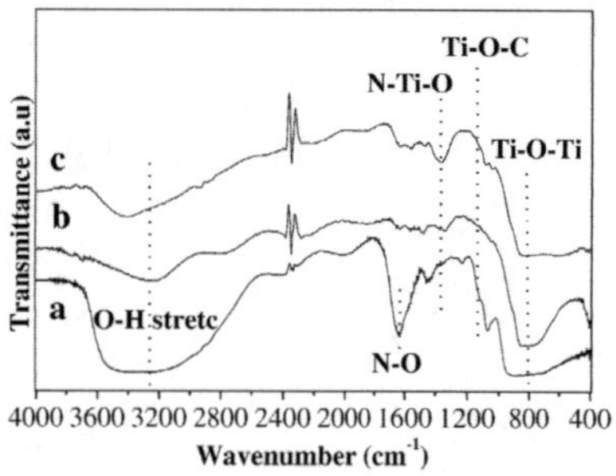

Figure 4. FTIR spectra of N-TNTAs-amorphous (a), TNTAs (b), and N-TNTAs (c)

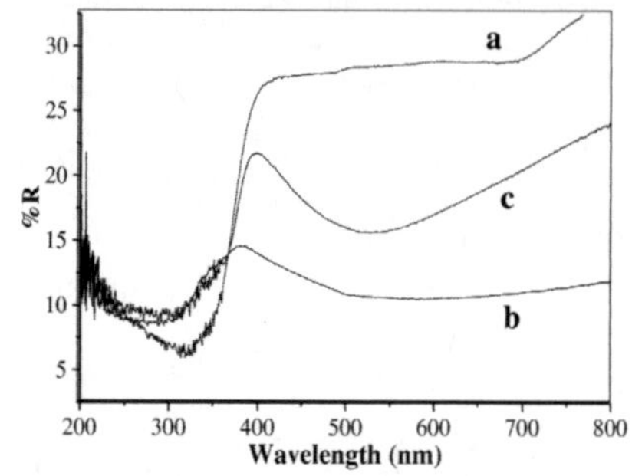

Figure 5. DRS UV-Visble reflectance spectra of TiO_2 P-25 (a), TNTAs (b), and N-TNTAs (c)

To verify the incorporation of N and C elements in the TiO_2 lattice, FTIR analysis was performed. The functional groups that present in the obtained samples of TNTAs and N-TNTAs are studied by characteristic FTIR spectra as shown in Fig.4. During the anodization process, the $C_6H_{15}N$ molecules as the nitrogen source were adsorbed on the surface of TiO_2 amorphous. Before calcination the peak at 1580 cm^{-1} could be assigned to the surface adsorbed of N-O (Fig.4.a). After calcination at 450°C under N_2 atmosphere, the peak at 1580 cm^{-1} was not appearance, but the new peak around 1460 cm^{-1} is attributed to N-Ti-O stretching vibration, as shown in Fig.4.c. Compared to FTIR spectrum of TNTAs, the peak around 1460 cm^{-1} is not observed. These results may be indicated that the incorporation of N atomic into TiO_2 lattice, cause shift the absorption band of O-Ti-O stretching to low wavenumber. Furthermore, a small peak at around 1100 cm^{-1} assigned to Ti-O-C stretching. It may be indicated that C atomic was incorporated to TiO_2 lattice. These results are also evidenced by XPS spectra analysis of C 1s core levels for N-TNTAs and TNTAs. The peaks observed at 410 and 805 cm^{-1} correspond to the Ti-O-Ti stretching vibrations of crystalline TiO_2 [22]. The strong absorption band at around 2500-3600 cm^{-1} is ascribed to the associated hydroxyls groups from either water molecules. In addition, after calcined at 450°C in N_2 atmosphere for 3 h, a small peak at around 3240 cm^{-1} which is assigned as O-H stretching of the hydroxyl group to Ti atoms for TNTAs and all the N-TNTAs was observed.

Figure 5. shows the diffuse UV-Vis reflectance spectra of the standard TiO_2 P 25 (curve a), TNTAs (curve b), and N-TNTAs (curve c). The band gap energy (Eg) of the photocatalysts are determined by applying Kubelka-Munk function F(R) [12] as follow:

$$[F(R).h\nu]^{\frac{1}{2}} = K(h\nu - Eg) \dots\dots\dots\dots\dots\dots\dots\dots (1)$$

where $F(R)=(1-R)^2/2R$; R is reflectance, hv is the photon energy and K is the constant characteristic of TiO_2. A plot of F(R) versus $h\nu$ should show a linear region just above the optical absorption edge [12,27]. Using the Tauc plot, the band gap energies for all samples are resulted. The band gap of N-TNTAs (2.8 eV; 420 nm), decrease from 3.19 eV (380 nm), is caused by incorporating N atomic to form N-Ti-O bonds [12]. The band gap of the N-TNTAs is lower than all photocatalysts. These results consistence with FTIR and XPS analysis. Since all photocatalysts undergo reducing the band gap

IOP Conf. Series: Materials Science and Engineering **172** (2017) 012005 doi:10.1088/1757-899X/172/1/012005

compared to TiO_2-P25 (3.22 eV), they might ascribe the extended photo-response to visible light region and could be activated by visible light as well.

3.2. Photoelectrochemical properties

The photocurrent response measurements were performed under visible light illumination to investigate the behavior of the transfer carriers. It can provide useful information of the photocatalytic process.

It is known that inter-band electron transition is a companied by relaxation and recombination in TiO_2, and consequently, only part of the photons absorbed by TiO_2 can contribute to generation of a photocurrent [8]. Therefore, it is necessary to take into account the photocurrent density-potential applied of the TNTAs and N-TNTAs photo-anode under visible light illumination. Figure 6 shows that the photocurrent density of TNTAs and N-TNTAs electrode in the dark and visible light irradiation. The photocurrent density of N-TNTAs and TNTAs photo-anode under darkness are negligible.

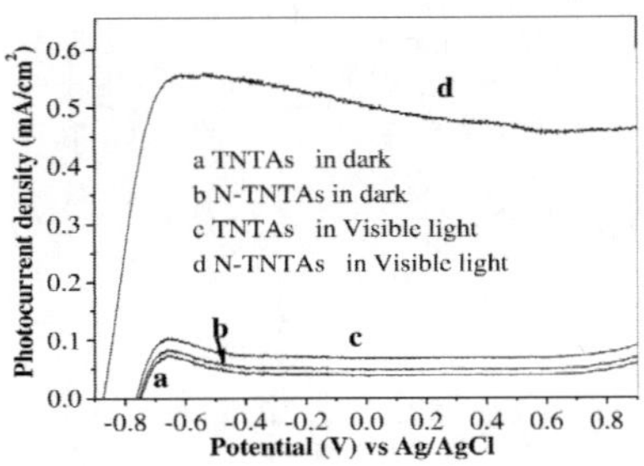

Figure 6. Linear sweep voltammograms (LSV) of the TNTAs and N-TNTAs : in dark and visible light illumination

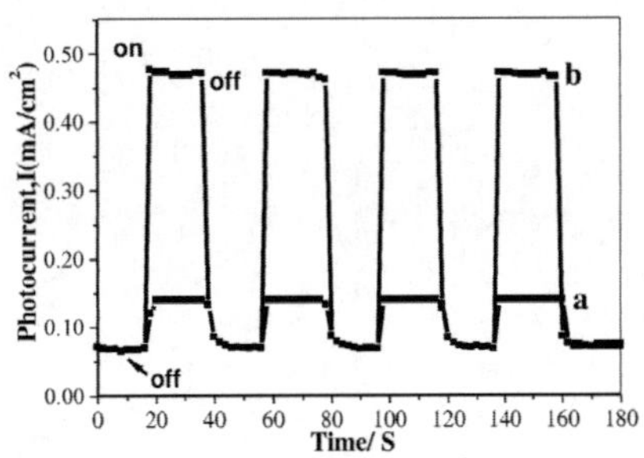

Figure 7. The amperometric *I-t* curves at fixed bias potential 0.2 V of TNTAs (a) and N-TNTAs electrode (b)

In visible light irradiation for the N-TNTAs, the photocurrent density is significantly higher than that of TNTAs at all bias potential, which is about 10 times than the TNTAs. The high photocurrent conversion efficiency may be contributed to improve the optical absorption capability caused by the N-doping [21]. The absorption shoulder up to visible region may have important roles in enhanced visible light photocurrent conversion efficiency of the N-TNTAs. For the undoped TNTAs electrode, no significant photocurrent density over the bias potential range because the energy of visible light illumination is smaller than the band gap energy of TNTAs. This result can be ascribed to N atoms incorporated in TiO_2 lattice which are able to narrow the band gap [6], promote the excitation of photoelectron-hole pairs by visible light [19]. It can be observed also from Fig.6, the zero-current potential (*E-zcp*) for the N-TNTAs and TNTAs were -0.91 and -0.76 V, versus Ag/AgCl, respectively. The negative shift of 0.15 V for the N-TNTAs compared to that the TNTAs can be ascribed to the N-doping need a lower potential(less energy) to extract the photoinduced electrons to external circuit, exhibited fast light response due to efficient charge transfer. Increasing bias potential applied lead to increase the generated photocurrent density then reach a relatively steady state within a certain range of potential (fixed potential 0.2 V). The photo-anode shift in photocurrent density is attributed to decrease in recombination on the surface of N-TNTAs, which could be determined by the surface structure.

The photoelectron-hole separation and electrode transfer interface, the photocurrent response of N-TNTAs electrode is evaluated by amperometric *I-t* curves, which were collected with visible light on or off cycles, at a fixed bias potential 0.2 V. As shown in Fig.7.the experiment run for 180 s firstly, then illuminating time was taken for 20 s each time. The photocurrent responds with the turn-on and turn-off the visible light quickly. When the visible light illumination is turn on, the photocurrent of

TNTAs quickly increases to 0.13 mA.cm^{-2} and remains constant until the irradiation is turned off while rapidly drop to the background photocurrent of steady state density (0.06 mA.cm^{-2}). The photocurrent arises from the photoelectron-hole separation in the surface of N-TNTAs, and the charge transfer process is very rapidly. For the N-TNTAs photo-anode, possess higher photocurrent than TNTAs. We can see in Fig.7. the photocurrent density of N-TNTAs showed is higher than TNTAs. Based on these results, the N-doping caused increased the oxygen vacancies and narrow band gap [6], indicating that broadening visible light region and hindering recombination of photoelectron-hole. The photocatalytic activity of TiO_2 usually depend on the transfer rate of surface charge carriers from the bulk to the surface and recombination rate of photogenerated electrons and holes, a faster photogenerated electrons and holes transfer leading to higher photocatalytic activity of TiO_2 [23-26].

3.3. Photo-electro-catalytic performance of N-TNTAs electrode
In the photoelectrochemical system, the MB may undergo decomposition driven by direct electrolysis (EC), photocatalysis (PC) and synergic photo-electro-catalysis (PEC).

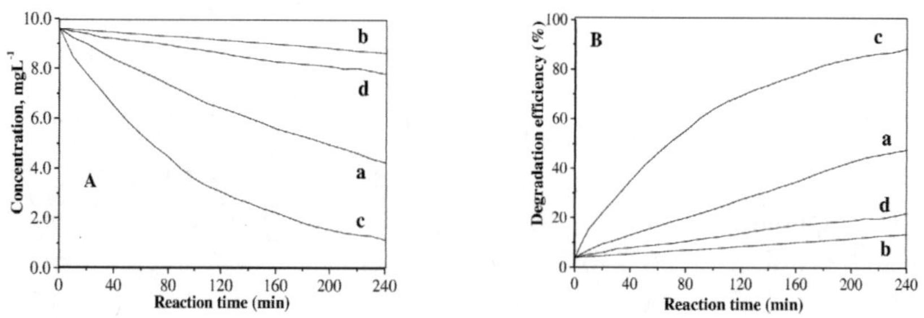

Figure 8.(A). Concentration of MB and (B) degradation efficiency (%) under different conditions : (a) PC under visible light; (b) direct EC at bias of 0.2 V in the dark; (c) PEC N-TNTAs under visible light and 0.2 V bias potential; and (d) PEC TNTAs under visible light and 0.2 V bias potential.

The EC, PC and PEC degradation of MB solution using N-TNTAs and TNTAs electrode was present in Fig.8. As shown in Fig.8.A that within 240 minutes, the EC process was obviously slower than PC and PEC process. The resulting concentration-reaction time curves show noticeable differences in the removal MB concentration among experiment. In order to quantitatively describe MB degradation in this work, can be applied by the Langmuir-Hinshelwood (L-H) equation, which is usually used to characterize the PC and PEC degradation process of organic compounds, as following the first order reaction [14,27] :

$$\ln\frac{Co}{Ct} = -k.t \quad \text{.................................(2)}$$

where Co is the initial concentration, Ct is the concentration of MB at the reaction time (t) and *k* is the rate constant (min^{-1}).

The degradation efficiency (%) under different conditions PC, EC and PEC reaction time for 240 min was shown in Fig.8.B. When no external potential was applied, the visible light illumination resulted 42 % of the MB was removed by photocatalysis on N-TNTAs electrode after 240 min (curve a). When the N-TNTAs is used as the photo-anode (0.2 V) without visible illumination, degradation of MB exhibited slow rate constant 1.0x10^{-3} min^{-1} (curve b) from electrolysis process. In contrast, when the N-TNTAs photo-anode apply external bias potential 0.2 V under visible illumination for 240 minutes, the MB was degraded about 89% and a much greater rate constant of 8.5x10^{-3} min was observed (curve c). Using the TNTAs as photo-anode (0.2 V) under visible light illumination only small the MB was degraded with a rate constant of about 1x10^{-3} min (curve d). The rate constant of N-TNTAs electrode was 8 times better compared to that the un-doped TNTAs electrode. These results were corresponding to the enhanced photocurrent density as given in Fig.6. Accordingly, there should be a synergistic effect between the electricity and radiation energy. The photocatalytic activity of TiO_2 usually depends on the charge transport ability [27,28] and a faster photoelectron leading to a higher photocatalytic activity [24].

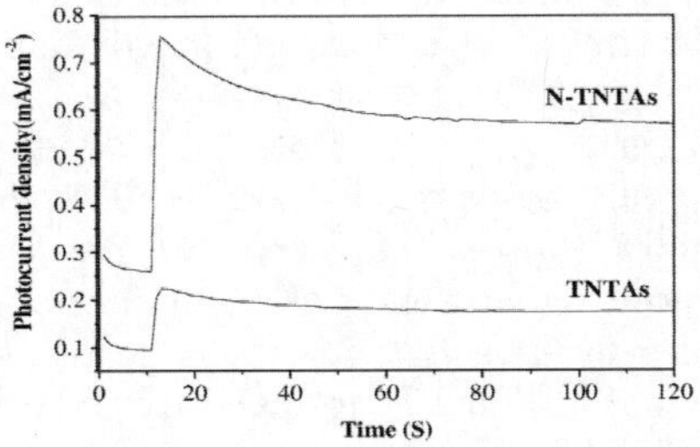

Figure 9. Comparison of photocurrent density-time profile for MB in KOH (10 mgL^{-1}) PEC degradation obtained fom N-TNTAs and TNTAs electrode (bias potential 0.2 V).

The typical photocurrent density versus time profiles obtained during of PEC degradation of MB with initial concentration of 10 mgL^{-1} in KOH solution under given condition were shown in Fig.9. The photocurrent density observed for N-TNTAs electrode is found much higher than TNTAs electrode. This result is consistent with the enhanced the LSV response, given in Fig.6, and improved the PEC activity given in Fig.8. of the N-TNTAs electrode. The photoelectron-hole could either recombine with electrons or react with OH^- ions or H_2O oxidizing them to form $OH^\bullet$ radicals. The $OH^\bullet$ radicals, are responsible for the decomposition of MB molecules. The photoelectron could reduce the adsorbed MB or react with O_2 adsorbed on the electrode surface or dissolved in water to form superoxide radical anion $O_2^{\bullet-}$, which can be transformed to $OH^\bullet$ radicals by further reactions [22]. The result is definitely consistent with the photocatalytic properties which can be attributed to the N atoms doped in TiO_2 can narrow the band gap, promote the excitation of electron-hole pairs by visible light and enhance the visible light photo-response [29]. The obviously enhanced rate constant in the PEC process can be attributed to the suppression of recombination between the photogenerated electron and hole pairs by external electric field and the special highly ordered nanotube array is favorable for the transport of the photogenerated electrons. In addition, such nanotube structures also show improved mechanical stability, since the TiO_2 is directly attached to the tube walls and there are more supporting points in the nanotube structure. The effect photo-stability of the electrode for the PEC degradation of MB was evaluated, there is no apparent deactivation of the N-TNTAs photocatalysts are observed after five cycles. This clearly shows that the photocatalysts are stable during the PEC process.

4. Conclusion

The highly ordered N-doped TiO_2 nanotube arrays (N-TNTAs) were applied as an photo-anode in the PEC degradation of MB aqueous solution, which the efficiency was degrade 89% compared with undoped TiO_2 electrode only 12%. Under visible light illumination, the N-TNTAs photo-anode showed higher photocurrent density and photo-electro-catalytic activity for degradation of MB solution than TNTAs electrode. The rate constant of N-TNTAs electrode about $8.5x10^{-3}$ min was 8 times better compared to that the TNTAs electrode. This improved the PEC performance can be attributed to the increased visible light absorption, the improved photoelectron-hole separation by external bias potential applied. These results suggest that the highly effective N-TNTAs electrode, with strong mechanical stability and superior transport of photoelectron and hole pairs, may serve well as a promising photo-electro-catalyst and may provide an easy and efficient route to remove pollutant organic such as methylene blue from wastewater.

References

[1] Jankulovska M, Barcelo I, Lana-Villarreal T and Gomez R 2013 Improving the photoelectrochemical response of TiO_2 nanotubes upon decoration with quantum-sized anatase nanowires *J. Phys Chem C.* **117** 4024

[2] Liu Y.Z, Zhang T.X, Nishimoto S, Jin M, Tryk D.A, Murakami T and Fujishima A 2008 Highly ordered TiO_2 nanotube arrays with controllable length for photoelectrocatalytic degradation of phenol *J. Phys. Chem. C.* **112** 253

[3] Wang X, Zhao M.H, Quan X, Y.Z. Zhao Z.Y and Chen S 2009 Visible light photoelectrocatalysis with salicylic acid-modified TiO_2 nanotube array electrode for p-nitrophenol degradation *J. Hazard. Mater* **166** 547

[4] Mor G.K, Shankar K, Paulose M, Varghese O.K and Grimes C.A 2006 Use of highly-ordered TiO_2 nanotubes arrays in dye sensitizer solar cells *Nano Lett.* **6** 215

[5] Chen S, Chu W, Huang Y.Y, Liu X and Tong D.G 2012 Preparation of porous nitrogen-doped titanium dioxide microspheres and a Study of their photocatalytic, antibacterial and electrochemical activities, *Materials Research Bulletin* **47** 4514

[6] Asahi R, Ohwaki T and Aoki K 2001 Visible-light photocatalysis in nitrogen-doped Titanium, *Science* **293** 269

[7] Li S, Lin S, Liao J, Pan L, Li D and Li J 2012 Nitrogen-doped TiO_2 nanotube arrays with enhanced photoelectrochemical property, *International Journal of Photoenergy* Article ID 794207 doi:10.1155/2012/794207

[8] Li J, Lu N, Quan X, Chen S and Zhao H 2008 Facile method for fabricating boron-doped TiO_2 nanotubes array with enhanced photoelectrocatalytic properties *Ind.Eng.Chem.Res.* **47** 3808

[9] Yan G, Zhang M, Hou J and Yang J 2011 Photoelectrochemical and photocatalytic properties of N + S co-doped TiO_2 nanotube array films under visible light irradiation *Materials Chemistry and Physics.* **129** 553

[10] Lindgren T, Mwabora J M, Avendano E, Jonsson J, Hoel A, Granvist C.G and Lindquist S.E 2003 Photoelectrochemical and optical properties of nitrogen doped titanium dioxide films prepared by reactive DC magnetron sputtering *J. Phys. Chem. B.* **107** 5709

[11] Zhang F, Pi Y, Cui J, Yang Y, Zhang X and Guan N 2007 Unexpected selective photocatalytic reduction of nitrite to nitrogen on silver-doped titanium dioxide *J. Phys. Chem. C.* **111** 3756

[12] Ratnawati, Gunlazuardi J, Dewi E L and Slamet 2014 Effect of $NaBF_4$ addition on the anodic synthesis of TiO_2 nanotube arrays photocatalyst for production of hydrogen from glycerol water solution *International Journal of Hydrogen Energy* **xxx** 1

[13] Yan G, Zhang M, Hou J and Yang J 2011 Photoelectrochemical and photocatalytic properties of N+S co-doped TiO_2 nanotube array films under visible light irradiation *Mater Chem Phys.* **129** 553

[14] Liu Y, Gan X, Zhou B, Xiong B, Li J, Dong C, Bai J and Cai W 2009 Photoelectrocatalytic degradation of tetracycline by highly effective TiO_2 nanopore arrays electrode *Journal of Hazardous Materials* **171** 678

[15] Zhang S, Peng F, Wang H, Yu H, Zhang S, Yang J and Zhao H 2011 Electro deposition preparation of Ag loaded N-doped TiO_2 nanotube arrays with enhanced visible light photocatalytic performance *Catalysis Communication* **12** 689

[16] Antony R P, Mathew T, Dash S and Tyagi A K 2013 Kinetics and Physicochemical process of photoinduced hydrophobic $\leftrightarrow$ superhydrophilic switching of pristine and N-dope TiO_2 nanotube arrays *J.Phys.Chem.C.* **117** 6851

[17] Shin S W, Lee J Y, Ahn K S, Kang S H and Kim J H 2015 Visible light absorbing TiO_2 nanotube arrays by sulfur treatment for photoelectrochemical water splitting, *J. Phys. Chem. C.* **119** 13375

[18] Xing M, Zhang J, and Chen F 2009 New approach to prepare nitrogen-doped TiO_2 photocatalysts and study on their photocatalytic activities in visible light *Applied Catalysis B. Environmental* **89** 563

[19] Jiao J, Tang J, Gao W, Kuang D, Tong Y, and Chen L 2015 Plasmonic silver nanoparticles matched with vertically aligned nitrogen-doped titanium dioxide nanotube arrays for enhanced photoelectrochemical activity *Journal of Power Sources.* **274** 464

[20] Sun H, Zhou G, Liu S, Ang M H, Tadé M O, and Wang S 2013 Visible light responsive titania photocatalysts codoped by nitrogen and metal (Fe, Ni, Ag, or Pt) for remediation of aquaeous pollutants *J.Chemical Engineering* **231** 18

[21] Lai K Y, Huang Y J, and Zhang X 2010 Nitrogen-doped TiO_2 nanotube array films with enhanced photocatalytic activity under various light sources *Journal of Hazardous Materials* **184** (3) 855

[22] Sakthivel S and Kisch H 2003 Daylight photocatalysis by carbon modified titanium dioxide *Angewandte Chemie* **42** (40) 4908

[23] Xi-ying Z, Jiajia S and Bin W 2013 A One-step electrochemical method for the production of TiO_2−xNx nanotubes *J.of The Electrochemical Society* **160** (6) 335

[24] Quan X, Yang S G, Ruan X L and Zhao H M 2005 Preparation of titania nanotubes and Their environmental application as electrode *Environ.Sci. Technol.* **39** 3770

[25] Mollavali M, Falamaki C and Rohani S 2015 Preparation of multiple-doped TiO_2 nanotube arrays with nitrogen, carbon and nickel with enhanced visible light photoelectrochemical activity via single-step anodization *International Journal of Hydrogen Energy* **40** 12239

[26] Hunge Y M, Mahadik M A, Mohite V S, Kumbhar S S, Deshpande N G, Rajpure K Y, Moholkar A V, Patil P S and Bhosale C H 2016 Photoelectrocatalytic degradation of methyl blue using sprayed WO_3 thin films *J Mater Sci: Mater Electron* **27** 1629

[27] Zhang X, Udagawa K, Liu Z, Nishimoto S, Xu C, Liu Y, Sakai H, Abe M, Murakami T and Fujishma A 2009 Photocatalytic and photoelectrochemical studies on N-doped TiO_2 photocatalyst *Journal of Photochemistry and Photobiology A: Chemistry.* **202** 39

[28] Xu J J, Ao Y H, and Chen M D 2010 Photoelectrochemical property and photocatalytic activity of N-doped TiO_2 nanotube arrays *Applied Surface Science* **256** (13) 4397

[29] Chen Y Y, Zhang S M, and Yu Y 2008 Synthesis, Characterization and Photocatalytic activity of N-doped TiO_2 nanotubes *Journal of Dispersion Science and Technology* **29** (2) 245

Acknowledgments
This research was partly funded by the cluster research scheme from University of Indonesia [Titania photo electro catalysis (TiPEC)], contract number: 1864/UN2.R12/HKP.05.00/2015), and JSPS-DGHE Bilateral Exchange.

IOP Conf. Series: Materials Science and Engineering **172** (2017) 012006 doi:10.1088/1757-899X/172/1/012006

Fabrication of hydroxyapatite from fish bones waste using reflux method

A Cahyanto[1], E Kosasih[2], D Aripin[3] and Z Hasratiningsih[1]

[1]Department of Dental Materials Science and Technology, Faculty of Dentistry, Universitas Padjadjaran, Sekeloa Selatan 1 Bandung, 40132, Indonesia
[2] Graduate Student, Department of Biomaterials, Faculty of Dental Science, Kyushu University, Maidashi 3-1-1, Higashi-ku, Fukuoka, Japan
[3]Department of Endodontology and Operative Dentistry, Faculty of Dentistry, Universitas Padjadjaran, Sekeloa Selatan 1 Bandung, 40132, Indonesia

Email: arief.cahyanto@fkg.unpad.ac.id

Abstract. The aim of this present study was to investigate the fabrication of hydroxyapatites, which were synthesized from fish bone wastes using reflux method. The fish bone wastes collected from the restaurant were brushed and boiled at 100°C for 10 minutes to remove debris and fat. After drying, the fish bones were crushed, and ball milled into a fine powder. The fish bone wastes were then processed by refluxing using KOH and H_3PO_4 solutions. The samples were calcined at 900°C and characterized by X-Ray Diffraction (XRD) and Fourier Transform Infrared Spectrometry (FT-IR). The XRD pattern of samples after treatment revealed that the peak of hydroxyapatite was observed and the bands of OH^- and PO_4^{3-} were observed by FT-IR. The scanning electron microscope evaluation of sample showed the entangled crystal and porous structure of hydroxyapatite. In conclusion, the hydroxyapatite was successfully synthesized from fish bone wastes using reflux method.

1. Introduction

Hydroxyapatite is one of the most common apatites used in dentistry and medicine. Hydroxyapatite and related calcium phosphate materials have been widely used for bone reconstruction or substitution due to their similarity to the bone. It is believed that natural hydroxyapatite is more biocompatible because it has similar physical-chemical characteristics with bone [1]. The chemical form of hydroxyapatite is $Ca_5(PO_4)_3OH$ but often written as $Ca_{10}(PO_4)_6(OH)_2$. Naturally, hydroxyapatite is an inorganic component that can be found in human hard tissue such as tooth and bone. These materials are commonly used as human body implant materials. Natural hydroxyapatite can be synthesized from eggshells, coral, fish bone, chicken bone, etc. [2].

Fish was consumed by humans because of their minerals and vitamins. However, the fish-bone wastes have not been used for anything. The fish-bone wastes have caused the environmental problem such as organic odor. The fish-bone contains calcium phosphate, collagen fiber, calcium carbonate, and hydroxyapatite. Therefore, the possibilities to synthesize hydroxyapatite using natural calcium sources such as fish bones are prominent [3].

Previously, the fabrication of natural hydroxyapatite was conducted using the fish bone of Brazilian river such as pintado (*Pseudoplatystomacorruscans*), jaú (*Paulicealutkeni*), and cachara

IOP Conf. Series: Materials Science and Engineering **172** (2017) 012006 doi:10.1088/1757-899X/172/1/012006

(*Pseudoplatystomafasciatum*) [4]. The fish bone was calcined at 900°C to achieve an optimal natural hydroxyapatite.

Most of Indonesia people have consumed both ocean fish and river fish. In West Java Province, especially in Bandung, people usually consume the river fish because it is easier to be obtained than the ocean fish. Minimum production cost, cheap and potential donor for calcium can be achieved by using fish bone [5]. Fabrication of hydroxyapatite from river fish bone required the simple methods. One of these methods is reflux. The reflux is a technique involving the condensation of vapors and the return of this condensate to the system. In the present study, the reflux method was employed using fish bone waste as Ca^{2+} sources. The Ca^{2+} could be bonded with PO_4^{3-} from the solution to form the hydroxyapatite.

2. Method and Materials

2.1 Preparation of fish bones waste

Fish bone wastes were obtained from the restaurant in local area. The collected fish bone wastes were brushed and boiled at 100°C water for 10 minutes to clean these samples from the debris and fat. Bones were dried using a microwave at 90°C for 100 minutes then crushed until they became powder. The powders were dried using dry-heat oven at the temperature of 60°C for 24 hours. Later, the powders were crushed using high-energy ball milling for 24 hours to obtain a fine powder. After milled, bones powder was dried using a microwave at a temperature of 60°C. Powder samples were put into a desiccator to maintain the sample in the dry condition.

2.2 Deproteinization

Powder samples were deproteinized by reflux method using 5% KOH solution. The purpose of 5% KOH addition is to create the strong base solution that can hydrolyse the calcium and protein in the fish bones. The samples and solution were boiled at 130°C for 18 hours.

$$(Ca^{2+}\ \text{Protein}) + KOH \rightarrow Ca^{2+} + \text{Protein}$$

$$Ca^{2+} + CO_3^{2-} \rightarrow CaCO_3\downarrow$$

2.3 Reflux using H_3PO_4 solution

The reflux used a weak acid of H_3PO_4 7N ≈ 2.33 M as a source of PO_4^{3-}. The PO_4^{3-} could act as a donor ion for synthesizing the hydroxyapatite. The solution was boiled at 130°C for 12 hours.

$$CaCO_3 + H_3PO_4 \rightarrow Ca_2PO_4 + H_2CO_3$$

$$H_2CO_3 \rightarrow H_2O \quad CO_3\uparrow$$

2.4 Calcination

Calcination was conducted using high-temperature furnace at 900°C for 8 hours. In this study, the samples were contaminated by excess of H_3PO_4. By calcination at 900°C for 8 hours, the color of contaminated samples changed from white to grey.

$$H_3PO_4 \rightarrow H_2O + P_2O_5$$

2.5 Characterization of the samples

The characterization of the samples was conducted using X-ray Diffraction (XRD), scanning electron microscopy (SEM) JEOL-JSM-6510LV, and Fourier Transform Infrared Spectrometry (FT-IR).

3. Results and Discussion

Table 1 summarized the comparison of hydroxyapatite at three highest peaks (2θ) in XRD between samples that calcined at 900°C and hydroxyapatite (Taihei Chemical Co. Ltd., Osaka, Japan) as a reference.

Table 1. Three highest peaks (2θ) of XRD in hydroxyapatite and samples calcined at 900°C.

Samples	Three highest peaks (2θ)		
	1	2	3
Hydroxyapatite (reference)	32.90	32.20	31.73
900°C	32.52	32.05	31.84

Table 1 and figure 1 showed the result of the sample that calcined at 900°C for three highest peaks (2θ) of XRD. The three highest peaks of hydroxyapatite as reference are 32.90; 32.20; 31.73 [6]. These values are quite similar to the sample calcined at 900°C.

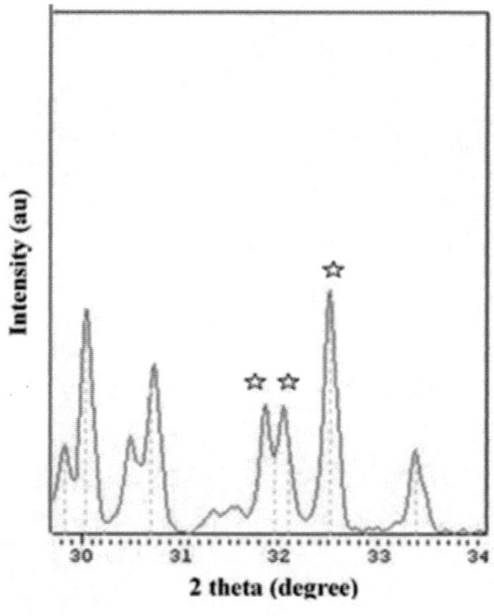

Figure 1. Three highest peaks of XRD pattern for the sample calcined at 900°C.

The XRD patterns of the sample calcined at 900°C showed that the hydroxyapatite could be observed and no peaks of $CaCO_3$ were detected. In other words, the reflux method that employed in this study was effective to fabricate the hydroxyapatite.

Figure 2 showed the functional groups of the samples that calcined at 900°C. The functional group of P-O and O-H was clearly observed, indicating that the hydroxyapatite was detected in the sample. For more details, table 2 summarized the comparison of infrared absorption spectroscopy of samples calcined at 900°C and hydroxyapatite as reference (Taihei Chemical Co. Ltd., Osaka, Japan).

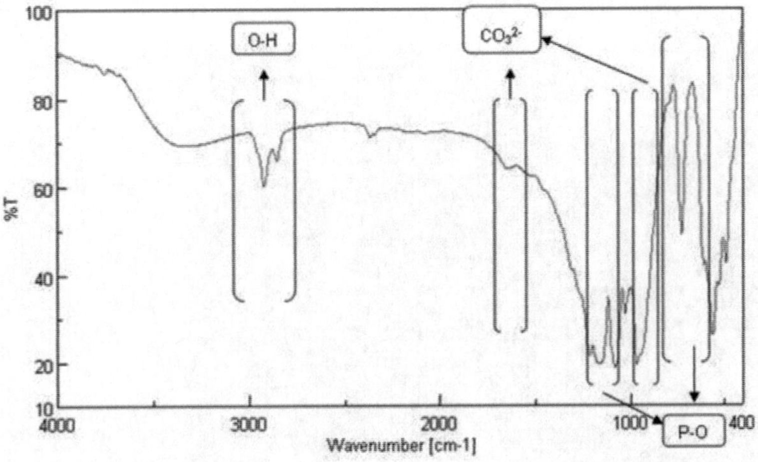

Figure 2. The peaks of P-O and O-H functional groups that appeared in the sample.

Table 2. The infrared absorption spectroscopy of sample calcined at 900°C compared to the reference of hydroxyapatite.

Sample	Infrared absorption spectroscopy (cm^{-1})	
	P-O	O-H
Hydroxyapatite (reference)	562-602;700-725;1096-1260	2700-3000
900°C	563; 725; 1160; 1214	2854; 2368

The CO_3^{2} functional groups were detected in this sample. The carbonate apatite was more useful compared to the hydroxyapatite. The carbonate apatite and hydroxyapatite were used in in vivo study [7]. The transformation of hydroxyapatite to carbonate apatite could also be possible under the low-temperature synthesis [8].

Table 3. The infrared absorption spectroscopy of CO_3^{2-} using FT-IR

Sample	Infrared absorption spectroscopy of CO_3^{2-} (cm^{-1})
900°C	1029.8; 1079.94; 1650.77

It should be noted that the absorption bands at 1650 cm^{-1} belonged to the vibration modes of C-O from the CO_3^{2-} group in A-type carbonate apatite, whereby the PO_4^{3-} was partially substituted by CO_3^{2-} [9].

The microstructure of sample observed by SEM can be seen in figure 3. In this image, the morphology of hydroxyapatite is porous. The pore size of sample is less than 1 μm in average and not homogeneous. One of the weaknesses of hydroxyapatite is brittle, but this porosity can be filled with a nano-filler like zirconia [10].

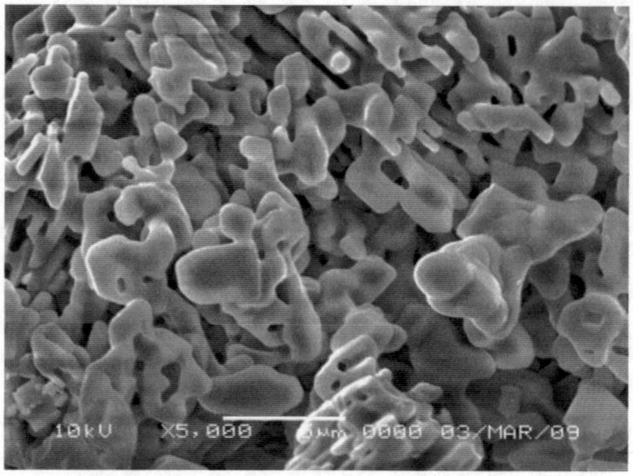

Figure 3. Scanning electron microscope image of sample calcined at 900°C.

4. Conclusions

The hydroxyapatite was successfully synthesized from the river fish bone wastes using a reflux method. The carbonate functional group (CO_3^{2-}) was found in the sample calcined at 900°C.

References

[1] Hench LL 1998 Biomaterials: a forecast for the future *Biomaterials* **19** 1419
[2] Afshar A, Ghorbani M, Ehsani N, Saeri MR and Sorrell CC 2003 Some important factors in the wet precipitation process of hydroxyapatite *Mater. design* **24** 197
[3] Prabakaran K and Rajeswari S 2006 Development of hydroxyapatite from natural fish bone through heat treatment *Trends Biomater. Artif. Organs* **20** 20
[4] Coelho TM, Nogueira ES, Weinand WR, Lima WM, Steimacher A, Medina AN, Baesso ML and Bento AC 2006 Thermal properties of natural nanostructured hydroxyapatite extracted from fish bone waste *J. Appl. Phys.* **101** 084701
[5] Chakraborty R and RoyChowdhury D 2013 Fish bone derived natural hydroxyapatite-supported copper acid catalyst: Taguchi optimization of semi batch oleic acid esterification *Chemical engineering journal* **215** 491
[6] Cahyanto A, Maruta M, Tsuru K, Matsuya S and Ishikawa K 2013 Basic properties of carbonate apatite cement consisting of vaterite and dicalcium phosphate anhydrous *Key Eng. Mater.* **529-530** 192
[7] Ishikawa K, Takagi S, Chow LC and Suzuki K 2003 Reaction of calcium phosphate cements with different amounts of tetracalcium phosphate and dicalcium phosphate anhydrous *J. Biomed. Mater. Res.* **46** 504
[8] Cahyanto A, Tsuru K and Ishikawa K 2016 Transformation of apatite cement to b-type carbonate apatite using different atmosphere *Key Eng. Mater.* **696** 9
[9] LeGeros RZ 1991 Calcium phosphates in oral biology and medicine *Monographs in oral science* **15** 15
[10] Cahyanto A, Gandasasmita S, Purwasasmita BS 2015 Fabrication of nano reinforced hydroxyapatite-zirconia used as Filler in Dental Cement Resin *Abs. Int. Conf. of 3rd Joint Scientific Meeting in Dentistry* p 35

Acknowledgements
The authors are grateful to Advanced Materials Processing Laboratory, Department of Engineering Physics, Faculty of Industrial Technology, Institute Technology of Bandung for supporting this study.

The influence of activating agents on the performance of rice husk-based carbon for sodium lauryl sulfate and chrome (Cr) metal adsorptions

Arneli, Z F Safitri, A W Pangestika, F Fauziah, V N Wahyuningrum and Y Astuti[*]

Department of Chemistry, Science and Mathematics Faculty, Diponegoro University, Central Java, Indonesia 50275

*Email: yayuk.astuti@live.undip.ac.id

Abstract. This research aims to study the influence of activating agents to produce rice husk based-carbon with high adsorption capacity and efficiency for either hazardous organic molecules or heavy metals which are unfriendly for the environment. Firstly, rice husk was burned by pyrolysis at different temperatures to produce rice husk-based carbon. To improve its ability as an adsorbent, carbon was treated with activating agents, namely, H_3PO_4 and KOH at room and high temperature (420 °C). The performance of carbon was then tested by contacting it with surfactant (SLS). Finally, the surfactant-modified active carbon was applied for chrome metal removal. The result shows that activation of carbon using phosphate acid (H_3PO_4) was more effective than potassium hydroxide (KOH) conducted at high temperature to adsorb sodium lauryl sulfate (SLS) and chrome metal with the adsorption capacity 1.50 mgg^{-1} and 0.375 mgg^{-1}, respectively.

1. Introduction
Rice husk is a by-product of rice milling. It is normally used for red rock production as a fuel, for cooking and sometimes it is thrown away as waste resulting in pollution to the environment. However, recently, rice husk has been applied as an adsorbent in the form of carbon for environmental treatments such as treatment of waste water caused by hazardous heavy metals [1, 2] and dye molecules [3]. To produce carbon, rice husk must be burned at a certain temperature. Furthermore, an abundant carbon can be generated in very limited oxygen condition since if oxygen is excess, carbon will not be produced perfectly, otherwise carbon monoxide and water are produced; this is called perfect combustion. After getting carbon, carbon is treated with activating agents to obtain activated carbon with the aim to produce carbon with high performance as an adsorbent [4]. The activating agents used for carbon activation are H_3PO_4, $ZnCl_2$, K_2CO_3, KOH, etc. [4, 5]. The use of activating agents is to eliminate the impurities such as alcohol, acids, aldehyde, etc. contained in the pores so that the pores of carbon become opened and the diameter of the pores and surface area of adsorbent increase. In addition to activation of carbon using activating agents, the performance of carbon can be improved by modifying the surface with the surfactant. This modification changes the surface properties of the activated carbon becoming more hydrophilic so that increasing its wettability and adsorption efficiency [6]. Therefore, this present work aims to investigate the influence of different activating agents to the performance of carbon as adsorbent for surfactant and chrome metal adsorptions.

2. Experimental Methods

2.1 Materials
Rice husk obtained from Semarang, Indonesia, phosphoric acid, potassium hydroxide, sodium lauryl sulfate (SLS) powder, sodium hydroxide, sulfuric acid, aquadest, chloroform, phenolphthalein,

methylene blue, monosodium phosphate monohydrate, and potassium dichromate. All the reagents were of analytical grade purchased from Sigma Aldrich.

2.2 Rice Husk Carbonization
Rice husk-based carbon was produced by several steps. Firstly, rice husk was washed using water and then dried under the sun. Subsequently, the dried rice husk had been burned by pyrolysis at different temperature 300, 350 and 400 °C for 10 minutes. The obtained carbon was then sifted using the top ending 100 mesh sieve.

2.3 Carbon Activation
20 gram rice husk-based carbon had been contacted with H_3PO_4 and KOH at different concentration for 1.5 hours at room temperature. The concentration used in each activating agent was 30, 40, 50, 60 and 70% for H_3PO_4 and 40, 50, 60, 70, 80 % for KOH. After that, the mixture was filtered. The residue, activated carbon, was then washed with aquadest until constant pH was achieved. Furthermore, the activated carbon had been dried in an oven at 105 °C for 1 hour and then being smoothed using a mortar and sifted using 100 mesh sieve. The optimum concentration of activating agents indicated by high adsorption capacity for SLS adsorption was then used to activate carbon at high temperature (420 °C) for 1 hour.

2.4 Determination of the Adsorption Efficiency and Capacity of Activated Carbon
1 gram rice husk-based activated carbon had been contacted with 100 mL surfactant (SLS) 60 ppm for 1.5 hours. Subsequently, the mixture was filtered. The obtained filtrate was then treated using MBAS method [7] to separate surfactant (SLS) unabsorbed by activated carbon. Finally, the obtained surfactant was then analyzed using UV-Vis spectrophotometer.

2.5 Characterization of Activated Carbon
Characterization of activated carbon was conducted using FTIR (Prestige 21 Shimadzu), SEM (JEOL-JSM-6510LV) and SAA (Quantochromenove - AsiQwin 1200e) to identify the functional groups on activated carbon surface, morphology, pore size, pore volume, surface area, which are of important on the adsorption process.

2.6 Adsorption of Chrome (Cr) by Activated Carbon Adsorbing Surfactant (SLS)
1 gram activated carbon adsorbing surfactant had been contacted with 25 mL of 20 ppm potassium dichromate (as a model pollutant) for 4 hours. Then, the mixtures were filtered and the obtained filtrate was analyzed using AAS (Shimadzu AA-640IF) to calculate the concentration of chrome metal not adsorbed by activated carbon.

3. Results and Discussion

3.1 Carbonation of Rice Husk
Carbonization process aims to eliminate the components of volatile organic compounds so that it can form carbon which also has pores [4]. Carbonization of rice husk was carried out using pyrolysis method. Using this method, oxygen was conditioned very limited, so that the carbon generated was optimal. Carbon generated is also influenced by temperature applied during pyrolysis.

Based on Figure 1, it can be observed that by increasing the carbonization temperature during the pyrolysis, the color of the resulting carbon is a mixture between black and gray. The difference of carbon in color was caused by the ash content. The presence of ash resulted from the oxidation process of carbon from oxygen coming from burning of organic molecules in the carbonization process. The black carbon was achieved at 300 °C with the carbon content of 63.51%.

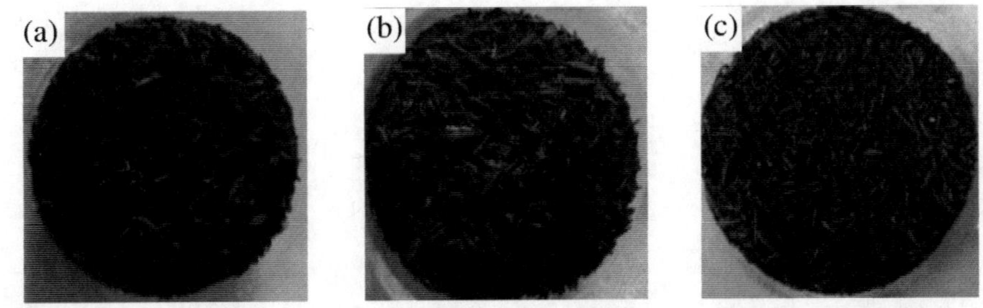

Figure 1. Carbon resulting from pyrolysis of rice husk at (a) 300 °C (b) 350 °C (c) 400 °C

3.2 Carbon Activation using H_3PO_4 and KOH

It can be seen from Table 1 that the adsorption capacity of activated carbon in H_3PO_4 - and KOH-activated carbons (CA-H_3PO_4 and CA-KOH, respectively) changed insignificantly when the concentration of activating agents increased. Furthermore, in comparison to the adsorption capacity of both samples, H_3PO_4 is little more effective than KOH to produce activated carbon applied to adsorb SLS. We suggest that H_3PO_4 is more effective to open up the layer of the carbon structure even though it has been reported that KOH as an activating agent enables to generate mesoporous and microporous carbons [8].

Table 1. Adsorption capacity of carbon after being activated using different concentration of H_3PO_4 and KOH at room temperature

[H_3PO_4]	Adsorption capacity (mgg^{-1})	[KOH]	Adsorption capacity (mgg^{-1})
30%	1,496	40%	1.432
40%	1,499	50%	1.493
50%	1,494	60%	1.491
60%	1,498	70%	1.485
70%	1,486	80%	1.482

Phosphoric acid (H_3PO_4) is an activating agent to open the pores and enlarge the surface area of activated carbon so that the adsorption power will increase. H_3PO_4 removes impurities contained in the carbon [9] as a result of carbonization process of biomass (rice husk) such as ketones, alcohols, acids, aldehydes. Moreover, H_3PO_4 also helps to form a bridge that connects the fragments of biopolymers (layer of carbon structure) by forming phosphate bonds that make pores more open and enlarge surface area. Further we used 60% H_3PO_4 and 50% KOH to activate carbon at 420 °C.

Figure 2 shows that carbons activated using H_3PO_4 both at room temperature and high temperature have higher % adsorption efficiency than that of using KOH. It can be said that H_3PO_4 is more effective than KOH as an activating agent even though at 300 °C, KOH is more effective to activate carbon but the difference is insignificant.

IOP Conf. Series: Materials Science and Engineering **172** (2017) 012007 doi:10.1088/1757-899X/172/1/012007

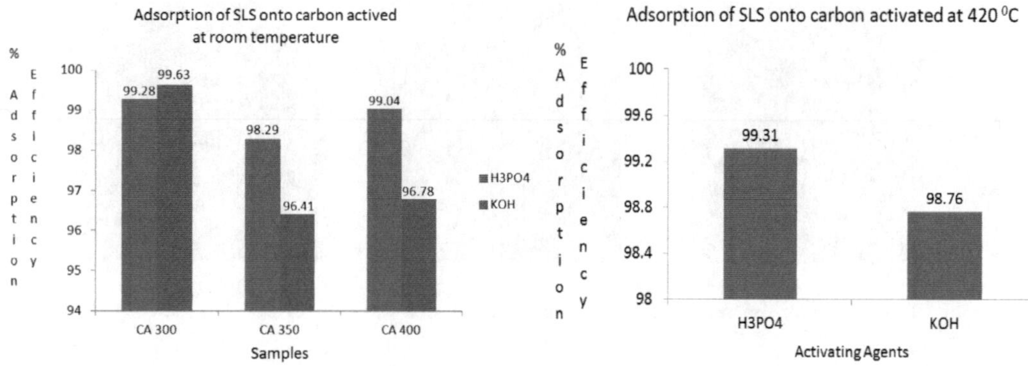

Figure 2. The adsorption of SLS surfactant onto carbon activated using H_3PO_4 and KOH at room temperature and high temperature (420 °C)

3.3 Characterization of Active Carbon and active carbon adsorbed SLS

As seen in figure 3, after activation using KOH (b), the vibration mode of –OH (3200-3800 cm^{-1}) [10] was higher than that of carbon. Si-O-Si vibration mode observed at 1097 cm^{-1} [10] almost disappeared after treating using KOH (b). Vibration mode of Si-H was also observed at 800 cm^{-1} [11]. The vibration mode at 1169-1173 cm^{-1} observed in CA-H_3PO_4 attributed to P=O vibration [12]. After adsorption of SLS surfactant, there is no significant change in the FTIR spectra of both samples as seen in figure 4 except for Ca-H_3PO_4. The appearance of S=O vibration mode was observed at 1110-1182 cm^{-1} [13]. It indicates that SLS has been well adsorbed on H_3PO_4-activated carbon.

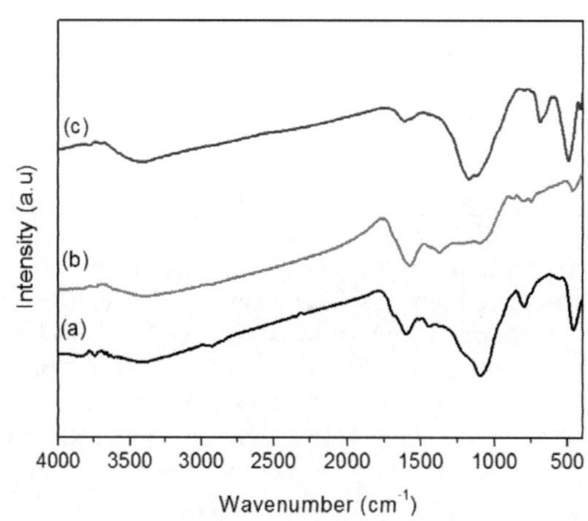

Figure 3. FTIR spectra of (a) carbon, (b) CA-KOH and (c) CA-H_3PO_4

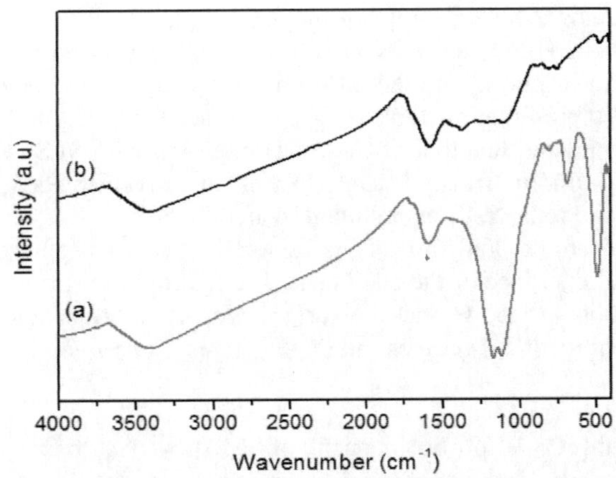

Figure 4. FTIR spectra of activated carbon adsorbed SLS coming from (a) CA-H_3PO_4 and (b) CA-KOH

SEM was applied to identify the morphology of resulting activated carbon. Because CA-H_3PO_4 gave better adsorption, SEM analysis was conducted for samples using H_3PO_4 as an activating agent. It can be seen in figure 5, before activation, carbon has heterogeneous pores (a) and after activation using H_3PO_4 at room temperature, the pores is more homogeneous and it looks brittle (b). After adsorption of SLS, the pores are still conserved. However, activation at high temperature made the carbon more rigid with the small pores below 1 μm.

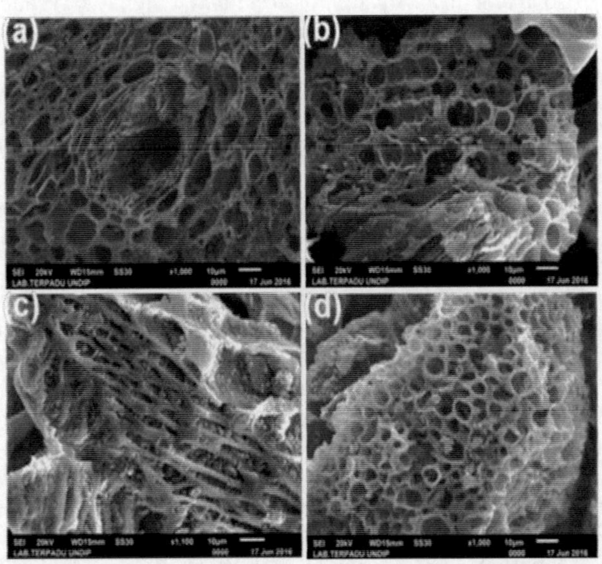

Figure 5. SEM images of (a) carbon, (b) activated carbon using H_3PO_4 at room temperature and high temperature and (d) H_3PO_4-activated carbon adsorbed SLS.

IOP Conf. Series: Materials Science and Engineering **172** (2017) 012007 doi:10.1088/1757-899X/172/1/012007

3.4 Adsorption of Cr^{2+} by Activated Carbon Adsorbing SLS

Carbons activated using both H_3PO_4 and KOH were then applied for synthetic chrome (Cr) waste metal removal with the aim to investigate the adsorption capacity of both samples. The result of Cr removal of both samples is presented in table 2. It can be seen that CA-H_3PO_4 modified with SLS adsorbed Cr (0.375 mgg^{-1}) higher than that of CA-KOH modified with SLS (0.113 mgg^{-1}). Therefore, it can be said that H_3PO_4 is more effective than KOH as an activating agent to produce surfactant-modified active carbon and to remove Cr as pollutant model.

The adsorption mechanism of ionic metals is generally due to exchange ions or electrostatic interaction on the hydrophilic surface of the adsorbent. The hydrophilic surface results from surfactant adsorbed on activated carbon surface so that the surface properties of activated carbon changes from hydrophobic into hydrophilic and becomes more negative. Consequently, the adsorption of Cr increases [6].

Table 2. Adsorption capacity of SMAC to adsorb Cr^{2+}

Activating Agent	Adsorption Capacity (mgg^{-1})
KOH	0,113
H_3PO_4	0,375

4. Conclusion

Carbon activated using H_3PO_4 is more effective to produce surfactant modified activated carbon than using KOH. Moreover, CA-H_3PO_4 adsorbing SLS is more effective for removal of Cr as pollutant model.

Acknowledgement

The authors would like to thank the Ministry of Research, Technology and Higher Education of Republic Indonesia for the financial support with the grant no. 022/SP2H/LT/DRPM/II/2016.

References

[1] Mehdinia S M, Khalilollah M, Tayyabeh R 2014 Rice husk silica adsorbent for removal of hexavalent chromium pollution from aquatic solutions *Iranica Journal of Energy & Environment* **5** 218-223

[2] Ali I O, Mostafa H A, Salah M S, Karam S S 2011 Synthesis and characterization of ZSM-5 zeolite from rice husk ash and their adsorption of Pb^{2+} onto unmodified and surfactant modified zeolite *Separation and Purification Technology* **83** 38-44

[3] Cheenmatchaya A, Kungwankunakorn S 2014 Preparation of activated carbon derived from rice husk by simple carbonization and chemical activation for using as gasoline adsorbent *International Journal of Environmental Science and Development* **5** 171-175.

[4] Soltani N, Bahrami A, Pech-Canul M I & González L A 2015 Review on the physicochemical treatments of rice husk for production of advanced materials *Chemical Engineering Journal* **264** 899-935

[5] Al-Latief D N, Arnelli and Astuti Y 2015 Synthesis of sodium lauryl sulfate (SLS)-modified activated carbon from risk husk for waste lead (Pb) removal *AIP Conference Proceedings* **1699** 060017

[6] Agustinus E T S, Mursito A T, Sembiring H 2013 Peningkatan daya serap karbon aktif terhadap ion logam hexavalent chromium (CrVI) melalui modifikasi dengan cationic surfactant (ethylinediamine) *Jurnal RISET Geologi dan Pertambangan* **23** 15-26.

[7] ASTM., Standart Test Method for Methylene Blue Active Substance, ASTM international, West Conshohocken, 2002

[8] Utama S, Kristianto H, Andreas A 2016 Adsorpsi ion logam kromium (Cr (Vi)) menggunakan karbon aktif dari bahan baku kulit salak *Prosiding Seminar Nasional Teknik Kimia "Kujangan"*, ISSN 1693-4393 1-6

[9] Yaman S 2004 Pyrolysis of biomass to produce fuels and chemical feedstocks *Energy Conversion and Management* **45** 651-671.

[10] Lambert J B, Herbert F S, David A L, Graham R C, *Introduction to Organic Spectroscopy*, 1987Macmillan Publ. New York

[11] Nakbanpote W, Goodman B A, Thiravetyan P 2007 Copper adsorption on rice husk derived materials studied by EPR and FTIR *Colloid Surface A* **304** 7-13

[12] Myglovets M, Poddubnaya O I, Sevastyanova O, Lindstrom M.E, Gawdzik B, Sobiesiak M, Tsyba M M, Sapsay V I, Klymchuk D O, Puziy A M 2014 Preparation of carbon adsorbents from lignosulfonate by phosphoric acid activation for the adsorption of metal ions *Carbon* **80** 771-783

[13] Abidin A Z, Lukman O W, Christian 2014 Konversi sampah plastic styrofoam menjadi bahan pembersih sulfur light diesel oil *Jurnal Sains Materi Indonesia* **16** 25-31

Neutron computed tomography investigation of the porosity on the titanium femoral knee investment casting

B T Prayoga[1], Suyitno[1], R Dharmastiti[1], F Akbar[2], Juliani[2]

[1] Departement of Mechanical Engineering, Gadjah Mada University, Yogyakarta Indonesia

[2] National Nuclear Energy Agency of Indonesia (BATAN), Serpong, Indonesia

Email: beni@ugm.ac.id; suyitno@ugm.ac.id; rini@ugm.ac.id

Abstract. Rotational velocity in the centrifugal casting is one of the factors that affect to the porosity of the femoral knee cast. The objective of this research is to investigate the internal porosity of as-cast femoral knee prostheses. CP-titanium femoral knee was cast using vertical centrifugal investment casting. The three rotational velocities parameter utilized in the vertical centrifugal casting were 45, 55 and 65 rpm. Neutron computed tomography characterization was conducted to investigate the size, percentage, and location of pores. The maximum size of pores in casting with the rotational velocity of 45, 55, and 65 rpm are 14.78 mm^3, 14.98 mm^3, and, 14.65 mm^3, respectively. The percentage of pores in the casting with the rotational velocity of 45, 55, and 65 rpm are 1.78%, 2.78% and, 1.78% respectively. The size and percentage of the pores for casting at the rotational velocity of 65 rpm are smaller than the size and percentage of the casting at 45 and 55 rpm.

1. Introduction

Total knee replacement has been known as one of the surgical procedures to recover the function of defective knee joint due to osteoarthritis and rheumatoid arthritis diseases. The demand for total knee replacement was predicted to increase by more than 607 % from 2005-2030 [1]. The materials which are often used as hip and knee joint prostheses are stainless steel 316L, cobalt-chromium alloys, titanium and titanium alloys because they are offering sufficient mechanical properties, those are non-toxic, corrosion resistance and wear resistance. Biomaterials for load-bearing joint prostheses should be acceptable to the body and able to withstand cyclic loading in a corrosive environment [2]. Today, titanium and titanium alloys have been widely used in biomedical application due to the biocompatible, corrosion resistance, high strength to weight ratio and low modulus [3-5].

Manufacture engineering is an important field in the production of joint prostheses especially to make a better quality and low-cost implant. The high quality and inexpensive implant from biomaterial are till interesting to study because it would have some impacts on the improvement of the quality of human life. The manufacturing of the joints prostheses can be done using casting technology. It is one of the competitive ways to reduce the production cost [6]. Investment casting or lost wax casting is known for the ability to produce a near-net shape workpiece with complex shape, dimensional accuracy, and good surface result. Unfortunately, the mechanical properties of castings are still inferior compared to the forgings because of the coarse of microstructure and porosity. Internal defects like shrinkage pore and porosity are the common problems which may cause deterioration of mechanical properties. The shrinkage pores are related to initiation sites of cracks[7].

To improve the quality of cast product, centrifugal investment casting has been developed recently [8–10]. The centrifugal force affects the improvement of fluidity, reduces pores, and increases dimension accuracy of casting [8].

Understanding the size, percentage, and distribution of porosity in the CP-titanium femoral knee casting is essential to characterize the casting soundness. X-ray and neutron computed tomography are the types of non-destructive tests which are increasingly used for industrial applications, specially for inspecting the inner defect or internal structure of the sample. The technique has been applied in numerous fields of research such as in cultural heritage for investigating the inner contour of clay sculpture [11]and the material for restoration of wooden-skull [12]. In civil engineering, this technique was used to study the water transports at fresh concrete that influencing in plastic shrinkage cracking [13]. In the metal casting, this method has been used to investigate the hidden defects in the Al casting [14] and titanium investment casting for aerospace industry [15,16]. There are still limited reports on the application of neutron computed tomography studies on the porosity in the centrifugal investment casting of CP-titanium. In this study, the effect of the rotational velocity of centrifugal investment casting on porosity size, percentage, and distribution was investigated using neutron computed tomography (NCT).

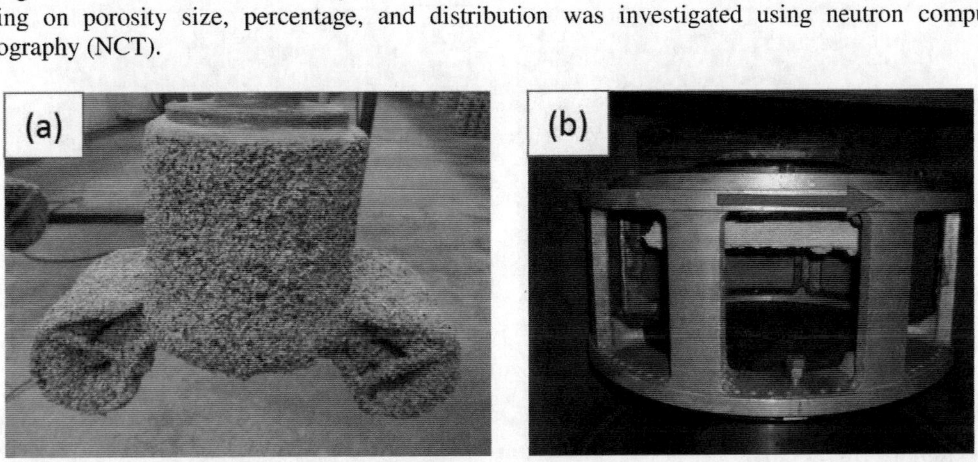

Figure 1. Centrifugal investment casting (a) assemblies tree and (b) rotational direction molding bracket in the casting chamber

2. Material and Methods

In this study, femoral knees were produced by using centrifugal investment casting method. The material used in this study was CP-titanium Grade 2. The dimensions of the as-cast CP-titanium femoral knee were 72 mm ×58 mm ×50 mm at its outermost points. Several steps were involved in this method.The first step was making the wax pattern of the femoral knee prosthesis. The wax femoral knee patterns were made by injecting the mushy wax into the aluminum die. Then, it was joined by wax sprue into the tree assemblies. After that, the tree assemblies were dipped into the ceramic slurry in eight layers and allowed it to dry and thus built a shell around the patterns (figure. 1a). Furthermore, the ceramic shells were steamed to remove the wax from the ceramic shell mold. Prior to the casting, the shell molds were baked in the oven to sinter and strengthen the green body of ceramic shell mold.

The melting process was performed in a vacuum electric-arc furnace equipment (Flash Caster, Japan). The ceramics molding was installed in mold bracket and then it was entered into the casting chamber (figure. 1b). Before the Argon gas (as shielding gas) was injected into the chamber, the casting chamber was set in vacuum condition. The CP-titanium was melted in a copper crucible that was installed in the middle and above the rotary mold bracket. In pouring the molten titanium into the ceramic mold, the crucible was tilted. The mold was turned vertically at anticlockwise direction (figure. 1b). The molten titanium was filled into the cavity due to the centrifugal force. This experiment used 45, 55, and 65 rpm rotational velocities. The melting, pouring, and solidifying processes were done in the argon environment. After the molten titanium was solidified, the shell was broken, and the sprues, gates and riser were cut off from the femoral knee casting. The as-cast femoral knee is shown in the figure 2.

The neutron radiography and tomography experiments were done in G.A. Siwabessy nuclear reactor Indonesia with 15 MW power. The femoral knee was rotated and bombarded by neutron beam. Neutron beam projection images were recorded at 200 s per image with 1° stepwise angle rotation of the sample. Octopus 8.5 software was used to reconstruct the 3D volume data, whereas the analysis was conducted using the software of Volume Graphics VG Studio Max 2.2 with additional modules to analyze the defect. The algorithm defect analysis was carried out using the default (v2.1) with the analysis mode voids, and material parameters were determined using determine surface. The parameters set of the software were the probability threshold of 0.5, the volume of the defect at the minimum of 2 mm^3, and the activated functions of neighborhood checking.

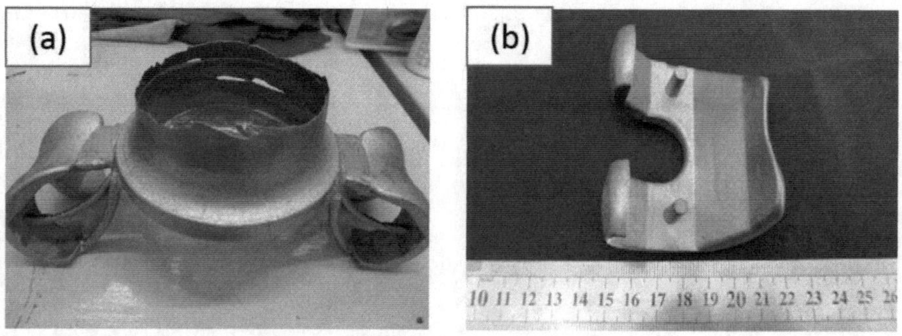

Figure 2. Femoral knee CP-titanium as-cast (a) with gating and (b) cut off from the sprue gating

3. Result and Discussion

Figure 3 shows the image neutron radiography of the as-cast femoral knee with different rotational velocities. In the case of neutron radiography images, solid areas are brighter than pores areas.There was no void on the surface and subsurface for the all casting parameter. The internal pores in the as-cast femoral knee were not visible while detected by the neutron radiography.

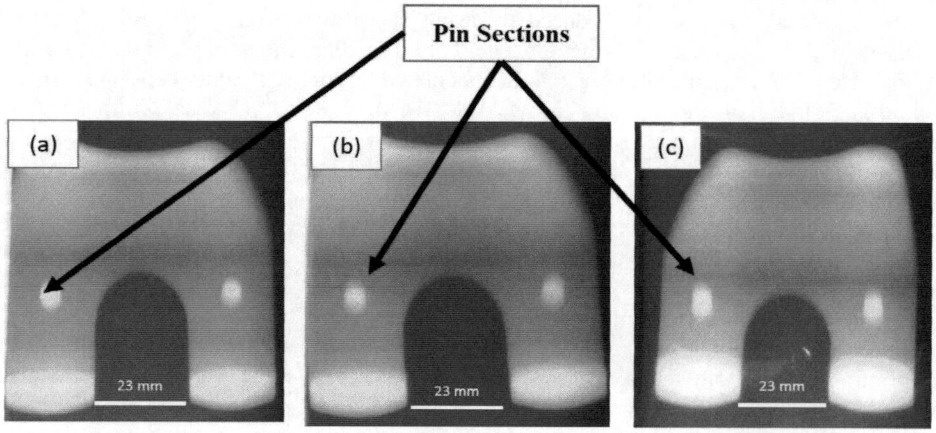

Figure 3. Neutron radiography of femoral knee investment casting with different rotational velocity, (a) 45 rpm, (b) 55 rpm and (c) 65 rpm

Figure 4 shows the defect volume of the internal porosity in femoral knee CP-titanium casting. The internal porosity was dominated by the small size of porosity (0-5 mm^3), while the large size porosity, (more than 5 mm^3), only occurred at several sites. The 3D images of the femoral knee were made full-transparent to view the size and location of the porosity. From the distribution and size of the defect shown in figure 4 (a-c), it can be seen that the location of the largest internal porosity in all the casting occurred under the pin section. The module of automatic defect analysis will be finding the pores and classifying the pores according to the volume size. The maximum size of pores in casting with the rotational velocity of 45, 55, and 65 rpm were 14.78 mm^3, 14,98 mm^3, and 14.65 mm^3, respectively. The percentage of voids in casting with the rotational velocity of 45, 55, and 65 rpm were 1.78, 2.78 and 1.78%, respectively. In particular, the casting with 55 rpm had the largest defects which were found in several sites of the casting, whereas the casting at 65 rpm had smaller volume and number of defects than those of the casting at 45 and 55 rpm.

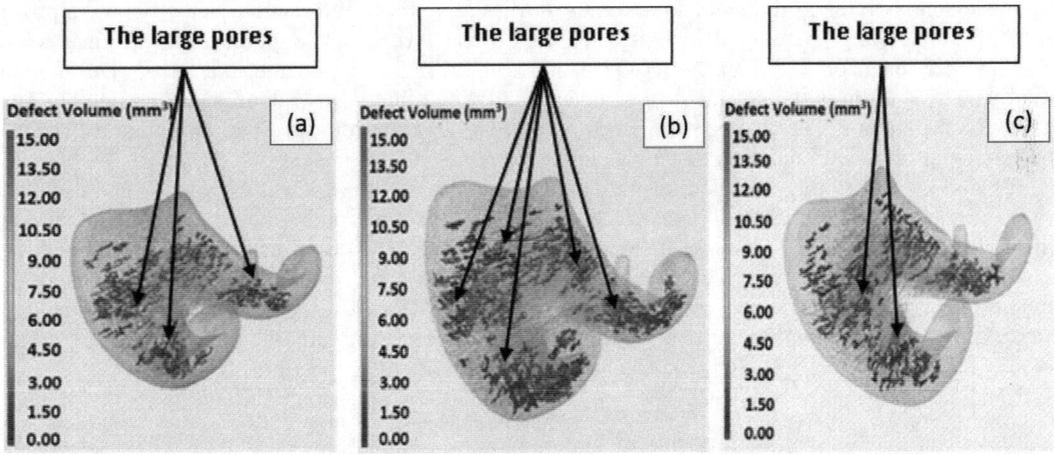

Figure 4. Defect volume of internal pores of as-cast femoral knee centrifugal investment casting, (a) 45 rpm, (b)55 rpm, (c) 65 rpm

Figure 5 shows the slice side view for the as-cast femoral knee with rotational velocity 45 rpm. It shows that the largest void was located beneath the pin section. Figure 6 shows the voids distributions which grouped in three range. The majority pores volume were (below 5 mm^3) for all the casting parameter.

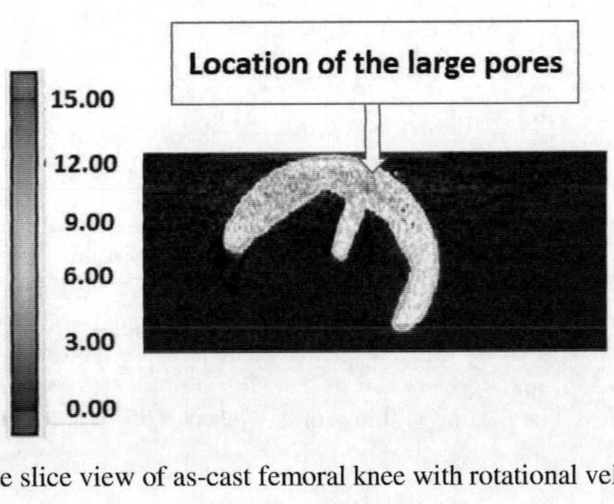

Figure 5. Side slice view of as-cast femoral knee with rotational velocity 45 rpm

In general, voids can be formed due to shrinkage during solidification and by the entrapped gas. The voids by the gas formed due to the process of releasing the gas from the face mold wall when the molten metal flows to fill it. The entrapped gasses could have been occurred due to the ceramic molds absorb air which contains moisture. The shape of pores has been grouped in three categorize by Wan as gas pores, gas-shrinkage pores, and shrinkage pores. The gas pores were near round shape. The shape of gas-shrinkage pore was presented as gas pore with many convexes and long tails. The shrinkage pore was very complex shape [17]. Figure 4 also shows gas-shrinkage and shrinkage porosity are dominant in the casting. In this study, the shape of porosity tends to the two type of pores are gas-shrinkage and shrinkage pores due to the shape sphere with a complex long tail and complex shapes pores.

The defects which are formed beneath the pin occur due to the shrinkage because the pin solidifies first before the thicker sections do. Shrinkage voids are the results from the reduction of the volume and contraction as the molten titanium solidifies [18]. Increased rotation velocity will improve the quality of castings because it increases the centrifugal force. Furthermore, the centrifugal force provides additional force so that the molten metal can fill the cavity in the mold [19]. The molten metal will flow towards the outside of the center of rotation radially. It produces casting without any defects on the surface. The casting parameters processes such as rotational direction, rotational radius, and rotational velocity play a significant role in the soundness of casting [19].

This study shows that the effect of the rotation velocity on the removal of pores is less due to a low rotation velocity. Withal, in this casting process, the mold is not preheated so that the cooling rate will be faster on the outside than the inside of the work piece. Therefore, the defects that formed are dominated by shrinkage pores. To reduce the internal pores, some researchers is using a design of complex gating system [20] and using hot isostatic pressing [8,21]. According to neutron computed tomography, the centrifugal investment casting of CP-titanium of the femoral knee with the rotational velocity of 65 rpm is the best result because it has fewest and smallest porosity.

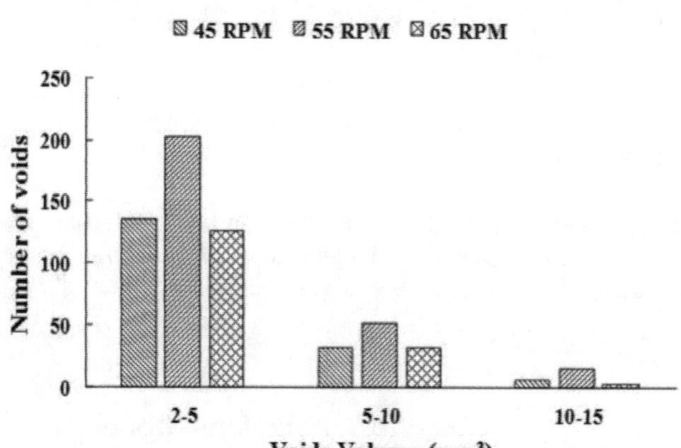

Figure 6. Voids distribution

4. Conclusions

In this study, the size and location of porosity of the as-cast CP titanium femoral knee which was made by the vertical centrifugal casting method with different rotational velocity were studied using neutron computed tomography. There are no defects were found in the surface and subsurface for all the casting. The maximum size of pores in casting with the rotational velocity of 45, 55, and 65 rpm are 14.78 mm^3, 14,98 mm^3 and 14.65 mm^3, respectively. The percentage of voids in casting with the rotational velocity of 45, 55, and 65 rpm are 1.78, 2.78 and 1.78%, respectively. The fewest and smallest porosity are achieved by casting with rotational velocity 65 rpm.

Acknowledgments
This work was financially funded by Ministry of Research, Technology and Higher Education of Indonesia through Doctoral Program Scholarship. The authors are grateful to Mr. Iwan and Mr. Baroto for his contribution and discussion about neutron radiography and the software.

References

[1] Kurtz S, Ong K, Lau E , Mowat F, Halpern M 2007 *J. bone Jt. Surg.* **89** 780–785.
[2] Mudali U K, Sridhar T M, Raj B 2003 *Sadhana.* **28 (3–4)** 601–637.
[3] Niinomi M 2008 *J. Mech. Behav. Biomed. Mater I.* **1** 30–42.
[4] Prasad S, Ehrensberger M, Gibson M P, Kim H, Monaco E A 2015 *J. Oral. Biosci.* **57 (4)** 192–199.
[5] Gepreel M A, Niinomi M 2013 *J. Mech. Behav. Biomed. Mater.* **20** 407–415.
[6] Lutjering G, Williams J C 2007 *Engineering Materials and Processes : Titanium.* 2nd ed, ed. Brain Derby Berlin Springer-Verlag Berlin Heidelberg p. 86
[7] Kuwazuru O, Murata Y, Hangai Y, Utsunomiya T, Kitahara S, Yoshikawa N 2008 *J. Solid. Mech. Mater. Eng.* **2 (9)** 1220–1231.
[8] Feng X, Qiu J, Ma Y, Lei J, Cui Y, Wu X, et al. 2016 *J. Mater. Sci. Technol.* **32** 362–371.
[9] Jia L, Xu D, Li M, Guo J, Fu H 2012 *Met. Mater. Int.* **18 (1)** 55–61.
[10] Rao A S, Tattimani M S, Rao S S 2015 *Metall. Mater. Trans B.* **46B** 793–799.
[11] Dinca M, Mandescu D 2015 *Phys. Procedia.* **69** 646–652.
[12] Osterloh K, Bellon C, Hohendorf S, Kolkoori S, Wrobel N, Nusser A, et al. 2015 *Phys. Procedia.* **69** 472–477.
[13] Wyrzykowski M, Trtik P, Munch B, Weiss, Vontobel P, Lura P 2015 *Cem. Concr. Res.* **73** 238–245.
[14] Balaskó M, Korösi F, Szalay Z 2004 *Appl. Radiat. Isot.* **61 (4)** 511–515.
[15] Richards W J, Barrett J R, Springgate M E, Shields K C 2004 *Appl. Radiat. Isot.* **61** 675–682.
[16] du Plessis A, Rossouw P 2015, *Case Stud. Nondestruct. Test Eval.* **3** 21–26.
[17] Wan Q, Zhao H, Zou C. 2014 *ISIJ Int.* **54 (3)** 511–515.
[18] Stefanescu D M 2009 *Macro-mass transport. Science and Engineering of Casting Solidification. 2nd ed.* New York Springer Science+Business Media LLC p. 68.
[19] Changyun L, Haiyan W, Lei X, Li, Changyun, Wang, Haiyan, Wu, Shiping, Xu, Lei., Wang K, Fu H 2010 *Rare Met. Mater. Eng.* **39 (3)** 388–392.
[20] Daming X, Limin J, Hengzhi F 2008 *China Foundry* **5 (4)** 249–257.
[21] du Plessis A, Rossouw P 2015 *J Mater Eng Perform.* **24 (8)** 3137–3141.

Coating of Sulfonic Silica onto Magnetite from Marina Beach Iron sand, Semarang, Indonesia

C Azmiyawati [a], L Suyati, Taslimah, R D Anggraeni

Department of Chemistry, Universitas Diponegoro, Semarang, Indonesia

[a] email: choiril.azmi@gmail.com, ph: +062 8157904691

Abstract. The mineral iron oxide is the main component of sand iron that are abundant in nature. Mineral iron oxide not yet widely applied into more useful products. The main component of iron ore is magnetite. Magnetite can be used as a basic ingredient in the manufacture of magnetite-modified silica adsorbent sulfonate. In this research, the adsorbent made from sulfonic functionalized silica-coated magnetic particle has been successfully produced, with the magnetite was obtained from iron sand at Marina Beach, Semarang Indonesia. This adsorbent was then used as a metal ion preconcentration media. From the research that it was found that the sulfonic has been bound to the silica marked by the emergence of element S on EDX. Whilst, the evidence that silica has coated on the magnetite could be seen from the SEM images which showed the morphology of sulfonic functionalized silica-coated magnetic particles were larger than the sulfonic functionalized silica without magnetite. From the DSC results showed that the addition of magnetite on sulfonic functionalized silica did not change the heat resistance of the sulfonic functionalized silica. Based on the XRD patterns show that magnetite sulfonate silica was formed.

1. Introduction

Indonesia is the largest archipelago in the world because it has 13 466 large and small islands with a fourth longest coastline in the world, which is 95.161 km [1]. The sand beaches in Indonesia with an iron content which is quite large and has not been used optimally. The main content of iron ore is iron oxide, one of which is magnetite. A magnetite nanoparticles are superparamagnetic [2] so that it can be used as a solid adsorbent support. The presence of magnetite in the adsorbent is expected to facilitate the desorption of the metal ions. However, magnetite has a poor stability so that the necessary modifications to the surface of the magnetite, one of which is the coating. Wang et al. [3] have done a magnetite coating with tetraethoxyisilane (TEOS) by sol-gel method at 50° C and pH 9. Magnetite coated silica has a better resistance to oxidation than magnetite without coating. Another source of silica as magnetite coating is sodium silicate (Na_2SiO_3). Sodium silicate can prevent the occurrence of corrosion reported that the magnetite coating using sodium silicate sol-gel method could improve resistance to thermal.

Research conducted Huang and Hu [4] using Mercaptopropyltrimethoxysilane (MPTMS) as a coating of silica coated magnetite by grafting generate magnetic with two layers, the first layer in the form of silica and a second layer MPTMS. Two of these layers increases the size of magnetite particles coated. The larger the size of the coated magnetite particles, the nature of magnetism is reduced so that it will complicate the process of separation of the adsorbent from the matrix. Efforts to overcome these problems can be done by changing the structure of the coated magnetite which has two layers into one layer, namely through the sol-gel process that is expected to form a single layer on the surface of magnetite. In general, this binding reaction depends on the ability of substitution between molecular modifier with the active side are distributed on the surface of a solid inorganic support. Of the few studies ever conducted showed that the metal adsorption by modified silica increases with increasing active group. Azmiyawati et al. [5] has studied the thermodynamics of adsorption of Mg (II), Cd (II), and Ni (II) on silica gel modified sulfonate to determine the type of bonding that occurs between the silica sulfonate with Mg (II), Cd (II), and Ni (II). Provided that the bond between the silica sulfonate with the metal ion is a chemical bond indicated by the value of metal ion adsorption adsorbent with greater than 20.9 kJ / mol. Other research is the use of modified silica sulphonate group as an adsorbent metal ions Ca (II), Cd (II), and Cu (II) which results from adsorption Ca (II) was better than Cd (II) and Cu (II) [6] In this research, sulfonate silica was coated on magnetite. The presence of silica sulfonate will increase against metal ion adsorption process. While the presence of magnetite as a support material of the compound will increase the metal ion desorption process. So expect the recovery of metal ions already desorption better.

2. Experimental

2.1. The Chemicals

Magnetite was collected from Marina Beach iron sand, Semarang, Indonesia. Chemicals for synthesis of MSS included 4-amino-5-hydroxy-naphtalene-2,7-disulfonic acid monosodium salt ($C_{10}H_8NNaO_7S_2$) (AHNSNa) as the sulfonato source, 3-chloroprophyltrimethoxysilane (CPTS), triethylamine ($C_6H_{15}N$), and hydrochloric acid. Sodium bicarbonate solution was used for the solvent of AHNSNa, diethyl ether and acetone for washing solvents. All chemicals were of analytical reagent grade (Merck) and used as receipted.

2.2. Procedure

Magnetite samples obtained by extracting magnetite from iron sand. The first step magnetic sample was grounded to obtain a powder with the particle size of 106 μm. The powder then was mixing and stirring with sodium silicate solution for two h. The second step of synthesis was the attachment of sulfonate groups on the CPTS by mixing 50 mL solutions of AHNSNa (4.08 %, w/v) in sodium carbonate solution 0.05 M, 4.85 mL triethylamine and six mL CPTS added dropwise and constantly stirring for 5 hours. The solution then was added to 100 mL magnetic sodium silicate solution and 3 M of HCl solution dropwise, and the suspension was then constantly stirred for 24 h at a room temperature (25 °C) until the gel was formed. This gel was aged for 2 h at 70°C, and dried in a vacuum for 24 h. The gel was filtered, and the residue was washed with distilled water till free of acid. The gel was dried at 70 °C for 2 h to obtain magnetite silica sulfonate (MSS). The functional group of the MSS was evaluated from FTIR spectra. The surface topology was identified with SEM-EDX. The heat resistance of MSS was evaluated from DSC.

3. Result and discussion

3.1. Magnetit Precursor from Iron sand

Iron sand is one of the abundant natural resource in Indonesia that haven't been used well. The main content of the iron sand is iron oxide, which is magnetite. Magnetite is an iron oxide that has the strongest magnetic properties so that it can be use as solid support for adsorbent. However, magnetite has a poor stability, so it's necessary to modify the surface of the magnetite by synthesis of magnetite-modified silica.

3.2. Chemical Structure of SS and MSS

The existence of functional groups of the Si-O-Si, Si-OH, Fe-O and S-O can be shown through FT IR spectra. Compared with silica sulfonate without magnetite , magnetite-silica resemblance pattern, followed by a decrease in the intensity of several bands, shifting some of the peak and the emergence of new bands in the absorption wavelength around 300 cm^{-1} which indicate the presence of iron oxide. The presence of magnetite particles indicated by the vibration absorption peak of Fe-O from bonding in the molecule Fe3O4 at wavelength 324-354 and 550-586 cm^{-1} are attributed to the formation of the ferrite phase. Different from magnetite, FT-IR spectra of silica coated magnetite there are pronounced changes, particularly at a region of 1300-700 cm^{-1}, indicating the presence of silica coating. The first change is the decrease of absorbance intensity that is probable doe to the magnetite coverage of magnetite by silica as well as AHNSNa. The presence of silica coated on magnetite is shown by a characteristic band at 450- 470 cm^{-1} from bending vibration of Si-O-Si and. Asymmetry bending vibration corresponding to Si-O-Si bonding is revealed with absorbance band at 787-802 cm^{-1} and 1072-1126 cm^{-1}. Absorbance bands in coated magnetite IR spectra around 1620-1640 cm^{-1} and 3400-3450 cm^{-1} come from bending and stretching vibration, respectively, of –OH groups from both Fe-OH and Si-OH. Stretching vibration of Si-O-H bonding results in absorbance band at 950-960 cm^{-1}. In IR spectra of magnetite coated only with silica, the absorbance of Si-OH vibration does not appear clearly due to overlap with the broad band of stretching vibration from Si-O-Si. In comparison to magnetite and silica coated magnetite, IR spectra of mercaptosilica coated magnetite gives the characteristic absorbance of propyl and mercapto groups from MPTMS. The presence of propyl groups can be indicated by bands at 1040-1050 cm^{-1} dan 1400-1480 cm^{-1} assigned to stretching vibration of –C-C- chain, bending of CH2-, respectively. Additionally, C-H bonding in propyl groups results in absorbance at 1240 cm^{-1} and 2920-2940 cm^{-1} corresponding to bending and asymmetric vibration of C-H. The increase of AHNSNa added during the coating process results in inclining the band intensity characteristics for propyl and sulfonato groups. In addition, a silica coating of magnetite indicated by the appearance of vibration corresponding to bond Si-O-Si at a wavelength of 1000-1200 and Si-O at 970 cm^{-1}. The FT-IR spectra of MSS compared to that of SS are shown in Fig. 1.

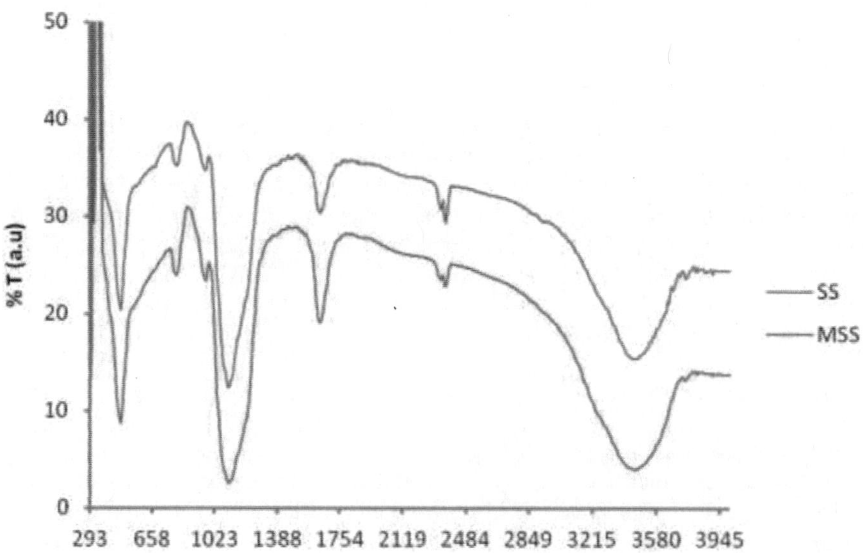

Figure 1. FT-IR spectra of SS and MSS

It can be seen from Fig. 1 that absorbance band for MSS at 972.12 cm^{-1} is broader than that of SS (at 956.69 cm^{-1}). It is probable due to overlaps between of Si-O stretching vibration from Si-OH and symmetric stretching of C-S at around 1,095 ~ 1,085 cm^{-1}. Also, the two sharp absorbance bands at 2931.80 and 2962.66 cm^{-1} for the SS and MSS indicate the presence of stretching vibration of C-H from methyl and methylene groups, respectively [7], [8]. Absorption bands for the stretching vibration of S = O characteristic at 1,028, 1,254, and 1,084 cm^{-1} can not be seen because those overlap with broad band of Si-O from siloxane (at around 1,072 cm^{-1}) [9], [10]. Broad absorption band from 3,100 to 3,450 cm^{-1} is predicted as an overlapping peak between the characteristic band for the O-H vibration (3,450 cm-1), and symmetry was stretching vibration of N-H (3,270 cm^{-1}) [11].

The success of the MSS synthesis can be observed from the composition of the hybrid material by an EDX analyzer (Table 1). If the material contains sulfur, can be expected to emerge successful synthesis SS and MSS. The presence of Fe in the MSS was indicated by the appearance of the element in EDX analyzer, MSS analysis results. This shows that the right of magnetite has been coated by silica sulfonate. The binding sulfur atom was supported by the emergence of other atoms as a supplement, such as the addition of oxygen.

Table 1. The average mass (in w/w) of the element S and Fe in SS and MSS based on data from EDX

Sample	**Element**	**Average mass (% w/w)**
Silika Sulfonat (SS)	sulfur, S	23,44
	Silika Oxide, SiO_2	76,56
Magnetit Silika Sulfonat (MSS)	Sulphur, S	21,02
	Silika Oxide, SiO_2	78,40
	iron oxide, (Fe_2O_3)	0,57

The presence of iron oxide indicates that magnetite compounds have been tacked on silica gel. While the presence of sulphur showed, a magnetit silica was attached by sulfonato groups. The above structure elucidation supports the reaction model proposed by Azmiyawati et al. [5] that the attachment of AHNSNa onto silica gel occurs in a sequence of two distinct steps. The first step is consisted of binding CTS onto AHNSNa to form disulfonato-aminopropyltrimethoxysilane (PTMDS), Eq. (1), and the second stage, APTMDS reacts with sodium silicate (from RHA) to produce disulfonato-silica hybrid (DSSH), Eq. (2).

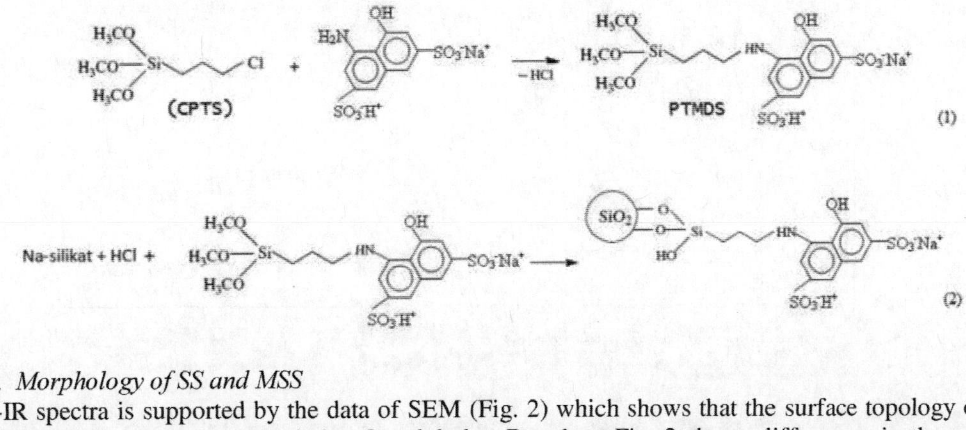

3.3. *Morphology of SS and MSS*

FT-IR spectra is supported by the data of SEM (Fig. 2) which shows that the surface topology of the SS looks and does not accumulation for globular. Based on Fig. 2 shows differences in the surface topology of the two materials. MSS looks to have larger grains than the SS. This size difference may

IOP Conf. Series: Materials Science and Engineering **172** (2017) 012009 doi:10.1088/1757-899X/172/1/012009

be result in the gel formation of those materials. SS formation bonding with a sulfonato compound, formation of siloxane framework stop when Si atom bonds to organic groups. Resulting the growth smaller ball polymer. MSS formation occurs continually and grow produce siloxane bonding, leading to formation of larger ball polymer [12]. Coating silica sulfonat onto magnetite, the larger ball polymer formed.

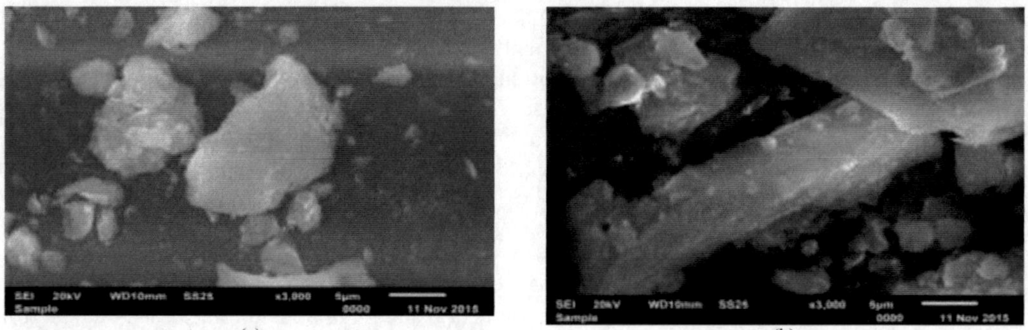

(a) (b)

Figure 2. Surface topology of (a) SS and (b) MSS from SEM to 3,000 times magnification.

3.4. Physical Properties of SS and MSS

The effect of magnetite on the formation of magnetite sufonate silica as shown by their XRD patterns, is illustrated in Fig.3. In the samples obtained by adding the amount of magnetite (b), the characteristic peak of the XRD patterns correspond to the magnetit (JCPDS 19-0629) [13]. And then there is no diffraction peak in the XRD pattern of (b). This shows that magnetite ssulfonate silica was formed.

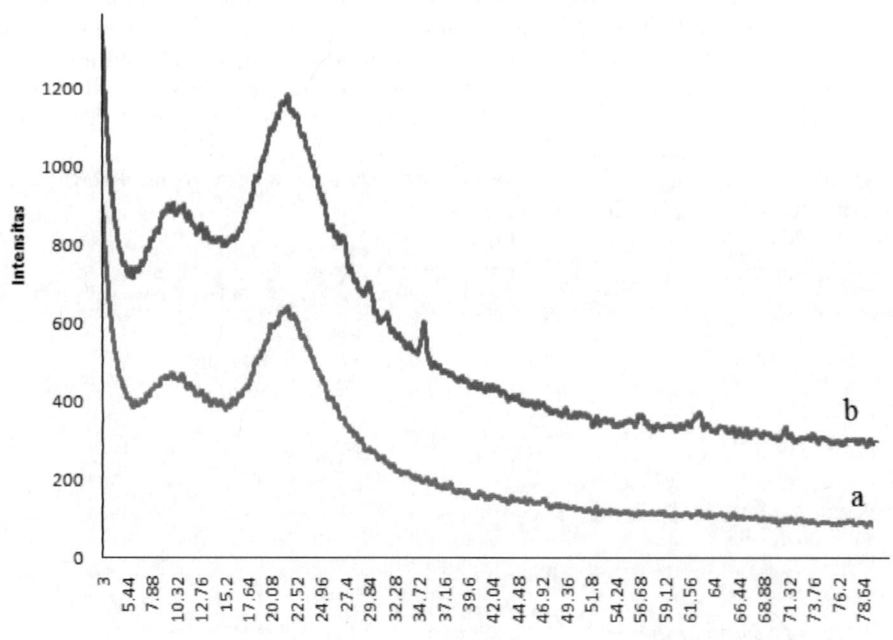

Figure 3. XRD patterns of (a) SS and (b) MSS

The physical properties of a material can be seen one of them through the material resistance to changes in temperature. The heat resistance of a material can be determined via DSC pattern (Fig. 3).

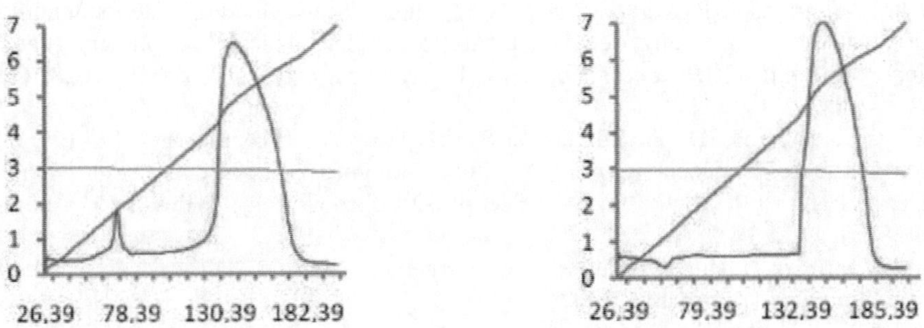

(a) (b)

Figure 3. DSC of (a) SS and (b) MSS

4. Conclusion

The iron sand can be used as material for the synthesis of magnetit-based material, such as magnetite silica sulfonate (MSS). The attached of magnetite on silica sulfonate will increase the size of ball polymer in the MSS and the emergence of iron oxide on FTIR spectra was obtained from MSS. In general sulfonate silica coating on magnetite has no effect on the thermal resistance of the MSS

References

[1] United Nations, United Nations Convention on The Law of The Sea, 10. December 1982

[2] H.A. Eivari and A. Rahdar, "Some Properties of Iron Oxide Nanoarticles Synthesized in Different Conditions", World Appl. Programming, 3, 52-55, 2013

[3] Wang Zhongzhu, Bi Hong, Liu J., Sun Tao, and Wu X., Magnetic and microwave absorbing properties of polyaniline/g-Fe_2O_3 nanocomposite. Journal of magnetism and magnetic materials 320 (2132-2139) Department of Chemistry and Department of Electronic Engineering, AnhuiUniversity, Hefei 230039, 2008, China.

[4] C. Huang and B. Hu, "Silica-Coated Magnetic Nanoparticles Modified with γ-Mercaptopropyltrimethoxysilane for Fast and Selective Solid Phase Extraction of Trace Amounts of Cd, Cu, Hg and Pb in Environmental and Biological Samples Prior to Their Determination by Inductively Coupled Plasma Mass Spectrometry", *Spectrochim. Acta*, 63B, 437-444, 2008.

[5] C. Azmiyawati, Nuryono, and Narsito, "Adsorption of Mg(II) and Ca(II) on Disulfonato- silica Hybrid," *Indo. J .Chem.* vol. 12, no. 3, pp. 223-228, November 2012.

[6] C. Azmiyawati, Nuryono, dan Narsito, "Synthesis of Disulfonato-Silica Hybrid from Rice Husk Ash", *Journal of Medical and Bioengineering (JOMB),* vol. 3, no 4, 301-305, December 2014.

[7] D. L. Pavia, G. M. Lampman, G. S. Kriz, and J. R. Vyvyan, "Introduction to Spectroscopy, fourth edition," *Brooks/Cole Cengage Learning*, Washington, pp. 33-37, 2007.

[8] A. G. S. Prado, J. A. A. Sales, R. M. Carvalho, J. C. Rubim, and C. Airoldi, "Immobilization of 5-amino-1,3,4-thiadiazole-thiol onto silica gel surface by heterogeneous and homogeneous routes," *Journal of Non-Crystalline Solid,* vol. 333, pp. 61-67, 2004.

[9] S. Shylesh, S. Sharma, S. P. Mirajkar, and A. P. Singh, "Silica functionalized sulphonic acid groups: synthesis, characterization and catalytic activity in acetalization and acetylation reactions," *Journal of Molecular Catalysis A: Chemical*, vol. 212, pp. 219-228, October 2003.

[10] V. Troitsky, T. Berzina, D. Shchukin, G. Sukhorukov, V. Erokhin, and M. P. Fontana, "Colloids surf. A: Physicochem," *Eng. Asp.*, vol. 245, pp. 163, 2004.

[11] H. Günzler and H-U. Gremlich, "IR spectroscopy and introduction," Wiley-VCH, Weinheim, 2002.

[12] U. Schubert and N. Hüsing, "Synthesis of inorganic materials," Wiley-VCH, Weinheim. (Federal Republic of Germany), 2000.

[13] H. Setyawan, F. Fajaroh, W. Widiyastuti, S. Winardi, I.W. Lenggono, N. Mufti, "One-step synthesis of silica coated magnetit nanoparticles by electrooxidation of iron in sodium silicate solution", J. Nanopart Res, 2012, 14:807, DOI 10.1007/s11051-012-0807-7

Acknowledgment

This study was financially supported in part by the Directorate General of Higher Education, Ministry of Riset, Technology, and Higher Education, Republic of Indonesia through *'Penelitian Fundamental'* Program, Number: 022/SP2H/LT/DRPM/II/2016, date: 17 Februari 2016.

IOP Conf. Series: Materials Science and Engineering **172** (2017) 012010 doi:10.1088/1757-899X/172/1/012010

The elevation effect on water-soluble polysaccharides and DPPH free radical scavenging activity of *Ganoderma lucidum* K

C Darsih[1], W Apriyana[1], S Nur Hayati[1], V Taufika Rosyida[1], Hernawan[1] and C Dewi Poeloengasih[1]

[1]Balai Penelitian Teknologi Bahan Alam, Lembaga Ilmu Pengetahuan Indonesia, Yogyakarta, Indonesia 55861

E-mail: cici001@lipi.go.id

Abstract. Water soluble polysaccharide is one of the important phytochemical in *Ganoderma lucidum* K. Phytochemicals in the plants, microorganisms, and plants were affected by internal and external factors. The objective of the research was to evaluate the effect of elevation on the water-soluble polysaccharides and its DPPH radical scavenging activity. We found that the water-polysaccharides in mushroom from Godean (elevation <100 mamsl) (35.28 ± 0.31%) higher than Kaliurang (elevation 800 mamsl) (25.17 ± 1.85%). The DPPH free radical scavenging activity of *Ganoderma lucidum* K from Godean (IC_{50} 11.5 ± 0.29 mg/mL) higher than Kaliurang (IC_{50} 14.4 ± 0.27%).

1. Introduction

Ganoderma lucidum K known as Lingzhi in Chinese and Reishi in Japanese [1]. *G. lucidum* is one the mushrooms that used as medicine in Asian countries. Several researches showed that metabolites contained in this mushroom exhibited specific biological activities, such as anti-diabetic effect, neuroprotective effects, immunomodulatory, antitumor, antioxidant, hepatoprotective, anti-hypertensive, anticancer, and antimicrobial [1-5]. Bioactive compounds found in *G. lucidum* include polysaccharides, triterpenes, protein, amino acids, mineral elements, vitamin, and phenols [6]. Two bioactive compounds, polysaccharides, and triterpenes are recently considered as the important constituent in this mushroom.

Several researches have been reported the important of polysaccharides from *G. lucidum.* Water-soluble polysaccharides from four strain *G. lucidum* showed antibacterial activity and the most sensitive bacteria toward polysaccharides was *Micrococcus luteus* ATCC 10240 with MIC of 0.63-1.25 mg/mL [7]. Water soluble polysaccharides of this mushroom demonstrated that polysaccharides reduced cerebral infarct area by oral administration in rats (100, 200, and 400 mg/kg) [8]. Polysaccharides of *G. lucidum* with a molecular weight of 5.2 kDa showed more effective in free radical scavenging and Fe^{2+} chelating than polysaccharides with a molecular weight of 15.4 kDa [9].

Metabolites in plants could be affected by corporal variations, ecological conditions, terrestrial variations, genetic factors and evolution, and political/social conditions [10, 11]. Gua *et al.* reported that secondary metabolites in *Scutellaria baicalensis* were influenced by internal and external environmental. Temperature affected the metabolites in *S. baicalensis* positively and chemical constituents high latitude lower than in high latitude [12]. Another research showed that production of secondary metabolites was influenced by environmental factors such as humidity, temperature, intensity of light, supply waters, minerals, and CO_2[13]. Elevation factor where the plant was grown also affected the biochemical composition in plants significantly. The research reported that increasing elevation effected significantly the variation of metabolites and increased the chloronogenic acid and fat concentrations on the traditional cultivar of coffee bean. Variation of chloronogenic acid and fat in Arabica hybrids were little effected by increasing elevation [14]. Those results indicated that the important environmental on metabolites production and encourage this research to know the effect of elevation factor on water-soluble polysaccharides and the antioxidant activity of *G. lucidum* extract.

2. Materials and methods

2.1. *Materials*

The *G. lucidum* was determined taxonomically at Pharmaceutical Biology Division, Faculty of Pharmacy, Universitas Gadjah Mada. The mushroom was cultivated in same media containing sawdust, bran, and dolomite in different elevation, Merapi mountain side, Kaliurang (800 mamsl) and lowland at Godean, Sleman (<100 mamsl). Furthermore, the mushroom was dried under direct sunlight and grounded to be powder.

2.2. *Water-Soluble Polysaccharides Analysis*

The water-soluble polysaccharides were determined based on the color reaction of polysaccharides and their derivatives with phenol and concentrated sulfuric acid [15]. Samples (100 mg) was extracted using 5 mL HCl 2 M at 100°C in the water bath (Memmert) for two hours. The filtrate and fruit bodies of *G. lucidum* were separated by filter paper, and then the filtrate was transferred into the centrifuge tube 5 mL. Further, 1 mL of filtrate was added with 5% phenol 0.5 mL and 2.5 mL of H_2SO_4 (98% v/v) (Merck). The mixture was then shaken for 2 minutes and incubated using water bath at 100°C for 15 minutes. The water-soluble polysaccharides concentration of *G. lucidum* was analyzed by UV-Visible spectroscopy (Dynamica Halo RB-10 spectrophotometer) at 490 nm. The blank solution contain 1 mL of aquadest, 0.5 mL of 5% phenol (Merck) and 2.5 mL of H_2SO_4 (98% v/v) (Merck). The standard glucose (Sigma, Milwaukee, WI, USA) was used as a standard solution.

2.3. *DPPH Radical Scavenging Assay*

The DPPH assay was conducted to find the radical scavenging activity of *G. lucidum* from two different elevations of land following the Shimada, *et.al* method [16] with some modification. The 2 mL methanol extracts of dried *G. lucidum* were added with 2 mL of DPPH solution at concentrations of 0.25, 0.5, 1.0, 1.5, and 2.0% w/v. The mixture was then shaken and incubated at room temperature in the dark for 30 minutes. The samples and control were analyzed by ELISA at 517 nm. The equation 1 was used to calculate the percent DPPH radical scavenging effect, where A0 is DPPH solution absorbance without sample and A1 is DPPH solution absorbance with sample *G. lucidum*.

$$\%\ \text{inhibition} = \left[\frac{A0-A1}{A0}\right] \times 100 \ \ldots\ldots\ldots\ldots\ldots 1$$

The antioxidant activities expressed as IC_{50} values for comparison. Effectiveness in antioxidant activities inversely corresponded with IC_{50} value. The IC_{50} value is the effective concentration at which the DPPH radicals were inhibited by 50%. The IC_{50} value was measured by linear regression of plots, where the abscissa (x) describe the concentration of tested methanolic extracts and the ordinate (y) describe the average percentage of inhibitory effect.

3. Results and discussions

The chemical compositions and concentrations in plants, fungi, and microorganisms were influenced by internal and external environmental. In this research, the mushroom *G. lucidum* was grown in different elevation and its effect on water-soluble polysaccharides was evaluated. Elevation difference related to the change of the temperature and intensity of light in the field [14]. The temperature in Kaliurang is 24-27 °C, lower than the temperature in Godean (29-33 °C). Relative humidity in Kaliurang (85-95%) higher than that Godean (75-85%), and the light intensity in Kaliurang (300-500 lux) lower than that Godean (400-600 lux). The physical geographic characteristics data are shown in Table 1.

Table 1. Physical geographic characteristics of each mushroom growing location

Physical geographic characteristics	Merapi mountainside, Kaliurang	Lowland at Godean, Sleman
Elevation	800 mamsl	<100 mamsl
Temperature	24 - 27 °C	29 - 33 °C
Relative humidity	85 - 95%	75 - 85%
Illuminance/light intensity	300 - 500 lux	400 - 600 lux

The results showed that water-soluble *G. lucidum* polysaccharides from Godean higher than *G. lucidum* mushroom from Kaliurang, shown in Table 2. The difference of polysaccharides concentration indicated that variation of elevation effected phytochemical production in mushrooms. The difference of phytochemical in elevated variations related to temperature, relative humidity, and light intensity. Variation of temperature can affect the regulation of metabolic in plant [17]. This results in line to previous research which the metabolites concentration were influenced by temperature and light. The research reported that biomass and ginsenoside production in *Panax ginseng* were influenced by variations of temperature and light quality [18]. Another research reported that leaf senescence and secondary metabolites of *Panax quinquefolius* root increased with elevation of temperatures. Increasing temperature to a 5°C affected the *Panax quinquefolius* [17].

Antioxidant activities of this mushrooms related to phytochemical in *G. lucidum*, mainly water-soluble polysaccharides. The result exhibited DPPH free radical activity of *G. lucidum* from Kaliurang and Godean with IC_{50} of 14.4 mg/ml and 11.5 mg/ml, respectively. This is suggested that concentration of metabolites for antioxidant in both mushrooms are proportional with water-soluble polysaccharides level.

Table 2. Water-soluble polysaccharides and DPPH free radical scavenging of *G. lucidum*

No	Samples	Water-soluble polysaccharides (%)	% inhibition at 10 mg/mL	IC_{50} value (mg/mL)
1.	GM	25.17 ± 1.85	34.81 ± 5.54	14.4 ± 0.21
2.	GG	35.28 ± 0.31	47.66 ± 3.94	11.5 ± 0.29

GM : *G. lucidum* was grown in Merapi mountain side, Kaliurang
GG : *G. lucidum* was grown in Lowland at Godean, Sleman

IOP Conf. Series: Materials Science and Engineering **172** (2017) 012010 doi:10.1088/1757-899X/172/1/012010

4. Conclusions

The chemical constituents on *G. lucidum* was influenced by external factor. Increasing of elevation decreased the water soluble-polysaccharides on *G. lucidum* as well as DPPH free radical scavenging activity of *G. lucidum*.

Acknowledgments

The authors acknowledge the Indonesian Institute of Sciences for providing financial support through Kegiatan Penguatan Kompetensi Inti Satuan Kerja 2015, Sanggar Tani Media Agro Merapi and SME Gubug Jamur Sitiram for supply fresh *G. lucidum*.

References

[1] Ma HT, Hsieh JF and Chen ST 2015 *Phytochem.* **114** 109-13
[2] Nie S, Zhang H, Li W and Xie M 2013 *Bioact. Carbohyd. Diet Fibre* **1(1)** 10-20
[3] Ferreira ICFR, Heleno SA, Reis FS, Stojkovic D, Queiroz MJRP, Vasconcelos MH and Sokovic M 2015 *Phytochem.* **114** 38-55
[4] Li Y-b, Wang J-l and Zhong J-J 2013 *Process Biochem.* **48(2)** 331-39
[5] Huang SQ and Ning ZX 2010 *Int. .J Biol. Macromol.* **47(3)** 336-41
[6] Batra P, Sharma AK and Khajuria R 2013 *Int. J. Med. Mushrooms* **15(2)** 127-43
[7] Skalicka-Wozniak K, Szypowski J, Los R, Siwulski M, Sobieralski K, Glowniak K and Malm A 2012 *Acta Soc. Bot. Pol.* **81(1)** 17-21
[8] Zhou ZY, Tang YP, Xiang J, Wua P, Jin HM, Wang Z, Mori M and Cai DF 2010 *J. Ethnopharmacol.* **131(1)** 154-64.
[9] Liu W, Wang H, Pang X, Yao W and GaoX 2010 *Int. J. Biol. Macromolec.* **46(4)** 451-57
[10] Pavarini DP, Pavarini SP, Niehues M and Lopes NP 2012 *Anim. Feed Sci. Technol.* **176(1- 4)** 5-16
[11] Figueiredo AC, Barroso JG, Pedro LG and Scheffer JJC 2008 *Flavour Frag. J.* **23(4)** 213- 26.
[12] Guo L, Wang S, Zhang J, Yang G, Zhao M, Ma W, Zhang X, Li X, Han B, Chen M and Huang L 2013 *Sci China Life Sci.* **56(11)** 1047-56.
[13] Ramakrishna A and Ravishankar GA 2011 *Plant Signal Behav* **6(11)**1720-31
[14] Bertrand B, Vaast P, Alpizar E, Etienne H, Davrieux F and Charmetant 2006 *Tree Physiol.* **26(9)** 1239-48.
[15] Sutherland IW 1990 *Biotechnology of microbial exopolysaccharides* (New York : Cambridge University press) p 1-151.
[16] Shimada K, Fujikawa K, Yahara K and Nakamura T 1992 *J. Agric. Food Chem.* **40** 945–48
[17] Jochum GM, Mudge KW and Thomas RB 2007 *Am. J. Bot* **94(5)** 819-26
[18] Yu K.-W, Murthy HN, Hahn E-I and Paek K-Y 2005 *Biochem. Eng. J.* **23(1)** 53-56

IOP Conf. Series: Materials Science and Engineering **172** (2017) 012011 doi:10.1088/1757-899X/172/1/012011

Development of high performance liquid chromatography method for miconazole analysis in powder sample

D Hermawan[1], Suwandri[1], U Sulaeman[1], A Istiqomah[1] and H Y Aboul-Enein[2]

[1] Department of Chemistry, Faculty of Mathematics and Natural Sciences, Jenderal Soedirman University, Purwokerto, Indonesia
[2] Department of Pharmaceutical and Medicinal Chemistry, Pharmaceutical and Drug Industries Research Division, National Research Centre, Dokki, Cairo, Egypt

Email: dadanphd@gmail.com

Abstract. A simple high performance liquid chromatography (HPLC) method has been developed in this study for the analysis of miconazole, an antifungal drug, in powder sample. The optimized HPLC system using C_8 column was achieved using mobile phase composition containing methanol:water (85:15, v/v), a flow rate of 0.8 mL/min, and UV detection at 220 nm. The calibration graph was linear in the range from 10 to 50 mg/L with r^2 of 0.9983. The limit of detection (LOD) and limit of quantitation (LOQ) obtained were 2.24 mg/L and 7.47 mg/L, respectively. The present HPLC method is applicable for the determination of miconazole in the powder sample with a recovery of 101.28 % (RSD = 0.96%, n = 3). The developed HPLC method provides short analysis time, high reproducibility and high sensitivity.

1. Introduction

Many types of drugs are produced synthetically and commercially available in dosage forms. One example of important drugs used as a medication is antifungal drugs. An antifungal drug is used to treat deep infections caused by a fungus. This drug can be divided into several classes, including triazoles and imidazoles. Miconazole is an imidazole antifungal agent (figure 1). It is commonly used to the skin or to mucous membrane to cure fungal infections, due to its high therapeutic properties. It has been extensively applied in the management of dermal, oral and vaginal mycosis. It is used in a variety of pharmaceutical formulations such as injections, tablets, oral gels, creams, ointments, topical powders and vaginal suppositories. The most usual application forms include creams, ointments or gels at 2.0 % concentration level, alone or associated with anti-inflammatory steroids, or other antimicrobials such as gentamicin for the treatment of dermatitis. Diaper dermatitis, a common dermatologic disorder in infancy, frequently associated with Candida albicans infections, is currently treated with ointments containing 0.25 % miconazole nitrate [1-2].

Chromatographic methods have achieved a great reputation in separation science; mainly using high performance liquid chromatography (HPLC) method [3-10]. The chromatographic process can be described as a separation technique that includes mass-transfer between stationary and mobile phase. HPLC uses a liquid mobile phase to separate the components in a mixture. The stationary phase can be in a liquid phase or a solid phase. These components are forced to flow through a chromatographic column under high pressure after being dissolved in a solvent. The mixture is then separated into its components. In our previous study, the HPLC method with ultraviolet (UV) detector has been successfully developed for the determination of miconazole nitrate in a cream sample using C_{18} column and ACN:MeOH (90:10 v/v) as mobile phase [11]. The inherent ability of HPLC for the quantification of drugs possessing good efficiency, reproducibility, selectivity and sensitivity combined with the ease of method development and rapid analysis make it a very good alternative for drugs analysis.

It is thus of interest in this study to develop the HPLC method for miconazole analysis in the different pharmaceutical sample. The developed HPLC method using C_8 column and MeOH:water as the mobile phase is effective, fast and reproducible for the determination of miconazole in the powder sample.

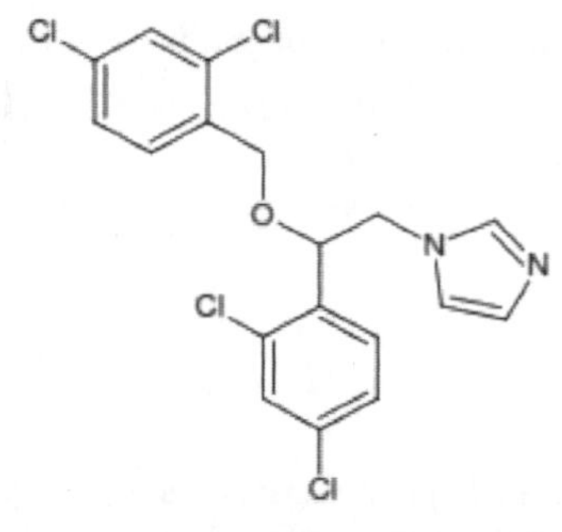

Figure 1. Chemical structure of miconazole

2. Method

2.1 Materials

Miconazole was obtained from Sigma (St. Louis, MO, USA). Methanol used is HPLC grade. Deionized water was taken from a Millipore Simplicity. All solvents were degassed prior to usage. Powder sample was obtained from local drug market. A stock solution of miconazole (1000 mg/L) was prepared by dissolving miconazole standard in methanol. A series of standard working solution with different concentrations ranging from 10 to 50 mg/L was prepared by further dilution of the stock solution. The stock and all standard working solutions were labeled and sealed with aluminum foil to avoid evaporation, and were stored in the refrigerator prior to use. The powder sample containing 2% of miconazole were weighed. It was sonicated in 10 mL methanol for 10 minutes. The extract was allowed to cool and then was filtered. The volume of the sample solution was adjusted to 25 mL with methanol. Finally, the sample was injected to the HPLC-UV method.

2.2 HPLC Instrument

The HPLC system used was Hitachi L-2000 series (Japan), equipped with a Model L-2130 pump, an on-line solvent vacuum degasser, an autosampler with 5 μL injection loop and a UV-Vis detector L-2420. The separation was carried out in a Phenomenex C8 column Luna 10μ (150 x 4,6 mm). The mobile phase consisted of methanol:water (85:15, v/v). The system was operated isocratically at flow rate 0.8 min/mL and UV wavelength 220 nm.

3. Results

3.1 Optimization of Mobile Phase

Effect of different composition of mobile phase (methanol:water) was first explored in this study to obtain the optimum HPLC condition. The mobile phase composition was varied in three different percentages of methanol-water (v/v): (85:15), (90:10), and (95:15). The typical result of HPLC chromatogram for miconazole with different composition of the mobile phase is shown in figure 1. It can be seen that the retention time of miconazole was gradually increased when the percentage of water was increased. Addition of water reduces the elution strength of the mobile phase; therefore the analyte was eluted slower. Methanol-water (85:15, v/v) has been chosen as the optimized mobile phase composition because it gave the best baseline of miconazole peak (base to base) with retention time less than 7 min.

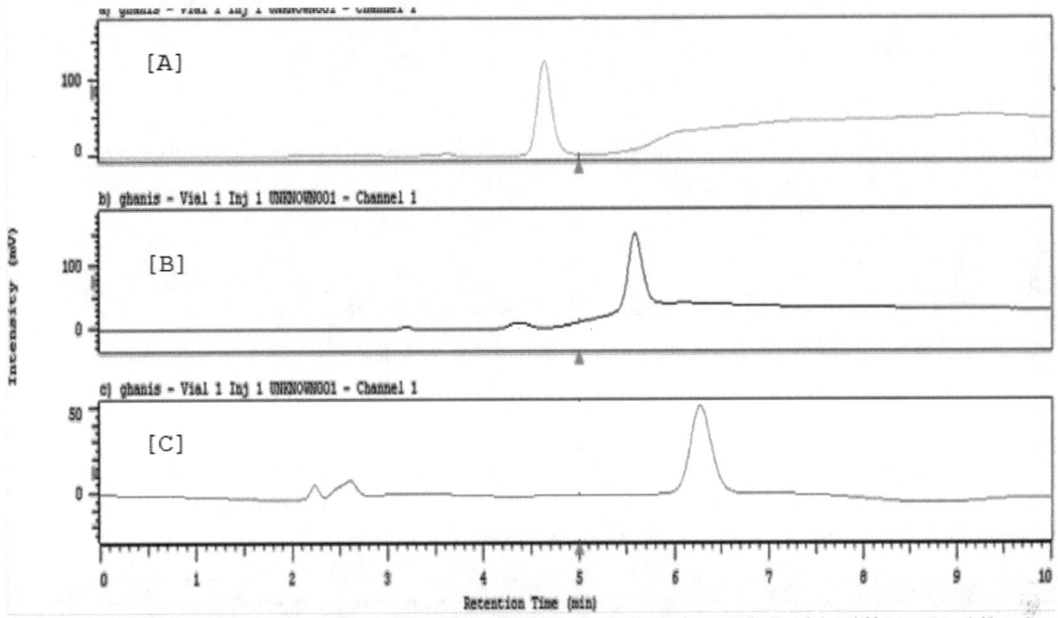

Figure 2. Chromatogram of miconazole (50 mg/L in methanol) by HPLC with different mobile phase composition. (A) methanol-water (95:5, v/v); (B) methanol-water (90:10, v/v); (C) methanol-water (85:15, v/v). HPLC conditions; flow rate: 0.8 mL/min, C_8 column (150 x 4,6 mm), UV detector (220 nm), injection volume: 5 μL

3.2 Optimization of Flow Rate

The effect of different flow rate of mobile phase was then investigated in this study. Flow rate was varied in three different flow rates: 0.7, 0.8 and 0.9 mL/min. The typical result of HPLC chromatogram for miconazole with a different flow rate of the mobile phase is shown in figure 2. It can be seen that retention time of miconazole was gradually increased when the flow rate of mobile phase was decreased. The 0.8 mL/min has been chosen as the optimized flow rate because it gave the best baseline of miconazole peak (base to base) with retention time less than 7 min.

3.3 Optimization of UV Detector Wavelength

The effect of different UV detector wavelength was also investigated in this study. UV detection was varied in three different wavelengths: 210, 220 and 230 nm. The typical result of HPLC chromatogram for miconazole with a different wavelength is shown in Fig 3. It can be seen that retention time of miconazole was not significantly different. UV detector wavelength 220 nm has been chosen as the optimized wavelength because it gave the best baseline of miconazole peak (base to base) with retention time less than 7 min.

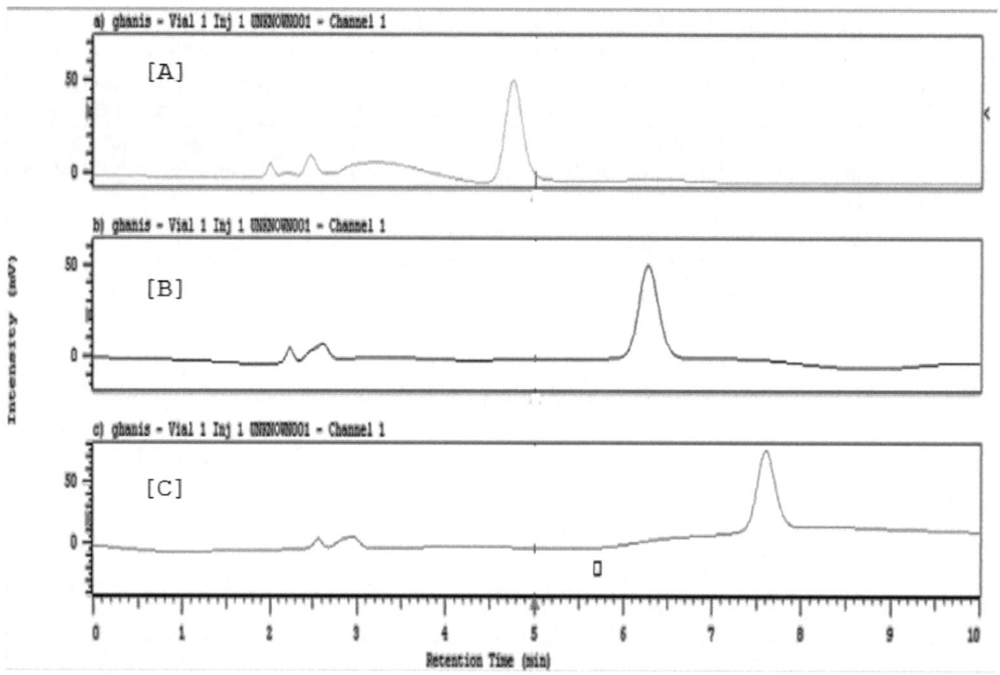

Figure 3. Chromatogram of miconazole (50 mg/L in methanol) by HPLC with different flow rate. (A) 0.9 mL/min; (B) 0.8 mL/min; (C) 0.7 mL/min. Mobile phase composition, methanol-water (85:15, v/v); other conditions: as shown in figure 2.

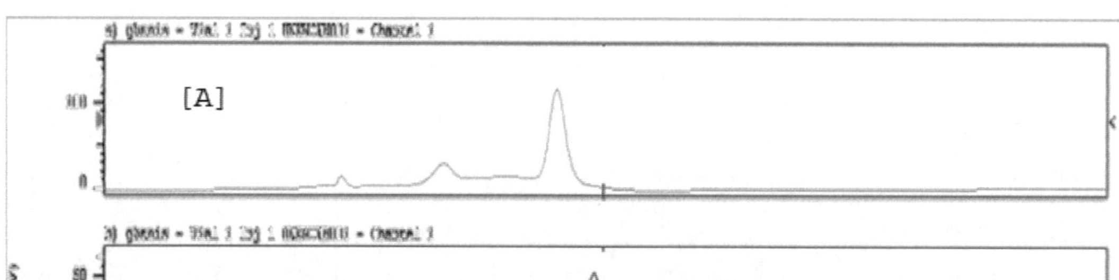

Figure 4. Chromatogram of miconazole (50 mg/L in methanol) by HPLC with different wavelengths. (A) 210 nm; (B) 220 nm; (C) 230 nm. Mobile phase composition, methanol-water (85:15, v/v); flow rate 0.8 mL/min. Other conditions: as shown in figure 2.

3.4 Analytical Performance of the HPLC Method

The performance of the developed HPLC method was examined in term of linearity, limit of detection (LOD), limit of quantitation (LOQ), recovery and repeatability (relative standard deviation/RSD). The calibration curve was linear for miconazole from 10 to 50 mg/L with $r^2 = 0.9983$. The LOD and LOQ of miconazole were 2.24 mg/L and 7.47 mg/L, respectively. The low LOD and LOQ showed that the developed HPLC method is sensitive and sufficient to determine miconazole in pharmaceutical samples for routine analysis. Average recoveries of miconazole in powder sample was calculated by comparing the peak height obtained from injections of miconazole standard with those obtained by injections of the samples with known concentration of miconazole drug. The percentage recovery obtained for analysis of miconazole in powder sample was 101.28% with RSD of 0.96% ($n = 3$).

4. Conclusions

In the present work, the high performance liquid chromatography (HPLC) method using C_8 column was successfully developed for the determination of miconazole in the pharmaceutical sample (powder). The optimum HPLC condition was achieved using mobile phase composition containing methanol:water (85:15, v/v), the flow rate of 0.80 mL/min, and UV detection at 220 nm. The HLPC condition gave retention time of miconazole within 7 minutes. The calibration graph was linear in the range from 10 to 50 mg/L with r^2 of 0.9983. The limit of detection (LOD) and limit of quantitation (LOQ) obtained were 2.24 mg/L and 7.47 mg/L, respectively. The optimized HPLC method was successfully applied to the analysis of miconazole in the pharmaceutical sample (powder). The percentage recovery obtained for analysis of miconazole in powder sample was 101.28 % with RSD of 0.96 % ($n = 3$).

Acknowledgments
Authors would like to thank Ministry of Research, Technology and Higher Education (RISTEKDIKTI) and Universitas Jenderal Soedirman (UNSOED), Indonesia for financial support through *International Research Collaboration and Scientific Publication* (KLN) Research Grant.

References

[1] De Zan MMD, Camara M, Robles JC, Kergaravat SV and Goicoechea H 2009 Development and validation of a simple stability-indicating high performance liquid chromatographic method for the determination of miconazole nitrate in bulk and cream formulations *Talanta* **79** 762-767.
[2] Cavrini V, Di Pietra AM and Gatti R 1989 Analysis of miconazole and econazole in pharmaceutical formulations by derivative UV spectroscopy and liquid chromatography (HPLC) *Journal of Pharmaceutical and Biomedical Analysis* **7** 1535-1543.
[3] Moety EMA, Khattab FI, Kelani KM and Alamein AMA 2002 Chromatographic determination of clotrimazole, ketoconazole and fluconazole in pharmaceutical formulations *II FARMACO* **57** 931-938.
[4] Low AS and Wangboonskul J 1999 An HPLC assay for the determination of ketoconazole in common pharmaceutical preparations *Analyst* **124** 1589-1593.
[5] Di Pietra AM, Cavrini V, Andrisano V and Gatti R 1992 HPLC analysis of imidazolic antimycotic drugs in pharmaceutical formulations *Journal of Pharmaceutical and Biomedical Analysis* **10** 873-879.
[6] Hermawan D, Ling LH, Wan Ibrahim WA and Sanagi MM 2013 Solid phase micro tip extraction coupled with high performance liquid chromatography for diclofenac sodium analysis in urine sample *Malaysian Journal of Fundamental and Applied Sciences* **9** 1-4.
[7] Hermawan D, Md Ali NA, Wan Ibrahim WA and Sanagi MM 2013 Analysis of fluconazole in human urine sample by high performance liquid chromatography method *Journal of Physics: Conference Series* **423** 012009.

[8] Velikinac I, Cudina O, Jankovic I, Agbaba D and Vladimirov S 2004 Comparison of capillary zone electrophoresis and high performance liquid chromatography methods for quantitative determination of ketoconazole in drug formulations *IL FARMACO* **59** 419-424.

[9] Mc Calley DV 1995 Influence of solvent organic modifier and solvent strength on peak shape of some basic compounds in high performance liquid chromatography using reversed-phase column *Journal of Chromatography A* **708** 185-194.

[10] Baranowska I, Markowski P and Baranowski J 2006 Simultaneous determination of 11 drugs belonging to four different groups in human urine samples by reversed-phase high performance liquid chromatography method *Analytica Chimica Acta* **570** 46-58.

[11] Hermawan D, Izdiani MY, Khaulah AR, Sanagi MM, Wan Ibrahim WA and Aboul-Enein HY 2013 Comparison of high performance liquid chromatography and microemulsion electrokinetic chromatography for miconazole nitrate determination in pharmaceutical formulation *Chromatographia* **76** 1527-1536.

IOP Conf. Series: Materials Science and Engineering **172** (2017) 012012 doi:10.1088/1757-899X/172/1/012012

Adsorption of Cd^{2+} using chitosan-linked p-t-buthylcalix[4]arene

D S Handayani[1], T Kusumaningsih[1], Pranoto[1], N Ihsaniati [1]

[1]Department of Chemistry, Faculty of Mathematics and Natural Sciences, Universitas Sebelas Maret, Jl. IrSutami 36A Surakarta Indonesia

Email: dsuci72@gmail.com

Abstract. This research aims to assess the use of chitosan -linked p-t-buthylcalix[4]arene as an adsorbent of Cd^{2+}. Adsorption was done with the batch method in the variation of acidity (pH), contact time and initial concentration of Cd^{2+}. The experimental results showed that the optimum pH of adsorption was pH 6 for the adsorbent chitosan and 8 for chitosan -linked p-t-buthylcalix[4]arene adsorbent. The optimum adsorption contact time was 240 minutes for the both adsorbent. Study of adsorption kinetics revealed that the adsorption of Cd for two adsorbents followed the kinetics model Ho, pseudo-order 2, with adsorption rate constant $1.037x10^{-2}$ and $1.777x10^{-2}$ g mg^{-1} min^{-1} respectively. Study of adsorption isotherm presented that the using chitosan adsorbent tends to follow the Langmuir isotherm, while the adsorbent chitosan- linked p-t-buthylcalix[4]arene tend to follow Freundlich isotherm. The maximum capacity of the adsorption using chitosan and chitosan- linked p-t-buthylcalix[4]arene adsorbent 30.19 and 21.67 mol/g , with the adsorption energy 41.459 and 28.627 KJ/ mol respectively.

1. Introduction

Chitosan is a polysaccharide amine resulted from deacetylation of chitin. This compound is an important natural polycationic biopolymer and can be applied in various fields such as metal and dyes adsorbent, materials for cosmetics and antibacterial agents [1]. Due to the ability to bind and to release back the metal ions by using certain solvents, thus chitosan is widely used as an adsorbent. Moreover, chitosan is able to be applied for continuous wastewater treatment [2]. The disadvantage of chitosan as the adsorbent is easily soluble in certain solvents such as acid, so it is necessary to improve the quality of chitosan by chemical modifications. Mahajan [3] found that the metal plating industry wastewater has a pH of 5 is still not meet the quality standards have permitted by the Government in accordance KEP-51 / MENLH / 10/1995.

Research on the use of chitosan and modified chitosan gradually increase, particularly in its application as a heavy metal adsorbent. Chitosan is available in various forms as an adsorbent, e.g. flakes [4] and powder [5], while the modified chitosan forms a crosslink chitosan [6]. Improving the ability of chitosan as an adsorbent can be conducted by modifying the chitosan to be crosslink chitosan as reported by Ngah *et al.* [7]. Other modifications can also be performed by crosslinking the crown ether as macromolecule on chitosan [8], or by cross-connecting between chitosan and Calix[4]arenes chelate to form the polymer. Previous research conducted by Restuti [9] also had modified chitosan with macromolecular compounds such as crown ether p-t-butilkaliks [4] arena.

The compound p-t-butylcalix[4]arene is one of the derivatives calixarene. Calixarene is a hollow cyclic molecule synthesized from the formaldehyde and phenols substituted by condensation under alkaline conditions [10]. Modifications compound p-t-butylcalix[4]arene on the fourth hydroxyl group with the addition of the carboxyl group a Quantitive Maming et al. [11] (2007) produce acid compound p-t-butilcalix[4]arene and the use of acid p-t-butylcalix[4]arene was made by Maming et al.[12] as an adsorbent ion phosphate. Modification of chitosan compound p-t-butylcalix[4]arene by agents optocoupler DIC (N, N'-diisopropylcarbodiimide) have been undertaken by Tabakci and Yilmaz [8]. Agents optocoupler DIC has a linear shape and is often used in the field of health and synthesis of chemical compounds and biological products. In this paper, we present the adsorption of Cd metal ion using chitosan-linked p-t-buthylcalix[4]arene.

2. Research Methods

2.2 Ingredients:

Metal solutions were prepared by diluting 1000 mg/L Cd standard solution in aqueous nitric acid to desired concentrations. The pH adjustment was carried out by adding slowly NaOH and/or HNO_3 solutions into the metal solutions followed by stirring until the desired pH was reached. All materials required were reagent grade from E Merck. The material of Chitosan-p-t-butylcalix[4]arene was synthesized according to Handayani, et al. [13].

2.3 Instruments:

Atomic Absorption Spectrophotometer (AAS; AA-F-6650 Shimadzu), pH meter (Hanna), Fourier Transform Infra Red (FTIR) Prestige-21 Shimadzu, Scanning Electron Microscopy (SEM) JEOL JSM 6360 - EDX (JED-2200) and adsorption stirrer (Intan Electroplating).

2.4 Procedure:

2.4.1 Effect of solution pH solution Cd

The experiment was conducted by adding 20 mg of Chitosan-p-t-butylcalix[4]arene into 25 mL of metal cation sample solution with the concentration of 10 mg/L. The adsorptions of Cd(II) was carried out in pH 4,0; 4,5; 5,0; 5,5; 6,0 and 6,5. The mixture was then stirred at Equal room treatment performed on chitosan as a comparison measured based on the concentration of metal cation before and after the adsorption using AAS. Equal treatment performed on chitosan as a comparison.

2.4.2 Effect of Stirring Time

As much as 60 mg of the calixarene polymer was added into 30 mL of metal cation sample solution with the concentration of 10 mg/L for each. The adsorption kinetics was studied by conducting the adsorption in various contact times, which were 5, 15, 45, 135 and 405 minutes. The concentration of metal ion before and after adsorption was then measured by AAS and corrected with the blank solution. Equal treatment performed on chitosan as a comparison.

2.4.3 Influence of Solution Concentration Cd

The adsorption processes were performed by stirring 20 mg of the calixarene polymer with 30 mL of metal cation sample solution having the concentration of 2, 4, 8, 16 and 20 mg/L at room temperature and the optimum conditions achieved before. Then, the adsorbed metal ion was measured based on the concentration of metal cation before and after the adsorption using AAS. Equal treatment performed on chitosan as a comparison.

3. Results and Discussion

3.1 Determination of Optimum pH

The degree of acidity (pH) solution is one important factor in determining the course of the adsorption process, as the pH can cause changes in the charge on the surface of the adsorbent and speciation of metal ions in solution. In the process of adsorption of metal ions, the acidity of the solution effect on two things. The first presence of protons in acidic conditions to the active site of the molecule will be protonated complexing. Both under alkaline conditions will be formed hydroxide and some metal ions will experience precipitation [14].

Effect of pH on the adsorption capacity of chitosan and chitosan- p-t-buthylcalix[4]arene on Cd^{2+} ion and the adsorption capacity of chitosan as a comparison is presented in Figure 1. Adsorption by chitosan-p-t-buthylcalix[4]arene in Figure 1, showed an increase in the number of Cd^{2+} concentration adsorbed of pH 2 to pH 4. The Concentration of Cd^{2+} adsorbed was relatively small at pH 2. Because at low pH, adsorption process can not be maximized [15]. At low pH, there is a high concentration of H^+, so the competition between H^+ with Cd^{2+}. Competition between H^+ with Cd^{2+} lead the interaction between Cd^{2+} with a hydroxyl group (OH) on chitosan p-t-buthylcalix[4]arene was reduced because of OH tend to interact with the H^+, consequently Cd^{2+} concentration was absorbed become small. The Concentration of Cd^{2+} was absorbed at pH 6 decreased again as adsorbent experiencing saturation point, consequently not capable of binding metal ions adsorbent. Adsorption of Cd^{2+} using chitosan-p-t-buthylcalix[4]arene reaches optimum at pH 8 with the adsorption capacity of 21.764 x 10^{-3} mmol/g.

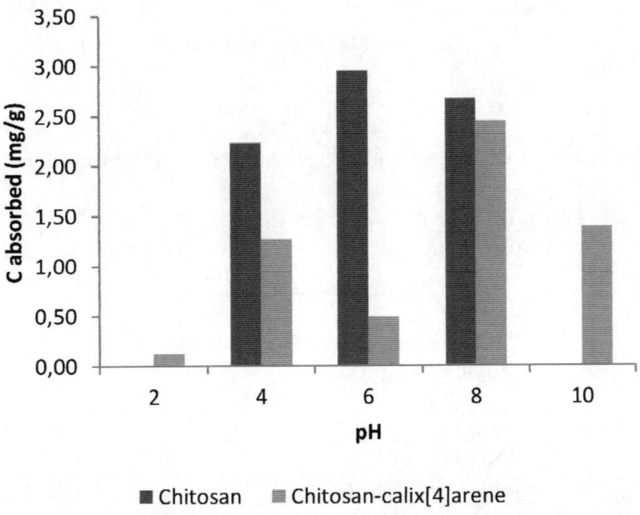

Figure 1. The concentration of Cd^{2+} adsorbed on the pH variation using chitosan and chitosan- p-t-buthylcalix[4]arene adsorbent

3.2 Effect of Contact Time and Parameter Kinetics

Adsorption kinetics are determined based on the adsorption contact time. Adsorption kinetics described the rate of binding of dissolved substances on contact time change of a reaction. The results of the determination of adsorption contact time were presented in Figure 2. The contact time optimum adsorption of Cd^{2+} on the minutes to 240 with a concentration of Cd^{2+} adsorbed was 32.895 x 10^{-3} mmol/g for Chitosan and 20.592 x 10^{-3} mmol/g for Chitosan-p-t-buthylcalix[4]arene. Adsorption kinetics was tested using two models of the adsorption kinetics pseudo first order (Lagergren) and pseudo second order (Ho), the results are presented in Table 1. Based on Table 1, using two adsorbents Cd adsorption kinetics followed the pseudo second order (Ho) with the adsorption rate constant of $1.037x10^{-2}$ g mg^{-1} $menit^{-1}$for chitosan and $1.777x10^{-2}$ g mg^{-1} $menit^{-1}$ for chitosan- *p-t-*buthylcalix[4]arene.

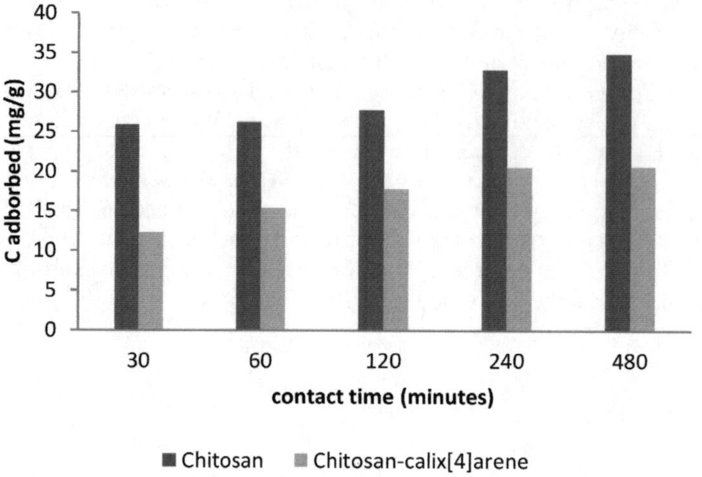

Figure 2. Adsorption of Cd^{2+} on variations of contact time

Table 1. Parameters of kinetics Adsorption of Cd^{2+}

Kinetic Model	Adsorbent	Parameter of Adsorption	
		R^2	K
Pseudo Order 1 (Lagergren) Log (qe-qt)=log qe-Kt	Chitosan	0.927	$3x10^{-3}$ minute^{-1}
	Chitosan- *p-t*-buthylcalix[4]arene	0.948	$10x10^{-3}$ minute^{-1}
Pseudo Order 2 (Ho) $\frac{t}{qt} = \frac{1}{Kqe^2} + \frac{1}{qe}t$	Chitosan	0.997	$1.037x10^{-2}$ g mg^{-1} minute^{-1}
	Chitosan- *p-t*-buthylcalix[4]arene	0.999	$1.777x10^{-2}$ g mg^{-1} minute^{-1}

3.3 Effect of Initial Concentration Dyes and Parameter Isotherm

Adsorption isotherms determined by the results of adsorption on the variation of the initial concentration of Cd^{2+}. Results of the adsorption by variation of the initial concentration of Cd^{2+} was presented in Figure 3.

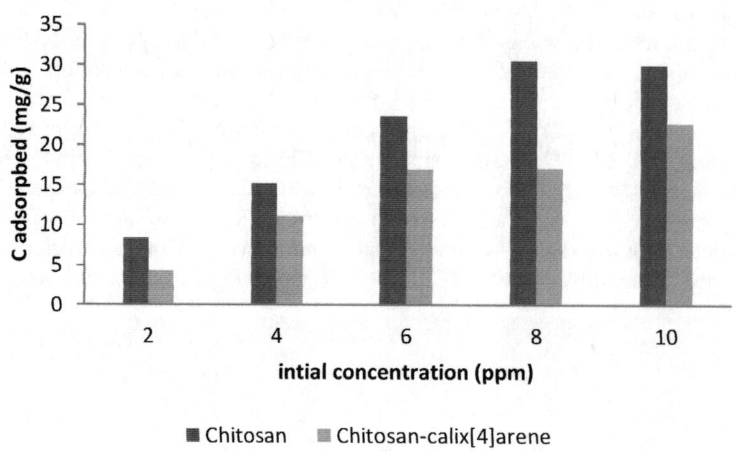

Figure 3. Initial concentration of Dyes variation Remazol Yellow FG on the initial concentration of dye

The results of adsorption on the variation of the initial concentration of Cd showed that the greater the initial concentration, the greater the concentration of Cd adsorbed. The study of the adsorption isotherm shown in Table 2.

Table 2. Parameters of Isotherm Adsorption of Cd^{2+}

Adsorbent	Parameter of Isotherm Adsorption						
	Langmuir				Freundlich		
	Xm (mol/g)	K x10^3 (L/mol)	ΔG (kJ/mol)	R^2	n	K(mol/g)	R^2
Chitosan	3.019 x 10^{-5}	16560	41.459	0.999	8.695	1.059 x 10^{-4}	0.822
Chitosan- *p-t-*buthylcalix[4]arene	2.167 x 10^{-5}	96.519	28.627	0.981	2.053	25.349 x 10^{-4}	0.995

According to the Table 2 chitosan tend to follow the Langmuir isotherm, while chitosan-Calix tend to follow the isotherm Freudelich. Cd adsorption using chitosan occurred monolayer, while using chitosan-p-t-calix[4]arene occurred multilayer, the adsorption capacity of each 3.019 x 10^{-5} and 2.167 x 10^{-5} mol/g. Adsorption energy of chitosan and chitosan-Calix was 41.459 kJ/mol and 28.627 kJ/mol. Modified chitosan adsorption capacity was smaller than chitosan, because due to steric factors of the adsorbent.

3.4 Adsorbent characterization With FTIR

Results of FTIR spectra of chitosan p-t-buthylcalix[4]arene before and after adsorption can be seen in Figure 4. Figure 4 shows a new peak at wave number 3152.78 cm^{-1} indicating the typical absorption group N-H stretching and weak-sharp absorption at wave number 1565.30 cm-1 which were the characteristic absorption of group N-H bending [16]. Thus occurred the Cd^{2+} Interaction with the group N-H [17]. In addition, there was a wide peak at 3383.29 cm^{-1} and 1600.39 cm^{-1}. These peaks are the absorption of the hydroxy group and physically adsorbed water [18], so there was interaction Cd^{2+} with an O-H group.

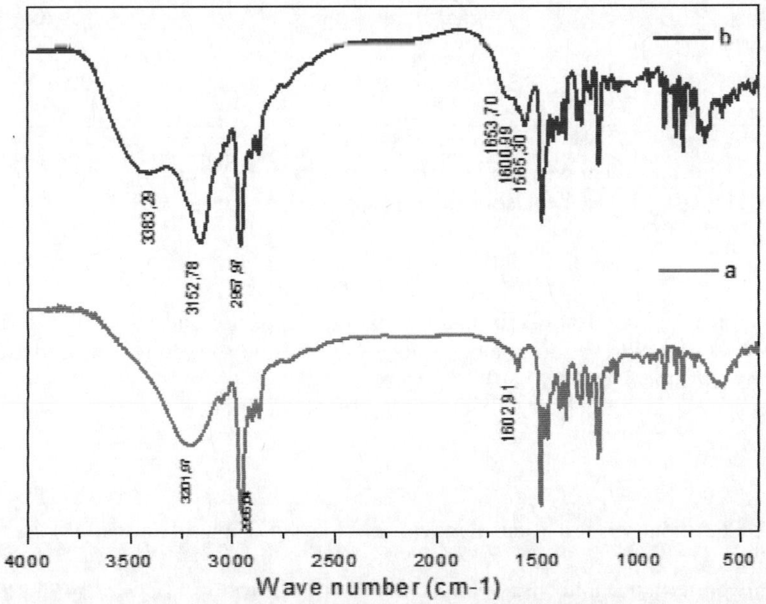

Figure 4. Spectra of FTIR chitosan p-t-buthylcalix[4]arene (a) before, (b) after adsorption

3.5 Adsorbent characterization by SEM – EDX

Morphology adsorbent chitosan p-t buthylcalix[4]arene before and after adsorption was characterized using Scanning Electron Microscopy (SEM) as shown in Figure 5.

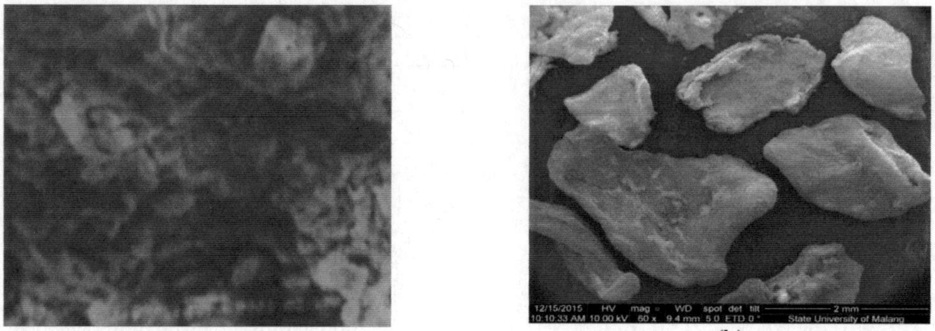

(a) (b)

Figure 5. Morphology adsorbent chitosan p-t- buthylcalix[4]arene (a) before, (b) after adsorption

Morphology adsorbent chitosan p-t-buthylcalix[4]arene before adsorption (Figure 5 (a)) showed the surface of the porous, so it was possible Cd can be adsorbed meet pores of the adsorbent. Morphology adsorbent chitosan p-t-buthylcalix[4]arene after the adsorption process (Figure 5b) surfaces look more meetings indicating that it happened adsorption adsorbate. Metals adsorbed in the adsorbent Chitosan p-t-butilkaliks [4] arena supported by the data of EDX. The composition of the adsorbent based on the EDX analysis as shown in Table 3.

Table 3. Result of EDX analysis

composition	p-t-butilkaliks [4] arena	
	Before adsorption (% mass)	After adsorption (% mass)
Atom C	53.95	61.57
Atom N	18.37	08.59
Atom O	7.55	28.50
Atom Cl	20.13	00.67
Atom Cd		00.67

Adsorption of Cd by the adsorbent was shown from the results of EDX analysis identified the existence of Cd, which did not exist prior to the adsorption process.

4. Conclusion

Ion of Cd adsorption using chitosan-Calix had a maximum capacity and adsorption energy of 21.67 mol/g and 28 627 KJ/mol, adsorption kinetics study followed the pseudo second-order kinetics (Ho) with adsorption rate constants for 1.777×10^{-2} g mg^{-1} min^{-1}.

References

[1] Bhuvana 2006 Studies on Frictional Behavior of chitosan-Coated Fabries, *Aux.Res. J.*, Vol 6(4): 123-130.

[2] Darjito, Purwonugroho D and Nisa SN 2006 Study on Adsorption of Cd(II) by Chitosan-Alumina. *Indo. J. Chem.*, 6 (3), hal.: 238-244.

[3] Mahajan SP 2002 Pollution Control in Process Industries, Tata McGraw-Hill Pub. Co. Ltd., New Delhi.

[4] Wu FC, Tseng RL, and Juang RS 2010 A review and experimental verification of using chitosan and its derivatives as adsorbents for selected heavy metals, Journal of Environmental Management, Elsevier, Vol. 91, Issue 4, Pages 798–806

[5] Airoldi C and Oryton AC 2000 Chitosan–organosilane hybrids—Syntheses, characterization, copper adsorption, and enzyme immobilization, Journal of Polymer Applied Science, volume 77. Issue 4, hal 797-804

[6] Ngah WSW, Endud CS, and Mayanar R 2002 Removal of copper(II) ions from aqueous solution onto chitosan and cross-linked chitosan beads, Reactive and Functional Polymers, Elsevier, Vol. 50, Issue 2, 181-190

[7] Ngah WSW, Teong LC, and Hanafiah M 2011 Adsorption of dyes and heavy metal ions by chitosan composites: A review, Carbohydrate Polymers, Elsevier, Vol.3, Issue 4, Pages 1446–1456

[8] Tabakci M, and Mustafa Y 2008 Synthesis of a Chitosan-linked Calix[4]arene chelating Polymer and Its Adsorption ability toward Heavy Metals and dichromate anions, Bioresour. Technol., 99: 6642-6645.

[9] Restuti A 2012 Sintesis dan Karakterisasi p-t-Butilkaliks[4]arena yang diikatkan pada kitosan, Skripsi, Surakarta : FMIPA Universitas Sebelas Maret.

[10] Gutsche CD 1998 Calixarene Revisited, Monograph in Supramolecular Chemistry, Royal Society of Chemistry, Cambridge.

[11] Maming 2008 Transpor Cr(III), Cd(II), Pb(II) dan Ag(I) melalui Membrane Cair Ruah mengandung Turunan Karboksilat, Ester, dan Amida p-t-butilkaliks[4]arena sebagai Pengemban Ion, Disertasi, PPS-UGM, Yogyakarta.

[12] Maming, Jumina, Siswanta D and Sastrohamidjojo H 2007 Transport of Cr3+, Cd2+, Pb^{2+}, and Ag+, Ions Trough Bulk Liquid Membrane Containing p-tert-Butylcalix[4]Arene-Tetracarboxilic Acid as Ion Carrier, Indo. J. Chem, 7(1), 172-179.

[13] Handayani D S, Kusumaningsih T and Restuti A 2014 Modification and Characterization a New Polymer Chitosan-p-t-b-calix[4]arenes, Prosiding ICOPIA part of series: Advance in Physic Research Atlantis Press 21-25

[14] Mollah. MYA, Schennach R, and Parga JR 2001 Electrocoagulation (EC)-science and applications, Journal of hazardous, Elsevier, Vol. 84, Issue 1, Pages 29–41

[15] Buhani, Narsito, Nuryono and Kunarti ES 2010 Chemical Stability of Cd(II) and Cu(II) Ionic Imprinted Amino-Silica Hybrid Material in Solution Media, Eksakta, 13 (1-2), hal.: 1-10.

[16] Sastruhamidjojo H 1991 Spektroskopi, Liberty Yogyakarta, Yogyakarta.

[17] Amri A, Supranto, and Fahrurozi M 2004 Kesetimbangan Adsorpsi Optional Campuran Biner Cd (II) dan Cr (III) dengan Zeolit Alam Terimpregnasi 2-merkaptobenzotiazol, Jurnal Natur Indonesia, 6 (2), hal.: 111-117.

[18] Taba P, Natsir H, Fauziah, and Ismail M 2010 Adsorpsi Ion Cd (II) oleh Kitosan-Silika Mesopori MCM-48, Marina Chimica Acta,11(1), hal.: 13-22.

Acknowledgment

The financial support from Indonesian Directorate General of Higher Education through the research project of Penelitian Unggulan Perguruan Tinggi (PUPT) 2015 is greatly appreciated.

IOP Conf. Series: Materials Science and Engineering **172** (2017) 012013 doi:10.1088/1757-899X/172/1/012013

Synthesis of Anti-Acne Ointment of Ethanol Extract of White Plumeria Leaves (*Plumeria Alba* L.)

D R Ningsih[1], Zusfahair[1], D Kartika[1], I T Lestari[1]
[1] Department of Chemistry, Faculty of Mathematics and Basic Sciences, Universitas Jenderal Soedirman Central Java Indonesia

Email: deeyan_bik@yahoo.com@gmail.com

Abstract. Acne is a chronic inflammatory skin disease pilosebaceous follicle, where the oil-producing glands are clogged and contaminated by bacteria. *Propionibacterium acne* is one of the bacteria that contributes to the pathogenesis of acne. Acne treatment was done by reducing the population of bacteria using an antibacterial. One of the plants that have antibacterial activity is white plumeria. The ethanol extract of white plumeria leaves contains antibacterial secondary metabolites, which are alkaloids and saponins. The aim of this study is to formulate white plumeria leaves extract into the water leached ointment base. Characteristics of the ointment were determined by evaluating the stability of the ointment including organoleptic, adhesion test, dispersive power test, determination of pH, and the antibacterial activity test. The results showed that the ointment of ethanol extract of white plumeria leaves has some characteristics, semisolid form, white, has distinctive smell of ointment, homogeneous but not protective, has a pH of 4.57 - 6.10, dispersive power of 5.10 - 6.06 cm, the adhesiveness of 1.67 - 3 seconds, and optimum antibacterial activity at concentrations of 5 ppm providing inhibition zone of 24.00 mm.

1. Introduction

Acne is a chronic inflammatory skin disease pilosebaceous follicle, where the oil-producing glands are clogged and contaminated by bacteria. *Propionibacterium acne* is one of the bacteria that contribute to the pathogenesis of acne. *P. acne* uses glycerol in sebum as a source of nutrients. *P. acne* will form free fatty of the sebum, causing a response of neutrophil cells to secrete enzymes that can damage the hair follicle wall. Thus it can cause inflammation [1]. Acne treatment was done by reducing the population of bacteria using an antibacterial. Acne treatment generally uses antibiotics, one of which is erythromycin. The use of Erythromycin antibiotic in some cases can give side effects on the body. This situation prompted scientists to develop new antibacterial compounds derived from natural resources [2].

The plant which is potential to be developed as an antibacterial is white plumeria. Ethanol extract of white plumeria leaves showed antimicrobial activity against *Staphylococcus aureus* and *E. coli* [3]. Antibacterial of white plumeria leaves is due to the existence of secondary metabolites. The ethanol extract of white plumeria leaves is known to contain alkaloids and saponins [3]. Alkaloid works as an antibacterial by disrupting components of the peptidoglycan in the bacterial cell, so the cell wall layers

IOP Conf. Series: Materials Science and Engineering **172** (2017) 012013 doi:10.1088/1757-899X/172/1/012013

are not fully formed and caused the death of these cells [4]. Saponins are substances that can interact with bacterial cell wall causing a bacterial cell lysis.

Increasing the effectivity of the use of white plumeria leaves to extract as a natural remedy was done using formulations in dosage forms. Anti-acne dosage forms have been widely circulated in the market, either in the form of gels, ointments, creams and lotions, but form is more suitable for acne. Ointment dosage is a dosage form that has suitable consistency for the treatment of skin diseases caused by bacteria. Based on its composition, ointment base was classified into hydrocarbon, absorption, water-soluble, and that can be leached by water [5]. Water-leached ointment base is considered to have advantages as a cosmetic, as it is easy to wash by water, the good release of the medicine because when used on the skin, there will be evaporation and increasing concentration of a drug that dissolves in water so as to encourage absorption into the skin tissue [6].

Accordingly, this research is aimed to formulate the ointment dosage of ethanol extract of white plumeria leaves and to test the activity against bacteria that cause acne which is *P. acne*. The effectivity of the release of active substances of ointment dosage was determined by evaluating the stability of the ointment dosage including organoleptic, adhesion test, dispersive power test, determination of pH, and the antibacterial activity test.

2. Methods

2.1 Tools and Materials

Tools used in this research are test tubes, petri dish, autoclave, analytical balance, oven, measuring cylinder, Erlenmeyer, beaker glass, micropipette, stirrer glass, aluminum foil, wrapping, volumetric flask, incubators, needle ose, stirrer, paper label, spirit burner, spectrophotometers, pipette volume, filler, cork drill, drugal sky, a pot of ointment.

Materials used in this research are ethanol extract of white plumeria leaves, aquades, *P. acne* bacteria, nutrient broth medium (NB), nutrient agar medium (NA), stearyl alcohol, white vaseline, nipasol, nipagin, polysorbate 80, glycerol, phenolphthalein indicator, KOH, alcohol 70%, solid paraffin.

2.2 Experimental Procedure

2.2.1 Water-leached basis of ointment dosage

Ointment dosage was based on compositions made by [7] with the addition of white plumeria leaves extracts as antimicrobial agents against *P. acne* bacteria in various concentrations. An ointment was made by fusion method. Ointment base was made with two preparations, namely preparation A and preparation B. Preparation A was made by fusing stearyl alcohol as much as 9.98% w/w and vaseline as much as 24.96% w/w at 70 °C while stirring and adding nipasol as much as 0.0025% w/w. Preparation B was made by heating distilled water as much as 50.035% w/w with glycerol as much as 9.98% w/w at 70 °C while stirring and adding polysorbate 80 as much as 5% w/w, nipagin as much as 0.005% w/w and extracts plumeria leave with concentrations of 0; 1; 5; and 10 ppm, as well as the positive control in the form of erythromycin ointment 10 ppm. Preparation A and B were mixed at a temperature of 70 °C while stirring in the mortar until cold and oil in water ointments were obtained. Ointment formulations of white plumeria leave extract was assembled as shown in Table 1.

The Ointment were evaluated in physical and chemical properties as well as the testing stability including the organoleptic, homogeneity, pH, dispersive power, and adhesion at room temperature on days 0, 5, 10 and 15. Antibacterial activity test of ointment was performed on the 15th day.

IOP Conf. Series: Materials Science and Engineering **172** (2017) 012013 doi:10.1088/1757-899X/172/1/012013

Table 1. Ointment formulation of white plumeria leaves extract.

Materials	Formula Salep				
	F1	F2	F3	F4	F5
Stearyl alcohol	9.98 g	9.98 g	9.98 g	9.98 g	9.98 g
White Vaseline	24.96 g	24.96 g	24.96 g	24.96 g	24.96 g
Nipasol	0.0025 g	0.0025 g	0.0025 g	0.0025 g	0.0025 g
Nipagin	0.005 g	0.005 g	0.005 g	0.005 g	0.005 g
Glycerol	9.98 g	9.98 g	9.98 g	9.98 g	9.98 g
Polysorbate 80	5 g	5 g	5 g	5 g	5 g
Aquades	50.035 g	50.035 g	50.035 g	50.035 g	50.035 g
Plumeria Extract	0 ppm	1 ppm	5 ppm	10 ppm	-
Erythromycin	-	-	-	-	10 ppm

2.2.2 Organoleptic test

The observations made in this test the form of dosage, smell and color of ointment dosage during storage to see the physical stability of ointment..

2.2.3 Homogeneity test

A total of 0.5 grams of ointment dosage was placed on the glass and then leveled and observed based on color uniformity and the absence of lumps and granules [8].

2.2.4 Protection power test

Filter paper (10x10 cm) was wetted with phenolphthalein and dried. The Ointment was weighed as much as one gram, then smeared on that paper. On another filter paper made an area (2,5x2,5 cm), then made a barrier at the edge of the area with the melted solid paraffin. This filter paper was attached on the filter paper before. A solution of KOH 0.1 N was dripped on these areas. The presence or absence of stains was observed at the time of 15; 30; 45; 60 seconds, 3 and 5 minutes, if no stains mean that ointments provide protection. Each type of ointment was repeated for three times [9].

2.2.5 pH test

Measurement of pH used a pH meter kit that dipped into 0.5 grams of ointment that has been diluted with ten mL of distilled water. The good pH value is 4.5-6.5 for ointment or appropiate with the pH value of human skin [10].

2.2.6 Dispersive power test

Dispersive power test was conducted by placing 0.5 grams of ointment between two transparent glass plates by a load of 100 g. Measurement of dispersive power was conducted after the ointment does not spread again or approximately 1 minute after loading. Ointment dosage which is good to use has a dispersive power of 5-7 cm [11].

2.2.7 Adhesion test

A total of 0.5 grams of ointment was weighed, then smeared on glass plates with an area of 2.5 cm^2. Both plates was affixed and pressed with a load of 1 kg for 5 minutes then the load was removed and given a load of 80 g for testing. Recorded the time until the two plates disjoint [12].

2.2.8 Antibacterial activity tests of ointment dosage

Antibacterial activity test of ointment dosage was conducted by pitting method. A total of one *ose* of *P. acne* bacteria was taken from the stock culture and incubated in 10 ml of liquid medium (Nutrient Broth) for 18-24 hours at a temperature of 37 °C. Some bacteria in a medium NB was taken and dispersed in the medium NA based on the results of absorbance at a wavelength of 600 nm. Based on the value of absorbance, if an absorbance value is less than or equal to 0.5 then taking 100 μL of bacterial culture, whereas if it is 0.6 to 1.0 then taking 50 μL of bacterial culture. The bacterial suspension on NA medium was streaked with spread plate using *drugalsky*, then dried for 15 minutes

IOP Conf. Series: Materials Science and Engineering **172** (2017) 012013 doi:10.1088/1757-899X/172/1/012013

at room temperature. Agar media was drilled using a *crockbor* with the diameter of ±7 mm. Ointment dosage of white plumeria leaves extracts was then put into the hole as much as 0.05 g. Comparators used for the negative control is ointment base without the addition of extracts and for positive control is erythromycin ointment. Then incubated at 37 ° C for 24 hours. The clear zone which is visible around the hole indicates the presence of antibacterial activity. Clear zone formed was measured using a caliper.

3. Results

3.1 Testing of Physical Properties of Ointment

3.1.1 Organoleptic test

Organoleptic testing is conducted to determine the shape, odor and color of ointment and to observe any changes that may occur during storage that last for 15 days. Organoleptic testing showed ointment dosage of ethanol extract of white plumeria leaves has a semisolid form and soft texture such as cream at concentrations of 0; 1; 5; and 10 ppm. [13] stated that water-leached ointment base has a soft creamy consistency, could be diluted and dissolved in the aqueous solution. The color test of dosage resulted the same color at concentrations of 0; 1; 5; and 10 ppm. The smell of ointment dosage by organoleptic test resulted that ointment dosage has a distinctive smell and no changes either in the concentrations of 0; 1; 5; and 10 ppm.

Testing of organoleptic stability of ointment resulted that no changes in color, smell, and shape of the four formulations during 15 days of storage indicating that the ointment is stable. Based on these results it can be said that the antiacne ointment dosage of white plumeria leaves extract has good physical stability based on its organoleptic properties. According to [14], one of the requirements for good quality of ointment was that the ointment should be stable by physical or chemical effect as long as the ointment was used to treat. Therefore, the ointment should be free from the occurrence of incompatibility and should be stable at room temperature.

3.1.2 Homogeneity test

Homogeneity test is conducted to analyze if during the synthesis process of the cream with active ingredient with its basic materials and other necessary additives are homogeneously mixed. Homogeneity test was conducted by applying 0.5 grams of ointment dosage on glass objects, and the ointment dosage is visually observed based on color uniformity and the absence of lumps and granules. The results of the homogeneity test, all creams of ethanol extract of white plumeria leaves, positive control in the form of erythromycin ointment, and negative controls in the form of ointment base without the addition of the extract, are the homogeneous ointment. This is indicated by the absence of clumps in certain material of the ointment. [15] stated that homogeneity of a dosage depends on the properties similarity of the material used, which were the base and the active substance. If there are differences in the properties of the base and the active substance, it will occur coagulation process resulting a larger particles of the dosage.

The results of this study, the ethanol extract of white plumeria leaves can be distributed evenly in a concentration of 0; 1; 5; and 10 ppm, and no effect of storage time for 15 days against homogeneity of dosage. Based on these results, it can be seen that the ointment dosage of ethanol extract of white plumeria leaves have the good physical properties based on its homogeneity. Homogeneous dosage will give good results because the drug substances are dispersed in the basic materials evenly so that in every part of dosage contain the same amount of drug. If the ingredients are not evenly dispersed in the basic materials, the drug will not achieve the expected therapeutic effect [9].

3.1.3 Adhesion test

An Adhesion test is one of the important parameters in the evaluation of the ointment because this test can analyze how far the ointment can be attached to the skin so that the expected therapeutic effect is achieved. According to [16] ointment should not impede the physiological functions of the skin, ointment base which has powerful adhesion will impede skin respiration, if it is too weak, then the therapeutic effect will not be achieved. The Adhesion test was conducted by applying 0.5 grams of ointment on a pair of glass plates. Both plates were taped and given a load of 1 kg for 5 minutes, the next load was released by the release load of 80 grams. The time required to detach both plates was recorded as the adhesion value of ointment dosage.

The test resulted that the concentration of 0 ppm has adhesion of 1.33 seconds, 1 ppm has adhesion of 2.33 second, and 5 ppm has adhesion from 3.07 to 2 seconds and 10 ppm has adhesion of 3.07 seconds. Figure 4.4 adhesion test results of ointment dosage, shows adhesion value of ointment dosage does not change at concentrations of 0; 1; and 10 ppm, but at a concentration of 5 ppm decreases on the 5th day and does not change until the 15th day. The decrease of adhesion does not affect on the stability because the value of adhesion is still suitable for the standards. According to [17], the adhesion of semisolid dosage should be more than 1 second.

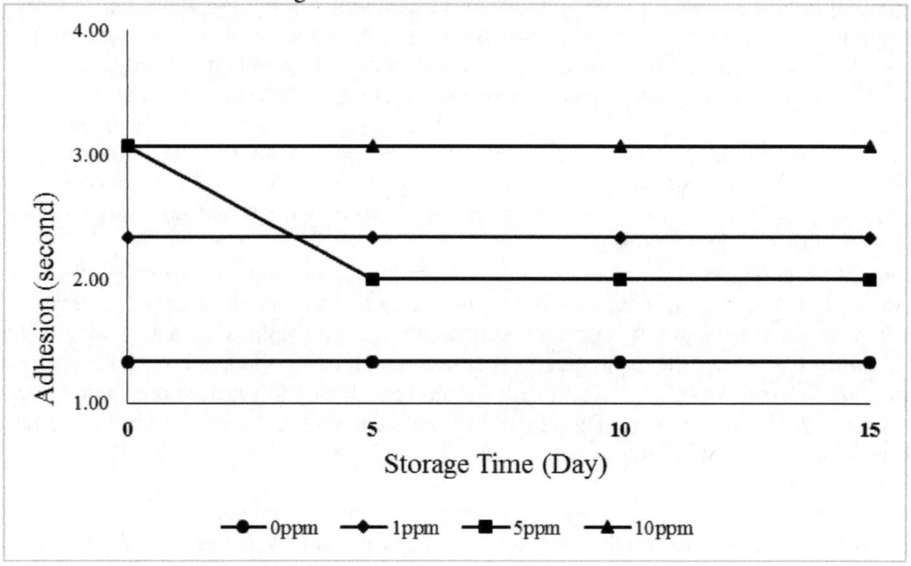

Figure 1 The relations between adhesion value of ointment dosage and storage time

3.1.4 Dispersive power test

Dispersive power test aims to analyze the dispersion of ointment dosage on treated skin. The greater dispersion of ointment, the greater medicine to be absorbed. Dispersive power test is done by placing 0,5 g of ointment between two plates of transparent objects by an additional load of 100 g. Measuring diameter of ointment that dispersal was conducted by taking the length of the average diameter of some sides. Picture of the dispersive power test with storage time can be seen in **Figure 2**.

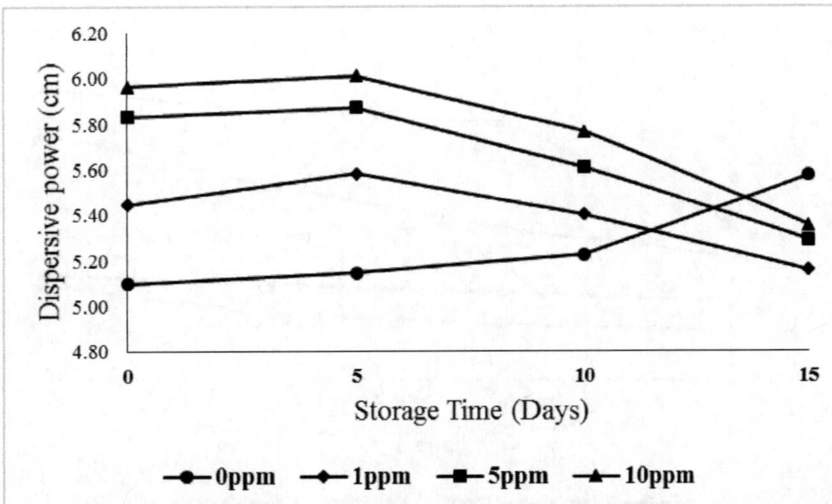

Figure 2 Relations between dispersion of ointment and storage time.

Test results showed that the dispersive power value with the addition of the ethanol extract, 1, 5, and 10 ppm, increased from day 0 to day 5, but decreased on day 10 and to 15, in contrast with the ointment without the addition of extracts continues increasing from day 0 to day 15. Despite the dispersion decreased on days 10 and 15, dispersion of ointment is still appropriate with the standard, which is good ointment dosage to use has dispersive power value of 5 s/d 7 cm [11]. The changes of dispersive power can be caused by the consistent changes of an ointment because storage containers are made of plastic. Different ways of stirring could also affect the particle size of dosage. Different particle sizes would change the value of dispersive power [18].

3.1.5 pH test

pH testing is a part of the physicochemical inspection criteria in predicting the stability of the ointment where the pH profile determines the stability of the active ingredient under acidic or alkaline condition [15]. [9] also stated that the suitability of the skin's pH with a pH of the topical dosage would also affect the skin response toward the ointment. The ideal topical dosage is not irritating to the skin. The possibility of skin irritation will be enormous if the ointment is too acidic or too alkaline. pH testing was conducted by dissolving 0.5 grams of ointment perfomed in 10 mL of aquades and then the pH was measured by using a pH meter. The resulting graph of pH test of dosage towards storage time can be seen in **Figure 3**.

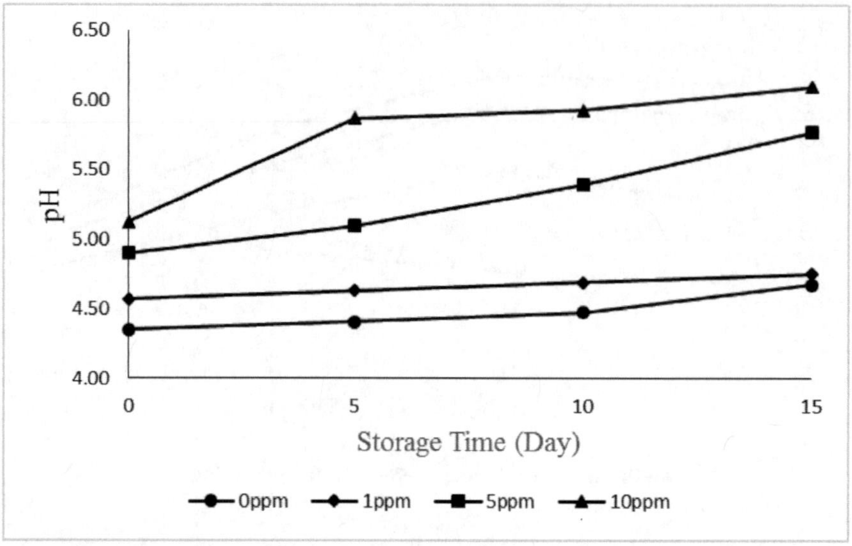

Figure 3 Relations between ointment pH and storage time.

Based on the test results is known that the pH of the ointment with the addition of white plumeria leaves extract at a concentration of 1; 5; and 10 ppm from day 0 to day 15 increased to more alkaline, it is thought to be caused by the active substances contained in the white plumeria leaves extract is alkaloid that has an alkaline pH. Alkaloids are nitrogen-containing compounds that are alkaline and have pharmacological activity [19]. Although the test results of pH ointment with the addition of white plumeria leaves extract on day 0 to day 15 increased but still achieve the Indonesia National Standard (SNI) for the topical dosage is 4.5-6.5. Ointment of extract ethanol of white plumeria leaves the concentration of 1 ppm has a pH of 4.57 to 4.76, a concentration of 5 ppm has a pH of 4.90 to 5.77, and the concentration of 10 ppm has a pH of 5.12 to 6.10. The ointment without adding the extract has a pH of 4.35 to 4.68 which shows the lowest pH value, the base does not achieve the standard of SNI. [8] stated that too acidic pH could irritate the skin; however, too alkaline pH could make a scaly skin.

3.1.6 Protection test

Protection test is conducted to analyze the ability of protection against foreign influences from outside which reduces the effectivity of the ointment. Protection test was performed by applying 0.5 grams of ointment dosage on filter paper that has been smeared by phenolphthalein indicator, then on another filter paper smeared by melted solid paraffin. Both papers are taped and the areas which subsequently spilled ointment smeared with 0.1 N KOH. A good ointment base can protect the skin from external influences such as acid-base dust and sunlight at the time of treatment, marked by the formation of red stains after addition of KOH [9]. Based on test results, the ointment of ethanol extract of Plumeria leaves did not have any protection. It is possibly caused by the ointment base used is the water-leached base that contains very little oil and easily removed by water, so it did not provide a protective effect. [20] suggested that the water-leached ointment base should be soluble and easily removed from the skin by washing with water.

3.2 The Ointment Antibacteril Activity Test to Bacteria

The test was conducted to analyze the effect of antibacterial activity after formulated into ointment dosage forms. The test was conducted by diffusion agar. Based on the test, the value of inhibition zone at the concentration of 1 ppm was 10.4 mm. At the concentration of 5 ppm was 24.00 mm and at the concentration of 10 ppm was 24.90 mm. The value of inhibition zone white plumeria leaves extract of ointment dosage forms has a greater value than in the form of the extract solution. The increasing of inhibition zone diameter of bacterial growth in dosage forms is possibly caused by the increasing penetration of the antibacterial compound that diffuses into the test object, resulting in larger inhibition zone diameter of bacterial growth. This is caused by the water-soluble cell wall of gram-positive bacteria that are more polar. The high water content of the water leached ointment base may increase the polarity so that it is easy to penetrate the polar peptidoglycan layer than the nonpolar lipid layer. [21] stated that the structure of the bacterial cell could determine the penetration, bond and antibacterial activity of the compounds. *P. acne* is a gram-positive bacteria whose cell wall structure has more peptidoglycan, a little lipid and cell wall contains polysaccharides (teichoic acid). Teichoic acid is a water-soluble polymer, which serves as a positive ion transport to leave or enter. The hydration effect of the preparation that is also used can lead to increased penetration of the drug into the skin [22].

The negative control of ointment base without adding the extract did not have inhibition zone. Positive controls such as erythromycin of ointment dosage have inhibition zone which is smaller than the ointment dosage of white plumeria leaves extract. This is caused by acidic ointment base affecting erythromycin work which is optimal under alkaline conditions. [23] stated that the antibacterial activity of erythromycin is bactericidal and increased at alkaline pH.

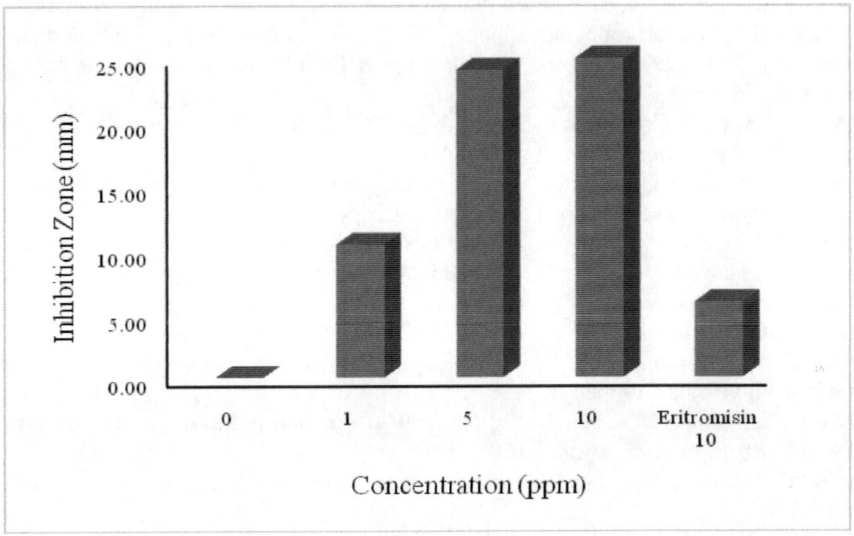

Figure 5 Inhibition zone obtained by the activity of ointment dosage of ethanol extract of white plumeria leaves

4. Conclusion

The evaluation of ointment dosage of ethanol extract of white plumeria leaves has some characteristics, semisolid form, white, distinctive smell, homogeneous but not providing power protection, has a pH of 4,57-6.10, dispersive power from 5.10 to 6.06 cm, adhesion of 1.67 to 3 seconds. The ethanol extract of white plumeria leaves in dosage forms can inhibit the activity of *P. acne* on the optimum concentration of 5 ppm, with the inhibition of 24.00 mm.

References

[1] Radji M 2011 Buku Ajar Mikrobiologi: Panduan mahasiswa farmasi dan kedokteran *EGC, Jakarta*

[2] Aliero A, Aliero B and Buhari U 2008 Preliminary phytochemical and antibacterial screening of Scadoxus multiflorus *Int J Pure Appl Sci* **2** 13-7

[3] Ningsih D R, Zusfahair Z and Purwati P 2014 Antibacterial activity cambodia leaf extract (Plumeria alba L.) to Staphylococcus aureus and identification of bioactive compound group of cambodia leaf extract *Molekul* **9** 101-9
[4] Heyne K 1987 Tanaman Berguna Indonesia *Penerjemah: Badan Litbang Kehutanan, Badan Litbang Kehutanan, Jakarta*
[5] Zulfa E, Prasetyo T B and Murukmihadi M 2015 Formulasi Salep Ekstrak Etanolik Daun Binahong (Anrederacordifolia (Ten.) Steenis) Dengan Variasi Basis Salep *e-Publikasi Ilmiah Fakultas Farmasi Unwahas Semarang* **12** 41-8
[6] Aulton M E 2002 *Pharmaceutics: The science of dosage form design*: Churchill livingstone)
[7] Himawati E R d E, T 2013 Pengembangan Formulasi Minyak Cengkeh sebagai Counter Irritant dalam Beberapa Sediaan Topikal,. In: *Majalah Farmasi Airlangga*
[8] Naibaho O H, Yamlean P V and Wiyono W 2013 Pengaruh basis salep terhadap formulasi sediaan salep ekstrak daun kemangi (Ocimum sanctum L.) pada kulit punggung kelinci yang dibuat infeksi Staphylococcus aureus *Pharmacon* **2**
[9] Susilowati E P 2013 Optimasi Sediaan Salep Yang Mengandung Eugenol Dari Isolasi Minyak Cengkeh (Eugenia caryophylatta Thunb.) *IJMS-Indonesian Journal on Medical Science* **1**
[10] Tranggono R I and Latifah F 2007 Buku Pegangan Ilmu Pengetahuan Kosmetik *Jakarta: PT. Gramedia Pustaka Utama*
[11] Garg A, Aggarwal D, Garg S and Singla A K 2002 Spreading of semisolid formulations: an update *Pharmaceutical technology* **26** 84-105
[12] Astuti I, Hartanti D and Aminiati A 2010 Peningkatan Aktivitas Antijamur Candida albicans Salep Minyak Atsiri Daun Sirih (Piperbettle LINN.) melalui Pembentukan Kompleks Inklusi dengan β-siklodekstrin *Majalah Obat Tradisional* **15** 94-9
[13] Ansel H 1989 Introduction to Pharmaceutical Dosage form, Terjemahan/; Farida Ibrahim *Pengantar Bentuk Sediaan Farmasi, Edisi Keempat, Edisi Keempat, Universitas Indonesia Press: Jakarta, Hal* 244-72
[14] Parrot E L 1971 *Pharmaceutical Technology Fundamental Pharmaceutic 3rd* (Mineapolis: Burgess Publishing Company)
[15] Lachman L, Lieberman H A and Kanig J L 1994 Teori dan praktek farmasi industri *Edisi III., diterjemahkan oleh Siti Suyatmi* **150** 658
[16] Voight R 1995 Buku Pelajaran Teknologi Farmasi (Terjemahan) *Universitas Gadjah Mada: Yogyakarta*
[17] Zats J L G P K, Gel, in Liebermen, H.A., Rieger, M.M., Banker, G.S 1996 *Pharmaceutical Dosage Forms: Disperse Systems 2* (New York: Marcel Dekker Inc)
[18] Widyaningrum N, Murrukmihadi M and Ekawati S K 2012 Pengaruh Konsentrasi Ekstrak Etanolik Daun Teh Hijau (Camellia sinesis L.) dalam Sediaan Krim terhadap Sifat Fisik dan Aktivitas Antibakteri *Sains Medika* **4** 147-56
[19] Lumbanraja L B 2009 Skrining Fitokimia dan Uji Efek Antiinflamasi Ekstrak Etanol Daun Tempuyung (Sonchus arvensis L.) terhadap Radang pada Tikus. *Skripsi* (unpublished) Faculty of Pharmacy. Universitas Sumatera Utara. Medan Indonesia.
[20] Anief M 1997 *Ilmu Meracik Obat* (Yogyakarta: Gajah Mada University Press).
[21] Jawetz E, Melnick J L, Adelberg E A, Brooks G, Butel J and Ornston L 1996 Mikrobiologi kedokteran *Alih bahasa: Nugroho E, Maulany. Edisi ke-20. Jakarta: EGC* 211-2
[22] Sulaiman T and Kuswahyuning R 2008 Teknologi dan Formulasi Sediaan Semipadat *Pustaka Laboratorium Teknologi Farmasi Fakultas Farmasi Universitas Gadjah Mada. Yogyakarta. Hal* **81** 101
[23] Katzung B G 2007 *Basic & Clinical Pharmacology* (United States: Lange Medical Publications)

IOP Conf. Series: Materials Science and Engineering **172** (2017) 012014 doi:10.1088/1757-899X/172/1/012014

Design and modeling balloon-expandable coronary stent for manufacturability

D Suryawan[1] **and Suyitno**[2]

[1] Department of Mechanical and Industrial Engineering, Gadjah Mada University, Yogyakarta, Indonesia

[2] Center for Innovation of Medical Equipments and Devices (CIMEDs) Department of Mechanical and Industrial Engineering, Gadjah Mada University, Yogyakarta, Indonesia

Email: donny_s@ymail.com, Suyitno@ugm.ac.id

Abstract. Coronary artery disease (CAD) is a disease that caused by narrowing of the coronary artery. The narrowing coronary artery is usually caused by cholesterol-containing deposit (plaque) which can cause a heart attack. CAD is the most common cause mortality in Indonesia. The commonly CAD treatment use the stent to opens or alleviate the narrowing coronary artery. In this study, the stent design is optimized for the manufacturability. Modeling is used to determine the free stent expansion due to applied pressure in the inner surface of the stent. The stress distribution, outer diameter change, and dogboning phenomena are investigated in the simulation. The result of modeling and simulating was analyzed and used to optimize the stent design before it is manufactured using EDM (Electric Discharge Machine) in the next research.

1. Introduction

Coronary artery disease (CAD) is a disease that caused by a plaque of cholesterol and lead the coronary artery to become narrow. CAD is the most deadly disease in Indonesia. More than 2 million people are indicated and diagnosed with CAD in 2013 [1]. The most common treatment to open or restore the narrowing coronary artery is stent implantation. A stent is a small tube that collapsed to a catheter balloon and then delivered and deployed in the narrowed coronary. The deployed stent will open the narrowing coronary and restore the blood flow.

Modeling is very important to provide research tools before a product designs are manufactured. It can be an alternative method to scenarios a various geometries or material and predict the failing of the design. In stent modeling, there are some strategies that used to simulate the deployment process of the stent. Negligible balloon, folded balloon, and hyper-elastic balloon are the most used strategy in stent modeling.

Several studies about stent deployment modeling are found in the literature. The negligible balloon is the most simple in stent deployment strategy. The pressure is applied directly to the inner surface of the stent and neglect the impact of balloon expansion [2-4]. The negligible balloon strategy needs a loading scenario to keep the expansion force equal as the accounted balloon strategy. This paper

IOP Conf. Series: Materials Science and Engineering **172** (2017) 012014 doi:10.1088/1757-899X/172/1/012014

presents a stent modeling with the negligible balloon to analyze and evaluate the deformation and stress distribution of the stent prior manufacturing.

2. Material and methods

The investigation about design and modeling of this research are based on 3D modeling and analyzing using a finite element method. A finite element model needs a design geometry, material properties, and numerical aspect.

2.1. Model Geometry

The model of the stent is designed for EDM manufacturability. The inner diameter and outer diameter of the stent are 2 mm and 2.6 mm. That dimension is obtained based on the availability the material and the turning machine capability to provide a raw material of the stent. The stent length of 10 mm with the cell length and width are 2 mm and 0.32 mm. The cell rounded to the stent with space between the other cells of 22.5^{o}. The stent design is shown in Figure 1.

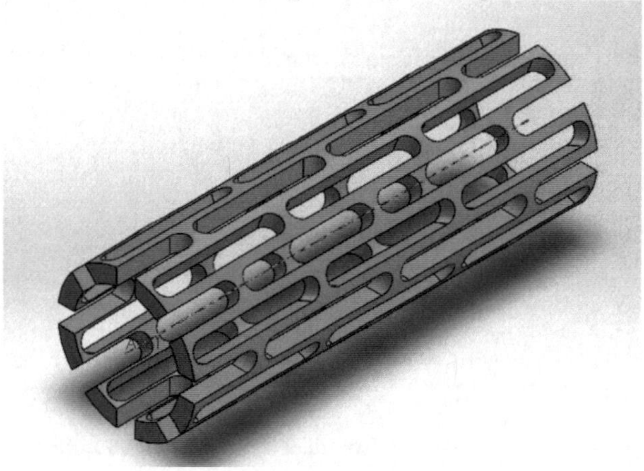

Figure 1. Stent Design

2.2. Material Properties

The stent material is 316L stainless steel. The specification of the material is the young modulus of 196000N/mm^2, the poison ratio of 0.3, the yield strength of 375N/mm^2 and the ultimate tensile strength (UTS) of 750 N/mm^2 [8].

2.3. Numerical Aspect

The modeling and simulation use ABAQUS with a dynamic explicit procedure. The stent is modeled with three-dimensional 8-node linear brick with are integration (C3D8R). The finite element mesh consists 23520 elements.

The four-step scenario is performed to simulate the stent deployment. The step scenario consists of initial step, crimping step, expansion step, and pressure release step. The crimping step is performed to simulate the crimping process or to fix the stent into the catheter. After crimped, a pressure applied directly to the inner surface of the stent until the stent deploy. Then the pressure is released. Loading scenarios are used to keep the expansion force equivalent to the balloon pressure. The loading scenarios are presented in Table 1. The contact interactions of the stent are defined as a self-contact.

2.4. Loading Scenario

The negligible balloon scenario causes a different expansion force to the stent. This expansion force definitely smaller than expansion force with the balloon. It happens because the inner stent surface with the hole of the cell is much smaller than the surface area of the balloon. The loading scenarios are created to keep the expansion force equal when the balloon is negligible. The loading scenario equation is derived from the Pascal law. Therefore, the equation is:

$$P_{wb} = \frac{A_{balloon}}{A_{stent}} P_{balloon}$$

Where P_{wb} is the pressure that applied in without balloon scenario. P_{ballon} is the pressure when the balloon is determined. $A_{balloon}$ is the surface area of the balloon and A_{stent} is the surface area of the stent.

The surface areas of the balloon are assumed as a stent inner surface without a cell. Therefore, the following surface area of the stent and the balloon respectively 11.62 mm^2 and 62.83mm^2. Table 1 summarized the loading scenarios that applied to the inner surface of the stent. It shows a comparison between the pressure of the balloon and the negligible balloon scenario.

Table 1. Loading Scenario.

With Balloon (MPa)	Without balloon (MPa)
0.3	1.62
0.4	2.16
0.5	2.70
0.6	3.24
0.7	3.78
0.8	4.32
0.9	4.86
1	5.40

3. Results and Discussion

This section presents the result of finite element analysis of the stent design and modeling. The result includes the stress distribution and the outer diameter change.

3.1. Deployment Result

The deployment process is depicted in figure 2. It shows the deployment process when a 5.40 MPa (1 MPa with balloon equivalent) pressure is applied.

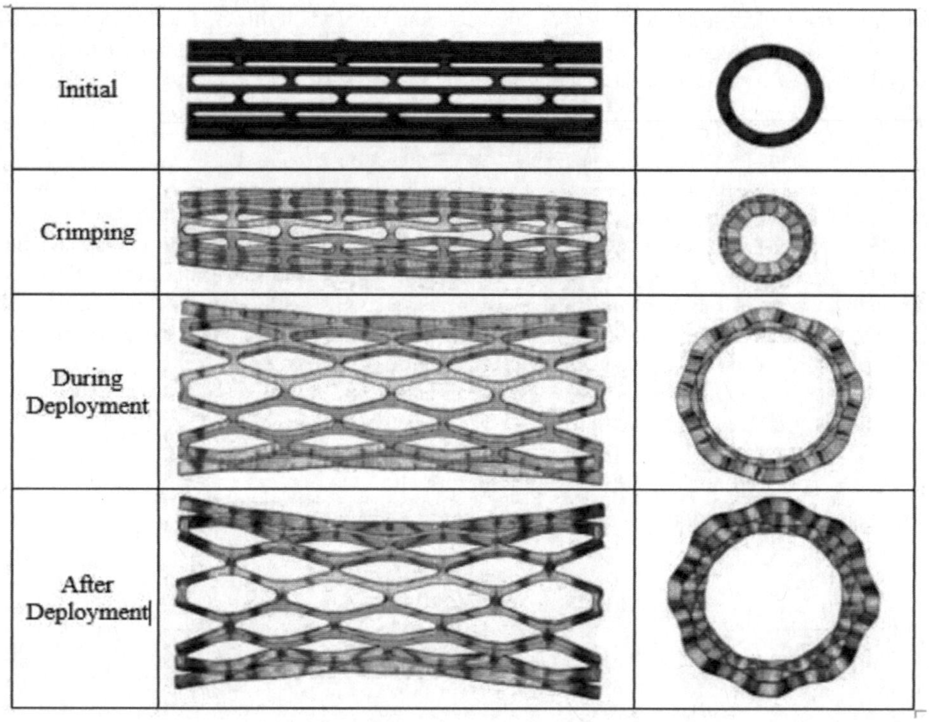

Figure 2. Deployment Process

3.2. Stress Distribution and Deformation

The deformations and the stress distributions when a 5.40 MPa (1MPa with balloon equivalent scenario) of pressure is applied is depicted in Figure 3. The stent deployment is designed in plastic deformation. So, the stress definitely higher than the yield strength of the material but it must under the ultimate tensile strength (UTS). The result shows that the maximum stress of crimping step and deployment process still below the UTS.

Two reference point (A and B) are used to measure the outer diameter change of the stent. The result of outer diameter change is depicted in figure 4.

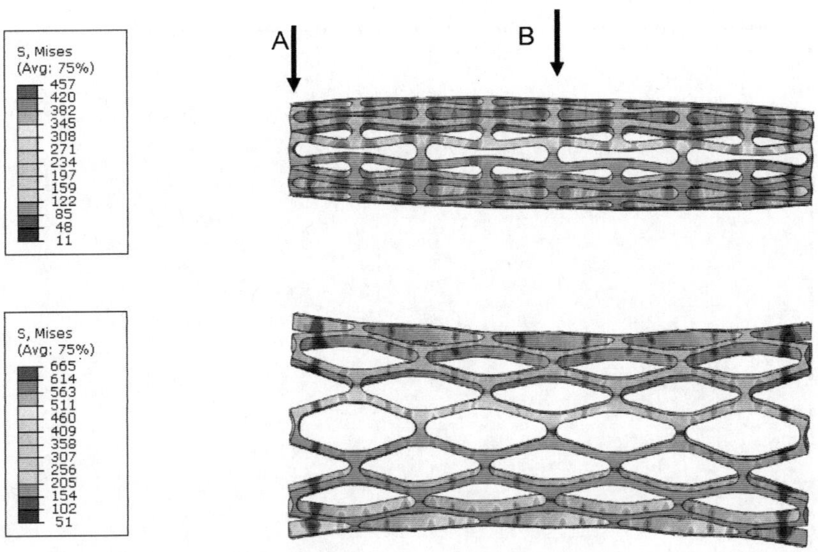

Figure 3. Stress Distribution

3.3. Outer Diameter Changes

The stent diameter changes against the pressure are shown in figure 4. The result shows that the diameter between middle and edge of the stent are different. The edge diameter often larger than the middle of the stent. This phenomenon is called dogbone.

Dogbone phenomena must be minimized because it can cause a vessel injuries. The dogbone phenomena are measured by comparing the ratio of the maximal displacement and the minimal displacement. The following equation is:

$$\boldsymbol{DB} = \frac{\boldsymbol{R}_{\mathbf{max}(\boldsymbol{edge})} - \boldsymbol{R}_{\boldsymbol{middle}}}{\boldsymbol{R}_{\mathbf{max}(\boldsymbol{edge})}}$$

Where $R_{max(edge)}$ is the maximum radius of the stent edge after deployment and R_{middle} is the middle stent radius after deployment. The percentage of dogbone against the pressure show in figure 5. The diameter reaches the initial diameter when the pressure is 2.16 MPa (0.4 with balloon equivalent) and the percentage of dogbone is 0.4 %.

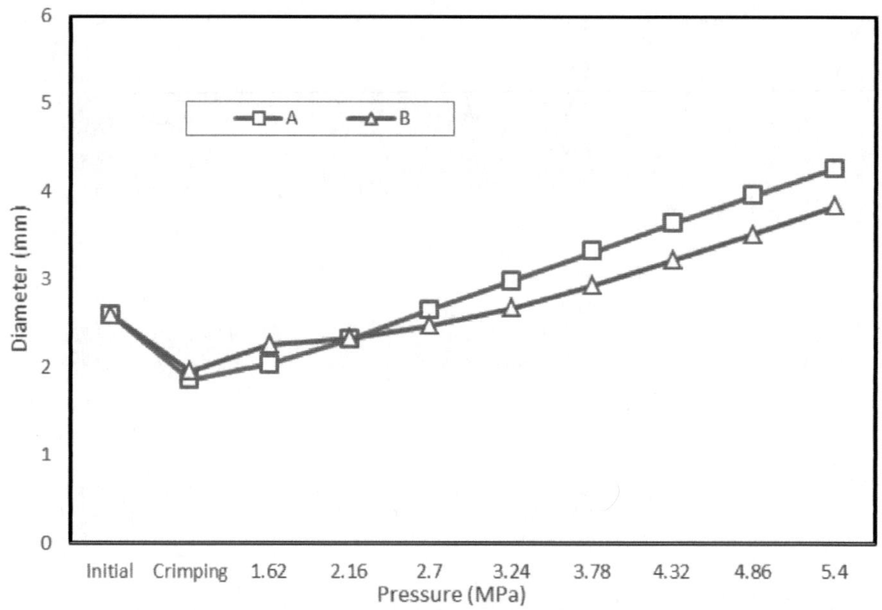

Figure 4. Outer Diameter Changes

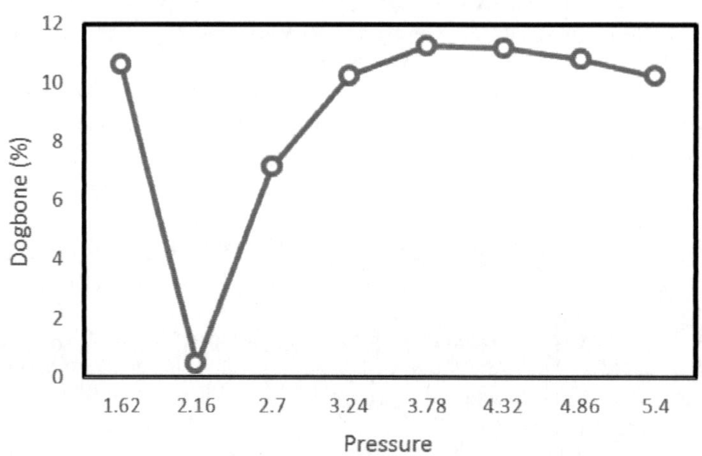

Figure 5. Percentage of Dogbone

4. Conclusion

This research is the preliminary step to investigate the stress distribution of stent expansion. The negligible balloon method is relative less accurate than included balloon method because it neglects the impact and contact of the balloon to the stent. However, this method can provide useful information about the stress distribution and deformation of the stent to evaluate the design. The result of the simulation shows that stent can deploy uniformly. The stent can reach the initial diameter with a small percentage of dogbone. The von misses stress also under the UTS of the material. It concludes that the stent design ready to be manufactured as a prototype.

IOP Conf. Series: Materials Science and Engineering **172** (2017) 012014 doi:10.1088/1757-899X/172/1/012014

5. References

[1] Data Centre and Information of Indonesian Health Ministry, 2013. The situation of heart health. Indonesian Health Ministry. Indonesia: Jakarta.

[2] Chua, S.N.D., Donald, B.J.M, dan Hasmi, M.S.J., 2002, Finite-element Simulation of Stent Expansion, Journal of Material Processing Technology, vol. 120, pp. 335-340.

[3] Auricchio, F., Di Loreto, M., Sacco, E., 2001. Finite-element analysis of a stenotic artery revascularization through a stent insertion. Computer Methods in Biomechanics and Biomedical Engineering 4, 249–263.

[4] Gu, L., Santra, S., Mericle, R.A., et al., 2005. Finite element analysis of covered microstents. Journal of Biomechanics 38, 1221–1227.

[5] Beule, M.D, Mortier, P., Carlier S.G., Verhegghe, B., Impe, R.V., Verdonck, P., 2008, Realistic Finite Element-based Stent Design: The Impact of Balloon Folding. J Biomech, vol. 41, pp. 383–389.

[6] Martin, D. dan Boyle, F., 2010, Drug-eluting Stents for Coronary Artery Disease: A Review. International Journal for Medical Engineering & Physics, vol. 33, pp. 148–163.

[7] Imani, S.M., Goundarzi, A.M., Valipour, P., Barzegar, M., Mahdinejad, J., Ghasemi, S.E., 2013, Application of Finite Element Method to Comparing the Nir Stent with the Multi-Link stent for Narrowings in Coronary Arteries. Acta Mechanica Solida Sinica, Vol. 28, No. 5.

[8] Murphy, B.P., Savage, P., McHugh, P.E, 2003. The Stress–strain Behavior of Coronary Stent Struts is Size Dependent. Annals of Biomedical Engineering, vol. 31, pp. 686–691.

IOP Conf. Series: Materials Science and Engineering **172** (2017) 012015 doi:10.1088/1757-899X/172/1/012015

Application of quantitative XRD on the precipitation of struvite from Brine Water

E Heraldy[1,2], F Rahmawati[1,2], Heryanto[1], and D P Putra[1]

[1] Department of Chemistry, Faculty of Mathematics and Natural Sciences, Sebelas Maret University
[2] Porous Materials Research Group
Jl. Ir. A. Sutami 36A, Kentingan, Surakarta 57126, Indonesia

Email: eheraldy@mipa.uns.ac.id

Abstract. The present studies have been conducted to quantify the varied phases in struvite formation from brine water as the magnesium source. The quantitative X-ray Diffraction (QXRD) method was performed to quantitatively determine the crystal phases and amorphous content of struvite samples. Substantial phase samples were employed quantitative analysis to calibrate against known phase composition information by Rietveld refinement on powder XRD data. The results showed that brine water could be considered as magnesium source the formation of struvite products. The study demonstrated that in general, the high N:P molar ratio (both pH 9 and 10) might lead to the significant formation of struvite.

1. Introduction

Struvite (magnesium ammonium phosphate hexahydrate) is a phosphate mineral with formula: $MgNH_4PO_4.6H_2O$. The struvite crystallization has been broadly used for simultaneous recovery of nutrients from human urine, swine wastewater, sludge supernatant and landfill leachate [1–5]. Its composition (i.e. Mg, N and P in equal molar concentrations) makes it a potentially marketable product for the fertilizer industry which can be further utilized as a valuable slow-release fertilizer [5-6].

In addition, precipitating struvite by recovery of phosphorus from wastewater has received much attention [7-10] due to struvite can provide both phosphorous and nitrogen nutrients simultaneously [11]. Examining the precipitation potential of struvite under different operational conditions [12-14] have been conducted.

Recently, Liu et al. [6] investigated the feasibility of using seawater and $MgCl_2$ as magnesium sources for phosphorus recovery from source-separated urine to struvite crystallization. The results demonstrated that both seawater and brine were effective magnesium sources to recover phosphorus from hydrolyzed urine to struvite formation. In addition, he was found that the presence of calcium in the magnesium sources could compromise struvite purity. Higher struvite purity could be obtained with higher Mg/Ca ratio in the magnesium source. In the previous study, as magnesium source, Heraldy et al. [15] have been utilized the brine water for Mg/Al hydrotalcite synthesis with the good performance. Therefore, the above mentioned challenges prompt us to investigate the feasibility of using brine water as magnesium sources for struvite crystallization.

The purity of struvite was investigated by various analytical methods, MAS-NMR spectroscopy [16], IC and ICP analysis [6, 17-18]. These methods were commonly used to determine the total Mg, P and N concentrations in precipitates. However, these methods may not be suitable for quantifying the relative phase abundances in precipitates because precipitates usually contain crystal and amorphous phases with similar elemental compositions. Although XRD methods have been used for struvite phase identification, the use has been limited to qualitative confirmation of the presence of phosphate phases in precipitates [17, 19]. However, quantitative phase analysis using XRD data can further be approached with Rietveld refinement [20], which fits a complete experimental diffraction pattern with a calculated profile [21]. Rietveld refinements for quantitative analysis of the phase compositions were processed with the RIETICA crystallographic program, where the pseudo-Voigt function was selected as the profile fitting function.

2. Materials and methods

The materials of H_3PO_4 and NH_3^+ were added to 200 mL brine water (as magnesium source) in stoichiometric proportion (Mg:N:P molar ratio) 1:1:1; 1:2:1 and 1:1:2 respectively at pH 9. 10 and 11. The mixture of raw materials was then agitated at room temperature for 2 h. After that, allow to separate the solids, and the solution is then decanted. The precipitate obtained were washed several times using free chlorine aquadest to remove the remaining impurities, separated by centrifuge 2500 rpm for 15 min and dried at 55°C for 24 h. Powder X-ray diffraction (XRD) data were collected by a Bruker D8 Advance X-ray diffractometer using monochromatized Cu-Kα radiation from $2\theta = 10°$ to 70°. XRD analysis was used to quantify the phase compositions of the samples using the Rietveld refinement method by fitting the full-profile experimental XRD data with a calculated diffraction pattern [22-23]. The Le Bail refinements were performed using RIETICA program [24]. Peak profiles were modeled using a pseudo-voigt peak shape. Experimental parameters refined were the instrument zero, scale factor, the lattice parameters and the peak shape parameters u, v, w, γ_0 and γ_1 [25].

3. Result and discussions

Rietveld refinement method was carried out for both samples using the lattice parameters from JCPDS standard as initial values. The refinement had been done as follow order zero, histogram parameter, lattice parameter and instrumental peak shape. For each step were used 30 cycles. In Fig. 1, the Struvite sample at pH 9 with molar ratio 1:1:1 was showed differently with JCPDS standard. The Rietveld analysis result which is shown in blue line is the differently between struvite and standard at 2 theta 33° and 16°. It is implied on the values of reability index parameters, i.e. the pattern factor (*Rp*), the weighted pattern factor (*Rwp*), and the goodness of fit ($\chi2$). The limitation of values of reability index parameters for *Rp* and *Rwp* closed to 8,0 [25] while for $\chi2$ closed to 4,0 [18, 27]. In Table 1, the struvite at pH 9 with molar ratio 1:1:1 have low reability values so as possible the high struvite content.

When the phosphor molar ratio in struvite was increased, appearing more new peaks nearby at 11°, 13° and around 30°, which are possible as the other phase of struvite like potassium cyclo-triphosphate ($K_3P_3O_9$) and hydroxylapatite [$Ca_5(PO_4)_3(OH)$]. Because of any other phase in struvite make it value of the reability index become higher (Table 1) and consequently can possible the content of struvite becoming lower. Therefore, the molar ratio Mg : P must be 1:1 to avoid the formation another phase in the formation of struvite. In addition, any other material will affect it. As is known in the brine water containing Ca^{2+} ions that will compete with Mg^{2+} ions to binding to P or N. Wang et al. [11] was stated the slightly amount of Ca^{2+} ions can afford reduces formation struvite. In according to Hao et al. [28], at pH higher than 10 will be a formation of Calcium compounds (Ca_3PO_4, $CaHPO_4$, etc.) so it will have an impact on the formation of struvite crystals.

Similarly, when the nitrogen mole was increased shows that the struvite crystallization was growing larger. It was proved in Fig. 1 and Fig. 2 which did not show new peaks and in the difference line no high peak as well as in Table 1 the *Rp* value closer to reliable value. Teddeo et al. [29] in his research was mentioned that the excessive nitrogen content are able to make newberyite [$MgHPO_4(H_2O)_3$] react to form struvite, so as to decrease the number of another phase than struvite formation. However, at pH 11 (Fig. 3) appears the new peaks in 11-13°. Hao et al. [28] were stated that when approaching pH 10.5 the content of NH_4^+ to struvite formation will be reduced from the deposition process and decreases rapidly when the pH exceeds 10.5 due to the equilibrium process to form NH_3. This is why in Fig. 3 at pH 11 struvite with molar ratio 1:1:2 tend to amorphous form. However when there is still excessive nitrogen content, is able to form struvite.

Molar ratio and pH is one of the parameters in struvite crystallization. When the pH is raised into pH 10 and pH 11, did not show significant changes in the XRD spectra. However, at pH 10 showed the formation of struvite higher than at pH 11. It can be seen in Fig. 2 where the formation of another phase besides struvite is lower than at pH 11 (Fig. 3) and also demonstrated in Table 1 at pH 10 is showed good quality in comparison to struvite at pH 11. Harrison et al. [30] were stated that the struvite crystallization could take place at a pH of between 7-11. Along with the increase of pH, struvite crystals was formed there will be more crystalline [30]. In accordance with Xia et al. [31] at pH 9, P total in solution tend to be in the form of HPO_4^{2-} that will produce a precipitate $MgNH_4PO_4.6H_2O$ and H^+, while at pH > 10 P total tend to be in the form of PO_3^{2-} that will produce $MgNH_4PO_4.6H_2O$ only. It is indicated that pH range 9-10 favorable for phosphate recovery. Similar trends were reported by Liu et al. [6].

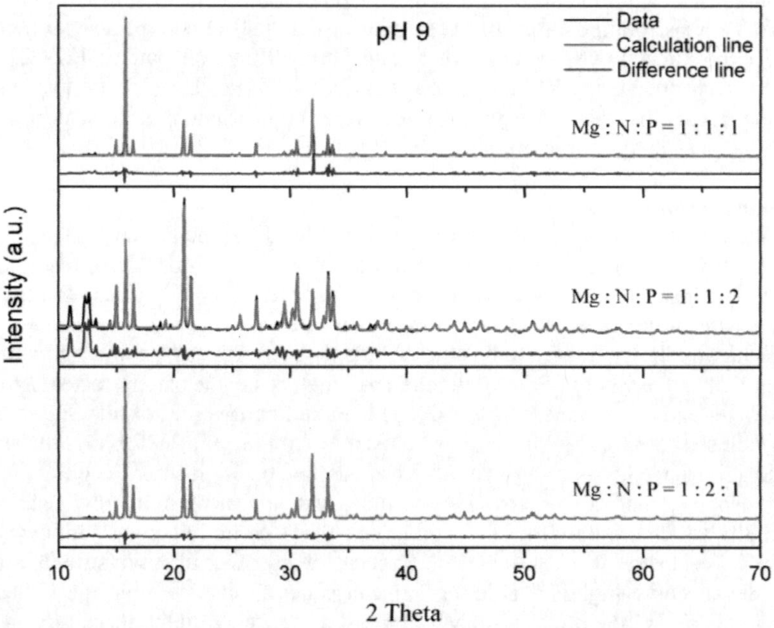

Figure 1. Graphic output of the Rietveld refinement of struvite at pH 9 in molar ratio Mg:N:P. The black line (---) represent XRD data; the red line (---) represents Rietveld calculated and the blue line (---) represent the difference line between XRD data and struvite standard

IOP Conf. Series: Materials Science and Engineering **172** (2017) 012015 doi:10.1088/1757-899X/172/1/012015

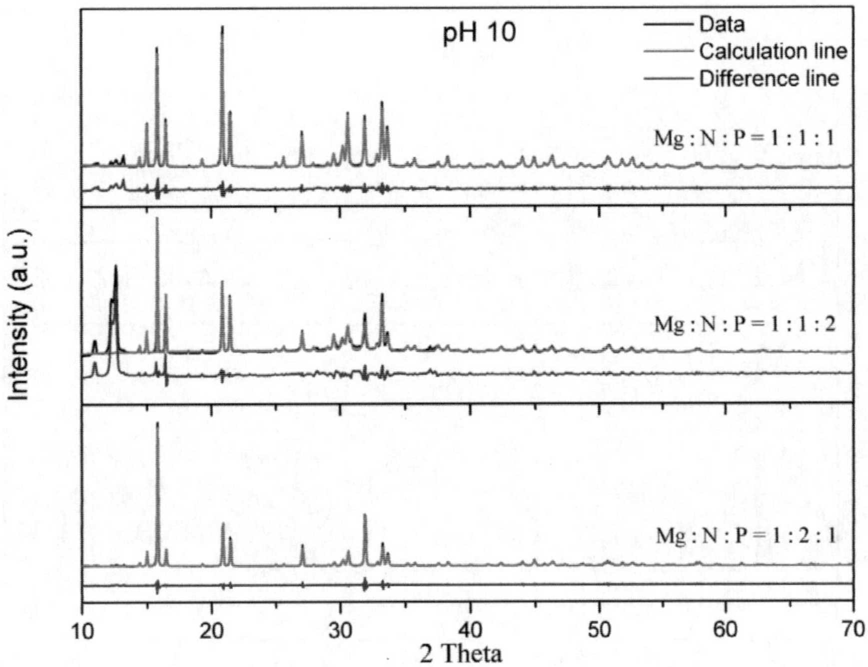

Figure 2. Graphic output of the Rietveld refinement of struvite at pH 10 in molar ratio Mg:N:P. The black line (---) represent XRD data; the red line (---) represents Rietveld calculated and the blue line (---) represent the difference line between XRD data and struvite standard

Table 1. Rietveld reliable factors, and phase composition of sample

Sample		Phase composition					Rietveld reliable factors		
pH	Molar ratio Mg : N : P						Rp	Rwp	χ2
9	1 : 1 : 1	A					17.66	26.62	134.5
	1 : 1 : 2	A	B	C	D	E	11.32	19.39	60.77
	1 : 2 : 1	A					7.71	11.55	25.56
10	1 : 1 : 1	A					8.46	12.16	22.59
	1 : 1 : 2	A	B	C	D	E	16.01	30.10	156.4
	1 : 2 : 1	A					6.97	9.52	12.45
11	1 : 1 : 2	A	B	C	D	E	15.50	25.78	91.45
	1 : 2 : 1	A	B	C			11.95	21.00	72.42

A = $MgNH_4PO_4.6H_2O$; B = $K_3(PO_3)_3$; C = $CaHPO_4$; D = $MgHPO_4(H_2O)_3$; E = $Ca_5(PO_4)_3(OH)$

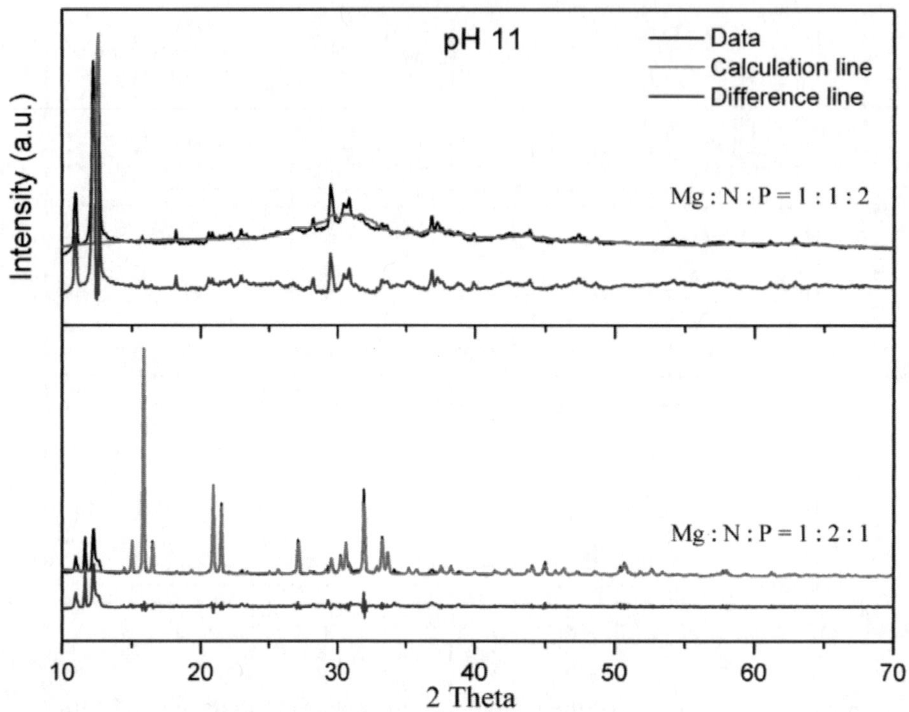

Figure 3. Graphic output of the Rietveld refinement of struvite at pH 11 in molar ratio Mg:N:P. The black line (---) represent XRD data; the red line (---) represents Rietveld calculated and the blue line (---) represent the difference line between XRD data and struvite standard

4. Conclusions

The present study showed that brine water could be considered as magnesium source in a formation of struvite products. The study demonstrated that in general, the high N:P molar ratio (both pH 9 and 10) might lead to the significant formation of struvite. The quantitative XRD technique is powerful and effective to determine the content of struvite in struvite products.

References

[1] Ganrot Z, Dave G, Nilsson E 2007 Recovery of N and P from human urine by freezing, struvite precipitation and adsorption to zeolite and active carbon, *Bioresour. Technol.* 98, 3112–3121.

[2] Huang H, Xu C, and Zhang W 2011 Removal of nutrients from piggery wastewater using struvite precipitation and pyrogenation technology, *Bioresour. Technol.* 102, 2523–2528.

[3] Shu L, Schneider P, Jegatheesan V and Johnson J 2006 An economic evaluation of phosphorus recovery as struvite from digester supernatant, *Bioresour.Technol*. 97, 2211–2216.

[4] Huang H, Xiao D, Zhang Q, and Ding L 2014 Removal of ammonia from landfill leachate by struvite precipitation with the use of low-cost phosphate and magnesium sources, *J. Environ. Manage*. 145, 191–198.

[5] Le Corre KS, Valsami-Jones E, Hobbs P, and Parsons SA 2009 Phosphorus recovery from wastewater by struvite crystallization: a review, *Crit. Rev. Environ. Sci.Technol.* 39, 433–477.

[6] Liu B, Giannis A, Zhang J, Chang VWC, and Wang J-Y 2013 Characterization of induced struvite

formation from source-separated urine using seawater and brine as magnesium sources, *Chemosphere* 93, 2738–2747.

[7] Battistoni P, Paci B, Fatone F, and Pavan P 2006 Phosphorus removal from anaerobic supernatants: start-up and steady-state conditions of a fluidized bed reactor full-scale plant, *Ind. Eng. Chem. Res.* 45 (2), 663-669.

[8] Karabegovic L, Uldal M, Werker A, Morgan-Sagastume F 2013 Phosphorus recovery potential from a waste stream with high organic and nutrient contents via struvite precipitation, *Environ. Technol.* 34 (7), 871-883.

[9] Lahav O, Telzhensky M, Zewuhn A, Gendel Y, Gerth J, Calmano W, and Birnhack L 2013 Struvite recovery from municipal-wastewater sludge centrifuge supernatant using seawater NF concentrate as a cheap Mg (II) source, *Sep. Purif. Technol.* 108, 103-110.

[10] Xie M, Nghiem LD, Price WE, and Elimelech M 2014 Toward resource recovery from wastewater: extraction of phosphorus from digested sludge using a hybrid forward osmosis-membrane distillation process, *Environ. Sci. Technol. Lett.* 1 (2), 191-195.

[11] Wang J, Burken JG, Zhand XJ, and Surampalli R 2005 Engineering struvite precipitation: impact of component-ion molar ratios and pH, *J. Environ. Eng.* 131, 1433-1440.

[12] Etter B, Tilley E, Khadka R, and Udert KM 2011 Low-cost struvite production using source-separated urine in Nepal, *Water Res.* 45 (2), 852-862.

[13] Lew B, Phalah S, Sheindorf C, Kummel M, Rebhun M, and Lahav O 2010 Favorable operating conditions for obtaining high-value struvite product from sludge dewatering filtrate, *Environ. Eng. Sci.* 27 (9), 733-741.

[14] Warmadewanthi, and Liu JC 2009 Recovery of phosphate and ammonium as struvite from semiconductor wastewater, *Sep. Purif. Technol.* 64 (3), 368-373.

[15] Heraldy E, Santosa SJ, Triyono, Wijaya K 2015 Anionic and cationic dyes removal from aqueous solutions by adsorption onto synthetic Mg/Al hydrotalcite-like compound, *Indones. J. Chem.* 15(3), 234-241.

[16] Scrimgeour SN, Chudek JA, Cowper GA, Lloyd CH 2007 ^{31}P solid-state MAS-NMR spectroscopy of the compounds that form in phosphate-bonded dental casting investment materials during setting, *Dent. Mater.* 23 (8), 934-943.

[17] Wang CC, Hao XD, Guo GS, van Loosdrecht MCM 2010 Formation of pure struvite at neutral pH by electrochemical deposition, *Chem. Eng. J.* 159 (1-3), 280-283.

[18] Yan H and Shih K 2016 Effects of calcium and ferric ions on struvite precipitation: A new assessment based on quantitative X-ray diffraction analysis, *Water Res.* 95, 310-318.

[19] Kruk DJ, Elektorowicz M, Oleszkiewicz JA 2014 Struvite precipitation and phosphorus removal using magnesium sacrificial anode, *Chemosphere* 101, 28-33.

[20] Nguyen DD, Devlin L, Koshy P, and Sorrell CC 2015 Quantitative X-ray diffraction analysis of anhydrous and hydrated portland cement-part 2: computer-based methods, *Adv. Mater. Res.* 1087, 498-503.

[21] Walenta G and Fullman T 2004 Advances in quantitative XRD analysis for clinker, cement, and cementitious additions, *Powder Diffr.* 19, 40-44.

[22] Wiles DT and Young R 1981 A new computer program for Rietveld analysis of X-ray powder diffraction patterns, *J. Appl. Crystallogr.* 14 (2), 149-151.

[23] Young R 1993 *The Rietveld Method*, IUCr. Monographs on Crystallography. Oxford University Press, Oxford.

[24] Hunter BA and Howard CJ 2000 *A Computer Program for Rietveld Analysis Of X-Ray and Neutron Powder Diffraction Patterns*, Lucas Heights Research Laboratories, NSW, Australia, 1-27.

[25] Peterson VK 2005 Lattice parameter measurement using Le Bail versus structural (Rietveld) refinement: A caution for complex, low symmetry systems, *Powder Diffr.* 20, 14–17.

[26] Fei L, Yanhuai L, Zhongxiao S, Kewei X, Zhi Z, and Hong C 2015 Rietveld Study on Evolution of Crystalline Structure of YSZ Nanoparticles during Co-Precipitation Synthesis, *Rare Met.*

Mater. Eng. 44, 2716–2720.

[27] Lu X, Shih K, Li X, Liu G, Zeng EY, and Wang F 2016 Accuracy and application of quantitative X-ray diffraction on the precipitation of struvite product, *Water Res.* 90, 9–14.

[28] Hao X, Wang C, Lan L, and Van Loosdrecht MCM 2008 Struvite formation, analytical methods and effects of pH and Ca^{2+}, *Water Sci. Technol.* 58, 1687–1692.

[29] Teddeo R, Kolppo K, and Lepisto R 2016 Sustainable nutrients recovery and recycling by optimizing the chemical addition sequence for struvite precipitation from raw swine slurries, *J Environ. Manage.* 180, 52-58.

[30] Harrison ML, Johns MR, White ET, and Mehta CM 2011 Growth rate kinetics for struvite crystallization, *Chem. Eng. Trans.* 25, 309–314.

[31] Xia P, Wang X, Song J, Wang H, Zhang J, and Zhao J 2016 Struvite crystallization combined adsorption of phosphate and ammonium from aqueous solutions by mesoporous MgO–loaded diatomite, *Colloids Surf., A.* 506, 220–227.

The Effect of Silane Addition on Chitosan-Fly Ash/CTAB as Electrolyte Membrane

E Kusumastuti[1], D Isnaeni[1], T Sulistyaningsih[1], F W Mahatmanti[1], Jumaeri[1], L Atmaja[2], and N Widiastuti[2]

[1] Chemistry Department, Faculty of Mathematics and Natural Sciences, State University of Semarang, D6 Building 2nd floor, Sekaran Unnes Campus, Gunungpati, Semarang, Indonesia
[2] Chemistry Department, Faculty of Mathematics and Natural Sciences, Tenth November Institute of Technology, Keputih ITS Campus, Sukolilo, Surabaya, Indonesia

Email : ella.kusuma@gmail.com

Abstract. Electrolyte membrane is an important component in fuel cell system, because it may influence fuel cell performance. Many efforts have been done to produce electrolyte membrane to replace comercial membrane. In this research, electrolyte membrane is composed of chitosan as an organic matrix and fly ash modified with CTAB and silane as inorganic filler. Fly ash is modified using silane as coupling agent to improve interfacial morphology between organic matrix and inorganic filler. This research aims to determine the best membrane performance based on its characteristics such as water uptake, mechanical properties, proton conductivity, and methanol permeability. The steps that have been done include silica preparation from fly ash, modification of silica surface with CTAB, silica coupling process with silane, synthesis of membranes with inversion phase method, and membrane characterization. The result shows that membrane C-FA/CTAB-Silane 10% (w/w) has the best performance with proton conductivity 8.00×10^{-4} $S.cm^{-1}$, methanol permeability 3.37×10^{-7} $cm.s^{-1}$, and selectivity 2.12×10^{3} $S.s.cm^{-3}$. The result of FTIR analysis on membrane C-FA/CTAB-Silane 10% shows that there is only physical interaction occured between chitosan, fly ash and silane, because there is no peak differences significantly at wave number 1000-1250 cm^{-1}, while morphology analysis on membrane with Scanning Electron Microscopy (SEM) shows good dispersion and there is no agglomeration on chitosan matrix.

1. Introduction

Electrolyte membrane is a critical component in the fuel cell because it serves as a proton conductor and also as a barrier between the two electrodes (anode and cathode) [1]. Currently, the wide use of membrane material is Nafion. The ability of Nafion as a proton conductor is very good with proton conductivity of 0.082 S / cm [2]. The main problem of Nafion if it is used at Direct Methanol Fuel Cell (DMFC) is there is methanol passing through the membrane (methanol crossover). The methanol permeability can cause the loss of fraction of fuel and cause the rate of reaction at the cathode to be slower, which means reducing the performance of cell voltage [1]. Many efforts have been done to produce the electrolyte membrane as Nafion replacement. One biopolymer that is intensively being investigated to replace Nafion is chitosan modified with inorganic fillers for example silica.

The use of inorganic filler with a smaller size can improve the performance of membranes for fuel cell applications [3]. This is according to research of [4] who did chitosan composite membrane/HZSM-5 with variations in particle size is 5, 2, and 0.4 lm. the results showed that there has been an increase in proton conductivity and methanol permeability decrease with decreasing particle size HZSM-5.

In this research, silica is coupled with silane, and then incorporated into chitosan to prepare C-FA/CTAB-Silane membrane. According to [5] the presence silane can improve the adhesion between the polymer matrix with filler, so that the interaction is stronger and can improve the electrolyte membrane properties. The proton conductivity, methanol permeability, and membrane selectivity are investigated and optimum membrane is characterized by FT-IR and SEM.

2. Experiment

2.1. Materials

Chitosan with the deacetylation degree of >80% was purchased from Bogor Agricultural Institute, Glycidoxypropyl-trimethoxysilane (GPTMS) and N, N-dimethylformamide (DMF) was purchased from CV. Wahana Hilab Indonesia (Yogjakarta), Fly ash was obtained from PLTU Tanjung Pati Jepara, CTAB, CH_3COOH, CH_3OH, NH_3, HCl, H_2SO_4 was purchased from CV. Chem-Mix Pratama (Yogjakarta).

2.2. Silica preparation from fly ash

Silica preparation from fly ash was done by washing method using HNO_3. The ratio of fly ash and HNO_3 was 1: 4, then stirred using magnetic stirrer for 3 hours at room temperature. After that, the solution was filtered and the residue was washed again with HNO_3. The residue obtained in this way was washed with distilled water to pH neutral and then was dried in an oven at 200°C for 6 hours [6].

2.3. Modification of fly ash surface with CTAB

First step, silica, CTAB and aquademin were mixed and stirred for 30 minutes at room temperature. The ammonia was added quickly until pH= 9, then the solution was stirred again for 2 hours at 60°C. After the solution was filtered, the residue was washed with aquademin to pH neutral and then was dried in an oven at 100°C for 24 hours and was dried in a furnace at 650°C for 3 hours [7].

2.4. Fly ash coupling process with silane

Fly ash and GPTMS were dissolved in 10 mL DMF at room temperature and stirred for 6 hours. Then the homogeneous solution was cast in a beaker glass and placed in an oven at 60°C, then at 100°C for 1 hour, and 120°C for 2 hours until the solvent removed [8]. After that, the residue was soaked with 0.1 M HCl solution at 80°C for 24 hours [9].

2.5. Synthesis of membranes

Chitosan powders 1.000 g were dissolved in 50 mL 2% (w/v) acetic acid and stirred for 4 hours at room temperature. Fly ash powders 0.0883 were dissolved in 50 mL 2% (w/v) acetic acid and stirred for 7 hours at room temperature. Then the two solutions were mixed and stirred for 4 hours at room temperature. The dope solution then was poured into acrylic mold and dried in oven at 60?C for 24 hours.

3. Result and Discussion

3.1. Characterization of silica preparation from fly ash

Tabel 1 shows comparison of fly ash chemical composition by *X-Ray Flourescence* (XRF) before and after being washed with HNO_3.

IOP Conf. Series: Materials Science and Engineering **172** (2017) 012016 doi:10.1088/1757-899X/172/1/012016

Tabel 1. The chemical composition of fly ash

No	Compound	Content before preparation (%)	Content after preparation (%)
1	SiO_2	44.94	56.61
2	Al_2O_3	19.10	17.58
3	Fe_2O_3	14.25	17.85
4	CaO	6.64	2.72
5	MgO	4.53	0.21
6	Na_2O	4.14	0.05
7	TiO_2	1.11	0.84

Table 1 shows the SiO_2 content increased by 11.67%. It shows that the nitric acid dissolve metal impurities better than Si in fly ash. The reaction between fly ash and nitric acid is shown in the following equation.

$$\text{Fly ash} + HNO_{3(excess)} \longrightarrow SiO_{2(s)} + M(NO_3)_{n\ (aq)}$$

where, M is a metal other than Si which contained in fly ash such as Al, Mg, Fe, Na, K, Ca, Ti, Mn and other metals [6].

3.2. Characterization of CTAB-modified fly ash particle

Figure 1 shows the morphology of fly ash modified with CTAB a) 0% and b) 4.10%. Modification process of silica with CTAB have been done successfully, it was supported by the particle size of fly ash from the beginning (1-5.5 μm) become slightly smaller than fly ash after modified with CTAB (0.75 – 3.33 μm) if were. This is accordance to[6] which states that the modification of CTAB on the surface of silica can reduce the size of original silica 240.38 nm to 107.89 nm.

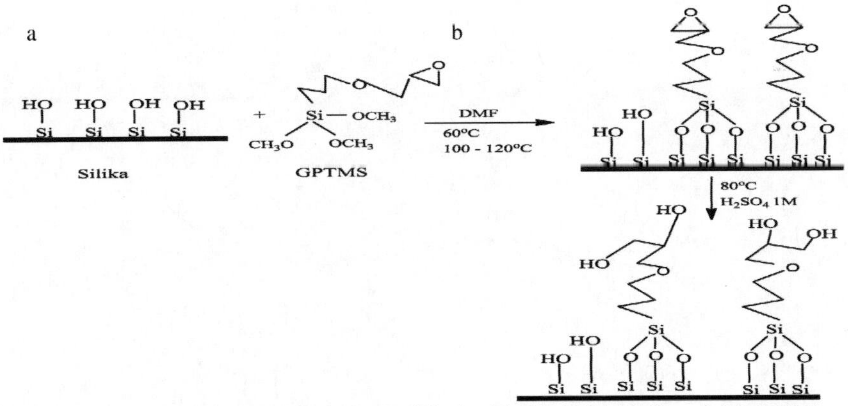

Figure 1. Morphology of fly ash modified with CTAB (a) 0% and (b) 4,10%.

In addition, agglomeration of silica particles in fly ash appeared to be smaller after they are modified with CTAB. According to [10], surfactant can reduce inter-particle agglomeration of silica, so it can facilitate silica joining the polymeric matrix.

3.3. Characterization of Fly ash coupling process with silane

Figure 2 shows the morphology of fly ash after being coupled with 10% of silane. After the fly ash is coupled with silane, it is obvious that there are silica particles that are spherical with a smaller size in fly ash. It indicates that fly ash coupling process with silane has been done successfully, because the silane that interacts with fly ash and cause coupled of silica which is contained in fly ash.

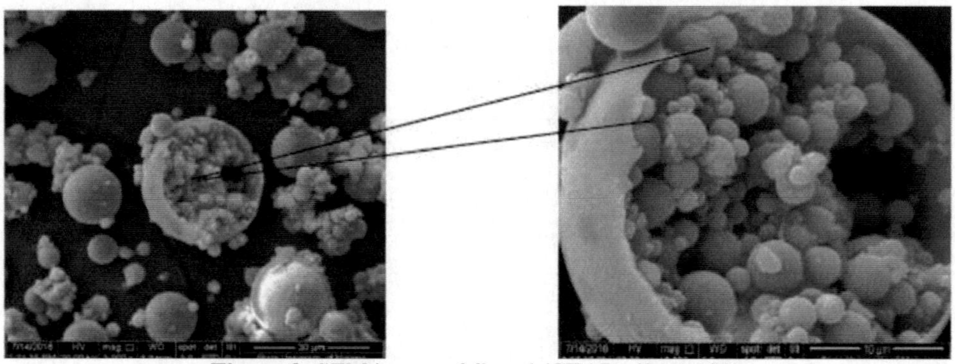

Figure 2. SEM image of fly ash/CTAB-silane 10%

Schematic representation of the coupling process of fly ash with silane can be seen in Figure 3. In the first stage, silane form silanol (Si-OH), then silanol from several silane molecules undergo condensation to form polisilane network through bonding Si-O-Si. The next process is incorporation of silane compound on the silica substrate. In this process, the hydroxyl group (OH) on polisilane is bonded to silanol (Si-OH) of the silica substrate through hydrogen bonds and releases water molecules, so it will form polysiloxane network between silica substrate with a silane compound.

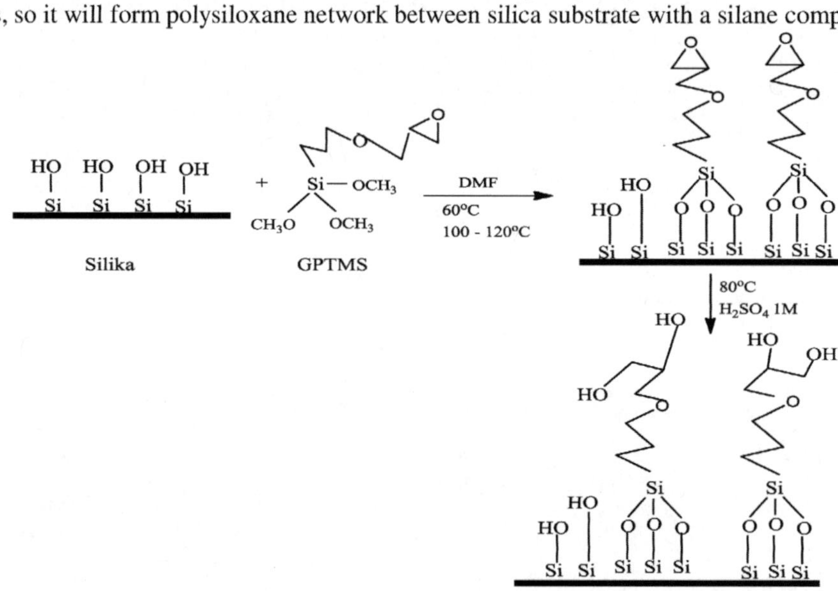

Figure 3. Schematic of the coupling process of fly ash with silane

The final stage of the fly ash coupling process with silane is the soaking powders with 0,1 M HCl solution at 80°C. It is according to [9] at this stage epoxy groups deformation (ring opening) of the silane compound will happen and produces hydroxyl (OH) that has the potential to react with the amine group on chitosan matrix.

3.4. Proton Conductivity

The proton conductivity of chitosan-fly ash (C-FA), C-FA/CTAB-Silane membranes that are in hydrated conditions at room temperature is presented in Figure 4. It is found that the proton conductivity increases after being coupled with silane, but in silane addition > 10% proton conductivity decreases. It is because the amount of added silane is strong enough to bind the chitosan amine and form a polysiloxane network with fly ash. Strong hydrogen bonds are formed between the amine group on the chitosan and epoxy silane, so it can facilitate the proton hopping.

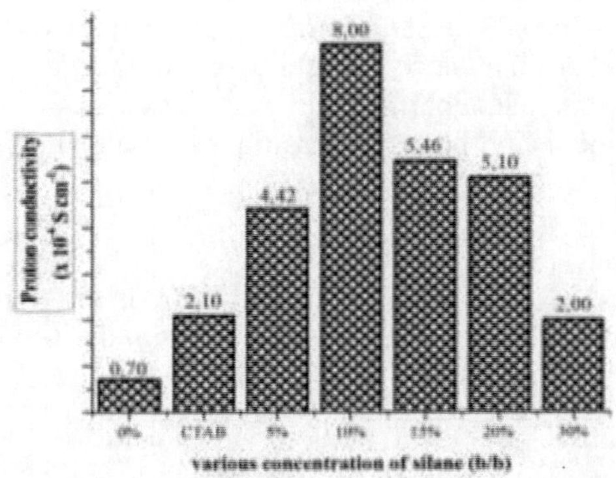

Figure 4. Proton conductivty membrane with various concentration of silane

It is consistent with the statement of [8] that the polysiloxane bond formed between the silica and silane fly ash can bind with water to facilitate the occurrence of proton hopping mechanism. In addition, silane at a concentration of >10%, the hydrogen bonds formed between the silane and chitosan and polysiloxane network become saturated and therefore can not facilitate proton hopping optimally and resulting lower proton conductivity.

All proton conductivity of membrane synthesis result shows a smaller value compared to Nafion, but this membrane can still be used as an electrolyte membrane for by [11] membrane having a proton conductivity above 1 x 10^{-5} S / cm can be used as an electrolyte membrane fuel cell. The result shows all membrane synthesis product can be used as an electrolyte membrane.

3.5. Methanol Permeability

The methanol permeability of chitosan-fly ash (C-FA), C-FA/CTAB-Silane membranes at 1 M methanol concentration is shown in Figure 5. All membranes show methanol permeability much lower in the range 1.77 x 10^{-6} to 0.337 x 10^{-6} $cm^2.s^{-1}$ than Nafion (5.6 x 10^{-6} cm2.s^{-1}). With the increase of silane addition, the methanol permeability first decreases then increases when silane addition >10% is done.

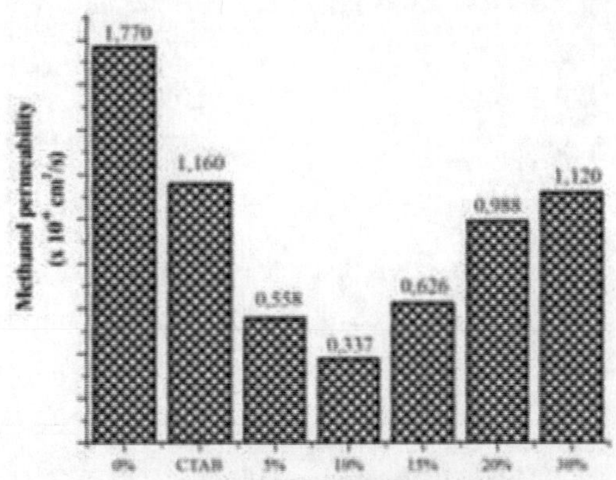

Figure 5. Methanol permeability membrane with various concentration of silane

Decreasing methanol permeability is caused by the power of a strong interface that causes empty space (void) contained on the membrane surface. It is according to [9] that decreasing methanol permeability can be attributed to better interfacial morphology between chitosan matrix and fly ash filler. The transitional phase between the silane and chitosan generated by the organic chains on the surface silica is favorable for reducing the non-selective voids, and the hydrogen bonds between fly ash and chitosan connect two phases closer. Thus, more methanol molecules would diffuse through the pores in fly ash particle, rather than the fly ash-chitosan interface, which is favorable for reducing the metanol crossover. Increase methanol permeability was caused by the aggregate pore on membrane. It can increase methanol permeability, because mass transfer (methanol) could occur through aggregate pore [2].

3.6. *Selectivity*

Membranes for the practical usage of PEM in fuel cell are required to posses high proton conductivity and low methanol permeability. Figure 6 shows the selectivity defined the ratio of proton conductivity and methanol permeability.

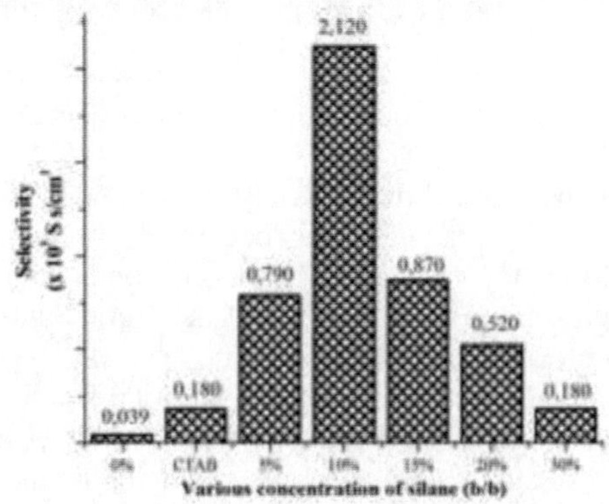

Figure 6. Selectivity membrane with various concentration of silane

The selectivity C-FA/CTAB-Silane 10% membrane closely resembled with Nafion (2.04 x 10^3 S.s.cm^{-3}), so the membrane can be used as electrolyte membrane.

3.7. *FT-IR Spectra of membrane*

The FT-IR spectra of C-FA, C-FA/CTAB, C-FA/CTAB-Silane 10 and 30% membrane are presented in Figure 7.

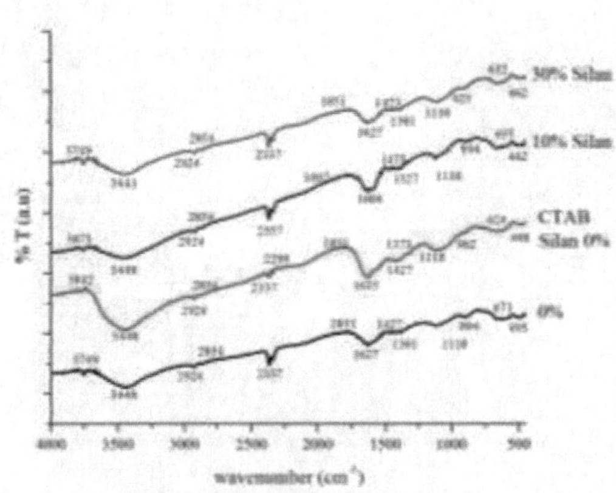

Figure 7. FT-IR spectra of membrane

The characteristic bands of chitosan including the –OH stretching, C-O stretching, -NH_2 bending, and –CH_2 stretching are located at 3448, 1118, 1627, and 2924 cm^{-1}, while the characteristic of fly ash is found at 671 and 493 cm^{-1} which shows Si-O and Al-O are stretching. The result of FT-IR analysis shows that there is physical interaction between chitosan, fly ash, and silane since there is no change singnificantly.

3.8. Morphology analysis of membrane

Figure 8 shows the typical surface-section SEM image of the C-FA, C-FA/CTAB, C-FA/CTAB-Silane 10 and 30% membrane.

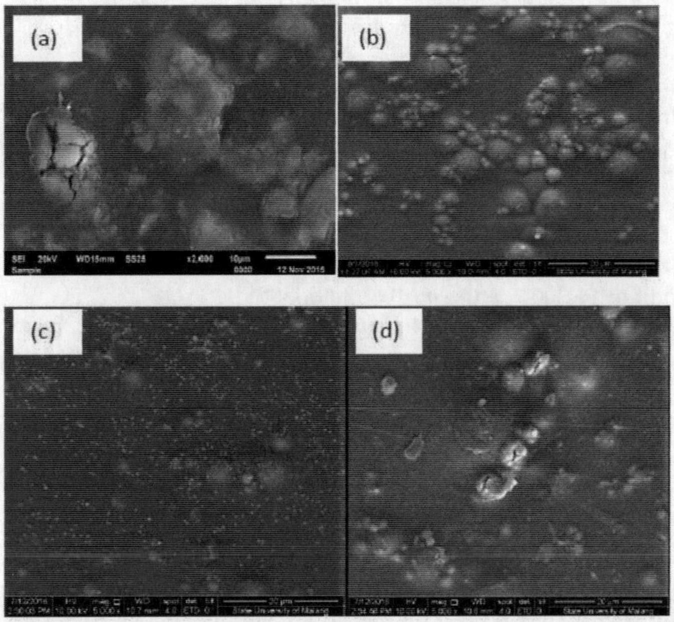

Figure 8. SEM image the (a) C-FA, (b) C-FA/CTAB, (c) C-FA/CTAB-Silane 10% and (d) C-FA/CTAB-Silane 30% membrane

As shown in Figure 8a) and 8b) there is obvious agglomeration of fly ash particle in the surface of membrane. If it is compared with C-FA membrane, the particle size of fly ash in C-FA/CTAB membrane is smaller than fly as in C-FA membrane, indicating that modification of surface of the fly ash with CTAB can reduce the size particle of fly ash. In addition, in Figure 8c) no abvious agglomeration and good dispersion of fly ash particle, while in Figure 8d) obvious agglomeration happenned. The result suggests that at C-FA/CTAB-Silane 10% membrane, interaction between chitosan and fly ash is good, so the membrane structure becomes compact, whereas at C-FA/CTAB-Silane 30% membrane, there is obvious agglomeration because the excess of silane addition, so interaction between chitosan and fly ash is not good.

4. Conclusion

C-FA/CTAB-Silane 10% membrane has the best performance with proton conductivity of 8.00 x 10^{-4} $S.cm^{-1}$, methanol permeability 3.37 x 10^{-7} $cm.s^{-1}$, and membrane selectivity 2.12 x 10^{3} $S.s.cm^{-3}$. The result of Fourier Transform Infra Red (FT-IR) analysis on membrane C-FA/CTAB-Silane 0% and 10% shows that there is only physical interaction occured between chitosan, fly ash and silane, while the result of morphology analysis membrane with Scanning Electron Microscopy (SEM) shows particle distribution was spread evenly and there is no agglomeration.

IOP Conf. Series: Materials Science and Engineering **172** (2017) 012016 doi:10.1088/1757-899X/172/1/012016

References

[1] Im M 2011 *Pembuatan dan Karakterisasi Komposit Membran PEEK Silika/Clay untuk Aplikasi Direct Metanol Fuel Cell (DMFC)* Thesis (Semarang: Program Pascasarjana Universitas Diponegoro)
[2] Handayani S and Dewi E L 2007 *Indo. J. Mater. Sci.* D **536**
[3] Wardani R K 2015 *Pengaruh Penambahan Surfaktan Kationik Pada Filler Terhadap Sifat dan Kinerja Membran Komposit Kitosan/Montmorilonit Termodifikasi Silan Untuk Aplikasi DMFC* Thesis (Surabaya: Institut Teknologi Sepuluh November)
[4] Wang J, Zheng X, Wu H, Zheng B, Jiang Z, Hao X and Wang B 2008 *J. Power. Sources.* **178**: 9-19
[5] Ismail H, Suhelmy S and Edyham M R 2002 *Eur. Polymer. J.* **38**: 39-49.
[6] Kim S J, Seo S G, and Jung S C 2010 *Korea. J. Chem. Eng.* **27**(6), 1901-1905
[7] Singh L P, Bhattacharyya S K, Mishra G and Ahalawat S 2011 *Appl. Nanosci.* 1:117–122
[8] Lin H, Chengji Z, Wenjia M, Shao K, Hongtao L, Yang Z and Hui N *J. Power. Sources.* 195: 762-768
[9] Wang Y, Jiang Z, Li H and Yang D 2010 *Chem. Eng. Process.* 49:278-285
[10] Zou H, Wu S and Shen J 2008 *J. Chem.* 108: 3893-3957
[11] Smitha B, Sridhar S and Khan A A 2004 *J. Macromolecules.* 37: 2233–2239.

IOP Conf. Series: Materials Science and Engineering **172** (2017) 012017 doi:10.1088/1757-899X/172/1/012017

In silico design of fragment-based drug targeting host processing α-glucosidase i for dengue fever

E P Toepak[1], **U S F Tambunan**[1]*

[1] Bioinformatics Research Group, Departement Of Chemistry, Faculty of Mathematics and Natural Science, Universitas Indonesia, Kampus UI Depok, West Java 16424, Indonesia

* Email:usman@ui.ac.id

Abstract. Dengue is a major health problem in the tropical and sub-tropical regions. The development of antiviral that targeting dengue's host enzyme can be more effective and efficient treatment than the viral enzyme. Host enzyme processing α-glucosidase I has an important role in the maturation process of dengue virus envelope glycoprotein. The inhibition of processing α-glucosidase I can become a promising target for dengue fever treatment. The antiviral approach using in silico fragment-based drug design can generate drug candidates with high binding affinity. In this research, 198.621 compounds were obtained from ZINC15 Biogenic Database. These compounds were screened to find the favorable fragments according to Rules of Three and pharmacological properties. The screening fragments were docked into the active site of processing α-glucosidase I. The potential fragment candidates from the molecular docking simulation were linked with castanospermine (CAST) to generate ligands with a better binding affinity. The Analysis of ligand - enzyme interaction showed ligands with code LRS 22, 28, and 47 have the better binding free energy than the standard ligand. Ligand LRS 28 (N-2-4-methyl-5-((1S,3S,6S,7R,8R,8aR)-1,6,7,8-tetrahydroxyoctahydroindolizin-3-yl) pentyl) indolin-1-yl) propionamide) itself among the other ligands has the lowest binding free energy. Pharmacological properties prediction also showed the ligands LRS 22, 28, and 47 can be promising as the dengue fever drug candidates.

1. Introduction

Dengue fever is a viral disease in humans caused by dengue virus (DENV)[1]. DENV belongs to the Flaviviridae family and because of that has a close relationship with other humans lethal viral pathogens such as Yellow Fever and West Nile virus. This virus has five serotypes, that are DENV-1, DENV-2, DENV-3, DENV-4, and DENV-5 [2,3]. Dengue virus is commonly found in tropical areas, especially in the regions of Asia and Africa. But recently, dengue infection has been reported in the Caribbean area, South America, and Europe [4,5]. DENV infection has spread around the world so that it has become one of the major health problems worldwide.

Each year, there are at least 40 to 100 million people are infected by DENV and more than half of the world's population at risk of infecting by this virus. Infections caused by the DENV can cause high fever and flu-like symptoms. These infections, in some people,may also develop into a more acute phase called dengue hemorrhagic fever (DHF) and dengue shock syndrome (DSS) [1,6,7]. DHF and DSS are responsible for causing death in major dengue cases, around 2.5 % from 500,000 clinical cases [6]. Therefore, an effective treatment is needed to repress the number of death caused by a dengue infection.

Antivirals studies based on virus factor have been widely reported. Thiazoles that has been modified by Li et al (2009) could use to inhibit envelope protein in the binding site of β-OG, Sousa et al (2015) reported that the compound agathisflavone and myricetin have noncompetitive inhibitor activity on NS2B-NS3 protease dengue virus serotype 2 and Tambunan et al (2015) through its research reported that 5-(3,4-dichlorophenyl)-N-[2-(p-tolyl) benzotriazol-5-yl]furan-2-carboxamidecan inhibit n-octyl-β-d-glucoside (β-OG) binding pocket on envelope protein of DENV-2 [8–10]. However, antiviral compounds that target the virus proteins usually only specific to a particular virus and sometimes fail due to the emergence of some resistant viruses. Antiviral targeting host protein that involved in the life cycle of the virus can be used as a therapy for various types of virus and the treatment of resistant viruses. There are a number of cellular proteins in the host cell that have been known to cause effects as anti-dengue when inhibited. Endoplasmic reticulum (ER)-resident α-glucosidase is one of the host proteins that can be used as a potential target [11–13].

Enzyme α-glucosidase is an enzyme that essential in the maturation process of the nascent glycoprotein. It involves in N-linked glycan part trimming process whereas their role is cutting glucose part on N-linked glycans that attached to immature glycoproteins. Envelope protein from some viruses have been known as glycoprotein and there is N-linked glycan attached to it. Therefore, inhibiting a-glucosidase in N-linked glycan part trimming process could interrupt the maturation of nascent envelope virus and hence it also disturb the function of viral envelope [12,14].

The α-glucosidase protein consists of processing α-glucosidase I and α-glucosidase II.Inhibiting processing α-glucosidase I by castanospermine (CAST) may cause an antiviral effect in the dengue infection [14–16]. Antiviral development with fragment-based drug approach could generate potential drug candidate. Fragment-based drug method aroused a larger and complex lead compound with higher binding affinity by joining two or more fragment [17].

In this study, we designed α-glucosidase I inhibitor using in silico fragments-based drug approach. Molecular docking is used to study the binding interaction between inhibitors and processing α-glucosidase I [18]. We used MOE 2014.09 software to perform molecular docking [19]. The objective of this study was to generate a lead compound using the silico fragment-based drug which showed inhibition activity toward processing α-glucosidase I and has good pharmacological properties. We expected the outcome from this study could provide an insight into a novel antiviral treatment for dengue infection.

2. Methods

2.1 Preparation of Processing α-glucosidase I

In this study, we used the three-dimensional structure of processing α-glucosidase I protein with PDB ID: 4J5T [16] that available in Protein Data Bank (http://www.rcsb.org/pdb). The other molecule that attaches in the enzyme, such as water (H_2O), were removed. LigX function in MOE 2014.09 was utilized to minimize and optimize the processing α-glucosidase I structure.

2.2 Fragment Preparation

Fragment database was downloaded from ZINC15 [20]. We used an abiogenic database that contains natural product and metabolite compound as fragment database. The rule of three (RO3) [21] was applied using Osiris Data Warrior software (http://www.openmolecules.org/datawarrior/) on the database to reduce the amount of the molecules and construct a database with the potential fragment. The CAST compound was drawn and convert to .mol type file structure using ACD/ChemSktech 12.01 (www.acdlabs.com). The energy minimized was applied to the fragment database and CAST using MOE 2014.09 and then the partial charge was also applied using the same software in an MMFF94x force field.

2.3 Molecular Docking of CAST and Fragment Database
Molecular docking simulation for the CAST and fragment database against processing α-glucosidase I was performed using MOE 214.09 software. We used Triangle Matcher as placements methods, London dG as scoring function and 30 times of placement retain as parameters. The other parameters were set according to MOE 2014.09 default parameters. The active site from Baker's (2013) processing α-glucosidase I model was chosen as docking area [16].

2.4 Hits Selection and Fragment linking
Potential fragments and CAST were linked using linker database in MOE 2014.09 to generated new ligands. The Lipinski Rules of Five (RO5) was applied in linking process. These new ligands were optimized using MOE 2014.09 and then were docked with processing α-glucosidase I. We choose several ligands that have low binding energies and good ligand-enzyme interaction to be continued into the next phase.

2.5 Analysis of Ligand Interactions and Drug Scan
The docked ligands with low binding energies were analyzed their interaction with the active site of processing α-glucosidase I. In this study, we also used some software such as Osiris DataWarrior and T.E.S.T (Toxicity Estimation Software Tool) (https://www.epa.gov/chemical-research/toxicity-estimation-software-tool-test) in drug scan to predicted the drug-likeness and the toxicity.

3. **Result and Discussion**

3.1 Preparation of processing α-glucosidase I
In this study, we used processing α-glucosidase I structure that has been reported by Barker and Rose (2013) with PDB ID: 4J5T in Protein Data Bank (PDB). This structure has similarity with processing α-glucosidase I structure in eukaryotes more than the other model structure of processing α-glucosidase I. They also reported that processing α-glucosidase I has catalytic residues that retained in Asp568 and Glu771 [16]. Therefore, the interaction of the ligand with one of this residue is important for a potential inhibitor candidates of processing α-glucosidase I. We used LigX in MOE 2014.09 to minimization and optimation to prepared the processing α-glucosidase I in Amber10:EHT force field. The minimization of processing α-glucosidase I structure was done when the RMS gradient of 0,05 kcal/mol/$Å^2$.

3.2 Creating and Preparation of the Fragment Libraries
The biogenic database was obtained from ZINC15 [20]. We used Osiris DataWarrior to screened and created the fragment libraries. The RO3 was applied to reduce the amount of the potential fragment. Moreover, The RO3 also helped to create libraries that only consist of fragment-like properties and simple molecules that easily to elaborated [21,22]. In general, RO3 would reduce the complexity of the molecule in fragment library in which have following criteria [17]:

- the compounds molecular weight are less than 300g/mol,
- have the number of hydrogen bond acceptor less than or equal with three,
- have less than or equal with three hydrogen bond donor, and
- have the number of cLogP 3

Furthermore, some rules are added in RO3 as well. These rules are:

- A number of rotatable bonds are less than or equal with three, and
- have 60 $Å^2$ surface area.

The RO3 screening results in 13.085 compounds from 198.621 compounds. Toxicity screening could also be used to avoiding reactive, unstable, or toxic fragments [21,23]. We used toxicity parameter such as mutagenic, tumorigenic, irritant and reproductive-effect compound as a parameter in Osiris DataWarrior to screen and create fragment libraries. After the toxicity screening process was conducted, we obtained 1.541 compounds in our fragment libraries.

The additional factor that must be considered to screened and creating libraries of compounds is drug-likeness. Some compounds fail to develop as a drug due of poor drug-likeness and pharmacokinetic properties [24]. Drug-likeness score could be used as a parameter to design fragment libraries. We used the drug-likeness score of the CAST as the standard ligand for screening the fragment database. This screening process resulted in 52 compounds in fragment libraries.

Fragment-based drug design method involves molecules that have a molecular weight less than 300 g/mol. Molecules such CAST which has molecule weight 189.210 g/mol could be involved in this method. CAST itself has inhibition activity towards processing α-glucosidase I [17,21]. Fragment linking between CAST and another potential fragment could build a bigger molecule with wider interaction area in the processing α-glucosidase I active site. Both energies minimized and partial charge were applied to both CAST and fragment libraries using an MMFF94x force field. The RMS gradient of the minimization was performed until it reaches 0,001 kcal/mol/Å.

3.2 Molecular docking Analysis of CAST and Fragment Libraries

Molecular docking is a method to predict and determine the favorable interaction between ligand and the receptor (e.g. enzyme or protein) [25]. Gibbs free energy binding energy (ΔG) was obtained from the result of molecular docking that used to predict which ligand-enzyme conformation that favourable. In this study, the molecular docking was performed using MOE 2014.09 software. First, we docked CAST towards processing α-glucosidase I to obtain the ΔG and their binding interaction. This simulation resulted that CAST has a binding interaction with three residues (Glu771, Asp392, and Trp391, as shown in figure 1.). The Glu771 itself is one of the catalytic residues in processing α-glucosidase I. The docking simulation also showed the ΔG of CAST is -28.7224 kcal/mol with an inhibition constant (pKi) of 20.9238.

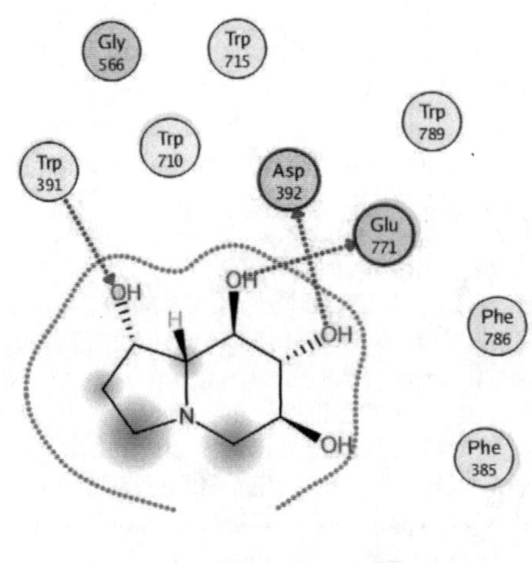

Figure 1. Binding interaction between CAST and the active site of processing α-glucosidase I protein.

Fragment libraries are used in second docking. There are 52 fragment molecules that docked towards processing α-glucosidase I. The potential fragment that chose have the small ΔG and have binding interaction around CAST's binding interaction [17,26]. The molecular docking simulation from 52 fragment result N-(2-methylindolin-1-yl)propionamide fragment is the primary candidate to be linked with CAST. This fragment has -26.271 kcal/mol with pKi 19.138 and distance 3.65 Å to CAST. Furthermore, the N-(2-methylindolin-1-yl)propionamide has interaction with Glu429 residue in the active site of processing α-glucosidase I. Interaction of processing α-glucosidase I with N-(2-methylindolin-1-yl) propionamide is showed in figure 2.

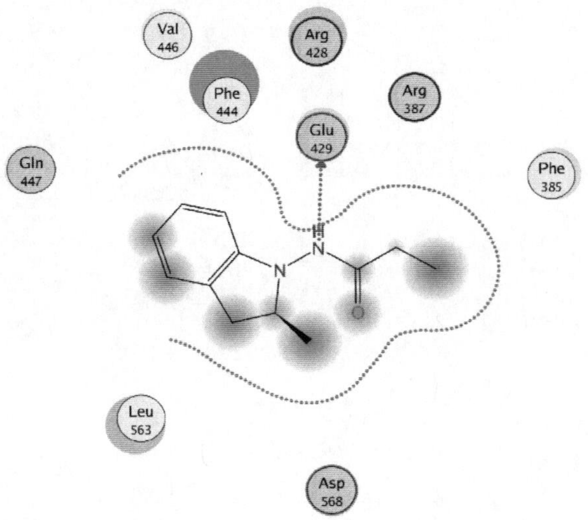

Figure 2. Binding interaction N-(2-methylindolin-1-yl) propionamide and α-glucosidase I protein.

3.3 Analysis of Fragment Linking and Molecular Docking of Ligand Formed from CAST and Potential Fragment

Fragment linking was used to linked two or more fragment with a linker to generated higher affinity binding compound with wider area interaction in the active site. The fragments should bind in the different area in binding site and close enough to be linked [17]. In this study, we linked CAST with N-(2-methylindolin-1-yl)propionamide using Link Multiple Fragments feature in MOE 2014.09. Fragment linking between CAST and another potential fragment could build a bigger molecule with wider area interaction in the active site of processing α-glucosidase I. To generated ligands with drug-likeness and good bioavailability properties, some parameter can be applied in Descriptor Filter such as molecular weight (less than 500g/mol, according to Lipinski's RO5), have total polar surface area (TPSA) less than 140 Å^2, and have rotatable bonds less than 10 [21,27]. This work resulted in 127 ligands. Each ligand was given code name LRS 1 to LRS 127.

All the ligands that were generated from fragment linking then docked towards active site of I. We chose five ligands out of 127 ligands that have small ΔG result than other ligands and have interaction with catalytic residues of processing α-glucosidase I. Table 1 showed the ΔG and inhibition constant (pKi) of the five best ligands and two standard ligands that used in our work. The docking result showed ligand LRS 28 (N-((S)-2-((S)-4-methyl-5-((1S,3S,6S,7R,8R,8aR)-1,6,7,8-tetrahydroxyoctahydroindolizin-3-yl)pentyl)indolin-1-yl)propionamide) has the lowest ΔG among the others. Ligand LRS 28 has -52.0572 kcal/mol with pKi 37.9229. The molecular interaction between LRS 28 and processing α-glucosidase I can be seen in figure 3.

IOP Conf. Series: Materials Science and Engineering **172** (2017) 012017 doi:10.1088/1757-899X/172/1/012017

Table 1. ΔG energy and pKi value from the selected linked compounds and standard ligands

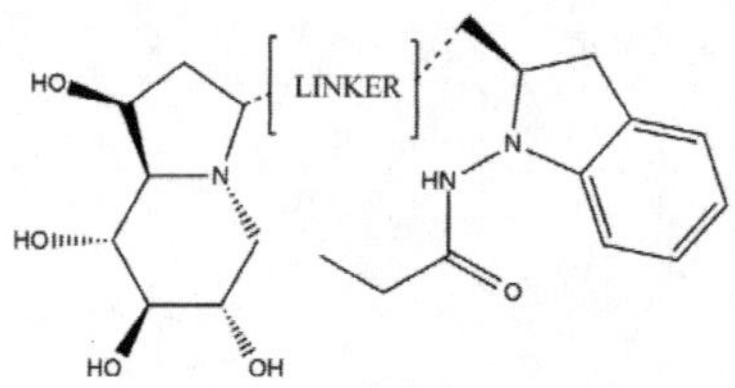

Ligands	Linker	ΔG (kcal/mol)	pKi
LRS 28	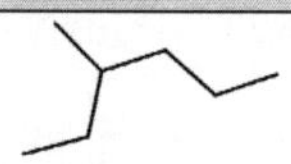	-52.0572	37.9229
LRS 85	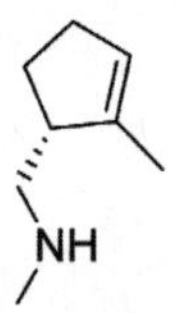	-52.0232	37.8981
LRS 24	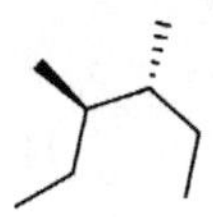	-51.181	37.2846
LRS 47	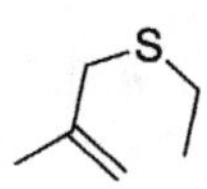	-49.5897	36.1254
LRS 22	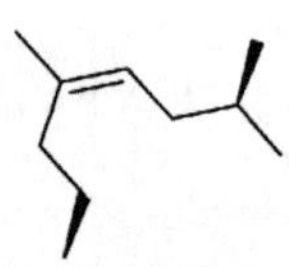	-49.5482	36.0951
*CAST	-	-28.7224	20.9238
*Bu-CAST	-	-34.3206	25.0020

Note : *Standard Ligands

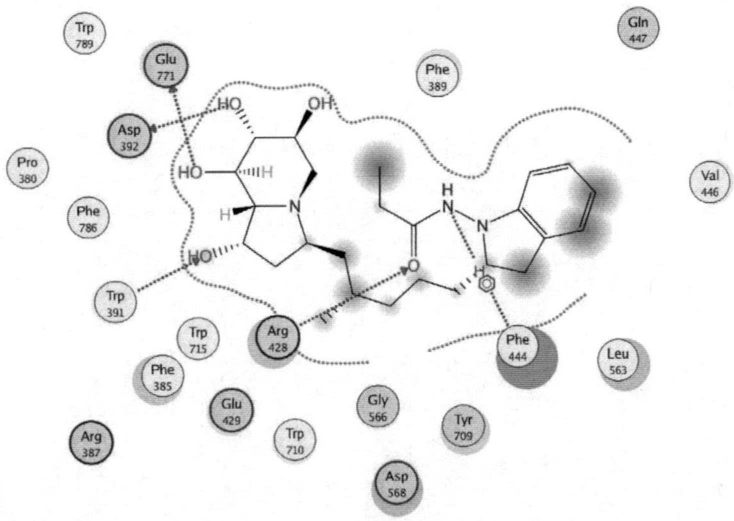

Figure 3. Interaction of LRS 28 with the active site of processing α-glucosidase I

3.4 Analysis of Drug Scan

The toxicity and drug-likeness analysis of the best ligands was performed using Osiris DataWarrior 4.2.2 and T.E.S.T (Toxicity Estimate Software Tool) [28,29]. Toxicity parameter such as mutagenicity, the tumorigenic and irritant effect was predicted using Osiris Data Warrior 4.2.2, while developmental toxicity was predicted using T.E.S.T. The toxicity prediction from best five ligands as well the standard ligands are shown in table 2.

Table 2. Toxicity prediction by Osiris DataWarrior and T.E.S.T Software

Ligands	Parameters			
	Mutagenic	Tumorigenic	Irritant	Developmental Toxicity
LRS 28	None	None	None	None
LRS 85	None	None	None	Yes
LRS 24	None	None	None	Yes
LRS 47	None	None	None	None
LRS 22	None	None	None	None
*CAST	None	None	None	None
*Bu-CAST	None	None	None	Yes

Note: *Standard Ligands

The drug-likeness properties are related to the Lipinski's RO5 and Veber rules. Ligands that have properties fit with these rules could be considered as potential drug candidates with good oral bioavailability [27,30,31]. RO5 states that drug candidates would easily absorb by the body if they have the following criteria:

- The molecular weight (MW) of the ligands is less than 500 g/mol
- clogP is less than 5.0
- Have less than 5 hydrogen bond donor and no more 10 hydrogen bond acceptor

Unlike RO5, Veber rules only state 2 criteria for candidate drugs to have good oral bioavailability. These criteria are had no more 10 rotatable bonds and the polar surface area equal to or less than 140 $Å^2$. We used Osiris DataWarrior to analysis the drug-likeness. The result drug-likeness prediction is shown in Table 3.

Table 3. The molecular properties prediction by Osiris DataWarrior

Ligands	Parameters					
	Molecular Weight	Number of H-Donor	Number of H-Acceptor	cLogP	Polar Surface Area ($Å^2$)	Number of Rotatable bonds
LRS 28	461.601	5	8	1.5144	116.5	8
LRS 85	486.611	6	9	-0.3757	128.53	7
LRS 24	475.628	5	8	1.7327	116.5	8
LRS 47	477.624	5	8	1	141.8	8
LRS 22	487.639	5	8	2.323	128.53	8
*CAST	189.210	4	5	-1.8813	84.16	0
*Bu-CAST	259.301	3	6	-0.4879	90.23	4

Note: *Standard Ligands

Table 2 has shown that one of two standard ligands and some of the modified ligands have no toxic properties. There is no risk of being mutagenic, tumorigenic and irritant for all modified ligands and the standard. However, the standard ligand of Bu-CAST and the modified ligand such as LRS 85 and LRS 24 have developmental toxicity properties. The result in table 3, all the standards have a good result and fit the criteria from both RO5 and Veber rules. Table 3 also indicates that among five modified ligands, LRS 47 violated one of Veber rules and LRS85 violate one of theRO5, for having six hydrogen bond donors and higher TPSA value than the stated rules (at 141 $Å^2$), respectively. Based on table 2 and table 3, among five modified ligands, LRS 28 and LRS 22 have the best result. These ligands showed to be negative on mutagenic, tumorigenic, irritant, and developmental toxicity properties. These ligands also fit the RO5 and also Veber rules. Ligand LRS 47, even though didn't fit with one of two criteria in Veber's rules, it has no mutagenic, tumorigenic, irritant, and developmental toxicity properties.

Conclusion

In this work, 127 ligands have been formed from fragment linking between CAST and N-(2-methylindolin-1-yl) propionamide. Molecular docking studies showed, best five ligands have shown their ΔG were smaller than the standard. Ligand LRS 28 showed the smallest ΔG at -52,0572 kcal/mol among all the modified ligands. Computational toxicity and drug-likeness test also showed the good result as well as the standard. Based on all the simulation that have been done, ligand LRS 28, LRS 22, and LRS 47 were the potential candidate to inhibit processing α-glucosidase I. We conclude these ligands are also potential ligands to developed as a drug. Therefore, in vitro and in vivo test is needed to further determine its inhibitory activity towards processing α-glucosidase I and bioactivity as dengue therapy.

References

[1] Lim S P, Wang Q Y, Noble C G, Chen Y L, Dong H, Zou B, Yokokawa F, Nilar S, Smith P, Beer D, Lescar J and Shi P Y 2013 Ten years of dengue drug discovery: Progress and prospects *Antiviral Res.* **100** 500–19
[2] Chen W, Kawano H, Men R, Clark D and Lai C J 1995 Construction of intertypic chimeric

dengue viruses exhibiting type 3 antigenicity and neurovirulence for mice. *J. Virol.* **69** 5186–90

[3] Mustafa M S, Rasotgi V, Jain S and Gupta V 2015 Discovery of fifth serotype of dengue virus (DENV-5): A new public health dilemma in dengue control *Med. J. Armed Forces India* **71** 67–70

[4] Frecer V and Miertus S 2010 Design, structure-based focusing and in silico screening of combinatorial library of peptidomimetic inhibitors of Dengue virus NS2B-NS3 protease. *J. Comput. Aided. Mol. Des.* **24** 195–212

[5] Kraemer M U G, Sinka M E, Duda K A, Mylne A, Shearer F M, Barker C M, Moore C G, Carvalho R G, Coelho G E, Van Bortel W, Hendrickx G, Schaffner F, Elyazar I R, Teng H-J, Brady O J, Messina J P, Pigott D M, Scott T W, Smith D L, Wint G W, Golding N and Hay S I 2015 The global distribution of the arbovirus vectors Aedes aegypti and Ae. albopictus *Elife* **4** e08347

[6] Nitsche C, Holloway S, Schirmeister T and Klein C D 2014 Biochemistry and Medicinal Chemistry of the Dengue Virus Protease *Chem. Rev.* **114** 11348–81

[7] Stevens A J, Gahan M E, Mahalingam S and Keller P A 2009 The medicinal chemistry of dengue fever *J. Med. Chem.* **52** 7911–26

[8] Li Z, Khaliq M, Zhou Z, Post C B, Kuhn R and Cushman M 2009 Design, Synthesis, and Biological Evaluation of Antiviral Agents Targeting Flavivirus Envelope Proteins *J. Med.* **51** 4660–71

[9] Sousa L R F de, Wu H, Nebo L, Fernandes J B, Da Silva M F D G F, Kiefer W, Kanitz M, Bodem J, Diederich W E, Schirmeister T and Vieira P C 2015 Flavonoids as noncompetitive inhibitors of Dengue virus NS2B-NS3 protease: Inhibition kinetics and docking studies *Bioorganic Med. Chem.* **23** 466–70

[10] Tambunan U S F, Zahroh H, Parikesit A A, Idrus S and Kerami D 2015 Screening analogs of β-OG pocket binder as fusion inhibitor of dengue virus 2 *Drug Target Insights* **9** 33–49

[11] Canard B 2011 Antiviral Research and Development Against Dengue Virus. *WHO Rep.* 1–101

[12] Chang J, Block T M and Guo J T 2013 Antiviral therapies targeting host ER alpha-glucosidases: Current status and future directions *Antiviral Res.* **99** 251–60

[13] De Chassey B, Meyniel-Schicklin L, Vonderscher J, André P and Lotteau V 2014 Virus-host interactomics: new insights and opportunities for antiviral drug discovery. *Genome Med.* **6** 115

[14] Sayce A C, Miller J L and Zitzmann N 2010 Targeting a host process as an antiviral approach against dengue virus *Trends Microbiol.* **18** 323–30

[15] Watanabe S, Chan K W K, Dow G, Ooi E E, Low J G and Vasudevan S G 2016 Optimizing celgosivir therapy in mouse models of dengue virus infection of serotypes 1 and 2: The search for a window for potential therapeutic efficacy *Antiviral Res.* **127** 10–9

[16] Barker M K and Rose D R 2013 Specificity of processing α-glucosidase I is guided by the substrate conformation: Crystallographic and in silico studies *J. Biol. Chem.* **288** 13563–74

[17] Rees D C, Congreve M, Murray C W and Carr R 2004 Fragment-based lead discovery *Nat. Rev. Drug Discov.* **3** 660–72

[18] Tambunan U S F, Parikesit A A, Unadi Y C and Kerami D 2015 Designing Cyclopentapeptide Inhibitor of Neuraminidase H5N1 Virus Through Molecular and Pharmacology Simulations *Tsinghua Sci. Technol.* **20** 431–40

[19] Vilar et al 2008 Medicinal chemistry and the molecular operating environment (MOE): application of QSAR and molecular docking to drug discovery. *Curr. Top. Med. Chem.* **8** 1555–72

[20] Sterling T and Irwin J J 2015 ZINC 15 - Ligand Discovery for Everyone *J. Chem. Inf. Model.* **55** 2324–37

[21] Scott D E, Coyne A G, Hudson S A and Abell C 2012 Fragment based approaches in drug discovery and chemical biology *Biochemistry*

[22] Jhoti H, Williams G, Rees D C and Murray C W 2013 The "rule of three" for fragment-based drug discovery: where are we now? *Nat. Rev. Drug Discov.* **12** 644–5
[23] Lyne P D 2002 Structure-based virtual screening: An overview *Drug Discov. Today* **7** 1047–55
[24] Chen G, Zheng S, Luo X, Shen J, Zhu W, Liu H, Gui C, Zhang J, Zheng M, Puah C M, Chen K and Jiang H 2005 {F}ocused combinatorial library design based on structural diversity, drug-likeness and binding affinity score *J Comb Chem* **7** 398–406
[25] Sousa S F, Ribeiro a J M, Coimbra J T S, Neves R P P, Martins S a, Moorthy N S H N, Fernandes P a and Ramos M J 2013 Protein-ligand docking in the new millennium--a retrospective of 10 years in the field. *Curr. Med. Chem.* **20** 2296–314
[26] Kobilka B and Schertler G F X 2008 New G-protein-coupled receptor crystal structures: insights and limitations *Trends Pharmacol. Sci.* **29** 79–83
[27] Veber D F, Johnson S R, Cheng H, Smith B R, Ward K W and Kopple K D 2002 Molecular Properties That Influence the Oral Bioavailability of Drug Candidates *J. Med. Chem.* **45** 2615–23
[28] El-saadi M W, Williams-hart T, Salvatore B A and Mahdavian E 2015 Use of in-silico assays to characterize the ADMET profile and identify potential therapeutic targets of fusarochromanone , a novel anti-cancer agent *Silico Pharmacol.* 1–20
[29] Ruiz P, Begluitti G, Tincher T, Wheeler J and Mumtaz M 2012 Prediction of acute mammalian toxicity using QSAR methods: A case study of sulfur mustard and its breakdown products *Molecules* **17** 8982–9001
[30] Leeson P 2012 Drug discovery: Chemical beauty contest *Nature* **481** 455–6
[31] Lipinski C A 2004 Lead- and drug-like compounds: The rule-of-five revolution *Drug Discov. Today Technol.* **1** 337–41

Conflict of Interest

We declare that there is no conflict of interest in this research nor publication

Acknowledgements

We would like to thank Directorate of Research and Community Engagement of Universitas Indonesia (DRPM UI) through Hibah Publikasi Internasional Terindeks Untuk Tugas Akhir Mahasiswa UI (PITTA) 2016 No:2049/UN2.R12/HKP.05.00/2016 for funding the research and publication of this article. USFT designed the study and supervised this research, while EPT was performed the experimental details. We also would like to thank Mr. Mochammad Arfin Fardiansyah Nasution for helping us to proofread this paper.

Utilization of composite membrane polyethyleneglycol-polystyrene-cellulose acetate from pineapple leaf fibers in lowering levels of methyl orange batik waste

E V Y Delsy[1], Irmanto[1] and F N Kazanah[1]

[1]Department of Chemistry, Faculty of Mathematic and Natural Sciences, Jenderal Soedirman University, Purwokerto

Email: vaulinaeva@gmail.com

Abstract. Pineapple leaves are agricultural waste from the pineapple that the fibers can be utilized as raw material in cellulose acetate membranes. First, made pineapple leaf fibers into pulp and then converted into cellulose acetate by acetylation process in four stages consisting of activation, acetylation, hydrolysis and purification. Cellulose acetate then used as the raw material to manufacture composite membrane with addition of polystyrene and poly (ethylene glycol) as porogen. Composite membrane is made using phase inversion method with dichloromethane-acetone as a solvent. The result of FTIR analysis (Fourier transform infra-red) showed that the absorption of the carbonyl group (C=O) is at 1643.10 cm^{-1} and acetyl group (C–O) at 1227.01 cm^{-1}, with a molecular weight of 8.05 x 10^4 g/mol and the contents (rate) of acetyl is 37.31%. PS-PEG-CA composite membrane had also been characterized by measuring the water flux values and its application to decrease methyl orange content (level) in batik waste. The results showed that the water flux value is of 25.62 L/(m^2.hour), and the decrease percentage of methyl orange content in batik waste is 71.53%.

1. Introduction

Membrane is thin layer, a border between two semipermeabel phases. Membrane technology is one of separation methods that is being developed for processing wastewater, the advantages are high separation efficiency, separation can take place continuously, requires no chemical additives, and the separation is more simple. Besides, the membrane technology is clean technology because it is environmentally friendly and does not generate waste [1]. According to Wenten [2], the success of separation process with membrane depends on the quality of the membrane.

Materials used in membrane preparation is cellulose acetate. Cellulose acetate can be synthesized by acetylation. Pineapple leaves (Ananas comosus) is one of the plants contain cellulose. This plant utilization is limited to the fruit, while the leaves are relatively not used widely.

Modifications to membrane base material is increasingly diverse, one of them is the addition of polyethylene glycol. The modification is affects only the flux and membrane rejection index, but it does not affect the strength. The added PEG affects the formation of pores in the membrane. In the presence of PEG, the number pores formed is increasing and the size is more even. A mixture of polymers that can be decomposed biologically and synthetic polymers can generate new physical properties. Besides the addition of polyethylene glycol, polystyrene can also be used as a mixture of natural polymers to increase the strength of the membrane obtained. According to Crowd [3], CA membrane with synthetic polymer addition, i.e. polystyrene (PS) have better mechanical properties compared with the CA. By adding polistyrene, the resulting membrane structure is getting tighter, and the crystallinity is getting higher, that improves the mechanical properties of the membrane. The manufacture of cellulose acetate composite membrane of pineapple leaf fibers has been made with the addition of PEG and polystyrene, as well as its application to decrease levels of the dye methyl orange on batik waste.The purpose of this research is: to determine the characteristics of the composite membrane PS-PEG-CA of pineapple leaf fibers and to determine the reduction in dye methyl orange percentage using a composite membrane PS-PEG-CA from pineapple leaf fibers.

2. Research Methods

2.1. Instruments

This study uses a set of reflux or extraction tools, analytical balance, filter paper, hot plate, shaker, measuring flask 25 mL and 100 mL, oven, one size pipette, measure pipette, Beaker glass 100 and 1000 mL, stirring rod, plastic containers, Viscometer Ostwald, filtration apparatus, desiccator, magnetic stirrer, glass plate, FTIR and UV-Vis Spectrophotometer.

2.2. Materials

This study uses materials pineapple leaves, distilled water, NaOH p.a., $Ca(OH)_2$, NaOCl, glacial acetic acid, KBr, phenolphthalein indicator, HCl, H_2SO_4 concentrated p.a., dichloromethane p.a., acetone p.a., acetic anhydride p.a., PEG (polyethylene glycol) 600, polystyrene, batik waste, dye methyl orange p.a., phosphate buffer, acetate buffer, pH indicators, pH meter, and aluminum sulfate.

2.3. Procedure

2.3.1. Preparation of pineapple leaf fibers pulp.

Added Ca (OH) 2 2.5% (w / v) of 150 mL to 20 grams Pineapple leaf fibers and soaked for three days. After washed with distilled water and put in a flat-bottom flask which had previously been filled with 300 ml of 17.5% NaOH (w/v), then refluxed for 4 hours. Once cool, pineapple leaf fibers are washed until it is free from NaOH, blended and molded into a pulp sheet and the pulp is then dried in an oven at 60 °C for 1 day.

2.3.2. Acetylation of cellulose pineapple leaf fiber[4]

Added 12 mL of glacial acetic acid to 5 grams Pulp that has been smoothed and stirred for 1 hour. After 1 hour the mixture was added 0.088 mL of concentrated sulfuric acid and 20 mL of glacial acetic acid and stirred again for 45 minutes. This mixture is cooled to a temperature of 18.3 °C then added 13.5 mL of acetic anhydride which has been cooled to a temperature of 15.6 °C and the mixture was added 0.612 mL of concentrated sulfuric acid with 20 mL of glacial acetic acid. The mixture is stirred for 20 hours until a solution is formed. The next stage is added 15 mL of 67% acetic acid dropwise over 1 hour while stirring at a temperature of 37.8 °C. The solution is left to stand for 20 hours. Furthermore, water is added while stirring until a precipitate is formed and allowed to stand for 10-15 minutes. The precipitate is filtered, then washed with water until neutral. The precipitate is then dried in an oven at a temperature of 50-60 °C.

2.3.3. Determination of Relative Molecular Weight of Cellulose Acetate[5]

Dissolved 0.15 grams of cellulose acetate in 100 mL acetone (called solution C). The resulting solution C, made into solution 0.2C; 0.4C; 0.6C; and 5 mL 0.8C Aceton put in viscometer and the flow time is measured at 25 °C. The same measurements performed on 0.2C; 0.4C; 0.6C; and 0.8C solution.

2.3.4 Preparation of Composite Membrane PS-PEG-CA[5]

Manufacture of membranes made by the method of phase inversion. The first stage began with the manufacture of the polymer solution CA 18% [w/v], PS (10% [w/v]), and PEG (10% [w/v]) in dichloromethane solvent: acetone (1:1). CA and SP polymer solution are mixed with a ratio of 9:1. Added 3 mL of PEG solution to each 100 mL CA: PS.The solution is stirred with a magnetic stirrer until it is homogeneous.Then the polymer solution is poured over a glass plate (18 × 18 cm^2) which has been given the duct tape on both sides to obtain the desired thickness, and then printed out by pressing and pushing the solution to obtain a thin layer. Furthermore, the polymer that sticks to the glass plate is left for 60 seconds to evaporate the solvent. Plate glass along the polymer is then soaked in the water to remove the polymer from the glass plate and eliminate PEG matrix trapped between CA and PS. PEG loss of the matrix leaving a cavity in the form of pores. Then, the membrane is soaked in a solution of sodium azide 1 ppm. This thin polymer is then used as a membrane.

2.3.5. Characterization of Composite Membrane PS-PEG-CA Flux[6]

Membrane samples are placed in a ultrafiltration cell. 150 mL of water put into the ultrafiltration cell and closed tightly, given the pressure of 2 kgf/cm^2. Flux value is determined as a function of time until they reached steady-state. Flux is expressed by the following equation :

$$J = \frac{V}{At}$$

J = fluks volume(L/m^2.hour)
V = permeate volume (L)
A = surface area (m^2)
t = time (hour)

permeation or solution across the membrane accommodated at specified intervals within 10-minute intervals and its volume is measured. Measurements are made until a time constant is obtained. The same treatment is conducted to measure the flux value of waste by replacing water with sewage.

2.3.6. Measurement of Methyl Orange Concentration.

Measurement of methyl orange concentrations in the samples is done by measuring the absorbance using a UV-Vis spectrophotometer before it is being discharged into the membrane and once flowed into the membrane. The absorbance of the sample solution is measured under the same conditions with a standard solution that is at the maximum wavelength. The calculation of methyl orange dye level in the samples is done by substituting the absorbance of the sample as "y" on a regression equation of standard solutions. The content of methyl orange is the "x" obtained.

3. Results and Discussion

3.1. Relative molecular weight cellulose acetate

Measurement of relative molecular weight cellulose acetate aims to determine the physical characteristics of pineapple leaf fibers' cellulose acetate which is a basic ingredient in the manufacture of cellulose acetate membranes. Membranes with a high average relative molecular mass will have a denser membrane structure and stronger chemical bonding than the membrane with a small average relative molecular mass [6]. Relative molecular weight is determined using Ostwald viscometer, where the principle of this method is the measurement of viscosity by comparing the flow time of a polymer solution with various concentrations and pure solvent flows through two boundary markers capillary tube at the viscometer [7].Measurements were performed by comparing the viscosity of the pure solvent with cellulose acetate solution at various concentrations, namely 0.03; 0.06; 0.09; and 0.12 g / dL. Pure solvent used is acetone because it can dissolve cellulose acetate well at room temperature

[8].The higher the concentration of the solution of cellulose acetate then the longer the flow time of the solution in the viscometer. Relative molecular weight cellulose acetate based on the results of the study was 8.05 x 10^4 g / mol. where relative molecular weight to commercial cellulose acetate are 7.5 x 10^4 g / mol [9]. The amount of the molecular weight of the cellulose acetate fibers is affected by the length of the cellulose material used as a basis for making membranes.

3.2. Composite membrane PS-PEG-CA Pineapple leaf Fibers

Composite means a blend of two or more materials to produce new materials with characteristics that more than origin. Manufacture of composite membrane PS-PEG-CA is done by phase inversion method. Phase inversion is a process of changing the form of the polymer from a liquid into a solid phase under controlled conditions. Cellulose acetate used in this study is the result of the synthesis of pineapple leaf fibers with acetyl level of 37.31%, soluble in acetone. The amount of the solvent effect on the acetyl content that will be used in the production of membrane [10].

3.3. Flux Membran

Flux measurements performed by flowing water into the filtration cells of dead-end system pressure 2 kgf/cm^2. Before measuring the flux values then firstly be compacted by passing distilled water on a membrane for 30 minutes. Compacting is done for conditioning the membrane so that the pores is equilibrium and constant flux data obtained. Compaction [1] is a mechanical change in the structure of the polymer membrane that occurs due to the thrust (pressure). This causes the cellulose acetate structure becomes more compact, and the membrane pores docked so it impacts the value of flux.

Flux measurements is done at 10 minute intervals until it reaches a constant permeate volume. Water flux value comparison chart on the membrane CA and membrane PS-PEG-CA can be seen in Figure 1

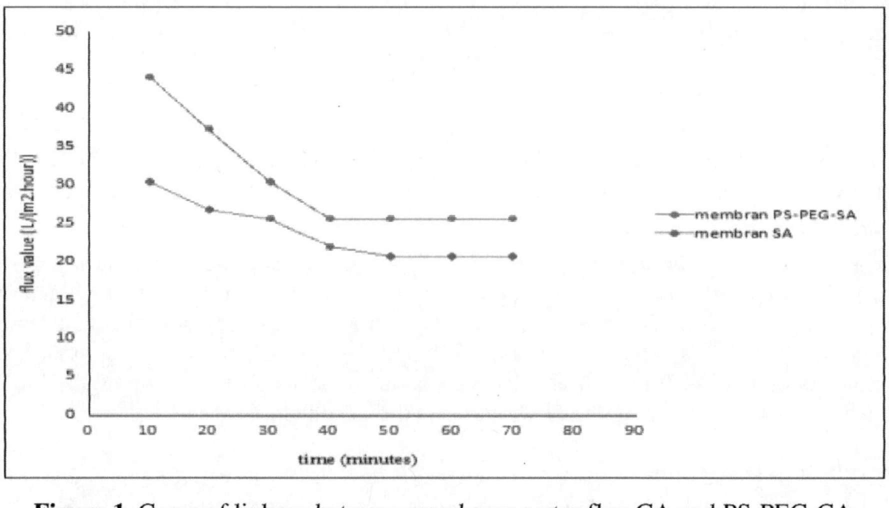

Figure 1. Curve of linkage between membrane water flux CA and PS-PEG-CA

Based on the chart above it shows that the longer membrane operation time, the flux value generated decreases. According [1], this is due to the occurrence of fouling, because the water used may still contain particles of the same size or larger than the pore size of the membrane, so that the particles will accumulate on the surface of the membrane. These conditions resulted in inhibition of the feed flow rate, so that the membrane's ability to parse the feed reduce. Based on the results of measurements using CA membrane and composite membrane PS-PEG-CA water flux values obtained successively at 20.74 L/(m^2.hour) and 25.62 L/(m^2.hour) Membrane water flux value PS-PEG-CA acquired is higher than the value of the water flux membranes CA. This is due to the addition of PEG. PEG is capable of forming pores in cellulose acetate membrane. The number of pores formed becomes more and more uniform in size, thereby increasing the value of water flux as it flows permeate will be easier with the increasing number of pores in the membrane.

3.5. Methyl Orange Level

Composite membrane PS-PEG-CA from pineapple leaf fibers is used to lower the levels of methyl orange dye in batik waste. Batik waste to be used previously undertaken the process of coagulation-flocculation. Coagulation-flocculation occurs at the fast and slow mixer unit. These coagulation-flocculation process through three phases, the addition of a coagulant, and then stirring rapidly during 5 min with a speed of 1500 rpm and slow stirring at 1000 rpm for 15 minutes. Rapid stirring serves to disperse the coagulant material in the water, resulting in sufficient contact between the coagulant with the particles suspended and will form a small flocs. Slow stirring serves to help establish a larger flocs so it will be easier to settle. The flocs formed then left to settle for 30 minutes. Coagulant used in this study is alum or $Al_2(SO_4)_3$.

Based on the results obtained by the equation y = 0,08415x + r2 of 0.0011 to 0.9998. The decreasing percentage methyl orange level of batik waste using CA membranes and membrane PS-PEG-CA from pineapple leaf fibers can be seen in Table 1.

Table 1. Decreasing Percentage of Methyl Orange levels On Batik Waste

Membrane	*Methyl Orange* Level Before Filtration	*Methyl Orange* Level After Filtration	Decreasing Percentage of *Methyl Orange* Level
Membrane CA	45.27 mg/L	17.52 mg/L	61.29 %
Membrane PS-PEG-CA	45.27 mg/L	12.885 mg/L	71.53 %

Based on Table 1 it can be seen that a decline in the concentration of methyl orange after filtration. Decreasing concentration of this methyl orange happens because methyl orange dye has a large molecular size so that when the filtration process the dye will be retained in the membrane surface and the resulting concentration of the dye will be reduced. The percentage decrease in the levels of methyl orange using CA membranes and membrane PS-PEG-CA amounted to 61.29% and 71.53%. Membrane PS-PEG-CA can degrade methyl orange dye is greater than the membrane CA. This is because the membrane PS-PEG-CA has a molecular structure that is denser than the membrane so that the macromolecules particles will be retained on the membrane [1].From the results of this study concluded that the composite membrane PS-PEG-CA from pineapple leaf fiber can degrade methyl orange dye in batik waste.

4. Conclusion

Based on the research described, it can be concluded that the composite membrane PS-PEG-CA from pineapple leaf fibers has a flux value of 25.62 L/(m^2.hour) with a relative molecular weight 8.05 x 10^4 g/mol and acetyl levels 37.31%. The concentration of methyl orange on batik waste before and after flow through the membrane PS-PEG-CA of fiber pineapple leaf is equal to 9.054 mg/L and 2.577 mg/L, so the percentage decrease in the levels of methyl orange on batik waste using a membrane PS-PEG-CA of pineapple leaf fiber is equal to 71.53%.

References

[1] Mulder M 1996 *Basic Principles of Membran Technology* (Netherland: Kluwer Academic Publisher)
[2] Wenten I G 2000 *Teknologi Membran Industrial* (Bandung: Penerbit ITB)
[3] Cowd, M A 1982 *Polymer chemistry* (London: J Murray)
[4] Masaoka S Ohe T and Sakota N 1993 Producting of cellulose from glucose by Acetobacter xylinum *Ferment Bio Eng* **75** 18

IOP Conf. Series: Materials Science and Engineering **172** (2017) 012018 doi:10.1088/1757-899X/172/1/012018

[5] Brandup J and Immergut E H 1999 *Polymer Handbook* 3rd Edition (New York : John Wiley&Sons)
[6] Doghieri F Nardella A Sarti G C and Valentini 1994 Pervaporation of Metano-MTBE Mixtures through Modified Poly-(phenylene oxide) Membranes *Journal of Membrane Science* **91** 283
[7] Stevens M P 2001 *Kimia Polimer* (Jakarta: Pradnya Paramita)
[8] Wafiroh S 2005 Aplikasi Membran Selulosa Asetat dari Pulp Abaka (Musa textilis) untuk Mengurangi Kadar Fe^{3+} dari Kekeruhan Air Sungai *Jurnal Kimia Lingkungan* **7** 1
[9] Radiman C L and Yuliani G 2000 Penggunaan Nata De Coco sebagai Bahan Membran Selulosa Asetat *Prosiding Simposium Nasional Polimer* 1410-8720.
[10] Kirk B E and Othmer D F 1993 *Encyclopedia of Chemical Technology* (New York : The interscience encyclopedia Inc)

Adsorption of Pb(II) ion from aqueous solution onto Chitosan/silica/polyethylene glycol (Ch/Si/P) composites membrane

F W Mahatmanti[1], W D P Rengga[2], E Kusumastuti[1], Nuryono[3]

[1] Department of Chemistry, Faculty of Mathematics and Natural Sciences, Universitas Negeri Semarang, Indonesia 50229
[2] Department of Chemical Engineering, Faculty of Engineering, Universitas Negeri Semarang, Indonesia 50229
[3] Department of Chemistry, Faculty of Mathematics and Natural Sciences, Universitas Gadjah Mada, Sekip Utara, Yogyakarta, Indonesia 55281

email: widhimahatmanti@rocketmail.com, fwidhi_kimia@mail.unnes.ac.id

Abstract. In this research, chitosan/silica/polyethylene glycol (Ch/Si/P) composites membrane was studied for selective adsorbent of Pb(II) from aqueous solution. This study started by preparation of Ch/Si/P composites membrane. The structure and surface morphology of Ch/Si/P composites membrane was characterized by X-ray diffraction (XRD), Fourier Transform Infra Red (FTIR) spectroscopy, and Scanning Electron Microscopy (SEM). Batch adsorption experiments were conducted with various contact time and Pb(II) concentrations to evaluate the adsorption kinetics and thermodynamics. Results show that adsorption fitted to Langmuir model and followed a kinetic model of pseudo-second-order. The adsorption capacity of Pb(II) onto the Ch/Si/P composites membrane is 0.16 mmol/g. The relative selective factor (α_r) value of Pb(II)/Rhodamine B is 2.38. This means that Pb(II) can be determined even in the presence of Rhodamine B interference.

1. Introduction

The textile industry is one of the industrial activities that are developing in Indonesia. One note on the textile industry is their products waste, which can pollute the environment in particular, the pollution of water. The textile industry waste will flow directly into sewers or rivers, so the wastewater can reduce potential natural cleaning power of water, can change the color of the river water, may even result in the death of aquatic organisms that are important for human life. The type of metal ions produce by the textile industry wastes mainly metal ions As, Cd, Cr, Pb, Cu, and Zn.

The intensity of the dye and the metal ion content in a textile industrial waste can be reduced by physics, chemistry or biology. In the chemical method, researchers used several methods to reduce pollution in the textile industry waste such as: coagulation-flocculation, electrochemical oxidation, adsorption and membrane filtration [1]. Membrane technology is considered an effective way to reduce chemical pollutants dye and metal ions in the waste. Membrane technology has several advantages compared with other processes, among others: (a) there are small, (b) low pressure,and c) high separation selectivity [2].

The manufacture of Chitosan-silica-polyethylene glycol (Ch/Si/P) membrane was mixed with silica isolated from rice husk ash (ASP) and polyethylene glycol (PEG) as plasticizer. The addition of silica in the manufacture of chitosan-silica membranes increases the tensile strength and Young's modulus, the pore size of the membrane, and the stability of the temperature rise [3]. In this paper studied the Ch/Si/P membrane to absorb Pb(II) ion in aqueous solution. Influencing factors that was studied were the effect of pH, contact time and initial concentration of the solution. The influence of the presence of Rhodamine B in aqueous solution was also studied.

2. Experimental

2.1. Material

Chitosan with the deacetylation degree of 83% was supplied from CV Ocean Fresh Bandung, Polyethylene-glycol (MW 400), sodium silicate solution from rice hull ash silica (25% SiO_2), sodium hydroxide pellet, HCl 37%, acetic acid (2% v/v), fumed HNO_3, $Pb(NO_3)_2$. All in analytical grade were purchased from Merck.

2.2. Adsorption experiment

The adsorption experiment was performed using a batch method at various pH (4.0 - 9.0). The solution pH was adjusted using HCl and NaOH solutions. An aqueous solution of Pb(II) ion 100 mg/L was shacked with 0.12 g membrane for 24 h at various pHs. The optimum pH was used to evaluate the kinetics parameter by varying the contact time (1, 2, 4, 8, 16, 24, and 26 hours) and thermodynamics parameter by varying the concentration (50 to 250 mg/L). Selectivity of adsorption was evaluated by competitive Pb(II) and Rhodamine B. The concentration of unabsorbed Pb(II) ion was detected by atomic absorption spectrophotometry. The concentration of unabsorbed Rhodamine B was detected by visible spectrophotometer .

3. Results dan Disccusions

3.1. pH Effect

The pH effect on the adsorption of Pb(II) ion onto Ch/Si/P composites membrane is shown in Figure 1. The results showed the highest adsorption of Pb(II) ion at pH 5. Increasing of pH lowers the level of $-NH_2$ protonation on the membrane surface, followed by decreasing in partial positive charge, consequently increasing of pH tends to increase adsorption of metal ions. At a higher pH, the metal ions that was adsorbed decreased. This was because of a decreasing in the adsorption of the adsorbate interaction is thought to occur due to the electrostatic repulsion between the active site NH_3^+ and metal ions have the same positive charge [4,5].

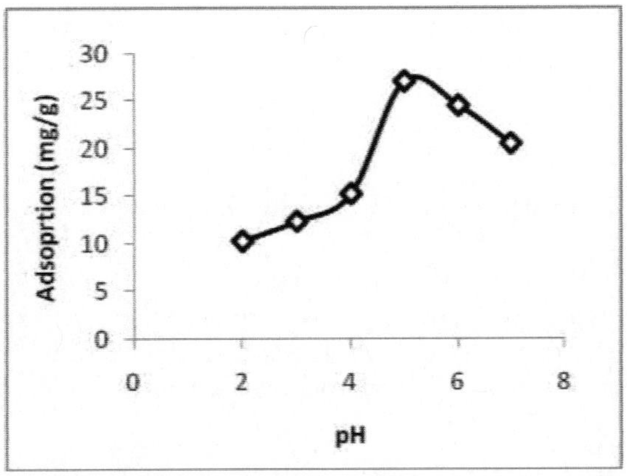

Figure 1. Effect of various pHs on adsorption of Pb(II) onto membrane

3.2. Kinetics Adsorption

In this study, the kinetic parameters obtained with the metal ion encounters at optimum pH and temperature variations in space and time of 1-26 hours. The observation of the effect of time to metal ions on the membrane is presented in Figure 2. In Figure 2, it appears that the adsorption of Pb(II) in the membranes Ch/Si/P has a high increasing at 4 hours among the contact time between 1-16 hours in the numbers of adsorbed metal ions, while the 16-26 hour of contact time has no increasing significantly. This indicates that there has been a balance between the amount of metal ions adsorbed by the number of metal ions remaining in the solution.

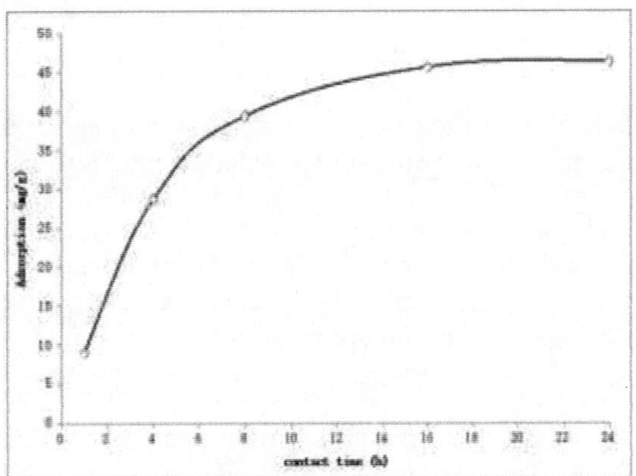

Figure 2. Adsorption curve of Pb(II) ion onto membrane at various contact times

Based on these phenomena, it can be predicted that the amount of adsorbate that is adsorbed is proportional to the number of active sites on the membrane that has not bonded with the adsorbate. The more the number of active sites on the membrane that is not binding adsorbate, adsorption occurs more easily and vice versa. At the beginning of the interaction, there are many empty active sites so there are many adsorbent are quickly bound to the membrane. At the time of the interaction time lasts longer then vacant positions provided the less so as the number of adsorbate bound decreases. Referring to Figure 2, the adsorption data at a different time of interaction can be processed using a pseudo-order kinetic model one proposed by Lagergren and pseudo-order kinetic model of two proposed by Ho. Some researchers have previously used equation kinetics pseudo first order Lagergren and equality kinetics pseudo-order 2 (Ho) to evaluate the tendency kinetic model that applies to the process of adsorption of metal ions by the adsorbent made of chitosan [6].

IOP Conf. Series: Materials Science and Engineering **172** (2017) 012019 doi:10.1088/1757-899X/172/1/012019

The kinetics of Pb(II) ion adsorption have been described on the basis of the pseudo-first-order and pseudo-second-order models. A simple kinetic model given by Lagergren is the pseudo-first order relation:

$$\frac{dQ}{dt} = k_1 (q_e\text{-}q) \quad (1)$$

And the pseudo-second-order model is given by the following equation:

$$\frac{dQ}{dt} = k_1 (q_e\text{-}q)^2 \quad (2)$$

Where q and q_e are the adsorbed Pb(II) ion onto the membrane at time t and equilibrium, respectively. The k_1 and k_2 are the kinetic rate constants of pseudo-first-order and pseudo-second-order models, respectively. We can obtain a linear equation for the pseudo-first-order and pseudo-second-order modes by integrating Eqs. (1) and (2), respectively :

$$\ln (q_e - q) = \ln q_e - k_1 t \quad (3)$$

$$\frac{t}{q} = \frac{1}{k2h} + \frac{t}{qe} \quad (4)$$

A fitting result for the kinetic models is given in Table 1.

Table 1. Fitting results of the pseudo-first-order and pseudo-second-order kinetic models for adsorption of Pb(II) ion onto membrane

Adsorbate	pseudo-first-order Lagergren model			pseudo-second-order kinetic model (Ho)		
	(h^{-1})	*(mmol/g)*	*R^2*	*($gmmol^{-1}h^{-1}$)*	*(mmol/g)*	*R^2*
Pb(II)	0.071	0.0089	0.951	0.0123	0.203	0.996

Table 1 obtained from research data were correlated according to an equation on kinetic models Lagergren pseudo- first order and pseudo-order kinetics equation 2 (Ho), for ions Pb (II). Data correlation factor R^2 values that is contained in Table 1, provides some significant data between the results obtained by pseudo first order kinetics model and the Lagergren pseudo second-order kinetic model. In general there is a tendency of correlation factor R^2 values higher pseudo second-order kinetic model [4,5]. In the pseudo second order kinetic model, the value q_e (adsorption equilibrium) more in line with the value q_e experimental results.

3.3. Adsorption isotherm models

The adsorption isotherm was studied by varying the initial concentration of Pb(II) ion (10 to 150 mg/L) at pH optimum. The effect of concentration Pb(II) ion on the adsorption is shown in the Figure 3. The observation of the ion adsorption Pb(II) in the membranes showed an increased number of metal ion adsorbed along with the increasing concentration of metal ion. At higher concentrations, the increase in the concentration of metal ion is not accompanied by an increase adsorption of metal ion significantly. Evaluation of adsorption data using the model of Langmuir isotherm and Freundlich isotherm models done by transforming the adsorption data to variables that correspond to the linear equation model of Langmuir and Freundlich isotherm. Data Langmuir and Freundlich isotherm parameters are presented in Table 2.

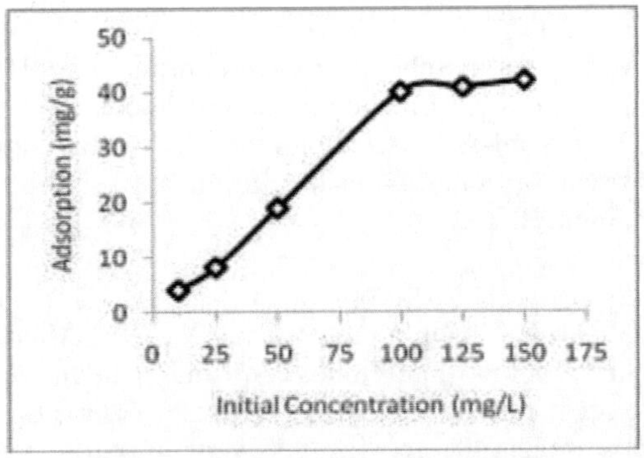

Figure 3 The effect of the initial concentration on adsorption of Pb(II) ion onto membrane

The R^2 obtained can be used to track the suitability of the process of adsorption by Langmuir and Freundlich isotherm models. In Table 2 shows that the Langmuir isotherm models provide conformity to the majority of the interaction between the metal ions with the membrane.

Table 2. Isotherm constant for adsorption of Pb(II) ion onto membrane

Adsorbat	Langmuir Isotherm			Freundlich Isotherm		
	q_e *(mmol/g)*	ΔG *(KJ/mol))*	R^2			R^2
Pb(II)	0,16	-25,97	0,995	2,040	1,573	0,984

The phenomenon that occurs in the ion Pb(II) with active sites on the membrane following the Langmuir isotherm models. Langmuir isotherm concept states that the increase in concentration followed by increasing the number of adsorbed substances so as to achieve equilibrium quantity. Langmuir adsorption theory suggests that at the adsorbent surface there are a certain number of active sites is proportional to the surface area of the adsorbent. In a state of active sites has not been saturated by adsorbate, there will be an increase in adsorption. Increasing the concentration of adsorbate that interaction does not increase the amount of adsorbate that is adsorbed adsorbent when active sites have been saturated by the adsorbate. If this trend occurs, then the interactions of metal ions involved in the surface of each adsorbent is monolayer / one layer. And vice versa, if the trend does not happen then the possibility of interaction between the metal ions and multilayer membrane occurs following the pattern Freundlich isotherm. Metal ion adsorption on an adsorbent is influenced by several factors, among others, the ability of the adsorbent, the size of the metal ion, and the environment in the process of adsorption. The ability of the adsorbent can be viewed through two different sides, namely active group and the pore size of the adsorbent.

The selectivity of the membrane against a mixture of dye Rhodamine B and Pb(II) ion. Adsorption of Rhodamine B dye mixture and ion Pb(II) is done by creating a mixture of Rhodamine B/Pb(II). The results of adsorption on a mixture of dye Rhodamine Band Pb(II) ion are presented in Table 3. That means that Pb(II) can't be determined even in the presence of Rhodamine B.

Table 3. Selective adsorption ion Pb(II)/Rhodamin B

Initial Concentration		D (L/g)		α
Rhodamine B	Pb(II)	Rhodamine B	Pb(II)	Pb(II)/ Rhodamine B
0.5	0.5	28.73	68.40	2.38
0.5	1	25.71	15.81	0.62

4. Conclusions

In the adsorption by a batch system for membrane obtained optimum at pH 5 for Pb(II). Adsorption of Pb(II) onto Ch/Si/P membrane follow the Langmuir isotherm model. Values ion adsorption capacity for Pb(II) was 0.16 mmol/g. In general the adsorption kinetics of data, there is a tendency of linearity coefficient R^2 values higher pseudo- second order kinetic model. The Pb(II) ion can't be determined even in the presence of Rhodamine B.

References

[1] Min M She L Hong G Zhu M Zhang Y and Wang X 2012 Micro-nano structure poly(ether sulfones)/poly(ethyleneimine) nanofibous affinity membranes for adsorption of anionic dyes and heavy metal ions in aqueous solution *Chem. Eng. J.* **197** 88-100

[2] Barakat M A 2011 New trends in removing heavy metals from industrial wastewater *Arab. J.Chem* ***4*** *361-377*

[3] Mahatmanti F W Nuryono Narsito 2014 Physical Characteristics of Chitosan based Film Modified with Silica and polyethylene glycol *Indones. J. Chem.* **14(2)** 131-137

[4] Salehi E and Madaeni S S 2014 Influence of poly(ethylene glycol) as pore-generator on morphology and performance of chitosan/poly(vinyl alcohol) membrane adsorbents *Appl. Surf. Sci.* **288** 537-541

[5] Madaeni E S S Rajabi V Vatanpour A A Derakhsan S Zinadini Sh Ghorabi and H A Monfared 2012 Novel chitosan/poly(vinyl) alcohol thin adsorptive membranes modified with amino functionalized multi-walled carbon nanotubes for Cu(II) removal from water : Preparation, characterization, adsorption kinetics and thermodynamics *Sep. Purif. Technol.* **89** 309-319

[6] Mahatmanti F W Nuryono Narsito 2016 Adsorption of Ca(II), Mg(II), Zn(II), and Cd(II) on Chitosan Membrane With Rice Hull Ash Silica and Polyethelyne Glycol *Indones. J. Chem.* **16(1)** 45-52

Acknowledgement

The authors highly appreciate the Institute for Research and Community Services, Universitas Negeri Semarang for the financial support through the Research Grant with Project No. SP DIPA-042.06.1.401516/2016 7th December 2015.

IOP Conf. Series: Materials Science and Engineering **172** (2017) 012020 doi:10.1088/1757-899X/172/1/012020

Characteristics of Iron Sand Magnetic Material from Bugel Beach, Kulon Progo, Yogyakarta

Fahmiati[1,2], Nuryono[3] and Suyanta[3]

[1] Department of Chemistry, Faculty of Mathematics and Natural Sciences, Universitas Halu Oleo, Anduonohu Kendari 93232, Indonesia
[2] Doctoral Student on Chemistry, Department of Chemistry, Faculty of Mathematics and Natural Sciences, Universitas Gadjah Mada, Sekip Utara, Yogyakarta 55281, Indonesia
[3] Department of Chemistry, Faculty of Mathematics and Natural Sciences Universitas Gadjah Mada, Sekip Utara, Yogyakarta 55281, Indonesia

fahmiati05@gmail.com

Abstract. Magnetic material (MM) of iron sands from Bugel Beach, Kulon Progo, Yogyakarta has been prepared and characterized. Magnetic material was separated from iron sands using a permanent magnet followed by treating with sodium hydroxide (NaOH) solution. The magnetic material product was characterized with X-ray Fluorescence, X-ray Diffraction, Fourrier Transform Infrared spectrophotometry, and Vibrating Sample Magnetometer to determine the chemical composition, crystallinity, presence of functional groups and the magnetization, respectively. Results showed that the investigated iron sand contained magnetic materials up to 89.47% (w/w). The main composition of MM included Fe_2O_3, TiO_2, and SiO_2, with percentages of 72.6, 7.0, and 10.0%, respectively, and the functional groups of material was dominated with Fe–OH and Fe–O. Treatment with NaOH 4M and NaOH 8M increased the content of Fe_2O_3 and TiO_2, otherwise reduced the concentration of SiO_2 and contributed to the improvement of the magnetization from 42.1 to 44.3 emu/g (with 4 M NaOH) and 64.0 emu/g (with 8 M NaOH). Additionally, MM was dominated with mineral of magnetite and contained functional groups of Fe–OH and Fe–O.

1. Introduction

The potential and distribution of iron sand in Indonesia were found on the western coast of Sumatera, Kalimantan, Sulawesi, Maluku islands, and the southern coast of Java [1-2]. Iron sand is generally only used as a building material, whereas it contains magnetic material (MM) constituting the basis for the development of devices in modern life. MM is generally composed of magnetic metals, such as iron, cobalt, nickel, or oxides such as magnetite, maghemite, nickel ferrite, cobalt ferrite, and others [3-5]. MM has been synthesized in different composition and phases such as pure metal (i.e. Fe, Co, Ni), alloys (i.e. FePt, $FePt_3$, $CoPt_3$), metal oxides or spinel ferrites type with general formula MFe_2O_4, where M is a divalent metal cation ($NiFe_2O_4$, $CoFe_2O_4$, $Zn_{0.35}Ni_{0.65}Fe_2O_4$, $MnFe_2O_4$, $MgFe_2O_4$) [3,5,6]. Although pure metals such as Fe, Co, Ni possess the highest magnetization saturation, they receive little interest due to their high characteristic toxicity and are extremely sensitive to oxidation [6-7]. In contrast, iron oxides are less sensitive to oxidation and are able to give a stable magnetic response [3].

The metal oxide of iron sand was found in two phases, magnetite and maghemite. Both are magnetic susceptibility that shows magnetic properties when in a magnetic field [4]. Iron oxide such as magnetite and maghemite last few years has been widely used in various applications, such as magnetic data storage devices with high density, magnetic resonance image formation, delivery system for medicines, cosmetics, dyes, inks as well as play a role in variation separation processes, including adsorption [7-9]. Several chemical methods can be used to synthesize MM iron oxide: co-precipitation reactions occurred in mixture solution of Fe^{2+} and Fe^{3+}, thermal decomposition of organic iron precursor, solvothermal, hydrothermal and microemulsion methods [8,10,11]. These methods were found to be suitable for synthesis of α-Fe_2O_3, γ-Fe_2O_3, and Fe_3O_4 with different proportions by controlling the reaction conditions. However, these methods have disadvantages such as difficult to avoid nucleation during reaction, require relatively high temperatures, complicated procedures, relatively slow kinetic due to the relatively low temperature used, and usage of multiple reagents.

Research on the use of natural MM only a few has been reported [12]. High magnetite ingredients in iron sands are certainly encouraging the utilization of iron sand for MM. Iron sands have high economic value, if it can be processed as MMs. Certainly the processing does not need a large scale mining, that the mining and processing can be done selectively [12]. The separation technique by magnetic separator can separate iron sands between magnetic materials and nonmagnetic materials, but MM may still contain minerals such as silica, magnesium oxide, aluminum oxide [13]. Improvement of MM magnetization with an acid solution such as hydrochloric acid, sulfuric acid and hydrofluoric acid can be used to reduce/eliminate the nonmagnetic oxide that may reduce the magnetic susceptibility [15]. However, treatment with acid solutions such as HCl and HF may reduce levels of metal oxide, which is soluble in acid, especially iron oxide. An alternative way to extract silica from the magnetic material is by using an alkaline solution such as sodium hydroxide (NaOH). Silica dissolves in basic condition, especially strong bases, such as alkali hydroxides. The extraction of silica with a solution of NaOH produce sodium silicate (Na_2SiO_3) solution.

This paper reports separation of MM from iron sand using an external magnetic field. To eliminate the nonmagnetic materials such as silica oxide, MM was extracted using a solution of NaOH. Effect of NaOH treatment on the MM characteristic was investigated and evaluated as well.

2. Materials and methods

2.1. Materials

Iron sand was collected from Bugel Beach, Kulon Progo, Yogyakarta. HF and NaOH was supplied from Merck.

2.2. Preparation of magnetic materials

Magnetic material (MM) of iron sand (30 grams) was separated with a permanent magnet and the MM obtained was grounded with a mortal and then sieved with a 200 mesh sieve size. MM size <200 mesh was washed with distilled water and dried in an oven at 95°C for 18 hours. Then, MM refluxed at 75-80°C using HF 10%, NaOH 4 M and NaOH 8M solutions for 2 hours, followed by washing using distilled water until neutral and dried in an oven at 95°C for 18 hours.

2.3. Characterization of magnetic materials

2.3.1. Analysis of oxides with X-ray fluorescence (XRF). The sample was grounded and put in a disposable plastic cell with mylar and analysed with PANalitycal type Minipal 4 (9 W X ray tube, max 30 kV, max 1 mA with high resolution silicon drift detector).

2.3.2. Characterization with Fourrier transform infrared (FT-IR) spectrophotometry. About 0.5 mg of sample was homogenized with 200 mg of KBr powder and was converted into a pellet form with 2000 psi of pressure. The pellet was put in a sample cell and the absorbance was measured using FTIR Spectrometer FTIR-PRESTIGE 21 at a wave number range 400-4000 cm^{-1}.

2.3.3. Identification of structure with X-ray diffraction (XRD). The sample was grounded and put in a sample cell and analyzed with Shimadzu XRD 6100 X-ray diffractometer. Cu kα radiation from 40 kV and 30 mA was applied to the sample with a 2θ range 0f 5-80° (scan speed of 5°/min).

2.3.4. Measurement of magnetization values. The magnetization values of the products were determined with vibrating sample magnetometer (VSM) Oxford types VSM 1.2H at the maximum external magnetic field of 1.2 Tesla at 25°C.

3. Results and discussion

3.1. Magnetic material of iron sand

Magnetic material (MM) was physically separated from iron sand Bugel Beach Kulon Progo Yogyakarta with magnet. Results showed that the investigated row iron sand contained magnetic materials (Figure 1) up to 89.47% (w/w). The content of MM depend on the location and this result is higher than that of iron sand magnetic material from the southern coast of Bantul Yogyakarta (66.32% (w/w)) [15].

Figure 1. Iron sand attracted to the permanent magnet.

3.2. Content of metal oxides

MM separated from iron sand produced material with black colour indicating high content of iron oxide, magnetite [8,10]. However, determination of the phase constituent of iron oxide magnetic material could not be conducted simply only by using visual observation. The oxide content of MM was determined by using X-ray fluorescence (XRF). Table 1 presents analysis results in oxides of iron sand, MM, and HF/NaOH treated MMs (MM-HF 10%, MM-NaOH 4M and MM-NaOH 8M). According to Table 1, iron sand contains dominant metal oxides of Fe_2O_3, SiO_2, TiO_2, Al_2O_3, and CaO, with percentages of 70.18%, 11.00%, 6.61%, 5.00%, and 4.26%, respectively. Moreover, some metal oxides with minor percentages also contain in the sample. The metal oxide content in a region varies depending on several factors such as geographic region, weather, and geologic condition. Iron sand deposits on the south coast of Yogyakarta comprise magnetite and maghemite mineral with levels of 59.97%, 11.7%, 6.48%, 4.34%, and 4.66% for Fe_2O_3, SiO_2, TiO_2, Al_2O_3, and CaO, respectively [15]. After being separated by permanent magnet, magnetic oxide such as iron oxide and titanium oxide content increased to 72.55% and 6.79%, respectively (table 1). However, non-magnetic oxide content such as SiO_2, Al_2O_3 and CaO did not show a significant decreased. This is due to the probability of non-magnetic oxides attached to the grain of iron sand magnetic material and also attracted by an external magnet field.

Non-magnetic oxide such as silica may be removed from magnetic material by extraction using an alkaline solution such as sodium hydroxide (NaOH). Silica reacts with a base to produce sodium silicate (Na_2SiO_3).

$$2NaOH_{(aq)} + SiO_{2(s)} \rightarrow Na_2SiO_{3(l)} + H_2O_{(l)}$$

Table 1. Metal oxide content in samples analyzed with XRF.

Oxides	Oxide Content (% weight) in sample				
	Iron sand	MM	MM-HF 10%	MM-NaOH 4M	MM-NaOH 8M
Fe_2O_3	70.18	72.55	78.77	80.56	82.73
SiO_2	11.00	10.00	4.00	3.00	2.00
TiO_2	6.61	6.79	7.56	7.39	7.68
Al_2O_3	5.00	5.00	3.00	3.00	4.00
CaO	4.26	3.31	2.86	1.83	1.07
Cr_2O_3	0,13	0.11	0.11	0.11	0.11
MnO	0.57	0.55	0.57	0.56	0.53
V_2O_5	0.48	0.49	0.53	0.55	0.56
NiO	0.12	0.11	0.14	0.09	0.13
P_2O_5	0.78	0.83	0.75	0.55	0.70
K_2O	0.38	0.37	-	0.10	0.07
ZnO	0.04	0.07	0.06	0.06	0.07
Rb_2O	0.23	0.23	0.25	0.26	-
Eu_2O_3	0.59	0.56	0.60	0.62	0.66
Re_2O_7	0.20	0.20	0.20	0.30	0.20
Bi_2O	0.47	0.48	0.53	0.59	0.66

Amorphous silica from rice hull ash (RHA) can be extracted using low temperature alkali extraction. The solubility of amorphous silica is very low at pH < 10 and increases sharply pH > 10. This unique solubility behaviour enables silica to be extracted in pure from RHA by solubilizing under alkaline conditions [16]. Alkali leaching of Kankara Kaolinitic clay is a function of concentration of the extracting agent. NaOH solution 40% w/w leached silica up to 65% [17]. All amorphous silica from soil samples were extracted at room temperature with 0.2M NaOH [18]. In this study, solutions of NaOH 4M and NaOH 8M are used as the leaching agents by refluxing the sample at 80°C for 2 hours. As a comparison, solution of HF 10% was also used for silica extraction. Silica leached from MM by NaOH 4M and NaOH 8M decreased the weight of MM up to 1.18% and 6.95%, respectively, while by HF 10% was 1.04%. XRF analysis results in Table 1 show that the iron oxide content after treated with HF 10%, NaOH 4M and NaOH 8M increased up to 78,77%, 80,56% and 82.73%, respectively. Furthermore, the titanium oxide content also increased significantly by activation. In contrast, silica oxide content decreased up to 4.00%, 3.00% and 2.00%, by HF 10%, NaOH 4M and NaOH 8M, respectively. Although hydrofluoric acid can be used to reduce the nonmagnetic oxide such as silica, treatment with hydrofluoric acid solutions may reduce levels of metal oxide, which is soluble in acid, especially iron oxide.

3.3. Functional groups

Analysis with infrared spectroscopy is aimed to identify the change of functional groups presence due to treatment of MM. FTIR spectra resulted from iron sand, MM, MM-HF 10%, MM-NaOH 4M and MM-NaOH 8M is presented in Figure 2. In FTIR spectra of iron sand (Figure 2(a)) is observed an absorbance band at 553.50 cm^{-1} corresponding Fe–O bond and it is attributed to the formation of ferrite phase. Similar with iron sand, FTIR spectra of MM, MM-HF 10%, MM-NaOH 4M and MM-NaOH 8M (Figure 2(b), (c), (d), (e)) Fe–O bond appears in absorbance band 570.93 cm^{-1} corresponding of magnetite mineral. The presence of silica was showed with characteristic bands at 493.78 cm^{-1} and 918.12-925.83 cm^{-1} coming from bending vibration of Si–O–Si and stretching vibration of Si–O–H (Figure 2(a) and (b)). The intensity of this spectrum decreased at Figure 2 (c), (d), and (e) indicating reduction of silica content by treating with HF and NaOH. FTIR spectra around 1627.92-1643.35 cm^{-1} and 3448.72 cm^{-1} come from bending and stretching vibration, respectively, of –OH groups from both Fe–OH and Si–OH [19].

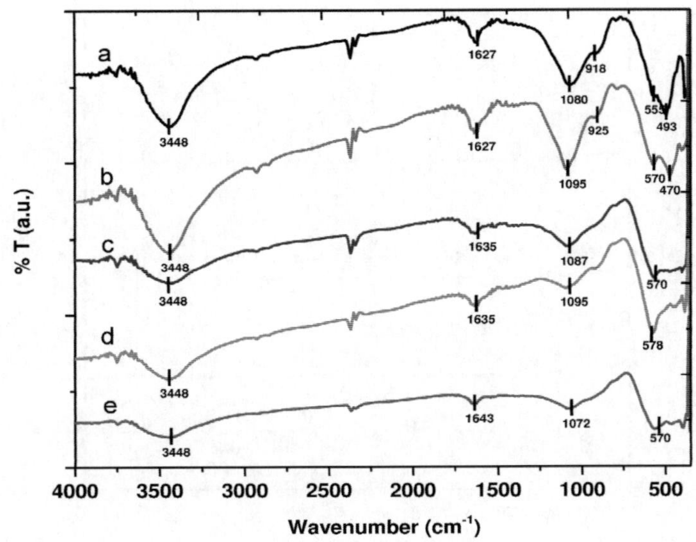

Figure 2. Infrared spectra of (a) Iron sand, (b) MM, (c) MM-HF 10%, (d) MM-NaOH 4M and (e) MM-NaOH 8M.

3.4. Structure of MM

XRD characterization was performed to obtain an X-ray diffraction pattern of the material indicating the crystallinity. Identification is based on the iron oxide phase matched with database. XRD measurements was performed in a range of $2\theta = 0\text{-}80°$. The XRD characterization results and characteristic peaks of iron sand, MM, MM-NaOH 4M and 8M-NaOH are presented in Figure 3 and Table 2, respectively. Almost the diffraction peaks can be indexed to an inverse spinel structure of magnetite (Fe_3O_4) particles (PDF file no 19-0629), except peak at $2\theta = 27.72°$ which is confirmed to quartz (SiO_2) particles [10]. This peak intensity decreased after separation with permanent magnet and treatment with NaOH. This suggests that treatment of MM with NaOH reduced the content of silica. The strong diffraction peak at $2\theta = 35.44\text{-}35.58°$ related to the spinel phase of Fe_3O_4 crystals. The sample appear high crystalline with sharp and intense peaks. The lattice parameters is about 8.32-8.39 Å, whereas crystal field spacing estimated using the Scherrer's formula at $\lambda = 1.54$ Å is 0.243-0.258 Å. It should be noted that the diffraction patterns of magnetite and maghemite (γ-Fe_2O_3) are almost identical due to their analogous inverse cubic spinel crystal structure [12].

Table 2. Crystal field spacing and lattice parameters of material sample.

Material sample	2θ (deg)	d_{311} (Å)	a (Å)
Magnetite	35.42	0.246	8.38
Iron sand	35.52	0.246	8.37
MM	35.58	0.258	8.36
MM-NaOH 4M	35.44	0.246	8.39
MM-NaOH 8M	35.50	0.243	8.32

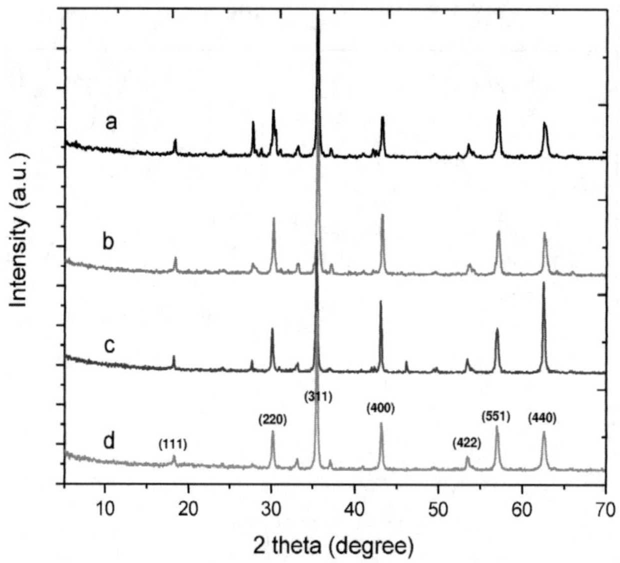

Figure 3. XRD spectra of (a) Iron sand, (b) MM, (c) (e) MM-NaOH 4M and (d) MM-NaOH 8M.

3.5. Magnetization values

Magnetic properties of the particles are studied using VSM at a room temperature and sweep width of external magnetic field from -1 to 1 T. VSM yields a magnetic hysteresis curve, which confirms relationship between magnetization, M (emu/g) and an external magnetic field, H (T). Important parameters in determining the magnetic properties based on the hysteresis curve are saturation magnetization, M_s, remanent magnetization, M_r and coercivity field, H_c. Measurements of magnetization versus applied external magnetic field of iron sand, MM, MM-NaOH 4M and MM-NaOH 8M are shown in Figure 4 and magnetic parameters of iron sand, MM, MM-NaOH 4M and MM-NaOH 8M are listed in Table 3. From Figure 4 seems that all samples give small curve area indicating low energy for magnetization and are classified as soft magnet. Magnet with low coercivity are expressed 'soft'. The term 'hard' refers to magnets having both a high permanent magnetization and a high coercivity [12]. This assumption is supported by the low H_c values of the samples. The value of $H_c \neq 0$ for iron sand, MM, MM-NaOH 4M and MM-NaOH 8M indicates ferrimagnetic properties of the samples. From Table 3, based on the M_s values, can be seen that treatment of MM with NaOH improves the magnetic property.

Table 3. Magnetic parameters of material samples.

Material samples	M_s (emu/g)	H_c (x 10^{-2} Tesla)	M_r (emu/g)
Iron sand	42,08	1.52	8.39
MM	43.95	1.8	10.39
MM-NaOH 4M	44.32	2.09	10.71
MM-NaOH 8M	64.28	1.76	12.18

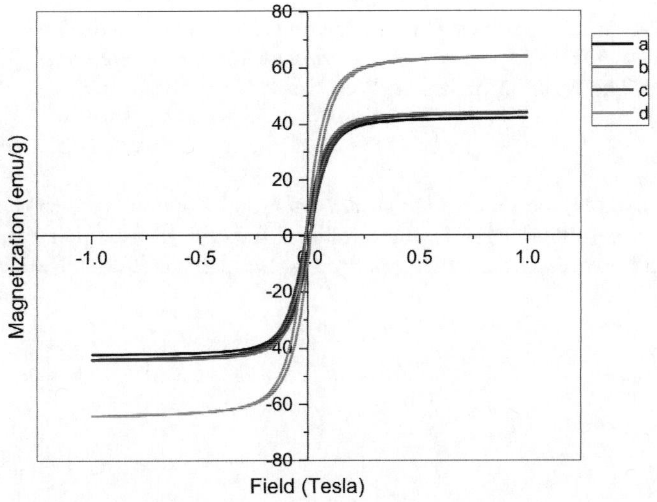

Figure 4. VSM hysteresis curve of (a) Iron sand, (b) MM, (c) MM-NaOH 4M and (d) MM-NaOH 8M.

4. Conclusions

In this research, iron sand collected from Bugel Beach, Kulon Progo, Yogyakarta contains 89.47% (w/w) magnetic material. Treatment with NaOH 4M and NaOH 8M increased the concentration of Fe_2O_3 and TiO_2, magnetization from 42.1 to 44.3 emu/g (with 4 M NaOH) and to 64.0 emu/g (8 M NaOH) and reduced the concentration of SiO_2. Additionally, MM was dominated with mineral of magnetite and contained functional groups of Fe–OH and Fe–O.

References

[1] Rusianto T, Wildan M W, Abraha K and Kusmono 2012 *J. Teknologi* **5** 62
[2] Bilalodin 2013 *J. Fisika Indonesia* **50** 29
[3] Indira T K and Lakshmi P K 2010 *Int. J. Pharm. Sci. Nanotech* **3** 1035
[4] Giakisikli G and Anthemidis A N 2013 *Anal. Chim. Acta* **789** 1
[5] Mazario E, Menendez N, Herrasti P, Cañ M, Connord V and Carrey J 2013 *J. Phys. Chem. C.* **117** 11405
[6] Pang Y L, Lim S, Ong H C and Chong W T *Ceramics Inter.* **42** 9
[7] Bagheri H, Afkhami A, Tehrani M S and Khoshsafar H 2012 *Talanta* **97** 87
[8] Ramimoghadam D, Bagheri S and Hamid S B A 2014 *J. Magn. Magn. Mater* **368** 207
[9] Laurent S, Forge D, Port M, Roch A, Robic L, Elst V and Muller R N 2008 *Chem. Rev.* **108** 2064
[10] Cornell R M and Schwertmann U 2003 *The Iron Oxide: Structure, Properties, Reaction, Occurences and Uses* 2 ed (New York: Wiley-VOH GmbH & KGaA) p 32
[11] Fajaroh F, Setyawan H, Widiyastuti W and Winardi S 2012 *Adv.d Powder. Technol* **23** 328
[12] Camacho C S, Villalobos M, Sanches M L, Alatorre J A, Cienfuegos J C and Riuz M G 2013 *Chem. Geology* **347** 233
[13] Zulfalina and Manaf A 2004 *Indo. J. Mater. Sci.* **5** 46
[14] Widihati I A G 2008 *J. Kim.* **2** 25
[15] Yulianto A, Bijaksana S and Loeksmanto W 2003 *Indo. J. Phys.* **14** 2

[16] Kalapathy U, Proctor A and Shultz J 2000 *Biosource Technol.* **73** 257
[17] Abdullahi Y, Ali E A and Lawal A O 2013 *J. Eng. Appl. Sci.* **8** 10
[18] Georgiadis A, Sauer D, Breuer J and Hermann L 2015 *Soil Research* **53** 392
[19] Stuart B 2004 *Infrared Spectroscopy: fundamentals and applications* (England John Wiley & Sons Ltd, The Atrium,Chichester, West Sussex PO19 8SQ) p 95

Acknowledgments
This work was funded by the Directorate of Research and Community Services, Directorate General of Higher Education (DP2M-DIKTI), Indonesian Ministry of Research, Technology and Higher Education through Research Grant Penelitian Produk Terapan No. 228v/UN29.20/PPM/2016.

IOP Conf. Series: Materials Science and Engineering **172** (2017) 012021 doi:10.1088/1757-899X/172/1/012021

Surface modifications of chalcopyrite $CuInS_2$ thin films for photochatodes in photoelectrochemical water splitting under sunlight irradiation

Gunawan,[1] A Haris,[1] H Widiyandari,[2] W Septina[3] and S Ikeda,[4]

[1]Chemistry Department, Faculty of Sciences and Mathematics, Diponegoro University, Semarang Indonesia
[2]Physics Department, Faculty of Sciences and Mathematics, Diponegoro University, Semarang Indonesia
[3]University of Zurich, Switzerland
[4]Konan University, Osaka, Japan

E-mail: gunawan_undip@yahoo.com

Abstract. Copper chalcopyrite semiconductors include a wide range of compounds that are of interest for photoelectrochemical water splitting which enables them to be used as photochatodes for H_2 generation. Among them, $CuInS_2$ is one of the most important materials due to its optimum band gap energy for sunlight absorption. In the present study, we investigated the application of $CuInS_2$ fabricated by electrodeposition as photochatodes for water splitting. Thin film of $CuInS_2$ chalcopyrite was formed on Mo-coated glass substrate by stacked electrodeposition of copper and indium followed by sulfurization under H_2S flow. The films worked as a H_2 liberation electrode under cathodic polarization from a solution containing Na_2SO_4 after loading Pt deposits on the film. Introduction of an n-type CdS layer by chemical bath deposition on the $CuInS_2$ surface before the Pt loading resulted appreciable improvements of H_2 liberation efficiency and a higher photocurrent onset potential. Moreover, the use of In_2S_3 layer as an alternative *n*-type layer to the CdS significantly improved the H_2 liberation performance: the $CuInS_2$ film modified with In_2S_3 and Pt deposits worked as an efficient photocathode for photoelectrochemical water splitting.

1. Introduction

The shortage of fossile sources has made an effort to get alternative energies such as hydrogen through its conversion to energy by direct combustion in conventional engines or reaction with pure oxygen in a fuel cell. Hydrogen can be produced by natural gas reforming, electrolysis, biomass, thermochemical, thermophysical, photoelectrochemical (PEC) methods, and photo-biological process [1]. Hydrogen (H_2) production by PEC water splitting is considered to be an attractive in view of energy and environmental issues.

Since the first report of a TiO_2 thin-film photoelectrode [2], a variety of semiconductor electrodes and devices have been investigated. To date, performances with conversion efficiency as high as 12.4% have been demonstrated for electrodes based *p*-$GaInP_2$/GaAs cells grown by atmospheric-pressure organometallic vapor-phase epitaxy [3]. However, due to limited corrosion resistance in aqueous electrolytes and expensive cost for production of these electrodes, these systems are not feasible for practical applications. Meanwhile, corrosion-resistant and inexpensive transition metal oxides are well-studied, but conversion efficiencies for these electrodes are not sufficiently high yet due to the lack of optical and photoelectrochemical properties required for realizing high photocurrents and H_2 evolution rates.

Cu-chalcopyrite p-type semiconductors such as $CuInSe_2$, $CuGaSe_2$, $CuInS_2$, $CuGaS_2$ and their mixed crystals are used as absorber layers in thin film solar cells. Due to their high absorption coefficient and tunable band gap values (1.0-2.4 eV) by changing the In/Ga and/or Se/S ratios, these materials are also attractive use as photocathodes for H_2 production [4-6]. Electrodeposition is one of the promising technologies for fabrication of low–cost Cu-chalcopyrite films because of the low cost equipment, scalability of large scale, and good control over film composition and morphology [7].

The insertion of n-type thin layers on photocathodes to improve the photo response had been done by reseachers using different absorbers, which are widely employed to fabricate *p-n* junctions for chalcophyrite and kseterite based solar cells. To date the surface modification of calcopyrite and kseterite [8] for photoelectrochemical water splitting had been done using n-type CdS by chemical bath deposition (CBD) and resulted apreciable current densities compared to the bare or Pt-thus films [9-15]. Meanwhile CdS is a toxic compound that should be avoided. Therefore in this reseach we tried to use In_2S_3 by CBD alternative to CdS as *n*-type thin layer on $CuInS_2$ photocathodes.

2. Experiment methods

Synthesis of the $CuInS_2$ thin film

The $CuInS_2$ thin film was synthesized by electrodeposition of Cu and In layers followed by sulfurization. Electrodeposition was carried out potentiostatically using a Hokuto Denko HSV-110 potentiostat-galvanostat under N_2 atomosphere without stirring. A vertical three-electrode setup consisting of an Ag/AgCl reference electrode, a Pt wire counter electrode, and an Mo-coated glass substrate (Mo/glass) as a working electrode (area of the working electrode being 0.7 cm^2) was employed. The electrolyte solution for Cu deposition consisted of 50 mM $CuSO_4$, 150 mM trisodium citrate, and 242 mM citric acid. The solution was adjusted to pH 2.4 using H_2SO_4. For In deposition, the electrolyte solution consisted of 30 mM $InCl_3$, 242 mM citric acid, and 36 mM trisodium citrate. Potentials used for Cu and In deposition were -0.2 V and -0.76 V (*vs.* Ag/AgCl), respectively. Electric charges of Cu and In deposition were fixed at 0.73 C and 0.84 C, respectively: this resulted in the composition ratio of Cu and In (Cu/In) in a Cu and In stacked layer of 1.3. The Cu/In stack as-deposited was then heated at 160 °C for 30 min under Ar flow (200 mL min^{-1}), followed by sulfurization at 560 °C under H_2S flow (5% H_2S in Ar, 200 mL min^{-1}) for 10 min in a glass tube furnace. Thus-obtained $CuInS_2$ films were then etched by immersion in an aqueous KCN solution (10%) for 2 min to remove excess Cu_xS components.

Surface modifications of the $CuInS_2$ thin film

A $CuInS_2$ thin film was added to an aqueous solution containing 12.5 mM $CdSO_4$, 0.22 M thiourea, and 11 M of ammonia; deposition of CdS was performed at 60 °C for 7 min. For In_2S_3 deposition, the $CuInS_2$ thin film was immersed in an aqueous solution containing 25 mM $In_2(SO_4)_3$, 0.1 M thioacetamide and 0.1 M acetic acid at 65 °C for 15 min. Loading of platinum deposits on bare $CuInS_2$ and these modified-$CuInS_2$ thin films was conducted by photoirradiation of the films in an N_2-saturated 0.1 M Na_2SO_4 solution containing 1 mM H_2PtCl_6 (with pH adjusted to 4.0) at -0.1 V (vs. Ag/AgCl) for 10 min. An Asahi Spectra HAL320 solar simulater was used as a light source.

PEC measurements

PEC water splitting was also performed using the three-electrode setup in an aqueous solution of 0.1 M Na_2SO_4 with pH adjusted to pH 9.0 under N_2 or Ar. Transient photocurrents were obtained by scanning applied potentials with a scan rate of 10 mV s^{-1} under chopped illumination of simulated AM 1.5 solar irradiation from the above solar simulator. Applied bias photon-to-current efficiency (ABPE) for photocathode was calculated from the data according to the following equation:

$$ABPE\ [\%] = J \times V_b \times 100 / P_{AM1.5},$$

where J (mA cm^{-2}) is the measured photocurrent density, V_b (V) is the bias voltage measured by the reversible hydrogen electrode (RHE) scale (E_{RHE}, see below), and $P_{AM1.5}$ is the photon flux of the AM 1.5 simulated sunlight (100 mW cm^{-2}). Incident photon-to-current conversion efficiency (IPCE) spectra of Pt-In_2S_3/$CuInS_2$ and Pt-CdS/$CuInS_2$ were also calculated using the following equation:
$\mathrm{IPCE}\ [\%] = 1240 \times J \times 100 / (\lambda \times P_i)$,
where J (mA cm^{-2}) is the measured photocurrent density, λ is the wavelength of incident photons and P_i is the incident light intensity (mW cm^{-2}). The measurement was performed at -0.73 V (vs. Ag/AgCl) by the lock-in technique using an NF LI5630 digital lock-in amplifier. For the measurement, photoirradiation was performed by chopping at 10 Hz of monochromatic light, which was obtained by passing light from the solar simulator through a monochromator. The number of incident photons was determined by an OPHIR Orion Laser power meter equipped with a photodiode. A PEC cell connected to an online gas chromatography system (INFICON 3000 Micro GC Gas Analyzer equipped with an MS-5 A column and a thermal conductivity detector) was used to detect H_2 and O_2 during the PEC water splitting. The PEC cell was immersed in a water bath to maintain the reaction temperature at 288 K, and photoirradiation was performed at -0.73 V (vs. Ag/AgCl) by using a Cermax LX-300F 300 W xenon lamp. The potentials in each measurement measured vursus Ag/AgCl electrode ($E_{Ag/AgCl}$) were converted into E_{RHE} using the following equation:
$E_{RHE} = E_{Ag/AgCl} + 0.059 \times \mathrm{pH} + 0.199$.

Characterizations
Crystal structures were measured by X-ray diffraction (XRD) using a Rigaku MiniFlex X-ray diffractometer (Cu Kα, Ni filter). Surface morphologies were examined using a Hitachi S-5000 FEG field emission scanning electron microscope (FE-SEM). Elemental compositions of thus-obtained films were determined using a Hitachi TM3000 scanning electron microscope equipped with a Swift ED3000 energy dispersive X-ray spectrometer (EDX).

2. Results and Discussion
In this study, a $CuInS_2$ was fabricated from electrochemically stacked layers of metallic Cu and In followed by sulfurization. The composition ratio of Cu and In (Cu/In) in metallic precursor was fixed at a Cu-rich composition (Cu/In = *ca.* 1.3) compared to the stoichiometric ratio of the final film. It is known that a high Cu/In ratio is desirable for obtaining a $CuInS_2$ film with high crystallinity and large grain sizes, due to the crucial role of Cu_xS impurity phase(s) to accelerate the growth of $CuInS_2$ crystallinity during the sulfurization process. The stacked metallic precursor film obtained on a Mo/glass substrate showed good uniformity and adherence. After sulfurization, the semi-glossy film with a white color became non reflective. A slightly bluish gray color of the resulting film suggests surface coverage of Cu_xS components that dissapeared after KCN etching, The elemental compositions of the film measured by EDX were 27.473, 25.087 and 47.439 % for copper, indium and sulfur, respectively (data not shown). The XRD pattern (Figure 1) of thus-obtained film showed typical difraction peaks assignable to the $CuInS_2$ crystal with a chalcopyrite structure, in addition to peaks due to the Mo substrate. The as-deposited film shows several diffraction peaks assignable to Cu, In and CuIn alloy. It is noted that we performed preheating under Ar to the as-deposited bilayer film before sulfurization to form the Cu-In alloy to suppress the formation of structural void on the fabricated $CuInS_2$ film [16]. XRD pattern of the sulfurized film showed typical diffraction pattern of the chalcopyrite $CuInS_2$ crystal without any impurity elements [17].

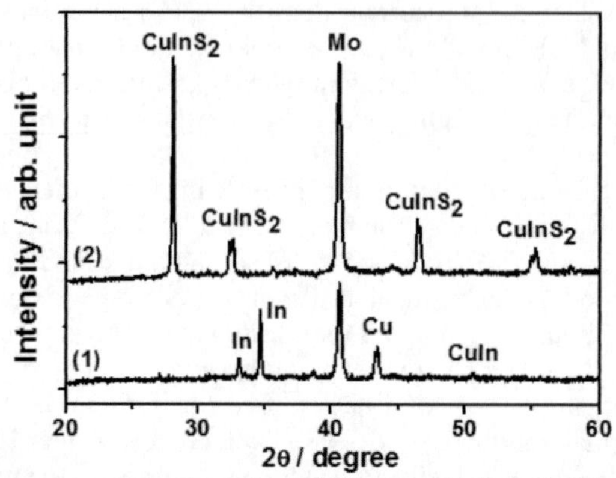

Figure 1. XRD patterns of as-deposited CuIn (1) and $CuInS_2$ (2) on Mo substrate.

Figure 2. SEM images of Sulfurized samples ($CuInS_2$).

Figure 2 Shows the scanning electron microscope (SEM) image of the $CuInS_2$ film after KCN etching. The top view SEM image indicates that the film consists of compact agglomerates of angular-shaped crystallite (Figure 2a). Cross sectional SEM image of the corresponding sample in Figure 2b also shows regularly shaped crystals with grain sizes ranging from 0.5 to 1.5 µm.

Figure 3 shows Current-potential curves for bare-$CuInS_2$, Pt-$CuInS_2$, Pt-In_2S_3/$CuInS_2$, and Pt-CdS/$CuInS_2$ under 0.1 M Na_2SO_4 at pH 9 under AM 1.5 chopped Xe lamp illumination. Figure 3A shows the bare-$CuInS_2$ film showed little photocurrents due to recombination and leakage [18], whereas the Pt-$CuInS_2$ film exhibited appreciable increases in photocurrents, indicating the effectiveness of loading Pt deposits as promoters for H_2 production and act as activation sites for evolution of hydrogen and improve the kientics of reaction, thereby reducing the bias voltage and enhanching IPCE at low bias [19]. The photocurrent values at 0 V *vs*. RHE were 0.51, 7.36, 15.16 and 15.12 mA cm^{-2} for bare-$CuInS_2$, Pt-$CuInS_2$, Pt-In_2S_3/$CuInS_2$, and Pt-CdS/$CuInS_2$ respectively. The open circuit potential (V_{oc}) or onset potensial of the Pt-$CuInS_2$ electrode was estimated to be 0.40 V *vs*. RHE (Figure 3A). The highly negative V_{oc} suggests that there is a significant potential barrier that hinders electron transfer to the surface loaded Pt deposits. In order to improve the photoresponse of the Pt-$CuInS_2$ electrode, before Pt deposition an n-type thin layer of In_2S_3 was inserted by using a CBD technique to form Pt- and In_2S_3-modified electrode (Pt-In_2S_3/$CuInS_2$). As a result, a significant improvement of cathodic photocurrents was observed. Moreover, these modifications resulted in appreciable positive shifts of V_{oc}, i.e. *ca*. 0.82 V (*vs*. RHE) indicating considerable facilitation of efficient charge separation by introduction of the p-n junction. This is due to good transparancy of In_2S_3 over the solar spectrum [20] and less absorption of blue photons [21], and may be the formation of *p-n* homojunction $CuInS_2$. We also compared In_2S_3 to CdS-modified it was found that introduction of a thin layer of In_2S_3 as an alternative n-type layer resulted comparable photocurrent than that of CdS and seems both had reached saturation at -0.2 V *vs* RHE, although CdS-modified one had lower

V_{oc} than In_2S_3 namely 0.61 V. To estimate solar-to-hydrogen conversion efficiency (STH) we used the applied bias photon-to-current efficiency (ABPE) [22]. The Pt-In_2S_3/$CuInS_2$ electrode gave maximum ABPE (Figure 3B) of 1.9% at 0.27 V (*vs.* RHE) higher than CdS that of 1.6% at 0.18V). Through the surface modifications, both the cathodic photocurrent and the V_{oc} of In_2S_3-modified $CuInS_2$ were increased significantly 29 times and 0.4 V respectively. This increase may be caused by formation of a built-in potential (V_{bi}) in the photoelectrode and asymetric distribution of V_{bi} across the junction which separates photogenerated charge carriers [23].

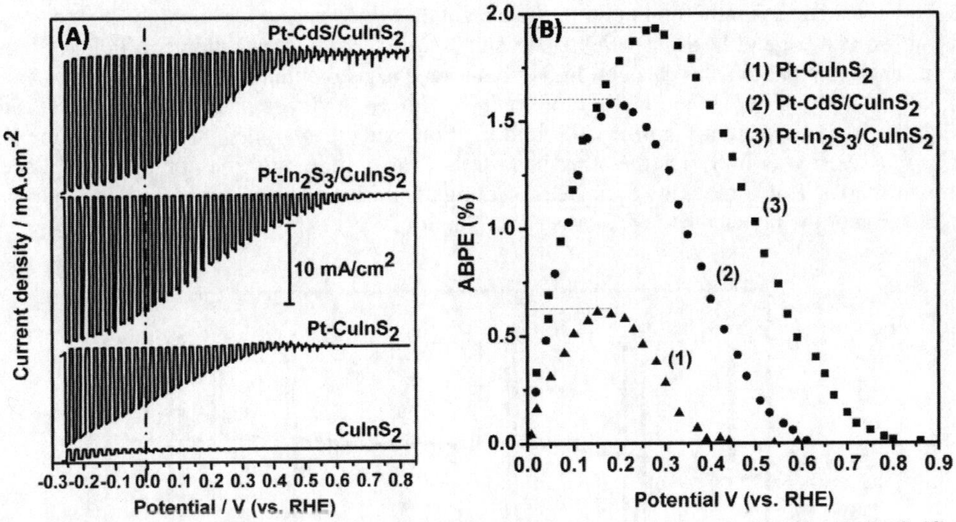

Figure 3. Current-potential curves (A) and APBE (B) of bare-$CuInS_2$, Pt-$CuInS_2$, Pt-In_2S_3/$CuInS_2$, and Pt-CdS/$CuInS_2$ electrodes in 0.1 M Na_2SO_4 at pH 9 under AM 1.5 G chopped Xe lamp illumination.

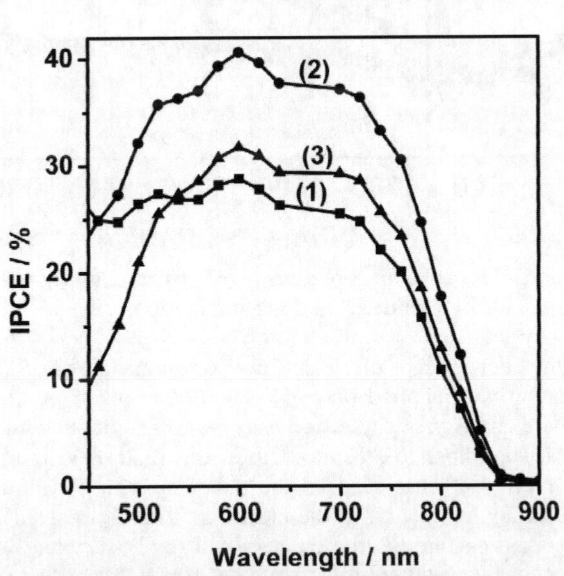

Figure 4. IPCE spectra of Pt-In_2S_3/$CuInS_2$ electrodes at In_2S_3 deposition of (1) 5, (2) 15 and (3) 30 min measured at 0 V (*vs.* RHE) in 0.1 M Na_2SO_4 at H 9.

For quantitative evaluation of photochatodes, the incident photon-to-current efficiency (IPCE) was measured (Figure 4). The measured IPCE values of Pt-In_2S_3/$CuInS_2$ photochatode at potentials of 0 V *vs* RHE. IPCE values were increased significantly under a negative electrode potential. All electrodes showed no photoresponse to the light with wavelength longer than 820 nm because the prepared $CuInS_2$ had a band gap around 1.5 eV, which corresponds to the onset of light absorption of 820 nm. The IPCE spectrum of Pt-In_2S_3/$CuInS_2$ at 15 min In_2S_3 deposition showed a rise at wavelengths ranging from 400 to 500 nm and it reached IPCE as high as 40.8% at wavelength 550-750 nm. Photocathode of Pt-In_2S_3/$CuInS_2$ obtained by In_2S_3 CBD for 15 min showed the higher IPCE than In_2S_3 deposited at 5 and 30 min at those wavelengths. However, at wavelength shorter than 560 nm (the absorption edge of In_2S_3) samples of 15 and 30 min CBD had much lower efficiency than that of 5 min. Therefore photon absorbed by In_2S_3 was forced to recombine [19].

Finally we examined the stability of Pt-In_2S_3/$CuInS_2$ and compared to Pt-$CuInS_2$ and Pt-CdS/$CuInS_2$ as can be seen in Figure 5 CdS modification tends to degrade, although it has the similar initial current density as In_2S_3. On the other hand In_2S_3 was a little bit stable for 1 h. And it seems it has similar trend as Pt-$CuInS_2$. However the protecting agent should be added to the surface to protect the degradation of the photocathode for longer examination.

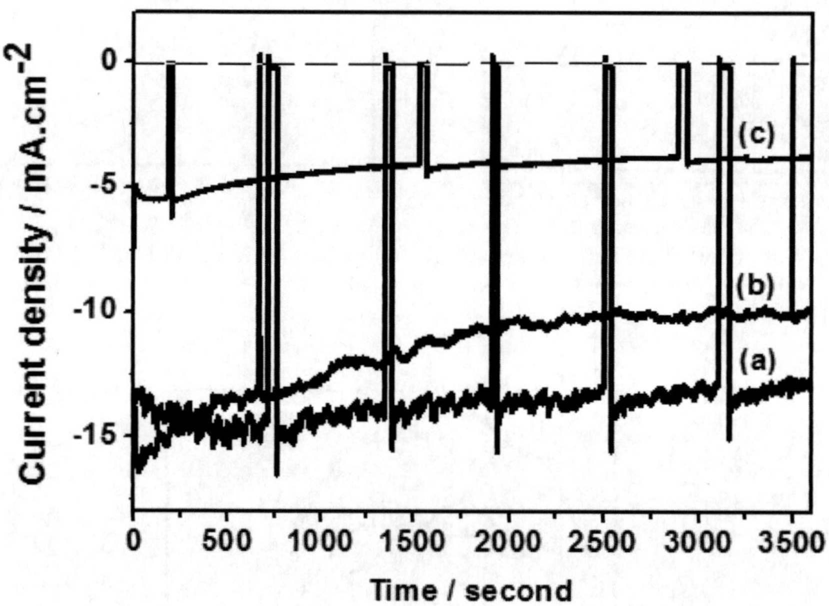

Figure 5. Current-time curves for (a) Pt- In_2S_3/$CuInS_2$, ((b) Pt-CdS/$CuInS_2$, (c) Pt-$CuInS_2$ electrodes in 0.1 M Na_2SO_4, pH 9 at 0 V *vs* RHE under Xe lamp illumination.

Figure 6 shows a time course curve of H_2 and O_2 evolution on Pt-In_2S_3/$CuInS_2$ electrode under 300 W Xe lamp illumination with applied bias of 0 vs RHE using three electrode system namely, photochatode, platinum wire and Ag/AgCl electrodes as working, counter and reference, respectively. Evolved H_2 and O_2 gases were close to stoichiometric ratio (2.2:1). Within 130 min was evolved 406.7 μmol H_2 and 185.1 μmol O_2 that exeeded the molar amount of $CuInS_2$ (ca. 1.6 μmol) these amount demonstrated the catalytic function of the electrode. The H_2 evolved was in agreement with with half of the electrons passing through the outer circuit ($e^-/2$), denoted by unfilled black square, showing that the efficiency of the water splitting is almost 100%. According to my knowledge the H_2 evolved at the same period gave the the highest result [9,23].

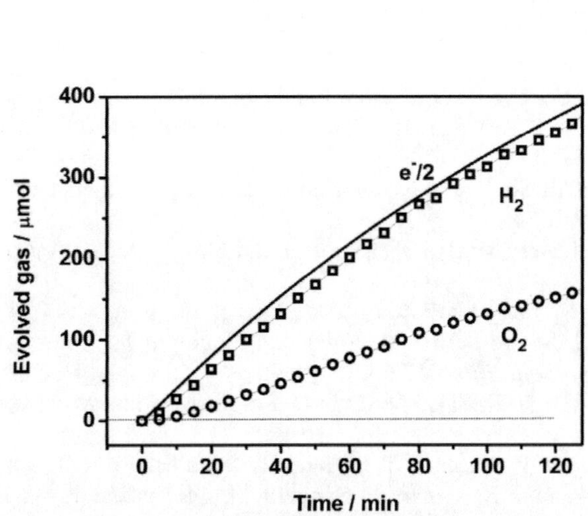

Figure 6. A time course curve of H_2 and O_2 evolution on Pt-In_2S_3/$CuInS_2$ electrode (0.7 cm^2) in 0.1 M Na_2SO_4, pH 9 at 0 V *vs* RHE under 300 W Xe lamp illumination. A black line denotes a time course of e^-_{half} during the H_2 evolution.

3. Conclusions

The formation of heterojunction by surface coverage of the $CuInS_2$ films with n-type buffer layers of In_2S_3 followed by loading of Pt deposits made them efficient H_2-evolving photoelectrodes. For the In_2S_3 modified sample, IPCE of more than 40.8% were obtained at wavelength ranging from 550 to 750 nm by applying a potential at 0 V vs RHE at pH 9. PEC water splitting under teresterial light was demonstrated using Pt-In_2S_3/$CuInS_2$ under external bias potential of 0 V vs RHE. Although the materials used in the present study were not examined for longer time with protecting agent to avoid photocathode corrosion. Thus, studies along these line are now in progress.

References

[1] Studer S, Stucki S and Speight J D in: Zuttel A, Borgschulte A and. Schlapbach L (Eds.), 2008 *Hydrogen as Future Energy Carrier* WILEY-VCH Verlag GmbH&Co.KgaA, ISBN:978-3-527-30817-0 Winheim p 49-54 59.

[2] Peter L M 2011 *Phil. Trans. R. Soc. A* **369** 1840 1856.

[3] Fujishima A and Honda K 1972 *Nature* **238** 37.

[4] Khaselev O and Turner J A 1998 *Science* **280** 425.

[5] Yokoyama D, Minegishi T, Maeda K, Katayama M, Kubota J, Yamada A, Konagai M and Domen K 2010 *Electrochem. Commun.* **12** 851.

[6] Yokoyama D, Minegishi T, Jimbo K, Hisatomi T, Ma G, Katayama M, Kubota J, Katagiri H and Domen K 2010 *Appl. Phys. Express* **3** 101202.

[7] Lincot D, Guillemoles J F, Taunier S, Guimard D, Sicx-Kurdi J, Chaumont A, Roussel O, Ramdani O, Hubert C, Fauvarque J P, Bodereau N, Parissi L, Panheleux P, Fanouillere Naghavi P N, Grand P P, Benfarah M, Mogensen P and O. Kerrec 2004 *Solar Energy* **77** 725–737.

[8] Rovelli L, Tilley S D and Sivula K 2013 *ACS Appl. Mater. Interface* **5**(16) 8018-8024.

[9] Ikeda S, Nakamura T, Lee S M, Yagi T, Harada T, Minegishi T and Matsumura M 2011 *ChemSusChem.* **4** 262.

[10] Ikeda S, Nakamura T, Lee S M, Yagi T, Harada T, Minegishi T and Matsumura M 2010 *ChemSusChem.* 0000,00 1-8.

[11] Tabata M, K Maeda, Ishihara T, Minegishi T, Takata T and Domen K 2010 *J. Phys. Chem. C* **114** 11215-11220.

[12] Ma G, Minegishi T, Yokoyama D, Kubota J and Domen K 2011 *Chem. Phys. Lett.* **501** 619-622.
[13] Moriya M, Menigishi T, Kumagai H, Katayama M, Kubota J and Domen K 2013 *J. Am. Chem. Soc.* **135** 3733-3735.
[14] Ikeda S, Nonogaki M, Septina W, Gunawan G, Harada T and Matsumura M 2013 *Catal. Sci. Tecnol.* **3** 18849-1854.
[15] Kumagai H, Minegishi T, Moriya Y, Kubota J and Domen K 2014 *J. Phys. Chem. C* 118 (30) 16386-16392.
[16] Li Z, Luo W, Zang M, Feng J and Zou Z 2013 *Energy Environ. Sci.* ***6*** 347-370.
[17] Sandoval-Paz M. G, Sotelo-Lerma M, Valenzuela-Jáuregui J J, Flores-Acosta M and Ramirez-Bon R 2005 *Thin Solid Films* **472** 5.
[18] Barkhouse D A R, Haight R, Sakai N, Hiroi H, Sugimoto H and Mitzi D B 2012 *Appl. Phys. Lett.* 100 193904.
[19] Cheng Z, Jaramillo T F, Deutsch T G, Kleiman-Swarsctein A K, Forman A J, Gaillard N, Garland R, Takanabe K, Heske C, Sunkara M, McFarland E W, Domen K, Miller E L, Turner J A and Dinh H N 2010 *J. Mater. Res.* **25** 3-16.
[10] Malizia M, Seger B J, Chorkendoff I and Vesborg P C K 2014 *J. Mater. Chem. A* **2** 6847-6853.
[21] Ho J C W, Batabya S K, Pramana S S, Lum J, Pham V T, Li D, Xiong Q, Tok A I Y and Wong L H 2012 *Mater. Express* **2** 4.
[22] Gopinath G R, Miles R W and Ramakrishna Reddy K T 2013 *Energy Procedia* **34** 399 – 406.
[23] Higashi M, Domen K and Abe R 2013 *J. Am. Chem. Soc.* **135** 10238-10241.

Acknowledgments
Financial support From Diponegoro University (No.: SP DIPA-042.01.2.400898/2016), Ministry of Research and Higher Education Republic of Indonesia is acknowledged.

IOP Conf. Series: Materials Science and Engineering **172** (2017) 012022 doi:10.1088/1757-899X/172/1/012022

Plant essential oils potency as natural antibiotic in Indonesian medicinal herb of " jamu "

H Soetjipto[1] and Y Martono[2]

[1,2]Satya Wacana Christian University, Salatiga, Indonesia

Email: hartati.sucipto@staff.uksw.edu

Abstract. The main purposes of this study are to compile antibacterial activity data of essential oils from Indonesian's plants in order which can be used as a natural antibiotic in "jamu" to increase potential Indonesian medicinal herb.By using Agar Diffusing method, Bioautography and Gas Chromatography Mass Spectrum, respectively, antibacterial activity and chemical compounds of 12 plants essential oils were studied in the Natural Product Chemistry Laboratory, Department of Chemistry, Faculty of Science and Mathematics, Satya Wacana Christian University, Salatiga since 2007 until 2015.The results of this studies showed that all of the essential oils have a medium to a strong antibacterial activity which are in the range of 30 – 2,500 μg and 80-5,000 μg. Further on, the essential oils analyzed by GCMS showed that each essential oils have different dominant compounds. These data can be used as basic doses in the usage of essential oils as natural antibiotics.

1. Introduction

One of Indonesian culture heritage is to drink "jamu" and another traditional herbal medicine to maintain health and disease preventive. Doctors also believed that herbal medicine has a good bioactivity and also safe. But one of the problems of Indonesian herbal medicine is they do not have enough clinical test to support medicinal treatment.

As a tropical country, Indonesia very close to many kinds of infectious diseases, so antibiotic using in medicine to be a basic need. Nowadays, around 96% medicine basic commodity were imported from China and India, and the large quantity is used for antibiotics; the amounts reached millions USD.

Medicinal and aromatic plants are used on a large scale in medicine against drug-resistant bacteria, which are considered one of the most important reasons for the lack of success of treatment in infectious diseases. Unresponsible antibiotic utilizing give the facility to form resistant super bacteria to antibiotic and will become a boomerang for the human shortly.

In 2007 a lot of countries decided to stop antibiotic production because the complicated process and the benefit which provide is not in proportion to the danger in the future. So what are we have to do to handle a lot of infections around us? Essential oils may be one of the alternatives to answer this problem, because of its ability as an antibacterial compound

As a tropical country, Indonesia has a lot of aromatic plants, in fact, 40-50 kinds of essential oils trading in the world came from Indonesia. [1] One of the important compounds in Indonesian herbal is essential oils, a volatile oil from plants, this oil is composed of many phenolic compounds which are responsible for the strong antibacterial effect. [2] This nature gives an opportunity to essential oils to be used as an antibiotic substitution, especially in "jamu" Indonesian traditional medicine utilizing.

The aims of this research are to equip data of essential oils dose which showed the antibacterial effect against Gram-positive and negative bacteria and to give information about dominant chemical compounds which responsible to antibacterial activity to be used optimally as an antibiotic substitution.

2. Material and method

2.1. Material

Essential oils of twelve species plants were used as samples, which are: Limo (Citrus amblycarpa (Hassk) Ochse), Lime (C. aurantifolia Swingle), Lemon (C. limon (L.) Burm.), Kefir lime (C. hystrix D.C.), Masoi (Cryptocarya massoia (Oken) Kosterm.), Surinam cherry/ Dewandaru (Eugenia uniflora L.), Anise (Foeniculum vulgare Mill.), Star anise (Illicium verum Hook. E.), Krangean (Litsea cubeba (Lour.) Pers.), Torch Ginger/Kecombrang (Nicolaea speciosa Horan), Basil (Ocimum basilicum Linn.) and Legetan (Spillanthes paniculata Wall), respectively.

Isolates of Gram-positive bacteria; Bacillus subtilis ATCC 6051, B. cereus FNCC 0057, Staphylococcus aureus FNCC 0047 and Gram-negative bacteria: Escherichia coli IFO 0091; Pseudomonas cepasia FNCC 0063; and P.aeruginosa FNCC 0063 were obtained from Laboratory of Microbiology, Faculty of Science and Mathematics, Satya Wacana Christian University, Salatiga.

All chemical reagents made by E-Merck, Jerman: toluene, ethyl acetate, ethanol, H2SO4, PEG, silica gel 60 F254, vanillin, NaCl, Nutrient Broth. Mueller- Hinton Agar (PA, Oxoid, Inggris), tetracyclin (PA, Oxoid, Inggris) and paper disc (PA, Whatman, Inggris),.
Steam distillation apparatus with clevenger, micropipette (Biohit), analytical balance (Mettler H80), oven (WIB-Binder), spectrophotometer (UV-Vis Mini Shimadzu U-1240), Gas Chromatography – Mass Spectrophotometer (Shimadzu QP2010S).

2.2.Method

2.2.1. Essential oil extraction [3]

Essential oil from fresh, clean, weighed aerial parts, fruit peel, bark, fruit, seed, flower and leaves and extracted by hydro-steam distillation using Clevenger apparatus were collected and store. The oil was filtered through anhydrous NA2SO4 to dry yielded essential oil, keep in tightened vials and stored in a refrigerator.

Physicochemistry test was done by rendement calculation, colour, transparent and aroma. GCMS analysis essential oil was done in Organic chemistry laboratory, Faculty of MIPA Universitas Gadjah Mada Yogyakarta.

2.2.2.Direct bioautographic test [4]

The essential oil was evaluated in vitro by Thin-layer Chromatography (TLC) plate, before using the plate was activated in 105°C 10 minutes. The plate of silica gel F254 4x10 cm (Merck) as a solid phase and toluene: ethyl acetate 93:7 as a mobile phase were used. Afterward, plate was sprayed with bacterial suspension in Mueller- Hinton Broth (MHB), and plate was stored in a water-vapor chamber at 37°C 24 hours. Iodonitrotetrazolium chloride 5 mg/ml was used to visualize the antibacterial spot.

2.2.3. Antimicrobial disk diffusion test [5]

One ose bacteria was inoculated to Nutrient Broth (NB), incubated at 30 oC for 24 hours. Dilute inoculum by using physiological solution (NaCl 0,9%) to match a 0,5 Mc Farland standard. The bacterial suspension was diluted and measured by UV-Vis Spectrophotometer to obtain Optical Density (OD) 0,4-0,5 at 550 nm.

A paper disc was dropped 20 µl essential oil in certain concentration and put the disc in a petri dish with medium content bacteria inside. The petri dish was incubated at 30 ° C 24 hours. Inhibition Area Diameter (IAD) was measured as a middle line start from the clear spot around the disc. The lowest concentration which shows the clear spot around the paper disc is the Minimum Inhibitory Concentration (MIC).

2.2.4. Isolation and identification antibacterial compound of eosential Oil
The essential oil was evaluated in vitro by using Thin Layer Chromatography (TLC) plate. 4x10 cm silica gel F254 sheets (Merck) as a solid phase and toluene: ethyl acetate 93:7 as a mobile phase. The isolated spot was analyzed by GCMS in Laboratory of Organic Chemistry, Fakultas MIPA Universitas Gadjah Mada Yogyakarta. The Mass Spectrum of every peak from the chromatogram was compared to database.

2.2.5. Data analysis [6]
Antibacterial activity data of each plant were analyzed by using Randomized Completely Block Design (RCBD) Sub-Sampling, five treatments, three subsamples, and five replications. As the blocks are the analysis time.

3. Result and discussion

Table 1 demonstrated that the rendement of each essential oil has variation in amount but in general, the amount of essential oil is relatively less than 3 %, and the oil content distributed in a different part of a plant. The small amounts of essential oil have already enough to do the function as a self-protecting to the prey, as a fascinating compound. The application essential oil in human life for example as a flavor in food, cosmetic, medicine, toiletries [2, 7, 8]

Table 1. Plants and their families, partly used and rendemen

Scientific name	Plant family	Part used	Rendement(%) w
Citrus amblycarpa (Hassk.) Ochse (Limo)[9]	Rutaceae	Peel	0.92
C. histryx D.C. (Kefir lime)[9]	Rutaceae	Peel	0.72
Cryptocarya massoiya (Oken)Kosterm.(Masoi) [10]	Lauraceae	Bark	1.40
*Illicium verum*Hook F.(Star anise) [11]	Illiciaceae	Flower	2.11
Litsea cubeba (Lour) Pers.(Krangean) [12]	Lauraceae	Bark	0.73
Nicolaia speciosa Horan (Torch Ginger) [13]	Zingiberaceae	Flower	1.34
Ocimum basilicum L.(Basil) [14]	Lamiaceae	Leaf	1.53

IOP Conf. Series: Materials Science and Engineering **172** (2017) 012022 doi:10.1088/1757-899X/172/1/012022

Table 2 demonstrated the strength of antibacterial activity at Minimum Inhibitory Concentration (MIC). Low (L) if the Inhibitory Area Diameter (IAD) less than 0.7 cm; Medium (M) the IAD 0.7-0.8 cm; and Strong (S) the IAD > 0,8 cm [15]. There are three big groups of essential oils have antibacterial activity in medium and strong levels, the first group has MIC less than 100 µg, the second group has MIC 100 - 1000 µg; and the last group with MIC > 1,000 µg. The higher the concentration of the essential oil, the lower the antibacterial activity obtained. Limo, kefir lime, star anise and torch ginger all these species usually used in food, they have antibacterial activity with MIC > 1,000 µg, whereas masoi and krangean usually used in traditional herb medicinal showed a medium to strong antibacterial activity with MIC less than 100 µg. All the essential oils samples indicated broad antibacterial spectrum because of its show antibacterial activity against Gram- positive and negative bacteria. Figure 1 showed the Inhibitory Area Diameter (IAD) as a clear spot around the paper disc.

Table 2. Minimum Inhibitory Concentration, Strength of Essential Oil Antibacterial Activity

		Minimum Inhibitory Concentration (MIC) µg.					
Scientific name	**Part used**	***B. cereus* FNCC 0057**	***B. subtilis* ATCC 6051**	***S. aureus* FNCC 0047**	***P. aeruginosa* FNCC 0063**	***P. cepasia* FNCC 0063**	***E. Coli* 0091 IFO**
Citrus amblycarpa(Hassk.) Ochse (JerukLimau)[9]	Peel	-	2,500(M)-5,000 (S)	-	2,500(M) -5,000(S)	-	2,500(M)-5,000(S)
*Citrus histryx*D.C (JerukPurut)[9]	Peel	-	2.500(M)-5.000 (S)	-	2.500(M)-5.000(S)	-	5.000(M)-7.500(S)
Cryptocarya massoiya (Oken) Kosterm (Mesoyi)[10]	Bark	-	30(M)-80(S)	-	-	30(M)-120(S)	30(M)-80(
*Illicium verum*Hook. F.(AdasBintang)[11]	Flower	-	3,000(M)-4,000(S)	3,000(M)-3,500(S)	4,000(M)-6,000(S)	-	3,500(M)-5,000(S)
Litsea cubeba(Lour) Pers.(Krangean)[12]	Bark	75(M)-100(S)	-	75(M)-100(S)	>200(S)	-	100(M)-150(S)
*Nicolaia speciosa*Horan (Kecombrang)[13]	Flower	-	1,700(M)-2,100(S)	-	-	-	1,700(M)-2,100(S)
*Ocimum basilicum*L. (Kemangi)[14]	Leaf	-	2,500(M)-5,000(S)	2,500(M) 5,000(S)	-	-	7,500(M)-10,000(S)

Note : – no test (M) : medium (IAD 0,7-0,8cm) (S) : Strong (IAD > 0,8 cm)

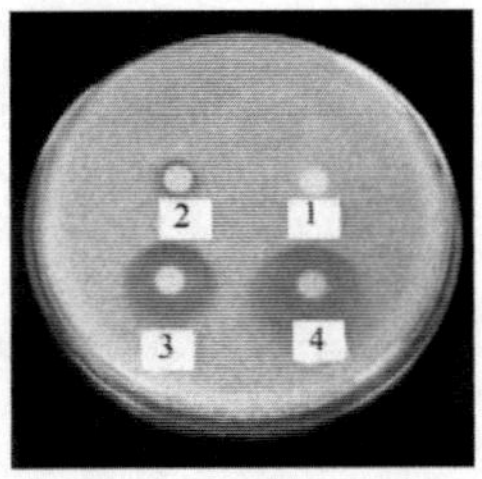

Figure1. MIC Essential oil of Limo peel Against *B. subtilis* ATCC 6051 : 1. Negative control, paper disc and solvent, 2. Limo essential oil 2,500 µg, 3. Limo essential oil 5,000 µg, 4. Tetracyclin

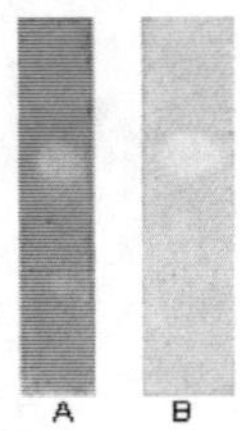

Figure 2. Bioautography test: A.Antibacterial compound in white spot. B.The isolated antibacterial compound was Rechecked.

Table 3 informed five main compounds of each essential oil.The essential oil of Citrus group has many similar compounds e.g.limonene, pinene, terpineol, however, the essential oil content is different. The main compound of essential oil and its amount are responsible for the characteristic aroma of the plant. Beside five main compounds, there are a lot of different compounds which is synergy to give antibacterial activity.

Table 3 : Five Main Compounds of each Essential Oil

No	Scientific name	5 main Compounds
1	*Citrus amblycarpa*(Haask.) Ochse (Limo), peel	limonena 26.85 %, β-pinena 20.52 %, Sitronellal 11.69 %, α-terpineol 10.68 %, Geranilasetat 6.91 %.
2	*C. histryx*D.C(kefir lime), peel	β – pinena 35.00 %, sitronelal 18.24 % , terpineol 8.11 % , limonena 6.29 %, geranilasetat 6.1 %.
3	*Cryptocarya massoia*(Oken) Kosterm (Mesoyi/masoi), bark	5,6-*dihydro-6-pentil-2H-pyran-2-one* 80.51 %,*1,2-benzene dicarboxylic acid, bis(2-ethylhexyl) ester* 6.99 %,*benzybenzoat*5.64%,*ethan, 1,1-dicyclopentyl* 1.52 %, nonana 1.34%.

IOP Conf. Series: Materials Science and Engineering **172** (2017) 012022 doi:10.1088/1757-899X/172/1/012022

4	*Illicium verum* Hook F. (star anise), flower	anetol 79.65 %, Metil kavikol 6.91% , fenikulin 4.01 %, limonene 3.83 %, linalool 1.10 %.
5	*Litseacubeba*(Lour.) Pers. (Krangean), bark	1-8 sineol 29.87 %, sitronelal 17.59 %, 1-limonen 10.74 %,Isopulegol 10.70 %, neoiso (iso) pulegol 5.12 %
6	*Nicolaia speciosa* Horan (Kecombrang), flower	1-dodecanol 30.26 %, dodecanal 26.17%, dodecanoic acid 18.96%. ester dodecyl 7.84 %, decanal 2.69 %,
7	*Ocimum basilicum* L. (Kemangi), leaf	Z-sitral 76.35 %, cis-α- bisabolene 5.14%, *trans-caryophyllene* 3.05 %, trans-caran 4,5-epoksi 2.09 %.

4. Conclusions

According to IAD all of seven essentials oil samples have antibacterial activity in the range of medium to strong level. Especially, Masoi and Krangean showed smaller doses in compared to another, and each essential oil has different dominant compound.

References

[1] Muhtadin AF, Ricky W, Pantjawarni P and Mahfud 2013 Pengambilan Minyak Atsiri dari Kulit Jeruk Segar dan Kering dengan Menggunakan Metode Steam Distillation *Jurnal Teknik Pomits* **2** **1** 2301-9271 ISSN 2337-3539

[2] Burt S. 2004 Essential oils: their antibacterial properties and potential application in foods – a review International. *Journal of Food Microbiology* **94** 223-253.

[3] Sastrohamidjojo H 2004 Kimia Minyak Atsiri 248 p

[4] Hamburger MO, Cordell JA. 1987 A Direct Bioautographic TLC Assay for Compounds Possessing Antibacterial activity *J. Nat. Prod* **50** **1** 19–22

[5] Gundidza M. 1993 Antifungal Activity of The Essential Oil from *Artemisia afra* Jacq *Cent. Afr. J. Med* **39** **7** 140-142

[6] Steel R G D, Torrie J H. 1981 Principles and Procedures of Statistic Biometrical Approach 2nd Japan Mc Graw Hill International Book Cop 633

[7] Elsner P, Maibach I H 2005 Cosmeceuticals and Active Cosmetics 2nd Ed Taylor and Francis New York 675

[8] Schrader K, Domsch A 2005 Cosmetology-Theory and Practice **3** 62 Verlag fur Chemische Industrie Augsburg

[9] Soetjipto H, Sulihingsih E and Dewi L 2010 Komposisi Kimia dan Efek Antibakteri Minyak Atsiri Kulit Buah Jeruk Purut (*Citrus hystrix* D.C) dan Jeruk Limau (*Citrus amblycarpa* Hassk) *Prosiding Seminar Nasional Kimia II Bervisi SETS* HKI Jawa Tengah Semarang 117-123

[10] Soetjipto H, Martono Y and Fitriani IL 2011 Identifikasi Senyawa dan Aktivitas Antibakteri Minyak Atsiri Kulit Masoi (*Cryptocarya massoia* (Oken) Kosterm) *Prosiding Seminar Nasional Kimia dan Pendidikan Kimia III* Fakutas Keguruan dan Ilmu Pendidikan Universitas Sebelas Maret Surakarta 582-588

IOP Conf. Series: Materials Science and Engineering **172** (2017) 012022 doi:10.1088/1757-899X/172/1/012022

[11] Widiastuti P, Soetjipto H and Hastuti SP 2007 Aktivitas Antibakteri dan Identifikasi Komponen Minyak Atsiri Adas Bintang (*Illicium verum* Hook F) *Prosiding Seminar Nasional Aplikasi Sains dan Matematika* Fakultas Sains dan Matematika Universitas Kristen Satya Wacana Salatiga 198-204

[12] Soetjipto H, Fajarwati D, Timotius KH 2007 Identifikasi Komponen dan Aktivitas Antibakteri Minyak Atsiri Kulit Batang Krangean (*Litsea cubeba* Pers.). *Prosiding Seminar Nasional Aplikasi Sains dan Matematika*. Fakultas Sains dan Matematika Universitas Kristen Satya Wacana Salatiga 190-197

[13] Soetjipto H, Hastuti SP and Kristanto O 2009 Identifikasi Senyawa Antibakteri Minyak Atsiri Bunga Kecombrang (Nicolaia Speciosa Horan). *Prosiding Seminar Nasional Sains dan Pendidikan Sains IV. Fakultas Sains dan Matematika* Universitas Kristen Satya Wacana Salatiga 640-655

[14] Ristanti WF, Soetjipto H and Timotius KH 2006 Identifikasi Citral dan Aktivitas Antibakteri Minyak Atsiri Daun Kemangi (*Ocimum basilicum* Linn) *Prosiding Seminar Nasional Kimia dan Pendidikan Kimia* Jurusan Kimia Fakultas MIPA- Universitas Negeri Semarang 67

[15] El Gayyar M, F A Draughon, D A Golden and J R Mount 2001 Antimicrobial Activity of Essential Oils Plants Against Selected Pathogenic and Saprophytic Microorganism *Journal of Food Protection* **64 7** 1019-1024

The Application of Red Pigments from *Streptomyces* K-4B and Dayak Onions (*Eleutherine palmifolia* (L.) Merr.) In Colouring Glycerine Soap

Herlina, A Asnani[*)], H Diastuti

Department of Chemistry, Faculty of Mathematic and Natural Sciences, Universitas Jenderal Soedirman, Jl. Dr. Soeparno, Karangwangkal 53123, Purwokerto, Central Java, Indonesia

[*]E-mail: ari.asnani@unsoed.ac.id

Abstract. Glycerin soap has been colored with red pigment from *Streptomyces* K-4 B and Dayak onion (*Eleutherine palmifolia* (L.) Merr). Both red pigments from *Streptomyces* K-4B and Dayak onion were extracted with ethanol by maceration method, followed with soxhlet extraction. The concentration of red pigment added was varied (0, 200, 300, 400 μL) to evaluate the best product. The resulted glycerine soaps were characterized and analyzed based on SNI 06-3532-1994. The research results indicated that the glycerine soap has water content ranged from 0.36% to 12.56%; the amount of fatty acid ranged from 14% to 36.75%; the amount of free fatty acids ranged from 0% to 0.37%; the non-saponifiable fat ranged from 0.001 to 0.019%; the pH ranged from 10.33 to 11.06; the foam stability ranged from 0.61% to 89.09%. The results of analysis of variance showed that the effect between treatments significantly different at 95% confidence level ($\alpha = 0.05$) on the characteristics of glycerine soap. The results of an organoleptic test with parameters observed were color, aroma, texture, foam, rough impression upon usage and rough impression after usage, gave "like to very like soap" with a maximum score of 4.67 (1 to 5 scale). Based on the color assessment, the organoleptic panelists preferred the glycerine soap of SK-4B3 (red pigment from *Streptomyces* K-4B, 200 μL) with the score of 4.30 (like to very like).

Keywords: Streptomyces, Dayak onion, glycerine soap, pigment,

1. Introduction

Glycerine soap is a translucent soap which has shiny appearance when compared with other regular soaps. Colouring glycerine soap becomes a necessity to increase consumer's interest for the product. The color itself can be originated from synthetic or natural dyes. Natural dyes are natural colors or pigments obtained naturally from microbes, plants, animals, or mineral sources. Asnani and Ryandini [1] have isolated 26 marine actinomycetes from the mangrove area in Segara Anakan Cilacap, Central Java. S*treptomyces* K-4B, as one of the isolated indigenous marine actinomycetes, appears to express red pigment produced by its mycelium. Furthermore, the ethanol extract of mycelium from *Streptomyces* K-4B recently has been reported to have antibacterial activity [2]. Interestingly, the ethanol extracts from Dayak onions (*Eleutherine palmifolia* (L.) Merr.), which is also an indigenous plant from Kalimantan, also contain beautiful red pigment. Harlita [3] has reported the various antibacterial activities from hexane, ethyl acetate, and ethanol extracts of Dayak onions. Based on those findings, both red pigments from indigenous *Streptomyces* K-4B and Dayak onions can be considered as potential natural pigments with antibacterial activities.

Due to the fact that glycerine soap needs colouring to attract consumers, thus the aims of the research were (1) to apply both red pigments from *Streptomyces* K-4B and Dayak onions to color glycerine soap, (2) to characterize the quality of glycerine soap colored with red pigments from *Streptomyces* K-4B and Dayak onions based on SNI 06-3532-1994; and (3) to evaluate the consumers preference towards the color of glycerine soap using red pigments from *Streptomyces* K-4B and Dayak onions. Characterizations of glycerine product were water content, a number of fatty acids, a number of free fatty acids, the nonsaponifiable fat, the pH, and the foam stability. The consumer preference was evaluated using organoleptic tests based on color, aroma, texture, and the impression of the glycerine soap before and after used application.

2. Materials and Methods

The main materials used in this study were *Streptomyces* K-4B, Dayak onions (*Eleutherine palmifolia* (L.) Merr.), and coconut oil (Barco) The cultivation of *Streptomyces* K-4B used Starch-casein nitrate media. The extraction of red pigments used 96% ethanol. The chemicals used for making glycerine soap were NaOH, white sugar (*Gulaku*), stearate acid, 70% ethanol, and glycerine. The chemicals used for soap characterizations were 20% H_2SO_4 (v/v), n-hexane, anhydrous sodium sulfate, KOH, 0.5N HCl, methyl red, and phenolphthalein indicators.

The research was conducted using Factorial Completely Randomized Design. Factors to be tested were two type of red pigments (W_1 = red pigment from *Streptomyces* K-4B; W_2 = red pigment from Dayak onions), and three concentrations of pigments (K_1 = 400 mL; K_2 = 300 mL; K_3 = 200 mL). The control group was treated without any addition of pigment. Each treatment was repeated three times, so there were 18 experimental units. The independent variables were types and concentrations of pigments. The dependent variable was the glycerine soaps produced using the pigments. The parameters measured were (1) water content, (2) a number of fatty acids, (3) a number of free fatty acids, (4) the nonsaponifiable fat, (5) pH value, (6) the stability of the foam, and (7) organoleptic preferences.

2.1. Pigment production from Streptomyces K-4B

The pigment production from *Streptomyces* K-4B was carried out following the procedure reported by Asnani, *et.al.* [4]. The pure culture of *Streptomyces* K-4B was cultivated in Starch-casein nitrate (SCN) medium for eight days which then served as an inoculum. As many as 10% of inoculum was subsequently cultivated in 100 mL of SCN, then incubated for 21 to 28 days. After incubation has completed, the pH was lowered to pH 2 with 2 N HCl, and the mycelium was separated using cold-centrifuge. The mycelium obtained was soxlet extracted with 96% ethanol, and then the red filtrate obtained was evaporated to yield concentrated red pigment.

2.2. Extraction of red pigment from Dayak onions

Extraction of red pigment from Dayak onions was performed using known maceration procedure [3]. A total of 100 grams of powder Dayak onions was macerated in 1.000 mL of 96% ethanol (1:10 w/v) for 3 x 24 hours with occasional stirring. After three days, the mixture was filtered, and the red filtrate obtained was evaporated to give concentrated red pigment from Dayak onions.

2.3. The making of glycerine soap

Production of glycerine soap was based on modified transparent soap formula from Tokosh and Mercer [5]. 60 grams of coconut oil was mixed with 20 grams of stearic acid then heated at a temperature of 70 °C until stearic acid completely dissolved. Saponification reaction was carried out by adding NaOH solution (13 g NaOH/ 27 g distilled water) into the solution of the oil-stearate acid. The mixture was stirred until saponification has completed as indicated by soap-trace formation.

After completion of saponification, a solution of sucrose (30 g sucrose/30 mL distilled water) was added with continuing heating until the mixture became homogeneous. Translucent appearance was obtained by adding around 6 grams of glycerin and 20 grams of ethanol with slow stirring. After that, the two types of red pigment with various concentrations were added to research treatments. The soap mixture was then poured into soap molds, covered with linen fabrics, and let them dried at room temperature.

2.4. Characterization of glycerine soap

The glycerine soaps produced were characterized using standard methods [6]. Water content was determined by weighing approximately 4 grams of soap (W) and placed in a weighed cup. The soap sample along with the cup (W_1) was heated in an oven with temperature of 100-105 °C until the weight remained constant (W_2). The calculation of water content was determined by equation (1).

$$\text{Water content} = \frac{(W_1\text{-}W_2)}{W} \times 100\% \qquad (1)$$

A number of fatty acids were determined by extraction method [6]. 10 grams of soap sample was diluted with 50 mL of distilled water, added three drops of methyl red indicator. 20% of H_2SO_4 solution was added until all fatty acids freed from sodium. The solution was then extracted with n-hexane. The extraction process was repeated until the solvent amounts reached approximately 100 mL. The n-hexane filtrates were combined, dried over anhydrous sodium sulfate, then filtered to remove anhydrous sodium sulfate. The filtrate obtained was evaporated in a weighed round bottom flask (W_1) at a temperature around 102-105 °C until the weigh remained constant (W_2). Fatty acid content was calculated by equation (2).

$$\text{Fatty acids content} = \frac{(W_2\text{-}W_1)}{\text{Sample weigh}} \times 100\% \qquad (2)$$

A number of free fatty acids were measured by titration method [6]. 100 mL of ethanol with 0.5 mL phenolphthalein indicator was boiled at a temperature of 70 °C, then neutralize with 0.1 N alcoholic KOH solution. After that, around 5 grams of soap sample (W) was added, then the mixture was refluxed for 30 minutes. After the mixture was no longer alkaline, the solution was titrated with 0.1 N alcoholic KOH solution (N) until the appearance of red color remained for at least 15 seconds (V in mL). Free fatty acid content was determined by equation (3). Multiplication by 0,205 is equivalent with the weight of lauric acid.

$$\text{Free fatty acids content (\%)} = \frac{\text{V} \times \text{N} \times 0{,}205}{\text{W}} \text{ x } 100\% \qquad (3)$$

The nonsaponifiable fat was measured by calculating the remaining solution from free fatty acids determination above. The remaining solution from the above procedure was added 5 mL of 0.5 N alcoholic KOH, then refluxed for 1 hour. After refluxing process, the solution was titrated with 0.5 N alcoholic HCl (N) until red color appeared (V_2). Titration was also carried out for blank with 0.5 N alcoholic KOH (V_1). The saponified fats can be calculated using equation (4). The value of 0.0561 is equivalent to the weight of KOH; 0.1981 is the saponification value of average oil.

$$\text{Nonsaponifiable fat content} = \frac{(\text{V2 - V1}) \times \text{N} \times 0{,}0561}{0{,}1981 \times \text{W}} \text{ x } 100\% \qquad (4)$$

The pH value was measured by dissolving 1 gram of soap into 10 mL of distilled water, then measured the pH of the soap solution [7].

The stability of foam was measured by dissolving 1 g of soap into 10 mL of distilled water, shook vigorously for 1 minute, then measured the foam height. The soap solution was allowed to stand for 60 minutes, and then the foam height was measured again [8]. The foam stability was calculated following the equation (5).

$$\text{Foam stability } (\%) = \frac{\text{foam height after 60 min}}{\text{initial foam height}} \times 100\% \qquad (5)$$

2.5. Organoleptic test of glycerine soap

An organoleptic test was conducted using 15 panelists who were randomly selected based on the difference between the sexes (male and female).The soap samples were blindly given to all panelists to know their preference based on color, aroma, texture and the impression of the glycerine soap before and after used application.

2.6. Data analysis

Resulted data from soap characterization were analyzed using two-ways ANOVA (Analysis of Varian) with a 95% confidence level ($\alpha = 0.05$). The resulted analysis which showed significant diversity was further analyzed using DMRT (Duncan's Multiple Range Test) with a 95% confidence level ($\alpha = 0.05$). The organoleptic tests were analyzed using data scoring from 1 to 5.

3. Results and Discussion

3.1. Pigment from Streptomyces K-4B

Actinomycetes are microorganisms that originally have the antibiotic potential. Excitingly, some of the actinobacteria express both extracellular and intracellular pigments with various colors. It was observed that *Streptomyces* K-4B has antibacterial activity and also produced red mycelium. Thus, in this study, the ability of *Streptomyces* K-4B (figure 1) to produce natural red pigment was thoroughly explored.

Streptomyces K-4B was cultivated in *starch-casein* nitrate agar supplemented with nystatin. The fermentation process for pigment production and ethanol extraction of pigment were done following the similar procedure reported by Asnani *et.al.* [4]. For mass extraction of mycelium, the soxhlet extraction method using ethanol 96% was applied. The ethanol extract was then evaporated to give concentrated red pigment (52.37%).

(a) (b) (c)

Figure 1. *Streptomyces* K-4B: (a) the colony, (b) ethanol extract, and (c) concentrated red pigment

3.2. Pigment from Dayak onions

Bawang Dayak (*Eleutherine palmifol*ia (L.) Merr) is one of the indigenous plants from Kalimantan [9]. In Indonesia, this plant is also known as the Mecca onions, garlic ghost, and Sabrang Arab garlic.

Local people in the area have used Dayak onions as traditional medicine. The part of Dayak onions that can be utilized is its bulb (figure 2a). Specific features of this plant are red bulb with a very smooth surface, leaves paired with double-finned leaf composition, and small white flowers. An extraction of red pigment from Dayak onions is easily done by maceration method using ethanol 96% as extraction solvent. The ethanol extract is then evaporated to obtained concentrated extract of red pigment which was readily used as the red pigment for colouring glycerine soap.

Figure 2. Dayak onions (*Eleutherine palmifolia* (L.) Merr), (a) the dried bulb, (b) the ethanol extract, and (c) the concentrated ethanol extracts.

3.3. Glycerine soap products

According to the Indonesian National Standard [6], bath soap is a compound of sodium or potassium fatty acids from vegetable oil or animal lipid solid, soft or liquid form, foaming used as a cleanser, and with the addition of fragrances or other non-harmful ingredients. The standard quality requirement of soap is in accordance with SNI 06-3532-1994. The term glycerine soap is based on the translucent apparent of the soap due to the addition of glycerine to the formula. The process of making glycerine soap is begun with heating the oil at a temperature of 70 °C and then added NaOH solution slowly. The mixture is stirred thoroughly until saponification process completed. The glycerine and alcohol are then added until the soap becomes transparent.

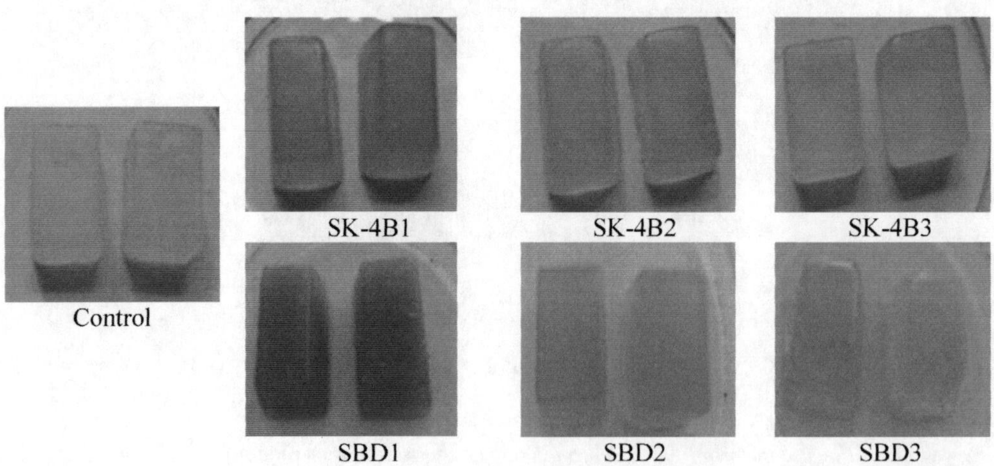

Figure 3. Glycerine soap products with red pigments from *Streptomyces* K-4B and Dayak onions

IOP Conf. Series: Materials Science and Engineering **172** (2017) 012023 doi:10.1088/1757-899X/172/1/012023

Color of glycerine soaps will depend on the choice of starting oil material. If the quality of oil used is poor, then it is likely to produce yellow soap [10]. In this research, the glycerine soaps were colored with various concentrations of natural red pigments from *Streptomyces* K-4B and Dayak onions. Control soap has a solid white to transparent, whereas soaps with the addition of pigments from *Streptomyces* K-4B and Dayak onions has increasingly reddish color with the increasing concentrations of red pigments (figure 3).

The coding used for glycerine soap products were as follow:

SK-4B1 = soap with 400 µL of pigment from Streptomyces K-4B
SK-4B2 = soap with 300 µL of pigment from Streptomyces K-4B
SK-4B3 = soap with 200 µL of pigment from Streptomyces K-4B
SBD1 = soap with 400 µL of pigment from Dayak onions
SBD2 = soap with 300 µL of pigment from Dayak onions
SBD3 = soap with 200 µL of pigment from Dayak onions

The characteristics evaluation of glycerine soaps is presented in Table 1. The water content indicates the amount of water contained in the product which will greatly affect the quality of the shelf life of the soap. Soap with high levels of water will have a lower shelf life because it gives an opportunity for the growth of microorganisms. In this research, the water content of glycerine soap products ranged from 0.36% to 12.56%. Those results indicated that all glycerine soaps products had met the requirement for water content when compared to SNI 06-3532-1994 (<15%).The lowest water content was SBD3 (0.36%) which contained 200 µL of pigment from Dayak onions. The highest water content was SK-4B2 (12.56%) which contained 200 µL of pigment from *Streptomyces* K-4B. The results from analysis of variance showed that the addition of pigment gave significant affected on water content at 95% confidence level ($\alpha = 0.05$). Further Duncan test showed that glycerine soaps SK-4B1, SK-4B2, SBD1, SBD2, and SBD3 were significantly different from that of SK-4B3.

A number of fatty acid is the total fatty acids contained in the soap. Both fatty acids occur as saponifiable or nonsaponifiable fats. Naturally, fatty acids are obtained by hydrolysis of the triglycerides [10]. The principle of measurement of a number of fatty acids is hydrolysing the bond between the fatty acids with Natrium (Na) on the soap using strong acids.

In this study, the average amount of fatty acids ranged from 14% to 38.5%. The lowest amount of fatty acid was SBD3 (14%), whereas the highest amount of fatty acids was SK-4B2 (38.5%) which was glycerine soap with 300 µL pigment from *Streptomyces* K-4B. However, those results have not met the standard requirement of SNI 06-3532-1994 (>70%), yet. The decreasing amount of fatty acid was probably due to the increasing amount of water used. The amount of fatty acid also has a tendency to decrease with the increasing stirring time and the ratio of water [11]. In addition, the amount of fatty acid in the soap product highly depended on the type of oils or fats used for soap production.

Table 1. The average characteristics value of glycerine soaps

Glycerine Soap	Water content (%)	Fatty Acids (%)	Free Fatty Acids (%)	Nonsaponifiable lipid (%)	pH	Foam Stability
Control	0.36	29.40	0.00	0.001	11.06	79.16
SK-4B1	0.59	35.00	0.37	0.015	10.02	85.87
SK-4B2	12.56	38,50	0.30	0.010	10.32	89.09
SK-4B3	0.87	32.25	0.16	0.010	10.35	82.16
SBD1	0.45	36.75	0.00	0.002	10.72	0.61
SBD2	0.36	17.50	0.14	0.002	10.74	0.70
SBD3	0.38	14.00	0.00	0.018	10.33	0.78
SNI	<15%	>64%	<2,5%	<2.5%	-	-

In glycerine soaps, there are possibilities that these soaps contain free fatty acids or free-alkaline. Free fatty acids are fatty acid contained in the soap but are not bound as soap or triglycerides. High contain of free fatty acids in soap will reduce the cleaning power of the soap against oil or grease. Alkaline-free is the amount of alkaline in the soap which is not bound to a compound. Excess alkaline in the soap can cause irritation to the skin. Analysis of free fatty acid aims to determine the amount of free fatty acid or free-alkaline contained in soaps.

The analysis of free fatty acid showed that glycerine soap with the addition of pigments from *Streptomyces* K-4B and Dayak onions contained free fatty acids ranged from 0% to 0.37%. These findings indicated that all soap products has free fatty acid less than 2.5%, which met the standard requirement based on SNI 06-3532-1994. The results from analysis of variance showed that the addition of pigment significantly affected the value of free fatty acids at 95% confidence level (α = 0.05). Further Duncan test showed that the type and concentration of pigment from *Streptomyces* K-4B (SK-4B1 and SK-4B2) were significantly different from SK-4B3 and SBD2.

Nonsaponifiable lipids are triglycerides or other fats in the soap that cannot be saponified by alkaline lye. The nonsaponifiable lipid will reduce the ability of soap to clean the dirt because the soap will tend to bind these nonsaponifiable lipids of the soap itself. The analysis gave the value for nonsaponifiable lipid ranged from 0.010% to 0.018%. The highest content of nonsaponifiable lipid was obtained from SBD3 (0.018%), whereas the lowest was 0.010% from both SK-4B2 and SK-4B3. All soap products had met the standard requirement for nonsaponifiable lipid (<2.5%) according to SNI 06-3532-1994. Based on the analysis of variance, the addition of the pigment concentration significantly affected the value of nonsaponifiable lipid at 95% confidence level (α = 0.05). Further Duncan test showed that SK-4B1, SK-4B2, SK-4B3, SBD2, and SBD3 were significantly different towards each other,

The pH value of the soap can be influenced by the content of free alkaline and other raw materials. The pH value of soap, whether it is too low or too high, can cause skin irritation. Normal skin has a pH of about 5 and it will increase if the skin is washed with soap solution. However, the increment in the pH value on the skin will not exceed to pH 7 [12]. In this research, the soap obtained has pH ranged from 10.02 to 10.74. The highest pH value of 10.74 was SBD2, and the lowest value were SK-4B1.Based on the results of analysis of variance, the addition of pigments significantly affected the pH on the 95% confidence level (α = 0.05). Further Duncan test showed that SBD3 was significantly different with SK-4B1 and SBD3, as well as with SBD1 and SBD2.

Foam is the dispersed gas in the fluid with the relatively stable structure consisting of air pockets encased in a thin layer [12]. Foam stability test is conducted by measuring the decrease in volume of foam after the soap solution is shaken and allowed to stand for 1 hour. Based on the measurement, the foam stability ranged from 0.61% to 89.09%. The highest value of the foam stability was SK-4B2 (89.09%), and the lowest value was SBD1 (0.61%). Based on the analysis of variance, the type of pigments significantly affect the foam stability on the 95% confidence level (α = 0.05). Further Duncan test showed that the increasing concentration of pigment has no effect, but the type of pigments (*Streptomyces* or Dayak onions) significantly affected the foam stability.

The tendency of the value of foam stability decreased with the increasing concentration of added pigments. The foaming capability is correlated with surface tension. The lower of the surface tension of soap solution, the greater the resulting foam. Low surface tension might occur because of the attraction of surface molecules are weak, and it will easily disturb the surface of the coating solution, hence increasing the ability for foam formation [7]. The addition of pigment eventually decreases the foam stability. This happens because of the presence of dissolved substances in the liquid will increase viscosity which in turn will increase surface tension and result in reduced foaming ability.

3.4. Consumers preference

Consumer preference towards a certain product begins with the assessment of the appearance, flavor, and texture of the product itself. In the end, the target is consumer acceptance to particular product. Consumer preference can be evaluated using an organoleptic test that uses human senses to measure the color, aroma, texture, appearance, and impression of soap products. To do so, the organoleptic test employs panelists who are asked to evaluate the characteristics of the product by the level of assessments that are very disliked (1), disliked (2), average (3), like (4), and very like (5).

In this research, the organoleptic test was a test for favorite or hedonic. The glycerine soap products to be evaluated were SK-4B1, SK-4B2, SK-4B3, SBD1, SBD2, SBD3, and control soap (without addition any pigment). The hedonic test was conducted to determine the level of consumer acceptance of products based on the characteristics of the standard soap (SNI 06-3532-1994). The test involved 15 panelists (eight women and seven men), who evaluated the soap characteristics which were color, aroma, texture, foam, and usage impression. The research results were presented in Table 2.

Color assessment was done visually. Increasing concentration of pigment gave significant effect on colour soap produced. Based on the color, panelists preferred SK-4B3 soap with an average value of 4.3 (like to very like). SK-4B3 soap was glycerine soap with the addition of 200 μL pigment from *Streptomyces* K-4B.

Aroma assessment was done by smelling the scent of soap products. Since the soap was originally made from coconut oil, the distinctive scent from coconut oil also added to the soap aroma. Apparently, an addition of pigment from Dayak onions could overcome the odor from coconut oil, whereas the pigment from *Streptomyces*-4B contributed no aroma at all. Hence, the panelists favored SK-4B3 and SBD1 products with the average value of 3.87 (average to like). SK-4B3 was glycerine soap with addition 0f 200 μL pigment from *Streptomyces*-4B, whereas SBD1 was added 400 μL of pigment from Dayak onions.

Texture assessment was done by touching and feeling the texture of the products. Evaluation results showed that panelists preferred SBD1 soap, which was glycerine soap with the addition of 400 μL of pigment from Dayak onions. The texture average value was 4.67 (like to very like).

Foam assessment was done by washing hands with the soap products. The amount of foam produced during the washing was observed by the panelists. The average ratings from panelists for resulted foam designated the SK-4B3 as the most favored soap with the value of 3.67 (average to like). SK-4B3 was glycerine soap with the addition of 200 μL pigment from *Streptomyces* K-4B.

Assessment of the utilization of glycerine soap was done by washing hands with soap products. The impression on the skin was observed during and after applying the products. The results indicated that panelists preferred SBD1 soap for impression during washing, but preferred SBD3 soap for impression after washing. SBD was the soap with the addition of pigment from Dayak onions. The application of glycerine soap SBD1 gave the average value of 3.97, whereas the application of SBD3 gave that of 4.37.

Table 2. The results from hedonic tests

glycerine soap	color average score	aroma average score	texture average score	foam average score	on usage average score	after usage average score
Kontrol	3.80	3.0	3.73	3.60	3.60	3.67
SK-4B1	3.53	3.53	4.06	3.60	3.80	3.87
SK-4B2	3.87	3.40	3.93	3.40	3.87	4.13
SK-4B3	4.30	3.87	4.06	3.67	3.73	4.13
SBD1	3.67	3.87	4.67	3.40	3.97	4.00
SBD2	3.86	3.64	3.86	3.36	3.50	4.00
SBD3	3.73	3.40	3.97	3.40	3.93	4.37

4. Conclusions

Red pigments from *Streptomyces* K-4B and Dayak onions can be used to color glycine soap. The soap products mostly met the standard requirement based on SNI 06-3532-1994. The water content ranged from 0.36% to 12.56% (SNI> 15%), the number of acid fatty acids ranged from 14.00% to 36.75% (SNI> 70%), the number of free fatty acid ranged from 0% to 0.37% (SNI <2.5%), the amount of nonsaponifiable fat ranged from 0.001% to 0.019% (SNI <2.5%), pH value ranged from 10.02 to 11.06, and the foam stability ranged from 0.61% to 89.09%. The organoleptic tests indicated that the most preferred color was SK-4B3, aromas were both SK-4B3 and SBD1, the texture was SK-4B3, the foam was SK-4B3, impression on usage was SBD1, and impression after usage was SBD3. Above all, the most preferred glycerine soap was SK-4B3 (addition of 400 µL pigment from *Streptomyces* K-4B) and SBD3 (addition of 100 µL pigment from Dayak onions).

Acknowledgment
The authors greatly acknowledge Indonesia Higher Education for providing research grant under *Strategis Nasional* Scheme in 2016. The authors also appreciate the Universitas Jenderal Soedirman, particularly the Faculty of Mathematics and Natural Sciences for kind supports and laboratories facilities.

References

[1] Asnani A and Ryandini D 2011 Screening of Marine Actinomycetes from Segara Anakan Indonesia for Antimicrobial Activity Proceeding ICONS 2011 Shaker Verlag Germany ISBN 978-38440-1403-7 ISSN 1434-5536.

[2] Prakoso R D 2016 Antibacterial Activities of Ethanol Extract from Mycellium actionmycetes K-2C and K-4B against *Escherichia coli* and *Staphylococcus aureus* on Different Incubation Time *Skripsi* (unpublished) Biology Faculty Universitas Jenderal Soedirman Purwokerto Indonesia.

[3] Harlita T D 2017 Antibacterial Activity of Dayak Onions (*Eleutherine palmifolia* (L.) Merr.) Againts Pathogenic Bacteria *Thesis* (unpublished) Biology Faculty Universitas Jenderal Soedirman Purwokerto Indonesia

[4] Asnani A, Ryandini D and Suwandri 2016 Screening of Marine Actinomycetes from Segara Anakan for Natural Pigment and Hydrolytic Activities In IOP Conference Series: Materials Science and Engineering **107 (1)** 012056 IOP Publishing.

[5] Tokosh R and Mercer A 1995 Transparent Soap Formulations and Methods of Making Same U.S. Patent No. 5,417,876 23 May 1995.

[6] Indonesia-SNI, S N 1994 Nomor 06-3532-1994 Tentang Sabun Mandi Padat. Badan Standarisasi Nasional. Jakarta.

[7] Retnowati D S, Kumoro A C and Budiyati C S 2014 Preparation and Characterization of Milk Soap with Cold Process *Jurnal Rekayasa Proses* 7 **(2)** 45-50.

[8] Hui Y H 1996 Industrial and Consumer NonEdible Products From Oils and Fats Bailey's Industrial Oil and Fat Products John Wiley & Sons Inc New York, NY.

[9] Krismawati A and Sabran M 2004 Management of Plant Genetic Resources for Specific Medicinal Plants in Central Kalimantan *Buletin Plasma Nutfah* **12 (1)** 16-23.

[10] Williams D F and Schmitt W H 2002 *Kimia dan Teknologi Industri Kosmetika dan Produk-Produk Perawatan Diri*. Translation FATETA Institut Pertanian Bogor.

[11] Wijana S, Soemarjo S and Harnawi T 2012 The Study on Liquid Soap Production from Recycled Frying Oil (The Effect of Mixing Time and Water: Soap Ratio on the Quality). *Jurnal Teknologi Pertanian* **10 (1)** 54-61.

[12] Hambali E, Bunasor T K, Suryani A and Kusumah G A 2005 Application of Diethanolamide from Lauric Acid of Palm Kernel Oil in The Production of Transparent Soap *Jurnal Teknologi Industri Pertanian* **15 (2)** 46-53.

Synthetic Studies of Bioactive Substances of 4-Hydroxybenzalhydantoin Derivatives

IW Hidayat[1*], D Sumiarsa[1], M Permatasari[1], AKania[1], Riska[1], L Priani[1]
[1]Department of Chemistry, University of Padjadjaran, Unpad Jatinangor 45363, Indonesia

[*]E-mail: i.wiani@unpad.ac.id or i.wiani@gmail.com

Abstract. 4-hydroxybenzalhydantoin derivatives were synthesized by the condensation reaction between benzaldehydes **12-13** and substituted hydantoins **14-16** under standard conditions of reflux in glacial acetic acid, in the present of sodium acetate and a little amount of acetic anhydride as a catalyst. All compounds were identified by spectral analysis to give 4-hydroxybenzalhydantoins **17-21**.

1. Introduction

Benzalhydantoin or 5-(benzylidene)imidazolidine-2,4-diones) derivatives are important and prevalent classes of biologically active natural and un-natural substances [1-10]. For example, (*Z*)-4-hydroxybenzalhydantoin **1** was isolated from the Red Sea sponge *Laxosubrites sp* but reported not to have any biological activity [11]. Recently, compound **1** was also isolated from the aqueous ethanolic extracts of the Red Sea sponge *Hemimycale arabica*, along with other substituted hydantoins, (*R*)-5-(4-hydroxybenzyl)hydantoin **2**, and (*Z*)-5-((6-bromo-1H-indol-3-yl)methylene)hydantoin **3** [12].

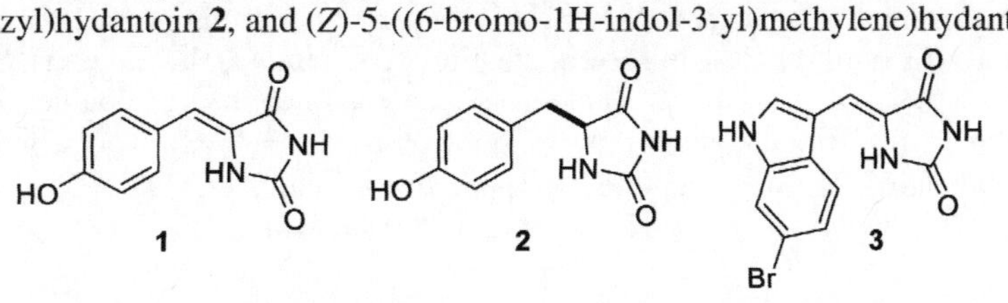

Interestingly, on this occasion, compound **1** was found to have potent *in vitro* anti-growth and anti-invasive properties against PC-3M prostate cancer cells [12]. As a result, thirty-nine other aryl-substituted benzalhydantoins **4-11** were synthesized and examined as novel antimetastatic agents with potential to control metastatic prostate cancer [12].

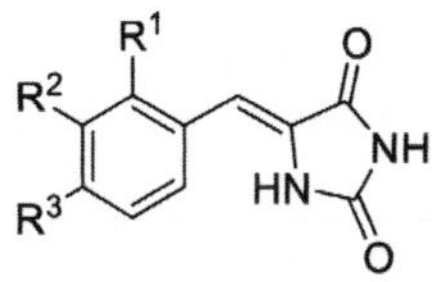

	R^1	R^2	R^3
4	H	H	Et
5	F	H	Me
6	H	Ph	H
7	H	CN	H
8	thiofen-3-yl	H	H
9	thiofen-2-yl	H	H
10	furan-2-yl	H	H
11	Ph	H	H

Benzalhydantoins substituted with alkyl, halogen, trifluoromethyl, and alkoxy groups on the phenyl ring were found to exhibit good anticonvulsant activity [2-12]. Researchers have found benzalhydantoins active against *Mycobacterium tuberculosis* **H37Rv** [14], while, others have reported that benzalhydantoins and 5-alkyl- and 5-arylmethyl-hydantoins, and their 2-thio analogues, have fungicidal and bacterial activity [8,15]. Many benzalhydantoins have been used for aromatic amino acid synthesis. Spectroscopic properties [16] and crystal structures of selected benzalhydantoin derivatives have been studied in order to understand the structure-activity relationships better, especially since *E*/*Z* geometrical isomers are possible because of the restricted rotation of the exocyclic methylidene double bond [17-18].

Our group has prepared a wide range of new [8-10], substituted benzalhydantoins **14-16** as part of a synthetic study toward the tandem approach to the molecules. It occurred to us that these can be viewed as spatial mimics of benzalhydantoin **1**.

2. Experimental

2.1. General

Melting points were measured on a Reichert hotstage microscope and are uncorrected. Ultraviolet spectra were measured on a Hitachi U-3200 spectrophotometer and refer to solutions in absolute MeOH, and data reported as wavelength (λ_{max}) in nm and absorption coefficient (ε_{max}) in $dm^3.mol^{-1}.cm^{-1}$. Infrared spectra were recorded on an FTIR Shimadzu 8400 spectrophotometer. The samples were prepared as neat thin films for liquids and KBr disks for solids. Proton NMR spectra were recorded in designated solvents on an Aligent Varian instrument. Proton spectra were recorded at 500 MHz and data were reported as chemical shift (δ) in parts per million (ppm) downfield from tetramethylsilane (TMS), multiplicity, observed coupling constant (*J*) in Hertz (Hz), and proton assignment. Multiplicities were reported as singlet (s), doublet (d), doublet of doublet (dd), triplet (t), quartet (q) or multiplet (m). Broad signals were reported as br. C-13 NMR spectra were recorded at 125 MHz. Chemical shifts (δ) were reported in ppm downfield from TMS. Mass spectra were recorded on High *Resolution Time of Flight Mass Spectroscopy* (HR-TOF-MS ES), Waters spectrometry.

2.2. Materials

4-hydroxybenzaldehyde **12**, 3-methoxy-4-hydroxybenzaldehyde **13**, imidazolidine-2,4-dione (hydantoin) **14** were obtained from commercial sources and used without purification.
The remaining compounds **15-21** were synthesized.

2.3. Preparation of Hydantoin Derivatives as precursor

2.3.1. Synthesis 3-methylhydantoin **15.**

Following the method of Orazi *et al* [22]. Hydantoin **14** (10.0 g, 0.10 mol) and potassium hydroxide (5.6 g, 0.1 mol) in ethanol (150 mL) and the mixture was stirred until gas evolution ceased (*ca.*15 min). Dimethyl sulphate (10.5mL) was added drop by drop and stirring was continued and the mixtures were refluxed for 5 h. The mixture was then cooled r.t and the precipitate was filtered and the filtrate was evaporated under reduced pressure till 20 mL and then extracted with ethyl acetate (6x20 mL), the combined ethyl acetate extracts were evaporated under reduced pressure and the residue were recrystallized from EtOAc to give 3-methylhydantoin **15** as white needles 44 %, m.p. 152 °C (lit. [22] m.p. 155 °C). Ultraviolet spectra have λ_{max} (MeOH) (ε_{max}): 220.4 (2.5x10), nm. ^{1}H NMR (DMSO-d_6, 500 MHz): δ(ppm) 10.6 (s, 1H, N1); 3.84 (s, 2H, C5); 2.8 (s, 1H, C3); ^{13}C-n.m.r (DMSO-d6)(125 MHz): δ (ppm) 158.3 (C2); 157.8 (C4); 46 (C5); 24(C6).

2.3.2. Synthesis 1-acetylhydantoin **16.**

Hydantoin **14** (1 g, 10 mmol) and acetic anhydride (1 mL, 10 mmol) and the mixture were stirred and refluxed for 5 h. The mixture was then cooled r.t and the precipitate was filtered and washed with ethanol. The residue was recrystallized from EtOAc to give 3-acetylhydantoin **1**as white needles 64 %, m.p. 143°C (lit.[22] m.p. 144 °C). HR-MS-ES$^-$ m/z: found for $C_5H_5N_2O_3$.^{1}H NMR (DMSO-d_6, 500 MHz): δ(ppm) 11.4 (*s*, 1H, N3); 4.1(*s*, 1H, C5); 2.5 (*s*, 3H, C7);^{13}C N (DMSO-d_6): δ(ppm) 169.8 (C2); 168.3 (C4); 154.4 (C6); 48.9 (C5); 24.1 (C7).

2.4. Preparative Experiments towards Benzalhydantoins **17-21**

2.4.1. General procedure. Following the method of Hoffman and Wheeler [23], hydantoins **14-16** (50 mmol) and benzaldehydes **12-13** (50-150 mmol), anhydrous NaOAc (50-150 mmol), Ac_2O (50 µL, 0.49 µmol) were dissolved together in glacial AcOH (3 mL) at r.t. and the mixture warmed to reflux (140 °C). Reactions were monitored by thin layer chromatography until all the aldehyde had reacted. Upon cooling, the mixture gave a precipitate that was collected and recrystallized.

2.4.2. Synthesis 5-(4-hydroxybenzylidene)imidazolidine-2,4-dione **17**

4-Hydroxybenzaldehyde **12** (5.3069 g, 50 mmol), hydantoin **14** (5.0184 g, 50 mmol), NaOAc (10.3583 g, 75 mmol), and Ac_2O (20 µL, 0.20 µmol) in AcOH (3 mL), were refluxed for 8 h, a white solid that was recrystallized from EtOH to give 5-(4-hydroxybenzylidene)imidazolidine-2,4-dione **17** as white needles (0.1506 g, 11%) m.p. 290 °C (lit.[21] m.p 311°C). HR-TOF-MS ES$^-$m/z: 203,0461 (Found: $C_{10}H_7N_2O_3$). Ultraviolet spectra have λ_{max} (MeOH) (ε_{max}): 203 (1.94 x 10^5) nm. ^{1}H NMR (DMSO-d_6, 500MHZ): δ (ppm) 11.1 (s, br, 1H, N3); 10.3 (s, br,1H, N1); 9.8 (s, 1H, OH); 7.4 (dd, 2H, C5' and C3'); 6.7 (dd, 2H; H6' and C2'); 6,3 (s, 1H, H7'). ^{13}C-NMR (DMSO-d_6, 125 MHz): δ (ppm) 165.6 (C2); 158.0 (C4); 155,6 (C4'); 131.2 (C2' and C6'); 125.3 (C5); 123.8 (C1'); 115.7 (C3' and C5'); 109.3 (C7').

2.4.3. Synthesis 5-(4-hydroxy-3-methoxybenzylidene)imidazolidine-2,4-dione **18**

Compound 3-Methoxy-4-hydroxybenzaldehyde **13** (1.0659 g, 7 mmol), hydantoin **14** (0.7005 g, 7 mmol), NaOAc (2.0007 g, 15mmol), and Ac_2O (20 µL, 0.20 µmol) in AcOH (3 mL), were refluxed for 8 h, a white solid that was recrystallized from EtOH to give 5-(4-hydroxy-3-methoxybenzylidene)imidazolidine-2,4-dione **18** as white needles (0.0731 g, 4 %) m.p. 245 °C (lit.[21]m.p 271°C). HR-TOF-MS ES$^-$ m/z: 233.0556 (Found: $C_{11}H_9N_2O_4$). Ultraviolet spectra have λ_{max} (MeOH) (ε_{max}): 345.4 (3.9 x 10) nm. (DMSO-d_6, 500MHZ): δ (ppm) 11,4 (s, br, 1H, N3); 10,4 (s, br, 1H, N1); 9,4 (s, 1H, OH); 7.0-7.1 (d, 1H, H5'); 6.7 (d, 1H, H6'); 6.3 (s, 1H, H7'); 3.8 (s, 3H, OCH_3). ^{13}C-NMR (DMSO-d_6): δ (ppm) 165.7(C4); 155.7 (C2); 147.7 (C3'); 147.6 (C4'); 125.4 (C5); 124.3 (C1'); 123.5 (C6'); 115.7 (C5'); 113.2 (C2'); 109.8 (C7'); 55.8 (OCH3).

2.4.4. Synthesis 5-(4-hydroxybenzylidene)-1-acetylimidazolidine-2,4-dione **19.**

Campound 4-hydroxybenzaldehyde **12** (1.0050 g, 7 mmol), acetyl hydantoin **16** (0.9942 g, 7 mmol), NaOAc (2.0293 g, 15mmol), and Ac_2O (10 µL, 0.10µmol) in AcOH (0.5 mL), were refluxed for 8 h, a yellow solid that was recrystallized from EtOH to give 5-(4-hydroxybenzylidene)-1-acetylimidazolidine-2,4-dione **19** as yellow needles (39 %) m.p. 165.7 °C (lit.[21] m.p 163.0 °C). HR-

TOF-MS ES$^-$ m/z: 245(Found: $C_{12}H_9N_2O_4$). Ultraviolet spectra have λ_{max} (MeOH) (ε_{max}): 334.8 (2.9 x 10^2) nm. ^{1}H-n.mr(DMSO-d$_6$, 500MHZ): δ (ppm) 11.10 (*s*, 1H, N3); 9.85 (*s*, 1H, OH); 7.46-7.48 (*d*, 2H, *J*8.5 Hz, C3' and C5'); 6.77-6.79 (*d*, 2H, *J*8.0 Hz,C2' and C6'); 6,40 (*s*, 1H,H7'); 2,27 (*s*, 3H, CH_3)

2.4.5. Synthesis 5-(4'-hydroxybenzylidene)-3-methylimidazolidine-2,4-dione **20**.

Compound 4-hydroxybenzaldehyde **12** (0.7179 g, 5 mmol), methyl hydantoin **15** (0.7102 g, 5 mmol), NaOAc (2.0293 g, 15mmol), and Ac_2O (10 μL, 0.10μmol) in AcOH (0.5 mL), were refluxed for 8 h, a white solid that was recrystallized from EtOH to give 5-(4'-hydroxybenzylidene)-3-methylimidazolidine-2,4-dione **20** as white needles (0.128 g, 12%) m.p. 302 ºC (lit.[21] m.p320ºC). HR-TOF-MS ES$^-$ *m/z:*217.0639(Found: $C_{11}H_9N_2O_3$). Ultraviolet spectra have λ_{max} (MeOH) (ε_{max}): 340.4(1.3 x 10) nm. υ_{maks}(KBr): 3265, 3020, 2954, 1705, 1654, 1029, 626 cm^{-1}. ^{1}H-NMR (DMSO-d$_6$, 500MHz): δ (ppm) 10.52 (s, 1H, N1); 9.88 (s, 1H, C4'); 7.5-7.45 (d, *J* 8.5, 2H, C2'-C3'); 6.8-6.7 (d, *J* 8.5, 2H, C5'-C6'); 6.4 (s, H, C7'); 3.1 (s, 3H, C6); ^{13}C-NMR (DMSO-d6)(125 MHz): δ (ppm) 164.4 (C2); 158.2 (C4); 155.3(C4'); 131.4(C2'-C3'); 124.1 (C5); 123.7 (C1'); 115.7 (C4'-C5'); 110.3 (C7'); 24.4 (C6).

2.4.6. Synthesis 5-(4'-hydroxy-5'-methoxybenzylidene)-3-methylimidazolidine-2,4-dione **21**

4-Hydroxy-5-methoxybenzaldehyde **13** (0.76 g, 5 mmol), methyl hydantoin **15** (0.57 g, 5 mmol), NaOAc (2.0293 g, 15 mmol), and Ac_2O (10 μL, 0.10 μmol) in AcOH (0.5 mL), were refluxed for 8 h, a white solid that was recrystallized from EtOH to give 5-(4'-hydroxy-3'-methoxybenzylidene)-3-methylimidazolidine-2,4-dione**21**as white needles (0.6267, 51%) m.p. 222,7-222,9°C(lit.[21] 222-223°C). Ultraviolet spectra have λ_{maks} (MeOH) (ε_{maks}): 346.5 (4.3x10)nm; ν_{maks}(KBr): 3554, 3231, 3100-3000, 1747, 1657, 1591-1469, 1233, 1165, 760-731 cm^{-1}; (ES-MS-E$^-$): *m/z* 247,5083 (Found: $C_{12}H_{11}N_2O_4$; ^{1}H-NMR (DMSO-d6, 500MHz): δ (ppm) 9.47 (s, H, OH); 10.61 (s, H, H1); 7.13 (s, H, H2'); 7.08-7.10 (d, *J* 8.0, H, H6'); 6.78-6.80 (d, *J* 8.0, H, H5'); 6.47 (s, H, H7'); 3.84 (s, H, OCH_3);2.95 (s, H, H6)

^{13}C-NMR (DMSO-d6, 125 MHz): δ (ppm) 164,4 (C4); 155.4 (C2); 147.8 (C3',C4'); 124.2 (C5,C1'); 123.7 (C6'); 115.8 (C5'); 113.3 (C7'); 110.8 (C2'); 55.8 (OCH_3); 24.3 (C6).

3. Result and Discussion

3.1. Synthesis of Benzalhydantoin Precursors

At the outset of this study, it was decided to adopt the potassium acetate modification [19-21] to synthesize a somewhat wider range of benzalhydantoins derived from hydantoin **14**, *N*(3)-methylhydantoin **15** and *N*(1)-acetyl **16** to more thoroughly explore the influence of substituents on the reaction. It was also decided to restrict the study to commercially available aldehydes **12-13**.

Hydantoin **14** is commercially available and very inexpensive, making it an attractive partner in this reaction. 3-Methylhydantoin **15** and 1-acetylhydantoin **16** are also well known [24], but not commercially available, and so were synthesized. Structures were confirmed by comparative melting point and by infrared, ^{1}H NMR and ^{13}C-NMRspectroscopy.

In hydantoin chemistry, it is known that the *N*-3 position of hydantoin is reactive towards electrophiles, presumably due to the most acidic position between the two activating carbonyl groups [24]. According to Pinner [24], hydantoin could be alkylated in the *N*-3 position by treatment with alkyl halides in alkaline solution. The reaction usually proceeds smoothly, especially for methylation reactions, while substitution at *N*-1 could not be achieved through direct alkylation [25]. Treatment of hydantoin **14** with dimethylsulphate in the presence of potassium hydroxide in ethanol gave 3-methylhydantoin **15** as white needles in 44 % yield. High resolution mass spectrometric analysis supported the molecular formula $C_6H_8N_2O_2$, and ^{1}H and ^{13}C NMR spectroscopy confirmed the presence of the *N3*-methyl group. Similar treatment of hydantoin **14** with acetic anhydride gave the corresponding *N1*-acetylhydantoin **16** as white needles in 64 % yield, again confirmed by high resolution mass spectrometry and appropriate ^{1}H and ^{13}C NMR spectroscopic signals.

3.2. Synthesis of Benzalhydantoins

With the necessary partner reactants in hand, the relative ease with which they reacted together was examined through the preparation of the compounds **17-21.** In this part of the research, reaction progress was monitored by comparative t.l.c. analysis. Products were isolated simply by allowing the reaction mixtures to cool, filtering off the desired crystalline material, and washing the crystals with cold ethanol. Examination of the isolated products by ^{1}H NMR spectroscopy indicated that all the products were impure. Pleasingly, they comprised at least 90% the derived benzalhydantoin. This was measured by integration of the olefinic methylenide signal compared with others. Notably, within the methylide region, in some cases *E/Z* isomers were evident. When this was observed, the highest chemical shift signal dominated and was assigned by literature analogy to the *Z* isomer while the minor singlet appeared at higher field and was assigned to the *E*-isomer.

Crude product yields were recorded at this stage, but in all cases the compounds were subsequently recrystallized to analytical purity for final characterization. The reaction times and yields, product melting points after recrystallization, and characteristic olefinic and N–H ^{1}H NMR chemical shifts were collected in Table 1.

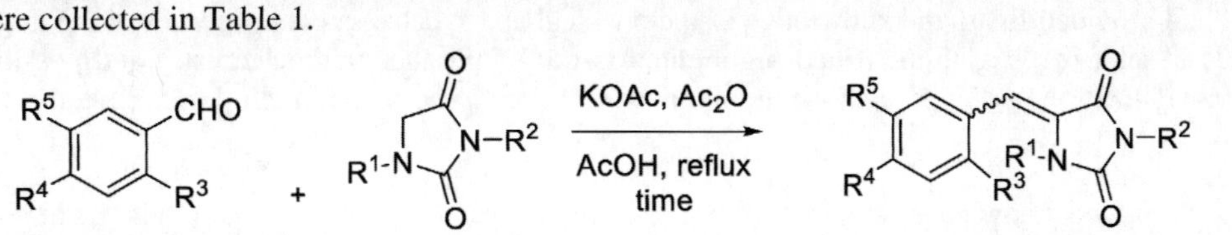

	R^3	R^4	R^5
12	H	OH	H
13	H	OH	OMe

	R^1	R^2
14	H	H
15	H	Me
16	Acetyl	H

	R^1	R^2	R^3	R^4	R^5
17	H	H	H	OH	H
18	H	H	H	OH	OMe
19	Acetyl	H	H	OH	H
20	H	Me	H	OH	H
21	H	Me	H	OH	OMe

Table 1. An initial study of the condensation of hydantoins **14-16** with aromatic aldehydes **12-13** in acetic acid in the presence of NaOAc and Ac_2O.

Entry	Structure No.	Reflux time (h)	Yields (%)	M.p. (°C)	Chemical shift (ppm) NH	Olefinic (H1') *Z*	Olefinic (H1') *E*	Integration ratio
1	**17**	8	11	290	10.3; 11.1	-	6.3	>99:<1
2	**18**	8	4	245	10.4; 11.4	-	6.3	>99:<1
3	**19**	8	39	165.7	11.1	6.4	-	>99:<1
4	**20**	8	12	302	10.5	6.4	-	>99:<1
5	**21**	8	51	222.8	10.6	6.5	-	>99:<1

All ^{1}H NMR spectra were recorded in DMSO-d_6 due to the low solubilities of the hydantoins in most non-polar solvents [26], including deuterochloroform. As mentioned earlier, geometrical isomers (*Z* and *E* isomers) were detected through the appearance of distinct signals for the olefinic proton on the exocyclic C=C double bond of the benzalhydantoins **17-21** [15].

Chui [13] reported that the olefinic proton signal, H1', in benzalhydantoins was diagnostic. It is deshielded in the *Z* form and the signal appears at δ 6.40-7.00, while in the *E* form the effect is less so and the signal appears at δ 6.20-6.30. The anisotropic effect exerted by the nearby C4 carbonyl group in the former is believed to be responsible [15]. This literature information was used in the assignment of geometric isomers to compounds listed in Table 1.

In keeping with earlier findings, the results presented in Table 1 indicated that the yields were dependent on both the identity of the aldehyde and the substituent at position N1 and N3 of the hydantoin. Compounds **17-18** resulted in (*E*)-isomer with low yields. Electron donating group (-OMe) on the aldehyde retard the initial addition process, while the same groups speed the dehydration reaction up by stabilizing the incipient cation charge at the benzylic center [21]. The lower yields of **17-18** may be an indication that the first step, addition, was rate limiting. Compounds **19-21** resulted in (*Z*)-isomer with varyingly higher yield, the product yields were dependent upon the type of hydantoin substituent at N1 or N3 position. However, an anomaly was observed with compound **21** where weak electron withdrawing (-OH) at R^4, strong electron donating (-OMe) at R^5 and substituent (-Me) at R^2 positions resulted in the highest yield.

A catalytic amount of acetic anhydride was used to increase the addition process of reaction to yield the intermediate alcohol, then elimination reaction was occurred to give benzalhydantoin.

4. Summary

This initial exploratory study had confirmed a number of claims from earlier studies in ours research group and had provided first-hand knowledge of the physical and spectroscopic properties of the target compounds of later investigations. It had highlighted issues of product stereochemistry and possible selective crystallization during product isolation that would need to be borne in mind in later studies.

Acknowledgements

Authors gratefully acknowledge to small funding support from Boeddi's Legacy via Karmini's Foundation, Bandung, West Java, Indonesia. A/Prof. Roger W Read and Prof. David St.C Black, The University of New South Wales, Sydney, Australia, for support and guidance's.

References

[1] Mehta N B, Diuguld C A R, Soroko F E 1981 *J. Med. Chem.* **24** 465-468
[2] Cortes S, Liao Z K, Watson D, Kohn H 1985 *J. Med. Chem.* **28** 601-606
[3] Poupaert J H, Vandervorst D, Gulot P, Mustafa M M M, Dumont P 1984 *J. Med. Chem.* **27** 76-78
[4] Krall R L, Penry J K, White B G, Kupferger H J, Swinyard E A 1978 *Epilepsia* **19** 409-428
[5] Brouillette W J, Jestkov V P, Brown M L, Akhtar M S, De Lorey T M, Brown G B 1994 *J. Med. Chem.* **37** 3289-3293
[6] Brown M L, Brown G B, Brouillette W J 1997 *J. Med. Chem.* **40** 602-607
[7] Kwon C H, Iqbal M T, Wurpel J N D 1991 *J. Med. Chem.* **34** 1845-1849
[8] Hidayat I W, Thu Y Y, Black D StC, Read R W 2016 *IOP Conf. Series: Material Science and Engineering* **107** 012058
[9] Hidayat I W, Bhadbhade M, Read R W 2015 *Procedia Chemistry* **17** 75-83
[10] Hidayat I W 2015 *Defining Protocol for the Synthesis of 3-Substituted-5-benzylideneimidazolidine-2,4-diones* PhD Thesis (Sydney NSW: The University of New South Wales)
[11] Perry T L 1998 Master's Thesis (Missisipi: The University of Mississippi) p 28-40
[12] Mudit M, Khafar M, Muralidharan A, Thomas S, Shah G V, Van Soest R W M, El Sayed K A 2009 *Bioorg. Med. Chem.* **17** 1731-1738
[13] Thenmozhiyal J C, Wong P T, Chui W K 2004 *J. Med. Chem.* **47** 1527-1535
[14] Szymanska E, Kiec-Kononowicz K 2002 *Il Farmaco.* **57** 355-362
[15] Marton J, Enisz J, Hosztafi S, Timar T 1993 *J. Agric. Food Chem.* **41** 148-152
[16] Tan S, Ang K, Fong Y J 1986 *Chem. Soc. Perkin Trans.* **II** 1941-1986.
[17] Menendez J C, Diaz M P, Bellver C, Solluber M M 1992 *Eur. J. Med. Chem.* **27** 61-66
[18] Drew M G B, Mok K F, Tan S F 1987 *Acta Crystallogr.* ***C43*** 745-748

[19] Akeng'a T A O 2000 Synthesis of Indoles from Benzalhydantoins and Benzalpiperazine-2,5-diones. PhD Thesis (Sydney NSW: The University of New South Wales)
[20] Akeng'a T O and Read R W 2005 *S. Afr. J. Chem.* **58** 93–97
[21] Akeng'a T O and Read R W 2007 *S. Afr. J. Chem.* **60** 11–16
[22] Orazi O O, Corral R A, Schuttenberg H 1974 *J. Chem. Soc. Perkin Trans.* **I** 219-221
[23] Wheeler H L and Hoffman C 1911 *Am. Chem. J.* **45** 368-383
[24] Ware E 1950 *Chem. Rev.* **46** 403-467
[25] Yang D H, Chen Z, Chen S, Zheng Q 2003 *Synth*esis 1891-1894

IOP Conf. Series: Materials Science and Engineering **172** (2017) 012025 doi:10.1088/1757-899X/172/1/012025

Preparation of TiO_2- SiO_2 via sol-gel method: Effect of Silica precursor on Catalytic and Photocatalytic properties

I Fatimah

Chemistry Department, Universitas Islam Indonesia, KampusTerpadu UII, Jl. Kaliurang Km 14, Sleman, Yogyakarta, Indonesia 55584.

Abstract. TiO_2- SiO_2have been synthesized by the sol-gel method from titanium isopropoxide and varied silica precursors: tetraethyl orthosilicate and tetra methyl ortho silicate. To study the effect of the precursor, prepared materials were characterized by X-ray diffraction, scanning electron microscopy, Diffuse Reflectance UV-vis optical absorption, and also gas sorption analysis. XRD patterns showed the formation of TiO_2 anatase in the TiO_2-SiO_2 composite with different crystallite size from different silica precursor as well as the different surface morphology. The DRUV-vis absorption spectra exhibit similar band gap energy correspond to 3.21eV value while the surface area, pore volume and pore radius of the materials seems to be affected by the precursor. The higher specific surface area contributes to give the enhanced activity in phenol hydroxylation and methylene blue photodegradation.

1. Introduction

Titanium dioxide is popular semiconductor material in some application including photocatalysis and dye-sensitized solar cell. To enhance its performance, some modifications have been investigated by such chemical and physical treatment[1][2]. Beside of doping and structural modification of titania nanoparticles, supporting titania has been reported to exhibit different photocatalytic performance from that titania itself. The contribution of support material has been explained to give better mechanism regarding the interaction between titanium oxide and the support[3][4]. By this scheme, titania-silica (TiO_2/SiO_2) composite is the well-known and intensively studied. Some studies revealed the relationship among surface titanium oxide structure, preparation method and loading amount, etc. with its photocatalytic activity.

As the different mechanism of preparation as well as synthesis route, TiO_2/SiO_2 material will give different performance related to its physicochemical character. Moreover, the porous structure of silica within the formation of the composite is also affected by some parameters, and one of these is the silica precursor. The composite formation between silica and titania precursors occurs between silanols or hydrolysis of titania from its hydroxides[5]. The condensation reaction between the titania precursor and the silica surface depends mostly on the hydroxyl groups of the silica since the rest of the silica is very inert and it is affected by the method used, and the amount of water and its pH used. In this present work the hydrolysis of tetra alkyl ortho silicate with two variations: tetraethyl ortho silicates (TEOS) and tetra meth kor to silicate (TMOS) is compared. Different hydrolysis rate is hypothesized to be the factor that controlling the physicochemical character and it may affect the photocatalytic properties. The physicochemical character is studied using UV/Vis diffuse reflectance, XRD, FTIR and also gas sorption analyzer; we discuss the relation between structure as a function of silica precursor and photocatalytic performance of TiO_2/SiO_2. Photocatalytic activity of the prepared materials was studied in methylene blue (MB) photodegradation and phenol oxidation reaction using H2O2 as oxidant.

2. Material and Method

2.1 Materials
Titanium (IV) isopropoxide, methylene blue, phenol, tetra ethyl ortho silicate (TEOS), tetra methylortho silicate (TMOS), and H_2O_2 were purchased from Merck.

2.2 Preparation of TiO_2-SiO_2
The materials were prepared with a sol-gel mechanism by mixing titania and silica precursors for two h under reflux condition. The molar ratio of Ti: Si is 4:9. The filtrate of each reflux mixture was evaporated and followed by calcination for four h at 773K. By varied precursor, the obtained material is encoded as TS-M and TS-E respect to the use of tetra methyl orthosilicate and tetra etylortho silicate.
The specific surface area (SSA) of the samples was determined using Quantachrome NOVA 1200e equipment. SSA analysis was based on Bruner-Emmet-Teller (BET) model of N2 low-temperature adsorption while the pore volume and pore radius were calculated using Barret-Jonna-Halenda (BJH) equation. Powder X-ray diffraction (XRD) patterns were collected using Shimadzu X6000 instrument. Ni-filtered CuKα radiation was used for collecting reflections in the 2θ range 15–80° with a step of 0.02°. Crystalline phases were identified by comparison with the JCPDS data files. Crystallites size of materials was estimated according to the Scherrer equation. FTIR spectra of prepared materials were obtained using Perkin-Elmer instrument.

2.3 Catalytic activity of TiO_2-SiO_2

For catalytic activity study of the prepared materials, phenol hydroxylation and methylene blue photodegradation were conducted. The hydroxylation reactor and photocatalytic reactor were set up as the scheme in Figure 1 and Figure 2. UV Lamp used in the reactor is Philips UVB Lamp with the power of 40W.

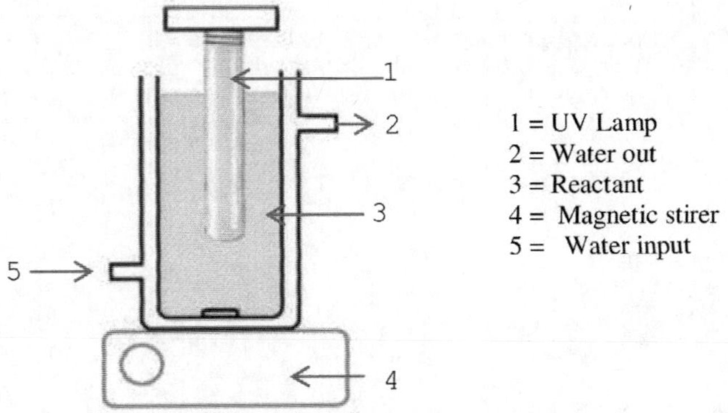

Figure 1. Scheme of photocatalytic reactor

IOP Conf. Series: Materials Science and Engineering **172** (2017) 012025 doi:10.1088/1757-899X/172/1/012025

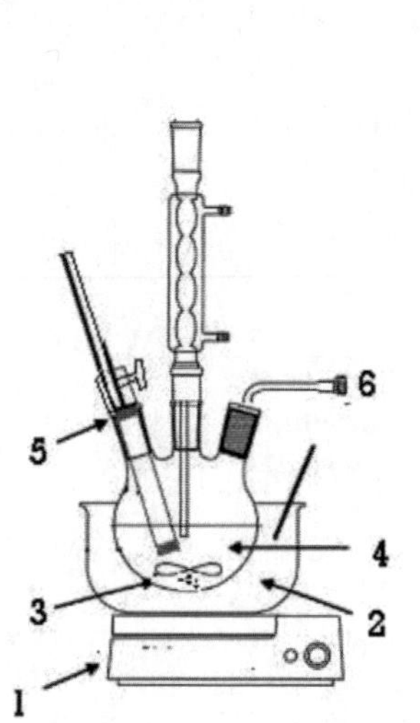

1 = Magnetic stirrer-heater
2 = Water bath
3 = Stirer
4 = Reactants
5 = H_2O_2 input
6 = Sampler

Figure 2. Scheme of phenol hydroxylation reactor

3. Results and Discussion

X-ray diffraction technique was performed to analyze and identify phase formation and crystallographic information of the samples. The XRD pattern of TiO_2/SiO_2 powder prepared from both precursors is given in **Figure 3**. The diffraction patterns indicate the formation of TiO_2 match very well with those of tetragonal TiO_2 anatase for both samples (JCPDS-21-1272). The presence of silica is not clearly indicated from patterns probably due to the high crystalline TiO_2 formed to cover amorphous silica which is also act as TiO_2 support. The patterns are similar with what obtained from the utilization of surfactant in sol-gel synthesis by using titanium isopropoxide precursor[6,7].

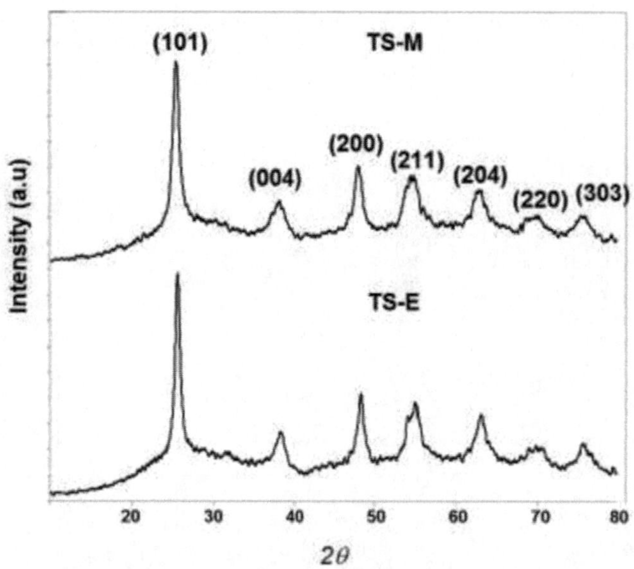

Figure 3. XRD pattern of prepared materials

The typical plot of N_2 adsorption-desorption isotherm and pore size distribution curves of the prepared TS samples are in Figure 4. Both samples exhibited type IV (BDDT classification) of isotherm with hysteresis loops of type H3 at relative pressure range of 0.65–1.0, representing the mesoporous structure. The pore size distributions indicated that TiO_2 presented a relatively narrow distribution ranging from 5 nm to 60Å. From the calculation it is found that the specific surface area of TS-E is higher compared to TS-M. It is also indicated from the pore distribution that TS-E gives the higher pore volume in all pore region. However the properties are not significantly appeared from the SEM profile analysis (Figure 5) in that both precursors give flaky morphology. The band gap energy values of TS-M and TS-E are also similar.

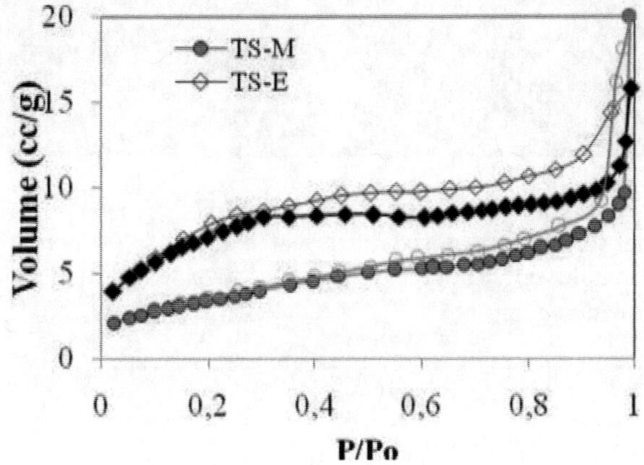

(a)

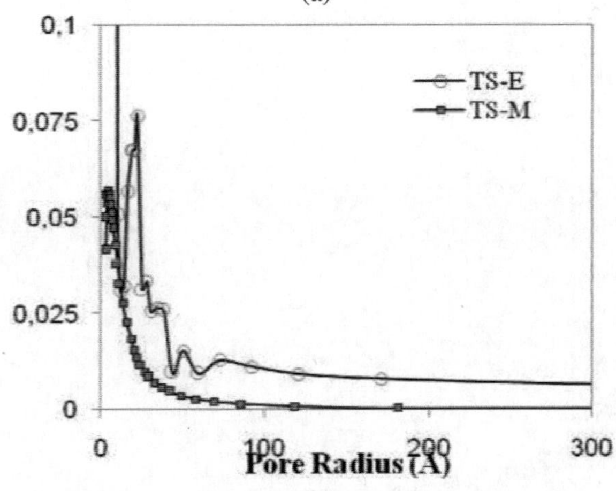

(b)

Figure 4. (a) Adsorption-desorption profile of materials (b) pore distribution curve of materials

(a) (b)

Figure 5. SEM profile of (a) TS-E (b) TS-M

The band gap energy of the materials are related to the photoactivity of materials and for anaysis DRUV-Vis was employed. The band gap energy is calculated as function of the edge wavelength (λ_{edge}) by following equation:

$$E_g = \frac{1240}{\lambda_{edge}} \quad (1)$$

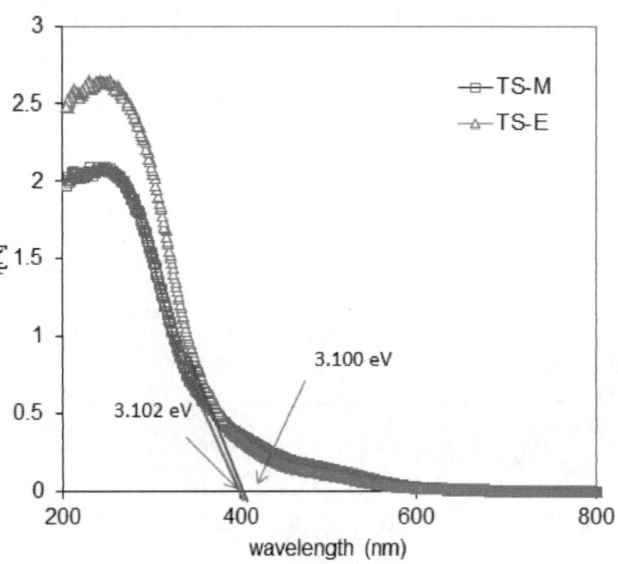

Figure 6.DRUV-Vis spectra of materials

Table 1: Surface profile data of prepared materials

Sample	Specific surface area (m^2/g)	Pore Volume (cc/g)	Pore radius (Å)
TS-E	235.71	0.2786	13.4
TS-M	128.22	0.1208	10.1

Photocatalytic activity of the materials is expressed from the kinetics of MB photooxidation in varied MB concentration presented in Figure 7. From the kinetics constant values (k) it is confirmed that TS-E gives higher rate respect to the faster MB degradation in the same pattern from varied time and concentration. The same pattern found that the higher MB concentration, the higher kinetic constant obtained. The data is fit with the theoretic concept in which the higher MB concentration creates more interaction from the more collision between catalyst surface and the reactants. Furthermore, related to the physicochemical properties of the materials the photocatalytic activity suggests that the specific surface area plays important role in the photodegradation mechanism. TS-E with the higher specific surface area provides more active sites to be effectively adsorb the MB molecules before oxidation with the presence of H2O2 as the oxidant in the system.

The similar reasons may apply for the kinetic of phenol hydroxylation data as presented in Figure 8. The trend of phenol conversion (Conv.(%)) as function of time of reaction represents that TS-E has higher activity than TS-M. Conv.is calculated with following equation:

$$Conv.(\%) = 100\left(\frac{[phenol]o-[phenol]t}{[phenol]o}\right) \qquad (2)$$

With [phenol]o and [phenol]t are initial phenol concentration and phenol concentration in time t of sampling.

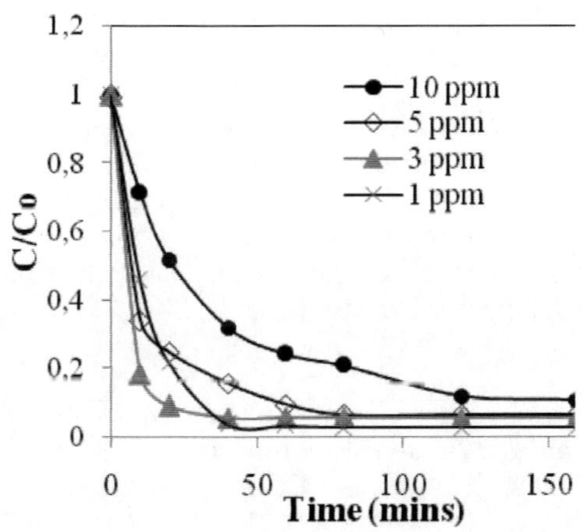

(a)

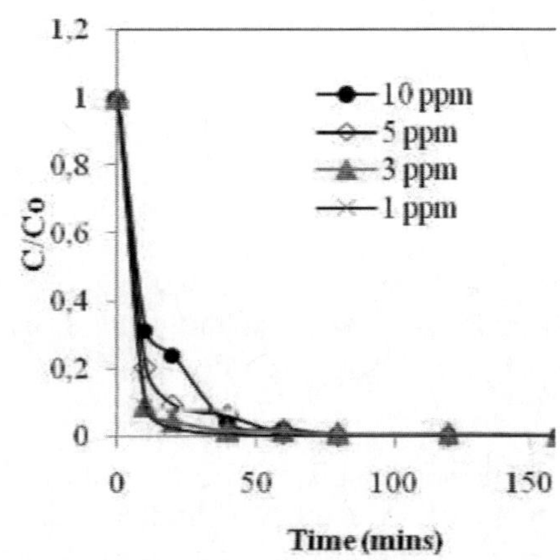

(b)

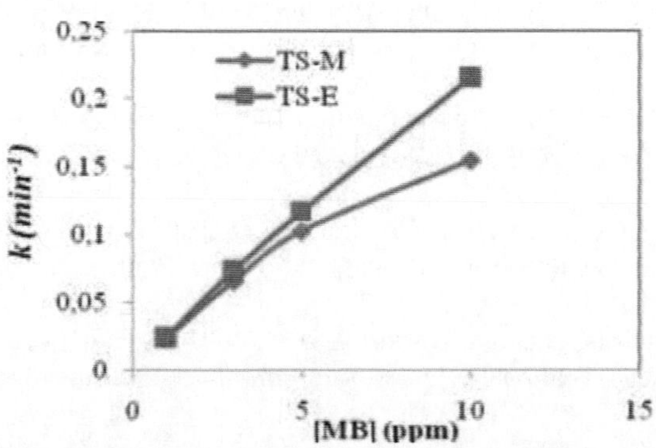

(c)

Figure 7. (a) Kinetic of MB photodegradation over TS-M (b) Kinetic of MB photodegradation over TS-E (c) Effect of MB concentration on kinetic constant

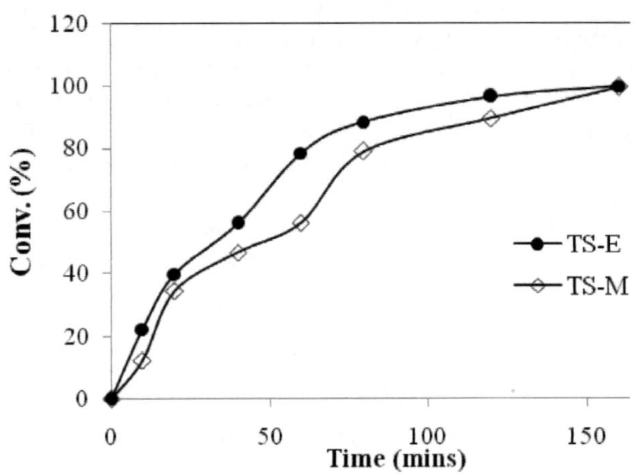

Figure 7. Kinetic of phenol hydroxylation by varied catalyst

4. Conclusion

In summary, the preparation of TiO_2/ SiO_2 composites with varied silica precursor has been successfully investigated. The physicochemical character of the materials reveals that the use of tetraethyl orthosilicate gives higher specific surface area and pore volume while the band gap energy and XRD analysis give similar pattern and the values of band gap energy as well as crystallite size. The parameter of specific surface area and pore volume of material facilitate the photocatalytic and catalytic activities resulting from the higher activity of TS-E compared to TS-M.

References

[1] Choi H, Al-abed S R, Dionysiou D D, Stathatos E and Lianos P 2010 *TiO2-Based Advanced Oxidation Nanotechnologies for Water Purification and Reuse* vol 2(Elsevier)

[2] Hashimoto K, Irie H and Fujishima A 2005 TiO 2 Photocatalysis: A Historical Overview and Future Prospects *Japanese Journal of Applied Physics***44** 8269–85

[3] Yao N, Cao S and Yeung K L 2009 Mesoporous TiO2–SiO2 aerogels with hierarchal pore structures *Microporous and Mesoporous Materials***117** 570–9

[4] Cetinkaya T, Neuwirthová L, Kutláková K M, Tomášek V and Akbulut H 2013 Synthesis of nanostructured TiO2/SiO2 as an effective photocatalyst for degradation of acid orange *Applied Surface Science***279** 384–90

[5] Aziz R A and Sopyan I 2009 Synthesis of TiO2-SiO2 powder and thin film photocatalysts by sol-gel method *Indian Journal of Chemistry - Section A Inorganic, Physical, Theoretical and Analytical Chemistry***48** 951–7

[6] Nandanwar R, Singh P, Syed F F and Haque F Z 2014 Preparation of TiO 2 / SiO 2 Nanocomposite with Non-ionic Surfactants via Sol-gel Process and their Photocatalytic Study

[7] Wei X, Zhu G, Fang J and Chen J 2013 Synthesis, characterization, and photocatalysis of well-dispersible phase-pure anatase TiO2 nanoparticles *International Journal of Photoenergy***2013**

IOP Conf. Series: Materials Science and Engineering **172** (2017) 012026 doi:10.1088/1757-899X/172/1/012026

Preparation Of KF-Modified Kaolinite As Green And Reusable Catalyst For Microwave Assisted Biodiesel Conversion

I Fatimah, R Z Andiena*, S P Yudha

Department of Chemistry , Faculty of Mathematics and Natural Sciences,Islamic University of Indonesia, Yogyakarta, Indonesia

*Email :restyzama@gmail.com, isfatimah@uii.ac.id

Abstract. Preparation of KF-modified kaolinite catalyst for microwave-assisted biodiesel conversion has been investigated. Kaolinite modification was conducted by solid-solid reaction between naturally occurring kaolinite mineral and KF salt followed by heating at 200oC for 2h. Prepared catalyst was characterized by using XRD, BET surface area analyzer, and SEM-EDX analysis and for catalytic activity tests, biodiesel conversion of jatropha oil was simulated. The comparison between microwave utilization and conventional method of biodiesel conversion were studied, moreover study on the catalyst reusability was performed. The results show that prepared catalyst gives the better physicochemical character of kaolinite as heterogeneous catalysts application as shown by the higher conversion and also reusability. Furthermore, the use of microwave irradiation exhibits the more time effectiveness. In general, the greener biodiesel conversion using presented methods is promising technique to be developed.

1. Introduction

Indonesia has huge world's natural resources particularly plants and minerals. In another side, fuel is a natural resource that could not be renewed, and one day it might be limited sources. Therefore the peoples try to initiate and develop alternative fuels. One of the alternative fuels is biomass based such as biodiesel. Biodiesel can be used as an alternative fuel diesel engines, and that is environmental friendly[1]. Biodiesel is produced from vegetable oils or animal fats over catalytic reaction. Vegetable oil is a raw material potential as a source of biodiesel because of its presence can be renewed. Vegetable oils used for biodiesel production including corn oil, cashew oil, coconut oil, sunflower oil, olive oil, soybean oil and castor oil have been reported[2]. The reaction requires a catalyst to accelerate the reaction for increasing the rate. The Homogenous catalyst of NaOH is usually applied, and it leads problems related to non-regenerable and consumable properties. Such condition affects to in sustainable process refer to green chemistry principles this research offers an environmentally friendly catalyst, i.e., heterogeneous catalyst[1–9]. Due to the very big potency of natural mineral in Indonesia, utilization of natural kaolinite as raw material for heterogeneous catalysts is an interesting topic. Refer to basic mechanism of biodiesel conversion and reaction a base catalyst plays an important role in the mechanism[6]. Some improvements have been attempted for increasing the effectivity of the solid catalyst in such biodiesel conversion, and an effort can be applied the modification with a salt consisting very basic anion[10]. Such modifications of solid basic catalysts were reported by previous papers. The addition and composite formation of a solid catalyst with potassium salt of KF, K_2CO_3 and KOH were attempted by the main mechanism of the improvement of solid basicity for increasing activity of the catalysis[11–14].

In this study, the modification of kaolinite with KF has been applied. Calcium Flouride (KF) was hypothesized to increase the basicity as well as biodiesel conversion rate. Another scheme for the benign environmental process the reaction will be accelerated over microwave assisted reaction. Some reports investigated the feasibility of microwave irradiation procedure in some organic conversion including biodiesel, and it gives some advantageous refer to time-effective and energy efficiency[15,16]. By a combination of the potency of kaolinite modification and the use of microwave irradiation, this study is aimed to evaluate the effect kaolinite modification onto physic-chemical character especially catalytic activity in biodiesel conversion over microwave utilization[1,4,17].

2. Materials and Method

2.1. Materials

Materials used in this research are castor oil supplied from National Company in Indonesia, natural kaolinite from Sukabumi West Java, KF, methanol, citric acid, NaOH were purchased from Merck, Germany.

2.2. Methods

KF/Kaolinite(KF/KAO) was prepared by grinding method. KF and kaolinite powder were mixed and grinded for 20 minutes, before heating at 200 C for 2 hours.

Prepared and raw material were characterized by XRD, Gas Sorption Analyzer, Scanning Electron Microscope – Energy Disspersive x-Ray (SEM-EDX). XRD X6000 Shimadzu was utilized for XRD measurement with Ni-filtered Cu-Kα radiation. For surface profile measurement, NOVA 1200e instrument was employed by using N_2 gas adsorption at the temperature of 77K, while for SEM-EDX a JEOL instrument was utilized. Surface basicity of the materials was determined by using back-titration method. The tested material was mixed with excess citric acid solution under stirring for a night. The unreacted citric acid was determined by titration using NaOH solution as a standard. The basicity was calculated from the sorbed citric acid per mass of the material (meq/g).
For catalytic activity test, 20 ml of castor oil was mixed with 60 ml of methanol, and 2 grams of catalyst (KAO and KF/KAO). Two variations of reaction system were applied: (1) microwave(MW) irradiation system (reactor set up is presented in Figure 1) for 30 minutes, (2) conventional method by refluxing the mixture for 2 hours. Results of the reactions were separated in separating funnel and the biodiesel fraction was analyzed by using GC-MS instrument. The calorific value of products was analyzed using Bomb-Calorimeter method.

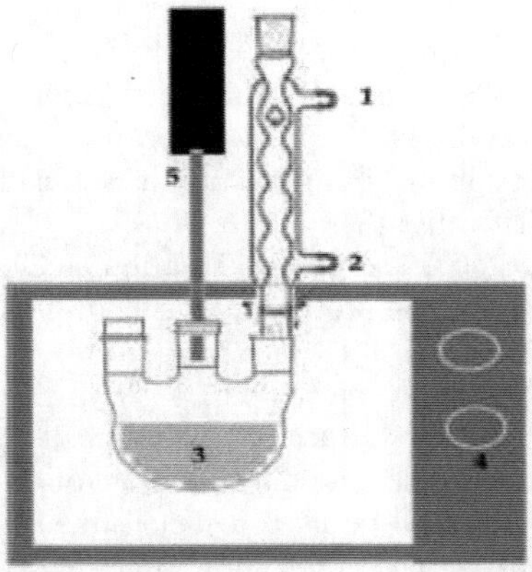

1: water out
2: water in
3: reaction mixture
4: power setting
5: stirrer

Figure 1. Microwave Reactor

3. Results And Discussion

3.1. Catalyst Preparation and Characterization

XRD analysis was used to identify and characterize the crystallinity of material. The XRD results can be displayed on the diffractogram in Figure 2.

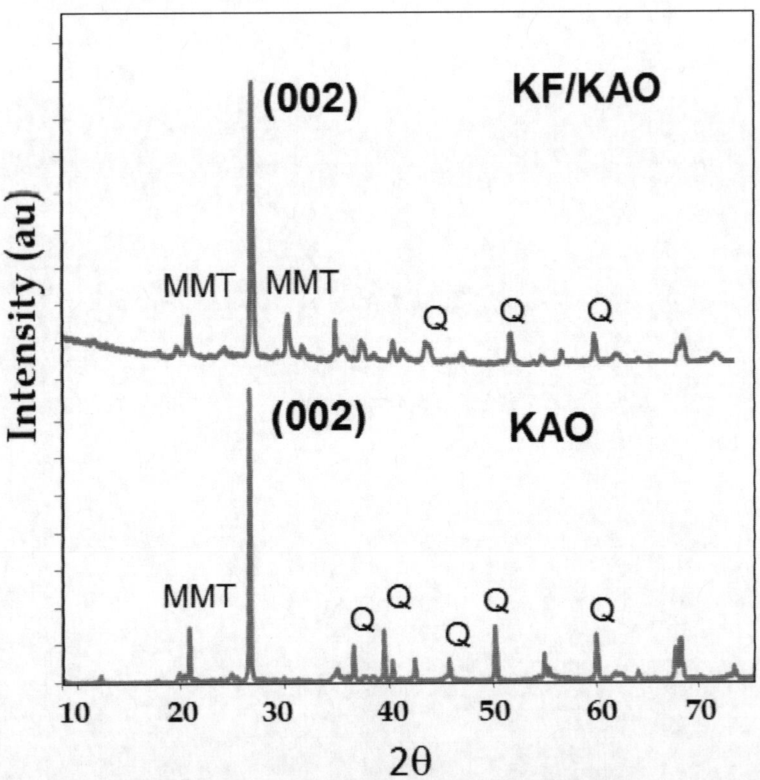

Figure 2. XRD pattern of KF/KAO and KAO

From the XRD pattern, it is showed that there some peaks in stare of kaolinite reflection at (002). Some impurities were detected i.e monmorilonite (MMT) and quartz (Q). It is because the materials used as raw material is natural kaolinite. From the comparison of both reflection it is concluded that there is no significant change of the reflection after the modification[8].

In order to evaluate the change of surface profile after KF modification, gas sorption analysis of the material was conducted by analyzing N_2 adsorption-desorption. The profile is depicted in Figure 3 and calculated parameters is listed in Table 1 .

Isotherm graph in Figure 3 confirm that adsorption isotherm obey type III [9]. It theoretically describes that the BET adsorption of gas molecules on the surface of solids will absorb in physisorption. In the adsorption process, nitrogen gas forms the multilayer region of molecules adsorption on the kaolinite surface and the monolayer region will be used to determine the specific surface area and pore size distribution that exist on the catalyst. The graph shows that KF/KAO gives higher volume of adsorption in all pressure region compared to KAO. From the pattern, specific surface area, pore volume and pore radius were calculated and the data are presented in Table 1. Both the curves and calculated data suggest that the modification of KF into the surface of KAO open the silica alumina surface. In addition the surface basicity of the materials which is the important parameter for trans esterification reaction also indicates the formation of basic interaction of attached F^- from KF to the surface as KF/KAO exhibits the higher value.

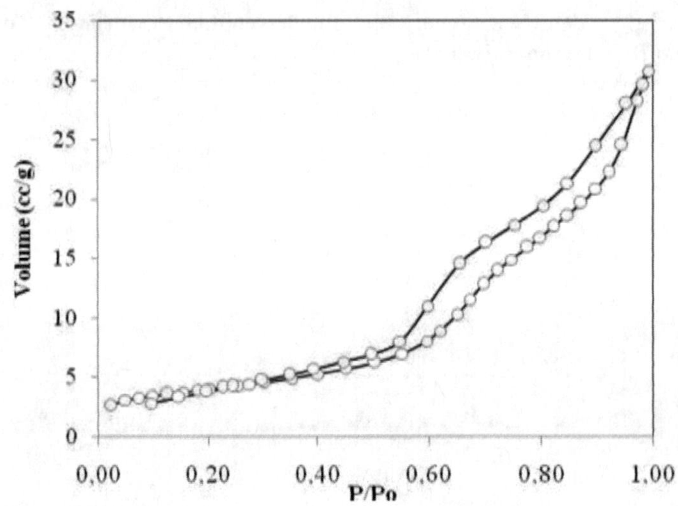

(a)

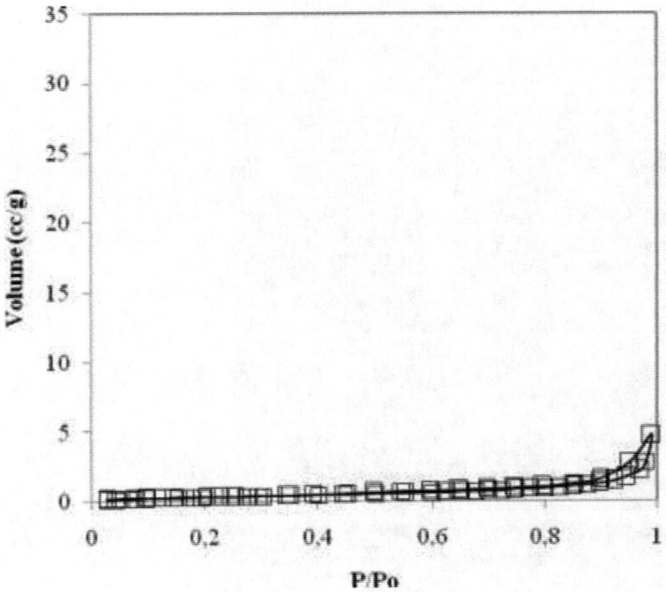

(b)

Figure 3. N_2 adsorption-desorption profile of (a) KF/KAO (b) KAO

Table. 1 Surface Parameters of the materials

Sample	Spesific Surface Area	Pore radius (Å)	Pore volume (cc/g)	Surface basicity (meq/g)
Kaolinite	11.303 m²/g	1.18. 10^2	6.68.10^{-2}	0.321
KF/Kao	27.878m²/g	1.36.10^2	9.504.10^{-2}	0.683

Data in Table 1 shows that after KF modification the surface area and pore volume were increased. The surface area is an important parameter which is referred to the texture of the surface design from heterogeneous catalysts. Related to the pore structure of the particles, pore shape and pore size distribution. The total surface area will be an important criterion to demonstrate the solid catalyst active sites that are directly related to the activity of the catalyst. Then the properties of the pore can control the movement of molecules related to selectivity. High selectivity becomes an important factor in biodiesel conversion.

Surface profile of KAO and KF/KAO was exhibited by SEM-EDX (Figure 4).

Element		(keV)	mass%
C K		0.277	24.04
O K		0.525	28.51
Na K	*	1.041	0.01
Mg K	*		
Al K		1.486	10.51
Si K		1.739	36.59
K K	*	3.312	0.33
Ca K	*	3.690	0.01
Total			100.00

Element		(keV)	mass%
O			49.04
Na K	*	1.041	0.10
Mg K	*	1.253	0.36
Al K		1.486	13.61
Si K		1.739	31.11
K K		3.312	5.58
Ca K	*	3.690	0.20
Total			100.00

Figure 5. SEM-EDX profile of KAO (above) and KF/KAO (bellow)

3.2. Catalytic Activity Test

For catalytic activity testing, biodiesel conversion offers prepared materials was conducted. To evaluate the effect of conversion method (Conventional and MW) and catalysts condition (fresh and reuse), the volume and methyl esters percentage are reported (Table. 2).

The mechanism of transesterification in the biodiesel conversion over heterogeneous catalyst can be schematically represented by the following reaction:

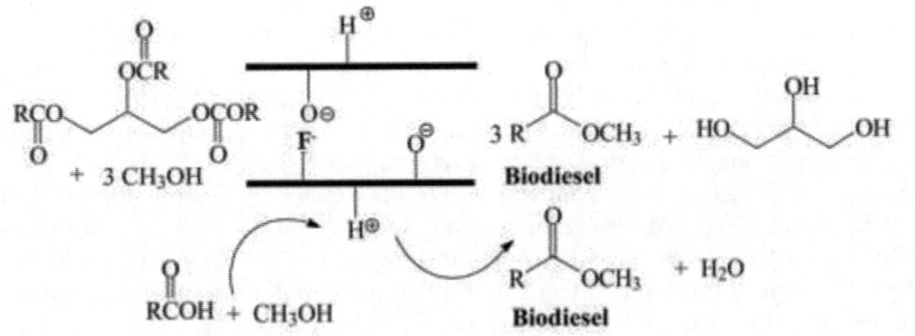

The yield (%) of biodiesel was calculated from GCMS data by identifying the percentage of methyl ester formed (5). The test is also conducted for reused catalyst. After was separated from the reaction mixture, used catalyst was washed with methanol followed by drying. Regenerated catalyst was further reused for the same reaction.

Table 2. Catalytic Activity Data of Materials by Variation Methods and Catalyts Condition

Catalyst	Reflux			MW		
	Volume (mL)	Yield (%)	Calorific value (Kal)	Volume (mL)	Yield (%) Ester	
KAO	16.4	45.67	3883.61	13.5	58.90	10922.90
KF/KAO	15.60	61.92	3391.49	13.20	72.15	8029.04
KAO-*reuse*	11.7	n.d*	4315.80	10.4	n.d	nd
KF/KAO--*reuse*	10.0	n.d	8552.95	9.8	15.10	3579.58

n.d= non detected

Table 2 shows that in general MW method gives higher yield compared to the conventional one. It is also confirmed that reusability of KF/KAO better than KAO as well as the utilization of MW gives better reusability from the produced methyl ester. The objective of sustainability in the proposed method is gained since the catalyst can be reused for further process.

4. Conclusions

Based on the results KF/KAO was successfully prepared with increasing parameters. The modification on KAO expressed the increasing activity in biodiesel conversion over microwave irradiation method. Compared with the conventional method, MW method gives the advantageous related to energy an efficient, time value of 12 hours which requires a very long time reaction is on a large scale. Production of biodiesel by microwave is effective and need to be developed so it can be a solution to the energy crisis currently.

References

[1] Kapilan N and Baykov B D 2014 A Review on New Methods Used for the Production of Biodiesel *Petroleum & Coal* **56** 62–73
[2] Ejikeme P M, Anyaogu I D, Ejikeme C L, Nwafor N P, Egbuonu C A C, Ukogu K, Ibemesi J

IOP Conf. Series: Materials Science and Engineering **172** (2017) 012026 doi:10.1088/1757-899X/172/1/012026

A, Chemistry I and Polytechnic F 2010 Catalysis in Biodiesel Production by Transesterification Process-An Insight *E-Journal of Chemistry* **7** 1120–32

[3] Weilong H, Xiang Y and Wenying C 2014 Prospective scenarios of ccs implementation in China's power sector: An analysis with China TIMES *Energy Procedia* **61** 937–40

[4] El Sherbiny S A, Refaat A A and El Sheltawy S T 2010 Production of biodiesel using the microwave technique *Journal of Advanced Research* **1** 309–14

[5] Chopade S G, Kulkarni K S, Kulkarni A D and Topare N S 2012 Biodiesel from trans-esterification of triglycerides with methanol : a review *Acta Chim. Pharm. Indica* **2** 8–14

[6] Zare M, Ghobadian B and Fayyazi, Ebrahim., Najafi G 2013 Microwave-assisted Biodiesel Fuel Production from Waste Cooking Oil *International Journal of Agriculture and Crop Sciences.* 1314–7

[7] Azcan N and Yilmaz O 2012 Microwave Irradiation Application in Biodiesel Production from Promising Biodiesel Feedstock : Microalgae (Chlorella protothecoides) *Would Congress on Enginerring and Computer Science* **II**

[8] Dang T H, Chen B H and Lee D J 2013 Application of kaolin-based catalysts in biodiesel production via transesterification of vegetable oils in excess methanol *Bioresource Technology* **145** 175–81

[9] Hassani M, Najafpour G D and Mohammadi M 2016 Transesterification of Waste Cooking Oil to Biodiesel using γ-alumina Coated.pdf *Journal of Materials and Environmental Science* 1193–203

[10] Liu H, Guo H shuang, Wang X jing, Jiang J zhong, Lin H, Han S and Pei S peng 2016 Mixed and ground KBr-impregnated calcined snail shell and kaolin as solid base catalysts for biodiesel production *Renewable Energy* **93** 648–57

[11] Hexin Zhang C H and Y Q 2011 Synthesis, Structural and Thermal Properties of Nano-porous SiO2-based Aerogels *Advances in Nanocomposites - Synthesis, Characterization and Industrial Applications* pp 40–60

[12] Soetaredjo F E, Ayucitra A, Ismadji S and Maukar A L 2011 KOH/bentonite catalysts for transesterification of palm oil to biodiesel *Applied Clay Science* **53** 341–6

[13] Jia L, Li Y, Chen J, Guo X, Lou S and Duan H 2016 Montmorillonite-supported KF/CaO: a new solid base catalyst for biodiesel production *Research on Chemical Intermediates* **42** 1791–1807

[14] Gao L, Teng G, Lv J and Xiao G 2010 Biodiesel Synthesis Catalyzed by the KF/Ca–Mg–Al Hydrotalcite Base Catalyst *Energy & Fuels* **24** 646–651

[15] Gude V, Patil P, Martinez-Guerra E, Deng S and Nirmalakhandan N 2013 Microwave energy potential for biodiesel production *Sustainable Chemical Processes* **1** 5

[16] Diamantopoulos N 2015 Comprehensive Review on the Biodiesel Production using Solid Acid Heterogeneous Catalysts *Journal of Thermodynamics & Catalysis* **6** 1–8

[17] Manco I, Giordani L, Vaccari V and Oddone M 2012 Microwave technology for the biodiesel production: Analytical assessments *Fuel* **95** 108–12

Synthesis of Furfural from Water Hyacinth *(Eichornia croassipes)*

Ismiyarto[1*], Ngadiwiyana[1], T windarti[1], Purbowatiningrum RS[1], M Hapsari[1], FH Rafi'ah[1], Suyanti[1], MS Haq[1]

Laboratorium of Organic Chemistry, Departement of Chemistry,
Faculty of Sciences and Mathematics Diponegoro University

Email: miyato_np@yahoo.com

Abstract. Furfural has been prepared from hydrolysis of dried biomass of water hyacinth *(Eichornia crassipes)* by using diluted hydrochloric acid and sulphuric acid as catalysts. This process involved the conversion of the pentosane fraction in water hyacinth into pentose, and then pentose was cyclodehydrated into furfural. The reaction was conducted in a distillation set with receiving the flask that contains chloroform. Furfural was identified by fehling test which was then characterized using Fourier Transform Infra Red (FTIR) and Proton Nuclear Magnetic Resonance (^{1}H-NMR), followed by Gas Chromatography with Mass Spectroscopy (GC-MS). The yield of furfural obtained using sulphuric acid catalyst was 0.38% and hydrochloric acid catalyst was 0.01% of dried biomass.

1. Introduction

Water hyacinth *(Eichornia crassipes)* is an aquatic plant which can grow fast especially in a rich nutrient water area. 140 millions of water hyacinth can increase every year and these plants cover 140 km^2 water area with a fresh biomass of 28.000 tons [4]. They cause an enormous ecological and economic disaster worldwide, and they must be removed every year [4, 11].

In Indonesia, water hyacinth biomass has been utilized as furniture and handicraft. Also, water hyacinth has been fermented to produce bioethanol [13]. Recently, research on the synthesis of Carboxymethyl Cellulose (CMC) from water hyacinth has been proposed. Carboxymethyl Cellulose (CMC) is a derivated cellulose that is used in food and non-food products for examples as a thickener agent, detergents, and paints [8]. However, thousand tons of this biomass still unutilized properly. A study [3] proven that water hyacinth biomass contains pentosans as much as 23.7%. It's mean that water hyacinth can be used as the promising precursor in the synthesis of furfural.

Furfural is an aldehyde compound that can be derived into furfuryl alcohol and tetrahydrofuran, solvents that widely used in industries. Furfural it self commonly used in the production of plastic resins, metal coatings industry, production of insecticide and food industry [9].

Furfural can be produced by hydrolysis of pentosan with the acid catalyst. Pentosan is one of the important fibre components of non-starch polysaccharides called hemicellulose that can be found in agricultural by products. Hemicellulose is one of the several heteropolymers that present in almost all plant cell walls [12]. The pentosan $(C_5H_8O_4)_n$ fraction of lignocelluloses can be converted into pentose $(C_5H_{10}O_5)$ by acid hydrolysis and continued by cyclodehydration process to produce furfural $(C_5H_4O_2)$. The reaction of this process is shown in **figure 1** [2].

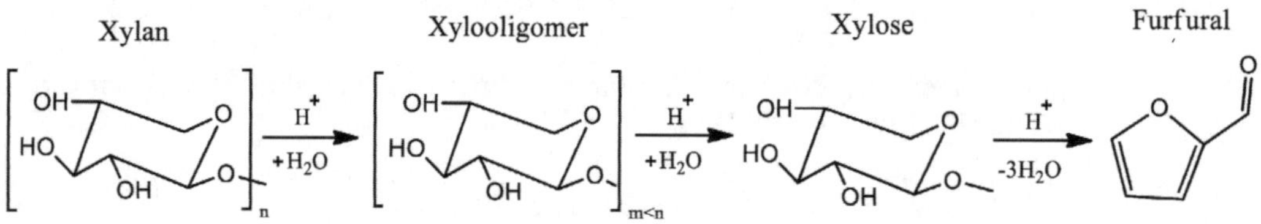

Figure 1. Formation of furfural from pentosan

2. Procedure

2.1 Materials and Instrumentation.

Water Hyacinth *(Eichornia crassipes)* was obtained from Pedalangan village, Sub-district of Banyumanik, Semarang City, Central Java, Indonesia. Freshwater hyacinth was thoroughly washed under tap water to remove sludge and then dried, milled, and stored as dried biomass [6]. The work was carried out in a distillation set as shown in **figure 2**. The apparatus was consisted of eight main parts: 1L round bottom flask as a batch reactor, an oil bath, a connector, thermometers to control the temperature in the round bottom flask and oil bath, a condenser, a receiving flask, and a magnetic stirrer. The condenser was attached to the connector which was also connected to round bottom flask and thermometer. The materials such as 50 g of dried water hyacinth, 700 mL of 1 M HCl, and 0.6 g of NaCl were placed in the round bottom flask and 13 mL chloroform is placed in the receiving flask.

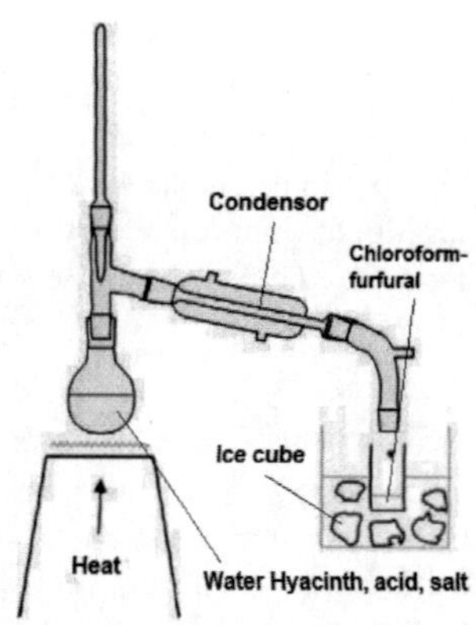

Figure 2. Distillation set up for preparation of furfural

2.2 Experimental procedure

Distillation process was carried out for approximately 7 hours, at 140°-180°C. The distillate containing 50 g of dried water hyacinth, 700 mL of 1 M HCl and 0.6 g of NaCl was set to flow into the receiving flask that containing of 13 mL chloroform. Two layers of aqueous and chloroform would be formed in which the aqueous layer would be at the top and the chloroform-furfural containing layer would be at the bottom of the flask. The chloroform-furfural containing layer was separated by heating in a water bath. The chloroform was recovered at 60°C while the furfural would have remained in the flask. The other experiment was conducted by replacing HCl with 1 M H_2SO_4 in order to compare which one would be more active.

2.3 Identification of furfural

Furfural was identified by Fehling test first before characterized using spectroscopic methods such as Fourier Transform Infra Red (FTIR) model Horizon MB300, ^{1}H-NMR model JEOL ECZ 500R at 500 MHz and Gas Chromatography with Mass Spectrometer (GC-MS) model Perkin Elmer GC Clarus 680 and MS Clarus SQ 8T. Elite- 5 MS Capillary Column (30 m x 0.25 mm I.D.) with 0.25 μm film thickness was used in this work, and Helium at a flow rate of 1 mL/min was used as the carrier gas. The injection port was maintained at 170°C while the column temperature was held at 50°C for 5 minutes to 150°C. The final temperature was held for 25 minutes. Scanning was done from m/z 50 to 300 in one scan. The mass spectral was carried out by comparing with the mass spectra library of National Institute of Standards and Technology (NIST) and Japan AIST/ NIMC Database. This GC-MS method is a modified method from GC-MS analysis furfural from Occupational Safety & Health Administration, US Department of Labor and GC-MS analysis furfural from Malaysian Agricultural Research and Development Institute [9].

3. Result and Discussion

This work is aimed to convert water hyacinth *(Eichornia crassipes)* as an agricultural waste biomass to furfural with diluted acid hydrolysis. The yield of furfural is depending on the percentage of pentosan. Water hyacinth has a composition of 40.71% cellulose, 23.7% pentosan or hemicellulose, and 15.42% lignin [3]. The yield is also affected by acid concentration, reaction time, acid-to-lignocellulosic mass ratio, and temperature of synthesis [7]. The most important parameter affected to sugar yield is acid concentration, while the highest impact on sugar degradation product is temperature [5].

The acid hydrolysis of water hyacinth was carried out using diluted hydrochloric acid and sulphuric acid at a concentration of 1 M. According to a research [1], 1 M is the best concentration to convert furfural from pentosan. The Furfural obtained was in a liquid form. It was colorless to yellowish. The product was confirmed as an aldehyde functional group by the red precipitate that appeared in Fehling test.

The yield of furfural that obtained from acid hydrolysis using 1 M H_2SO_4 and 1 M HCl were 0.19 g (0.38%) and 0.05 g (0.01%) of dried water hyacinth, respectively. It means that H_2SO_4 is more active than HCl in converting furfural from pentosan. As compared to HCl, H_2SO_4 tends to be more selective and stable in the batch reactor. In addition H_2SO_4 have higher boiling point than HCl. Residual sulfates are more easily separable and had less impact than residual chlorides. Besides, H_2SO_4 can reduce corrosiveness, improve selectivity, and has a lower cost [2].

The presence of furfural (aldehyde) was proven by IR spectrum (**Figure 3**) with the existence of four peaks gained at 3132.17 cm^{-1} (sp^{2}C-H), 2812.01 cm^{-1} and 2842.87 cm^{-1} (C-H aldehyde), 1670.25 cm^{-1} (-C=O, carbonyl) and 1461,1 cm^{-1} (-C=C-).

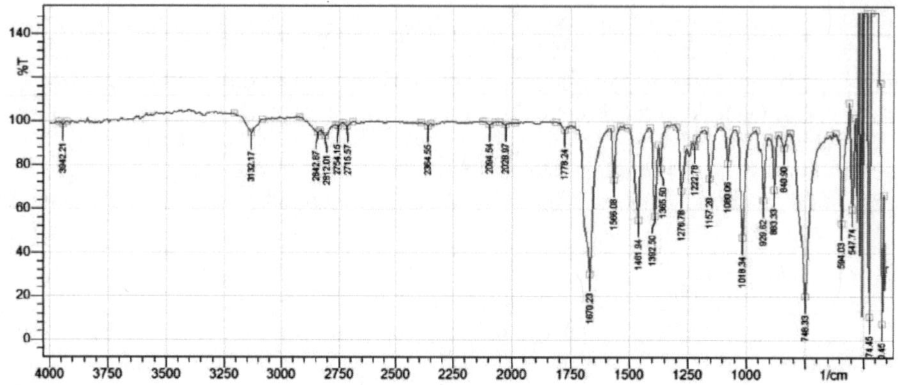

Figure 3. FTIR spectrum of furfural

A typical ^{1}H-NMR spectrum of furfural can be seen in **figure 4**. The aldehyde proton signal appears in the normal low-field position at δ 9.633 (1H). The ring protons appear due to the asymmetry as three groups of signals in a region 6.57 - 7.67 ppm (3H).

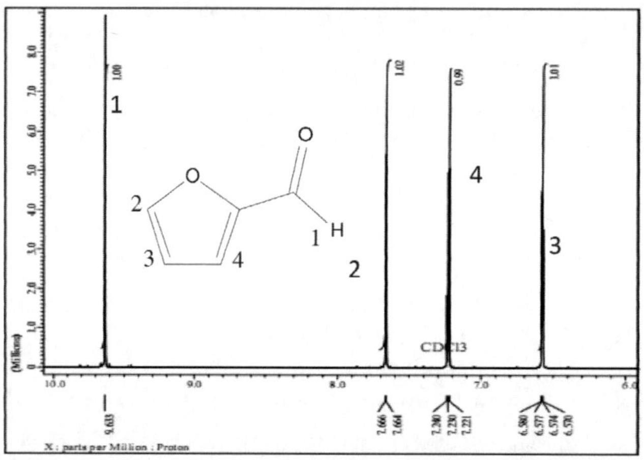

Figure 4. ^{1}H-NMR (500 MHz, $CDCl_3$) spectrum of furfural

Chromatogram in **Figure 5** showed a dominant peak which was indicating the presence of only one constituent. Furfural has a retention time of 5.604 minutes with probability or relative abundance of 100%. On comparison with the mass spectra of the constituent with NIST and Japan AIST/ NIMC, that constituent was characterized and identified as Furfural.

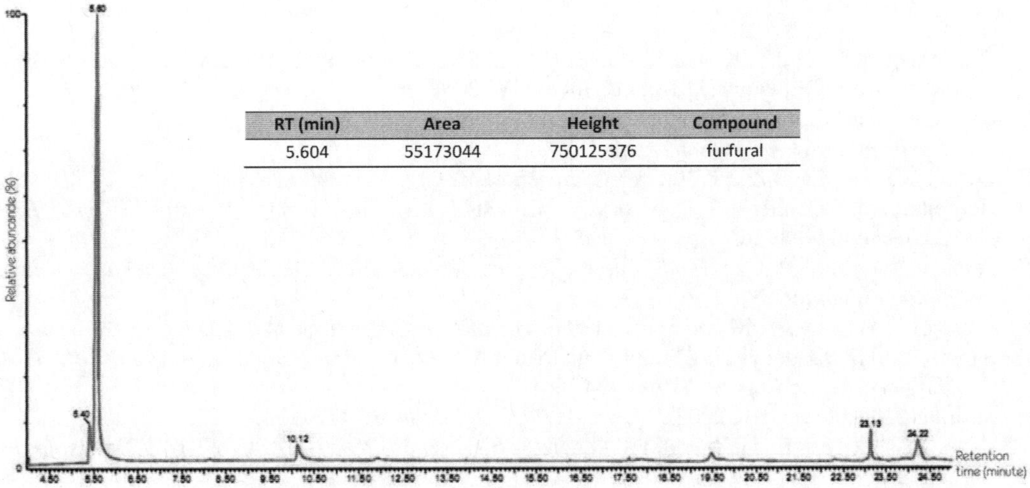

RT (min)	Area	Height	Compound
5.604	55173044	750125376	furfural

Figure 5. Chromatogram of furfural

In comparison with the mass spectra of the constituent with NIST and Japan AIST/ NIMC, that constituent was characterized and identified as Furfural (**Figure 6**). The mass spectrum (**Figure 6**) showed a molecular ion peak at m/z 96 which correlates to a molecular formula of Furfural ($C_5H_4O_2$).

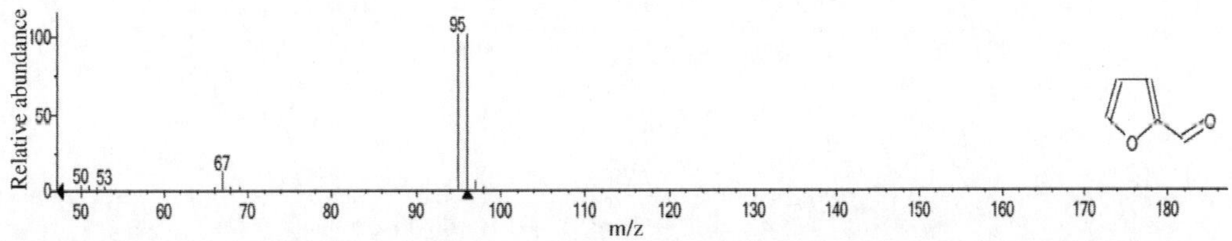

Figure 6. Mass spectrum of furfural (on 5.604 min)

4. Conclusion

Furfural has been successfully synthesized from water hyacinth *(Eichornia crassipes)* by acid hydrolysis method. The yield of furfural obtained using sulphuric acid catalyst was 0.19 g (0.38%) and hydrochloric acid catalyst was 0.05 g (0.01%) of dried biomass. The furfural was colorless to yellowish, with the chemical formula of $C_5H_4O_2$ and molecular weight of 96.

Acknowledgments
We are grateful to the financial support from Diponegoro University (No: SP DIPA-042.01.2.400898/2016, December, 7th 2015).

References

[1] Ambalka, VU and MI Talib 2012 *International Journal of Engineering and Sciences.* **1** 30-36.
[2] Bond JQ, Upadhye AA, Olcay H, Tompsett GA, Jae J, Xing R, Alonso DM,

IOP Conf. Series: Materials Science and Engineering **172** (2017) 012027 doi:10.1088/1757-899X/172/1/012027

Wang D, Zhang T, Kumar R., Foster A, Sen SM, Maravelias CT, Malina R., Barret SRH, Lobo R., Wyman CE, Dumesic JA and Huber GW 2014 *Energy Environ. Sci* . **7** 1500-1523.

[3] Ganguly A, Das S, Bhattacharya A, Singh P, Chatterjee PK and Dey A 2013 *Advances in Chemical Science*. **2** 1

[4] Lu J, Wu J, Fu, Z and Zhu L 2007 *Environmental Management*. **40** 823-830.

[5] Neureiter M, Danner H, Thomasser C, Saidi B and Braun R 2002 *Appl. Biochem. Biotechnol*. **98** 49-58.

[6] Paterson D 2009 Furfural–Fuel and Flavour Molecule of the Future. *Chemistry Education in New Zealand.*

[7] Pessoa JR, Mancilka AIM and Sato S 1997 *Brazillian J. Chem. Eng.* **14** 1-10.

[8] Saputra AH, Qadhayna L and Pitaloka AB 2014 *International Journal of Chemical Engineering and Applications*. **5** 347.

[9] Sashikala M and Ong HK 2007 *J. Trop. Agric. and Fd. Sc.* **35** 165–172.

[10] Tao R, Zeng R, Yin XY, Zhang SX and Yang ZH 2016 *BioResources* **11** 2372-2380.

[11] Villamagna A and Murphy B 2010 *Freshwater Biology*. **55** 282-298.

[11] Wankasi D and Naidoo EB 2012 *American Journal of Food and Nutrition*. **2.2** 47-50.

[12] Zhang Q, Weng C, Huang H, Achal V and Wang D 2016 *Journal Frontiers in Microbiology*. **6** 1411.

IOP Conf. Series: Materials Science and Engineering **172** (2017) 012028 doi:10.1088/1757-899X/172/1/012028

Adsorption of Crystal Violet Dye Using Zeolite A Synthesized From Coal Fly Ash

Jumaeri[1,a), E Kusumastuti, [1,b) S J Santosa[2,c) and Sutarno[2,d)

[1)Department of Chemistry, Faculty of Mathematics and Natural Sciences, Semarang State University, Semarang, Indonesia.
[2)Department of Chemistry, Faculty of Mathematics and Natural Sciences, Universitas Gadjah Mada, Yogyakarta, Indonesia.

Email: [a)jumaeri.unnes@gmail.com, [b) ella.kusuma@gmail.com, [c) sjuari@yahoo.com, [d) sutarno@ugm.ac.id

Abstract. Adsorption of Crystal Violet (CV) dye using zeolite A synthesized from coal fly ash (ZA) has been done. Effect of pH, contact time, and the initial concentration of dye adsorption was studied in this adsorption. Model experimental of adsorption isotherms and adsorption kinetics were also studied. The adsorption is done in a batch reactor at room temperature. A total of 0.01 g of zeolite A was added to the Erlenmeyer flask 50 mL containing 20 mL of the dye solution of Crystal Violet in a variety of conditions of pH, contact time and initial concentration. Furthermore, Erlenmeyer flask and its contents were shaken using an orbital shaker at a speed of 200 rpm. After a specified period of adsorption, the solution was centrifuged for 2 minutes so that the solids separated from the solution. The concentration of the dye after adsorption determined using Genesis-20 Spectrophotometer. The results showed that the Zeolite A synthesized from coal fly ash could be used as an effective adsorbent for Crystal Violet dye. The optimum adsorption occurs at pH 6, and contact time 45 minutes. At the initial concentration of 2 to 6 mg/L, adsorption is reduced from 79 to 62.8%. Crystal Violet dye adsorption in zeolite A fulfilled kinetic model of pseudo-order 2 and model of Freundlich adsorption isotherm.

1. Introduction

The use of coal as a fuel in steam power plants continue to grow, as long with the increasing demand for electricity for various purposes. The burning of the coal not only to produce energy but also produce solid wastes such as fly ash and bottom ash. The amount of the generated fly ash from the coal burning to approximately 7.87% of the coal used [1]. The total coal consumption in Indonesia (new and existing power plant power plant) will reach 87.7 million tons in the year 2015 and 166.2 million tons in the year 2019. The fly ash and bottom ash production will attain 4.8 million tons in the year 2015 and 8.31 million tons in the year 2019 (assumptions 5% from fuel consumption) [2]. If the accumulated fly ash is not used, the waste would cause problems for the surrounding environment and have no economic value. That's why the need for efforts to cultivate and utilize the fly ash in the various fields. The content of SiO_2 and Al_2O_3 in the fly ash, allowing fly ash can be used as raw materials in the synthesis of zeolite A.

Meanwhile, the use of synthetics dyes for industry application continuously increases. The disposal of dye waste into the stream of water cause not only high color intensity but also disrupt aquatic life water accepts it. Waste water containing dye is very difficult to be processed because the dye is an organic compound that is stable, resistant to aerobic decomposition and stable to light [3]. The most synthetic dye is an aromatic organic compound, soluble and dispersible in water [4]. The presence of dyes in the wastewater is undesirable and potentially inhibiting photosynthesis process [5].

The cationic dyes of Crystal Violet (CV), with C.I 42555, are widely used in the textile dyeing. The dye has a brightness and high color intensity, even at very low concentrations [6]. The disposal of CV into the hydrosphere can lead to environmental degradation. The CV is easily adsorbed fish tissue from the exposure water and reduced metabolism of fish due to leucomoiety, leucocrystal violet [7]. Some studies exhibited carcinogenic and mutagenic effects of the CV in rodents. Therefore the waste water containing dyestuff CV needs to be treated before it is discharged into water bodies.

The various methods have been used for removal of dyes in aqueous solution, either through ion exchange, activated carbon adsorption, membrane technology and coagulation, flocculation, and so on. Although the use of activated carbon obtained the most effective for the removal of dyes, but its use is limited due to high operating costs [8]. Adsorption is considered as the most effective method for the treatment of waste water. The effectiveness of alternative adsorbent to remove the dye has been studied. Zeolite is a once alternative material used as an adsorbent in treatment waste water of the heavy metals, dyes and other organic pollutants. Regarding with isomorphic substitution of Si by Al, the overally surface of the zeolite is negatively charged. The lower of Si/Al ratio of the zeolite, the higher negative charge of the zeolite surface. Zeolite A has a low Si/Al ratio, so that suitable for absorbing the cationic dye, CV. In this research, the adsorption of CV in zeolite A from coal fly ash has been studied. Adsorption was carried out in batch system with a variety of adsorption conditions, namely pH, contact time and initial concentration of the solution. Adsorption isotherms and kinetics models were also studied based on the experimental data obtained.

2. Materials and Methods

In this study, the entire chemical used: the dye Crystal Violet, NaOH, and HCl, are analytical reagent grade, produced E Merck. The tools used include a set of glassware, analytical balance, orbital shaker, centrifuge, universal pH indicator sticks, and Genesis Spectrophotometer.

2.1. Adsorbent

The used adsorbent in this study is zeolite A from coal fly ash of PLTU Suralaya Banten Indonesia. The zeolite A (ZA) was hydrothermally synthesized through a modified alkaline fusion. The synthesis process and characterization of zeolite A has been described in our previous research. The synthesized zeolite A exhibited surface area of 18.998 m^2/g and showed the similarity with the commercially zeolite A Wako produced Japan, in the diffraction patterns, SEM image, and IR spectra [9]. Figure 1 showed the diffraction patterns and SEM image of the zeolite A synthesized from fly ash.

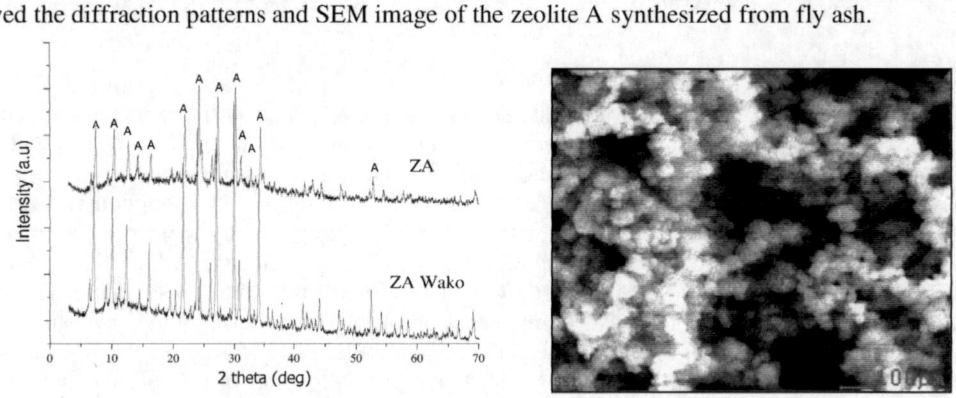

(a) (b)

Figure 1. Diffraction patterns (a) and SEM image (b) of the synthesized zeolite A [9].
ZA: zeolite A synthesized from fly ash, ZA Wako: zeolite A commercial Wako

2.2 Adsorbate

The used adsorbate in this study is the cationic dye of *Crystal Violet* (CV), having molecular formula $C_{25}H_{30}N_3Cl$ and molecular weight of 407.99 g/mol. Figure 2 shows the molecular structure of the *Crystal Violet* dye. The stock solution of CV was prepared by dissolving 1 g of CV in 1 liter of demineralized water and stored in glass jars for the further use.

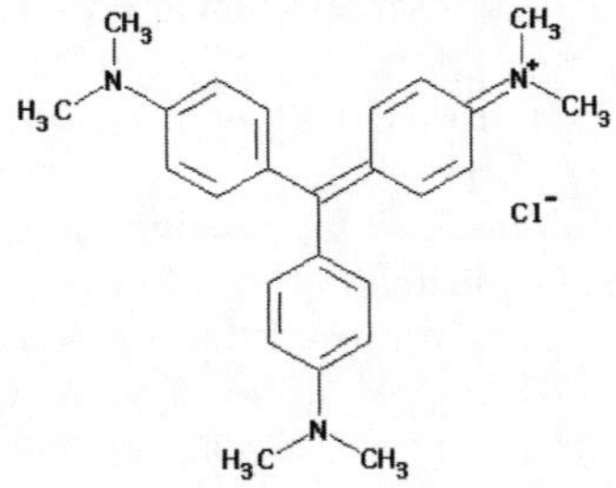

Figure 2. Molecular structure of Crystal Violet [10].

2.3 Adsorption of dye Crystal Violet (CV)

The Crystal Violet adsorption on zeolite A (ZA) was conducted in a batch reactor at room temperature. A total of 0.01 g ZA was added to 20 ml of the dye solution of Crystal Violet in various conditions (pH, contact time and initial concentration) and shaken by using an orbital shaker at a speed of 200 rpm. After the adsorption in the given period is carried out, the solution was centrifuged for 2 minutes at 1000 rpm, so the solids separated from the solution. The concentration of the dye after adsorption was determined using Genesis-20 spectrophotometer, at wavelengths corresponding to the pH condition of each solution.

3. Result and Discussion

3.1 Effect of pH

Previously the adsorption is carried out, firstly determined the maximum wavelength of the CV solution at the different levels of acidity (pH). The results of the measurement of CV absorbance solution show the maximum absorbance occurred at pH 6 and wavelength (λ) 490 nm.

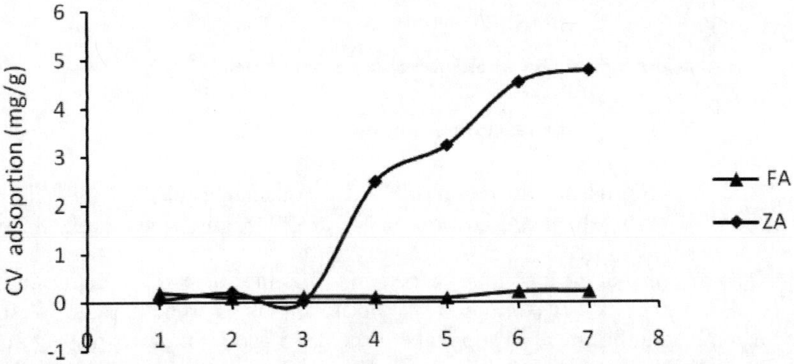

Figure 3. Adsorption of CV dye at various pH.
FA: fly ash, ZA: zeolite A synthesized from fly ash

The adsorption of CV dye in aqueous solution is performed in orbital shaker 200 rpm during 60 minutes, with the initial concentration of 3 mg/L. The adsorption used the adsorbents of fly ash (FA) 0.1 g and 0.01 g zeolite A (ZA). After the adsorption process, the CV concentration was determined using UV-VIS Spectrophotometer at wavelength 590 nm. The amount of CV dye adsorption at various pH is presented in figure 3. In this figure, it appears that the number of CV adsorbed by zeolite A increases with pH. The maximum adsorption of CV occurred at pH 6. This is in line with the maximum CV adsorption on acacia leaves adsorbent which occurs at pH 6 [7]. Figure 3 also shows that the amount of CV adsorption at the ZA adsorbent is greater than FA adsorbent. This occurs because of the zeolite A have greater electrostatic attraction than fly ash in their interaction with CV cationic dye.

The increase in pH, which means an increase in the concentration of OH^-, causing the surface of the zeolite A more negatively charged and increasing the electrostatic interaction with cationic dye so that the adsorption of the dye CV increases. In general, for a cationic dye, the adsorption capacity and rate constants tend to increase with increasing pH of the solution [7]. For the subsequent experiments were performed at the optimum adsorption namely at pH 6 and the maximum wavelength of CV 490 nm.

3.2 Effect of contact time

Contact time is an important factor in the adsorption process. In general, the adsorption process runs quickly at the start of contact time and gradually decreases with increasing time of adsorption. In physical adsorption, the adsorbate species are adsorbed in a short contact time. Nevertheless, the adsorption equilibrium can often be achieved within a longer time. The adsorption of CV dye was performed on contact time of 0, 5, 10, 20, 30, 45, 60, 75 and 90 minutes, in the orbital shaker at 200 rpm. The adsorption was carried out at pH 6, the initial concentration of 3 mg/L and using adsorbent 0.1 g FA, 0.01 g ZA. The amount of the adsorption of CV at a variety of contact time (qt) was presented in figure 4.

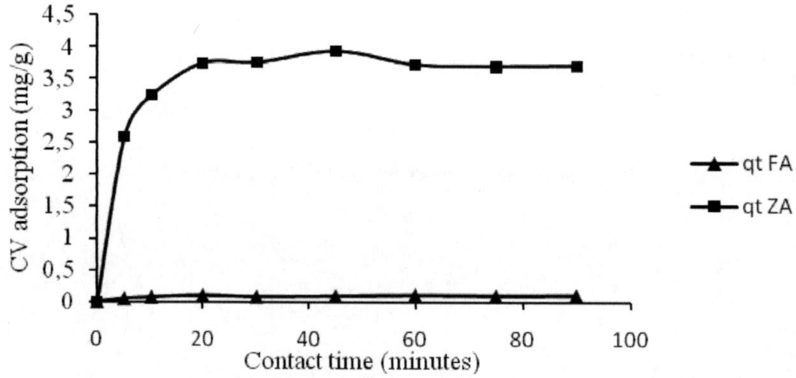

Figure 4. Adsorption of CV dye at contact times.
qt FA: adsorption CV at fly ash, qt ZA: adsorption CV at ZA: zeolite A synthesized from fly ash

From figure 4, it appears that the amount of adsorbed CV is increasing with increasing contact time and gradually fixed and reaches a maximum at 60 and 45 minutes respectively to FA and ZA. The result exhibited that the equilibrium adsorption take place at 45 and 60 minutes respectively for zeolite A and fly ash. In the early research, the CV adsorption equilibrium time on BTEA-CTMA-bentonite and bentonite was obtained at 70 minutes [5], while the adsorption CV on acacia leaves at 120 minutes [7]. From figure 4 also appears that the maximum adsorption of Crystal Violet to the adsorbent FA and ZA each other 0.1134 mg/g (17.35%), and 4.2927 mg/g (65.68%). Adsorption CV runs quickly at the start of contact time and gradually reduced and the adsorption rate is relatively constant at the optimum time (equilibrium). With increasing contact time the active sites present in adsorbent ZA were filled adsorbate CV and diminishing with the passage of the adsorption process. The CV dye adsorption in zeolite A is higher than in the fly ash adsorbent. This happens because the zeolite A have the higher electrostatics attraction than fly ash, due to the surface of the zeolite A more negatively charged.

3.3 Effect of initial concentration

To study the effect of the initial concentration, the adsorption of CV was done in orbital shaker at 200 rpm, pH 6, the contact time of 60 and 45 minutes respectively to adsorbent FA and ZA. The initial concentration solution of CV is 1, 2, 3, 4, 5 and 6 mg/L. Adsorption CV at the different initial concentrations was presented in figure 5.

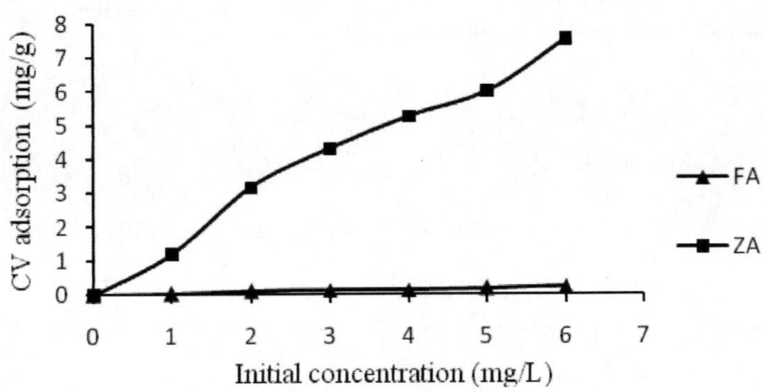

Figure 5. Adsorption of CV dye at various initial concentration.
FA: fly ash, ZA: ZA: zeolite A synthesized from fly ash

Adsorption CV per unit mass of adsorbent increases with the initial concentration, although the percentage of adsorption decreases with increasing initial concentration. At the initial concentration of 1 to 6 mg/L adsorption CV increased from 0.043 becomes 0.2276, and 1.2045 becomes 7.5359 mg/g, respectively to adsorbent FA and ZA. This happens because there is an increase of the concentration gradient driving force with the increase in the initial concentration of the dye [11]. Similar results were obtained that the adsorption CV increased from 7 to 23 mg/g with increasing initial concentrations of 50 to 200 mg/L [7].

3.4 Adsorption isotherm of CV

Study of the adsorption isotherm is obtained by variation of the initial concentration of dye CV from 2 to 6 mg/L, at pH 6 and the optimum contact time for adsorbent FA and ZA. Langmuir and Freundlich isotherm models used to describe the color characteristics of adsorption equilibrium CV. Langmuir and Freundlich isotherm constants respectively can be calculated from the slope and intercept the plot between Ce/qe vs Ce and log qe vs log Ce , each other from equation (1) and (2).

$$q_e = \frac{q_{m}K_L C_e}{1 + K_L C_e} \qquad (1)$$

$$q_e = K_F C_e^{1/n} \qquad (2)$$

The linear curve of Langmuir and Freundlich isotherm models on dye adsorption CV in ZA were presented in Figure 6. Based on the value of slope and intercept in Figure 6, the adsorption isotherm parameters value CV in ZA were obtained, as listed in Table 1.

By comparing the correlation coefficient between the Langmuir and Freundlich isotherm models can be concluded that the Freundlich isotherm models more suitable for adsorption CV in ZA. The maximum monolayer adsorption (qm) obtained is 10.753 mg/g and K_L 7.75 L/mg.

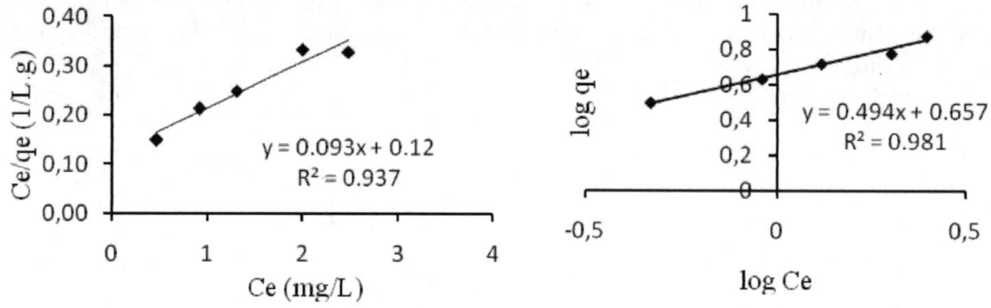

(a) (b)

Figure 6. Model of adsorption isotherm CV in ZA: Langmuir (a) and Freundlich (b)

Table 1. Parameter of adsorption isotherm model of CV.

Adsorbent	Adsorption Isotherm					
	Langmuir			Freundlich		
	q_m (mg/g)	K_L(L/mg)	R^2	n	K_f (mg/g)	R^2
ZA	10.672	7.808	0.937	2.02	4.5	0.981

The obtained equilibrium factor value, R_L, is 0.06 to 0.02 in range, indicating a favorable adsorption of CV dyes on the adsorbent ZA. In the model of Freundlich isotherm parameters obtained K_f 4.5 and 1/n is 0.494 or n = 2.02. The quantity of 1/n indicate that the adsorption involves a heterogeneous surface and chemical adsorption.

3.5 Kinetics of Adsorption of Dyes CV

Kinetics of adsorption of the dye CV solution on the adsorbent zeolite A of fly ash were tested using kinetic model of pseudo- order first Lagergren, as in equation (3) and the pseudo-order second of Ho, as in equation (4).

$$\log(q_e - q_t) = \log(q_e) - \frac{k_1}{2{,}303} t \qquad (3)$$

$$\frac{t}{q_t} = \left(\frac{1}{k_2\, q_e^{2}}\right) + \left(\frac{1}{q_e}\right) t \qquad (4)$$

The kinetic constants in the linear equation (3 and 4) can be computed from the slope and intercept of plot log (qe-qt) vs. t and t/qt vs. t respectively. The resulted adsorption kinetics curves of CV in ZA were presented in figure 7. Figure 7 showed plot of ln(qe-qt) againts t and t/qt against t for adsorption CV in zeolite A, which resulted in a straight line, with R^2 0.851 and 0.997 respectively. Of the value of the slope and intercept of figure 7 can be obtained constanta of k1, k2 and qe as listed in table 2.

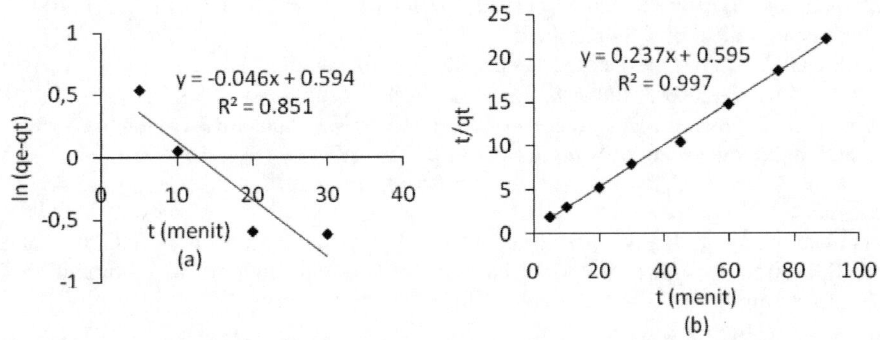

Figure 7. Kinetics of pseudo-first-order (a) and pseudo-second order (b) of adsorption CV on ZA.

Table 2. Parameter of adsorption kinetics CV on ZA.

Adsorbent	Pseuodo-orde 1			Pseudo-orde 2		
	qe(mmol/g)	k_1 (min^{-1})	R^2	qe (mmol/g)	k_2(g/mmol.min)	R^2
ZA	4.4×10^{-3}	0.046	0.851	10.3×10^{-3}	38.51	0.997

Based on the obtained R^2, it can be concluded that the kinetic model of pseudo-order two is more appropriate to describe the process of adsorption CV dye in ZA. Similar results were obtained that the CV adsorption is best described by kinetics model of pseudo-second order [5, 7]. Although the mechanism of adsorption on zeolites are complex due to the pore structure, surface charge on inside and outside, heterogeneity and the crystal imperfect on its surface, but the zeolite adsorption properties mainly related to its ability as an exchanger ion [12]

4. Conclusion

Regarding on the adsorption of Crystal Violet, zeolite A synthesized from fly ash showed adsorption capacity much greater than the fly ash. The adsorption of Crystal Violet depending on the pH and the maximum adsorption occurred at pH 6. The adsorption process is best described by an equation pseudo-orde 2 kinetics. Furthermore, the adsorption equilibrium data fulfilled the Freundlich adsorption isotherm models. Based on these results, can be concluded that zeolite A synthesized from fly ash is an effective adsorbent for Crystal Violet dye in aqueous solution.

References

[1] Prijatama H 1993 *Fly ash and Utilization* National Seminar Papers of Coal Indonesia (Universitas Gadjah Mada, Yogyakarta 7- 8 September 1993)
[2] Ministry of Energy and Mineral Resources Republic Indonesia 2016 *Indonesia Electricity Development Plan And Indonesia Coal-Ash anagement Implementation* International Coal Based Power Conferences (New Delhi, 17th March 2016)
[3] Garg V K, Gupta R, Yadav A, and Kumar R 2003 *Bioresour. Technol.* **89** 121–124
[4] Hazrat A and Muhammad SK 2008 *Environ. Sci.Technol.* **1(3)** 143-150
[5] Jian-min R., Si-wei W and Wei J 2010 *World Acad. Sci. Eng. And Technol.* **65**, 790-795

IOP Conf. Series: Materials Science and Engineering **172** (2017) 012028 doi:10.1088/1757-899X/172/1/012028

[6] Chu K H and Chen K M 2002 *Process Biochem.* **37** 595–600.
[7] Prasad A L and Santhi T 2012 *Sustain. Environ. Res.* **22** 113-122
[8] Nasuha N, Zurainan H Z, Maarof H I, Zubir N A and Amri N 2011 *Int.Conf. Environ. Sci and En.IPCBEE* **8**
[9] Jumaeri, Santosa SJ, Sutarno and Kunarti E S 2014 *Advanced Materials Research* **1043** 198-203 Trans Tech Publications, Switzerland
[10] © Merck KGaA, Darmstadt, Germany, chemicals(at)merck.co.id, 2012
[11] Ashtoukhy E S Z E 2009 *J. Environ. Manage.,* **90(8)** 2755-2761
[12] Calzaferri G, Brühwiler D, Megelski S, Pfenniger M, Pauchard M, Hennessy B, Maas H, Devaux and Graf A 2000 *Solid States Sci* **2** 421-447

Acknowledgment
The authors would like to thank Directorate General of Higher Education (DGHE), Department of National Education Republic of Indonesia for the financial support in Competitive Research Program (Contract Number: 1.5.5/UN37/PPK.3.112014).

IOP Conf. Series: Materials Science and Engineering **172** (2017) 012029 doi:10.1088/1757-899X/172/1/012029

The Acute Toxicity Test of Methanolic Extract of *Hyptis pectinata* Poit on Liver Balb/c Mice

M Suzery[1], B Cahyono[1] and P Astuti[2]
[1]Department of chemistry, Faculty of Science and Mathematics Diponegoro University, Semarang, Indonesia
[2]Physiology and Microanatomy Division, Faculty of Veterinary Medicine, Gadjah Mada University, Jogjakarta, Indonesia

Email: meiny_suzery@yahoo.com

Abstract. Plants from Lamiaceae family has almost entirely reported having physiological activities. One of them is *Hyptis pectinata* Poit plant. Research on the toxicity of *Hyptis pectinata* needs to be done to protect people from the possibility of its harmful effects. This study aim to know the acute toxicity effects of *Hyptis pectinata* extract (HPE) on liver of Balb/c mice. This research was a laboratory experimental study using the post test only controlled group design. Balb/c mice were randomly divided into 4 groups. K (control group), P1, P2 and P3 (treatment groups; given HPE 200mg/kgBW, 1000 mg/kgBW, and 5000 mg/kgBW, respectively). The extract was orally given with gastric tube on the first day and the mice were terminated at the 8th day then the livers were observed. The Kruskal-Wallis test for macroscopic morphological and volume of the liver showed there were no significant difference with p=0.406 and p=0.054. The highest liver histopathological score was in P3 group. The Kruskal-Wallis test showed significantly difference (p=0.000). Continued with Mann-Withney test that showed a significant difference in K-P1 (p=0.009), K-P2 (p=0.009), K-P3 (p=0.009), P1-P2 (p=0.028), and P1-P3 (p=0.009). In conclusion, the HPE is safe to use which has no complication with liver of mice.

1. Introduction

There are at least 295 genera and 7775 species of plants from Lamiaceae family and most widely found in Mexico, Senegal, Brazil, Spain and in Asia such as India, Malaysia and Indonesia. Some species of this genus of plants, namely *Hyptis pectinata* Poit, *Hyptis suaveolens*, *Hyptis oblingofolia*, *Hyptis tumentosa*, *Hyptis verticillata*, *Hyptis capitata*, *Hyptis brevives* Poit, *Hyptis fruktikosa* and *Hyptis urticoide*, almost entirely reported to have physiological activities. The activities shown by this plant: anti-inflammatory, antipyretic, cancer [1, 2], ease breathing, skin diseases [3], disorders of the stomach, fever [4], and infections by bacteria [5].

In India, *Hyptis pectinata* Poit leaves are made in the form of tea which used as a beverage prevent abdominal pain [2]. Water extract of this plant has also been reported to increase liver regeneration [6].

Currently, in society the issue back to nature is growing and thus influence use of traditional medicine tends to increase. Meanwhile, many people assume that the use of herbs or traditional medicines are relatively safer than synthetic drugs. However it does not mean herbs or traditional medicines do not have any adverse side effects when the use is not in the right path. The accuracy concerns to the right dose, the method and time of usage and also the selection of appropriate ingredients in accordance to indications of its use. Traditional medicine is usually used to maintain health, prevent disease, cure disease and restore health.

Research on the toxicity of *Hyptis pectinata* need to be done to protect people from the possibility of its harmful effects. Toxic effects of drugs are often seen in the liver because the liver has a central role in metabolizing a substance that goes into our bodies. The liver will change the structure of drugs that are lipophilic to be hydrophilic thus can easily removed from the body through urine or bile. Biliary excretion of a substance allows the accumulation of xenobiotics in the liver, causing hepatotoxic effects. Research on the toxicity and adverse reactions of *Hyptis pectinata*, both the single and multi herbal usage, has not been done. This makes the knowledge of the therapeutic dose and toxic dose of *Hyptis pectinata* herbal, especially to the liver, becomes ambiguous. Therefore, this study purpose to determine the safety and effects of this plant extract acute toxicity to the liver as the organ that functions in drug metabolism.

2. Material and method

Materials: *Hyptis pectinata* Poit plants, Liver materials tissue in mice for histopathological examination routine, standard feed mice, aquadest, buffered formalin. Additional material for immunohistochemical staining. Equipment: Tool making preparations for histological, mice cages, stomach probes, minor surgical tools, beakers, microscopes and digital cameras.

2.1. Colection anfd extraction of the plant

Hyptis pectinata poit plants was obtained from the village Kanayakan East Dago, Bandung. The specimens were stored at the herbarium of Department of Biology, Faculty of Science and Mathematica, Diponegoro University, Semarang. *Hyptis pectinata* leaves (1 kg) was dried under the sun for 8 days, and obtained 650g of crude simplicia (62%). The dried material result was then determined the moisture content by using the method of distillation (moisture content 7.6%) and smoothed by squeezing to obtain a powder. Furthermore, macerated using methanol (3x24 hours), then filtered and the solvent is evaporated using vacuum rotary evaporator (Buchi) until obtained thick extract (HPE) and ready for the next study.

2.2. Animal

Balb/c male mice (weight 25-35g) were obtained from Integrated Research and Testing Laboratory (LPPT) Unit IV, Gadjah Mada University. All the animals were maintained under standard condition, with temperature at 25±2 0C, 12 h dark-light cycle and relative humidity. The rat received food and water ad libitum. This study was approved by the Ethics Committee for Health Research and Medical University School of Medicine and Kariadi Hospital Semarang, Indonesia (Medical Letter No: 611/EC/FK/RSDK).

2.3. Acute toxicity study

This study uses a post-test only controlled group design using adult Balb/c mice males (age 2-3 months), in health physical condition and not disabled. The study was conducted in May-July 2016 in microanatomy laboratory of Veterinary Medicine Faculty, University of Gadjah Mada, Yogyakarta, Indonesia. Research carried out for two weeks with each of the seven days for acclimatization and seven days for a treatment time of *Hyptis pectinata* Poit extract (HPE).

Sample 20 mice and were divided into four treatment groups:

1. The control group (K) which is only given aquadest
2. P1 group is group that exposed HPE with dose of 200 mg/kg
3. P2 group is group that exposed HPE with dose of 1000 mg/kg
4. P3 group is group that exposed HPE with dose of 5000 mg/kg

Then observed for 7 (seven) days and terminated on the eighth day for observation of its liver. After that, the picture of macroscopic morphology, volume, and liver histology of Balb/c mice are observed.

2.4. Statistical Analysis

The data were analyzed and the results are presented in tables and boxplot. To determine the normality of the data normality test using the Shapiro-Wilk test. When the data were normally distributed and homogeneous, the parametric analysis will do. Where the conditions are not met then the parametric analysis performed in non-parametric analysis. Limit when significance level was $P \leq 0.05$ with 95% confidence interval. The data analysis done using software SPSS Ver. 15.0 for Windows.

3. Results

3.1. The Liver Macroscopic Image

Table 1. Results of liver volume measurement

Groups	*Liver volume (ml)*		
	Median	**Minimum**	**Maximum**
Control	4	4	4
P1	3.5	3.0	4.5
P2	3.5	3.0	4.0
P3	2.5	2.5	3.5

*difference test non-parametric Kruskal-Wallis $p = 0.054$

$p > 0.05$: there are no differences in liver volume significantly.

3.2. The Liver Microscopic Image

The highest mean score of histopathological changes in the treatment groups of three (P3). Scores were assessed include changes in the form parenkimatosa degeneration, hydropic degeneration and necrosis. The mean value of changes in the liver histological structure in each treatment group are shown in Table 2 as follows:

Table 2. The mean score of histopathological changes image of liver cells of Balb/c mice

Groups	***Mean score of histopathological changes image of liver cells of Balb/c mice***	
	Median	**Standard deviation**
Control	314.20	66.53
P1	430.80	70.06
P2	651.20	33.38
P3	757.60	32.11

* Kruskal-Wallis test: $p=0.000$

$p < 0.05$: there are no significant differences in the structure of the histopathology of liver cells in two groups. Furthermore, Mann-Whitney test was used to assess the differences between groups. The results of the Mann-Whitney test can be seen in Table 3 below:

Table 3. The p-value at the Mann-Whitney test between groups Group K, P1, P2 and P3

Groups	*K*	*P1*	*P2*	*P3*
K		0.175	0.009	0.009
P1	0.175		0.009	0.009
P2	0.009	0.009		0.028
P3	0.009	0.009	0.028	

4. Discussion

Results of macroscopic observation and analysis of the liver showed that the *Hyptis pectinata* extract (HPE) acutely not have a significant effect on the macroscopic image of the liver, in this case assessed morphology and the volume of the liver. Kinds of substances contained in a material, the administered dose, and duration of exposure are factors that affect the liver damage [7]. The duration of the exposure can be categorized as acute (≤ 2 weeks), sub-chronic (2 weeks-3months) and chronic (>3months). Small degeneration often does not cause macroscopic changes in the liver and these changes usually occur in chronic administration [8]. Besides the liver also have high regeneration ability, tissue loss due to toxic substances or due to surgery can stimulate the regeneration of liver cells so that the mass of the liver tissue back to normal [9].

Results of statistical analysis of the microscopic image showed that the control group (K) and the treatment group one (P1) was not significantly different ($p = 0.175$), whereas the control group (K) with treatment group two (P2) and the treatment of three (P3) there significant differences with each value of $p = 0.009$. The control group (K) compared with the treatment group (P1) was not significantly different it is because the administered dose is the smallest dose and still a safe dose thus the resulting effect is also small. The control group compared to the treatment group two (P2) and the treatment of three (P3) had a significant difference, it indicates that the concomitant increase in the dose administered, the effects are also getting bigger then offset by the increasing number of cells undergoing necrosis.

The score changes in the microscopic image of liver cells increases in accordance with the increasing of dose given of *Hyptis pectinata*. This is consistent with the dose-response theory, where the higher the dose given, the higher the response generated and a considerable range will be found therapeutic doses, toxic and lethal doses [10].

Changes of histopathologic image seems increase with the increasing of doses. These changes are also consistent with a dose-response theory. Large doses can cause liver cell damage also increase further.

5. Conclusion

In the acute treatment of HPE, the macroscopic appearance of liver between the control group compared with the treatment groups did not significantly difference and the microscopic appearance of the liver between the control group which compared with treatment groups was significantly difference. In conclusion, the HPE is safe to use which has no complication with liver of mice.

6. Acknowledgements

The authors thank the Directorate of Research and Community Services Directorate General for Strengthening Research and Development of the Ministry of Research Technology and Higher Education through a National Strategy program (STRAGNAS) in 2016.

7. References

[1] Bispo MD, Mourao RHV, Franzotti EM, Bomfim KBR, Arrigoni-Blank MF, Moreno MPN, Marchioro M and Antoniolli A 2001 Antinociceptive and antiedematogenic effects of the aqueous extract of Hyptis pectinata leaves in experimental animals *J. Ethnopharmacol* **76** 81–86

[2] Boalino DM, Connolly JD, McLean S, Reynolds WF, and Tinto WF 2003 A-pyrones and a 2(5H)-furanone from *Hyptis pectinata Phytochemistry* **64** 1303–1307
[3] Malan K, Pelissier Y, Marion C, Blaise A and Bessiere, JM 1988 The essential oil of *Hyptis pectinata Planta Medica* **54** 531–532
[4] Martinez, M 1989 *Las Plantas Medicinales de Me´xico* Editorial Botas Mexico p. 508
[5] Rojas A, Hernandez L, Pereda-Miranda R and Mata, R 1992 Screening for antimicrobial activity of crude drug extracts and pure natural products from Mexican medicinal plants *Journal of Ethnopharmacology* **35** 275–283
[6] Melo GB, Silva RL, Melo VA, Antonioll AR, Jordani Souza ME, Jordani MC, and Castro-e-Silva Jr O 2005 Effects of aqueous extract of Hyptis pectinata on liver mitochondrial respiration *Phytomedicine* **12** 359-362
[7] Darmansjah I, Wiria MSS. Dasar Toksikologi. In: Gunawan SG, Setiabudy R, Nafrialdi, Elysabeth 2007 *Farmakologi dan Terapi*. 5th ed. Jakarta: Departemen Farmakologi dan Terapeutik Fakultas Kedokteran Universitas Indonesia pp 820-825
[8] Robins SL and Kumar V 1995 *Buku Ajar Patologi II* 4th ed. Alih bahasa: Pendit BU Jakarta: EGC; 3 p 18
[9] Junqueira LC and Carneiro J. 1997 *Histologi Dasar* 3rd ed. Alih bahasa Hartono A. Jakarta: EGC; p 354
[10] Suzery M and B cahyono 2014 Evaluation of Cytotoxicity Effect of *Hyptis pectinata* Poit (Lamiaceae) extracts using BSLT and MTT methods *Jurnal Sains dan Matematika*. **22** 82-86

Non-invasive Self-Care Anemia Detection during Pregnancy Using a Smartphone Camera

M D Anggraeni[1*], A Fatoni[2]

[1]Department of Nursing, Faculty of Health Sciences, Universitas Jenderal Soedirman, Purwokerto, Indonesia
[2]Department of Chemistry, Faculty of Mathematics and Natural Sciences, Universitas Jenderal Soedirman, Purwokerto, Indonesia.

*E-mail: mekar.dwi@gmail.com

Abstract. Indonesian maternal mortality rate is the highest in South East Asia. Postpartum hemorrhage is the major causes of maternal mortality in Indonesia. Anemia during pregnancy contributes significantly to postpartum hemorrhage. Early detection of anemia during pregnancy may save mothers from maternal death. This research aim to develop a non-invasive self-care anemia detection based on the palpebral color observation and using a smartphone camera. The color intensity (Red, Green, and Blue) was then measured using a Colorgrab software (Loomatix) and analyzed compared to the hemoglobin concentration of the samples, measured using standard Spectrophotometer method. The result showed that the red color intensity had a high correlation (R^2=0.814) with a linear regression of y=14.486x + 50.228. This preliminary study may be used as anemia early detection which more objective compared to visual assessment usually performed.

1. Introduction

The maternal mortality rate (MMR) among Indonesian mothers is the highest in ASEAN country[1]. The MMR in Indonesia was 228 per 100,000 live births in 2007 and it increased up to 359 per 100,000 live births in 2012 [2, 3]. The MMR in Indonesia is still far from the Millennium Development Goal's target which stated that the MMR should be 102 per 100,000 live births in 2015 [4]. Several factors contribute to the high MMR in Indonesia. According to Khan [5], the principal cause of MMR in Asia was a hemorrhage. The Indonesian Ministry of Health [4] also reported that the major cause of MMR during 2010-2013 remained the same, it was a hemorrhage. In 2013, 30.3% of Indonesian mothers died due to hemorrhage during the perinatal period. There was a significant association between perinatal hemorrhage with anemia status [5, 6]. The lower hemoglobin level during pregnancy caused the higher risk of postpartum hemorrhage [7]. The combination between low hemoglobin level during pregnancy and postpartum hemorrhage contributed significantly to 40 - 43% of maternal death in Africa and Asia [8].

Anemia during pregnancy is a public health problem in developing countries. More than 50% of pregnant mothers in developing countries were anemic [9]. In addition, Indonesia Basic Health Research [4] reported that 37.1% of Indonesian women suffered from anemia during pregnancy. Anemia early detection may prevent Indonesian mothers from postpartum hemorrhage [10]. Several strategies are conducted by the Indonesian government to reduce anemia during pregnancy. It includes anemia early detection which is conducted during trimester 1 and 3 of pregnancy at the health care facilities. Then, the pregnant mothers are categorized into certain anemia status. The anemia cut-offs among pregnant mothers are divided into normal: >11g/dL, mild: 10 - 10.9g/dL, moderate: 7 - 9.9g/dL, and severe: < 7g/dL [11]. The Indonesian health care providers use several methods to detect anemia among pregnant mothers.

Several methods have been developed to detect hemoglobin levels, such as immunoassay, electrochemical, chemiluminescence, high performance liquid chromatography and mass spectroscopy. Those methods require expensive equipment, complicated sample preparation, and trained analysts. In the other hand, a physical examination has long been used to detect anemia status among pregnant mothers by health care providers. The anemia status is evaluated by looking at the reddish color of the inside lower eyelid. A pale red color shows an anemia status but a pink to clear red color shows a non-anemia status [12]. This method has long been used by health care providers, particularly who have limitation to access modern testing instruments. However, this method requires examiner's experience and has a high subjectivity. Develop a simple instrument to detect the anemia status by using the reddish color of the lower eyelid may increase the result's objectivity. Pregnant mothers also may use the simple instrument to detect the anemia status by themselves.

The use of simple equipment for colorimetric analysis (color comparison) pioneered by Liebhafsky and Winslow in 1950 that using a light bulb and a cylindrical graded to determine the levels of iron and copper in the water[13]. Furthermore, various studies has been reported in the use of simple equipment as an analytical tool, such as using a scanner to determine iron (III) [14], camera to detect complex starch-iodine and biological markers (biomarkers)[15] and chromium - iron [16].

Nowadays, almost everyone has a mobile phone with camera-equipped. This phone camera could be an opportunity as a self-tool for camera-based detection. We reported the use of a mobile phone application to detect anemia among pregnant women, which would replace the conventional method of anemia early detection by visual observation of palpebral conjunctiva color. The method was based on the capturing of a digital image of the inferior palpebral conjunctiva using a mobile phone. The color of the palpebral conjunctiva was then correlated with the participant's hemoglobin levels obtained by the standard method performed in the hospital. In this way, the evaluation of the inferior palpebral conjunctiva would more objective compared to the visual assessment. Another factor which may affect the hemoglobin level to predict the anemia were also collected, including the parity status, age, body mass index, blood type and geographical altitude of the participants. This self-care early anemia prediction would easy, fast, accurate as an important anemia monitoring status among pregnant women. Furthermore, the application will be used widely by the health professionals and the society as a self-care method as early detection of anemia during pregnancy especially in the location which the absence of modern instruments to detect hemoglobin.

2. Materials and Methods

2.1. Study design

Blood samples were collected from a local hospital. The palpebral image was captured using smartphone of Asus Zenfone and the image was analyzed using ImageJ software. Blood hemoglobin was measured using a Shimadzu Biospec 1601 UV-Vis spectrophotometer (Shimadzu, Japan). Blood sample was collected from the participants using standard method used in the hospital, continued by analyzing their hemoglobin level

2.2. Palpebral conjunctiva imaging

Palpebral conjunctiva digital image was taken using smartphone camera (Asus ZenFone 2 Laser, rear camera, 13 MP, laser autofocus, f/2.0 aperture). Participants asked to hold the lower eyelid with their right hand to easily observe the palpebral conjunctiva, while their right-hand holds a white paper as the white correction (Fig 1). Photograph of palpebral conjunctiva was taken in ambient lighting, without flash, and at a distance about 30-40 cm.

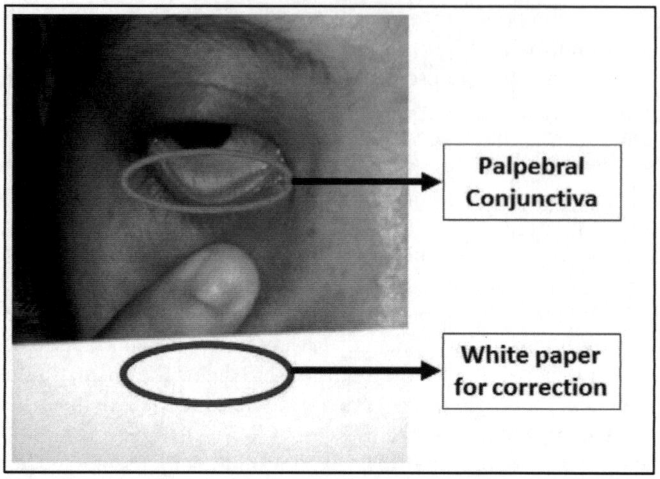

Figure 1. Palpebral conjunctiva image (marked with red oval) with a white paper as standard white color (marked with blue oval) taken with smartphone camera at about 30-40 cm.

2.3. Image analysis
Images were transferred from the smartphone to a laptop running Windows 8.0 (Microsoft) with a free software image analysis of ImageJ version 1.48v (National Institute of Health, USA). The image was recorded their color intensity of Red, Green, and Blue for both palpebral conjunctiva and white paper as a standard white color. The white color standardized was performed by assuming white paper would have color intensity for both R, G and B of 255. The real color intensity of white when they were less than 255, then added to the sample color intensity.

2.4. Data analysis
Raw data of color intensity (Red, Green, and Blue) of palpebral conjunctiva were collected. In another side, hemoglobin (Hb) level of the participants was analyzed using the standard method of spectrophotometric. Color intensity as the dependent variable was then analyzed using simple regression (Microsoft Excel, Data Analysis) related to Hb level as the independent variable. This regression analysis would result in a regression equation, linear regression chart, and correlation coefficient.

3. Results and Discussion

3.1. Study Participants
Twenty pregnant patients were asked as participants during the research. The age of participants were 22 – 36 years with an average of 28 years. The blood type of participants including A, B, AB and O. Body mass index (BMI) were 19 – 32.5 with an average of 25.6. Geographical altitude based on their address were 25 – 230 with an average of 84.4 m above sea level. All participants have consumed Fe tablet during their pregnancy.

3.2. Palpebral conjunctiva imaging and data processing
All participants were asked to take their palpebral conjunctiva photograph by the research assistant. The image capturing may be repeated in case of low light condition or sufficient view that it should be included the white calibrator paper and palpebral conjunctiva easily observed. The color intensity of palpebral conjunctiva images was corrected their white balance using paper as a reference color. For example, when white paper intensity (R,G,B) was 200,200,200, then the color intensity of palpebral conjunctiva was added their correction based on the maximum color intensity of 255, which was in this

example of 55 (255-200). Many strategies have been reported to correct the digital image color intensity as detection tools, such as software-based white balance [17], red label based white balance [18], standardized color palette [19] and white paper. The use of white paper as the standard white color was simple and easy to apply in wide condition and location, including rural application. The color intensity after correction was then used for further study, analyzing using simple regression related to a hemoglobin level of each participant.

3.3. Regression analysis

The results showed a significant relationship between palpebral conjunctiva image color intensity and hemoglobin concentration of the participants. Among three color intensity (RGB), the red color intensity showed the highest relationship with r^2 of 0.8139 (Fig. 2). The correlation coefficient (r) of 0.92 was showed a high correlation [20] between red color intensity and hemoglobin concentration. However, the highest sensitivity of this color intensity versus Hb level was the blue color, with the slope of 17.89 with an r^2 of 0.568. Therefore, we could conclude the best color intensity to predict the Hb level using palpebral conjunctiva image was the red color intensity, considering the highest coefficient of determination.

The limit of detection (LOD) was calculated based on the regression equation of red color intensity. The calculated LOD was 3.79 g/dL, which was lower than the cut of severe anemia during pregnancy of < 7g/dL. Thus, this imaging based anemia prediction could be used as early detection of anemia during pregnancy.

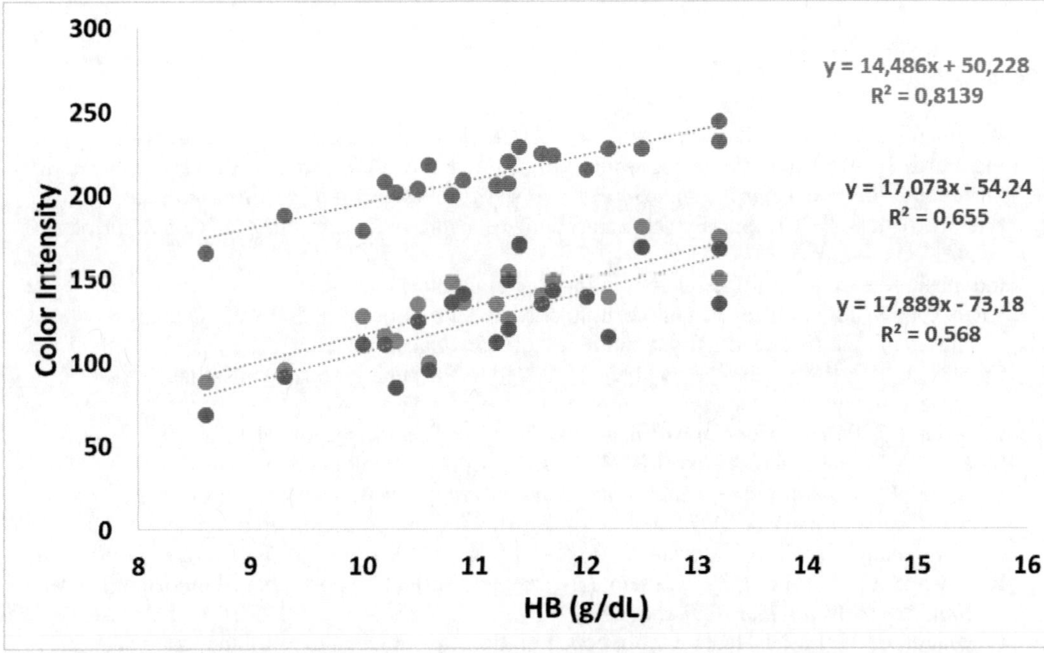

Figure 2. Relationship between palpebral conjunctive color intensity and measured hemoglobin (n=20). Dotted line represents best fit by linear regression.

3.4. Factor affecting hemoglobin level

Many factors have been reported to make a variation of the hemoglobin level, such as body mass index (BMI) [21], parity [22] and geographical altitude [23]. We also analyzed this parameter that may would influence in the proposed anemia prediction. The result showed a weak correlation [24] between BMI and hemoglobin level, with a correlation coefficient of 0.27 which was similar to the previous report [21]. The altitude of the participant was analyzed using online map based on Google map (www.daftlogic.com/sandbox-google-maps-find-altitude.htm) by looking for their address, and mapping on the online altitude map assistant. The result showed no correlation between altitude hemoglobin levels with a correlation coefficient of 0.07 ($P > 0.05$). The similar hemoglobin level may due to the participants were from 25 to 230 m above sea level, with an average of 84.4 m above sea level. The previous study reported that the altitude affected the hemoglobin level in the order of 3000 ft (914 m) above sea level [23], thus, below the 914 m such in this research would not significantly different. The blood type was no correlation between hemoglobin level and blood type with a correlation coefficient of 0.21 ($P > 0.05$). The parity status of the participants was nullipara (p0) of 12%, primipara (p1) of 65% and multipara (p2) of 24%. A previous study [22] reported that the parity was directly related to the prevalence of anemia (low hemoglobin level), however, in this research, the parity was also showed no correlation with the hemoglobin level (r of 0.05, $P > 0.05$), which may due to all participants had consume Fe tablet during their pregnancy.

4. Conclusion

Smartphone camera could be used as non-invasive and self-assessment of anemia prediction, with a high correlation between red color intensity of palpebral conjunctiva and blood hemoglobin concentration and calculated LOD of 3.79 g/dL. The use of digital image would more objective compared to conventional visual assessment usually perform in the rural region as early diagnostic of anemia.

References

[1] Organization W H 2015 *World health statistics 2015*: World Health Organization)
[2] Indonesia S 2008 Indonesia demographic and health survey 2007 *Jakarta/Calverton: National Family Planning Coordinating Board, Ministry of Health and Macro International*
[3] Nasional B K K B 2013 Survey demografi dan kesehatan Indonesia (SDKI) 2012 *Jakarta: BKKBN*
[4] Indonesia K K 2015 Profil kesehatan Indonesia tahun 2011
[5] Khan K S, Wojdyla D, Say L, Gülmezoglu A M and Van Look P F 2006 WHO analysis of causes of maternal death: a systematic review *The lancet* **367** 1066-74
[6] Frass K A 2015 Postpartum hemorrhage is related to the hemoglobin levels at labor: Observational study *Alexandria Journal of Medicine* **51** 333-7
[7] Christian P 2008 *Handbook of Nutrition and Pregnancy*: Springer) pp 319-36
[8] Sloan N L, Jordan E and Winikoff B 2002 Effects of iron supplementation on maternal hematologic status in pregnancy *American Journal of Public Health* **92** 288-93
[9] Candio F and Hofmeyr G 2007 Treatments for iron deficiency anemia in pregnancy: RHL commentary *The WHO Reproductive Health Library. Geneva: World Health Organization*
[10] Risnawati I and SN A H P 2015 Anemia in pregnancy effect on postpartum hemorrhage *Jurnal Ilmu Keperawatan dan Kebidanan* **6** 57-67
[11] Organization W H 2011 Haemoglobin concentrations for the diagnosis of anaemia and assessment of severity
[12] McKinney E S, Murray S S, James S R and Nelson K 2013 *Maternal-child nursing*: Elsevier Health Sciences)
[13] Kehoe E and Penn R L 2013 Introducing colorimetric analysis with camera phones and digital cameras: an activity for high school or general chemistry *J. Chem. Educ.* **90** 1191-5
[14] Kompany-Zareh M, Mansourian M and Ravaee F 2002 Simple method for colorimetric spot-test quantitative analysis of Fe (III) using a computer controlled hand-scanner *Anal. Chim. Acta* **471** 97-104

[15] Oncescu V, O'Dell D and Erickson D 2013 Smartphone based health accessory for colorimetric detection of biomarkers in sweat and saliva *Lab Chip* **13** 3232-8
[16] Firdaus M L, Alwi W, Trinoveldi F, Rahayu I, Rahmidar L and Warsito K 2014 Determination of chromium and iron using digital image-based colorimetry *Procedia Environmental Sciences* **20** 298-304
[17] Seo J-H, Park Y-B and Park Y-J 2014 Reliable facial color analysis using a digital camera and its relationship with pathological patterns: a pilot study *European Journal of Integrative Medicine* **6** 322-7
[18] Zhao Y, Tao J and Tu P 2013 Quantitative evaluation of efficacy of photodynamic therapy for port-wine stains using erythema index image analysis *Photodiagnosis and photodynamic therapy* **10** 96-102
[19] Collings S, Thompson O, Hirst E, Goossens L, George A and Weinkove R 2016 Non-Invasive Detection of Anaemia Using Digital Photographs of the Conjunctiva *PloS one* **11** e0153286
[20] Asuero A, Sayago A and Gonzalez A 2006 The correlation coefficient: An overview *Crit. Rev. Anal. Chem.* **36** 41-59
[21] Ghadiri-Anari A, Nazemian N and Vahedian-Ardakani H-A 2014 Association of body mass index with hemoglobin concentration and iron parameters in Iranian population *ISRN hematology* **2014**
[22] Uche-Nwachi E, Odekunle A, Jacinto S, Burnett M, Clapperton M, David Y, Durga S, Greene K, Jarvis J and Nixon C 2010 Anaemia in pregnancy: associations with parity, abortions and child spacing in primary healthcare clinic attendees in Trinidad and Tobago
[23] Ruíz-Argüelles G J 2006 Altitude above sea level as a variable for definition of anemia *Blood* **108** 2131-2
[24] Munro B H 2005 *Statistical methods for health care research* vol 1: Lippincott Williams & Wilkins)

IOP Conf. Series: Materials Science and Engineering **172** (2017) 012031 doi:10.1088/1757-899X/172/1/012031

Comparison on mechanical properties of single layered and bilayered chitosan-gelatin coated porous hydroxyapatite scaffold prepared through freeze drying method

M D Effendi, D Gustiono, Lukmana, D Ayu and F Kurniawati

Agency for The Assessment and Application of Technolgy, Serpong, Indonesia

Email: dachyar.effendy@bppt.go.id

Abstract. Biopolymer coated porous hydroxyapatite (HA) scaffolds were prepared for tissue engineering trough freeze drying method and impregnation. in this study, to mimic the mineral and organic component of natural bone, synthetic hydroxapatite (HA) scaffolds coated by polymer were prepared.Highly porous Hap scaffolds, fabricated by synthetic HA impregnation method on polyurethane foam, were coated with polymer coating solution, consisting of chitosan, Gelatin, and bilayered chitosan-gelatin prepared by aging and impregnating technique. For the purpose of comparison,The bare scaffolds without polymer coating layer were investigated. The Bare scaffolds were highly porous and interconnected with a pore size of around 150 μm–714 μm, has porosity at around 67,7% -85,7%, and has mechanical strength at around 0.06 Mpa - 0.071 Mpa,which is suitable for osteoblast cell Proliferation. Chitosan coated porous HA scaffold and gelatin coated porous HA scaffold had mechanical strength at around 0.81-0.85 Mpa, and 1.32-1.34 Mpa, respectively, with weight ratio of biopolymer and Hap was around 18%-22%. To compare these results, the coating on the bare scaffold with gelatin and chitosan had been conducted. Based on the result of FTIR, it could be concluded that coating procedure applied on porous hydroxy apatite (HA) coated by gelatin,chitosan coated HA scaffold, and bilayered Gelatin-chitosan coated porous HA scaffold, confirming that for allsampleshad no significant chemical effect on the coating structure. The compressive strength of bilayered Gelatin-chitosan coated HA scaffold had middle values between the rest, at around 1,06-1.2 Mpa for the samples at the same weight ratio of biopolymer: HA (around 18% - 22%). These results also confirming that coating by gelatin on porous hydroxyapatite was highest compresive strength and can be applied to improve mechanical properties of porous hydroxyapatite bare scaffold

1. Introduction

Hydroxyapatite ($Ca_{10}(PO_4)_6(OH)_2$, HA) has been widely used in medicine and dentistry because it is biocompatible, osteoconductive, and has excellent chemical and biological affinity with bony tissue [1]. As a result, HAp is accepted as a bioactive scaffold material for guided bone regeneration.hydroxyapatite (HA) ceramics also have been used mainly as non-load-bearing parts in the form of powders and granules [2]. Compared to dense bodies or granules, the porous scaffolds are highly attractive for their biological benefits, such as osteoconductivity and fast bone ingrowth, due to high surface area and sufficient blood circulation [1][3]. However, the intrinsically poor mechanical properties of HA, such as low compressive strength and fracture toughness restrict its application only to small sizes of granules and powders or non-load bearing implant [2]. In order to expand its applicability in hard tissue applications, the brittleness of HA needs to be overcome. In addition to the requirements for the chemical composition of the scaffold material, an interconnected porous structure is necessary to allow cell attachment, proliferation, and differentiation, and to provide pathways for biofluids. With this approach, the porous scaffold serves an important role in the manipulation of the functions of osteoblasts and a central role in the guidance of new bone formation into desired shapes. Therefore, the scaffold materials must be biocompatible, osteoconductive, and osteointegrative, and have enough mechanical strength to provide structural support.

Chitosan is a natural cationic polymer that is biologically renewable, biodegradable, biocompatible, non-antigenic non-toxic, and biofunctional [4]. Gelatin is a natural protein derived from the organic constituent of bone (collagentype I). Therefore its combination with the natural mineral constituent of bone (HA) is supposed to provide closer properties to the natural bone. Gelatin is readily assimilated by the body [5]. A composite scaffold of HA and Gelatin is therefore expected to show increased osteoconductivity and biodegradation together with sufficient mechanical strength [6]. beside that, gelatin, a natural protein derived from the organic phase of bone is much cheaper and more easily obtainable in solutions than collagen.In this study, a coating design of the HA porous scaffold was proposed to optimize these requirements.The polymer layer is expected to improve the brittleness of the porous scaffold, and the HA powder to enhance biocompatibility and hydrophilicity of the polymer coating layer, it assumed close to the mineral and organic component of natural bone.

2. Experimental details

2.1.Fabrication of Porous Scaffolds

Porous scaffolds were prepared by mixing hydroxyapatite powder (BPPT, synthesized from limestone with characteristic are : > Purity 99%, crystallinity : > 99%,Density 1.865 (gram/cc), Particle Size (nm) : 50 – 100) were mixed with distilled water (DW). and suspension agent (Duramax D3005, Aldrich) was added to make a viscous HA slurry (HA/DW = 0.5 mg/mL). A polyurethane foam template (45 ppi, 3M) was replicated by immersing into the slurry, drying, and heat-treating at 600 °C for 3 h. Porosity was calculated by measuring the dimension and weight of specimen as well as by the Archimedes method (n =3).

2.2. Biopolymer coated porous hydroxy apatite(HA) scaffold procedure

The scaffold, prepared in a dimension of (diameter x height) 10 mm × 20 mm, was immersed into the biopolymer containing solutions (with concentration was 2% chitosan solution and 20% Gelatin solution) for 24 h at room temperature. For samples containing double immerse of biopolymers, the scaffold was immersed into 20% gelatin solution and was applied to freeze drying method for 24 hours, and then immersed into 2% chitosan solution and was applied to freeze drying for 24 hours to obtain bilayered gelatin-chitosan coated HA scaffold.

2.3.Characterization and compressive strength test

The morphology of the coated scaffolds was evaluated using optical micrscope (Olympus BH-2) and polarized light microscope (Olympus CX31).The coating structure was analyzed using Fourier transformed infrared (FT-IR; System 2000, Perkin Elmer, USA) spectroscopy. Compressive mechanical tests were performed on scaffolds of a dimension (diameter x height) 10 mm × 20 mm using a Shimadzu TrapeziumX at a crosshead speed of 0.1 mm/min. The stress-strain curve was obtained, and the compressive strength and elastic modulus were determined from the maximum load recorded and the initial slope of the curve (<2% strain). The capacity of energy absorption (Wab) of the coated and the bare scaffolds was defined as the energy necessary to deform a specimen to a strain (ε), and was calculated from the area under the stress-strain curve at a given strain, as follows [7]

$$W_{ab}(\varepsilon) = \int_0^{\varepsilon} \sigma(\varepsilon')\, d\varepsilon' \quad (1)$$

Six samples were tested for each condition, and data were analyzed statistically using Student t test and significance was considered at $p < 0.05$.

Water absorption. Water absorption of bare scaffold, gelatin coated porous scaffold, chitosan coated porous scaffold and gelatin/chitosan coated porous Hap scaffold were studied to evaluate the effect of HA content on the size and stability of material. The ratio of water absorption (Wa) at time t was calculated using the following equation:

$$W_a\% = \frac{W_t - W_0}{W_0} \times 100\% \quad (2)$$

Where Wt and W0 are the weights of sample at time t and the dry state at room temperature, respectively

Measurement of porosity and density. The density measurements provided information about pore size and distribution, permeability, and the presence of structural faults in sintered ceramic structures [8]. A scaffold of weight W was immersed in a graded cylinder containing a known volume (V1) of ethanol. The cylinder was placed in a vacuum to force the ethanol into the pores of the scaffold until no air bubble emerged from the scaffold. The total volume of the ethanol and scaffold was then recorded as V2. The volume difference (V2 – V1) was the volume of the skeleton of the scaffold. The scaffold was removed from the ethanol and the residual ethanol volume was measured as V3. The total volume of the scaffold, V, was then

$$V = V_2 - V_3 \quad (3)$$

The apparent density of the scaffold, ρ was evaluated using,

$$\rho = \frac{W}{(V_2 - V_3)} \quad (4)$$

The porosity of the open pores in the scaffold, Ɛ was evaluated using [8],

$$\varepsilon = \frac{(V_1 - V_3)}{(V_2 - V_3)} \quad (5)$$

3. Results and discussion

3.1. Bare scaffold morphology

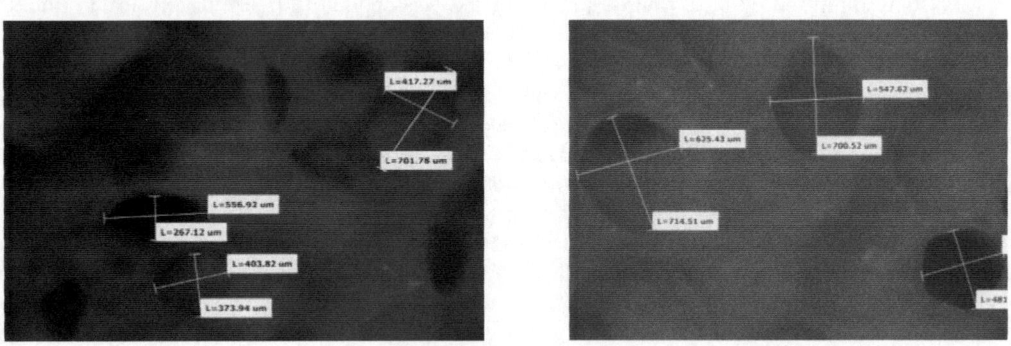

Figure 1. Optical microscopy morphology with 4x magnification of bare scaffold on the a) X axis cross section, b) middle cross section

The microscopy morphologies of the porous scaffold are represented in figure 1 a) and b). The bare HA scaffold, obtained by polyurethane reticulated foam method, exhibited a well-developed open pore structure. The porosity and average pore size were approximately 84% and 200–700 μm, respectively. It occurs caused by replication of morphology of polyurethane foam applied

3.2. Coated scaffold morphology

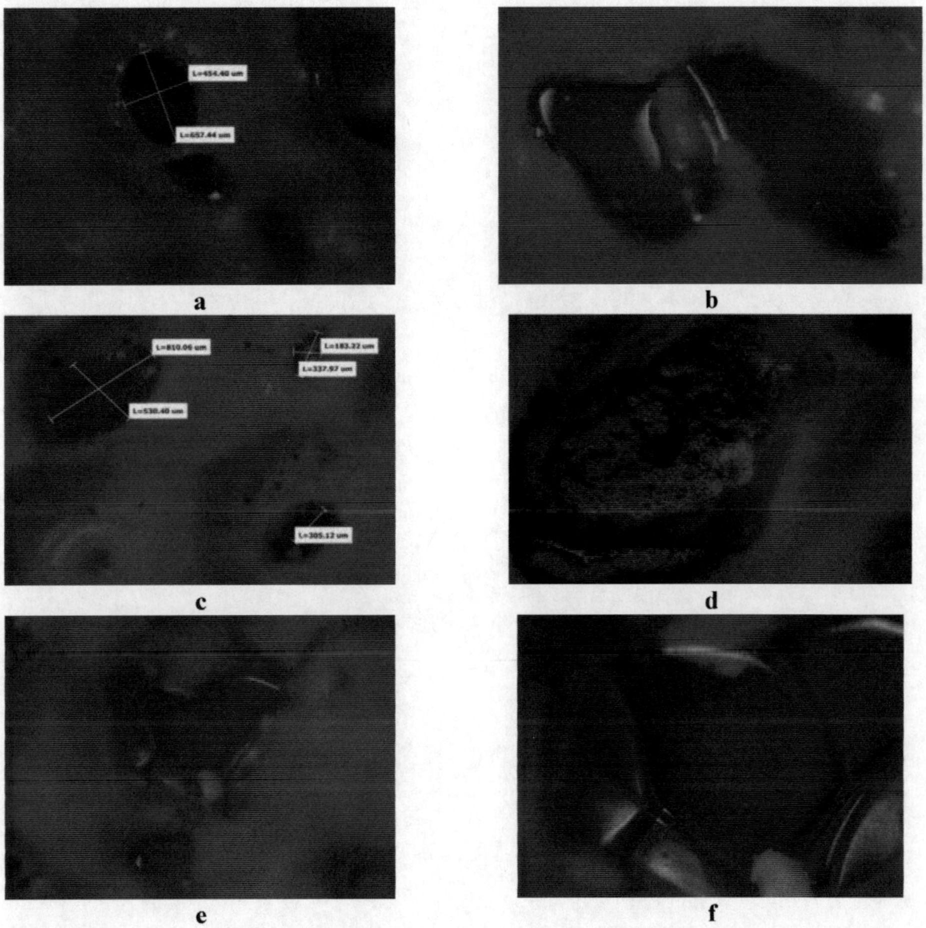

Figure 2. Optical microscopy with 4x magnification of Gelatin coated HA scaffold, and b)10x magnification ;c) 4x magnification of chitosan coating layer covered scaffold surface and d) 10x magnification, e) 4x magnification of bilayer gelatin-chitosan coating layer covered scaffold surface and f) 10x magnification

When a chitosan and gelatin was coated as single layered and bilayered applied on the scaffold, the pore structure changed slightly. The pore wall became thicker and some pores were partially closed (figure 2 a), 2b), 2c) and 2e). The porosity of the coated scaffold decreased slightly (around 78%). At high magnification, the coating layer was somewhat rough but uniformly covered the scaffold surface throughout (figure 2 b,2d and 2 f). Figure 2e) and 2.ef) showed that bilayered coated gelatin (dark side) and chitosan (bright side) had covered scaffold surface.

IOP Conf. Series: Materials Science and Engineering **172** (2017) 012031 doi:10.1088/1757-899X/172/1/012031

3.3. Coated scaffold structure

The coating structure was evaluated with FTIR spectroscopy and is shown in figure 3 and figure 4. Data on pure 20% Gelatin, 2% chitosan and HA powders are represented as references as below :

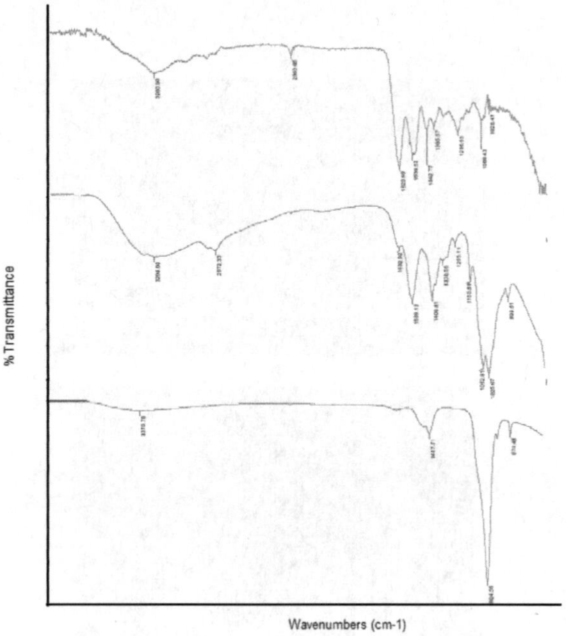

Figure 3. FTIR spectroscopies of a) pure 2% solution chitosan, b) pure 20% gelatin solution, pure hap powder c) were represented as references;

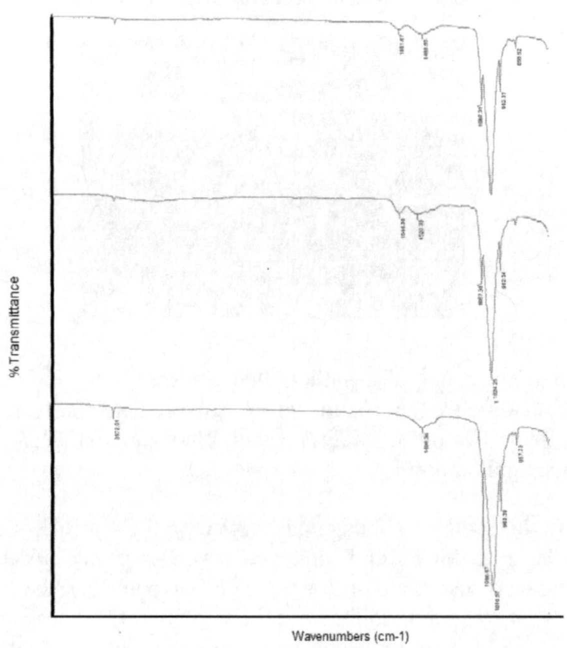

Figure 4. FTIR spectroscopies of d) chitosan coated porous Hap scaffold d) the chitosan/gelatin coated porous Hap scaffold e) and gelatin coated porous Hap scaffold

Figure 2a) represented mixed bands typical of Gelatin. The absorption at 1628 cm^{-1} and 1539 cm^{-1} are attributed to ((C=O)) and ((C-N)) vibrations of amide I and amide II, which is representative of the collagen secondary structure. According to *Yakimet & al* [9], the bond at 1628 cm^{-1} gives information on the helicity of the protein. A broad absorption band at 3390 cm^{-1} of OH groups corresponding to absorbed water, secondary alcohols (CMC) and (intramolecular/ intermolecular) hydrogen bonding. [10] [11]. Figure 2b represented mixed bands typical of HA (P-O at 827,963–1026 cm−1 and O H at 3370 cm−1(broad)). Figure 2c) represented The spectrum of chitosan depicts characteristic absorption bands at 3280 cm^{-1}, 2872 cm^{-1}, attributed to the -OH and -CH_3 groups. Furthermore, bands were identified at 1539 and 1406 cm^{-1} typical of the N-H group bending vibration and vibrations of -OH group of the primary alcoholic group, respectively. The bands at 1406 cm^{-1} and 1025 cm^{-1} correspond to the stretching of C-O-N and C-O groups. The bands at 1062 and 898 cm^{-1} are attributed to the glycosidic bondings. The shoulder at 1632 cm^{-1} represents the stretching of C=O. According to those IR absorption references, it could be concluded that process applied on porous HA coated by gelatin,chitosan coated HA scaffold and mixed Gelatin/chitosan coated porous HA scaffold, confirming that for all samples had no significant chemical effect on the coating structure as shown in figure 2d), figure 2e), and figure 2f) respectively.

3.4. Mechanical properties

The mechanical properties of a scaffold used for tissueengineering are very important due to the need for the structural stabilityto oppose the various stresses incurred during culture in vitro or implantationin vivo. The effects of the content of different biopolymer in the coated scaffolds on the mechanical properties are shown inTable 1, which were remarkably higher that those of the bare HA scaffolds (0.06 Mpa - 0.071 Mpa).

Table 1. Compressive strength

Sample(s)	Bare scaffold (BS) weight, gr	Coated scaffold weight after freeze drying	Biopolymer fraction, (%)	Max. Compresive strenght (Mpa)
BS	1,812- 1.855	-	-	0.067-0.071
G20	1.825- 2.188	2.145 -2.54	18.09% - 17,53%	2.95-3.27
K2	1.33-1.52	1.455- 1.76	15.79% - 16.17%	0.41-0.45
GK	1.779 – 1.915	2.16 - 2.275	18,80%-21,42%	2.02-2.77

The mechanical properties of the different concentrations of these samples were evaluated by compressive strength measurements.The compressive strength obtained in this study were quite low and within the margin of natural trabecular bones, presumably due to the high porosities (at least over 80%) where the ultimate compressive strength of the trabecular bone ranged from 0.22 to 10.44 Mpa with a mean value of 3.9 Mpa [12]. Practically, the trabecular bones had mechanical properties in a wide range depending on the porosity, position, and direction, as well as testing methods. [13-15]. The energy-absorption values in the coated scaffolds had higher values than those reported HA scaffolds (4.5–13 N * cm) with much lower porosities (∞60–70%).[15][17].

The thicker stems and consequently reduced porosity with cyclic coating also played some role in improving the mechanical properties. In this respect, the polymer component is quite beneficial in terms of providing mechanical flexibility and efficient energy absorption.where the gelatin coated porous hydroxyapatite has shown better value of compressive strength than other

4. Conclusions

A hydroxyapatite (HA) porous scaffold was coated with gelatin, chitosan and immersed gelatin and chitosan respectively shown the improvement of compressive strength related to the bare scaffold (without coating applied). The process applied on porous hydroxyapatite confirming that the mix-polymer had no significant chemical effect on the coating structure. It means that the individual characteristics of components such as biocompatible, osteoconductive, and osteointegrative were no change.Besides, the polymer layer could be improve the brittleness of the porous scaffold. Compared to trabecular bone, The compressive strength of all types of biopolymer coated HA scaffolds were quite low and within the margin of natural trabecular bones

Acknowledgment
This work was supported by Project of Inovation of biocompatible materials for implant, BPPT, 2016

References

[1] L.L. Hench 1991 *J. Am. Ceram. Soc.* **74** 1487–1510
[2] Y. Kimura 1991 *J. Amer. Ceram. Soc.* **74** 1487
[3] H. W. Kim, Y. J . Noh, Y. H. Koh And H. E. Kim 2003 *J. Mater. Sci.: Mater. Med.* **14** 899
[4] R.A.A. Muzzarelli 2011 *Carbohydr. Polym.* **83** (4) 1433–1445
[5] Cooper, C.W. and Falb, R.D. 1971 *Ann. NY. Acad. Sci.* ***146*** *(1)* 218-224
[6] Bigi, A., Panzavolta, S. and Roveri, N. *1988 Biomaterials* ***19*** *739-744.*
[7] P . Kenesei, Cs. Kadar, Zs. Rajkovits And J. Lendvai, 2004 *Script. Mater.* **50** 295
[8] Hodgkinson, R. and Curry, J.D. 2002 *Proc. Inst. Mech. Eng. [H]* **204** 115-121
[9] I. Yakimets, N. Wellner, A. Smith, R. Wilson, I. Farhat, J. Mitchell, 2005 *Polymer* **46** 12577
[10] D.R. Biswal, R.P. Singh, 2004 *Carbohydrate Polymer* **57** 379
[11] L. I. Kutsenko, A. M. Bochek, E. N. Vlasova, B. Z. Volchek, 2005 *Russian Journal of Applied Chemistry* **78** 2014
[12] Misch CE,Qu z,et.al. 1999 *J Oral MaxillofacSurg*. 700-6, discussion 706-8
[13] Hayes WC, Bouxsein ML. *Biomechanics of cortical and trabecular bone: Implications for assessment of fracture risk*. In: Mow VC, Hayes WC, editors. Basic orthopedic biomechanics. New York: Lippincott-Raven, (1997) 69–111
[14] Keaveny TM, Guo XE, Wachtel EF, McMahon TA, Hayes WC.1994 *J Biomech* **27** (9) 1127–1136
[15] Keaveny, TM. 2001 *Strength of trabecular bone*. In: Cowin SC, editor. Bone mechanics handbook. New York: CRC Press, 6-1–16-41
[16] Kim HW, Deng Y, Miranda P, Pajares A, Kim DK, Kim HE,Lawn BR. 2001 *J Am Ceram Soc* **84** 2377–2384
[17] Roach DH, Lathabai S, Lawn BR. 1988 *J Am Ceram Soc* **71** 97–105

IOP Conf. Series: Materials Science and Engineering **172** (2017) 012032 doi:10.1088/1757-899X/172/1/012032

Separation of Cu^{2+}, Cd^{2+} and Cr^{3+} in a Mixture Solution Using a Novel Carrier Poly(Methyl Thiazoleethyl Eugenoxy Acetate) with BLM (Bulk Liquid Membrane)

M C Djunaidi, Khabibi and I Ulumudin

Departement of Chemistry, Diponegoro University, Indonesia

Email: cholidjunaidi@live.undip.ac.id

Abstract. The separation process using a novel carrier polyeugenol has active groups N and S has been done with the technique BLM. Polyeugenol has groups active N and S was synthesized from eugenol which is then polymerized into polyeugenol. This polymeric compounds was then acidified become acidic poly (eugenoksi acetate). After the acid formed , then the synthesis was continued by add 4-methyl-5-tiazoleetanol to form esters poly (methyl thiazole eugenoxy ethyl acetate) (PMTEEA). The result of the synthesis was analyzed by FTIR and 1H NMR. This polyester product synthesis was applied as a carrier for separating metal ions Cu^{2+}, Cd^{2+} and Cr^{3+} with variations in feed phase pH = 5 and pH = 7 in the membrane of chloroform using techniques BLM. Receiving phase after 24 hours was analyzed by AAS. In variations of feed pH = 5 ions was obtained 66.21% Cd^{2+}, 28.83% Cu^{2+} and 10.92% of Cr^{3+}, at pH = 7 was obtained 70.77% Cd^{2+}, 30.14% Cu^{2+}, and 3.72% of Cr^{3+}.

1. Introduction

Due to the discharge of large amounts of metal-contaminated wastewater, industries bearing heavy metals, such as Cd, Cr, Cu, Ni, As, Pb, and Zn, are the most hazardous among the chemical-intensive industries. Because of their high solubility in the aquatic environments, heavy metals can be absorbed by living organisms. Once they enter the food chain, large concentrations of heavy metals may accumulate in the human body. If the metals are ingested beyond the permitted concentration, they can cause serious health disorders [1].

One method for separating metal ions are liquid separation membrane [2]. In the liquid membrane technique, carrier compounds play an important function. Compounds carrier as a facilitator is determinant in the separation performance from the feed phase. A carrier compounds should has high ability to extraction via stable complex formation within the membrane, having high selective separation towards certain species, has good solubility as well as diffusion coefficient in the suitable organic solvent (membrane) and can be used in relatively a small amounts [3]. Besides that, a selectivity compounds carrier towards certain metal ions determined by the active groups on the carrier compound.

Carrier compounds will form complexes with metal ions through chemical bonds between the active group with a metal ion, this establishment is based on the theory HSAB (Hard and Soft Acids Bases), this theory states that in general the metal acid hard ions (such as alkali, alkali earth, and Cr^{3+})

will form stronger complex with the bases hard (such as RO-), ion acid soft metal (such as Cd^{2+}) will form a stronger complex with a soft metal (such as RS), and metal acid borderline ions such as Cu^{2+} will form strongwe complex with a bases borderline (such as pyridine) [4].

Carrier compounds used in this study was poly (methyl thiazole eugenoksi ethyl acetate) (PMTEEA). PMTEEA has thiazole groups containing atoms of N and S was selective towards certain metal ions. PMTEEA was a derived compounds of polyeugenol. Polyeugenol is polymerised of eugenol. Eugenol contained in clove oil, has three active groups: allyl, methoxy and hydroxyl. From the hydroxyl groups can be substituted by more selective groups. make eugenol has potential as a selective compound carrier. Eugenol has been widely used for the separation of heavy metals, including eugenol polymer (polyeugenol) used to adsorb ion Fe (III) [5,6], as a functional polymer in membrane in situ [7] and in particles membrane [8,9]. Polyeugenol that has many OH groups (bases hard) has been researced the possibility of using as the carrier for separating metal ions Cr^{3+} [10] and the carrier compound that has active group N derived from polyeugenol has selective towards metal ion Cu^{2+} with percent transport 87, 54% [11]. In this study, we expected that the active group N which is borderline base ligand will selective binds towards Cu^{2+}(bordrline acid metal), while the active ion/group S which is a soft base ligand will selective binds towards Cd^{2+} (soft acidic metal ion).

2. Methodology

Polymers synthesized used as a carrier substance in heavy metal recovery used Bulk Liquid Membrane (BLM) techniques. Metal solution containing each 30 ppm Cr^{3+}, Cu^{2+} and Cd^{2+} with variations in pH = 5 and pH = 7 as feed phase and HCl as the receiving phase (stripping phase). pH was kept constant, pH = 1. The remaining heavy metals in feed phase and transported in receiving phase were determined by AAS.

2.1. Materials.

The materials were purchased from SIGMA-Aldrich, Eugenol, BF_3-diethylether while other reagents were purchased from E Merck, Germany: $SOCl_2$, 4-methyl-5-thiazoleethanol, NaOH, chloroacetic acid , $CrCl_3$. $6H_2O$, $CuCl_2.2H_2O$, $CdCl_2.H_2O$. Chloroform, Methanol, Diethylether and demineralized water were purchased from Bratachem.

2.2. Instrumentation

The instruments used to characterization of PMTEEA in this study were FTIR Spectrophotometer (Shimadzu 8201PC), and GC MS (Shimadzu QP2010S), NMR ^{1}H JEOL-MY60, analytical balance (Mettler Tolendo AB54-S), Atomic Absorption Spectrophotometer (Perkin Elmer), pH meter (HACH E C20).

2.2.3. Equipments. Laboratory Glassware and a set of tools BLM (U tube, Fig 1).

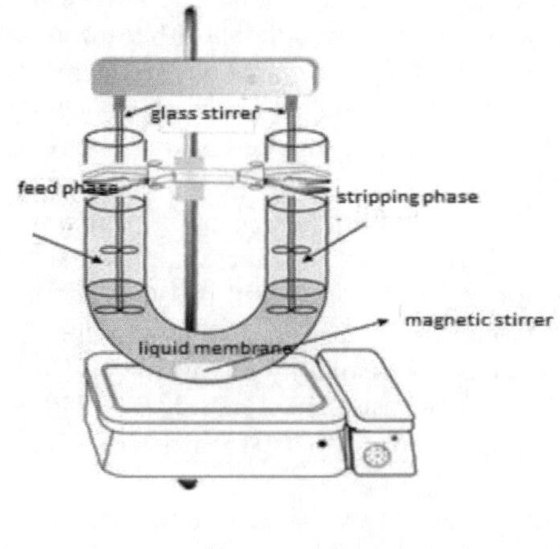

Figure 1. U Tube BLM

3. Procedure

3.1. Liquid Membrane Preparation
0.7 g PMTEEA was dissolved in 30 mL of chloroform.

3.2. Recovery process
A solution of 30 mL PMTEEA was poured in the U tube was placed between the feed phase and phase receiver each 13 mL, then stirred for 24 hours.

3.3. Measurement of pH
After the feed phase and stripping phase through the mixing process for 24 hours, the pH was measured by pH meter on both side.

3.4. Analysis by AAS
Analysis of the metal ion content in the feed phase and phase receiver after the separation process was done by atomic absorption spectrometer.

3.5. Synthesis of Compounds Carrier PMTEEA
3.5.1. Synthesis Polyeugenol. Eugenol (5.8 g) was put in a 3-necked flask, then 0.25 mL of boron trifluoride diethyl ether, $BF_3O(C_2H_5)_2$ was added as catalyst. The addition was done 4 times every hour while stirring with magnetic stirrer at room temperature. The occurrence of a reaction can be characterized by the color change of the solution into red. After the last addition of the catalyst, the polymerization was allowed to continue up to 12-16 hours, after which 1 mL of methanol was added to stop the reaction. The gel produced was dissolved in chloroform and put into a separating funnel and then washed repeatedly with distilled water until neutral. The organic layer was transferred into a 50 mL erlenmeyer flask and added with anhydrous Na_2SO_4. The liquid was separated by decantation. Afterwards, the solvent was evaporated in rotary evaporator at 40 °C. The residue obtained was further dried in the desiccator, and was subsequently weighed and characterized using FT-IR.

3.5.2. Synthesis Poly Eugenoxi Acetic Acid . A total of 5 g of polyeugenol was added to the boiling flask size of 100 ml, and then added a solution of 17.5 mL NaOH 33% (33 g NaOH in 100 mL) of. Subsequently the mixture was stirred for approximately for 30 minutes, and added 12.5 mL of a 50% solution of chloroacetic acid (50 g in 100 mL water) slowly with a pipette and stirred constantly. The mixture was heated in a water bath with a temperature of 80-90° C. Heating was carried out for 2 hours, then cooled and acidified with 6 M HCl until pH reached 1. Step further was extracted with diethylether for 3 times, each 50 mL. Ether extracts were combined and extracted with sodium bicarbonate 5% w / v for 3 times each 30 mL, and then the water layer was acidified with 6 M HCl to pH = 1. The subsequent steps were filtered, dried and weighted. The results were analyzed by FTIR and ^{1}H NMR.

3.5.3. Synthesis PMTEEA . Amount of 3 g of polyeugenoxi acetic acid was added to three neck flask of 100 mL size with additional equipment (an addition funnel, reflux) and .added by 3 ml of thionyl chloride dropwise. Then the mixture was refluxed for 150 minutes in a warm water bath (40 ° C), and then allowed to cool. Subsequently added to the mixture was added 2.5 mL tiazoletanol dropwise and refluxed again in the warm water bath (40° C) for 6 hours. After coolled, the results obtained were dissolved in chloroform and washed with water. The extraction was dried with anhydrous sodium sulphate, filtered and then evaporated to remove the remaining solvent. The results obtained were analyzed by FTIR and ^{1}H NMR and also GC MS.

4. Results and Discussion

4.1. Synthesis Polyeugenol

The result of polyeugenol synthesis was reported [9,by Djunaidi, et al, 2010 and 2015. Polyeugenol synthesis then was used for synthesised polyeugenoxy acetic acid.

4.2. Synthesis Poly (Eugenoksi Acetic) Acid

Polyeugenol synthesised has a phenol group, allyl and methoxy. It makes polyeugenol can be synthesized into other compounds such as in the form of poly eugenoxi acetic acid. Polyeugenol has hydroxy groups which can react with bases to form polyeugenolat salts. Proton in OH is easily separated because the shape of the anion stabilized by the resonance of the benzene ring. The addition of NaOH excess were intended to obtain salts as much as possible. This polyeugenolat sodium salt was reacted with chloroacetic acid to form polyeugenoxy acetic acids and then purified with diethyl ether to remove nonpolar impurities and extracted with sodium carbonate to remove impurities- which is polar (figure 2). The results obtained acid synthesis of poly (eugenoxi acetate) with a yield of 4.53 g of 90.6%.

The following reaction mechanisms that occur:

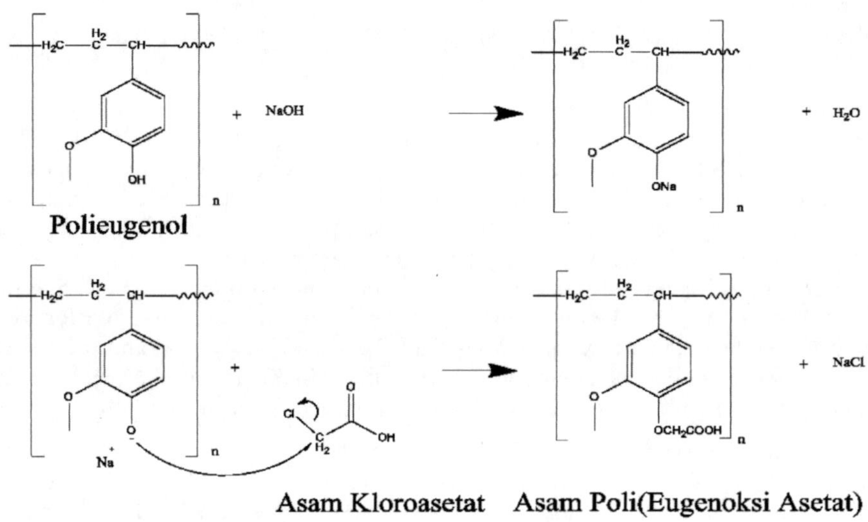

Figure 2: Mechanism acid synthesis reaction of poly (eugenoxi acetate)

Whereas the FTIR spectra was obtained as follows:

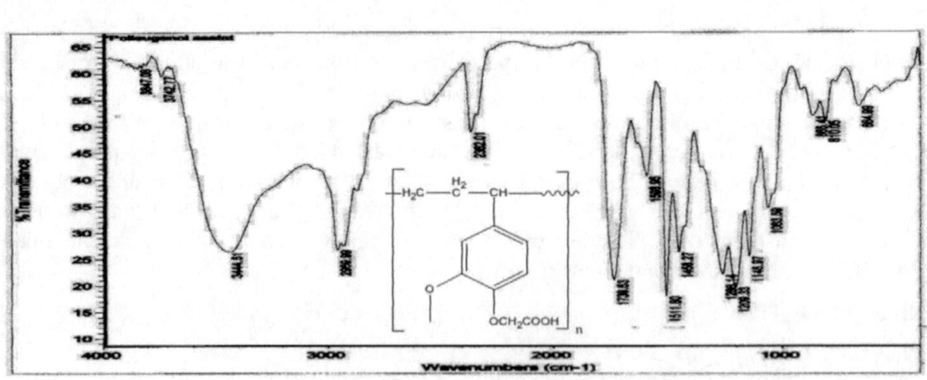

Figure 3. FTIR spectra of compounds polyeugenoxy acetic acid

FTIR spectra appear (figure 3) on acid carbonyl group shown in the bands 1739 cm^{-1}. This indicates there has been a carboxylation reaction in the presence of acetate groups on polyeugenol. While the ^1H NMR spectra was as follows:

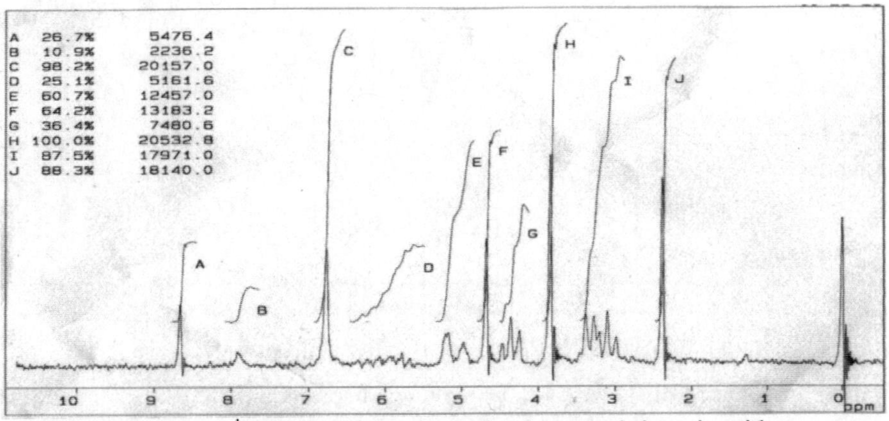

Figure 4. ^1H NMR acid compound polyeugenoksi acetic acid

Table 1. Correlation of ^1H NMR signals (figure 4) to the type of proton polyeugenoxy acetic acid

Chemistry Shift δ (ppm)	Integration	Information
6.7-7	3 H	$-C_6H_3$, s
4.6	2 H	$-OCH_2-$, s
3.8	3 H	$-OCH_2-$, s
0.5-1.5	3 H	$-CH_2-CH_2-$, m

Spectrum -IR of poly eugenoxi acetic acid (figure 4) was similar to the spectra poly eugenoxi acetic acid through the monomer way synthesis [12].

4.3. Synthesis of PMTEEA

This compound was esterified use thionyl chloride to converse the polyeugenoxi acetic acid into hydrochloric acid. The hydrochloric acid formed then was reacted with an alcohol (4-Methyl-5-Thiazoletanol). The result was 2.9361 g blackish brown solid. The yield results was 97.9%. The following reaction mechanisms that occur:

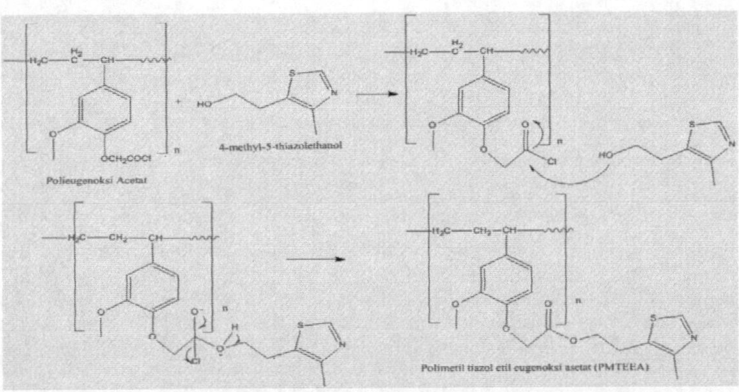

Figure 5. The mechanism of the synthesis reaction PMTEEA

FTIR analysis of the product was presented in the figure 6 :

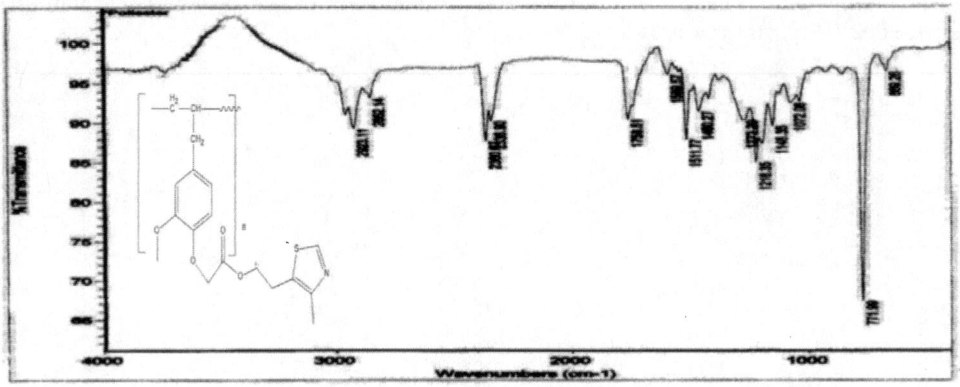

Figure 6. FTIR spectra PMTEEA (with solvents)

The FTIR spectra (figure 6) showed loss of the OH group (3400-3500 cm^{-1}) which was replaced with a methyl group thiazole ethanol and the appearance of an absorption band in the region ester carbonyl group (1758 cm^{-1}) which showed an esterification reaction. (Table 3.2 distinguishes acid FTIR results poly (eugenoksi acetate) with PMTEEA).

Table 2. Results of the analysis of the infrared spectra PMTEEA.

Group	Absorption (cm^{-1})	
	Polyeugenoxy acetic acid	PMTEEA
Hydroxyl	3400-3500	-
Carbon saturated	2960	2923
Carbonyl ester	-	1758
Carbonyl acid	1739	-
Metylen	1458	1460
Eter (C-O)	1300-1000	1300-1000

PMTEEA compounds was tested by ^{1}H NMR instrument for strengthening that the esterification reaction occurred. 1H NMR spectra following (figure 7):

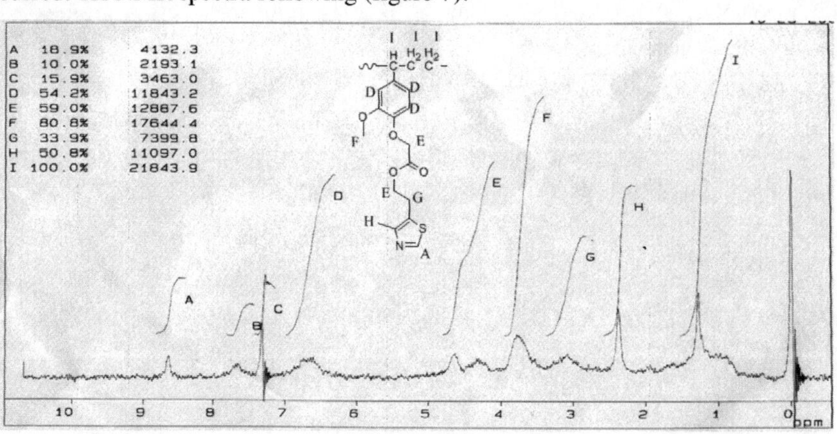

Figure 7. ^{1}H NMR PMTEEA

Peak at 8.5 δ (figure 7) indicates that the hydrogen atoms of the thiazole group has been entered in the acid poly eugenoxi acetic acid. The loss of OH peak at 5.5 δ on the compound 4-methyl-5-

tiazoletanol showed that the formation of esters. While the peak at 7.3 δ showed that the H atom of chloroform which has not evaporated.

4.4. Transport Mixed Metal Ion BLM Technique Using Active Compounds Carrier PMTEEA.
In this application, 0.7 g PMTEEA as compound carrier was dissolved in 30 mL of chloroform, with the feed phase was a mixture of metal ions Cu^{2+}, Cd^{2+} and Cr^{3+} in buffers pH = 5 and pH = 7, while stripping phase was a solution HCl pH = 1. The recovery process using BLM method was based on the formation of a stable complex between the ligand to the central atom. BLM process was carried out for 24 hours with continuous stirring.

Table 3 indicated that after stirring for 24 hours there was decreased pH in the feed phase and the increase in the pH in the receiving phase. This was because at the time of contact between the feed and the membrane phase, carrier compounds would form complexe with metal ions, will then be taken to the interface layer membrane-phase stripping. In this layer carrier compound released metal ions bounded and replaced with H^+ and then migrate to the interface layer membrane- feed phase to be released and replaced with a metal ion. This process occurs repeatedly until no metal ions can be exchanged. The mechanism of transport of metal ions from the feed phase to receiving phase through the chloroform membrane as shown in the picture below (figure 8).

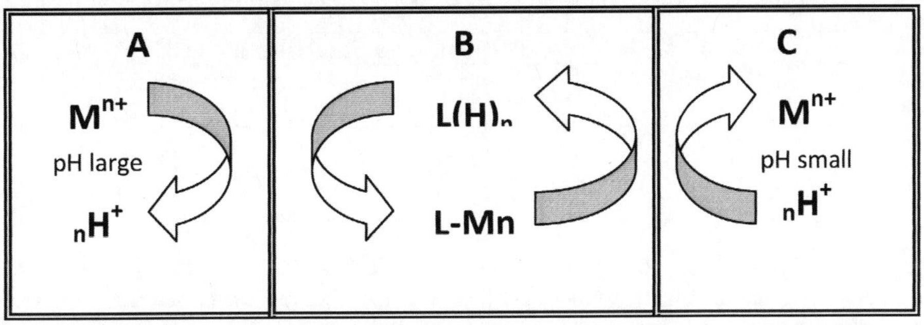

Figure 8. The mechanism of transport of metal ions from the receiving phase to phase feed through the membrane chloroform
A: feed phase B: organic phases C: phase receiver
L: carrier M: metal ion

Table 3. Changes in pH before and after stirring at BLM

pH *buffer*	pH Feed phase		pH stripping phase	
	$pH_{t=0}$	$pH_{t=24}$	$pH_{t=0}$	$pH_{t=24}$
5[a]	5.01	4.2	1.0	3.06
7[b]	7.4	4.3	1.0	2.91

[a] pH buffer 5: mixture of metal ion (Cr^{3+}, Cd^{2+} and Cu^{2+}) in a buffer solution with pH = 5
[b] pH 7 buffer: : mixture of metal ion (Cr^{3+}, Cd^{2+} and Cu^{2+}) in a buffer solution with pH = 7

4.5 Metal Ion Transport In the feed phase pH = 5 and pH = 7
In this experiment, 0.7 g PMTEEA was dissolved in chloroform as membrane phase and feed phase adjusted at various pH = 5 and pH = 7 to test the selectivity and effectiveness of the transport of metal ions on the effect of feed concentration. pH = 5 and pH = 7 was used because all of metal ions can be extracted at a pH near neutral (Hiratani, et al, 1992). In addition, according to a study [13] most of the reaction complex formation required a very low level of acidity or slightly alkaline as a condition to get a perfect extraction. Results transport mix metal ion was shown in Figure 9 and Table 4.

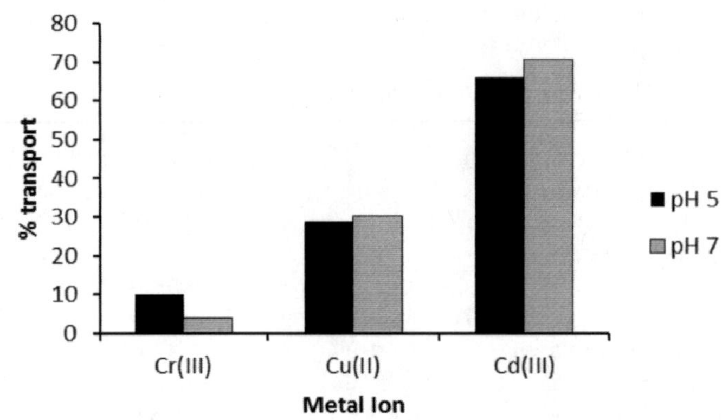

Figure 9. Diagram of metal ion transport in feed phase in feed pH = 5 and pH = 7

Table 4. Percent transport of metal ions in the receiving phase

Metal Ion	% Transport at receiving phase	
	pH (Feed phase) = 5	pH (receiving phase) = 7
Cr^{3+}	10.92	3.72
Cu^{2+}	28.83	30.14
Cd^{2+}	66.21	70.77

Ligands PMTEEA has active group S and N, based on the theory HSAB Pearson (1963) [4], which states that in general ions hard metals (such as alkali metals, alkaline earth, and Cr^{3+}) form stronger complex with donor atoms hard (such as RO-), ion soft metals (such as Cd^{2+}) form a stronger complex with donor atoms of soft (such as S), and the metal ions borderline as Cu^{2+} with donor atoms borderline such as pyridine (N), then the active group S binds to complex strongly with Cd^{2+} whereas the active group N complex binds strongly with Cu^{2+}. Cahyono [11] has been conducting research using eugenol has N as a carrier, his study indicated that the carrier more selective towards Cu^{2+} followed next by Cd^{2+} and Cr^{3+}. Similarly, HSAB theory was worked in the selective transport OH group of polyeugenol toward metal ions Cr^{3+} [10].

Cd^{2+} was transported greatest at pH = 5 and pH = 7 because the S atom in the thiazole group has a great affinity towards Cd^{2+} compared with the N atom to Cu^{2+}. In addition, the percent metal ion transport at pH = 7 was slightly larger than at pH = 5. This was because most of the reactions of complex formation, required a very low level of acidity or slightly alkaline as a condition to get a perfect extraction [13].

5. Conclusion

Poly methyl thiazole eugenoksi ethyl acetate) (PMTEEA) could be synthesized from eugenol (especially from polyeugenol). PMTEEA could be used as a carrier for the recovery Cd^{2+}, Cu^{2+} and Cr^{3+} with Bulk Liquid Membrane (BLM) Technique. At pH 5 and 7 feed phase the sequences of transport in BLM were Cd^{2+}, Cu^{2+}, and Cr^{3+} that proved HSAB theory works.

References

[1]. Babel, T.A. Kurniawan, 2004. *Chemosphere,* **54** (7), pp. 951-96
[2]. Cleij,M.C., Scrimin, P., Tecilla, P., and Tonellato, U. 1997 *J. Org. Chem.*, **62** (16), 5592-5599•

[3]. Bartsch, R.A., dan Way, J.D., 1996, *Chemical Separation With Liquid Membrane : An Overview, American Chemical Society,* Washington, page 1-6
[4]. Peterson, R. T., 1996, *Design of Macrocyclic Carriers for Liquid Membrane, In Chemical Separations with Liquid Membranes,* American Chemical Society, Washington DC
[5]. Djunaidi, M.C, Jumina, Siswanta, D., Ulbricht, M., 2015. *Indones. J. Chem.* **15** (3). *305 – 314*
[6]. Djunaidi, M.C, Jumina, Siswanta, D., 2015. *Oriental Journal of Chemistry*. **31** (4). 2223-2229
[7]. Djunaidi, M.C, Jumina, Siswanta, D., Ulbricht, M., 2015. *Asian J. Chem* **27** *(12) p. 4553-4562*
[8]. Djunaidi, M.C, Jumina, Siswanta, D., Ulbricht, M., 2015. *Proceeding International Conference of Chemical and Material Engineering (ICCME)*
[9]. Djunaidi, M.C, Jumina, Siswanta, D., 2016. *Oriental Journal of Chemistry*. **32** (1). 77-84
[10]. Djunaidi, M.C., Lusiana, R.A., Kartikawati, N.G., 2010. *Conference Chemistry National and Paedegogig Chemistry 2010*
[11]. Cahyono, H, Djunaidi, M.C., Ismiyarto, 2010, *Conference Chemistry National and Paedegogig Chemistry 2010.*
[12]. Umi, H.T., 2002, Thesis. Department of Chemistry. University Gadjah Mada, Yogyakarta
[13]. Boon, S.T., 2006, *Selective Liquid-Liquid Extraction Of Precious Metals From Semiconductor Wastes*, Department of Chemical Engineering, Universiti Teknologi Malaysia, Malaysia, page 98

IOP Conf. Series: Materials Science and Engineering **172** (2017) 012033 doi:10.1088/1757-899X/172/1/012033

Finite Element Analysis for Prediction of Shear and Stress Concentration & Distribution in Femoral Bone

N Suhendra, D Gustiono, E A Nugroho, Masmui, H Yuliani

Agency for the Assessment and Application of Technology, Puspiptek Distric 224nd Bld, Serpong – South Tangerang – Banten Indonesia

E-mail : nandang.suhendra@.bppt.go.id

Abstract. The effect of micromotion on the shear shielding and size of yielding region in the bone asperity in contact with metal of femoral stem was investigated. The main objective of this work was to gain an understanding of bone wear particleformation mechanism from the two-dimensional finite element model of cementless femoral stem type. To assess the influence of the parameters of interest, different friction coefficients and sliding distance (micromotion)were used in the numerical simulations. Results from the finite element analysis showed that the increase ofthe yielding region is strongly influenced by the rise in sliding distance (micromotion), which is related to the generation of bone wear particle formations. Finite element bone wearparticle formation model, based on strain discontinuities, was therefore proposed for further works. The results obtained in this study can lead to the development of an accurate finite element wearparticle formation mechanism model that would be of use in the assessment of an artificial implant performance and their development.

1. Introduction

Normal human joint performance is facilitated by low-friction articular cartilage bearing surfaces, which are conforming and self-regenerating[1-4]. Severely damage of natural joints, e.g. due to osteoarthritis, they are often replaced by artificial implants. The most common material combinations used for articulating surfaces in a total hip joint replacement are: metal - metal, metal - ultra high molecular weight polyethylene (UHMWPE), ceramic – ceramic and ceramic – UHMWPE[1, 3, 5-7]. Meanwhile, materials used for femoral stem are stainless steel, Titanium Alloys and recently polymeric composites.

Commonly, artificial hip joint designs utilizing a Titanium Alloys femoral stem, can last 20 years in the human body without failure[8]. However, there are several factors which affect the durability of the total hip joint prosthesis, one of which is implant – femoral bone contacting surface wear (and related phenomena)[9]. It is postulated that micromotion occurring within the interface of the implant and femoral bone premature failure of hip joint prostheses [8, 10, 11]. Micromotion of femoral stem from the bone surface was found to initiate aseptic loosening and early failure of the prosthesis[10, 12-16]. After reviewing the literature, one can conclude that, in some cases, roughened stems have been shown to fail earlier than polished stems of the same implant[16].

Herberts et al[17]analyzed the in-vivo failure behavior of a stem with a surface roughness of Ra=2 μm. They concluded that the stem was not stable and generated excessive osteolysis, particularly around defects in the cement mantle. Based on this observation, it was suggested that under similar conditions such as number of cycles, load, stop to stop sequence and activity level, the harder counter-face roughness is the most important factor in both wear volume and the number of wear particles generated from the softer bone surfaces[18].

It was suggested that migration of prosthesis is defined as a change inposition of prosthesis, cement mantle or both andis thought to indicate implant failure and representloosening[19]. Once migration has begun,stability is lost and periprosthetic particles maymodulate latter stages of loosening[19]. Mechanismsby which migration occurs are not fully understood.It could be due to fatigue failure ofcancellous bone surrounding the prosthesis[20] leadingto loss of osteo-integration of a stableprosthesis, or it could be attributed to surgicaltechniques—for example, reaming which disturbs capillarycirculation of periprosthetic bone, leading to necrosis. Theinitial use of cement with first generation cementing techniqueallowed defects and stresses to occur within the cement whichresulted in a weaker bone–cement interface and permitted theingress of polyethylene particles. thus resulting in loosening[20].With improved cementing techniques which include the use ofa medullary plug, a cement gun, lavage of the canal,pressurisation, centralisation of the stem, and reduction inporosity in the cement, the incidence of femoral lysis has beenreduced[20].

Attempts to reduce wear in order to prevent bone resorption have been made by incorporating hard bearing surfaces, i.e. metal-on-metal and ceramic-on-ceramic prostheses [21]. Apart from the improvement in the lubricating quality of the joint, a reduction in sliding distance was also expected to be beneficial in minimizing the rate of volumetric wear. The use of a smaller, up to a certain value, femoral head radius was another solution suggested [22].

Processes contributing to wear of cancellous bone in total hip arthroplasty (THA) are associated with surface features and the overall geometrical conformity of the mating components [3, 23, 24]. Therefore, it is important to understand the relationship between the microscopic wear mechanism, material properties such as: contact loading, surface roughness, etc. The loaded contact during micro-motion with sliding between two asperities can result in local elastic-plastic deformation, development and extension of microscopic cracks[25].

In general, there are two approaches that are used in simulation studies of wear. Damage arising from a single, isolated asperity contact can be studied[25-27], or alternatively the wear resulting from contacts between multiple asperities is considered[18, 28]. The first approach involves numerical modelling the movement of a hard asperity of known geometry over a specimen surface, resulting in wear, while the second approach is mainly experimental, using specially prepared specimens and a tribometer, such as a joint simulator.

Information about wear mechanisms involved in generation of wear particles of particular sizes during articulation has been well documented[5, 28-31]. However, details of the mechanisms by which particles separate from the parent material, caused by sliding contact in THA, are still being investigated. These processes are difficult to study in a laboratory environment, and consequently computer modelling must be used to advance our knowledge of such systems. The interlocking asperities of the cancellous bone surface and its harder counter-face (Titanium Alloys surface) are modelled using a simple geometry so that deformation and stresses distribution, and their effects on failure and wear mechanisms of cancellous bone in THA, can be identified.

In this work possible wear particle generation mechanisms in a Cancellous bone in a total hip joint prosthesis under sliding conditions were investigated using a 2-D finite element computational model. The aim of the study conducted was to develop the model that could be used to predict the effects of load, geometry and material properties on the stress and strain distributions occurring within the structure of hip joint replacement components. Prediction of mechanisms involved in the formation of cancellous bone wear particles is of the main objective of this study.

2. Experimental method

2.1. Finite Element Geometrical Model

IOP Conf. Series: Materials Science and Engineering **172** (2017) 012033 doi:10.1088/1757-899X/172/1/012033

A finite element model of the contact between two asperities has been developed. A contact between a single Titanium Alloy asperity (representing the femoral stem) and a single Cancellous bone asperity (representing the femoral bone) in the interlocking position was modelled. The scale of the model is of the order of micrometer size. Illustration of asperity geometrical scale is shown in Figure 1.

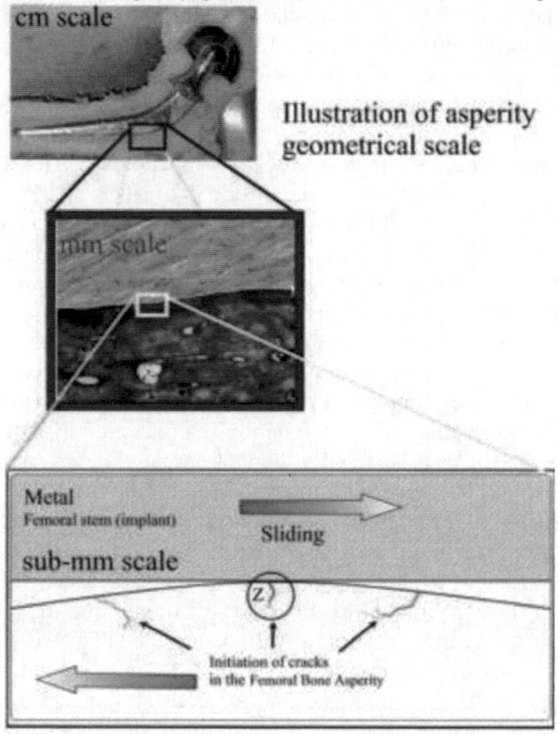

Figure 1: Illustration of Asperity Geomterical Scale

It was suggested by Fisher et al [32] that the real peak-to-valley height, Rt , of asperities on femoral stems are about 0.07 – 0.08 μm. However, average asperity size of the well finished stem components is about 0.02 – 0.05 μm in height[3], with the average asperity slopes of less than one degree, while the size of Cancellous bone asperities range from 0.1 μm, measured from peak to valley, to several tens of micrometers[3]. In this work, both contacting surfaces, i.e. Cancellous bone and Titanium Alloy, initially were of the same surface roughness. Surface topography of both the harder femoral stem and the softer cancellous bone components was modelled by triangular-shape asperities, as illustrated in Figure 2. The assumption of similar surface roughness allowed full interlocking contact between the two asperities to be modelled.

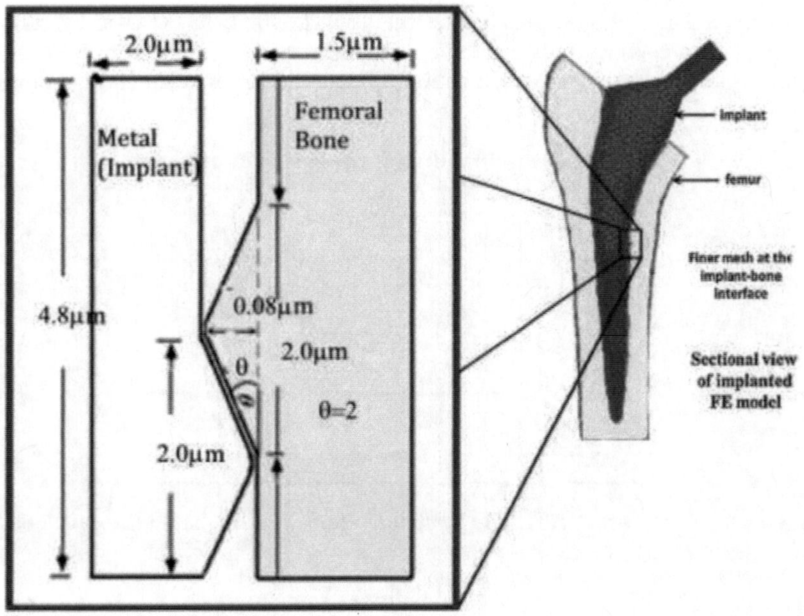

Figure 2: Schematic illustration of the contacting asperities in total hip joint replacement model (*the scale does not represent the correct ratio of vertical to horizontal dimensions*)

As shown in Figure 2, it was assumed that each asperity side-face is in contact with a side-face of the matching asperity, i.e. forming a micro-scale asperity contact. The height of the asperities modelled was 0.08 μm with base angle of $\theta = 2°$ (the asperity height to width ratio is 0.035). The values of the asperity slope, height and width were chosen with the aim to study the failure mechanisms affected by stresses and deformations during sliding contact, with geometrical dimensions being within the size range of the real head and cup asperities. Asperity base angle of $\theta = 2°$ was used instead of $\theta = 1°$ or less in order to study the interlocking effect on the stress and strain changes in the softer asperity.

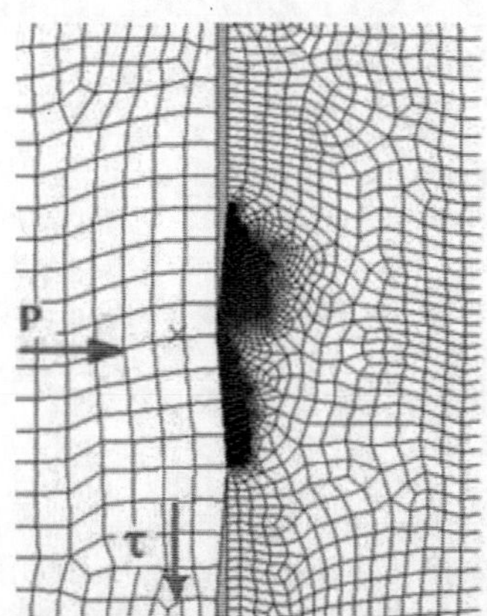

Figure 3: 2-D finite element geometrical model of contacting asperities

For simplicity, a 2-D finite element model was used to simulate the asperity contact. The model was developed using a commercially available FEM package ABAQUS (version 6.3, HKS Inc. U.S.). The contacting asperities were modelled with four-noded plane strain elements as shown in Figure 3.

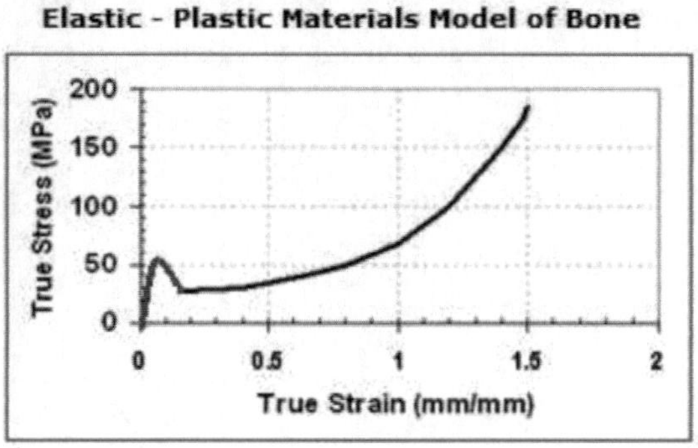

Figure 4: Stress-strain curve of Cancellous Bone used in the model developed [33, 34]

2.2. Material Model

Elastic-plastic material behaviour was proposed. It was assumed that the material starts to deform plastically when its yield point reached, i.e. $\frac{12}{7}\sigma_Y$, as the material starts to flow plastically when its flow stress is reached, i.e. $\sigma_f = \frac{12}{7}\sigma_Y$ [35]. In the model proposed the values of plastic yield stress, true stress and true strain at the cancellous bone breaking point were set as 19 MPa, 185 MPa and 1.5 mm/mm, respectively [33]. The stress-strain curve for cancellous bone used in the finite element model is shown in Figure 3 [33], while the material characteristics are listed in Table 1 [34, 36].

Table 1: Mechanical characteristics of materials used in the finite element analysis[34, 36]

Material	Properties		
	Elastic modulus (GPa)	Poisson ratio	Density (kg/m3)
Cancellous (femoral bone)[36]	1.0	0.40	950
Titanium Alloy(stem)[34]	208.0	0.30	7800

2.3. Loading and Boundary Conditions

The contact between the two surfaces was established first by moving the harder femoral head asperity in the vertical direction onto the fixed softer bone surface to give an initial load. The Titanium AlloysFemoral stem asperity was then moved in the vertical direction, cyclic sliding of micro-motion along the Cancellous bone asperity with a speed of 0.2 ms-1. This sliding speed was used in the frictional characterization of explanted Charnley hip prostheses[32]. The vertical sliding results in the higher contact stress at the asperities shown in Figure 3, causing the softer asperity to change its behaviour from elastic through elastic-plastic to fully plastic. The verticalmicro-motion or micro-sliding distance, used in the simulations, was chosen to ensure that the harder sliding asperity (Titanium Alloy stem) had completely passed over the softer fixed asperity (Cancellous bone). The loading and boundary conditions in this model are set in such a manner that the tangential load intensifies as the micro-sliding distance increases. The graph of the contact force changes occurring within the sliding contact is shown in Figure 5.

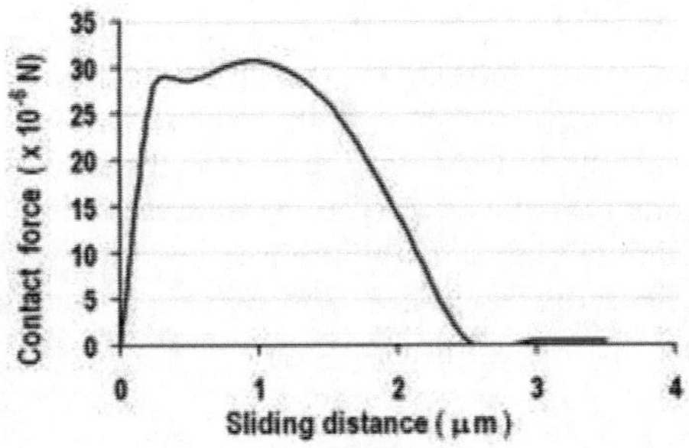

Figure 5: Plot of contact force vs. sliding distance

The effect of friction coefficient was examined with values of 0.8 used in this study. A friction coefficient of 0.8 was assumed to represent a sliding contact between Titanium Alloy and cancellous bone in a total hip joint arthroplasty.

2.4. Material failure in a strain discontinuity approach

The concepts of material failure in discontinuity settings have been reported in several studies[37-40]. The aim of these studies was to identify the mechanisms of material failure and to quantify corresponding critical structural response. It was suggested that material failure (i.e. crack development) usually occurs when the displacement discontinuities take place[25, 37]. Material fracture processes can be categorized into three distinct zones – diffuse failure, weak discontinuity and strong discontinuity zones[37]. The diffuse failure zone is denoted by an increase in strain concentration in a continuous fashion, without the appearance of material discontinuities. The weak discontinuities zone is characterized by a rapid increase in the strain field, resulting in a narrow strain band which exhibits strain discontinuities, however, the displacement field remains continuous. The strong discontinuities zone is defined as the development of the weak discontinuity zone into a band, where the displacement experiences a real jump (discontinuity) and unbounded strain[37].

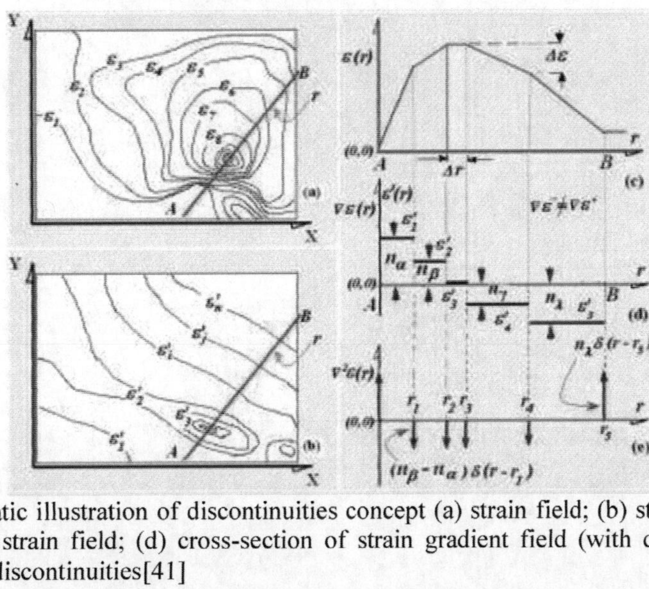

Figure 6: Schematic illustration of discontinuities concept (a) strain field; (b) strain gradient field; (c) cross-section of strain field; (d) cross-section of strain gradient field (with discontinuities); (e) magnitude of strain discontinuities[41]

IOP Conf. Series: Materials Science and Engineering **172** (2017) 012033 doi:10.1088/1757-899X/172/1/012033

The discontinuity concept used in this work is based on the discontinuity state of the equivalent plastic strain, which is derived from the rate of change of the equivalent plastic strain with respect to geometrical position. The magnitude of the strain discontinuity itself is obtained from the second derivative of equivalent plastic strain field with respect to the spatial position, as illustrated in Figure 6. Since no prior assumptions regarding the type of material failure (in terms of strain discontinuity settings) are made in modelling of the contact between the two surfaces, the material can be categorized into both diffuse failure zones and weak discontinuity zones. Therefore, a fully conventional finite element method could be implemented in this study in calculating the strain discontinuity fields.

In the narrow band of the continuous strain field zone, the derivation of the strain field with respect to the geometrical position becomes discontinuous across the limits of a narrow band (strain-gradient localization band). In order to observe the magnitude of finite plastic strain discontinuity jumps, it is necessary to examine the point-wise behaviour of the finite plastic strain discontinuity derivation. It can be seen from Figure 5, that n_α , n_β or n_γ are constants or the gradients of $\varepsilon(r)$ and that the delta functions approach the origin from either side [42]. The significance of the sequence $\nabla^2\varepsilon(r)$ is relatively easy to interpret. If $f(r)=\nabla\varepsilon(r)$ and $f'(r)=\nabla^2\varepsilon(r)$, then mathematical descriptions for the strain discontinuity derivative along section AB (Figure 6c), between r0 and r5 are:

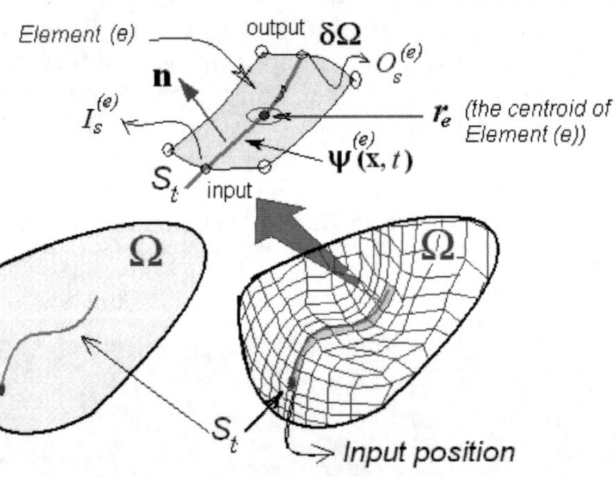

Figure 7: Schematic illustration of 2-D domain with a discontinuity tracking[41]

$$f(r)=\begin{cases} n_\alpha & \text{for} \quad r_0 \le r < r_1 \\ n_\beta & \text{for} \quad r_1 \le r < r_2 \\ \dots & \dots \quad \dots \\ n_\gamma & \text{for} \quad r_4 \le r < r_5 \\ n_\lambda & \text{for} \quad r_5 \le r < r_6 \end{cases} \tag{1}$$

$$f'(r) = \begin{cases} n_\alpha [\delta(r-r_0) - \delta(r-r_1)] \\ n_\beta [\delta(r-r_1) - \delta(r-r_2)] \\ \ldots\ldots \\ n_\gamma [\delta(r-r_4) - \delta(r-r_3)] \\ n_\lambda [\delta(r-r_5) - \delta(r-r_4)] \end{cases} \quad (2)$$

Equation (2) can also be presented in the following form:

$$f'(r) = n_\alpha \delta(r-r_0) + (n_\beta - n_\alpha)\delta(r-r_1) + \ldots. + (n_\gamma - n_\lambda)\delta(r-r_4) + \ldots. \quad (3)$$

The equivalent plastic strain discontinuities are identified based on the calculation of invariants or components at elements common to two or more other elements; then they are compared to determine the greatest difference (discontinuity jump). The elements that are receiving equivalent values of plastic strain from all nearest elements will exhibit a zero value in a plot of the equivalent plastic strain discontinuities $f(r) = \nabla\varepsilon(r) = 0$.

The strain discontinuity, in simple cases, can be used as the criterion of a local failure. The onset of discontinuity also relates to the end of the elastic regime in the bulk material through the fulfilment of some yield or damage criterion[40]. The propagation direction of material failure (on empirical basis) is made orthogonal to the maximum tensile stress[43].

Based on strain discontinuities, it is assumed that a 2-D body Ω experiences discontinuities along a discontinuity path, St, which is characterized by its root r, i.e. the first nodal point (in the finite element context this is the first element) as schematically illustrated in Figure 6. It is also assumed that for each x of Ω (i.e. every element in the finite element mesh) a propagation direction $Æ^{(e)}(\mathbf{x},t)$, orthogonal to the normal n to the discontinuity path, is available at every time of the analysis $Æ^{(e)}(t)$ [37].

Prediction of the root of discontinuity propagation location is based on the geometrical region where large-magnitude of finite strain discontinuity jumps (the difference between strain gradients, e.g. $(n_\alpha - n_\beta)$ etc.) are accumulated. Estimation of the material failure propagation path is made by identifying the tracking line of the displacement of the high finite strain discontinuity jumps to new geometrical positions[41].Detailed displacement track of accumulating finite strain discontinuity jumps is shown in Suhendra and Stachowiak work[41].

Since the asperities were arranged in an interlocking position, it was assumed that material failure could originate at the base of the softer asperity of the cancellous bone part. Material failure has been associated with a local specific behaviour of the nodal points involving the progressive softening of the material response that spreads across the body and leads to the structural failure. The variation of tensile fracture direction due to loads imposed on the asperity of the softer material, i.e. cancellous bone, is the key issue in this approach aiming to predict the crack initiation and wear particles generated. The direction of tensile fracture occurring in the softer material was examined by the standard continuum mechanics approach, i.e. strain vs. stress non-linear constitutive equations. It is well known that the inclusion of strain-softening features in this approach leads to the strain-

IOP Conf. Series: Materials Science and Engineering **172** (2017) 012033 doi:10.1088/1757-899X/172/1/012033

localization phenomenon, in which the strains concentration is confined to a narrow band[37, 38]. This approach, based on the implementation of the plastic strain discontinuities in the model, might be helpful in predicting the mechanisms involved in wear particle formation. The strain discontinuities are understood here as jumps in the strain field experienced by solid materials during deformation (strain) processes[41].

Characterization of material failure has usually been performed by means of local strain with high discontinuity jump concepts. Local failure at a nodal point or at a discrete set of points progresses through the body along failure surfaces in the form of cracks or shear bands [37]. The accumulation of strain discontinuities in a narrow band is responsible for an increasing dissipation of strain energy, and can then be treated as an initial local failure which eventually manifests itself in a form of physically observable propagating surface failure exhibiting a discontinuity in the displacement field [38]. Therefore, strain discontinuities accumulation in the surface layer of contacting bodies can be also treated as an initial process of wear particle formation. The narrow bands of strain concentration, addressed in this study, can prove to be a useful tool in the prediction of wear particle generation mechanisms occurring in joint prostheses.

3. Results and discussion
Results obtained from the model developed consist of equivalent (von Mises) stress, equivalent plastic strain, strain energy and plastic strain discontinuity. Based on these results, the possibility of predicting the characteristics of the wear particles generated (e.g. size and shape) is discussed.

3.1. Von Mises stress
The distribution of equivalent von Mises stress field in the subsurface of cancellous bone asperity is shown in Figure 8. The friction coefficient and sliding speed applied in the simulation were 0.8 and 0.2 ms-1, respectively.

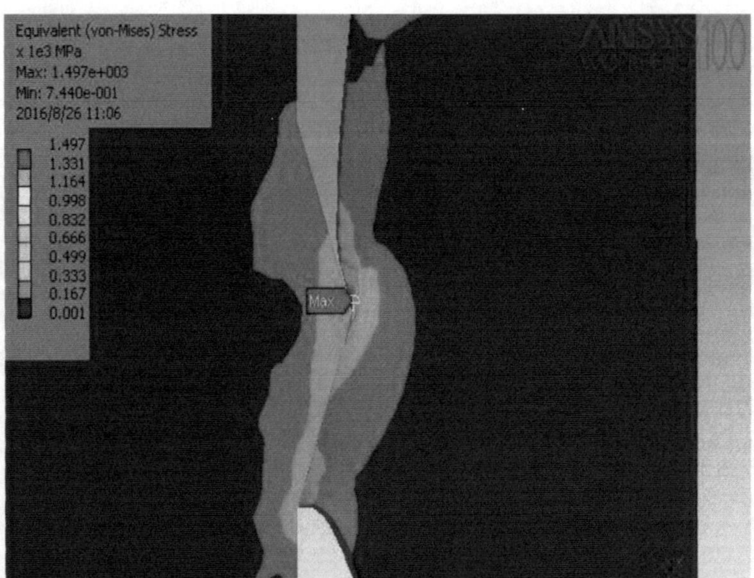

Figure 8: Distribution of equivalent von Mises Stress

The von Mises stress distribution, shown in Figure 8, illustrates how the stress pattern changes as the sliding distance increases. It can be seen from Figure 8 that as the harder Femoral stem asperity slides along the cancellous bone surface, the geometrical position of the highest value of von Mises stress in the cancellous bone subsurface also moves. In addition, as the sliding distance increases, the von Mises stress value raises from 26.3 MPa at the initial position to 33.1 MPa at a sliding distance of 3.0μm. As the tangential force increases with the displacement, the maximum shear stress value repositions itself closer to the interface/surface. It might be assumed that the possible degradation of Cancellous bone asperity, leading to crack development, may take place in the region where the highest values of von Mises equivalent stress occur.

3.2. Equivalent plastic strain

The surface and subsurface deformation of the softer asperity during loaded sliding contact was also examined. The surface stress change and internal equivalent plastic strain field in the deformed cancellous bone asperity occurring during contact under the friction coefficient of 0.8 are shown in Figure 9. It can be seen from Figure 9 that the geometrical position of the maximum value of equivalent plastic strain in the cancellous bone asperity moves during sliding.

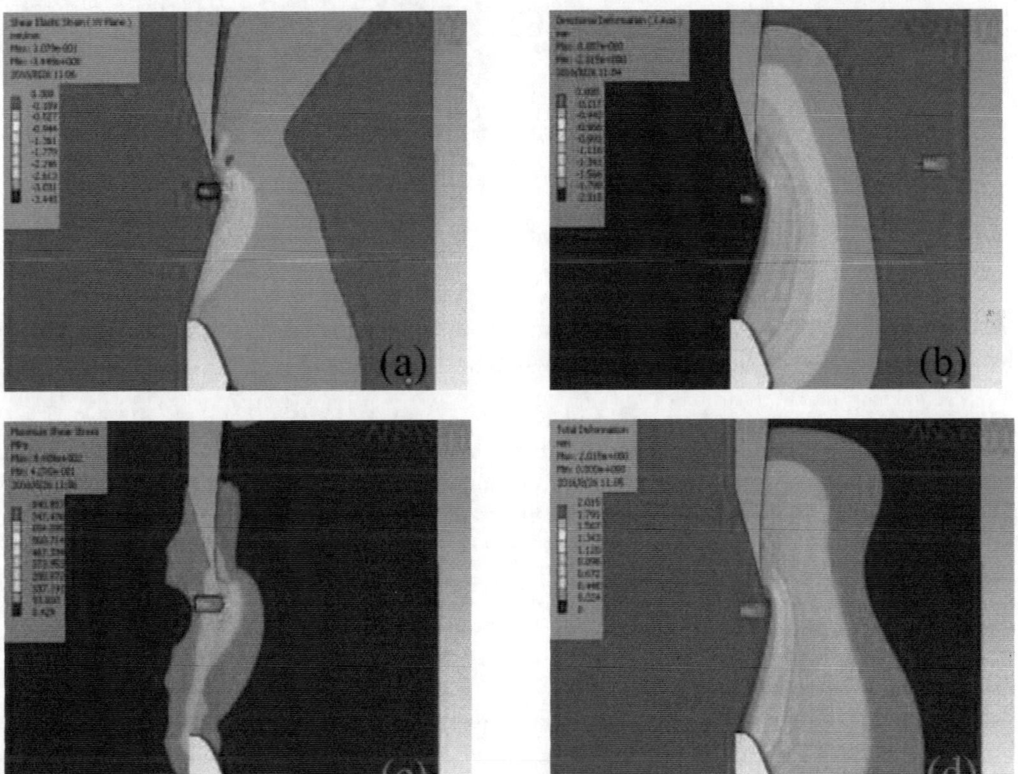

Figure 9: (a) Shear Elastic Strain; (b) Directional Deformation; (c) MaximumShear Stress; (d) Total Deformation

As the sliding contact moves across the surface of the THR (between the Stainless steel Femoral stem and cancellous bone) much of the damage done to the softer material occurs beneath the micro-sliding surface. Cyclic plastic deformation occurs across the softer asperity surface area as the harder

IOP Conf. Series: Materials Science and Engineering **172** (2017) 012033 doi:10.1088/1757-899X/172/1/012033

asperity passes over the surface and the frictional traction forces increase the plastic deformation[44]. Although internal strain in the softer material may eventually attain extremely high levels, it does not directly contribute to crack growth unless it is accompanied by an internal triaxial compressive stress field that occurs directly beneath a contacting asperity [45].

3.3. Plastic strain discontinuity

The location where the initial crack (material's failure) may occur in the softer asperity has been examined based on the von Mises stress field behaviour (Figure 7). Real wear processes are complex, as many factors are involved in the formation of a wear particle occurring between the articulating surfaces of THR [33]. Nevertheless, plastic strain discontinuity examination gives information that might prove to be useful in the prediction of wear particle detachment and formation.

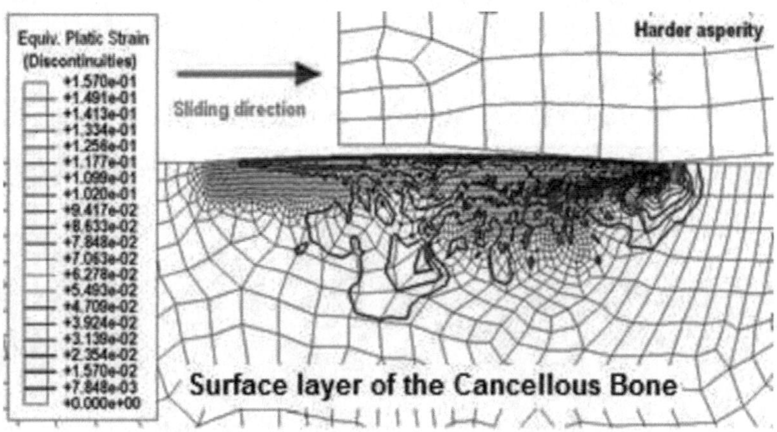

Figure 9: Fields of equivalent plastic strain discontinuities in the cancellous bone asperity subsurface with friction coefficient of 0.8 and sliding speed of 0.2 ms-1

Some of the continuum mechanical models of solids allow for the formation of surface strain discontinuities when the body is deformed. Across such a deforming surface, the stress and strain fields suffer finite discontinuity jumps even though the traction and displacement fields remain continuous. The fields of equivalent plastic strain discontinuity in the cancellous bone asperity subsurface with friction coefficient of 0.8 and sliding speed of 0.2 ms-1 are shown in Figure 9. It can be seen from Figure 9 that plastic strain discontinuity occurs in the subsurface of cancellous bone asperity close to its surface. The strain discontinuity was observed at sliding distance of 1.35 μm when the strain energy reaches its highest value.

4. Conclusions

From the study conducted the following conclusions can be drawn:

- The model of a loaded sliding contact between two asperities has been developed. The calculation results confirmed that the positions of maximum plastic strain and von Mises stress depend on the friction coefficient and the magnitude of load applied.
- As the friction coefficient and sliding contact force increase, the maximum von Mises stress and plastic strain are moving closer to the surface.
- The evolution of equivalent von Mises stress and equivalent plastic strain within the polyethylene, resulting from a frictional sliding contact, might be useful in interpretation of wear particle detachment mechanisms in tough polymers, like cancellous bone.

- The strain energy as well as the strain discontinuity generated might be of use in the prediction of wear particle formation.
- The friction coefficient has a significant effect on the stress and strain state occurring between the contacting asperities, as well as on the area subjected to higher stress and strain states.
- Strain discontinuity approach might be a useful tool in prediction of wear particle formation.

References

[1] Mow, V.C., A. Ratcliffe, and S.L.Y. Woo, Bio-tribology of synovial joints. Vol. 1. 1990, New York: Springer-Verlag. 323 - 345

[2] Podsiadlo, P., M. Kuster, and G.W. Stachowiak, Numerical analysis of wear particles from non-arthritic and osteoarthritic human knee joints. Wear, 1997. 210(1-2): p. 318-325

[3] Dowson, D., A comparative study of the performance of metallic and ceramic femoral head components in total replacement hip joints. Wear, 1995. 190(2): p. 171-183

[4] Podsiadlo, P. and G.W. Stachowiak, 3-D imaging of surface topography of wear particles found in synovial joints. Wear, 1999. 230(2): p. 184-193.

[5] Firkins, P.J., et al., A novel low wearing differential hardness, ceramic-on-metal hip joint prosthesis. J. Biomechanics, 2001. 34(10): p. 1291-1298.

[6] Maloney, W.J., et al., Fibroblast response to metallic wear debris in vitro. J. Bone Jt. Surgery, 1993. 75A: p. 835 - 844.

[7] Wirganowicz, P.Z. and B.J. Thomas, Massive ostelysis after ceramic on ceramic total hip arthoplasty. Clinical Orthop., 1997. 338: p. 100 - 104

[8] Catelas, I., et al., Induction of macrophage apoptosis by ceramic and polyethylene particles in vitro. Biomaterials, 1999. 20(7): p. 625-630.

[9] Panichkul, P., et al., New Approach and Stem Increased Femoral Revision Rate in Total Hip Arthroplasty. Orthopedics, 2016(1): p. 86 - 92.

[10] Goodman, S.B., Huie, P., Song, Y. P., Schurman, D., Maloney, W., Woolson, S., Sibley, P., Cellular profile and cytokine production at prosthetic interfaces. J. Bone Jt. Surgery, 1998. 80: p. 531 - 539.

[11] Verdonschot, N., E. Tanck, and R. Huiskes, Effects of prosthesis surface roughness on the failure process of cemented hip implants after stem-cement debonding. J Biomed Mater Res, 1998. 42: p. 554 - 559.

[12] Chiba, J., et al., Biochemical analyses of human macrophages activated by polyethylene particles retrieved from interface membranes after failed total hip arthoplasty. J. Arthroplasty, 2001. 16(8): p. 101 - 105

[13] Kobayashi, A., Freeman, M. M., Bonfield, W., Number of polyethylene particles and ostelysis in total joint replacements: Quantitative study using a tissue-deggistion method. J. Bone Jt. Surgery, 1997. 14: p. 840 - 873.

[14] Green, T.R., Fisher, J., Stone, M., Wroblewski, B. M. and Ingham, E., Polyethylene particles of 'critical size' are necessary for the induction of cytokine by macrophages in vitro. Biomaterials, 1998. 19(24): p. 2297 - 2302.

[15] Oparaugo, P.C., et al., Correlation of wear debris-induced osteolysis and revision with volumetric wear-rates of polyethylene: A Surver of 8 reports in the literature. Acta Orthop. Scand., 2001. 72(1): p. 22 - 28.

[16] Collis, D. and C. Mohler, Comparison of clinical outcomes in total hip arthroplasty using rough and polished cemented stems with essentially the same geometry. J Bone Joint Surg Am, 2002(84): p. 586 - 592.

[17] Herberts, P., H. Malchau, and G. Garellick, Annual report 2003. The Swedish National Hip Arthroplasty Register, 2004.

[18] Lancaster, J.G., et al., The wear of ultra-high molecular weight polyethylene sliding on metallic and ceramic counterfaces representative of current femoral surfaces in joint replacement.

Proc. Instn. Mech. Engrs. (H) J. Eng, Med., 1996. 211(Part H): p. 17 - 24.
[19] 19. Harris, W., J. McCarthy, and D. O'Neill, Femoral component loosening using contemporary techniques of femoral cement fixation. J Bone Joint Surg Am, 1982. 64: p. 1063 - 1067.
[20] Taylor, M. and T. KE, Fatigue failure of cancellous bone: a possible cause of implant migration and loosening. J Bone Joint Surg Br, 1997. 79: p. 181 - 182.
[21] Tipper, J.L., et al., Characterisation of wear debris from UHMWPE on zirconia ceramic, metal-on-metal and alumina ceramic-on-ceramic hip prostheses generated in a physiological anatomical hip joint simulator. Wear, 2001. 250(1-12): p. 120-128.
[22] Hall, R.M., et al., The association between rates of wear in retrieved acetabular components and the radius of the femoral head. Proc. Instn. Mech. Engrs. (H) J. Eng. Med., 1998. 211: p. 321 - 326.
[23] Learmonth, I.D. and J.L. Cunningham, Factors contributing to the wear of polyethylene in clinical practice. Proc. Instn. Mech. Engrs. (H) J. Eng, Med., 1997. 211(Part H): p. 49 - 57.
[24] Hall, R.M., et al., The effect of surface topography of retrieved femoral heads on the wear of UHMWPE sockets. Medical Eng. & Physics, 1997. 19(8): p. 711-719.
[25] Suhendra, N., Computational analysis for prediction of loaded sliding contact effects in total hip joint prosthesis, in Proceeding of the 3th International Biomaterials Conference, I.N. Jujur, Editor. 2014, BIOMAT 2014: Surabaya - Indonesia.
[26] McNie, C., et al., Prediction of plastic strain in ultra-high molecular polyethylene due to microscopic asperity interactions during sliding wear. Proc. Instn. Mech. Engrs. (H) J. Eng, Med., 1998. 212(Part H): p. 49 - 56.
[27] Adams, G.G. and M. Nosonovsky, Contact modeling -- forces. Tribology Intl., 2000. 33(5-6): p. 431-442.
[28] Stachowiak, G.W., Friction and Wear of Polymers, Ceramics and Composites in Biomedical Applications, in Advances in Composites Tribology, K. Friedrich, Editor. 1993, Elsevier: Amsterdam. p. Chapter 14: 509-557.
[29] Cooper, J.R., D. Dowson, and J. Fisher, Birefringent studies of polyethylene wear specimens and acetabular cups. Wear, 1991. 151(2): p. 391-402.
[30] Scholes, S.C., S.M. Green, and A. Unsworth, The wear of metal-on-metal total hip prostheses measured in a hip simulator. Proc. Instn. Mech. Engrs. (H) J. Eng, Med., 2003. 215(Part H): p. 523 - 530.
[31] Vassiliou, K. and A. Unsworth, Is the wear factor in total joint replacements dependent on the nominal contact stress in ultra-high molecular weight polyethylene contacts? Proc. Inst. Mech. Engr. (H) J. Eng. Med., 2004. 218: p. 101 -- 107.
[32] Fisher, J., Dowson, D., Hamdzah, H. and Lee, H. L., The effect of sliding velocity on the friction and wear of UHMWPE for use in total artificial joints. Wear, 1994. 175(1-2): p. 219-225.
[33] Kurtz, S.M., Pruitt, L., Jewett, C. W., Crawford, R. P., Crane, D. J., Edidin, A. A., The yielding, plastic flow, and fracture behavior of ultra-high molecular polyethylene used in total joint replacements. Biomaterials, 1998. 19: p. 1989 - 2003.
[34] Anoname. In: Matweb (Material Property Data), BioDur Carpenter CCM Alloy. Data alloy [online]. Available from: http://www.matweb.com/search/ SpecificMaterial.asp?bassnum=NCAR40. [cited [Accessed 25 July 2004].BioDur Carpenter CCM Alloy. Data alloy [online].
[35] Kaliszky, S., Plasticity: Theory and Engineering Applications. Studies in Appl. Mech. Vol. 21. 1989, New York: Elsevier Science Publishing Company, Inc. 505.
[36] Jin, Z.-M., A general axisymmetric contact mechanics model for layered surfaces, with particular reference to artificial hip joint replacement. Proc. Instn. Mech. Engrs. (H) J. Eng, Med., 2000. 214(Part H): p. 425 - 435.
[37] Oliver, J., On the discrete constitutive models induced by strong discontinuity kinematics and continuum constitutive equations. Intl. J. Solids Struct., 2000. 37(48-50): p. 7207-7229.

[38] Sukumar, N. and J.-H. Prevost, Modeling quasi-static crack growth with the extended finite element method Part I: Computer implementation. Intl. J. Solids Struct., 2003. 40(26): p. 7513-7537.
[39] Wells, G.N. and L.J. Sluys, Application of embedded discontinuities for softening solids. Eng. Fract. Mech., 2000. 65(2-3): p. 263-281.
[40] Wells, G.N. and L.J. Sluys, A new method for modelling cohesive cracks finite element. Intl. J. Numer. Method. Eng., 2001. 50: p. 2667 - 2682.
[41] Suhendra, N. and G.W. Stachowiak, Computational model of asperity contact for the prediction of UHMWPE mechanical and wear behaviour in total hip joint replacements. Tribology Letters, 2007. 25(1): p. 9 - 22.
[42] Hoskins, R.F., Generalised Functions. Mathematics and its applications, ed. G.M. Bell. 1979, London: Ellis Horwood Limited (Div. John Willey & Sons).
[43] Pradeilles - Duval, R.-M. and C. Stolz, Mechanical transformations and discontinuities along a moving surface. J. Mech. Phys. Solids, 1995. 43(1): p. 91 - 121.
[44] Edidin, A.A., et al., Plasticity-induced damage layer is a precursor to wear in radiation-cross-linked UHMWPE acetabular components for total hip replacements. J. Arthroplasty, 1999. 14(5): p. 616 - 627.
[45] Krzypow, D.J. and C.M. Rimnac, Cyclic steady state stress-strain behavior of UHMW polyethylene. Biomaterials, 2000. 21(20): p. 2081-2087.
[46] Oliver, J. and A.E. Huespe, Continuum approach to material failure in strong discontinuity settings. Comp. Meth. Appl. Mech. Eng., 2004. 193(30-32): p. 3195-3220.

Acknowledgments
The author wish to thank Ministry for Research and Technology in the scheme of Incentive Programme of Applied Research for their financial support and Agency for the Assessment and Application of Technology, Centre for Materials Technology for its help during the preparation of this paper.Thanks to THA team members for the very good collaboration as a team work. Thanks also to Professor GwidonStachowiak for his inspiration from beginning the writer worked in Tribology Laboratory at UWA – Australia until presently.

Oyster mushroom's lipase enzyme entrapment on calcium alginate as biocatalyst in the synthesis of lauryl diethanolamide

N Wijayati[1], K Masubah[1] and Supartono[1]

[1] Department of Chemistry, Semarang State University, Indonesia

Email: nanikanang@gmail.com

Abstract. Lipase is an enzyme with large biotechnology applications, such as hydrolysis in the food industry, applications in chemical industry, synthesis of polymers and surfactants. Lipase was isolated from oyster mushroom with activity 0,93 U/mg and protein content 1,1234 mg/mL. Lipase was immobilized by entrapment method in a matrix of Ca-alginate. This report describes that we have developed for the synthesis of lauryl diethanolamide The result showed that the optimum condition of lipase immobilization was achieved on 3% Na-alginate solution with protein content 0,84 mg/mL and the activity 3,33 U/mg. An amide (22.911%) formed from the amidation of lauric acid and diethanolamine.

1. Introduction

Surfactants are widely used in various industrial detergent, softener, paints, inks, emulsifiers (emulsifier), insecticides and others. In general the surfactant synthesis using renewable and non-renewable raw materials. Surfactants are generally synthesized from petroleum derivatives, such as linear alkyl benzene sulphonate (LAS) and alkyl sulphonate (AS) [1]. However, these surfactants can cause pollution to the environment as waste after use will be difficult degraded and can not be updated (non-renewable). At this time with the limited non renewable materials, there should be efforts to replace non-renewable raw materials.

The potential of Indonesia as a producer of surfactants which are synthesized from fatty acids of coconut oil is huge, considering Indonesia's palm oil production has increased from year to year. High lauric acid content which is 50.50% pure coconut oil has the potential to make a diethanolamide surfactant. Lauril nonionic surfactant derived from alkanolamides. These surfactants derived from lauric acid and diethanolamine through enzymatic amidation. diethanolamide surfactants are more effective both as a foam stabilizer, thickener and foam booster [2]. Diethanolamide can be found in the manufacture of detergents, emulsifiers agents, and cosmetics.

An important alternative route for the production of diethanolamide involves the use of a biocatalyst in organic media. The use of enzymes as biocatalysts have a very important role in the chemical and pharmaceutical industries. Lipase (EC 3.1.1.3) is an enzyme with biotechnology applications is very broad, such as hydrolysis of the milk fat in the food industry, the oleochemical industry applications, the synthesis of triglyceride structure and the synthesis of polymers and surfactants. Mushrooms was a source of enzymes of interest in biotechnology due to the availability and high stability. However, the use of lipase as a biocatalyst has several weaknesses, which are only applicable to one reaction. One way to overcome this drawback is to apply the enzyme immobilization techniques to be used. Enzyme immobilization aims to improve the stability and productivity of these enzymes so that the lipase can be reused [3-5].

Methods of immobilizing enzymes entrapment in calcium alginate is one important method of immobilization. One commercially available alginate is sodium alginate has been used for more than 65 years in the food and pharmaceutical industries as coatings, emulsifying and film-forming agent. Entrapment using calcium alginate gel that does not dissolve quickly known, non-toxic, inexpensive, and versatile method of immobilization of an enzyme [6].

Surfactant synthesis of renewable materials and environmentally friendly is very important as to minimize the impact of pollution. Lauryl diethanolamide is one surfactant made from natural easily degraded and contaminant. Free lipase is an environmentally friendly biocatalyst which catalyzes the synthesis of surfactant. This report describes that we have developed for the synthesis of lauryl diethanolamide. The optimum concentration of Na alginate in the process of entrapment lipase oyster mushrooms and know entrapped lipase stability for repeated use in the synthesis of lauryl diethanolamide.

2. Methods

Please follow these instructions as carefully as possible so all articles within a conference have the same style to the title page. This paragraph follows a section title so it should not be indented.

2.1. Material

Oyster mushrooms, n-hexane, diethanolamine (*E Merck*, 99%), NaOH (*E-Merck)*, KOH (*E-Merck))*, Phenolphthalein (*E-Merck)*, KH_2PO_4 (*E-Merck)*, $KH_2PO_4.2H_2O$, lauric acid, H_2SO_4 (*E Merck, 96%)*, VCO, distilled water, ethanol (Brataco Chem), sodium alginate (Shandong Jiejing Group Corporation), bovine serum albumin (*E-Merck).*

2.2. Isolation of Lipase

Isolation lipase of oyster mushrooms was done by washing 200 g of oyster mushrooms until completely clean and sterile. Oysters mushrooms and 100 mL aquabidest were blended until crushed oyster mushrooms and the slurry obtained centrifuged at 1000 rpm for 20 min.

2.3. Protein assay

Determination of protein content was carried by Biuret method. A total of 1 mL sample, four mL Biuret reagent and five mL of distilled water, then shaken and allowed to stand for 10 minutes until a purple solution. The wavelength of maximum absorbance at Bovine Serum Albumin (λmax = 540 nm). The protein content was known by converting the absorbance value of the standard curve regression equation Bovine Serum Albumin.

2.4. Lipase Activity

Lipase activity test with a volumetric method using the VCO as a substrate. A total of 1 mL of lipase was added to 4 ml of n-hexane and incubated at 50°C for 30 minutes. The results formed extracted using ethanol solution for 5 minutes. Followed by titration of the extract using a 0.05 M NaOH solution by adding 2-3 drops of PP indicator.

2.5. Immobilization Lipase

Immobilization of lipase in Ca-alginate is done by adding 30 mL of lipase in 20 mL Na-alginate with the weight ratio between the enzyme and the Na-alginate (2:3). The mixture is stirred with a magnetic stirrer at a room temperature. Then, Na-alginate solution mixture and lipase put in 10 mL of 0.1 M $CaCl_2$ little by a little drop of 2 mL using a pipette while stirring slowly with a magnetic stirrer to form an immobile enzyme granules. A gel containing the enzyme removed from the $CaCl_2$ solution by filtration. Then the gel is stored at 4°C until to be used. To obtain optimum process conditions immobilized lipase was determined of the percentage concentration of the immobilized enzyme with Na-alginate concentration of 1%, 2%, 3% and 4% (w/v).

2.6. Synthesis of Lauryl Diethanolamide
Diethanolamine was mixed with lauric acid in the ratio 2/1 and entrapped lipase at a concentration of 12.5% (w/w of lauric acid), then dissolved in n-hexane with a ratio of 2/1 (v/b lauric acid). The mixture was stirred using a shaker with a speed of 150 rpm at 50°C for 24 h.

3. Result and Discussion

3.1. Immobilization of Lipase from oyster mushrooms

The initial protein content was obtained from the isolation of 1.12 mg / mL and a specific activity of 4.03 U mg. Lipase enzyme immobilization process was carried out by the method of oyster mushrooms entrapment. Alginate entrapment conducted by dripping a suspension of lipase and Na-alginate into a solution of $CaCl_2$ that causes intermolecular crosslinking of alginate with Ca^{2+}. The crosslinking causes the formation of alginate gel in less than 1 minute. The gel is formed shortly after hatch is soft. After 20 minutes, alginate gel becomes harder allegedly due to crosslinking between the alginate and Ca^{2+} has been going perfectly. Calcium alginate formed is not soluble in water. That's because the calcium to form a bi-cations Ca^{2+} and forming a chelate with oxygen guluronic [3].
The initial steps to immobilize it entrapment reaction was mixed the enzyme into each Na-alginate solution of 1%, 2%, 3%, and 4% in comparison with the lipase alginate is 2: 3. Beads obtained is not the same at each concentration of Na-alginate, alginate beads were obtained small and fragile (Figure 1).

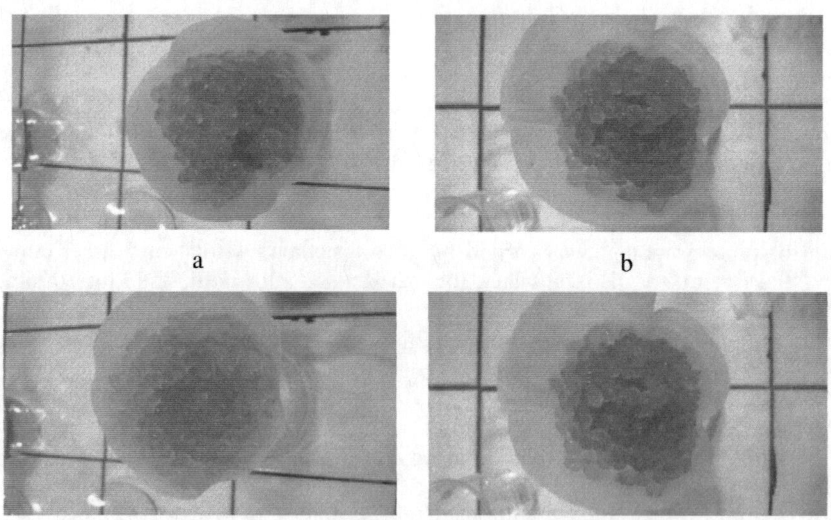

a b c d

Figure 1. Bead Lipase entrapped resulting from a concentration of Na alginate. 1%, b. 2%, c. 3%, and d. 4%.

Entrapment method using Ca-alginate gel pore density is affected by the alginate. The greater the concentration of Na-alginate crosslinks has formed pores closer. The pore density can prevent lipase which has been stuck for detaching thereby increasing the amount of lipase trapped in Ca-alginate beads. Lipase activity entrapped in Ca-alginate increased to a concentration of 3% and at a concentration of 4% Na-alginate decreased activity. This is because the pores are formed more tightly so that the process of diffusion substrate passes Ca-alginate to be blocked so that products are more difficult to form lead to decreased activity. The optimum concentration of Na-alginate obtained at concentrations of Na-alginate 3% of total lipase trapped by 0.84 mg / mL (Figure 2) and activities at 3:33 U/mL (Figure 3).

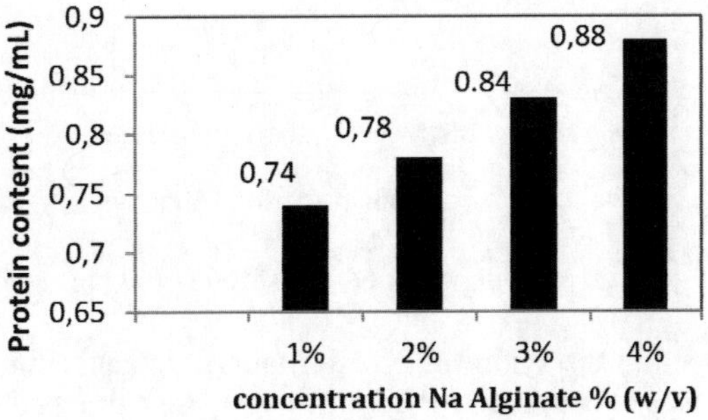

Figure 2. The relationship between the concentration of Na-alginate Protein levels Trapped

Specific enzyme activity is expressed as the number/unit of enzyme activity per mg protein enzymes. Determination of enzyme activity performed by titrimetric methods.

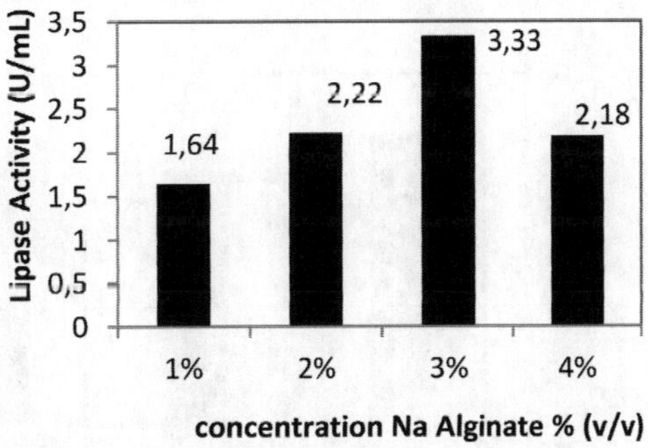

Figure 3. The relationship between the concentration of Na-alginate Protein levels Trapped

3.2. Activity of Lipase entrapped in Synthesis lauryl diethanolamine

Lauryl diethanolamide obtained from amidation reaction between lauric acid and diethanolamine. Numbers ester synthesis are relevant to the biosurfactant alkanolamide and fatty acid esters [7].

IOP Conf. Series: Materials Science and Engineering **172** (2017) 012034 doi:10.1088/1757-899X/172/1/012034

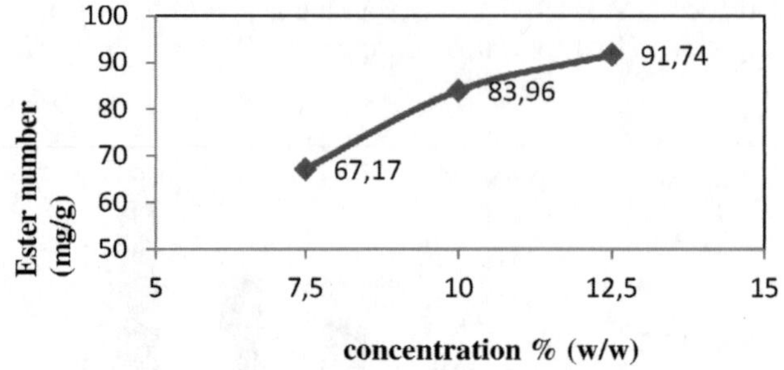

Figure 4. Relations concentration variation entrapped lipase% (w / w lauric acid) with numbers ester

The addition of lipase entrapped in the synthesis of ester lauryl diethanolamide then the numbers are increasing. The ester number was obtained on the addition of lipase entrapped concentration of 12.5% (w / w of lauric acid). The addition of concentration is then used as the optimum concentration in the stability test repeated use of lipase entrapped (Figure 4).

3.3. Stability and Reusability of Lipase immobilization

One enzyme immobilization goal was to determine whether the immobilized enzyme can be reused after being used in a reaction. Lipase has been immobilized by entrapment used as a biocatalyst in the amidation reaction between lauric acid and diethanolamine to produce lauryl diethanolamide. In this research note that entrapped lipase can still be used up to three times a reaction cycle (Figure. 5).

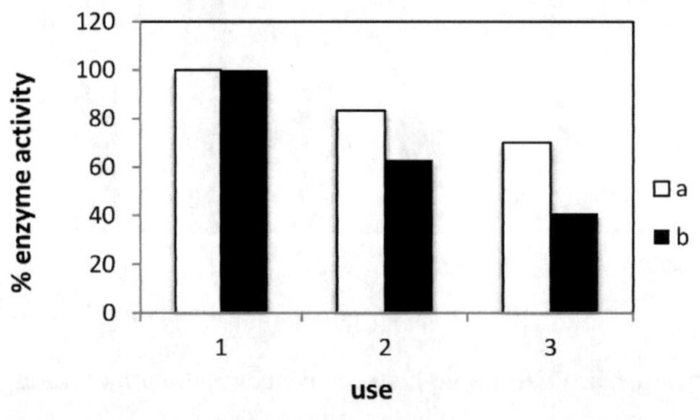

Figure 5. Relations with the repeated use of % of enzyme activity a) recovery using NaCl, b). recovery without NaCl

Entrapped lipase activity was tested for three times to determine the stability of the repeated use in the synthesis of lauryl diethanolamide. Figure 5a shows 83.25% activity during second reuse and 70.11% activity on its third use. While in Figure. 5b shows 62.86% activity during second reuse and 40.90% activity on its third use.

3.4. Characterization of product Synthesis

The product of a reaction between acid and lauric diethanolamide was then characterized using FTIR and HPLC.

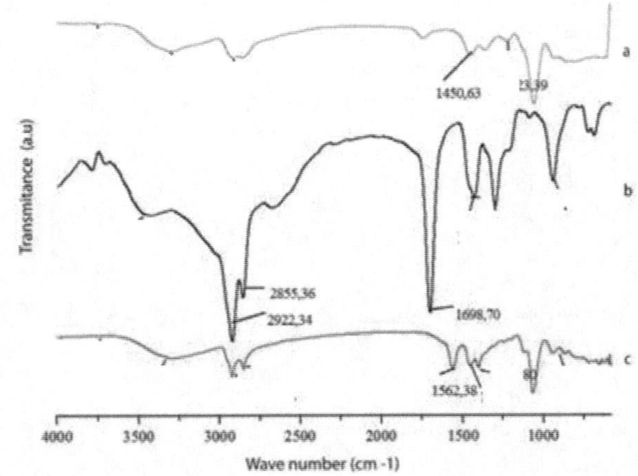

Figure 6. IR spectra a. diethanolamine, b. lauric acid, and c. product of synthesis

Analysis of a reaction between lauric acid and diethanolamine using FT-IR (Figure. 6) was suspected there is some absorption, the peak of 3748 cm^{-1} indicate the presence of strain NH, 3302 cm^{-1} indicate the presence of OH groups. The peak of 2924 cm^{-1} and 2853 cm^{-1} indicates an aliphatic CH group, the peak of 1562 cm^{-1} declared an amide group CO-N, 1456 cm^{-1} indicate the presence of CN bond stretching vibration, 1403 cm^{-1} shows the CH_2 stretching vibration, 942 cm^{-1} indicate the presence of CO stretching vibration.

Amidation diethanolamine with lauric acid was analyzed using HPLC to determine the percentage of reaction products formed. HPLC chromatogram of the results can be seen in Fig 7 and Table 1.

Table 1. The product of amidation reaction using HPLC analysis

No.	Retention Time (min)	Consentration (%)
1	2,43	24,59
2	2,63	49,01
3	5,85	0,29
4	9,22	22,91
5	12,52	1,59
6	16,29	1,22
7	19,26	0,39

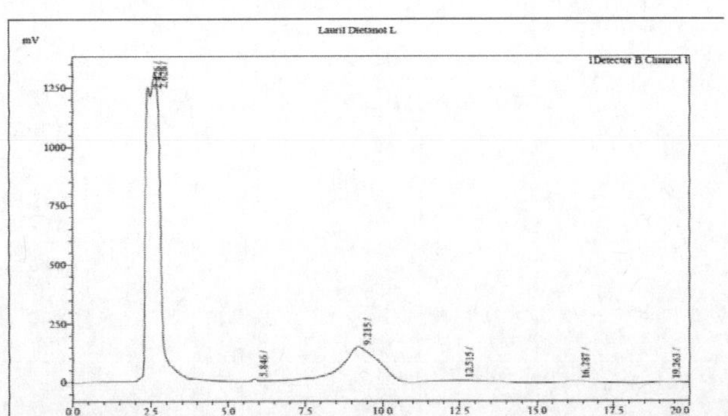

Figure 7. HPLC Chromatogram of product synthesis

The product of the amidation reaction between lauric acid and diethanolamine was characterized using HPLC [8]. The peak of lauryl diethanolamide appears at a retention time of 9.22 min with a concentration 22.91%. The peak may be suspected as an amide formed from the reaction of lauric acid amidation and diethanolamida (Figure. 8).

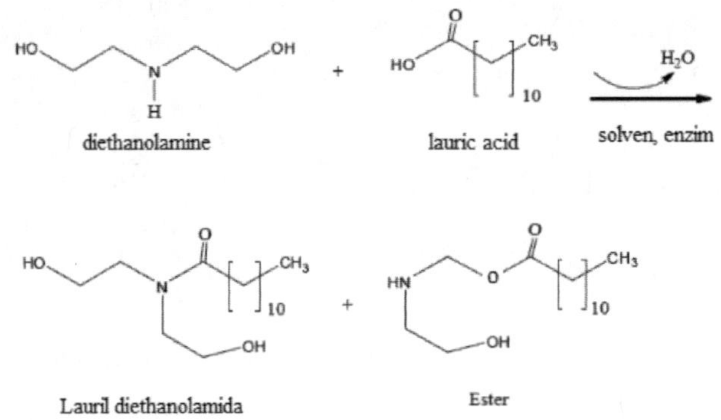

Figure 8. Amidation Reaction of Lauryl acid and diethanolamine

4. Conclusion

The result showed that the optimum condition of immobilization lipase were 3% of Na-alginate solution with protein content 0,84 mg/mL and the activity is 3,33 U/mL. Stability of entrapped lipase for the synthesis of lauryl diethanolamide showed that entrapped lipase could be used for three cycles. An amide (22.91%) was formed from the reaction amidation of lauric acid and diethanolamine.

5. References

[1] Hendra, Rahman, and Nurhaeni. 2013. Online Journal of Natural Science, 2 **3** 55.
[2] Yabur, R., Bashan, Y., and Carmmona, G. H. 2007, *J. Appl. Phycol.* **19** 43.
[3] Gulay S. 2009. Immobilization of Thermophilic Recombinant Esterase Enzyme by Entrapment in Coated Ca-Alginate Beads, **Thesis**, Izmir Institute of Technology, Izmir
[4] Yang XM, Yu W, Ou ZP, Ma HL, Liu WM, Ji XL. 2009., *fruit, plant food Hum Nutr* **64** 167.
[5] Cao, L. 2005.. Weinherm, Jerman : WILEY-VCH Verlag GmbH & Co. KgaA.
[6] Anwar A., Shah Ali U. Q., Samina I., dan Abid A. 2009. *World Applied Science Journal* 7 **10** 1281
[7] Fernandez-Perez, M dan Otero, C. 2003. *Enzyme and Microbial Technology* **33** 650.
[8] Gaber, Y., Akerman, C. O., and Hatti-kaul, R. 2014, *Chemistry Central Journal*, 8 **1** 33.

IOP Conf. Series: Materials Science and Engineering **172** (2017) 012035 doi:10.1088/1757-899X/172/1/012035

Cinnamomum casia Extract Encapsulated Nanochitosan as Antihypercholesterol

Ngadiwiyana[1], Purbowatiningrum[1], Enny Fachriyah[1], Ismiyarto[1]
[1]Chemistry Department, Science and Mathematics Faculty, Diponegoro University
Jl. Prof. H. Soedarto, SH, Tembalang, Semarang, Indonesia, Postcode 50275

Email: ngadiwiyana@gmail.com

Abstract. Atherosclerosis vascular disease with clinical manifestations such as cardiovascular disease and stroke are the leading cause of death in Indonesia. One solution to these problems is a natural antihypercholesterol medicine by utilizing *Cinnamomum casia* extract. However, the use of natural extracts to lower blood cholesterol levels do not provide optimal results because it is possible that the active components of extract have been degraded /damaged during the absorption process. So that, we need to do the research to get a combination of chitosan nanoparticles-*Cinnamomum casia*. extract as a compound which has an antihypercholesterol activity through the in vitro study. Modification of natural extracts encapsulated nanochitosan be a freshness in this study, which were conducted using the method of inclusion. The combination of both has the dual function of protecting the natural extracts from degradation and deliver the natural extracts to the target site. Analysis of nanochitosan using the Particle Size Analyzer (PSA) shows the particle size of synthesis product that is equal to 64.9 nm. Encapsulation efficiency of *Cinnamomum casia* extract-Chitosan Nanoparticles known through UV-VIS spectrophotometry test and obtained the efficiency encapsulation percentage of 84.93%. Zeta Potential at 193,3 mv that chitosan appropriate for a delivery drug. Antihypercholesterol activity tested in vitro assay that showed the extract-nanoparticle chitosan in concentration 150 ppm gave the highest cholesterol decreasing level in the amount of 49.66% w/v. So it can be concluded that *Cinnamomum casia* extract can be encapsulated in nanoparticles of chitosan and proved that it has a cholesterol-lowering effect through the in vitro study.

1. Introduction

According to data from the Ministry of Health in 2013 the number of patients with cardiovascular disease, stroke and heart attack showed an upward trend and that the disease has become one of the major causes of death in Indonesia. Cardiovascular disease caused by hypercholesterolemia, one of them is a medical condition characterized by blood cholesterol levels more than normal. Many factors cause hypercholesterolemia among others caused by genetic factors such as familial hypercholesterolemia and polygenichypercholesterolemia, that habit, obesity and lack exercise [1]. The main purpose of treatment of hypercholesterolemia is to maintain normal range of cholesterol levels. Treatment of hypercholesterolemia may include weight control, exercise and diet. But it does not help much to lower blood cholesterol that usually needs to be given cholesterol- lowering drug therapy. Anti-cholesterol drugs have been widely available in the market, but almost all is synthetic compounds would pose a risk to use in a long time so it needs to be developed natural anti-cholesterol drugs by utilizing several Indonesia's plant extracts.

Cinnamomum casia is a popular plant Indonesian and most widely used as a traditional medicine. Cinnamomum is believed to have very large benefits including increased appetite,anti-cholesterol, anti- inflammatory, anemia, antioxidant, cancer prevention, and anti-microbial. Cinnamomum bark is composed of three fractions, namely fraction of cinnamic acid, cinnamaldehyde and essential oil [2]. The use of natural extracts to reduce blood cholesterol levels still give unsatisfactory results when compared with the use of synthetic drugs [3]. The efficiency of absorption of the body is less because of the relatively large particle size, low solubility and possible active components of the extract has been degraded/damaged during the absorption process. This research will be a combination of Cinnamomum extract with nanochitosan. The cinnamomum will be encapsulated in nanoparticles of chitosan. The purpose of this are protecting the natural extracts from degradation and it delivering to the target side [4]. Starting from the above problems, it appeared an innovation to improve and explore types of antihypercholesterol compounds with high activity so as to reduce, and even be able to cope with demand antihypercholesterol synthetic drugs, and replace them with cheap raw materials and with a simple method. Overall these activities are divided into three groups: the extraction of bioactive compound from cinnamomum and synthesize of nanochitosan, a modified nanoparticle of chitosan-cinnamomum extract as antihypercholesterol drug and in vitro assay of antihypercholesterol. The advantages from cinnamomum extract encapsulated in nanoparticles of chitosan are natural extract protected from degradation and it can be released slowly and continuously over a certain period, so the delivery toward the target cells occurs precisely as well as the use of natural extracts to be optimal. Nano chitosan - cinnamomum extract is an innovation in the field of nano herb that has a synergistic effect. That is mutually reinforcing in order to produce high nutritious as an antihypercholesterol drug.

2. Experimental

2.1 Materials and procedure

Cinnamomum casia were collected in the surroundings of Semarang, Central Java province (Indonesia), in February 2016, shrimp shells werr collected from Traditional Market at Semarang, Central Java province (Indonesia), Cholesterol was purchased from Sigma Chemicals Company, St. Louis, Mo. USA. All other and reagents were of analytical grade and purchased from Organic Laboratories, Diponegoro University.

Research tools used are glass that commonly used in the laboratory, Blender, Oven, Hot Plate Stirrer (IKA C-MAG HS 7), Centrifuge (Corning LSE Compact), Freeze drying (LLI500), PSA (Microtrac-Particle Size Analyzer), FTIR (Shimadzu-Fourier Transform Infrared) and Spectrophotometer Ultraviolet-Visible (Spectroquant).

2.2 Active Components Extraction of Cinnamomum casia

The active components of plants extracted with ethanol for 24 hours. The next stage is filtered to collect the active component. The solvent is evaporated by using rotary evaporator.

2.3 Phytochemical screening Extract

Phytochemical screening was conducted to determine the active component from The obtained extract. Screening/qualitative tests were performed such as alkaloids, saponin, flavonoids, phenolics, tannins, steroids and terpenoids tests.

2.4 Nanochitosan Synthesis of Shrimp Shell

Demineralization, shell shrimp or crab shell has been mashed into a powder which then it was added by HCl. The mixture was heated at 80 °C for 4 hours while stirred with a stirrer of 50 rpm and filtered. The solids washed by distilled water to remove HCl solution. Floured solids dried in the oven at 70 °C for 24 hours. The shrimp shell powder or dried crab shell from demineralization stage added NaOH and it stirred at 90 ^{O}C for 1 hour. Furthermore the result obtained is filtered then washed with distilled water until neutral pH.The result is a chitin then added it with 70% ethanol. The precipitate washed in hot distilled water and acetone to remove the color.It is a chitosan. Chitosan nanoparticles

IOP Conf. Series: Materials Science and Engineering **172** (2017) 012035 doi:10.1088/1757-899X/172/1/012035

were prepared based on the ionic gelation method. Chitosan was dissolved in acetic acid. Then stirred it until homogeneous. Forward it added by sodium tripolyphosphate slowly. Gel suspension centrifuged and dried it with freeze dryer. Fathermore nanochitosan charachterized by FT-IR and SEM.

2.5 Encapsulation of cinnamomum in chitosan nanoparticles

Encapsulation cinnmomum in chitosan nanoparticles were prepared based on inclusion complexation method. First, 0,1% of cinnamomum was added chitosan nanoparticles. Gel suspension was centrifuged and dried using freeze dryer, then determined of encapsulation efficiency.

2.6 Chemical and physical properties tests

Morphology of natural extracts encapsulated nano chitosan characterized using Particle Size Analyzer (PSA) to analyzing nanoparticles of chitosan size. The efficiency of the extract encapsulation process in nano chitosan was evaluated by calculating the value of encapsulation efficiency (Encapsulation Efficiency, EE), and loading capacity (Loading Capacity, LC) extracts in nanoparticles of chitosan. Levels of natural extracts in the supernatant was analyzed by ultraviolet-visible spectrophotometry. EE value and LC is calculated using the formula below:

Where, $$EE\ (\%) = \frac{m_0 - m_s}{m_0} \times 100 \qquad (1)$$

$$LC\ (\%) = \frac{m_0 - m_s}{w_{np}} \times 100 \qquad (2)$$

m_0 = initial mass of natural extracts
M_s = mass of natural extracts in the supernatant
w_{np} = total weight of the naturally obtained extract of nanoparticles [5]

2.7 Anticholesterol assay

Anticholesterol assay conducted by cholesterol Lieberman-Burchard reaction. It determines the amount of free cholesterol in the sample which is reacted in green. Furthermore, its absorbance measured by UV-Vis spectrophotometer. If the concentration of free cholesterol in the sample increased then the solution color become lighter. [6]

3. Results and Discussion

3.1 Phytochemical study

The purpose of the phytochemical assay was determine the secondary metabolite that content in cinnamomum extract. The result of phytochemical of cinnamomum extract is presented in table 3.1.

Table 3.1 Phytochemical screening results

No	Group	Result
1	Flavonoids	(+)
2	Saponins	(+)
3	Alkaloid	(-)
4	Steroids/Triterpenids	(+)
5	Tannin	(+)
6	Phenolic	(+)

Cinnamomum extract is dark brown that was successfully obtained from cinnamomum maceration in ethanol. Based on the phytochemical screening which are flavonoids, triterpenoids and phenolic compounds.

3.2 FTIR spectra analysis of Chitosan.

Analysis of FTIR spectra provided information about the functional groups of products. The spectra showed that the compound has a functional group as expected compound.

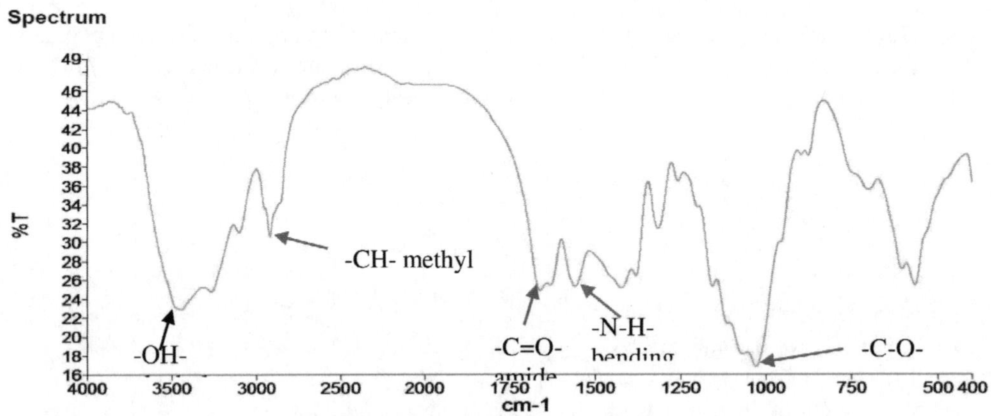

Figure 3.1. FTIR Spectra of Chitosan

FTIR spectra of chitosan residues showed a broad absorption band at 3490 cm^{-1} indicating hydrogen bond of OH stretching vibration. The absorption band at 2950 $cm^{-1}$and 1050 cm^{-1} are responsible for stretching vibration of C-H methylene group and C-O group, respectively. Chitosan absorption band are similar to that of chitin. The differences occur after deacetylation step, in which there are changes in the absorption spectrum at 1690 cm^{-1} from stretching of C=O.The spectra (Fig 3.3) showed a shift of C=O absorption from 1650 cm^{-1} to 1690 cm^{-1} and a decrease of N-H absorption band in CONH group at 1560 cm^{-1} in chitosan. It also showed a new appearance of a weak absorption at 1555 cm^{-1}.

3.3 Data analysis of PSA

Particle size analyzer phase aims to determine the size of particles produces from chitosan become nanoparticles-chitosan.

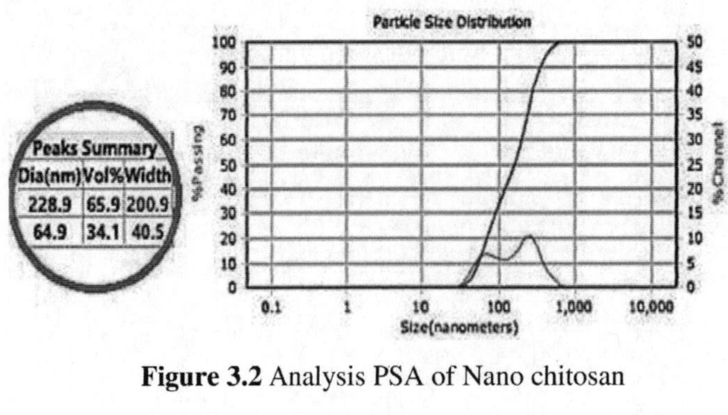

Figure 3.2 Analysis PSA of Nano chitosan

Chitosan nanoparticles of the PSA test is known particles size of nano chitosan synthesized is 64.9 nm. Based on the result of PSA test that chitosan naoparticle synthesis has been successful.

3.4 Percentage of Encapsulation efficiency from cinnamomum Extract-Chitosan Nanoparticles

Encapsulation of cinnamomum extract in chitosan was added crosslink agent Na-TPP. The interaction between chitosan and Na-TPP will optimal cause the distance between chitosan chains stretched so that the amine group is protonated form $-NH_3^+$ that will interact with cinnamomum. Percentage of encapsulation efficiency suggests how much cinnamomum coated in nanoparticles of chitosan.

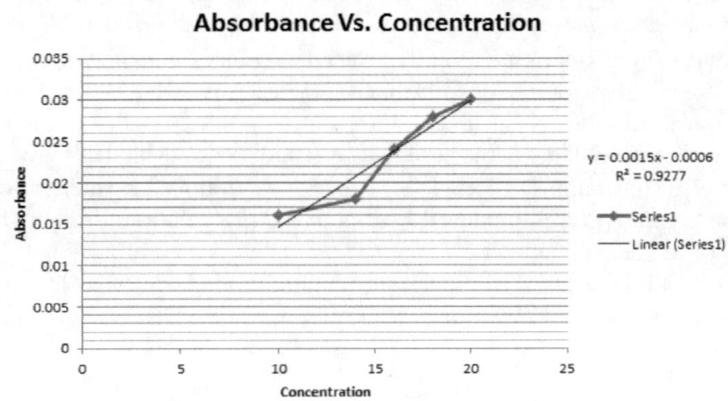

Figure 3.3 Calibration curve of Standard cinnamomum Extract

The percentage of encapsulation efficiency is 84.93%. This showed that as much as 84.93 % of cinnamomum are successfully encapsulated by chitosan nanoparticles.

3.5 Anticholesterol assay

Anticholesterol assay aims to determine encapsulation nanoparticles-cinnamomum extract as in vitro anticholesterol activity.

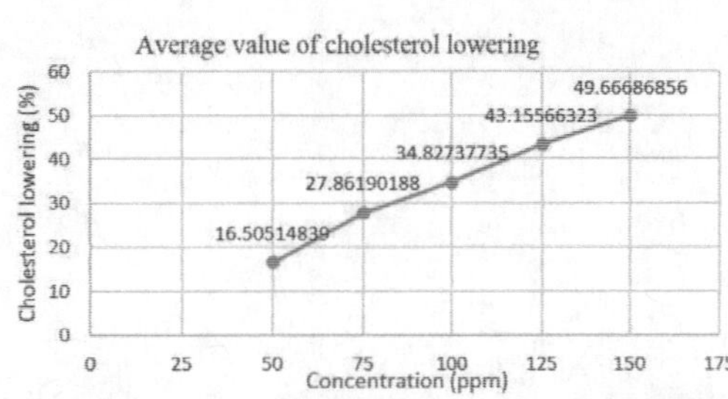

Figure 3.4. Reduction of cholesterol concentration

Based on the percent reduction obtained that ethanol extract at 150 ppm decreased by 49.666% w/v. It showed that *cinnamomum* extract encapsulated by nanochitosan have hypercholsterol activity. On the Figure. 3.4 also showed that the larger sample concentration of ethanol extract of *cinnamomum* encapsulated by nanochitosan provides anti-cholesterol activity.

4. Conclusion

Cinnamomum casia extract can be encapsulated by nanoparticles of chitosan and proved that it has a cholesterol-lowering effect through the invitro study.

Acknowledgment

The author would like to thank the Directorate of Research and Community Service (Ditlitabmas), Directorate General of Higher Education Ministry of Education and Culture, which has provided research grant.

References

[1] Brown, Carol T, 2003, Penyakit Aterosklerotik Koroner dalam: Price. Sylvia A and Wilson Lorrain M. Patofisiologi: Konsep Klinis Proses-Proses Penyakit Brahm U. Pendit, Huriawati H Hartanto, Pita Wulansari, dan Dewi Asih Mahanani, Penerjemah). (Ed Ke-6, Vol I) Jakarta: Penerbit Buku Kedokteran EGC, p580-588.

[2] Prasetya, NBA and Ngadiwiyana 2006 Identifikasi Senyawa Penyusun Minyak Kayu Manis (*Cinnamomun cassia*) Menggunakan GC-MS, Jurnal Sains dan Matematika **14** 2.

[3] Kammona, O and Kiparissides C 2012 Recent advances in nanocarrier-based mucosal delivery of biomolecules. Journal of Controlled Release. **161** 781-794.

[4] Ngadiwiyana, Prasetya, N Basid and Purbowatiningrum 2011 Potensi Sinamaldehid Hasil Isolasi Minyak Kayu Manis Sebagai Senyawa Antidiabetes. Majalah Farmasi Indonesia **22** 9-14

[5] Sarmento B, Ferreira D, Veiga F, Riberio A 2006 Characterization of Insulin-Loaded alginate nanoparticles produced by ionotropic pre-gelation through DSC and FTIR studies. Carbohydrate Polimers **66** 1-7.

[6] Rudel LL and Morris MD 1973 Determination of cholesterol using o-phthalaldehyde, Journal of Lipid Research **14** 364-366.

IOP Conf. Series: Materials Science and Engineering **172** (2017) 012036 doi:10.1088/1757-899X/172/1/012036

Synthesis CuO-ZnO nanocomposite and its application as an antibacterial agent

N Widiarti, J K Sae and S Wahyuni

Department of Chemistry, Semarang State of University
D6 Campus of Sekaran Gunungpati Tel. (024) 8508112 Semarang 50229

E-mail: wiwid_mgl_78@yahoo.co.id

Abstract. The synthesis of CuO-ZnO composites and their application as anti-bacterial have been conducted. Nanocomposite CuO-ZnO was synthesized using sol-gel method. The nanocomposite products were characterized by XRD, DR-UV, SAA, FTIR, SEM-EDX. The results of the XRD analysis showed that the CuO-ZnO composite has a nanometer size with the average of 15.99 nanometer. The DR-UV analysis showed that the CuO-ZnO composite has a band-gap of 2.28 eV in the average. The analysis of SAA showed that the CuO-ZnO has a surface area of 23.20 m^2/g in average. FTIR investigation revealed that the vibration of ZnO was observed at 432.05 cm^{-1} whereas CuO at 524.64 cm^{-1} and 594, 06 cm^{-1}. The SEM-EDX analysis showed that the ZnO has a hexagonal structure whereas the CuO has a monoclinic structure. The CuO-ZnO nanocomposite has the ability as an antibacterial against *S. aureus* as gram-positive and *E. coli* as Gram-negative bacteria.

1. Introduction

Nanomaterial is a material with the size of nanoscale (1-100 nm). This material has a unique properties and high value for commercial applications. The key factors of nanoparticles are small particle size, narrow size distribution, low agglomeration and high dispersion [1]. Nanomaterial can be applied in various fields such as cosmetics, paints, displays, batteries, medicine, catalysis, gas sensor, food engineering (production, processing, safety and packaging), agriculture, energy (storage and conversion) and construction [2].

ZnO is an excellent antibacterial agent because it has photocatalytic properties. This material can be activated by UV and visible light to form the electron-hole pairs. These holes can split the H_2O molecule (from suspension from ZnO) to OH^- and H^+. Dissolved oxygen molecules on could be converted to superoxide radical anions ($\bullet O_2^-$) which react with H^+ to produce ($HO_2\bullet$) radicals. The collisions of ($HO_2\bullet$) with electrons produce the hydrogen peroxide anion (HO_2-). These species react with hydrogen ions to produce H_2O_2 molecules that can penetrate the cell membranes and kill the bacteria [3]

The advantage of ZnO is their photocatalytic ability that can produce radical compounds. This material can be used as an antibacterial agent [3]. However, the ZnO material has large band gap as reported by Vijayakumar *et al.* [4] which can decrease the photocatalytic properties. Other researchers reported that the ZnO have the band gap of 3.3 eV [5].

The addition of CuO to ZnO could form the CuO-ZnO composite that increased particle size and decrease the band gap energy [6]. The higher concentration of CuO in composites, the smaller band gap energy could be obtained. The CuO percentage of 0.5%, 1%, 3%, 5% and 10 % on the CuO-ZnO composite resulted in the band gap energy of 3.1, 3.0, 2.95, 2.75 and 2.65 eV respectively. The ZnO band gap energy before impregnated with CuO was 3.20 eV [5]. Other researchers reported [4] that the addition of CuO to ZnO lowered the band gap energy from 3.3 eV to 3.1 eV. It increased the stability of photocatalytic reaction and increased radical species such as superoxide anion radical ($\bullet O_2^-$), ($HO_2\bullet$) and (HO_2-) which can inhibit the growth of bacteria.

Sol-gel is a simple method for nanomaterial synthesis. This method involves two stages of sol and gel formation. Many techniques could be applied in the sol-gel technology such as the changing of initial precursor, time of gel formation, type of catalyst, rate of solution formation, gel formation conditions and gel physical processing. Thus, the sol-gel process enables the formation of solid material through gelation from a solution [7].

The main difference between gram-positive and gram-negative bacteria is that the protein Gram-positive bacteria have thicker cell wall containing many layers (consisting of peptidoglycan and acid teichoic) while protein Gram-negative bacteria have thinner cell wall that contain many layers (consisting only of peptidoglycan). These cause the difference between the gram-positive and gram-negative reaction to an antibiotic. Although the cell walls are thicker, Gram-negative bacteria are more resistant to antibiotics than Gram-positive bacteria because the bacteria can penetrate the lipid layer on the outer membrane of the bacteria [8].

Based on the above description, the authors synthesized the CuO-ZnO nanocomposite using sol-gel method to be applied as antibacterial for the *Staphylococcus aureus* as gram-positive and *Escherichia coli* as Gram-negative bacteria.

2. Experimental

2.1 Materials

Zinc(II) acetate dihydrate ($Zn(CH_3COO)_2.2H_2O$) 98% (Merck), copper(II) sulphate pentahydrate ($CuSO_4.5H_2O$) 98% (Merck), ethanol p.a (Merck), NaOH p.a, ethylene glycol p.a (Merck) ρ= 1.11 g/mL (Merck), 99% citric acid (Merck), NaCl, Nutrient agar, *Staphylococcus aureus* and *Escherichia coli*, and distilled water.

2.2 Synthesis of ZnO nanoparticles

Synthesis of ZnO nanoparticles used the sol-gel method [9]. Typically, 4.6751 g of $Zn(CH_3COO)_2.2H_2O$ was dissolved in 20 mL ethanol and stirred with a magnetic stirrer for an hour. The mixture was kept for 48 hours to form a gel and washed with distilled water several times. Then, the gel was dried in an oven at 120 °C to constant weight and calcined at 500 °C for 4 hours to obtain the ZnO nanoparticles. These materials were characterized by Diffuse Reflectance-UV (DR-UV) and X-ray Diffraction (X-Ray Diffraction), SAA (BET), FTIR, and SEM.

2.3 Synthesis of CuO nanoparticles

Synthesis of CuO nanoparticles used the sol-gel method [10]. The 20 mL $CuSO_4.5H_2O$ 0.1M was added to 10 mL of 0.1 M citric acid and stirred with a magnetic stirrer until homogenous. The solution was added by 10 mL of 1 M NaOH drop by drop and stirred with a magnetic stirrer for 1 hour to dark blue. Then, it was stored (aging) for 48 hours and washed with distilled water several times. The gel was dried in an oven at 120 °C to constant weight and calcined at 500 °C for 4 hours to obtain the ZnO nanoparticles. These materials were characterized by Diffuse Reflectance-UV (DR-UV) and X-ray Diffraction (X-Ray Diffraction), SAA (BET), FTIR, and SEM.

2.4 Synthesis of CuO-ZnO nanocomposite

Synthesis of CuO-ZnO nanocomposite used the sol-gel method [11]. Firstly, 25 mL of ethylene glycol was mixed with 5.3247 g of $Zn(CH_3COO)_2.2H_2O$, 25 mL of distilled water and 6.3 g of citric acid with vigorous stirring for an hour. After an hour, 0.4244; 0.6741; 0.9488; 1.2733 g of $CuSO_4.5H_2O$ were added to the suspension at 60 °C for 3 hours and mixed quickly. The solution was kept in the dark for 48 hours and washed with distilled water several times. The obtained gel was dried at 120 °C to constant weight and calcined at 500 °C for 4 hours to produce the CuO-ZnO nanocomposites. These materials were characterized by Diffuse Reflectance-UV (DR-UV), X-ray Diffraction (X-Ray Diffraction), SAA (BET), FTIR, and SEM-EDX.

2.5 Antibacterial activity test

2.5.1 Agar medium oblique and bacterial inoculation. All the tools were sterilized by washing them, then wrapped in paper and heated in an oven at a temperature of 170 °C for 2 hours. Agar medium was made by weighing 0.23 g Nutrient agar. All the ingredients were put into the Erlenmeyer and dissolved in 10 mL of distilled water to a boil. Poured each five mL into two test tubes. Then, the solutions were sterilized in an autoclave at 121 °C for 15 minutes and waited for into a warm solution for ±30 minutes to condense at a slope of 30°. On an agar medium oblique, a bacterial suspension was stroked using a needle ose. The medium oblique was incubated at 37 °C for 24 hours for breeding bacteria.

2.5.2 Media Test. Media test was made by Nutrient Agar (NA) 2.3 gram dissolved in 100 mL of distilled water (23g/1000 mL) using Erlenmeyer. The medium was homogenized with the stirrer to boil. The medium that has been homogenized, was sterilized in an autoclave at 121 °C for 15 minutes, and chilled to a temperature of ±45-50 °C.

2.5.3 Preparation of bacterial suspensions. The inoculated test bacteria were taken with a wire loop and suspended in a sterile tube containing 2 mL of 0.9% NaCl solution. Then, the concentration of bacteria was measured using Denis check to obtain Mc Farland standard, 0.5. The same treatment was carried out on any kind of test bacteria.

2.5.4 Testing Antibacterial. Poured 25 mL essential media into a sterile Petri dish. It was left until freeze. It was added the bacterial suspension with a soft cotton until evenly distributed to the media. Then, the paper discs were immersed in the sample solution 10 mg/mL, lifted and aerated briefly. The paper disc was placed on the test medium with a predetermined distance. The petri dish was incubated in an incubator at 37 °C for 24 hours. After that, the clear zone was measured as obstruction growth of bacteria.

3.Results and Discussion

3.1 The synthesis of ZnO, CuO, and CuO-ZnO

The synthesis of ZnO nanoparticles by modifying the sol-gel method was performed by Mallick *et al.* [9], by mixing Zinc (II) acetate dihydrate and ethanol. They were stirred using a magnetic stirrer for an hour. Ethanol is used as a solvent. The mixture was obtained and stored (aging) for 48 hours to obtain a gel formation. The gel was dried in an oven at a temperature of 120 °C to constant weight to evaporate the solvent ethanol, and calcined at temperatures of 500 °C for 4 hours.

The synthesis of CuO nanoparticles by modifying the sol-gel method was performed by Radhakrishnan*et al.* [10]. Solution Copper(II) sulfate pentahydrate and citric acid (as a capping agent) were stirred using a magnetic stirrer for 30 minutes, and NaOH was added dropwise and stirred for an hour with the aim of CuO nanoparticles suspension formation [12]. NaOH was used to maintain the solution in a bases condition during the synthesis process. The mixture was stored (aging) for 48 hours to obtain a gel formation. The gel was washed with distilled water to remove impurities from the gel. The gel was dried in the oven at a temperature of 120 °C to constant weight to remove water and calcined at a temperature of 500 °C for 4 hours.

The synthesis of CuO-ZnO nanocomposite by modifying the sol-gel method was conducted by other researchers [11] by mixing up first zinc (II) acetate dihydrate, ethylene glycol, distilled water, and citric acid with a magnetic stirrer for 1 hour. After an hour, copper(II) sulfate pentahydrate solution was added and stirred rapidly at 60 °C for 3 hours to accelerate the reaction. The mixture was stored in dark room (aging) for 48 hours. The gel was washed with distilled water and dried at a temperature at 120 °C to constant weight. Calcination at temperature of 500 °C for 4 hours was to eliminate the organic compound, to decompose $CuSO_4.5H_2O$ into CuO and $Zn(CH_3COO)_2.2H_2O$ into

ZnO, as a result the CuO-ZnO structure was formed. In this synthesis, the polymerization reaction occurred. The nanoparticles can be grown in the polymer matrix and the formation of particle agglomeration can be prevented by maintaining distribution good spatial in a polymer matrix [13].

The function of citric acid is to form the complex compounds with zinc citrate before reacting with ethylene glycol to form a polymerization. The use of citric acid is intended for formation the metal complex compound citrate [14].

3.2 Characterization of ZnO, CuO, and CuO-ZnO

3.2.1. X-Ray Diffraction (X-RD). XRD was used to analyze the size of ZnO, CuO, and CuO-ZnO nanocomposite. The instrument used was XRD (Shimadzu) with Cu radiation at wavelength of 1.54060 Å, 40.0 kV, and a strong current of 30.0 mA. The XRD analysis results of the compound ZnO, CuO, CuO-ZnO nanocomposite can be seen in figure 1.

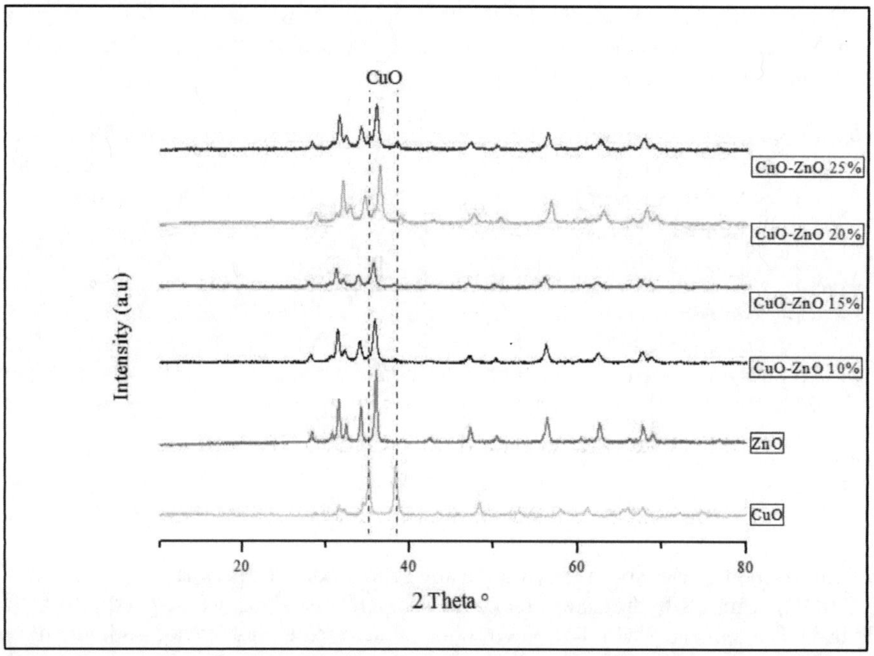

Figure 1.X-ray diffractograms of ZnO, CuO, 10%CuO-ZnO, 15%CuO-ZnO, 20%CuO-ZnO, and 25%CuO-ZnO.

The result of XRD analysis showed that ZnO diffraction pattern appears in the peak 2θ of 31.47, 34.12 and 35.96. The CuO appears at the peak 2θ of 35.10, 38.34 and 48.36. The XCuO-ZnO nanocomposite (X=10, 15, 20, and 25%) has high peak 2θ at 31, 34, 35, 47, 56 and 67, according to JCPDS 36-1451 that has the ZnO peak at 2θ= 31.7; 34.4; 36.2; 47.5; 56.5; 62.8; 67.9; 69.05 and JCPDS 05-0661 that has the CuO peak at 2θ= 30 to 50 [5].

Based on the diffraction pattern of the nanocomposite XCuO-ZnO, the addition of CuO resulted in a decreased crystallinity of ZnO. The results can be seen in table 1.

Table 1. Intensity of CuO and ZnO in CuO-ZnO.

Samples	ZnO intensity around 2θ= 67	CuO intensity around 2θ = 38
10%CuO-ZnO	275	42
15%CuO-ZnO	171	45
20%CuO-ZnO	49	136
25%CuO-ZnO	47	155

The particle size could be calculated using the Debye-Scherer equation. The results of these calculations can be seen in table 2.

Table 2. Particle size of ZnO, CuO, and X CuO-ZnO XRD analysis.

Samples	Particle size (nm)
ZnO	26,83
CuO	19,83
10%CuO-ZnO	16,23
15%CuO-ZnO	15,56
20%CuO-ZnO	16,43
25%CuO-ZnO	15,74

Based on table 2, it was observed that the CuO had been successfully composited on ZnO. After synthesis, the particle size decreased. The particle size of 16.23 nm, 15.56 nm, 16.43 nm and 15.74 nm was observed in the sample of 10% CuO-ZnO, 15% CuO-ZnO, 20% CuO-ZnO, and 25% CuO-ZnO, respectively. This result is almost similar with other report [11] which has the particle size of 19 nm for 10% CuO-ZnO nanocomposite.

The result of the smaller size might be due to the longer aging process of 48 hours. Moreover, the smaller particle size may be formed as a result of CuO covering the surface of ZnO. Thus, the particle size of CuO-ZnO composite followed the CuO that had smaller size compared to the ZnO.

3.2.2. Band Gap (Spektrofotometer UV-Vis Diffusion Reflectance). The characterization of the band gap energy determination was performed by using UV-Vis diffusion reflectance spectrophotometer. The calculation was performed using the Kubelka-Munk by plotting the $(F(R).hv)^{1/2}$ against h.v(eV). The band gap energy of ZnO, CuO and CuO-ZnO nanocomposite can be seen in table 3 and figure 2.

Table 3. The band gap of ZnO, CuO and XCuO-ZnO DR-UV analysis.

Samples	Band gap(eV)
ZnO	3,39
CuO	2,04
10%CuO-ZnO	2,48
15%CuO-ZnO	2,30
20%CuO-ZnO	2,23
25%CuO-ZnO	2,12

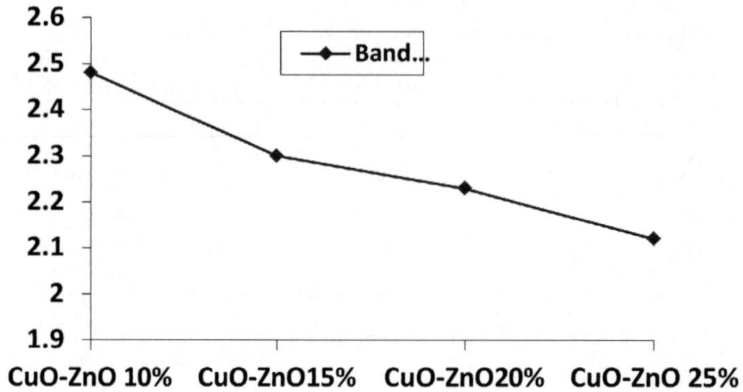

Figure 2. Graph the band gap of ZnO as a result of the addition of CuO.

The addition of CuO decreased the band gap energy of composite. According to another report [5], the smaller band gap energy was caused by the combined transition that enhanced from O_2 (2p) to Zn^{2+} ($3d^{10}$- 4s) and to Cu^{2+} ($3d^9$). Other researchers [16] reported that the swift heavy ion irradiation could decrease in the band gap with an increase in the ion fluence. Due to smaller band gap energy, the CuO-ZnO has a better photocatalytic properties compared to the ZnO or CuO a [15]. A decreased band gap energy can facilitate stepping electrons from the valence band to the conduction band and increase the photocatalytic activity through enhancing the radical species such as superoxide anion radical (•O_2-), (HO_2•) and (HO_2^-) [16].

3.2.3. Surface area (BET). The surface area of ZnO, CuO and CuO-ZnO nanocomposite can be determined using the BET surface area. The method works based on a single gas adsorption at a constant temperature. Nitrogen gas was used as an adsorbate on the surface of ZnO, CuO, and CuO-ZnO.

Table 4. The surface area of ZnO, CuO, and XCuO-ZnO BET method.

Samples	Surface area (m^2/g)
ZnO	9,63
CuO	21,61
10%CuO-ZnO	22,48
15%CuO-ZnO	22,07
20%CuO-ZnO	19,52
25%CuO-ZnO	28,74

The addition of CuO on ZnO produced a larger surface area of CuO-ZnO composite. The larger surface area of composite possessed better photocatalytic activity compared to the smaller one. According to another report [17], the large surface area of CuO-ZnO more effectively reduced the concentration of Rodhamin B compared to the smaller surface area of ZnO. It was caused by the higher adsorption of CuO-ZnO composite.

The surface area also affected the antibacterial properties. The large surface area of composites can increase the antibacterial properties due to easier to absorb the UV light [17] and generating the radical species. The CuO-ZnO composite could effectively kill the bacterial cells due to increased production of Zn^{2+} ions and reactive oxygen species (ROS)[18].

3.2.4. Analysis FTIR spectrophotometer. FTIR spectrophotometer was used to analyze the functional group of ZnO, CuO and CuO-ZnO nanocomposite. The wavelength of 4000-400 $cm^{-1}$was used for the analysis of functional groups of samples, and the results can be seen in figure 3.

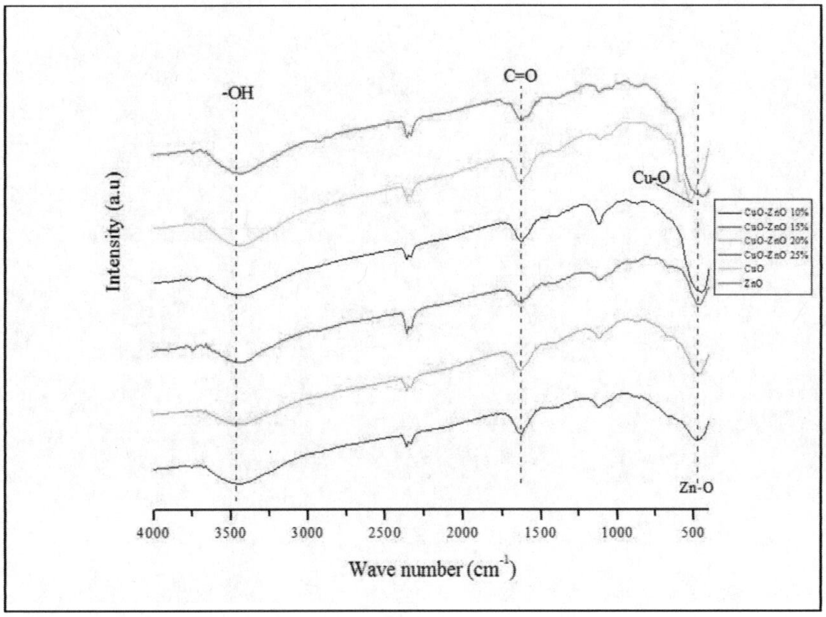

Figure 3. FTIR spectra of ZnO, CuO, and XCuO-ZnO nanocomposite.

Functional groups of ZnO, CuO and CuO-ZnO nanocomposite with FTIR spectrophotometer can be seen in figure 3. The -OH functional groups could be observed at the frequency of 3500-3000 cm^{-1}, the peak at 3449 cm^{-1} could be observed at the sample of ZnO, CuO, and CuO-ZnO (10 %, 15%, 20%, and 25%). The C=O functional group could be observed at the wavenumber of 1627.92 cm^{-1} for ZnO and 10% CuO-ZnO, and 1635.64 cm^{-1} for CuO, 15% CuO-ZnO, 20% CuO-ZnO and 25% CuO-ZnO. The wavenumber of 2854.65 cm^{-1}, 2862.36 cm^{-1}, 2924.09 cm^{-1} and 2931.80 cm^{-1} are the C-H functional group according to the reference [19] that reported the C-H group appeared at wavenumber of 2855.05 $cm^{-1}$and 2922.59 cm^{-1}. The C=O and C-H groups appeared because the synthesis used the precursor of Zn $(CH_3COO)_2.2H_2O$ that was not perfectly decomposed to ZnO.

The wavenumber of 432.05 cm^{-1} is assigned to Zn-O bond similar to another report [19] that the vibration Zn-O appeared at wavenumber of 424.26 cm^{-1}, 471.36 cm^{-1} and 542.86 cm^{-1}. The wavenumber of 524.64 and 594.06 cm^{-1} are assigned to Cu-O bond according to reference [20] that the vibrations of the Cu-O are in the range frequency of 700-400 cm^{-1}. By adding the CuO to ZnO, the shift wavenumber at the frequency of 750-400 cm^{-1} could be observed. The wavenumber of 447.49 cm^{-1}, 462.92 cm^{-1}, 462.92 cm^{-1} and 478.35 cm^{-1} was found in the 10% CuO-ZnO, 15%CuO-ZnO, 20% CuO-ZnO and 25%CuO-ZnO respectively. The shift of wavenumber indicated that the addition of CuO may change the structure of ZnO.

3.2.5. SEM-EDX. SEM is a scanning electron microscope that illustrates the sample surface by scanning with a beam of high-energy electrons. X-rays in the SEM can be used to identify the elemental composition of a sample by a technique known as energy dispersive x-ray (EDX) [21]

Figure 4. SEM images of ZnO 20000x magnification (a), CuO magnification 20000x (b) and 25% CuO-ZnO with the magnification of 15000x (c) and magnification 20000x (d).

The surface morphology of ZnO, CuO and 25% CuO-ZnO nanocomposite synthesized by the sol-gel method can be seen in figure 4. The ZnO, CuO and 25% CuO-ZnO have a uniform particle shape and size with low level of agglomeration because they did not form a large cluster. The agglomeration can usually be demonstrated by the formation of large cluster [22].

EDX analysis of 25% CuO-ZnO nanocomposite is expected to having the elements composition similar with the desired composition in the synthesis process.

Table 5. The composition of 25% CuO-ZnO nanocomposite using EDX analysis

Sample	Mass (%)				
	CuO	ZnO	SO_3	C	O
25% CuO-ZnO	15,53	68,50	1,22	14,75	-

Table 5 showed that the composition of CuO measured by EDX in the composite was 15.53 %. It is lower than that in the composition of synthesis (25%). It may be due to an inhomogeneous process of composite synthesis such as stirring treatment.

3.3. Antibacterials

The antibacterial test was conducted by the method of the disc. It can be seen that the CuO-ZnO nanocomposite inhibited the growth of gram-positive bacteria (*S. aureus*) and gram-negative bacteria (*E. coli.*). The experiment used the S.O.S floor cleaner as the positive control because it was proven to inhibiting the growth of bacteria *S. aureus* and *E. coli*. The distilled water was used as negative control because it can be used as a solvent of CuO-ZnO nanocomposite.

Table 6. Diameter of clear zone antibacterial.

Samples	Clear zone diameter (mm)	
	Staphylococcus aureus	*Escherichia coli*
ZnO	1,9	Resistant
CuO	1,7	1,4
10%CuO-ZnO	1,3	1
CuO-ZnO 15%	1,5	2
CuO-ZnO 20%	1,6	2,1
CuO-ZnO 25%	2,1	2,3
Control positive (S.O.S)	7,5	14,5
Control negative (distilled water)	-	-

In this experiment, the sample was irradiated using UV light for only 15 minutes to avoid the bacterial death. When the ZnO, CuO and CuO-ZnO exposed by UV light, the superoxide anion radicals (O_2^{-}•) might be formed then reacted with H^+ to produce HO_2•. The electrons will produce hydrogen peroxide anion (HO_2-). Then, the HO_2- anion reacted with hydrogen ions to form the H_2O_2 molecules that can serve to kill bacteria.

Other researcher [15] reported that the ZnO-CuO nanocomposite had higher photocatalytic properties than the respective oxides. It can increase the radical compounds and the H_2O_2 molecule to effectively kill the bacteria. The higher CuO was added, the higher antibacterial activity could be produced. The antibacterial activity of CuO-ZnO nanocomposite is shown in figure 3.5.

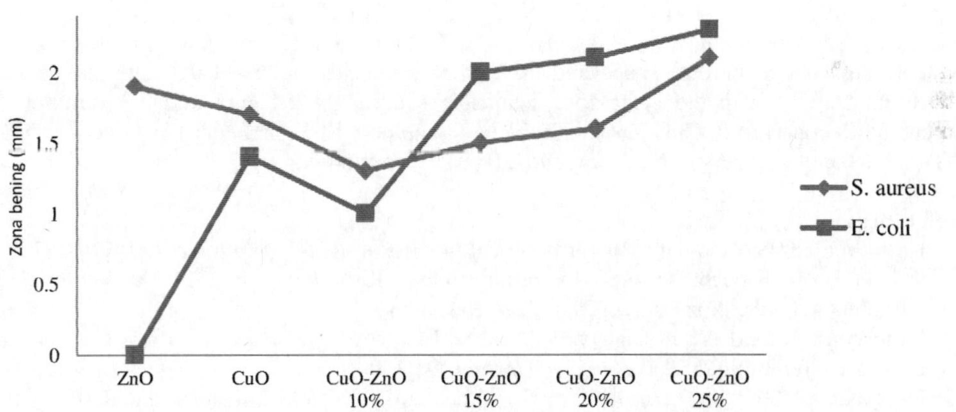

Figure 5. The diameter clear zone antibacterial of ZnO, CuO and CuO-ZnO composites against *S. aureus* and *E. coli.*

Figure 5 showed that the ZnO is resistant as antibacterial against *E. coli* and better at inhibiting the growth of bacteria *S. aureus*. The inhibitory of CuO was not so much different between the bacteria *E. coli* and *S. aureus*. Therefore, the CuO-ZnO nanocomposite can be used as inhibitors of the bacterial growth of *E. coli* and *S. aureus*. The antibacterial activity increased by the increase of CuO on the composites, indicating that the CuO increases the photocatalytic properties of composite.

The size of CuO-ZnO nanocomposite of 10%, 15% and 25% CuO were not significantly different. Therefore, the composition has the important role in antibacterial activity. In this experiment, 25% CuO on the CuO-ZnO composite showed the highest antibacterial activity to *S. aureus* and *E. coli*, it is better than CuO and ZnO. The ZnO was better antibacterial properties against S. aureus whereas the CuO was better antibacterial properties against *E. coli.*

The surface area of CuO-ZnO nanocomposite may influence the antibacterial activity. In this research the highest surface area of composite showed the highest antibacterial activity. Other researchers [18] reported that the larger surface area of CuO-ZnO will easily enter to the cell and kill the bacteria due to increase of Zn^{2+} ions and reactive oxygen species (ROS).

The smaller band gap energy could also enhance the antibacterial activity because the electrons can be easily exited from the valence band to the conduction band. Therefore, the photocatalytic activity could be more efficient due to high species of superoxide radical formation as inhibitors of bacterial growth.

The size of CuO-ZnO nanocomposite with the CuO content of 10%, 15%, 10% and 25% were not significantly different. However, the antibacterial activity of 25% CuO-ZnO nanocomposite to S. aureus was better than ZnO, CuO and other CuO contents of CuO-ZnO nanocomposites. The CuO-ZnO composite of 15-25% CuO was better antibacterial activity than ZnO and CuO to *E. Coli*. The ZnO is better antibacterial properties against *S. aureus*, whereas the CuO is better antibacterial properties against E. coli.

The larger surface area of CuO-ZnO nanocomposite, the higher antibacterial activity was observed. The 25% CuO-ZnO nanocomposite showed the highest antibacterial activity against *S. Aureus* due to high surface area. Other researchers [18] reported that the CuO-ZnO with large surface area could be easier to get bacterial cells and kill them due to increase of Zn^{2+} ions and reactive oxygen species formation.

4. Conclusion

The composites of CuO-ZnO was successfully synthesized using sol-gel method. The particle size of 16.23 nm, 15.56 nm, 16.43 nm and 15.74 nm were observed in composite for the CuO content of 10%, 15%, 20% and 25% respectively. Addition of CuO reduced the value of the band gap energy of ZnO i.e.: 2.48 eV (10%), 2.30 eV (15%), 2.23 eV (20%) and 2.12 eV (25%). The synthesis of composite increased the surface area up to 28.74 m^2/g for 25% CuO-ZnO. The morphology of ZnO, CuO and 25% CuO-ZnO has a uniform shape and size. The composite of 25% CuO-ZnO showed the best antibacterial activity with the clear zone against *S. aureus* of 2.1 mm and 2.3 mm for *E. coli*. Photocatalytic properties of CuO-ZnO with the band gap of 2.12 eV is better than ZnO or CuO which is having the band gap energy of 3.39 eV and 2.04 eV, respectively.

5. References

[1] Lines M G 2008 Nanomaterials for practical functional uses *J. Alloys Compd.* **449** 242
[2] Arivalagan K, Ravichandran S, Rangasamy K and Karthikeyan E 2011 Nanomaterials and its potential applications *Int. J. ChemTech Res.* **3** 534
[3] Padmavath N and Vijayaraghavan R 2008 Enhanced Bioactive of ZnO Nanoparticles an Antimicrobial Study *Sci. Tech. Adv. Mater.* **9** 035004
[4] Vijayakumar G N S, Deva Shankar S, Rathnakumari M and Sureshkumar 2010 Synthesis of Electrospun ZnO/CuO Nanocomposite Fibers and Their Dielectric and Non linear Optic Studies *J. Alloy. Compd.* **507** 225
[5] Qamar M T, Aslam M, Ismail I M I, Salah N and Hameed A 2015 Synthesis, Characterization and Sunlight Mediated Photocatalytic Activity of CuO Coated ZnO for the Removal of Nitrophenols. *ACS Appl. Mater. Interfaces***7** 8757
[6] Barokah, S. &Widiarti, N. 2015. Photocatalytic activity CuO-ZnO on oxidation reaction of Phenol*Indo. J. Chem Sci.* **4** 88
[7] Owens G J, Singh R K, Foroutan F, Alqaysi M, Han C M, Mahapatra C, Kim H W, Knowles J K 2016 Sol-gel based materials for biomedical applications *Prog. Mater. Sci.***77** 1.
[8] Dehzangi A, Sohrabi S, Heffernan R, Sharma A, Lyons J, Paliwal K and Sattar A 2014 Gram-positive and gram-negative subcellular localization using rotation forest and physicochemical-based features *BMC Bioinformatics***16(Suppl 4)** S1.

[9] Mallick P and Sahu S 2012 Structure, Microstructure and Optical Absorption Analysis of CuO Nanoparticles Synthesis by Sol-Gel Route *Nanosci. Nanotechnol.* **2** 71

[10] Radhakrishnan A and Beena B 2014 Structural and Optical Absorption Analysis of CuO Nanoparticles.*Indian J Adv. Chem. Sci.* **2** 158

[11] Giahi M, Badalpoor N, Habii and Taghavi H 2013 Synthesis of CuO/ZnO Nanoparticles and Their Application for Photocatalytic Degradation of Lidocaine HCl by the Trial-and-error and Taguchi Methods *Bull. Korean Chem. Soc.* **34** 2176

[12] Shahmiri M, Ibrahim N A, Zainuddin N, Asim N, Bakhtyar B, Zaharim A and Sopian K 2013 Effect of pH on the Synthesis of CuO Nanosheets by Quick Precipitation Method *WSEAS Trans. Environ. Develop.***9** 243

[13] Guo Q, Ghadiri R, Weigel T, Aumann A, Gurevich E L, Esen C, Medenbach O, Cheng W, Chichkov B and Ostendorf A 2014 Comparison of *in situ* and *ex-situ* Methods for Synthesis of Two-Photon Polymerization Polymer Nanocomposite *Polymers***6** 2037

[14] Danks A E, Hall S R and Schnepp Z 2016 The evolution of Sol-Gel Chemistry as a Technique for Materials Synthesis *Mater. Horiz.***3** 91

[15] Benxia and Yanfen 2010 Facile synthesis and photocatalytic activity of ZnO-CuO nanocomposite *Superlattices Microstruc.* **47** 615

[16] Kuriakose S, Avasthi D K, Mohapatra S 2015 Effects of Swift Heavy Ion Irradiation on Structural, Optical and Photocatalytic Properties of ZnO-CuO Nanocomposites Prepared by Carbothermal Evaporation Method *Beilstein J. Nanotechnol.* **6** 928

[17] Liu Z, Bai H and Sun D D 2011 Hierarchy CuO/ZnO Membranes for Environmental Applications under the Irradiation of Visible Light *Int. J. Photoenergy***2012** 804840

[18] Sirelkhatim A, Mahmud S, Seeni A, Kaus N H M, Ann L C, Bakhori S K M, Hasan H and Mohamad D 2015 Review on Zinc Oxide Nanoparticles: Antibacterial Activity and Toxicity Mechanism *Nano-Micro Lett.* **7** 219

[19] Chitra K and Annandurai G 2013 Antimicrobial activity of wet chemically engineered spherical shaped nanoparticles on the foodborne pathogen *Int. Food Res. J.***20** 59

[20] Karthik K, Jaya N V, Kanagaraj M, Arumugam S 2011 Temperature-dependent magnetic anomalies of CuO nanoparticles *Solid State Commun.* **151** 564

[21] Joshi M, Bhattacharyya A and Ali S W 2008 Characterization Techniques for Nanotechnology Application in Textiles *Indian J. Fibre Text.* **33** 304

[22] Phiwdang K, Phensaijai M and Pecharapa W 2013 Study of Antifungal Activities of CuO/ZnO Nanocomposites Synthesized by Co-precipitation Method *Adv. Mat. Res.* **802** 89

Acknowledgments
Through this article the author would like to thank the Department of Chemistry who has been given the opportunity to study and Jalud Kumba Sae as a working partner in this study

Synthesis and Enhanced Photocatalytic Activity of Ce-Doped Zinc Oxide Nanorods by Hydrothermal Method

N Aisah[1], D Gustiono[1], V Fauzia[2], I Sugihartono[3] and R Nuryadi[1]

[1]Agency for the Assessment and Application of Technology, Serpong, Indonesia
[2]Department of Physic, Universitas Indonesia, Depok, Indonesia
[3]Department of Physic, State University of Jakarta, Jakarta, Indonesia

Email: nuning.aisah@bppt.go.id

Abstract. Zinc oxide (ZnO) is a n-type semiconductor material which has a wide direct band gap energy of ~ 3.3 eV, and other interesting optical properties, hence it's potentially applied to various fields such as electronics, optoelectronics, sensors, photonic devices, and also photocatalyst. Dopant in ZnO nanostructures is an effective way to improve ZnO's structural properties in various applications. In this study, undoped and Ce doped ZnO nanorods were synthesized on ITO coated glass substrates by ultrasonic spray pyrolysis for seeding deposition and hydrothermal methods at a temperature of 95 ^{0}C for 2 hours for growth. X-ray diffraction, field emission scanning electron microscopy (FESEM), UV-VIS and Photoluminescence spectroscopy were used to characterize the crystal structure, surface morphology and optical properties of ZnO nanorods and the photocatalytic activity test for methylene blue degradation. The experimental results showed that 3% Cerium dopant has produced hexagonal morphology ZnO nanorod growing more uniform on (002) crystal planes, increased the intensity of ultraviolet absorbance thereby increase the degradation speed of methylene blue.

1. Introduction

Zinc oxide is n-type semiconductor which has wide and direct band gap energy of ~ 3.3 eV at room temperature, large exciton binding energy of 60 meV, excellent thermal stability, high electron mobility and specific electrical and optical properties [1], [2]. ZnO can be used for some application such as electronics, optoelectronics, sensors, photonic devices and photocatalyst.

ZnO nanorod is one of the nanostructure materials most widely synthesized due to its large surface area. ZnO nanostructures were fabricated by various methods such as thermal evaporation, vapor phase deposition, pulsed laser deposition that requires the complex processes, sophisticated equipment and high vacuum [3]. Then, the chemical based approaches such as the ultrasonic spray pyrolysis (USP) [4], [5], sol-gel [6] and hydrothermal methods [2] show the advantages i.e. it can be operated at low temperature, low pressure with simple operation.

Doping of ZnO nanostructures is one effective way to improve the structural properties of ZnO for various applications. In particular, doping ZnO with rare earth materials is of interest in tailoring its optical properties [7]. One of the rare earth metal elements that commonly known as the lanthanide group of the periodic system is cerium. The previous researches such as Lang *et al.* [8] have observed that the Ce dopant has a strong effect on the structure and PL characteristics of ZnO nanorods. Murugadoss *et al.* also have investigated the various doping concentrations of Ce that showed strong absorption and emission in a visible region of UV-vis and PL spectra [9]. While Nidhi Sinha *et al.* reveal that Ce doping resulted in a shift of XRD peaks to higher angles, decreased size of nanorods, lowered the band gap of ZnO and increased in the intensity of green emission peak in PL [10]. As can be seen from extensive searching that there is only limited literature available about the utilization of Ce doped ZnO nanorods for the treatment of organic contaminants via photocatalysis. One promising approach is to utilize a photocatalytic process from semiconductors nanomaterial such as TiO_2 and ZnO. Selection of ZnO as a photocatalyst due to the wider energy band gap so it can absorb more light [11].

In this work, we synthesized and characterized the undoped and Ce doped ZnO nanorods on Indium Tin Oxide coated glass substrates with USP and hydrothermal method and then investigated the optical properties and the photocatalytic activity. The ultrasonic spray pyrolysis (USP) has many advantages such as low cost, deposition equipment that is simple and easy fabrication of large area films. USP is a method of fabricating a thin layer based on the thermal decomposition of the precursor solution is sprayed through an ultrasonic nozzle above the surface of the substrate which has been heated [4], [12], [13]. Nanorod growth by hydrothermal method is expected to produce nanorod with minimal defects, high crystallinity, good homogeneity as well as having a special morphology and metastable phase thermodynamics [14].

2. Experimental details

2.1. Synthesis of ZnO Nanorods

Firstly, ZnO seed layers were deposited on ITO-coated glass substrates with ultrasonic spray pyrolysis technique. The seed solution was 0.2 M zinc acetate dihydrate $Zn(CH_3COO)_2.2H_2O$ (Merck) that was diluted in distilled water and placed into a nebulizer. The fine droplets were atomized with ultrasonic wave 1.65 MHz and sprayed onto the substrate at a distance of 10 cm. During the spraying process for 10 minutes, the substrates were heated at a temperature of 450 ^{0}C on the hotplate and then cooled down to room temperature. The second step is the growth of nanorods by the hydrothermal method. The growth solution consists of 0.1M Zinc nitrate tetrahydrate $Zn(NO_3)_2.4H_2O$ (Merck) and 0.1M hexamethylenetetramine (HMT) $C_6H_{12}N_4$ (Sigma Aldrich) with the molar ratio 1:1 and then mixed using a magnetic stirrer for 10 minutes at room temperature. Then, 3mol% of cerium nitrate hexahydrate $Ce(NO_3)_3.6H_2O$ (Sigma Aldrich) was added into the growth solution and kept on stirring until the temperature reached 60 ^{0}C. In order to grow the Ce-doped ZnO nanorods, the substrates were then immersed in this solution and heated at a temperature of 95 ^{0}C for 2 hours in oven. After that, the samples were rinsed with distilled water several times and dried at 80 ^{0}C for 30 minutes.

2.2. Characterization

The surface morphology of ZnO nanorods was determined by field emission scanning electron microscopy (FESEM) INSPECT F50. The crystalline structures of the samples were characterized by Shimadzu X-ray Diffraction 7000 Maxima X with CuKα (1.54060 Å) X-ray source using a current of 30 mA and a voltage of 40 kV. The UV-Vis absorption spectrum analysis was performed by Thermo Fisher Scientific GENESYS 10S UV-vis Spectrometer and the Hitachi UV-vis Spectrometer U-3900H for the diffuse reflectance mode. The room-temperature photoluminescence (PL) measurements were carried out using a FLS920 (Edinburg Instruments).

2.3. Photocatalytic Experiments

The photocatalytic activity was investigated by means of the degradation of methylene blue (MB) in an aqueous solution. The ZnO nanorods were placed in a beaker that contained 3mL of 1×10^{-5}MMB and they were then irradiated with backlight lamps. After irradiating for 0, 10, 20 and 30 min, the concentration of residual MB was determined by a UV–vis spectrophotometer.

3. Results and discussion

3.1. Surface Morphology of ZnO Nanorods

Figure 1 (a) and (b) shows the FESEM images of undoped ZnO and Ce-doped ZnO nanorods. It can be seen that the undoped ZnO nanorod has a sharp tip-shaped like a needle. Furthermore, the rod diameter is small with an average of 125 nm. The ZnO nanorod with Cerium dopant is generally hexagonal in shape with an average diameter is large enough about 340 nm. This phenomenon shows that the cerium dopant influences the direction and shape of the nanorods growth. This result was slightly bigger than the result obtained from the same methods in previous researches which the diameter rod produced were in the range of 100-200 nm [1], [8], [15], [16].

Figure 1. FESEM images of (a) undoped and (b) Ce-doped ZnO nanorods.

3.2. XRD analysis

The X-ray diffraction patterns of undoped Zno and Ce-doped ZnO nanorods are shown in figure 2. The diffraction peaks, i.e. (100), (002), (101), (102) and (103) are similar to bulk ZnO and correspond to hexagonal wurtzite ZnO (ICSD No. 98-005-7478). In addition to ZnO phase, there are other peaks at angles of 30.27, 35.17 and 50.98 which indicate the indium oxide phase derived from the ITO substrate. Preferred orientation of ZnO crystal along (002) and (013) indicates the direction of the crystal growth perpendicular to the substrate surface which is in agreement with the results of SEM morphology in figure 1. For Ce-doped ZnO, there is no diffraction peaks of the phases containing Ce were found in the samples which indicated that Ce^{3+} ions substituted the Zn^{2+} sites or interstitial sites in ZnO lattice.

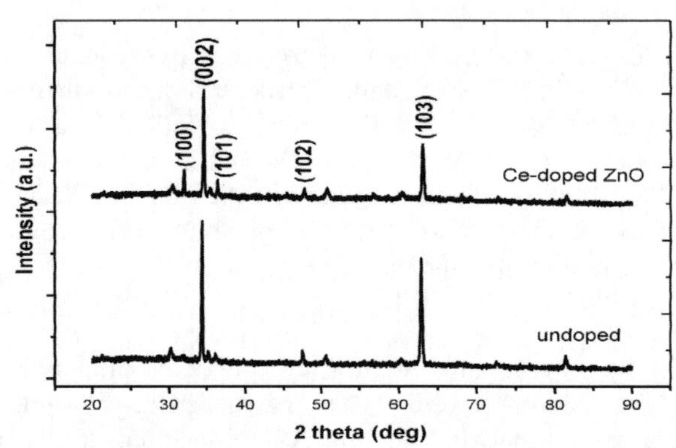

Figure 2. XRD patterns of undoped and Ce doped ZnO.

The diffraction peaks (002) have shifted from 34.40^0 to 34.60^0 after the doping of 3 mol% Ce. The addition of Cerium decreased the intensity of (002) and (103) planes, the full width at half maximum

(FWHM) also widened from 0.14 into 0.15 and the crystallite size also decreased from 84 nm into 74 nm. Decreasing the intensity and widening of FWHM indicates a slight decrease in crystallinity of ZnO nanorods.

The lattice parameters of Ce-doped ZnO nanorods (a = 3.2530, c = 5.2132) increased compared with the undoped one (a = 3.2492, c = 5.2027). It indicates that Ce ions have been incorporated into the ZnO lattice because the ionic radius of Ce^{3+} (1.034 Å) is much bigger than that of Zn^{2+} (0.74 Å).

3.3. Optical properties of ZnO nanorods

Absorption and transmittance spectra of ZnO nanorod can be seen in figure 3. From the absorption spectra, the curve dropped sharply in a range of 375 nm, known as the absorption edge relating to band gap energy of ZnO nanorod samples. Figure 3 (a) exhibits the absorption edge of ZnO nanorod at 373 nm then slightly shifted to 365 nm for Cerium doped. In the sample of 3% Cerium doped ZnO, the shift is followed by an increase in the UV absorption intensity. In general, an increase in the concentration of Cerium dopant will be decreasing the intensity of the absorbance in the visible region. Increasing the intensity of absorption in the UV region is also obtained by N.Sinha et. al. which used 1mol% cerium doped [10].

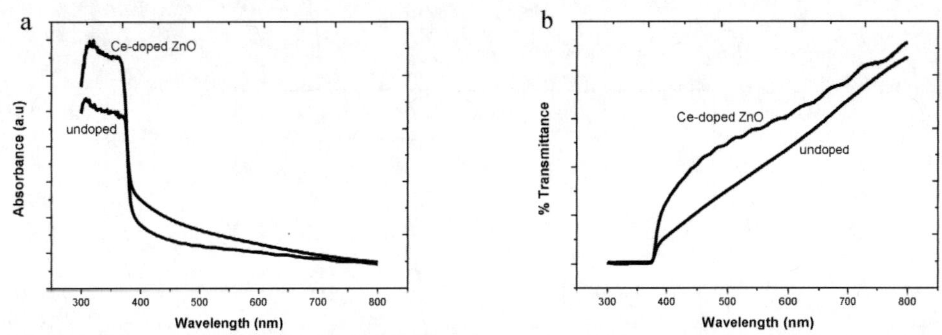

Figure 3. UV-vis absorption spectra of undoped ZnO and Ce-doped ZnO (a), transmittance of undoped ZnO and Ce-doped ZnO (b).

The transmittance spectra of ZnO nanorods are shown in figure 3(b). The transmittance in the visible light region was increased for Ce-doped ZnO nanorods, while the transmittance in UV region was decreased.

3.4. Diffuse reflectance spectra

The direct band gap energy was calculated by the diffuse reflectance spectrum using Kubelka-Munk formulation and then transferred into Tauc plot as shown figure 4 and figure 5.

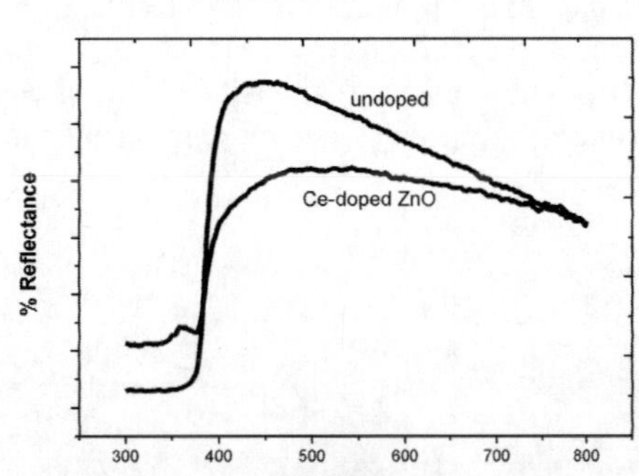

Figure 4. UV-vis diffuse reflectance spectra of ZnO and Ce-doped ZnO

IOP Conf. Series: Materials Science and Engineering **172** (2017) 012037 doi:10.1088/1757-899X/172/1/012037

The band gap energy of undoped ZnO nanorods was 3.23 eV and decreased into 3.18 eV after Ce doping. The change of band gap energy may be due to the new energy levels of cerium in ZnO lattice.

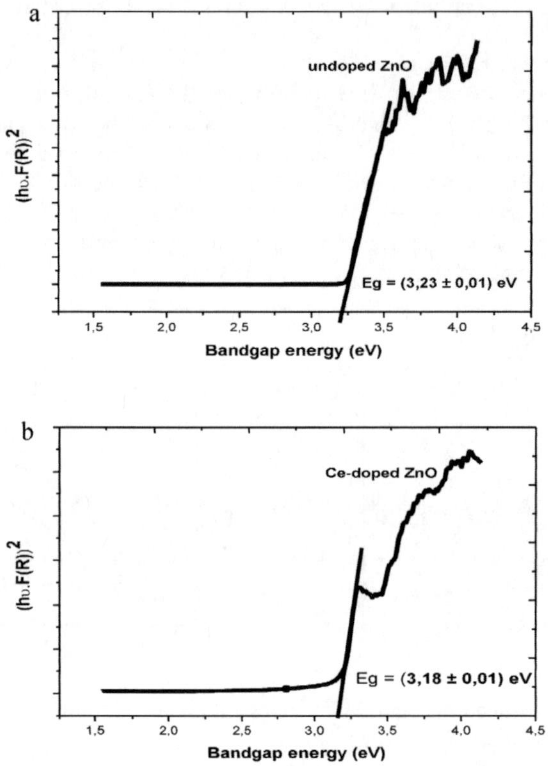

Figure 5. Tauc plot of (a) undoped ZnO. (b) Ce-doped ZnO.

3.5. Photoluminescence spectra

The room temperature PL spectra of undoped ZnO and Ce-doped ZnO nanorods is shown in figure 6. The PL spectra of undoped ZnO exhibits stronger near-band-edge (NBE) emission centered at a wavelength of 398 nm. The UV emission results from the recombination of free or bound exciton indicate the highest crystal quality of the material. It is also observed there is a broad emission band centered at a wavelength of 620 nm. This broad green emission peak is usually associated with the presence of defect levels in the band gap such as oxygen vacancies, zinc vacancies and interstitial zinc ion vacancies [5], [10], [17], [18]. The enhanced intensity of the green emission attributed to the increase of oxygen vacancy introduced by the existence of ce impurities in ZnO nanorods [8].

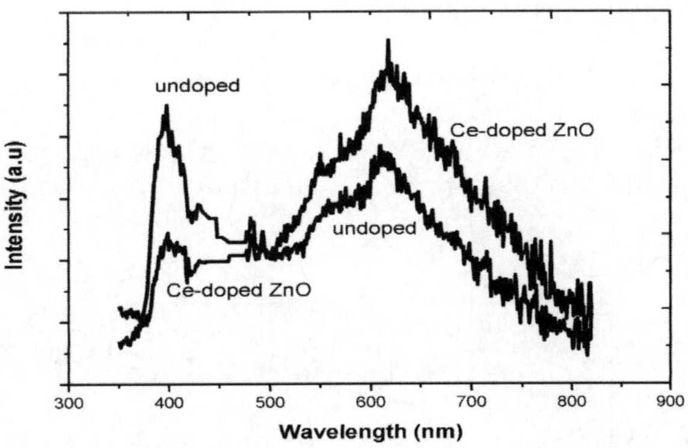

Figure 6. PL spectra of undoped ZnO and Ce-doped ZnO

3.6. *Photocatalytic degradation of methylene blue*

Figure 7 shows absorption spectra of the MB solution under a UV light irradiation using the ZnO nanorods. Quantitative degradation of methylene blue is shown at 597 nm as absorption peak.

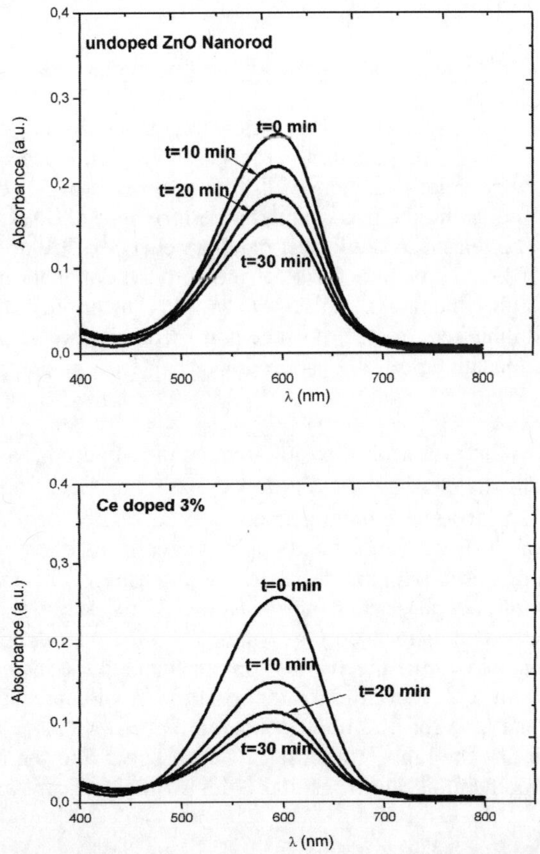

Figure 7. Absorbance spectra of MB by ZnO nanorods photocatalytic activity

The absorbance decreases by the increase of the irradiation time. The degradation of MB was calculated by the following equation (1):

$$\text{Degradation (\%)} = [(A_0\text{-}A_t)/A_0] \times 100 \quad (1)$$

where A_t is the absorbance at 597 nm of the MB solutions after irradiation in selected time, and A_0 is the absorbance at 597 nm of the MB solutions before irradiation.

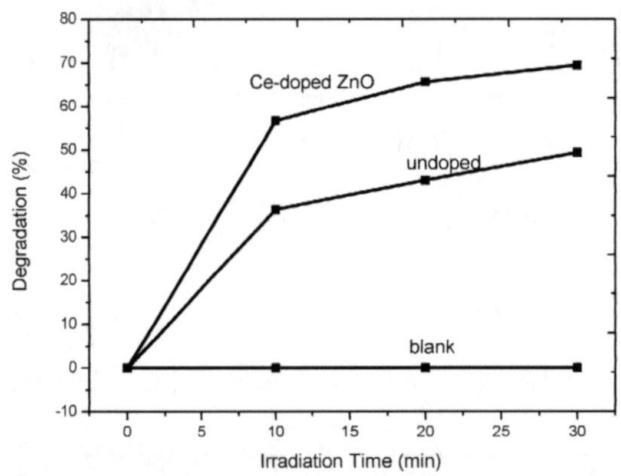

Figure 8. Degradation of MB aqueous solution with ZnO nanorods under UV light irradiation

Figure 8 shows the degradation spectra of MB when treated in the presence of ZnO nanorods photocatalysts. It can be seen that the degradation percentage under undoped Ce is 49% after 30 min of irradiation whereas 69% degradation was achieved in the presence of 3% ce doped ZnO after 30 min of exposure to UV light indicating the photocatalytic performance of 3% Ce doped ZnO has been running well.This result may be related to the lowest bandgap energy of Ce doped ZnO nanorod, thus widening the spectrum of UV light can excite the electrons into the conduction band which generates electrons and holes with higher numbers. Faisal *et al.* [19] mentioned that cerium acts as a "warehouse" of electrons and inhibitors of electron-hole pair recombination in producing light so as to increase the ability of the photocatalytic of ZnO nanorod.

4. Conclusions

The Ce-doped ZnO nanorods were synthesized by low-temperature hydrothermal process. In FESEM, the size increase of nanorods was obtained for Ce doped ZnO and Cerium dopant influence the direction and shape of the nanorods. Ce doping resulted in decreasing of the intensity of (002) and (103) planes and widening of FWHM, it indicate a slight decrease in crystallinity of ZnO nanorods. Increasing the concentration of Cerium dopant decreases the intensity of the absorbance in the visible region. The band gap energy of undoped ZnO nanorods was 3.23 eV that was decreased into 3.18 eV after Ce doping. The change of band gap energy may be due to the new energy levels of Cerium in ZnO lattice. The enhanced intensity of the green emission attributed to the increase of oxygen vacancy introduced by the existence of Ce impurities in ZnO nanorods. Photocatalytic experiment ZnO nanorods to degrade methylene blue for 30 minutes showed that 3% Ce doped ZnO have the fastest degradation ability of MB due to the highest intensity of absorbance and the lowest bandgap energy making it more effective for generating electrons and holes with significant amount that required to degrade methylene blue.

References

[1] Chang C, Lin C, Chen J, and Hsu M 2014 Ce-doped ZnO nanorods based low operation

temperature NO_2 gas sensors *Ceram Int.* **40** 10867
[2] Liu Y and Gao W 2015 Growth process, crystal size and alignment of ZnO nanorods synthesized under neutral and acid conditions *J. Alloys Compd.* **629** 84
[3] Yang L, Zhao Q, Willander M and Yang J H 2009 Effective way to control the size of well-aligned ZnO nanorod arrays with two-step chemical bath deposition *J. Cryst. Growth* **311** 1046
[4] Ma H L, Liu Z W, Zeng D C, Zhong M L, Yu H Y and Mikmekova E 2013 Nanostructured ZnO films with various morphologies prepared by ultrasonic spray pyrolysis and its growing process *Appl. Surf. Sci.* **283** 1006
[5] Sri P 2014 Structural and Optical Properties of Cerium doped Zinc Oxide thin films using spray pyrolysis *J. NanoSci. NanoTechnol.* **2** 34
[6] Benramache S, Rahal A, and Benhaoua B 2014 The effects of solvent nature on spray-deposited ZnO thin film prepared from Zn $(CH_3COO)_2$, $2H_2O$ *Opt. - Int. J. Light Electron Opt.* **125** 663
[7] Yang J, Gao M, Yang L, Zhang Y, Lang J, Wang D, Wang Y, Liu H, and Fan H 2008 Low-temperature growth and optical properties of Ce-doped ZnO nanorods *Appl. Surf. Sci.* **255** 2646
[8] J. Lang, Q. Han, J. Yang, C. Li, X. Li, L. Yang, Y. Zhang, M. Gao, D. Wang, and J. Cao 2010 Fabrication and optical properties of Ce-doped ZnO nanorods *J. Appl. Phys.* **107** 2
[9] G. Murugadoss, R. Jayavel, and M. Rajesh Kumar 2015 Structural and optical properties of highly crystalline Ce, Eu and co-doped ZnO nanorods *Superlattices Microst.* **82** 538
[10] N. Sinha, G. Ray, S. Bhandari, S. Godara, and B. Kumar 2014 Synthesis and enhanced properties of cerium doped ZnO nanorods *Ceram. Int.* **40** 12337
[11] N. Febiana Djaja, Rosari Saleh 2013 Characteristics and photocatalytics activities of Ce-Doped ZnO nanoparticles *Mat. Sci. Appl.* **4** 145
[12] A. Bedia, F. Z. Bedia, M. Aillerie, N. Maloufi, and B. Benyoucef 2015 Morphological and optical properties of ZnO thin films prepared by spray pyrolysis on glass substrates at various temperatures for integration in solar cell *Energy Procedia.* **74** 529
[13] S. Benramache, B. Benhaoua, F. Chabane, and A. Guettaf 2013 A comparative study on the nanocrystalline ZnO thin films prepared by ultrasonic spray and sol–gel method *Opt. - Int. J. Light Electron Opt.* **124** 3221
[14] T. T. Y. Tan 2012 *Rare Earth Nanotecnhology*. Singapore: Pan Stanford Publishing Pte.Ltd.
[15] X. Y. Chen, F. Fang, A. M. C. Ng, A. B. Djurisic, W. K. Chan, H. F. Lui, P. W. K. Fong, C. Surya, and K. W. Cheah 2011 Effect of doping precursors on the optical properties of Ce-doped ZnO nanorods *Thin Solid Films* **520** 1125
[16] G. N. Dar, A. Umar, S. A. Zaidi, A. A. Ibrahim, M. Abaker, S. Baskoutas, and M. S. Al-Assiri 2012 Ce-doped ZnO nanorods for the detection of hazardous chemical *Sensors Actuators, B Chem.* **173** 72
[17] Y.-I. Jung, B.-Y. Noh, Y.-S. Lee, S.-H. Baek, J. H. Kim, and I.-K. Park 2012 Visible emission from Ce-doped ZnO nanorods grown by hydrothermal method without a post thermal annealing process *Nanoscale Res. Lett.* **7** 43
[18] S. Y. Kuo, W. C. Chen, F. I. Lai, C. P. Cheng, H. C. Kuo, S. C. Wang, and W. F. Hsieh 2006 Effects of doping concentration and annealing temperature on properties of highly-oriented Al-doped ZnO films *J. Cryst. Growth* **287** 78
[19] M. Faisal, A. A. Ismail, A. A. Ibrahim, H. Bouzid, and S. A. Al-Sayari 2013 Highly efficient photocatalyst based on Ce doped ZnO nanorods: Controllable synthesis and enhanced photocatalytic activity *Chem. Eng. J.* **229** 225

Acknowledgments
This work was supported by Hibah Publikasi International Terindeks untuk Tugas Akhir Mahasiswa (PITTA) 2016 Universitas Indonesia and grants from the Ministry of Research, Technology and Higher Education of the Republic of Indonesia (Insinas 2016).

IOP Conf. Series: Materials Science and Engineering **172** (2017) 012038 doi:10.1088/1757-899X/172/1/012038

Colorimetric chemosensor of symmetrical benzoylthiourea derivatives as for detection of Cu^{2+} in aqueous solution

N A Hamedan[1], S Hasan[1], H M Zaki[1] and N Z Alias[1]

[1]Faculty of Applied Sciences, Universiti Teknologi MARA, 40450 Shah Alam, Selangor, Malaysia.

E-mail: sharizal187@perlis.uitm.edu.my

Abstract. A novel receptor, designed with a combination of oxygen (O), nitrogen (N) and sulfur (S) -binding sites for metal ions was synthesized. Ortho (**A**), meta (**B**) and para (**C**) bearing benzoyl thiourea were designed and synthesized with triamine group to apply as colorimetric chemosensors for detection of Cu^{2+}. The structure was confirmed by characterized the compound using Elemental analysis, Fourier Infrared (FTIR) and proton Nuclear Magnetic Resonance (^1H NMR) spectroscopy. Functional groups of C=O, N-H, C=N and C=S were found at 1677 cm^{-1}, 3240 cm^{-1}, 1591 cm^{-1}, 1024 cm^{-1} respectively while ^1H NMR shows peaks of alkane (CH_2), benzene (Ar-H), CONH, CSNH at 3.68 – 4.14, 7.16 – 7.86, 8.74, and 9.2 respectively. Elemental analysis for **A**, **B** and **C** $C_{20}H_{21}N_5O_2S_2Br_2$ found was compatible with the expected theoretical calculation. For an application, all of these three sensors showed excellent colorimetric specific selectivity and high sensitivity for Cu^{2+} in acetonitrile/water binary solutions, so only **A** was selected for further studies towards sensitivity. When Cu^{2+} was added to the solution of **A**, a dramatic color change from yellow to green, while other cations Fe^{2+}, Zn^{2+}, Ni^{2+}, Co^{2+}, Cr^{3+} and Mn^{2+} did not interfere with the recognition process for Cu^{2+}. The detection limit of the sensor **C** toward Cu^{2+} was 1.15 x 10^{-5} M, which is less sensitive that sensor **A** and **B** with a detection limit of 6.2 x 10^{-6} M and 1.5 x 10^{-6} M respectively. This indicated that the sensor **A** and **B** might be useful as an efficient chemical sensor.

1. Introduction

Thiourea derivatives are organosulfur compounds with the general formula $(R^1R^2N)(R^3R^4N)C{=}S$. In organic synthesis, thiourea derivatives are continuously increasing in numbers, due to their unique properties and synthetic simplicity. They have been extensively studied in coordination chemistry, and are known for various applications in the field of material science, agriculture, biological studies and pharmaceutical. The ability of thiourea derivatives to bind with metal ions via thiono group (C=S) make them applicable as ion sensors and transition metals extractors. In recent years, there were different types of sensor designed and synthesized to sense different types of metals such as multisubstituted hydrazone based Schiff bases as a sensor to detect F^- and Cu^{2+} [1], thiosemicarbazone-appended 3-hydroxynaphthalimide detect Cu^{2+} [2], naphthalimide-based fluorescent chemosensor that detect Zn^{2+} [3] and calix[4]arenes to detect Hg^{2+} ion [4]. Copper (Cu) is an essential trace metal that is used as a catalytic cofactor for many enzymes. Copper is also a vital dietary nutrient, although only small amounts of the metal are needed for well-being. Although copper is the third most abundant trace metal in the body [behind iron and zinc], the total amount of copper in the body is only 75-100 milligrams. Copper is present in every tissue of the body, but is stored primarily in the liver, with fewer amounts found in the brain, heart, kidney, and muscles [5]. Despite this, Cu is dangerous when it is present in excess, mainly because it can participate in the Fenton reaction and produce free radical species. Many studies have reported that copper overload leads to the overproduction of reactive oxygen species (ROS) and the establishment of an oxidative stress condi tion that is responsible for the pro-oxidative damage of key proteins, lipids and DNA structure [6].The present study aimed at the determination of the isomerization effect of the methyl group at ortho, meta and para positions of the three newly synthesized bromobenzoyl thiourea isomers, namely 1-(2-Bromo-benzoyl)-3-(2-{2-[3-(2-bromo-benzoyl)-thioureido]-ethylamino}-ethyl)-thiourea **A**, 1-(3-Bromo-benzoyl)-3-(2-{2-[3-(3-bromo-benzoyl)-thioureido]-ethylamino}-ethyl)-thiourea **B** and 1-(4-Bromo-benzoyl)-3-(2-{2-[3-(4-bromo-benzoyl)-thioureido]-ethylamino}-ethyl)-thiourea **C** on the spectroscopic data (UV-vis, NMR, FTIR and EA).

2. Methods

2.1. Reagents and physical measurements

The inorganic salts $Fe(SO_4).6H_2O$, $Mn(SO_4)$, $Cr(SO_4)XH_2O$, $Co(Cl_2)_2.6H_2O$, $Ni(NO_3)_2.6H_2O$, $Cu(NO_3)_2.3H_2O$ and $Zn(Cl_2)_2$ were purchased from Sigma-Aldrich. All solvents and other reagents were of analytical grade. ^1H NMR spectra were recorded on a Bruker Avance III 300 Spectrometer at room temperature. ^1H chemical shifts were reported in ppm downfield from tetramethylsilane (TMS), d scale with the solvent resonances as internal standards). UV-vis spectra of compounds were recorded on a Perkin Elmer LAMBDA 25 UV-Vis spectrometer. Melting points were measured using BÜCHI Melting Point B-545 and the elemental analyses were conducted using CHNS Analyzer Flash EA 1112 series. The infrared spectra were performed using FTIR Perkin Elmer 100 Spectrophotometer in the spectral range of 4500-600 cm^{-1}.

2.2 General Procedure for UV-Vis spectroscopy

All the UV-Vis experiments were carried out in Acetonitrile solution on a Perkin Elmer LAMBDA 25 UV-Vis spectrometer. Any changes in the UV-vis spectra of the synthesized compound were recorded on the addition of sulfate, chlorate, and nitrate metal salts while the ligand concentration was kept constant in all experiments.

2.3 Binding constant

The binding constant values for receptor towards Cu^{2+} were calculated from the UV–Vis titration results using Benesi–Hildebrand (B–H) Eq. (1).

$$\frac{1}{\Delta A} = \frac{1}{b\Delta\varepsilon[G_0][H_0]K_a} + \frac{1}{b\Delta\varepsilon[G_0]} \quad (1)$$

where ΔA is the change in absorbance, $[G_0]$ is the concentration of the guest molecule, $[H_0]$ is the concentration of the host molecule and K_a is binding constant. The plot can be made with $1/\Delta A$ as a function of $1/[H_0]$. $\Delta\varepsilon$ can be derived from the intercept while K_a can be calculated from the slope.

2.4 Synthesis and characterization of sensors A, B and C.

Preparation of 1-(2,3,4-Bromo-benzoyl)-3-(2-{2-[3-(2,3,4-bromo-benzoyl)-thioureido]-ethylamino}-ethyl)-thiourea **(A-C)**. The 2,3,4-bromobenzoyl chloride (1.2 mL, ten mmol) was added dropwise to a stirring acetone solution (40 mL) with ammonium thiocyanate (0.76 g, 10 mol). A white precipitate of

NH_4Cl was formed immediately and was filtered. The filtrate then was taken to be added dropwise into diethylenetriamine (0.5 mL, five mmol) in 40 mL acetone then put at reflux with constant stirring for 4 hours. The solution was poured into a beaker containing some ice blocks. Then, the resulting yellow precipitate was filtered and washed with little cold methanol, then dried in a vacuum desiccator. Finally, the crude product was purified by recrystallization from methanol to afford the title compound as colorless single crystalline solids.

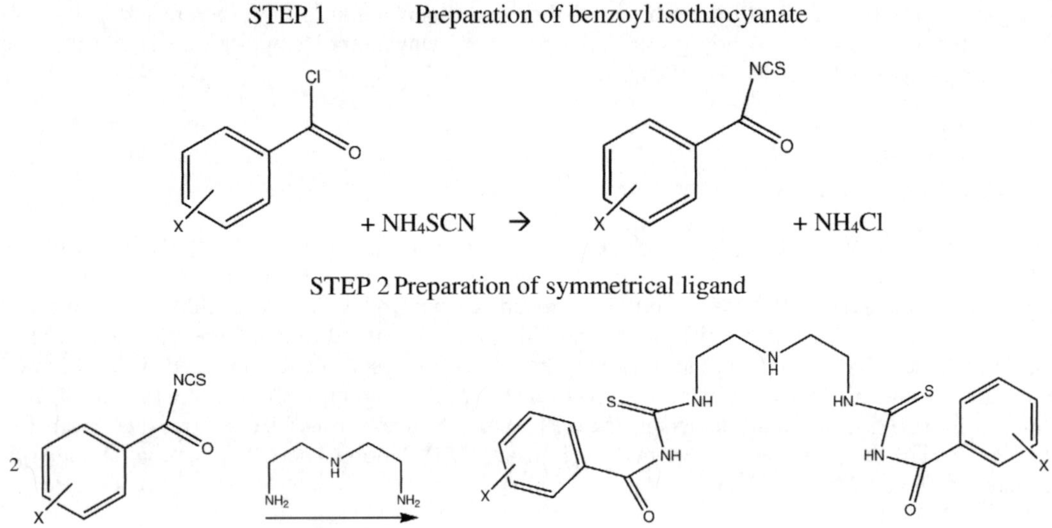

X = o/m/p- Br

Scheme 1 Synthesis of the symmetrical benzoylthiourea derivatives

A: Yield 52.8%; pale yellow solid, m.p 80°C. IR (cm^{-1}): υ(C=O) 1673, υ(N-H) 3230, υ(C=C) 1589, υ(C=S) 1025. ^{1}H NMR (600 MHz, $CDCl_3$-d_6) δ 10.65 (s, CSNH, 1H), 8.98 (s, CONH, 1H), 7.14-7.66 (m, Ar-H, 6H). Anal. Calc. for $C_{21}H_{26}N_5OSBr$: C, 40.90; H, 3.60; N, 11.92. Found: C, 41.42; H, 3.39; N, 11.83.

B: Yield 61.4%; pale yellow solid, m.p 158°C. IR (cm^{-1}): υ(C=O) 1713, υ(N-H) 3343, υ(C=C) 1622, υ(C=S) 1065. ^{1}H NMR (600 MHz, $CDCl_3$-d_6) δ 10.85 (s, CSNH, 1H), 8.95 (s, CONH, 1H), 7.19-8.32 (m, Ar-H, 6H). Anal. Calc. for $C_{21}H_{26}N_5OSBr$: C, 40.90; H, 3.60; N, 11.92. Found: C, 40.46; H, 3.62; N, 10.02.

C: Yield 54.2%; pale yellow solid, m.p 95°C. IR (cm^{-1}): υ(C=O) 1668, υ(N-H) 3263, υ(C=C) 1589, υ(C=S) 1008. ^{1}H NMR (600 MHz, $CDCl_3$-d_6) δ 10.87 (s, CSNH, 1H), 8.98 (s, CONH, 1H), 7.28-8.06 (m, Ar-H, 6H). Anal. Calc. for $C_{21}H_{26}N_5OSBr$: C, 40.90; H, 3.60; N, 11.92. Found: C, 41.12; H, 3.99; N, 12.26.

3. Results and Discussions

*3.1 Absorption studies of receptor **A**, **B** and **C**.*

The experimental maximum absorption bands of benzoylthiourea isomers **A, B** and **C** are showed in Figure 1 and Table 1. The experimental spectra were recorded in acetonitrile solvent (c 5 x 10^{-5}M). The maximum absorption bands at 250, 250 and 246 nm in **A**, **B** and **C** respectively, correspond to $\pi \rightarrow \pi^*$ electronic transition. The solvent effect induces a hypsochromic shift (blue shift) of λ_{MAX} ($\pi \rightarrow \pi^*$) from *ortho* and *meta* (250nm) to *para* (246nm), which was shifted to shorter wavelength and higher energies. According to Woodward-Fieser Rules for Aromatic, it stated that a substituent will have a different effect depending on whether it is *ortho, meta* or *para* substituents. This rules also noted that the strongest effect is observed for *para* substituent. According to Chetana and co-workers, the electronic absorption spectra of compounds N, N'- disubstituted thiourea show two intense broad bands with maxima in the range of ~270 and ~325 nm; the high energy band can be attributed to ligand $\pi \rightarrow \pi^*$ transitions on the pyridine and benzene aromatic rings, where as the lower energy band can be considered as a thione originating calf thymus (CT) transition at the C=S bond [7].

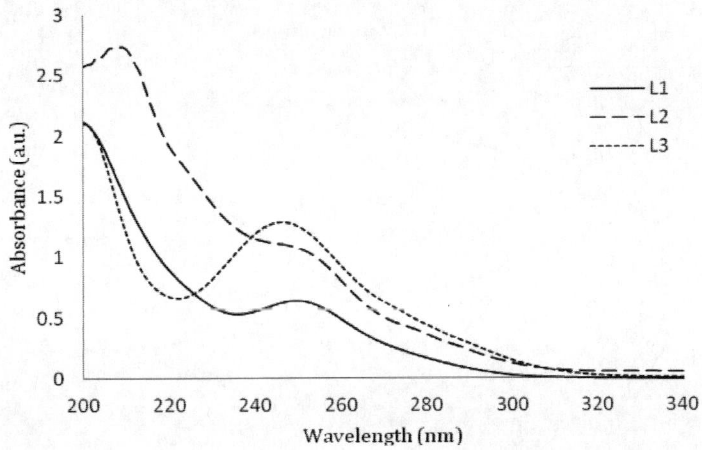

Figure 1 UV spectra of A, B and C; c= 5 x 10^{-5}M in acetonitrile, cell length = 1mm.

Table 1. Experimental λ_{MAX} (nm) of $\pi \rightarrow \pi^*$ transition of **A**, **B** and **C** benzoylthiourea derivatives.

Sensor	λ_{MAX} (nm)
A	250
B	250
C	246

*3.2 Absorption studies of receptors **A B** and **C** toward various cations.*

All ligands **A, B** and **C** are more sensitive sensor towards Cu^{2+} ion than other metals such as Mn^{2+}, Cr^{3+}, Fe^{2+}, Co^{2+}, $Ni^{2+}$and Zn^{2+}. The cation binding abilities of the different receptors were investigated by UV-vis spectroscopic titration in ACN/ H_2O (10:1/v:v) at pH 7.0. The colorimetric sensing abilities were primarily investigated by adding various cations (Cu^{2+}, Mn^{2+}, Cr^{3+}, Fe^{2+}, Co^{2+}, $Ni^{2+}$and Zn^{2+} to the ACN/ H_2O (10:1/v:v, pH 7.0) solutions of sensors. Addition of Cu^{2+} to the solutions (5.0 x 10^{-5} M), was responded sensors with dramatic color changes from yellow to green while other cation did not induce any changes (Fig. 3) which is compatible with the UV-Vis results indicated that only Cu^{2+} ion can be detected by the sensors **A**, **B** and **C**. In the corresponding UV-vis spectrum (Fig. 2), the absorption peak of Cu^{2+} were shifted from 250nm (**A**), 250nm (**B**) and 246nm (**C**) to a new peak appeared at visible spectra at 800 nm of Cu^{2+} complexes formed, which attributed to the color change from yellow to green.

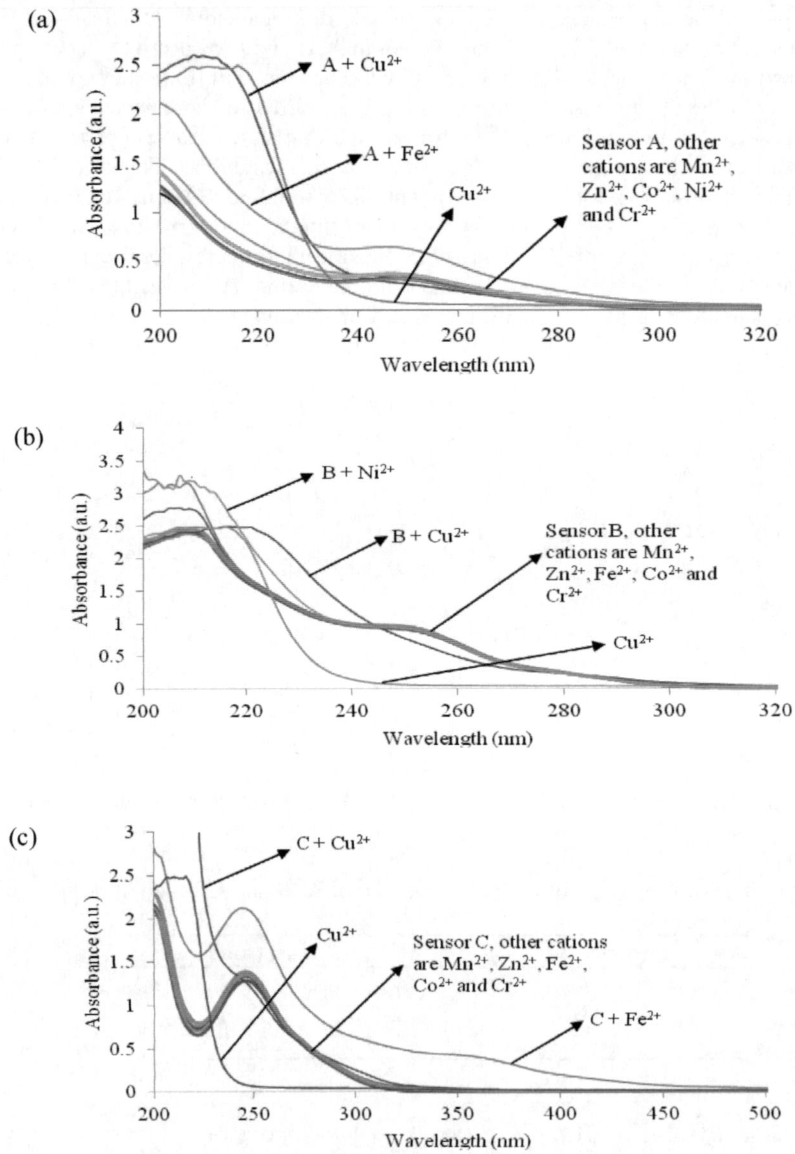

Fig. 2. UV-vis absorption spectra of **A** (a), **B** (b) and **C** (c) in the presence of of various cations in ACN/H_2O (10:1, v:v, pH 7.0) solution at room temperature.

In order to exclude the possibility of these results being due to Cu^{2+} self-absorption, a blank test were carried out via adding the same amount of Cu^{2+} to blank ACN/ H_2O solution (without containing **A, B or C**) as a result, no color change were observed. In corresponding UV-vis spectra, there is no absorption peak appeared in the visible region (Fig. 2). It was confirmed that all sensor **A, B** and **C**

IOP Conf. Series: Materials Science and Engineering **172** (2017) 012038 doi:10.1088/1757-899X/172/1/012038

could colorimetrically detect Cu^{2+} in ACN/ H_2O binary solution [8]. **B** and **C** show same result as **A** so only **A** was selected for further studies towards sensitivity to Cu^{2+}. A series of experiments was carried out to investigate the Cu^{2+} recognition capability and mechanism of **A**.

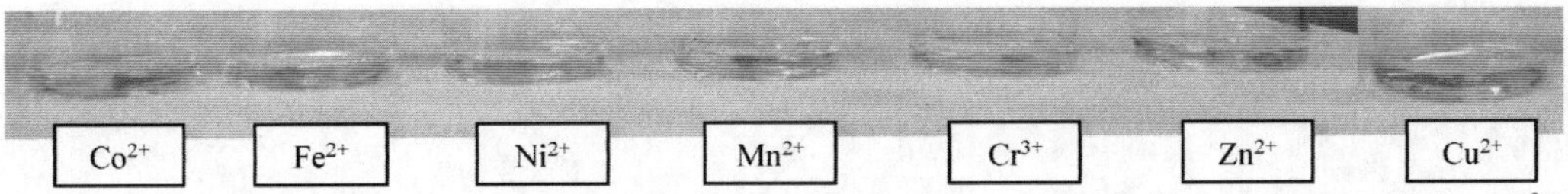

Fig. 3. Color changes observed upon the addition of various cations to the solutions of sensor **A** (1.5 x 10^{-3} M) in ACN/H_2O (10:1, v:v, pH 7.0) solutions. Left to right : free **A**, Co^{2+}, Fe^{2+}, Cr^{3+}, Mn^{2+}, Zn^{2+}, Ni^{2+} and Cu^{2+}.

Job's plot studies (Fig. 3) were done to find out the stoichiometry of the complex formed between the receptors and cations and it was found that the complex formed was in 2:1 stoichiometries. The molar fraction found of the complex **A**-Cu^{2+} was 0.7. Moreover, the recognition mechanism of the sensor **A** with Cu^{2+} was investigated by IR spectra titration methods. Based on IR spectra of **A**, the stretching vibration absorption peaks of cyanide C=N, acyl C=O and thiourea C=S appeared at 1589, 1673 and 1025 cm^{-1} respectively. However, when **A** coordinated with Cu^{2+}, the stretching vibration absorption peaks of acyl C=O were shifted to 1698 cm^{-1} while stretching vibration absorption peaks of cyanide C=N and thiourea C=S didn't take place any shift. According to IR spectrum, it shows left shifted (higher wavenumber) with the higher energy of C=O that indicated **A** complexed with Cu^{2+} via Cu^{2+}- N coordination bond only near to C=O as shown in Scheme 2. The secondary amine near to C=O will bind to the Cu^{2+} because it may affect the C=O through back-bonding theory. The addition of metal should form a lower energy but for peak C=O, the higher energy of C=O formed might be affected by the back bonding of Cu^{2+}- N coordination bond that may shorten the C=O bond, lower the vibration or frequency of the bond. Due to the equation of $E=h\upsilon$, the frequency is directly proportional to the energy. The higher the frequency, the higher the energy present.

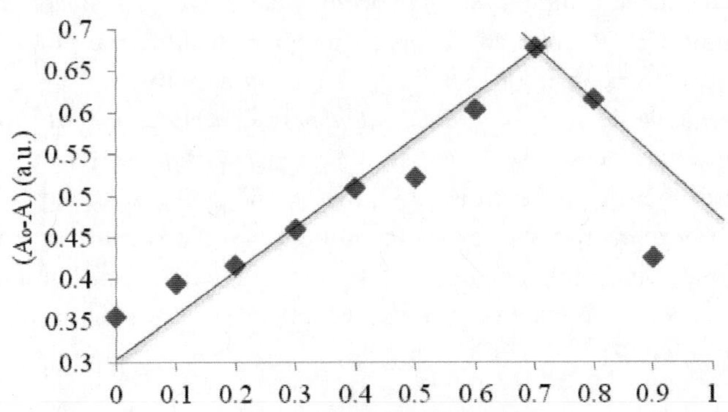

Fig. 3. A Job plot of **A** and Cu^{2+}, which indicated that the stoichiometry of **A**-Cu^{2+} complex was 2:1.

IOP Conf. Series: Materials Science and Engineering **172** (2017) 012038 doi:10.1088/1757-899X/172/1/012038

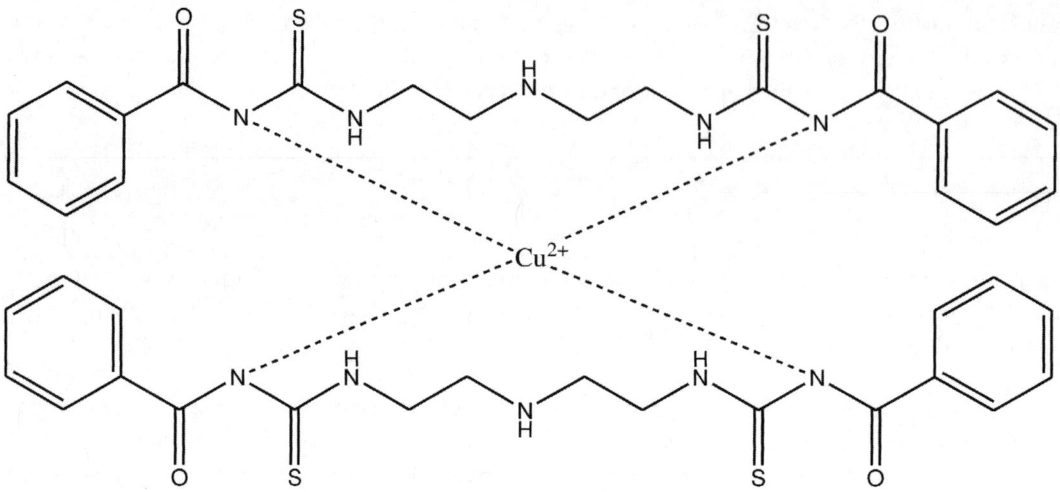

Scheme 2. The proposed reaction mechanism of the sensor **A** with Cu^{2+}.

From the UV-Vis absorption measurements, the binding constant (K_a) of the copper complex of the receptor is calculated from the variation in the absorbance at 800nm, respectively and is found to be 1.04 x 10^4 M. This may be due to the presence of electron withdrawing group (–Br) in it. This suggests that the receptor forms strong binding with Cu^{2+} ion [1].

The detection limit for **C** (para) is lower than **A** and **B** that correspond to the maximum absorption bands of **C** which is shifted to shorter wavelength and higher energy. Higher energy may have a lower detection limit towards metal. The detection limit of metal ions by **A, B** and **C** was calculated from the plot of UV–vis absorbance versus metal ions concentration. From the plot, detection limit was calculated using the formula 3σ/m. Where 'σ' is the standard deviation and 'm' is the slope of the straight line. A calibration curve was obtained from the plot of UV-Vis absorbance with the addition of Cu^{2+} concentration. The curve equation as shown in Fig. 5 was R^2=0.9987. The linear range of quantitative detection for Cu^{2+} was determined as 2..5 x 10^{-4} to 1.95x10^{-2} M with a detection limit of 6.2 x 10^{-6} M (sensor **A**), 1.5 x 10^{-6} M (sensor **B**) and 1.15 x 10^{-5} M (sensor **C**) for Cu^{2+} from the curve in the used aqueous solution system. Thus all **A, B** and **C** can be utilized as an efficient Cu^{2+} ion sensor at micromolar level. According to World Health Organization (WHO-1984), 1 mg/l of copper present in drinking water. Our sensor **A** detects copper in micromolar levels (10^{-6} M/L) which is lower or equal to maximum permitted amount of copper in drinking water defined by World Health Organization [9].

IOP Conf. Series: Materials Science and Engineering **172** (2017) 012038 doi:10.1088/1757-899X/172/1/012038

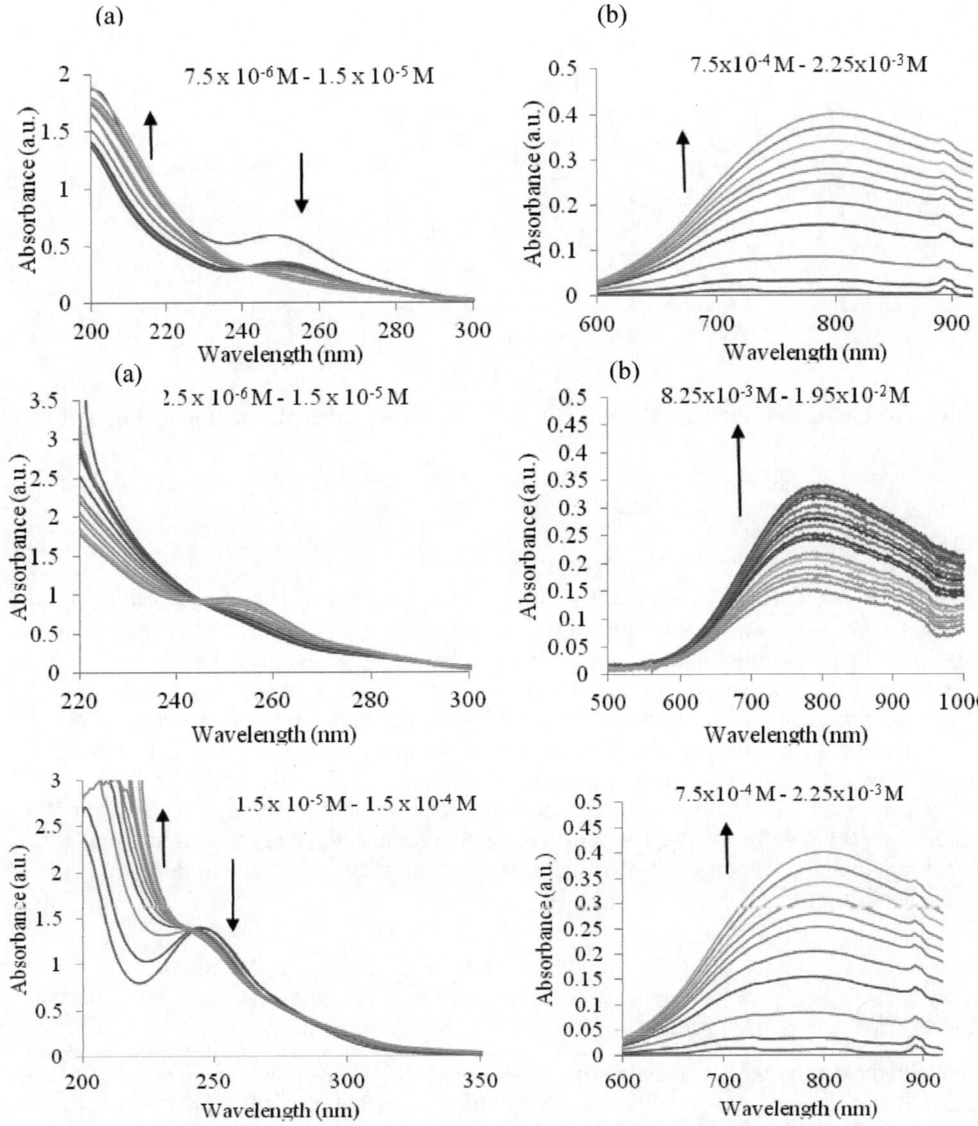

Fig. 4. (a) UV spectra titration of sensor **A, B** and **C** (5×10^{-5} M) with Cu^{2+} (2.5×10^{-6} M - 1.5×10^{-4} M) in 10:1 ACN/H_2O (v/v) solution. (b) Visible spectral titration of sensor **A, B** and **C** (5×10^{-5} M) with Cu^{2+} (7.5×10^{-4} M – 1.95×10^{-2} M) in 10:1 ACN/H_2O (v/v) solution.

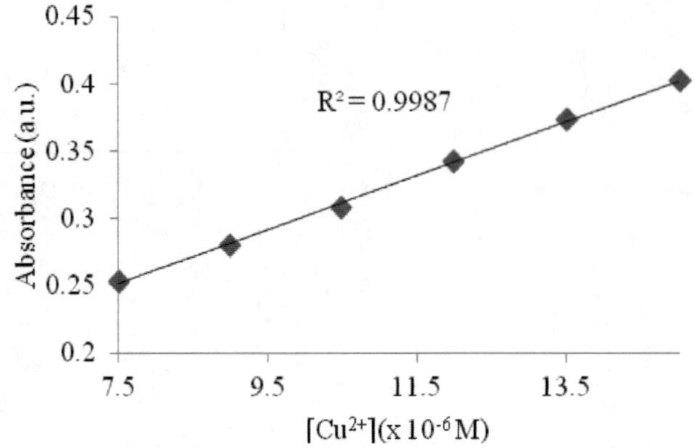

Fig. 5. Calibration curve was obtained from the plot of UV-Vis absorbance with the addition of Cu^{2+} concentration.

4. Conclusion

In conclusion, **A**, **B** and **C** benzoylthiourea derivatives were designed and synthesized to apply as a colorimetric sensor for Cu^{2+} ion. This sensor showed specific selectivity for Cu^{2+} in ACN/H_2O binary solutions indicated that benzoylthiourea also may act as a chemosensor and played a crucial role in the process of colorimetric recognition. Only sensor **A** was chosen for further studies due to the same selectivity towards metal of Cu^{2+} ion. Investigation of the recognition mechanism indicated that the sensor **A** recognized Cu^{2+} by forming a stable 2:1 **A-** Cu^{2+}complex. The coexistence of other cations did not interfere with the Cu^{2+} recognition process. Moreover, the detection limit of the sensor **C** toward Cu^{2+} was 1.15 x 10^{-5} M, which is less sensitive that sensor **A** and **B** with a detection limit of 6.2 x 10^{-6} M and 1.5 x 10^{-6} M respectively. The sensitive detection indicated that the sensor **A** and **B** may be useful for preliminary detection of Cu^{2+} ions by simple solvent extraction in chemical and environmental applications.

Acknowlegdements

We would like to thanks the Malaysian Ministry of Education (KPM) for providing generous financial support under RAGS grants: 600-RMI/RAGS 5/3 (2/2013) in conducting this study and Universiti Teknologi Mara (UiTM) for providing all the needed facilities.

References

[1] A. Sundar, M. Prabhu, N. I. Gandhi, M. Marappan, and G. Rajagopal, "Spectrochimica Acta Part A : Molecular and Biomolecular Spectroscopy Synthesis , characterization , structural analysis of metal (II) complexes of N' - [(E) -3-Bromo-5-Chloro-2-hydroxybenzidene] -4-hydroxy benzohydrazide — Multisubstituted Schiff ba," vol. 129, pp. 509–518, 2014.
[2] H. Park and S. Chang, "Sensors and Actuators B : Chemical Selective colorimetric and ratiometric signaling of Cu^{2+} ions by," vol. 220, pp. 376–380, 2015.
[3] Z. Xu, X. Qian, J. Cui, and R. Zhang, "Exploiting the deprotonation mechanism for the design of ratiometric and colorimetric Zn^{2+} fluorescent chemosensor with a large red-shift in

emission," *Tetrahedron*, vol. 62, no. 43, pp. 10117–10122, 2006.

[4] T. Kao, C. Wang, Y. Pan, Y. Shiao, J. Yen, C. Shu, G. Lee, S. Peng, and W. Chung, "Upper Rim Allyl- and Arylazo-Coupled Calix [4] arenes as Highly Sensitive Chromogenic Sensors for Hg^{2+} Ion," pp. 2912–2920, 2005.

[5] J. Osredkar and N. Sustar, "Copper and Zinc , Biological Role and Significance of Copper / Zinc Imbalance," 2011.

[6] N. Arnal, M. J. T. De Alaniz, and C. a. Marra, "Carnosine and neocuproine as neutralizing agents for copper overload-induced damages in cultured human cells," *Chem. Biol. Interact.*, vol. 192, no. 3, pp. 257–263, 2011.

[7] P. R. Chetana, B. S. Srinatha, M. N. Somashekar, and R. S. Policegoudra, "Synthesis , spectroscopic characterisation , thermal analysis , DNA interaction and antibacterial activity of copper (I) complexes with N , N' - disubstituted thiourea," vol. 1106, pp. 352–365, 2016.

[8] Q. Lin, P. Chen, J. Liu, Y. Fu, Y. Zhang, and T. Wei, "Dyes and Pigments Colorimetric chemosensor and test kit for detection copper (II) cations in aqueous solution with speci fi c selectivity and high sensitivity," vol. 98, pp. 100–105, 2013.

[9] K. B. Narayanan and H. H. Park, "Spectrochimica Acta Part A : Molecular and Biomolecular Spectroscopy Colorimetric detection of manganese (II) ions using gold / dopa nanoparticles," vol. 131, pp. 132–137, 2014.

IOP Conf. Series: Materials Science and Engineering **172** (2017) 012039 doi:10.1088/1757-899X/172/1/012039

Preparation, Characterization and Methylene Blue Dye Adsorption Ability of Acid Activated-Natural Zeolite

O A Saputra, M D Prameswari, V T D Kinanti, O D Mayasari, Y D Sutarni, K Apriany and W W Lestari

Chemistry Department, Sebelas Maret University, Faculty of Mathematic and Natural Sciences, Jl. Ir Sutami 36A Surakarta, 57126 Indonesia.

Email: oziadisaputra@student.uns.ac.id

Abstract. The aim of this research was to prepare an acid-activated natural zeolite (Ac-Zeo) as a low-cost adsorbent material and to investigate their ability on methylene blue dye removal in aqueous solution. The natural zeolite was activated using hydrochloric acid and the final product was characterized using Fourier transform infra-red (FTIR), X-ray diffraction (XRD) and scanning electron microscopy (SEM). The adsorption process was carried out using the batch method. Some parameters like pH condition, contact time and varied dye initial concentration were studied to determine the adsorption ability of Ac-Zeo. In this study, kinetic adsorption was evaluated using pseudo-second order model approach and found that the kinetic adsorption rate constanta (k) and adsorption capacity at equilibrium are 0.1872 $mg.g^{-1}.min^{-1}$ and 14.94 $mg.g^{-1}$, respectively. Moreover, Langmuir, Freundlich, and Dubinin–Kaganer–Radushkevich isotherm adsorption models as well as sorption mechanism were studied in this research.

1. Introduction

Zeolites are a class of microporous crystalline aluminosilicate material, widely available in Indonesia [1]. Zeolites have three-dimensional frameworks, consist of $[SiO_4]^{-4}$ and $[AlO_4]^{-5}$ tetrahedral [2, 3]. Beside of that, the zeolites also contain cations (i.e. Na^+, Ca^+ and K^+) to balance the negative charge in its structure [2, 4]. The presence of negative charge in zeolite frameworks makes them have an adsorption capacity to some cationic compound like metal ion and basic dye [5-7]. In the past work, numerous researchers have studied about the applicability of zeolite in various purposes, i.e. catalyst [8], ion metal adsorption [9], gas purification [10], environmental engineering [11], and dye adsorption [7, 12]. However, the commercial synthetic zeolites with high purity are high cost [5]. Thus the using of natural zeolite is very necessary to utilize these minerals also to reduce the operating cost in waste water management in textile industry.

Synthetic dyes are commonly used in dyeing process than natural dyes due to its low-cost production, brighter colors, high resistance to the environment and application, thus the consumption of synthetic dye is higher than natural dyes even though highly toxic and carcinogenic [13]. The dye compound consists of chromophores groups which influence the dye color, and auxochromes group which determines the color intensity [14]. The rapid development of industries like textile, paper and pulp, leather, food, pharmaceutical and cosmetic, causes the increasing of colored effluent and generates a considerable amount of polluted waste water [15, 16]. This pollution becomes a major concern in environmental problems, human health and marine lives [13].

It was noted that the removal dye in aqueous solution is difficult to be achieved due to the inert properties of synthetic dye and their low concentration in aqueous solution [17]. During the past decade, several methods were used for dye removal in aqueous solution, including photocatalytic degradation [18], ion exchange [19], electrochemical [20], coagulation [21], physicochemical treatment [22, 23] and adsorption [7, 12, 24]. Among all of these methods, the adsorption process is very suitable for dye removal due to its advantages such as low operational cost to remove the organic dye molecules, easy and relatively effective in dye removal and can be operated at low concentration of dyes [13, 25, 26].

Lately, a low-cost adsorbent by utilizing the natural resources, such as chitosan [27], cellulose from waste product [24], clay minerals [28], and natural zeolite [7, 29-31] has been developed. Due to their porosity, the zeolite has a good ability to be used as adsorbent material in dye removal. Han et al. [7, 29] has been successfully utilized natural zeolite in particle form for methylene blue dye removal in aqueous solution. Alpat et al. [30] also used Turkish natural zeolite with the structure like-clinoptilolite for cationic dye removal, Toluidine Blue O, and gave the best adsorption condition at pH 11 with an adsorption maximum capacity of 2.1×10^{-4} mol.g^{-1}. Another work has been reported that an enhancement of maximum adsorption capacity could be achieved by modifying the natural zeolite [32]. The adsorption ability of zeolite is determined by their porous structure. Elaiopoulus et al. [2] reported that by the dealuminated process using acid treatment (HCl) was able to reform the pore structure and removed the amorphous Al-species. This caused the crystallinity of zeolite increased as well as influenced in their porosity. Based on the literature review, in this research was tested the effect of acid treatment in natural zeolite on the methylene blue dye removal. Furthermore, the characteristic of natural zeolite after treatment was analyzed through FTIR (Fourier transform infrared), XRD (X-ray diffraction) and SEM (scanning electron microscopy).

2. Experimental

2.1. Materials
Natural zeolite was collected from Klaten, Indonesia. Hydrochloric acid (HCl), sodium hydroxide (NaOH) and ammonium chloride (NH_4Cl) were purchased from Merck. The methylene blue dye was commercially available from Surakarta, Indonesia. The distilled water was supplied from Chemistry Laboratory of Sebelas Maret University.

2.2. Preparation of Acid-Activated Natural Zeolite (Ac-Zeo)
Natural zeolite was crushed into powder form and sieved to obtain ± 200 mesh of particle size. The zeolite powder was refluxed with hydrochloric acid 50% at 50°C for 1 hour. Afterward, the material was filtered and washed with distilled water until pH 7. The sample was recorded as acid activated-natural zeolite (Ac-Zeo). The Ac-Zeo was then immersed in 1 M ammonium chloride solution for five days. Then, the Ac-Zeo was washed several times using distilled water and dried at 60 °C for 24 hours. The Ac-Zeo was stored in vacuum desiccator before used.

2.3. Characterization of Ac-Zeo
The Ac-Zeo was characterized using Fourier transform infra-red (FTIR) spectroscopy IR Prestige 21 SHIMADZU. The sample was prepared in pellet form and recorded in the range of 400-4000 cm^{-1} wave number. The scan number was set up to 45 times with resolution 2 cm^{-1}. The crystallinity of Ac-Zeo was studied using X-Ray Diffraction (XRD) model Shimadzu XRD Lab-X 6000. The morphology of Ac-Zeolite was observed using Scanning Electron Microscopy (SEM) model Phenom FEI.

2.4. Methylene Blue Adsorption

IOP Conf. Series: Materials Science and Engineering **172** (2017) 012039 doi:10.1088/1757-899X/172/1/012039

2.4.1. Effect of pH. A Methylene blue solution was prepared into 30 mg.L^{-1} of concentration by dissolving 30 mg dye with 1 L distilled water. About 0.02 g of Ac-Zeo was added into 10 mL methylene blue solution with various pH (3, 5, 7, 9, and 11). Then, the mixture was shaken about 30 minutes at room temperature. The dye solution was filtered immediately and analyzed using UV-Vis Perkin Elmer Lambda 25 Spectroscopy.

2.4.2. Effect of Contact Time. Methylene blue solution was adjusted at optimum pH. About 0.02 g of Ac-Zeo was added into methylene blue solution (10 mL, 30 mg.L^{-1}) and shaken for 10, 20, 30, 40 and 60 minutes. After adsorption process, the dye solution was filtered immediately. The Adsorbed dye was calculated using UV-Vis PerkinElmer Lambda 25 spectroscopy.

2.4.3. Effect of Initial Dye Concentration. Methylene blue solution was prepared at 10, 20, 30, 40 and 50 mg.L^{-1} and conditioned at optimum pH condition. About 10 mL of dye solution was added into an erlenmeyer containing 0.02 g of Ac-Zeo and shaken for 30 minutes. The solution was filtered and analyzed using UV-Vis PerkinElmer Lambda 25 spectroscopy.

3. Results and Discussion

3.1. Material Characterization

Quantitative analysis of natural zeolite and the activated product was conducted using XRD. The result showed that natural zeolite was classified as zeolite mordenite. Figure 1 describes the comparison of natural zeolite and Ac-zeo with the standard diffraction pattern of zeolite mordenite (Zeo-mordenite) from the International Zeolite Association [40]. As we can see in Figure 1, there is no significant effect on the pattern of natural zeolite after the acid activation. However, natural zeolite has broad peak at 2θ 5.4o which indicates the amorphous phase. After acid treatment, the pattern shows that broad peak has vanished. According to Elaiopoulus et al. [2], the acid treatment will eliminate the amorphous Al^{3+}-species in natural zeolite structure. The eliminating of amorphous Al-species (dealumination) could increase the crystallinity of zeolite. Consequently, it will increase the pore in zeolite and it can be used as an active site for organic dye molecules binding.

Chemical analysis of zeolite and its modification was conducted using FTIR. This technique is very important to analyze the functional group of each material. Moreover, it was also used by the researcher for proving the chemical interaction in compounding, to analysis the material structure and to obtain information about the ions substitution in the material like zeolite [2, 33]. The FTIR spectra of the natural zeolite and acid-treated zeolite are shown in figure 2. It showed that natural zeolite which has classified as mordinite generally consists of Si-O tetrahedral layer and Al-OH octahedral layer. It was proven by the absorption band appeared at 1042 cm^{-1} which correspond to the stretching vibration band of Si (Al)-O [34, 35]. After acid treatment, the effect can be seen as a shift in the vibration position at 1042 cm^{-1} to 1086 cm^{-1}. The shifting was due to dealumination process [8]. It also confirmed by X-ray diffractogram (figure 1). The adsorption peak at 798 cm^{-1} is characteristic vibration of –OH deformation in Al-OH. The vibration of Si-O-Al deformation observed as a band at 530 cm^{-1}, while the vibration of Si-O-Si deformation is also found at 456 cm^{-1}. A broad peak appearing at 3438 cm^{-1} is assigned to hydrogen bonding of OH in zeolite framework. Moreover, the peaks at 3630 cm^{-1} and 1637 cm^{-1} are typical adsorption band of the water molecules in the zeolite. As we know, the presence of water could poison the active sites of zeolite. Adding acid into zeolite could make the pore wide open and clean from impurities. It was proven after acid treatment intensity at 3630 cm^{-1} and 1637 cm^{-1} decreased. The Infrared spectrum of Ac-Zeo has absorbance band at 3156 cm^{-1} and 1400 cm^{-1}, which characteristic of NH stretching and bending vibration which generated from NH_4^+ ion exchange. The presence of NH_4^+ ions could increase the acid site of zeolite.

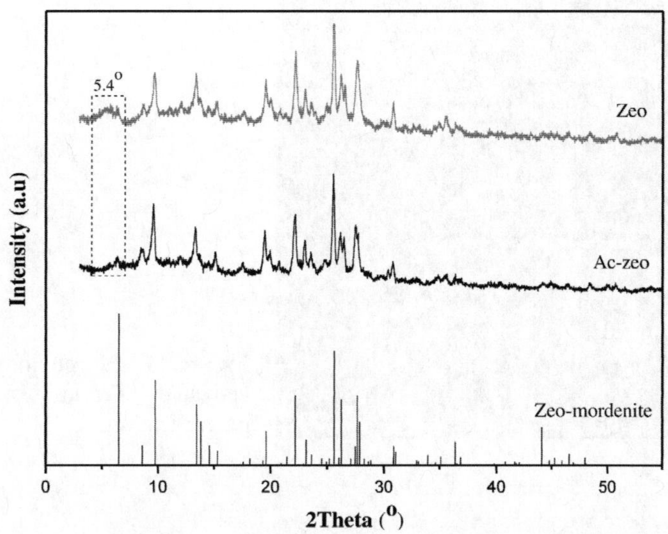

Figure 1. XRD pattern of Zeolite and Ac-Zeo.

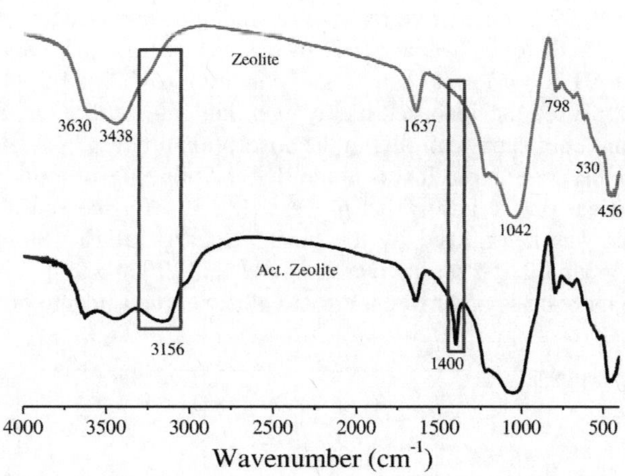

Figure 2. Infra-red spectrum of zeolite and Ac-Zeo.

Morphological study of Ac-Zeo surface was examined using scanning electron microscopy (SEM). Figure 3 showed the morphology of the Ac-Zeo with 40000 times magnification. Based on the SEM image, it was clearly seen that the physical structure of the Ac-Zeo like a tubular rod.

Figure 3. Morphology of Acid-activated Zeolite with 40000x magnification

3.2. Effect of pH

The pH played an important role in the adsorption process [34]. However, in our study, the dye removal percentages at various pH condition were not significantly different, where at acid or basic condition, it has high dye removal percentages which are more than 90%. Figure 4 shows the dye removal in various pH condition with the best condition is achieved in acid condition (pH 3). In the pH range of 3-5, the methylene blue dye removal slightly decreases, where at pH 3 the dye removal is 97.9% while at pH 5 reach 97.6%. The dye removal of methylene blue decreases gradually at pH 7, which is 15.6% compared to the condition at pH 3. At basic condition (pH 9 and 11), the dye removal increases compared to at pH 7, but lower than at acid condition (pH 3 and 5). The other investigation [7, 36], the increasing of pH led the adsorbed methylene blue onto zeolite increased because at lower pH the presence of H^+ ions competed with MB in the adsorption process. However, based on this data, the lower of pH gave the best condition. It was noted that the presence of a porous structure of zeolite also played an important part in the adsorption process [31, 37]. By the acid treatment, the negative charge in zeolite surface was neutralized by the positive charge of H^+. Moreover, the addition of ammonium chloride compound in the zeolite increased the acid site of zeolite. Thus, at acid condition, the MB is possible to be more basic and make it easy to attach at the acid site of zeolite.

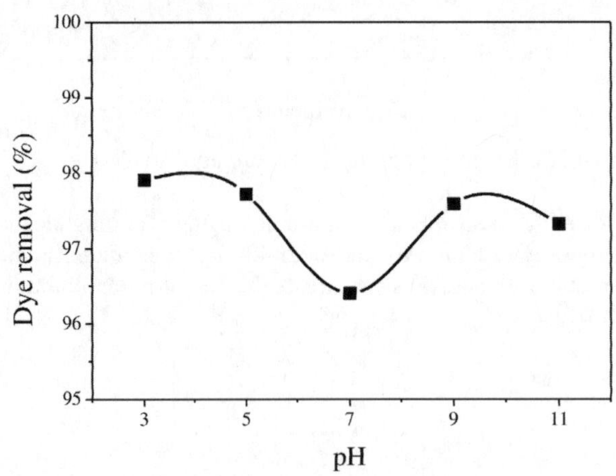

Figure 4. Effect of pH condition on dye removal.

IOP Conf. Series: Materials Science and Engineering **172** (2017) 012039 doi:10.1088/1757-899X/172/1/012039

3.3. Adsorption Kinetic

The effect of contact time on methylene blue adsorption onto zeolite can be seen in figure 5. In the beginning, the dye removal reaches more than 90%, and it quickly increases after 20 minutes of applied contact time by 98.6%. The mechanism of methylene blue adsorption follows two phases. At the first stage, the methylene blue adsorption is fast due to high chemical and physical contact through the zeolite pores. At second phase, the adsorption is slow because the methylene blue adsorbed onto zeolite reached an equilibrium or the concentration of methylene blue was relatively small [7,38]. The adsorption of methylene blue onto zeolite has reached an equilibrium state at 40-60 minutes. By acid treatment, the ability of zeolite to adsorb methylene blue increased, compared to Han et al. [7] work, which reached equilibrium state at 820 minutes.

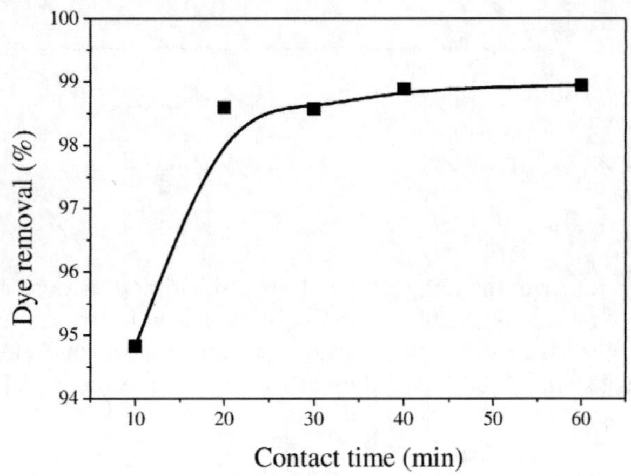

Figure 5. Effect of contact time on dye removal.

A simple kinetic adsorption can be described using a linear form of pseudo-second order using equation 1 [31]:

$$\frac{t}{Qt} = \frac{1}{k.Qet^2} + \frac{1}{Qet}t \tag{1}$$

where, t is the contact time (min), Qt is the adsorption capacity at t time ($mg.g^{-1}$), k is the rate constant of pseudo-second order ($g.mg^{-1}.min^{-1}$) and the Qet is adsorption capacity at equilibrium ($mg.g^{-1}$).

Figure 6 shows the linear form of the second-order kinetic of methylene blue dye adsorbed onto zeolite. From t versus t/Qt plot pseudo second order model, it can be obtained the kinetic rate constant (k) and the adsorption capacity at equilibrium (Qet). The k and Qet of methylene blue adsorbed onto Ac-Zeo are 0.1872 $mg.g^{-1}min^{-1}$ and 14.94 $mg.g^{-1}$, respectively. The correlation coefficients (R2) obtained by using the linear form of pseudo-second order model is closed in 0.99995, indicating the models is applicable for the adsorption. A similar result has been found by Wang *et al.* [31] on methylene blue dye adsorbed onto zeolite.

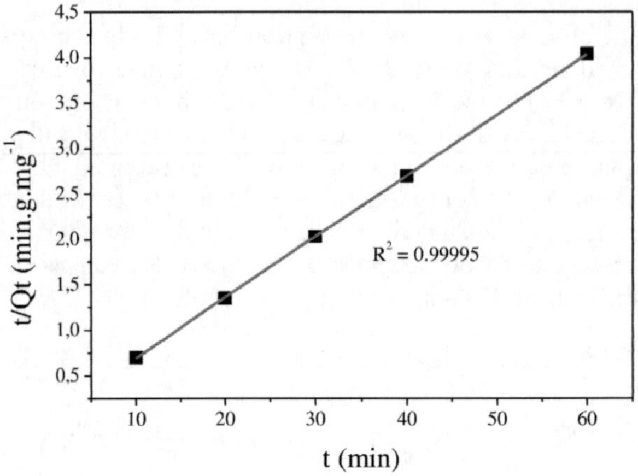

Figure 6. Effect of contact time on dye removal.

3.4. Adsorption Isotherm

To study the adsorption isotherm, the Langmuir and Freundlich models were applied in this research. Moreover, the Dubinin–Kaganer–Radushkevich (DKR) model was also studied to clarify the energy sorption in methylene blue adsorbed onto Ac-Zeo. The linear form of Langmuir, Freundlich, and Dubinin–Kaganer–Radushkevich Models equation are described in equation 2 [7], 3 [31], and 4 [39].

$$\frac{Co}{qe} = \frac{1}{K_L.Qm} + \frac{1}{Qm}Co \qquad (2)$$

$$\ln qe = \ln K_F + \frac{1}{n}\ln Co \qquad (3)$$

$$\ln Qe = \ln Qs - \beta\epsilon^2 \qquad (4)$$

Where, Co is initial dye concentration (mg.L^{-1}), qe and Qm is adsorption capacity (mg.g^{-1}) at equilibrium and maximum, respectively. Qe is the amount of methylene blue dye adsorbed in mmol.Kg^{-1}. Qs is the DKR monolayer capacity (mmol.Kg^{-1}), β is energy sorption constant (mol^2.J^{-2}), and ε is Polanyi potential which calculated using Equation 5.

$$\epsilon = RT\ln\frac{1}{Co} \qquad (5)$$

Energy sorption can be calculated following equation 6.

$$E = \frac{1}{\sqrt{2\beta}} \qquad (6)$$

Figure 7, 8 and 9 shows the linearized Langmuir, Freundlich, and Dubinin–Kaganer–Radushkevich adsorption isotherms of methylene blue on Ac-Zeo. The parameters of all models, such as the adsorption maximum capacity, n value, energy sorption, the isotherm constants and correlation coefficients are presented in Table 1. The R^2 of Freundlich model is higher than other models, which is 0.999. It means that the adsorption process well fit to Freundlich model. Thus it followed the Freundlich mechanism, similar to Alpat *et al.* [30] works.

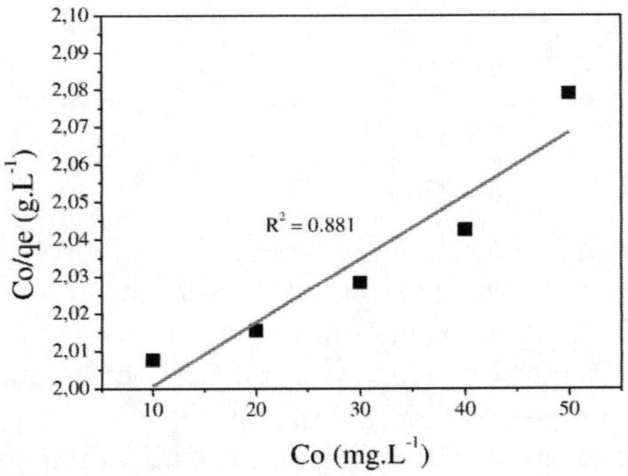

Figure 7. Langmuir isotherm model Co versus Ce/qe.

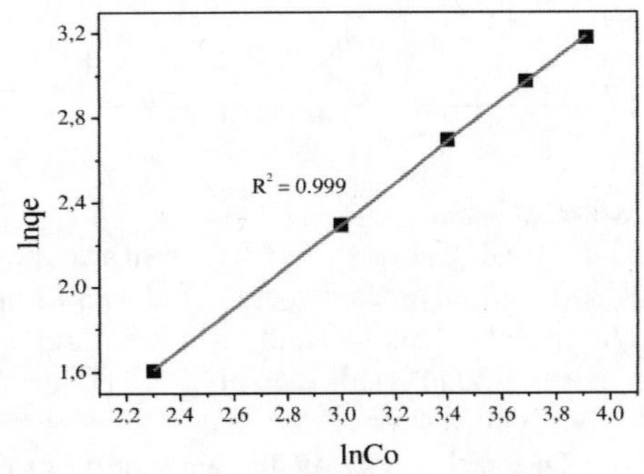

Figure 8. Freundlich isotherm model lnCo versus lnqe.

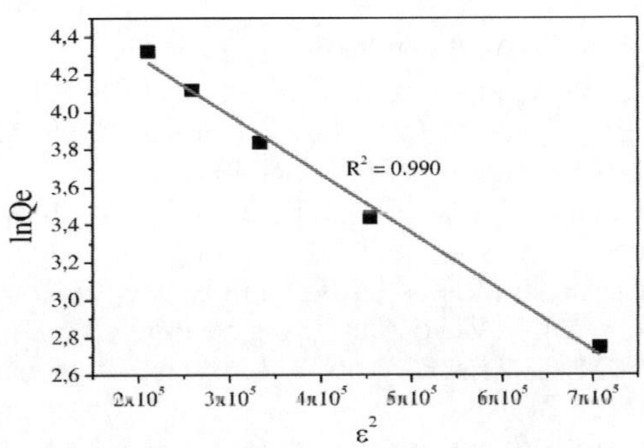

Figure 9. Dubinin–Kaganer–Radushkevich isotherm model ε^2 versus lnQe.

From the Langmuir model, it can be obtained the adsorption maximum capacity by plotting Co versus Co/qe [7, 30, 31]. The adsorption capacity maximum is 588.24 mg.g^{-1}. The Freundlich model was generated by the heterogeneity factor of adsorption sites (n). The 1/n value of methylene blue adsorbed onto Ac-Zeo is 0.98 ($0<1/n<1$), indicating favorability and higher adsorption ability of methylene blue on zeolite [7]. The DKR monolayer adsorption capacity is 137.36 mmol.Kg^{-1}, while the energy sorption of methylene blue is 0.399 Kj.mol^{-1}, indicating physical sorption [39].

Table 1. Some parameters on Isotherm adsorption.

Isotherm	Parameters	Value
Langmuir	Qm (mg.g^{-1})	588.24
	K_L (L.mg^{-1})	8.57 x 10^{-4}
	R^2	0.881
Freundlich	1/n	0.98

	K_F ((mg.g^{-1})(mg.L^{-1})^n)	0.52
	R^2	0.999
Dubinin–Kaganer–Radushkevich	Qs (mmol.Kg^{-1})	137.36
	β (mol^2.J^{-2})	3.13 X 10^{-6}
	E (kJ.mol^{-1})	0.399
	R^2	0.990

4. Conclusions

The acid treatment of zeolite has been conducted and caused an increasing of zeolite crystallinity. The Ac-Zeo has the porous structure and acts as an active site for methylene blue dye adsorption. The pH condition did not significantly influence the methylene blue dye removal. However, the best condition was given at acid condition (pH 3). The adsorption equilibrium was reached at range 40-60 minutes with dye removal was more than 90%. The rate constant of pseudo-second order (k) is 0.1872 mg.g^{-1}min^{-1}. The isotherm study and adsorption mechanism followed the Freundlich model with a favorable mechanism. The maximum and DKR monolayer adsorption capacities are 588.24 mg.g^{-1} and 137.36 mmol.Kg^{-1}, respectively. The adsorption mechanism is physical adsorption with energy sorption of 0.399 kJ.mol^{-1}.

References

[1] Hernawan, Wahono S K, Maryana R and Pratiwi D 2015 *Energy Proced.* **65** 116
[2] Elaiopoulos K, Perraki T and Grigoropoulou E 2010 *Micropor. Mesopor. Mater.* **134** 29
[3] Castaldi P, Santona L, Enzo S and Melis P 2008 *J. Hazard. Mater.* **156** 428
[4] Bowman R S 2003 *Micropor. Mesopor. Mater.* **61** 43
[5] Johan E, Yamada T, Munthali M W, Corner P K, Aono H and Matsue N 2015 *Proced. Environ. Sci.* **28** 52
[6] Erdem E, Karapan N and Donat R 2004 *J. Colloid Interface Sci.* **280** 309
[7] Han R, Zhang J, Han P, Wang Y, Zhao Z and Tang M 2009 *Chem. Eng. J.* **145** 496
[8] Sriningsih W, Saerodji M G, Trisunaryanti W, Triyono, Armunanto R and Falah I I 2014 *Proced. Environ. Sci.* **20** 215
[9] Ren H, Jiang J, Wu D, Gao Z, Sun Y and Luo C 2016 *Water Air Soil Pollut.* **227** 101
[10] Hung H and Lin T F 2006 *J. Hazard. Mater.* **135** 210
[11] Wang Y, Kmiya Y and Okuhara T 2007 *Water Res.* **41** 269
[12] Shen Y, Fan C C, Wei Y Z, Du J, Zhu H B and Zhao Y 2016 *Dalton Trans.* **45** 10909
[13] Ngah W S, Teong L C and Hanafiah M A K M 2011 *Carbohydrate Polym.* **83** 1446
[14] Moussavi G and Mahmoudi M 2009 *J. Hazard. Mater.* **163** 806
[15] Srinivasan A and Viraraghavan T 2010 *J. Environ. Management.* **91** 1915
[16] Yao Y, Xu F, Chen M, Xu Z and Zhu Z 2010 *Bioseour. Technol.* **101** 3040
[17] Crini G and Badot P M 2008 *Prog. Polym. Sci.* **33** 399
[18] Lenzy G G, Evangelista R F, Duarte E R, Colpini L S M, Fornari A C, Neto R M, Jorge L M M and Santos O A A 2016 *Desalination and Water Treatment* **57(30)** 14132
[19] Labanda J, Sabate J and Liorens J 2009 *J. Membrane Sci.* **340** 234
[20] Mittal A, Jain R, Mittal J, Varshney S and Sikarwar S 2010 *Int. J. Environ. Pollut.* **43** 308
[21] Shi B, Li G, Wang D, Feng H and Tang H 2007 *J. Hazard. Mater.* **143** 567
[22] Akbari A, Remigy J C and Aptel P 2002 *Chem. Eng. Process.* **41** 601
[23] Pia A B, Roca J A M, Miranda M I A, Clar A I and Clar M I I 2003 *Desalination* **157** 73
[24] Lazim Z M, Mzuin E, Hadibarata T and Yusop Z 2015 *J. Teknologi* **74(11)** 129
[25] Mahmoodi N M and Arami M 2009 *Chem. Eng. J.* **146** 189
[26] Kiakhani M S, Arami M and Gharanjig K 2013 *J. Environ. Chem. Eng.* **1** 406
[27] Esquerdo V M, Cadaval Jr T R S, Dotto G L and Pinto L A A 2014 *J Colloid Interface Sci.* **424** 7

[28] Gurses A, Dogar C, Yalcin M, Acikyildiz M, Bayrak R and Karaca S 2006 *J. Hazard. Mater.* **131** 217
[29] Han R, Wang Y, Zou W, Wang Y and Shi J 2007 *J. Hazard. Mater.* **145** 331
[30] Alpat S K, Ozbayrak O, Alpat S and Akcay H 2008 *J. Hazard. Mater.* **151** 213
[31] Wang S and Zhu Z H 2006 *J. Hazard. Mater.* **B136** 946
[32] Ren H, Jiang J, Wu D, Gao Z, Sun Y and Luo C 2016 *Water Air Soil Pollut.* **227** 101
[33] Favvas E P, Tsanaktsidis C G, Sapalidis A A, Tzilantonis G T, Papageorgiou S K and Mitropoulos A A 2016 *Microporous Mesoporous Mater.* **225** 385
[34] Alshameri A, Yan C and Lei X 2014 *Microporous Mesoporous Mater.* **196** 145
[35] Perraki T and Orfanoudaki A 2004 *Appl. Clay. Sci.* **25** 9
[36] Jin X, Jiang M Q, Shan X Q, Pei Z G and Chen Z 2008 *J. Colloid Interface Sci.* **328** 243
[37] Rafatullah M, Sulaiman O, Hashim R and Ahmad A 2010 *J. Hazard. Mater.* **177** 70
[38] Heraldy E, Santosa S J, Triyono and Wijaya K 2015 *Indo. J. Chem.* **15(3)** 234
[39] Dada A O, Olalekan A P, Olatunya A M and Dada O 2012 *IOSR J. Appl. Chem.* **3(1)** 38
[40] Treacy M M J and Higgins J B 2001 Collection of Simulated XRD Powder Patterns fo Zeolites 4th edition Elsevier

Acknowledgement
Author(s) would like to acknowledge to Indonesian General of Higher Education for financial support through PKM-Dikti grants 2016.

IOP Conf. Series: Materials Science and Engineering **172** (2017) 012040 doi:10.1088/1757-899X/172/1/012040

Ab initio computational study of reaction mechanism of peptide bond formation on HF/6-31G(d,p) level

P Siahaan[1*], M N T Lalita[1], B Cahyono[1], M D Laksitorini[2], S Z Hildayani[1]

[1] Diponegoro University, Semarang, Indonesia;
[2] Gadjah Mada University, Yogyakarta, Indonesia.

Email: siahaan.parsaoran@live.undip.ac.id; siahaan_parsaoran@yahoo.com

Abstract. Peptide plays an important role in modulation of various cell functions. Therefore, formation reaction of the peptide is important for chemical reactions. One way to probe the reaction of peptide synthesis is a computational method. The purpose of this research is to determine the reaction mechanism for peptide bond formation on Ac-PV-NH_2 and Ac-VP-NH_2 synthesis from amino acid proline and valine by ab initio computational approach. The calculations were carried out by theory and basis set HF/6-31G(d,p) for four mechanisms (path 1 to 4) that proposed in this research. The results show that the highest of the rate determining step between reactant and transition state (TS) for path 1, 2, 3, and 4 are 163.06 kJ.mol^{-1}, 1868 kJ.mol^{-1}, 5685 kJ.mol^{-1}, and 1837 kJ.mol^{-1}. The calculation shows that the most preferred reaction of Ac-PV-NH_2 and Ac-VP-NH_2 synthesis from amino acid proline and valine are on the path 1 (initiated with the termination of H^+ in proline amino acid) that produce Ac-PV-NH_2.

1. Introduction

The application of peptide as a drug to treat brain diseases is hampered by the difficulty in the delivery system of peptide past the Blood-Brain Barrier (BBB). [1,2] In general, there is only one way that can be passed by peptide molecules that is a paracellular pathway. However, this pathway is blocked by a tight junction which is the result of cadherin-cadherin interactions between cells. One way to open this pathway is to modulate cadherin protein by their peptide derivatives synthesis. Peptide syntheses that is used to modulate these interactions are ADT (Ac-QGADTPPVGV-NH2) and HAV (Ac-LFSHAVSSNG-NH2) which are derived from the bulge and groove region on EC1 domains. [3,4] The sequence of amino acids in the peptide domain EC1 or ADT contained amino acids proline (P) and valine (V) with proline-valine (Ac-PV-NH2) sequence are tend to be formed, while the sequence valine-proline (Ac-VP-NH2) is not formed. Based on this fact, there is a preferred reaction mechanism in peptide synthesis from proline and valine amino acid. So it is important to learn that mechanisms on the thermodynamics and kinetics aspect. The reaction of proline and valine peptide synthesis is a reaction between two active groups on the amino acids, the amine group, and the carboxyl group. [5-10] Therefore there are several possible products that can be formed, that are Ac-PV-NH2and Ac-VP-NH2. Each product will be through different mechanism reactions. In this research, there are four reaction mechanisms (path 1 to 4) which allow for the synthesis of peptides from proline and valine amino acid (Scheme 1).

One method that can be used to study the reaction mechanism of peptide synthesis is ab initio computational methods. [11-18] Computational method has a high level of accuracy. In the present study, we used the level of theory and basis set HF / 6-31G (d, p) to calculate the molecular energy of the reactants, products, intermediates and transition state (TS). Data of energies that obtained were used to determining the reaction mechanism are preferred.

2. Computational Method

This research is done using ab initio method on HF/SCF string theory and 6-31G(d,p) basis set. Calculating software that used are Gaussian03 (Linux operating system) Meanwhile, Gauss view05, Chemcraft, Avogadro, and Jmol are used as visualization software. File input construction are done using notepad++.

The calculation is done on each single and transition state molecules. The directive on determining transition state is QST3 and the command of "freq" is used to obtain the vibration frequencies from each pertinent molecules. [19-23]

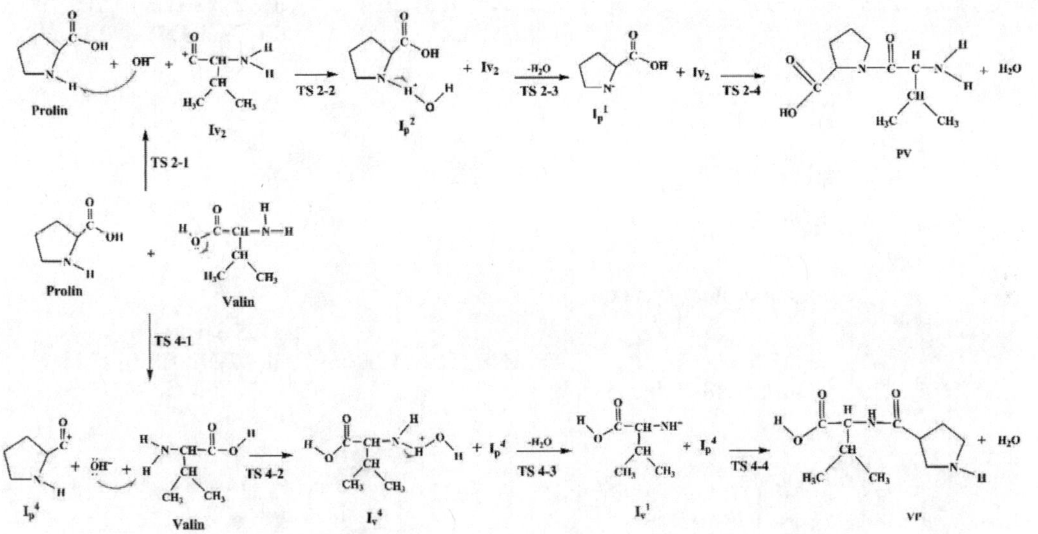

Scheme 1. Mechanism reaction of Ac-PV-NH_2 and Ac-VP-NH_2 peptide in various pathway

3. Result And Discussion

Path 1 and 3 is a reaction pathway that begins with the termination of hydrogen bonds in the amine group. The difference between them are situated on compounds that initiate the reaction, and then resulting product is different. Termination of hydrogen bonds on path 1 generate an intermediate I_p^1 with the enthalpy is 414 kcal.mol^{-1}, while the enthalpy formation of the molecular path 3 generates the intermediate I_v^3 at 416 kcal.mol^{-1}. The thermodynamic data shows that the intermediate I_p^1 is more easily formed than I_v^3. The stability of the molecule is evidenced by the distribution of the charge on each constituent atom of the compound.[5,6]

Termination of hydrogen bonding to the amine group can be shown by changes in some parameters of the molecular structure that involved in this reaction stage (figure 1). The most significant changes can be seen in the bond length in the reaction area (N-H). The N-H bond length on the proline molecule is 0.997 Å while in the transition state structure (TS 1-1) was obtained 4.101 Å. Similarity occurs on molecule valine which the N-H bond from 1 Å to 1.009 Å on its transition state. This condition strengthened by the imaginary vibrational frequencies on the center reaction area (υ=-40.88 cm^{-1}). The molecular structure of TS 3-1 also indicates the imaginary vibrational frequency in the central area of the reaction(υ=-191.41cm^{-1}).

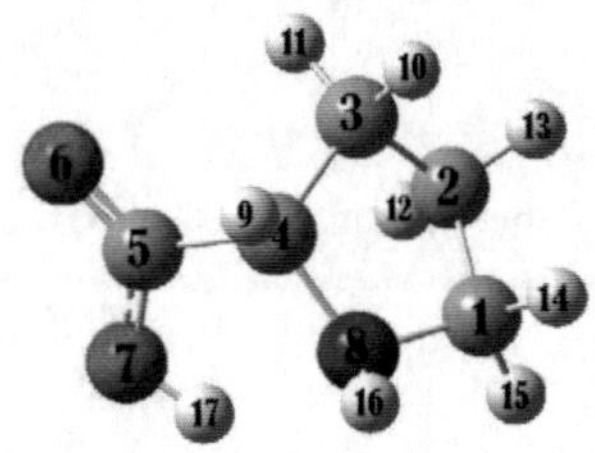

a) Proline
r(N8-H16) 0.997 Å α(C1-N8-H16) 113°
r(C5-O7) 1.317 Å α(C4-C5-O7) 115°
r(C5-O6) 1.185 Å

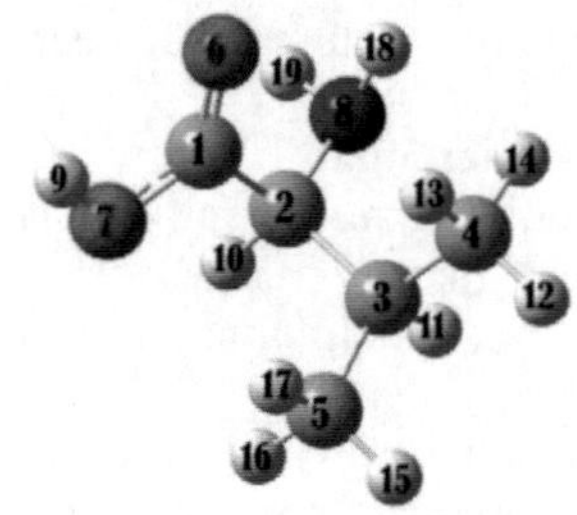

a) Valine
r(N8-H19) 1 Å r(C1-O7) 1.328 Å
r(N8-H18) 1 Å α(C2-N8-H19) 110°
r(C1-O6) 1.189 Å α(C2-C1-O7) 113°

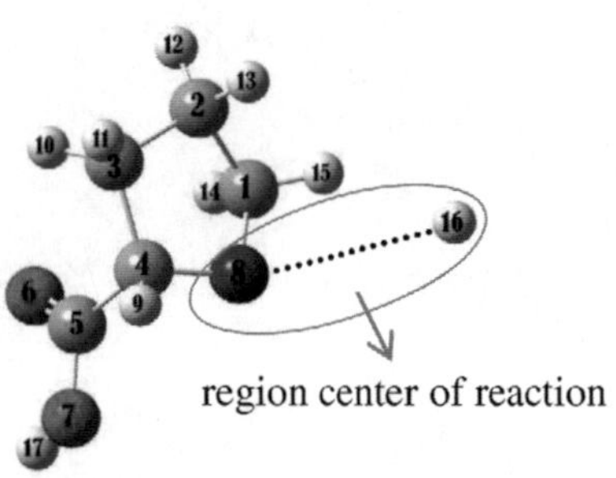

b) TS 1-1
r(N8-H16) 4.101 Å α(C1-N8-H16) 98°

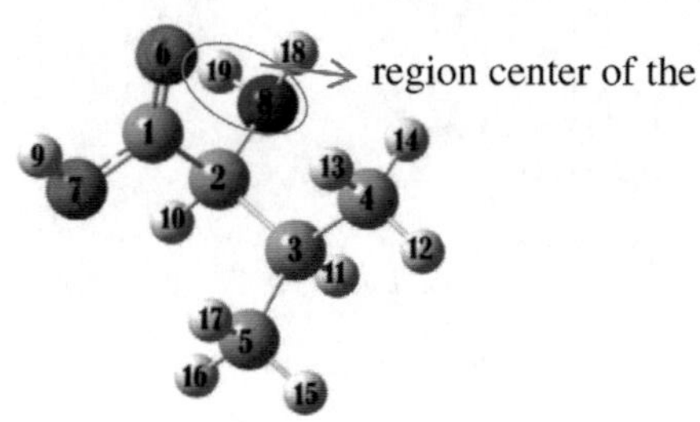

c) TS 3-1
r(N8-H19) 1.009 Å α(C2-N8-H19) 122°

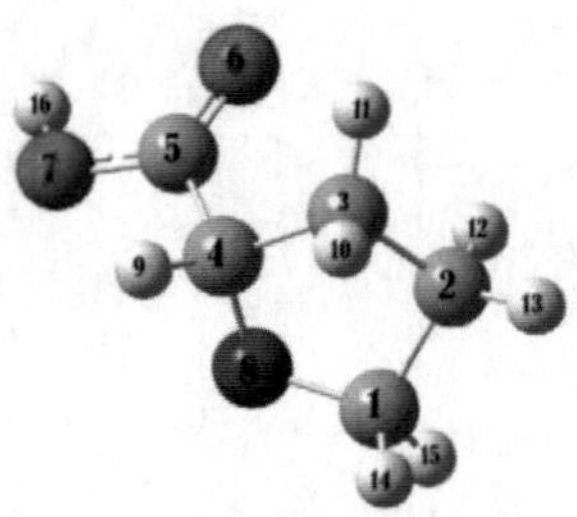

e) I_p^1
r(C1-N8) 1.426 Å α(C1-N8-C4) 106°
r(N8-C4) 1.441 Å α(N8-C4-C5) 104°

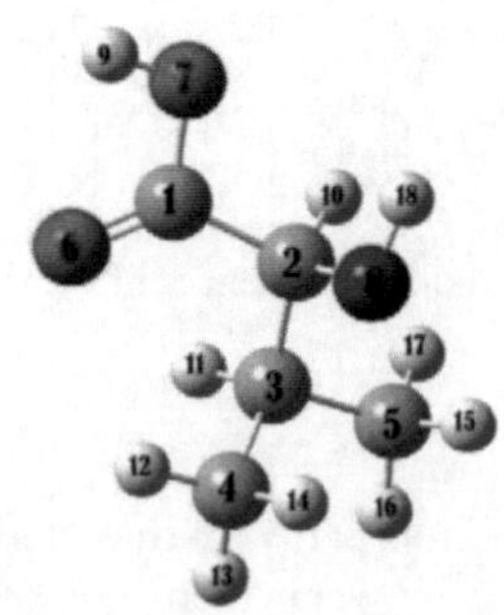

f) I_v^3
r(C2-N8) 1.428 Å α(C1-C2-N8) 109°
r(N8-H18) 1.011 Å τ(C1-C2-N8-H18) -75°

Figure 1. Calculated structures of pertinent stationary points along the first step

In the insertion reaction of H^+ on the path 1 produces intermediate I_v^1, while on the path 3 produces I_p^3. The calculations show the energy of I_v^1 is -1.05×10^6 kJ.mol^{-1} and the enthalpy is -174.9 kcal.mol^{-1}. It also refers to the energy of I_p^3 that is -1.05×10^6 kJ.mol^{-1} with the enthalpy is -150.7 kcal.mol^{-1}. The value of negative enthalpy indicates an exothermic reaction, which is the energy to be used to break the ties.

Both of this intermediates contain H_2O group which is a good leaving group, so this reaction step tended to release the H_2O group. It is proved that the intermediates have high reactivity. This second reaction step also involves the large activation energy (Ea). The activation energy that has passed by valine to become I_v^1 is 398.83 kJ.mol^{-1} and the Ea of proline to become I_p^3 is 397.74 kJ.mol^{-1}. The transition state in the second reaction step (TS 1-2) was found an imaginary vibrational frequency on the center reaction area, υ = -1096.69 cm^{-1}. It also occurs on proline, the reaction between H^+ ions with carboxylate groups on valine shows the imaginary vibration frequencies, υ = -246.89cm^{-1}).

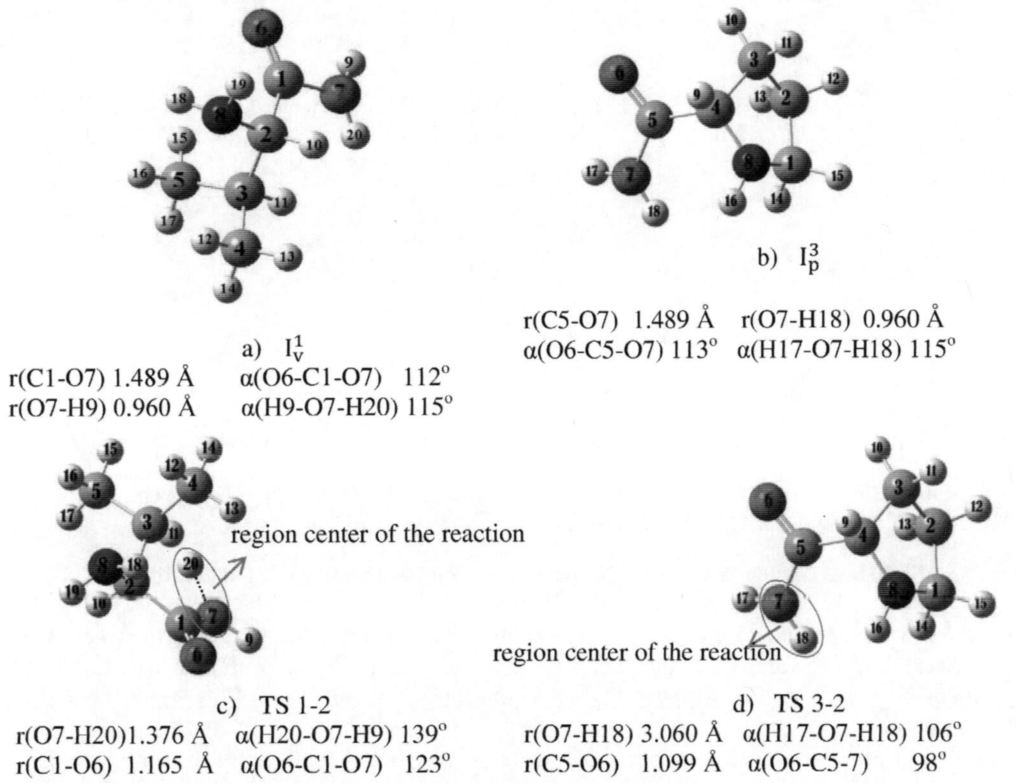

Figure 2. Calculated structures of pertinent stationary points along the second step

Furthermore, in the reaction step of releasing H_2O produces intermediates that are I_v^2 in the path 1, while on the path 3 produces I_p^4. The enthalpy of I_v^1 to become I_v^2 is 47,730 kcal.mol^{-1}, while the enthalpy of I_p^3 to become I_p^4 is 47,690 kcal.mol^{-1}. The high enthalpy value in intermediates is caused the molecular energy of this intermediates are high. It indicated that the intermediates are unstable which the molecular energy of I_v^2 and I_p^4 are -851,000 kJ.mol^{-1} and -848,000 kJ.mol^{-1}. The activation energy which is traversed by this reaction steps higher than the previous reaction step.

In the process of bond termination conduces change of some parameters of intermediates structure with each transition state (figure 3). The bond length of C-O in the transition state structure of I_p^3 to become I_p^4 (TS 3-3) is 2.07 Å, while the transition state structure of I_v^1 to become I_v^2 (TS1-3) not be found. The transition state structure of I_p^3 to become I_p^4 has an imaginary frequency that characterized by the value of wavenumber (υ=-85.941 cm^{-1}).

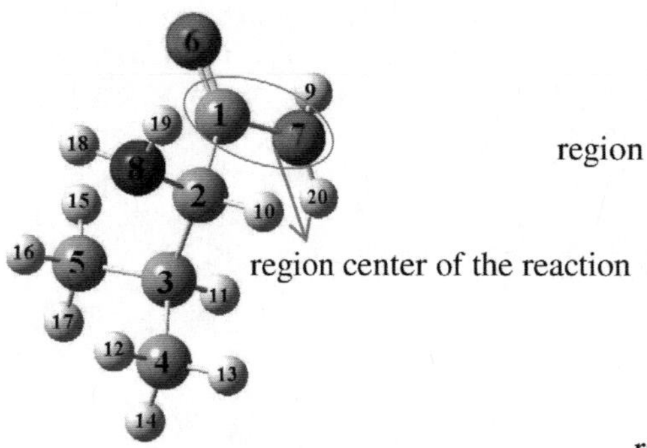

a) TS 1-3

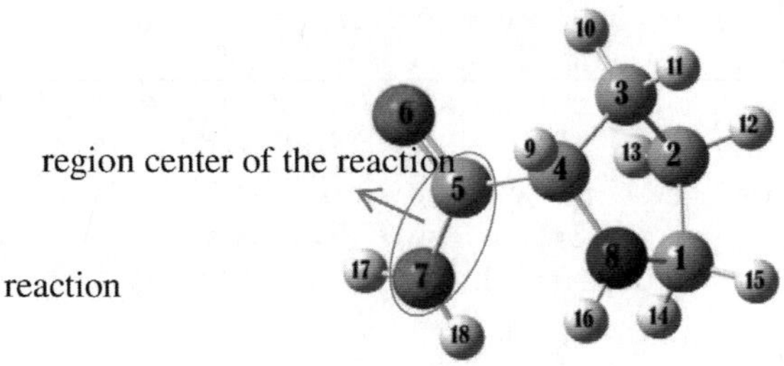

b) TS 3-3

r(C5-O7) 2.076 Å α(C4-O5-O7) 93°
r(C5-O6) 1.121 Å α(O6-C5-O7) 104°

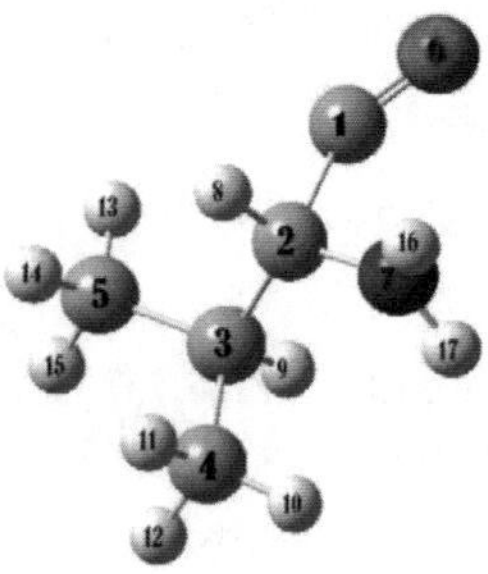

c) I_v^2

r(C1-O6) 1.2 Å α(C2-C1-O6) 140°
r(C2-C1) 1.616 Å τ(C3-C2-C1-O6) -143°

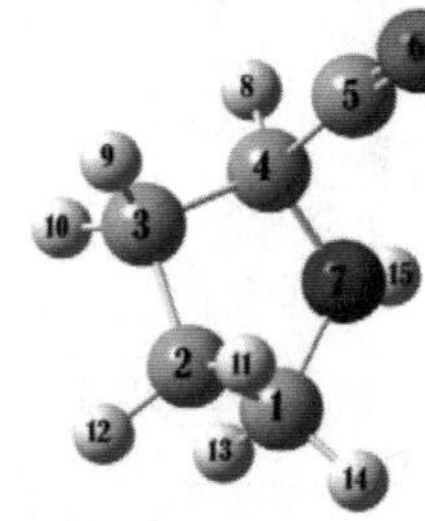

b) I_p^4

r(C5-O6) 1,2 Å α(C4-C5-O6) 174°
r(C4-C5) 1,611 Å τ(C3-C4-C5-O6) 68°

Figure 3. Calculated structures of pertinent stationary points along the third step

On the last step, peptides are formed by an establishment of C-N bond from intermediates I_v^2 with I_p^1 in the path 1 and I_p^4 with I_v^3 in the path 3. The product that produces from path 1 is Ac-PV-NH2, while on path 3 produce Ac-VP-NH2. The enthalpy of Ac-PV-NH2 formation on the path 1 is -257.58 kcal.mol^{-1}, while the enthalpy of Ac-VP-NH2 formation on path 3 is -253.50 kcal.mol-1. In the reaction of peptide formation is found an imaginary vibrational frequency that can be seen by the value of the wave number. The imaginary vibration frequency in the Ac-PV-NH2 formation is -211.997cm^{-1}, while in the Ac-VP-NH2 is -6.18 cm^{-1}.

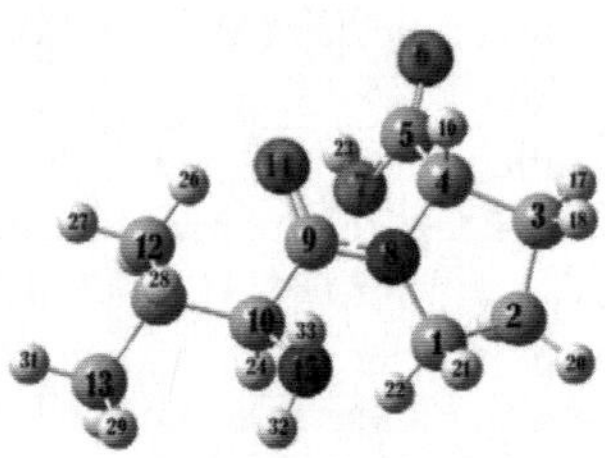

a) Ac-PV-NH_2
r(C9-N8) 1,352 Å α(N8-C9-C10) 118°

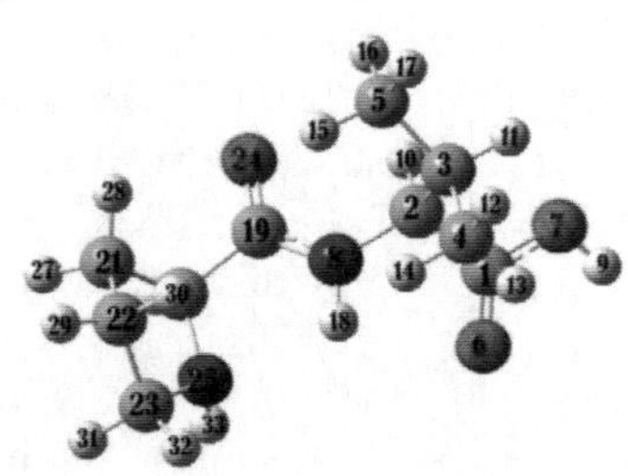

b) Ac-VP-NH_2
r(C19-N8) 1.345 Å α(N8-C19-O24)124°

Figure 4. Calculated structure of stationary points

Overall, the reaction on path 1 needed an enthalpy 47,708 kcal.mol^{-1}. It is indicated endothermic reaction occurs. In the path 3 the entirety of enthalpy is 47,702 kcal.mol^{-1}. It also indicated that the reaction in path 3 is an endothermic reaction. However, the enthalpy that is traversed by the path 1 is lower than path 3. The lowest of activation energy in path 1 are on step 3. Similarly, it is obtained at step 3 in the path 3. According to the Arrhenius equation, $k = Ae^{-E_a/RT}$, getting smaller of activation energy will generate the greater of the rate constant, which v=k[A][B] so the rate become faster, according to esterification reaction [5,6].

Based on the computational calculation, the value of each step on the path 1 and 3 (Figure 5) shows that the path 1 is a reaction pathway that is more likely than the path 3 because on the third step in path 3 obtained elevated activation energy (Ea) that is 5,685 kJ.mol^{-1}. Therefore, by following the reaction in path 1, the products tend to be stable and easier to obtain Ac-PV-NH2.

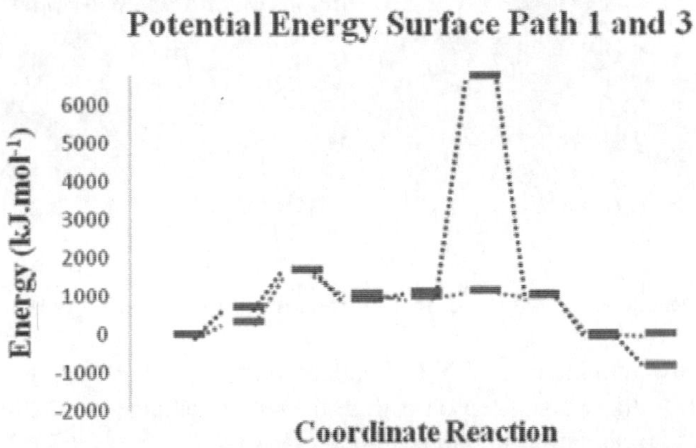

Figure 5. Calculated reaction path 1 and 3

▬ path 1 ▬path 3

In the path 2 and 4 is begins with the termination C-O bonds in the carboxylic group. Dissolution of C-O bond in path 2 produce intermediate I_p^2 with the enthalpy is 47,550 kcal.mol^{-1}, while in the path 4 produce intermediate I_p^4 with the enthalpy is 47,540 kcal.mol^{-1}. Based on the enthalpy that is obtained in this reaction, it requires large energy to break the C-O bond. The termination reaction can be verified through the vibrational transition state structure (figure 6) which have negative value. The transition state structure of step 1 in the path 2 (TS 2-1) was given the frequency -87.055 cm^{-1}, while in the path 4 (TS 4-1) was given frequency -40.677 cm^{-1}.

The next step is an insertion of OH^-, which in the path 2 occurs in proline molecule while in the path 4 it occurs on valine. Insertion of OH^- in path 2 produce intermediate I_p^2 with the enthalpy is -46,995 kcal.mol^{-1}, while in the path 4 produce intermediate I_v^4 with the enthalpy -47,003 kcal.mol^{-1}. The energy value is proved by the high reactivity of this intermediates due to the presence of H_2O as a good leaving group. Furthermore, the transition state structure could be proven through imaginary vibrational frequency. The imaginary frequency of transition state structure in path 2 is -2,183.96 cm^{-1}, whereas in the path 4 is-2,241.371cm^{-1}.

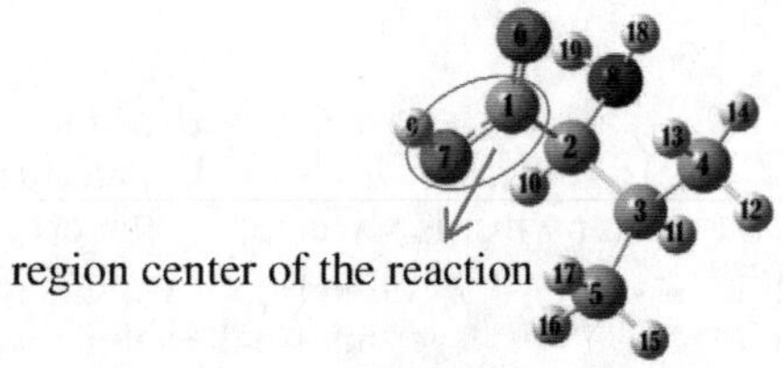

a) TS 2-1
r(C1-O7) 1.302 Å r(C1-O6) 1.187 Å
r(O7-H9) 0.952 Å α(C2-C1-O7) 113°

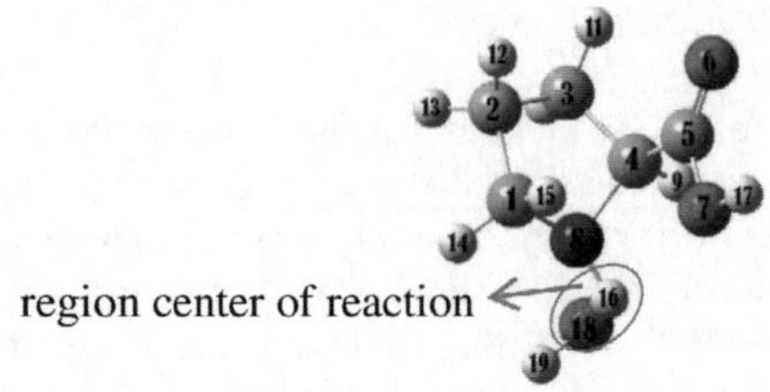

a) TS 2-2
r(H16-O18) 1.128 Å r(N8-H16) 1,24 Å

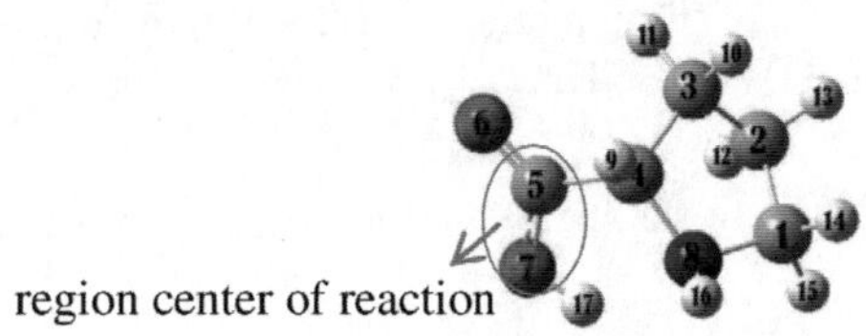

b) TS 4-1
r(C5-O7) 1.319 Å α(C4-C5-O7) 120°
r(O7-H17) 0.942 Å α(C4-C5-O6) 117°

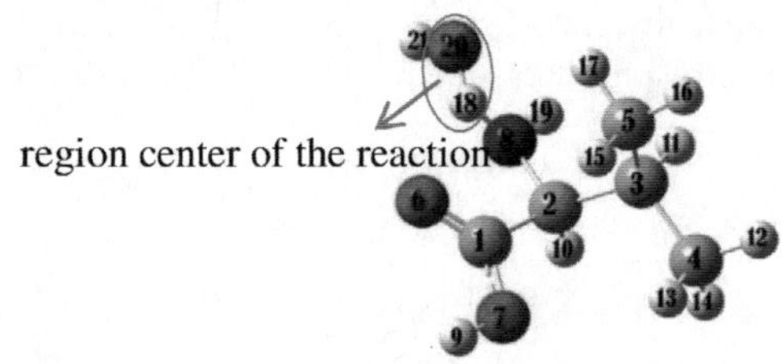

b) TS 4-2
(H18-O20) 1.142 Å α(N8-H18-O20) 78°

Figure 6. Calculated structures of transition in first and second step

On the next step, the termination of N-H bond in intermediates I_p^2 and I_v^4 occur and H_2O molecules are released. In the second step on path 2, the results obtained enthalpy 47,400 kcal.mol^{-1}, while on the path 4 give an enthalpy 548,000 kcal.mol^{-1}.

The transition state structure in this reaction step can be proved by the molecular vibration, which is the transition state of I_p^2 to become I_p^1 (TS 2-3) was obtained the imaginary frequency that is -1420,437 cm^{-1}, while the transition state of I_v^4 to become I_v^3 is not found because the reaction run rapidly.

The last step of this reaction pathway much the same with the reaction in path 1 and 3, which is a product formation. The product that produces from path 2 is Ac-PV-NH2 and on path 4 is Ac-VP-NH2. Based on computational calculations, the mechanism more likely is through the reaction path 4 with the Ac-VP-NH2 product. However, between path 1 and 4 the mechanism is preferred to the path 1 because the lower activation energy (Ea) that be passed in this reaction.

IOP Conf. Series: Materials Science and Engineering **172** (2017) 012040 doi:10.1088/1757-899X/172/1/012040

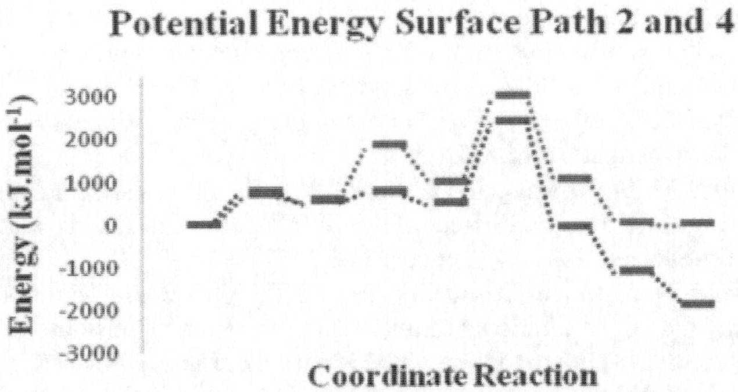

Figure 8. Calculated reaction path 2 and 4.

path 2 path 4

4. Conclusion

Peptide formation reaction from proline and valine amino acid can produce Ac-PV-NH2 and follow the reaction path 1. It is also can produceAc-VP-NH2with the reaction that occurred is on path 4. However, followed by the activation energy, the reaction on path 1 have a faster reaction rate so it is enabled to prefer label more to pass by the reaction of peptide bond forming. The rate determining step on path 1 is obtained in step 3 which H_2O molecules are released from valine with the activation energy is163,06 kJ.mol^{-1}.

5. References

[1] Sun L 2013 *Mod. Chem. Appl.* Peptide-Based Drug Development **I** 1-2.

[2] Kiptoo P, Laksitorini M D and Siahaan T J 2013 *Handbook of Biologically Active Peptides (Second Edition)* Chapter 233-Peptide Delivery, In A.J. Kastin, (Boston: Academic Press) 1702-10.

[3] Laksitorini M D, Kiptoo P K, ON H Ngoc, Thliveris J A, Miller D W, Siahaan T J 2015 *J. Pharm. Sci* Modulation of Intercellular Junction by Cyclic-ADT Peptides as a Method to Reversibly Increase Blood-Brain Barrier Permeability **104(3)** 1065-75.

[4] Pattabiraman V R and Bode J W 2011 *Rev. Nature* Rethinking amide bond synthesis **480** 471-79.

[5] Arlinghaus R, Shaeffer J and Schweet 1964 *Biochemistry* Mechanism of Peptide Bond Formation in Polypeptide Synthesis **51** 1291-9.

[6] Bucher Götz 2013 *Beilstein J. Org. Chem* New reactive intermediates in organic chemistry **9** 613-14.

[7] Chandrudu S, Simerska P and Toth I 2013 *Molecules* Chemical Methods for Peptide and Protein Production **18** 4373-88.

[8] Kulkarni J A and Asgaonkar K D 2012 *Int. J. Res in Pharm. And Bion. Sci* Study of Various Reaction Intermediates **3(1)** 325-51.

[9] Morgan P E, Pattison D I, Davies M J 2011 *Biology & Medicine* Quantification of hydroxyl radical-derived oxidation products in peptides containing glycine, alanine, valine, and proline, Free Radical **52** 328-39.

[10] Wang Q, Gao J, Zhang D, Liu C, 2015 *Chemical Physics* A theoretical model investigation of peptide bond formation involving two water molecules in ribosome supports the two-step and eight-membered ring mechanism 450-51, 1-11.

[11] Besora M, Braga A C and Sameera W M C 2014 *J Organometallic Chem* A Computational view on the reactions of hydrocarbons with coinage metal complexes **784** 2-12.
[12] Fessenden R J, Fessenden J S 1982 Kimia Organik Edisi Ketiga (Jakarta: Erlangga).
[13] Kwan E E and Evans D A 2010 *Organic Letters* Intermolecular Michael Reaction: A Computational Investigation **12** 5124-5127.
[14] Nielsen R J, Keith J M, Stoltz B M, and Goddard W A 2004 *Jacs Articles* A Computational Model Relating Structure and Reactivity in Enantioselective Oxidations of Secondary Alcohols by (-)-Sparteine-Pd^{II} Complexes **126** 7967-7974.
[15] Shafei H, Haqgu M, Nematollahi D and Gholami M R 2008 *Int. J. Electrochem. Sci* An Experimental and Computational Study on the Rate Constant of Electrochemically Generated N-Acetyl-p-Quinoneimine with Dimethylamine **3** 1092-1107.
[16] Sultana N and Fabian W M F 2013 *Beilstein Journal of Organic Chemistry* A Computational study of base-catalyzed reactions of cyclic 1,2-diones: cyclobutane-1,2-dione **9** 594-601.
[17] Thiel Indre and Hapke Marko 2013 *Angwandte Chemistry International* Computational Studies and Experimental Results-An Example of Excellent Teamwork in Studying Carbocyclization **52** 2-5.
[18] Ujaque G, Cooper A C, Maseras F, Eisenstein O, and Caulton K G 1998 *J. Am. Chem. Soc.* Computational Evidence of the Importance of Substituent Bulk on Agostic Interactions in $Ir(H)_2(PtBu_2Ph)_2^+$ **1998** 120 361-65.
[19] Valiev M, Bylaska E J, Govind N, Kowalski K, Straatsma T P, Van Dam H J J, Wang D, Niepolcha J, Apra E, Windus T L, and de Jong W A 2010 *Journal of Computer Physics Communication* NWChem: A comprehensive and scalable open-source solution for large scale molecular simulation **181** 1477-89.
[20] Cramer C J 2004 Essentials of Computational Chemistry: Theory and Models (London: John Wiley and Sons), 1-579.
[21] Foresman J B, Frisch, Æ, Exploring Chemistry with Electronic Structure Methods *second edition*, Pittsburgh, 3-297.
[22] Quinn C M 2002 Computational Quantum Chemistry *California: Academic Press,* 1-231.
[23] Siahaan P, Windarti T 2009 Struktur Molekular-Mikro Material *Semarang: Diponegoro University,* 19-27

Acknowledgement
Thanks to Prof. Teruna J. Siahaan, Ph.D. (Pharmacy Department University of Kansas, US) for the valuable direction and discussion of the cells and drug delivery system. Thanks to Prof Krzysztof Kuczera, Ph.D. (Chemistry Department, University of Kansas, US) for in-depth discussion of a computational modeling. Last but not least, thanks to the Indonesian Directorate General of Higher Education, which has funded this research and technology Funding Scheme 2013.

IOP Conf. Series: Materials Science and Engineering **172** (2017) 012041 doi:10.1088/1757-899X/172/1/012041

Partial biochemical characterization of crude extract extracellular chitinase enzyme from *Bacillus subtilis* B 298

P Lestari[1], N Prihatiningsih[2] and H A Djatmiko[2]

[1] Department of Chemistry, Faculty of Mathematics and Natural Sciences, University of Jenderal Soedirman, Purwokerto, Indonesia
[2] Department of Agrotechnology, Faculty of Agricultural University of Jenderal Soedirman, Purwokerto, Indonesia

E-mail: pujilestari_06@yahoo.co.id

Abstract. Extraction and characterization of extracellular chitinase from *Bacillus subtilis* B 298 have been done. Growth curve determination of *B. subtilis* B 298, production curve determination of crude extract chitinase from *B. subtilis* B 298, and partial biochemical characterization of crude extract chitinase have been achieved in this study. Optimum growth of *B. subtilis* B 298 was achieved at logarithmic phase within 9 hours incubation time, so it was used as inoculum for enzyme production. According to production curve of the enzyme, it was known that incubation time which gave the highest chitinase activity of 15 hours with activity of 6.937 U/mL respectively. Effect of various temperatures on chitinase activity showed that optimum activity was achieved at 40°C with an activity of 5.764 U/mL respectively. Meanwhile, the optimum pH for chitinase activity was achieved at pH of 5.0 with an activity of 6.813 U/mL respectively. This enzyme was then classified as metalloenzyme due to the decline of the activity by EDTA addition. All divalent cations tested acted as inhibitors.

1. Introduction

Chitin (β-1,4-linked polymer of N-acetyl-D-glucosamine) is one of the most abundant polysaccharides in nature, after cellulose. It is a major cell wall constituent of higher fungi belonging to Chytridiomycetes, Ascomycetes, Basidiomycetes and Deuteromycetes, insect exoskeletons and crustacean shells [1]. Approximately, 75% of total weight of shellfish such as shrimp, crabs, and krill are considered as waste and chitin comprises 20 to 58% of dry weight. Microbial degradation of insoluble macromolecules such as lignin, cellulose, and chitin depend on the production of extracellular enzymes that have the ability to degrade these components [2].

One kind of the enzyme that plays an important role in degrading these components is chitinase. Chitinase is the hydrolytic enzyme that spesifically degrade chitin. According to [1], chitinases are produced by several bacteria, actinomycetes, fungi and also by higher plants. Microorganisms produce the chitinase primarily for the assimilation of chitin as a carbon and nitrogen source. Chitinases have been isolated from a variety of bacteria including Bacillus spp., and some of them are reported to produce multiple forms of chitinases with different molecular masses [3]. Chitinase enzymes are used in various applications such as biological control of fungal pathogens [4], preparation of oligosaccharides and N-acetyl-D-glucosamine [5], and protoplast from filamentous fungi [6].

IOP Conf. Series: Materials Science and Engineering **172** (2017) 012041 doi:10.1088/1757-899X/172/1/012041

In this study, a chitinolytic bacterium *Bacillus subtilis* B 298 was investigated for the production and characterization of crude extract extracellular chitinase. The investigation included (i) bacteria growth curve determination, (ii) chitinase production curve determination, (iii) partial biochemical characterization of crude extract chitinase.

2. Materials and methods

2.1. Materials

B. subtilis B 298 isolate, Nutrient Agar (Merck), Nutrient Broth (Merck), Chitin powder (Sigma), DNS (sigma), NAG standard solutions, NaOH solution 1% (w/v), Sodium acetate buffer 50 mM pH 4.0, Sodium acetate buffer 50 mM pH5.0, Sodium phosphate buffer 50 mM pH 6.0, Sodium phosphate buffer 50 mM pH 7.0,Tris-HCl buffer 50 mM pH 8.0, Tris-HCl buffer 50 mM pH 9.0, $CaCl_2$ 1.0 mM, $MgCl_2$ 1.0 mM, $CuCl_2$ 1.0 mM, $ZnCl_2$ 1.0 mM, $HgCl_2$, EDTA 1.0 mM.

2.2. Determination of bacterial growth curve

Fresh isolate from solid Nutrient Agar (NA) medium was transferred to 100 mL of liquid Nutrient Broth (NB) medium pH 7.0 and was incubated using shaker bath at room temperature. Sampling was done every 3 hour, and the absorbance of the culture was measured using UV spectrophotometer at λ 600 nm. Growth curve was made by plotting the absorbance at Y axis and incubation time at X axis. Growth at logarithmic phase was used as inoculum for the production of chitinase enzyme.

2.3. Determination of chitinase production curve

Chitinase production curve was made by transferred inoculum to the 200 mL production medium (NB) fortified with 0.5% colloidal chitin (w/v) as an inducer, and incubated at room temperature. Sampling was done every 3 hours, and the culture was then measured the absorbance at λ 600 nm. After that the culture was centrifuged (5000 rpm, 4°C, 20 minutes). The supernatant is the crude extract enzyme and was determined the chitinase activity. Production curve was made by plotting the chitinase activity at Y axis and incubation time at X axis. The incubation time that gave the highest chitinase activity was used to produce the enzyme.

2.4. Production of crude extract chitinase

Production of crude extract chitinase was done according to the growth curve and chitinase production curve obtained. *B. subtilis* B 298 isolate was grown in instant Nutrient Agar medium and incubated for 48 hours at room temperature. After that, growing colonies were transferred to 200 mL inoculum medium (Nutrient Broth) in erlenmeyer flask and incubated for 9 hours at room temperature in a shaker bath. The inoculum (20% v/v) was then transferred into 1 L production medium (Nutrient Broth) in erlenmeyer flask, fortified with 0.5% (w/v) colloidal chitin as inducer, and allowed to grow at room temperature for the time of incubation that gave the highest chitinase activity from production curve, in shaker bath. The medium was then centrifuged (5000 rpm, 4°C, 20 minutes). The cell-free supernatant obtained was crude extract chitinase.

2.4.1. Assay of chitinase activity. Chitinase activity was determined spectrophotometrically by estimating the amount of free reducing groups formed after colloidal chitin hydrolysis. The reaction mixture was composed of 1 ml of 0.5% colloidal chitin suspended in 0.2 M sodium phosphate buffer pH 7.0 and 1 ml of the enzyme solution. After 30 minutes incubation at 37° C, the reaction mixture was stopped by adding 1 ml of 1% NaOH solution, and then boiled for 5 minutes. The mixture was then centrifuged (4000 rpm, 5-10 minutes). One ml of supernatant was added with 1 ml of 1% DNS solution, then incubated in boiled water for 5 minutes. The supernatant absorbance at 535 nm was measured. A standard curve was obtained using N-acetylglucosamine as a standard. One unit of chitinase activity was defined as the amount of the release of 1 μmol product (NAG) per minute.

2.5. Characterization of crude extract extracellular chitinase

2.5.1. Determination of various temperatures effect towards chitinase activity. Optimum temperature was determined over a temperature range of 30-50°C in Sodium phosphate buffer 50 mM pH 7.0. The activity was measured using the method described before.

2.5.2. Determination of various pH effect towards chitinase activity. Optimum pH was determined over pH values range from 3.0 to 9.0 at the temperature of 37°C. Sodium acetate buffer 50 mM was used for pH 3.0 to 5.0, Sodium phosphate buffer 50 mM for pH 6.0 to 7.0, and Tris-HCl buffer 50 mM for pH 8.0-9.0. The activity was measured using the method described before.

2.5.3. Determination the effect of EDTA and divalent cations towards chitinase activity. The effect of EDTA was examined by estimating activity in the presence of 1.0 mM EDTA solution. The effect of various divalent cations (Ca^{2+}, Mg^{2+}, Zn^{2+}, Cu^{2+}, Hg^{2+}) was examined by estimating activity in the presence of 1.0 mM solution of chloride metal salts. 250 μL of enzyme solution was preincubated in the presence of 250 μL of EDTA or divalent cations for 30 minutes at 30°C, followed by measuring activity using colloidal chitin as substrate at pH 7.0 and temperature of 37°C as described before. Control was done by measuring the chitinase activity without addition of either EDTA or divalent cations.

3. Results and discussion

3.1. Determination of B. subtilis B298 growth curve

The bacterial growth curve was made to determine the time of incubation that reaches logarithmic phase. The result was shown at figure 1.

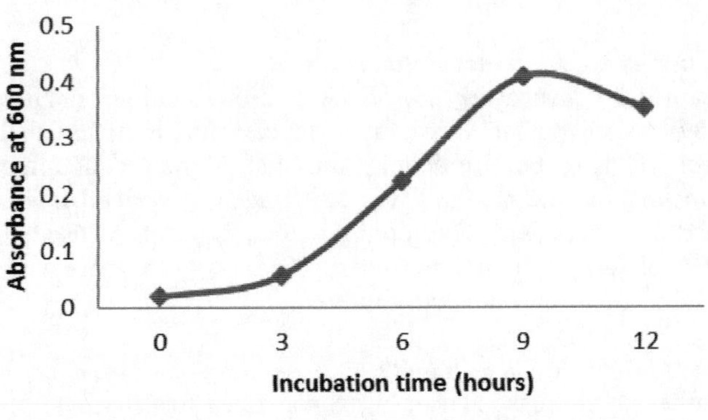

Figure 1. *B.subtilis* B 298 growth curve

According to the result shown at figure 1, the bacterial growth was reached optimum within 9 hours incubation with the absorbance value is 0.408. It's mean that the logarithmic phase of this bacterial growth was achieved at 9 hours incubation time. So, the inoculum that is used as the source for chitinase production is *B.subtilis* B 298 isolate that is incubated for 9 hours.

3.2. Determination of chitinase production curve

Chitinase production curve was made to determine the time of incubation that give the highest chitinase activity. The result was shown at figure 2.

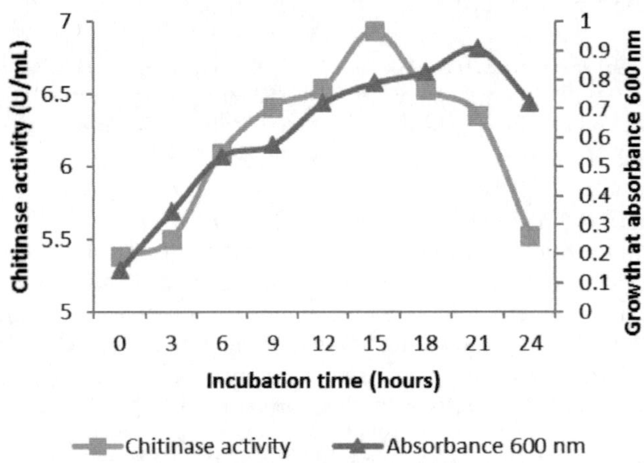

Figure 2. Chitinase production curve from *B. subtilis* B 298

Based on the result shown at figure 2, the chitinase enzyme from *B. subtilis* B 298 was produced maximumly within 15 hours incubation of the inoculum. That was indicated by the highest chitinase activity with its value is 6.937 U/mL. It's mean that for the production of crude extract chitinase, the inoculum should be incubated for 15 hours. Contrary, the chitinase enzyme from *B. licheniformis* MB-2 was produced maximumly within 72 hours incubation of the inoculum [7]. It's mean that *B. subtilis* B 298 produced this enzyme maximumly faster than *B. licheniformis* MB-2.

3.3. Production and characterization of crude extract chitinase

Crude extract chitinase was obtained by incubating *B.subtilis* B 298 inoculum in Nutrient Broth medium fortified with 0.5% (w/v) of colloidal chitin as the inducer for 15 hours that has been determined before. The addition of colloidal chitin in the medium induced the production of chitinase enzyme in bacteria cells. After incubating, the medium was centrifuged to get cell-free supernatant that is considered as crude extract chitinase. The crude extract was then used for further characterization including the effect of various temperatures and pHs towards chitinase activity as well as the effect of the addition of EDTA and various divalent cations.

3.4. Effect of various temperatures toward chitinase activity

Assay of various temperatures effect on chitinase activity was conducted to determine the optimum temperature that give the highest chitinase. The result was shown at figure 3. Based on the result shown at figure 3, the enzyme was active at the temperature range observed. The optimum temperature for chitinase activity from *B.subtilis* B 298 was achieved at 40°C with activity value is 5.764 U/ mL respectively, meanwhile the activity decreased slightly above 40°C. So, chitinase from *B.subtilis* B 298 is classified as the mesophilic enzyme. Similar study was described by other workers. Two chitinases isolated from *Pseudomonas aeruginosa* K-187 has the optimum temperature of 40°C and 50°C [8]. Some other workers obtained different results. Chitinase from *Bacillus amyloliquefaciens* SAHA 12.07 had optimum temperature 50°C, from *Serratia marcescens* KAHN 15.12 had optimum temperature 60°C [9], chitinase from *Bacillus cereus* 108 had optimum temperature 65°C [1]. Increasing the incubation temperature of an enzyme could cause the increase of enzyme activity. This is due to the increased speed of thermal motion of enzyme molecules, increased the numbers of molecules that have enough energy to reach the transition state [10].

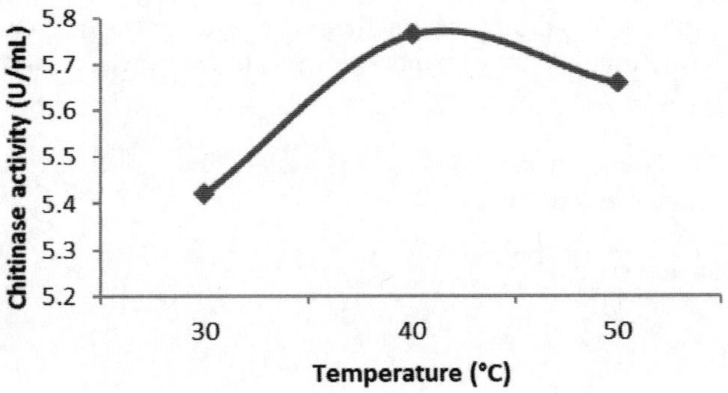

Figure 3. Effect of various temperatures on chitinase activity

3.5. Effect of various pH toward chitinase activity

Effect of various pH is examined to determine the optimum pH for chitinase activity. The result was shown at figure 4.

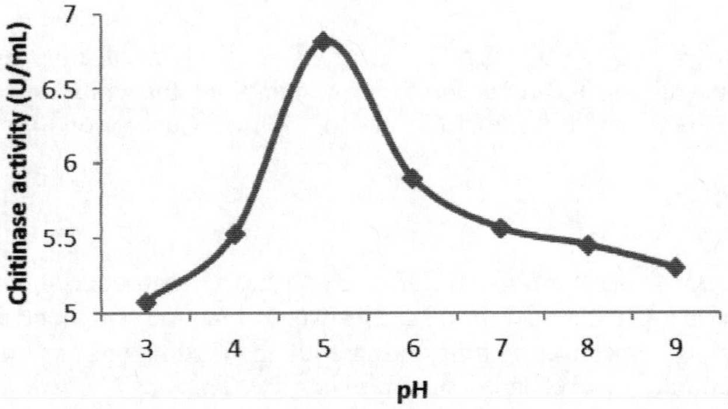

Figure 4. Effect of various pH on chitinase activity

Based on the result exhibited at figure 4, the enzyme was active in the pH range of 3.0 to 9.0 with optimum pH for chitinase activity from *B.subtilis* B 298 was achieved at pH 5.0, with its activity is 6.813 U/mL. The activity decreased sharply above the pH of 5.0. Similar study has been reported by other researchers. Crude extract chitinase from *Bacillus amyloliquefaciens* SAHA 12.07 had optimum activity at pH 5.0 with its activity is 1.158 U/mL [9]. Meanwhile, chitinase from *B. amyloliquefaciens*

V656 had optimum activity at pH 6.0 and 7.0 [11], chitinase from *S. marcescens* GEI and *S. marcescens* MO 1 had optimum activity at pH 7.0 [12, 13]. Each enzyme has an optimum pH that causes maximum activity. If the pH of the reaction is below or above the maximum pH, the enzyme activity decreases [14]. Bacterial chitinases generally have a pH range from acidic to alkaline [9].

3.6. Effect of EDTA and divalent cations addition towards chitinase activity

Several enzymes are known to require certain metal ions for their catalytic activity. Metal ions are required as a component of the active site of the enzyme to maintain its stability. Test of the addition of EDTA and divalent cations is conducted to determine whether the enzyme includes metalloenzyme or not and what kind of cations that are required for enzyme activity. The result was displayed at table 1.

Table 1. Effect of EDTA and divalent cations addition on chitinase activity

Treatment	Relative activity (%)
Control	100
EDTA	51.225
$CaCl_2$	53.088
$MgCl_2$	82.118
$ZnCl_2$	54.584
$CuCl_2$	63.826
$HgCl_2$	86.695

Based on table 1, the relative activity of enzyme decreased with the addition of 1.0 mM EDTA compared to control. The decline in activity reached about 51%. This means that chitinase from *B. subtilis* 298 bacteria is a metalloenzyme. According to [15], metalloenzyme is an enzyme that requires metal ions in the active site for catalytic ability. If addition of EDTA decreased enzyme activity compared to control, the enzyme including metalloenzyme. EDTA inhibits enzyme activity by chelating metal in the active site to alter the tertiary structure of enzyme and enzyme lose it's catalyzing ability.

Addition of 1.0 mM Ca^{2+}, Mg^{2+}, Zn^{2+}, Cu^{2+} and Hg^{2+} cations showed the decline of chitinase activity. This means that all the divalent cations tested were inhibited the chitinase activity of *B.subtilis* B 298. The largest decline in activity occured in the addition of Ca^{2+} cation that is equal to 53%.

4. Conclusion

Crude extract chitinase extracellular from *B.subtilis* B 298 bacteria has been produced and partially characterized the biochemical properties. Production of crude extract enzyme was conducted using inoculum within 9 hours incubation and production time within 15 hours incubation. The crude extract chitinase has the optimum temperature at 40°C and optimum pH at 5.0. This enzyme was categorized as metalloenzyme.

Acknowledgements

We thank The Ministry of Research, Technology and Higher Education The Republic of Indonesia for supporting the funding for this research.

References

[1] Hammami I, Siala R, Jridi M, Ktari N, Nasri M and Triki M A 2013 Partial purification and characterization of chilO8, a novel antifungal chitinase produced by *Bacillus cereus* IO8 *J. Appl. Microbiol.* **115** 358-66

[2] Farag M A and Al-Nusarie S T 2014 Production, optimization, characterization and antifungal activity of chitinase produced by *Aspergillus terrus Afr. J. Biotechnol.* **13** 1567-78

[3] Ajit N S, Verma R and Shanmugam V 2006 Extracellular chitinases of fluorescent pseudomonas antifungal to *Fusarium oxysporum* f. Sp *Curr. Microbiol.* **52** 310-16

[4] Someya N, Nakajima M, Hirayae K, Hibi T and Akutsu K 2001 Synergistic antifungal activity of chitinolytic enzymes and prodigiosin produced by the biocontrol bacterium *Serratia marcescens* strain B2 against the grey mold pathogen, *Botrytis cinerea J. Gen. Plant. Pathol.* **67** 312-17

[5] Makino A, Kurosaki K, Ohmae M and Kobayashi S 2006 Chitinase catalyzed synthesis of alternatingly N-deacetylated chitin: a chitin-chitosan hybrid polysaccharide *Biomacromolecules* **7** 950-57

[6] Yano S, Wakayama M and Tachiki T 2006 Cloning and expression of an alpha-1,3-glucanase gene from Bacillus circulans KA-304: the enzyme participates in protoplast formation of *Schizophyllum commune Biosci Biotechnol Biochem* **70** 1754-63

[7] Toharisman A, Suhartono M T, Spindler B M, Hwang J K and Pyun Y R 2005 Purification and characterization of a thermostable chitinase from *Bacillus licheniformis* MB-2 *World J. Microbiol. Biotechnol* **21 (5)** 730-38

[8] Wang S L and Chang W T 1997 Purification and characterization of two bifunctional chitinase/lysozymes extracellularly produced by *Pseudomonas aeruginosa* K-187 in a shrimp and crab shell powder medium *J. Applied. Environ. Microbiol* **63 (2)** 380-86

[9] Azizah S N, Mubarik N R and Sudirman L I 2015 Potential of chitinolytic *Bacillus amyloliquefaciens* SAHA 12.07 and *Serratia marcescens* KAHN 15.12 as biocontrol agents of *Ganoderma boninense Res. J. Microbiol* **10 (10)** 452

[10] Bollag D M and Edelstein S J 1991 *Protein Methods* (USA: Willey-Liss) p230

[11] Wang S L, Shih I L, Liang T W and Wang C H 2002 Purification and characterization of two antifungal chitinases extracellularly produced by *Bacillus amyloliquefaciens* V656 in a shrimp and crab shell powder medium *J. Agric. Food Chem.* **50** 2241

[12] Tu S, Qiu X, Cao L, Han R, Zhang Y and Liu X 2010 Expression and characterization of the chitinases from *Serratia marcescens* GEI strain for the control of Varroa destructor, a honey bee parasite *J. Invertebrate Pathol.* **104** 75

[13] Okay S, Ozdal M and Kurbanoglu E B 2013 Characterization, antifungal activity and cell immobilization of a chitinase from *Serratia marcescens* MO-1 *Turk. J. Biol.* **37** 639

[14] Nelson D L and Cox M M 2004 *Lehninger principles of biochemistry* (Electronic Material fourth edition) (University of Wisconsin-Madison) chapter 6 pp 190-237

[15] Lestari P, Raharjo T J, Matsjeh S and Haryadi W 2016 *AIP Conf. Proc.*vol 1755 080003 (USA: AIP Publishing)

Antidiabetic Activity from Gallic Acid Encapsulated Nanochitosan

Purbowatiningrum[1)], Ngadiwiyana[1)], Ismiyarto[1)], E Fachriyah[1)], I Eviana[1)], O Eldiana[1)], N Amaliyah[1)], A N Sektianingrum[1)]

[1)] Chemistry Department, Science and Mathematics Faculty, Diponegoro University
Jl. Prof. H. Soedarto, SH, Tembalang, Semarang, Indonesia, Post code 50275

Email: purbowatining@gmail.com

Abstract. Diabetes mellitus (DM) has become a health problem in the world because it causes death. One of the phenolic compounds that have antidiabetic activity is gallic acid. However, the use of this compound still provides unsatisfactory results due to its degradation during the absorption process. The solution offered to solve the problem is by encapsulated it within chitosan nanoparticles that serve to protect the bioactive compound from degradation, increases of solubility and delivery of a bioactive compound to the target site by using freeze-drying technique. The result of chitosan nanoparticle's Scanning Electron Microscopy (SEM) showed that chitosan nanoparticle's size is uniform and it is smaller than chitosan. The value of encapsulation efficiency (EE) of gallic acid which encapsulated within chitosan nanoparticles is about 50.76%. Inhibition test result showed that gallic acid-chitosan nanoparticles at 50 ppm could inhibite α-glucosidase activity in 28.87% with 54.94 in IC_{50}. So it can be concluded that gallic acid can be encapsulated in nanoparticles of chitosan and proved that it could inhibit α-glucosidase.

1. Introduction

Diabetes mellitus (DM) is a chronic metabolic disorder caused by dysfunction of insulin that causes hyperglycemia [1]. It caused long-term damage, dysfunction, and failure of various organs. The World Health estimates that in 2030 about 366 million people will suffer *diabetes mellitus* [2]. One of therapeutic to reduce hyperglycemia after a meal is prevented adsorption of glucose by inhibiting the enzyme hydrolysis on carbohydrate, such as α-glucosidase enzyme. The denial of α - glucosidase enzyme, glucose in the blood can be returned within normal limits [3]. Recently, some of the natural phenolic compounds and polysaccharides (such as tea, grapefruit, strawberries) have been shown to inhibit the α - glucosidase enzyme and potential as the natural therapeutic agent for the treatment of *diabetes mellitus* [4].

Chitosan is a polysaccharide derived from deacetylation of chitin which is found in the shells of crabs, shrimp, and squid. Chitosan has been shown afford to biodegradable, biocompatible, non-toxic, non-antigenic, and antimicrobial, so chitosan could be applicated in biomedical. Chitosan nanoparticles have activity as anti-diabetic [5].

Gallic acid is known as a phenolic compound that has drug activity for some disease, one of them as anti-diabetic. The use of gallic acid as *diabetes mellitus* (DM) results unsatisfactory because it is not stable at extreme temperatures or presence of oxygen or light [6] and is suspected because less of absorption by the human body the affected by particle's size is large, poor solubility during the absorption process. It is encouraging to modify for optimized the properties of the active compound by performing encapsulation in nanoparticles that serve to improve the solubility and *bioavailability. Encapsulation method is done to protect the bioactive compounds packaged in a* secondary ingredient (polysaccharide, protein, or lipid) to protect them from the effects of oxygen, heat and light that can affect the stability of compound [7].

This research investigated the characterization of chitosan structure, and anti-diabetic potential of encapsulation gallic acid-chitosan nanoparticles. Firstly, synthesized chitosan nanoparticles, next encapsulated gallic acid in chitosan nanoparticles. To determine the encapsulation of gallic acid-chitosan nanoparticles as anti-diabetic was test by inhibition α – glucosidase enzyme.

2. Procedure

2.1 Materials and Instrumentation

Materials; Chitosan with 80-85% deacetylation degree that obtains from deacetylation of chitin, acetic acid p.a (Merck), sodium tripolyphosphate (Na-TPP) p.a (Sigma-Aldrich), water de-ionization, gallic acid (Sigma Aldrich), 4-nitrophenol, phosphate buffer, α-glucosidase enzyme, bovine albumine serum, dimethyl sulfoxide (DMSO), *p*-nitrophenyl, α-D-glucopiranoside as a substrate.

Research tools used are glass that commonly used in the laboratory, Blender, Oven, Hot Plate Stirrer (IKA C-MAG HS 7), Centrifuge (Corning LSE Compact), Freeze drying (LLI500), PSA (Microtrac-Particle Size Analyzer), FTIR (Shimadzu-Fourier Transform Infrared) and Spectrophotometer Ultraviolet-Visible (Spectroquant).

2.2 *Eksperimental*

2.2.1. Preparation of chitosan nanoparticles. Chitosan nanoparticles were prepared based on the ionic gelation method: Chitosan was synthesized by chitin deacetylation isolated from shrimp shells. Chitosan was dissolved with acetic acid, then stirred until homogeneous. Hereafter, it added by solution of sodium tripolyphosphate slowly. Gel suspension was centrifuged and dried using freeze dryer, then charachterized by FT-IR, and SEM.

2.2.2. Encapsulation of gallic acid in chitosan nanoparticles. Encapsulation of gallic acid in chitosan nanoparticles were prepared based on inclusion complexation method. First, gallic acid 0,1% was added chitosan nanoparticles. Gel suspension was centrifuged and dried using freeze dryer, to be followed by determination of encapsulation efficiency.

2.2.3. Alpha-glucosidase inhibition test. The enzyme inhibition assay was done based on the methods from Lee and Lee, 2001; and Prashanth et al., 2001. Enzyme solution was prepared by dissolving 1.0 mg of α-glucosidase in 100 mL of phosphate buffer (pH 7.0) containing 200 mg of bovine albumin serum. Prior to use, as much as 1 mL of that solution was diluted 25 times using phosphate buffer (pH 7.0) which then referred as the enzyme solution. A total of 250 μL 20 mM *p*-nitrophenyl α-D-glucopyranoside as a substrate, then added 490 μL 100 mM phosphate buffer (pH 7.0) and 10 μL solution of synthesized compounds in DMSO. The reaction mixture was incubated at the optimum temperature for 5 minutes, then 250 μL of enzyme solution was added and then it was incubated. Enzymatic reaction was stopped by the addition of 1000 μL of 200 mM sodium carbonate. The product of the enzymatic reaction was *p*-nitrophenol which can be measured by using UV-visible spectrophotometer at 400 nm. The concentration of sample were 6.25 ppm, 12.5 ppm, 25 ppm, and 50 ppm in DMSO solvent. Percent inhibition can be calculated from the equation: [(C - S) / C] x 100; with S = absorbance of the sample (obtained from the S1 - So, where S1 = absorbance of the sample with enzyme addition and So = the absorbance of the sample without enzyme addition) and C = absorbance of control (DMSO), without sample (control-blank). IC_{50} values was obtained from: a linear regression equation Y = a + bx.

IC50 → Y = 50, then it was substituted into the linear regression equation becomes:

$$IC_{50} = (50-a)/b$$

3. Results and Discussion

3.1 Characterization

Analysis of FTIR spectra provided information about the functional groups of products. The spectra showed that the compound has a functional group as expected compound. FTIR spectra of chitin (**Fig.1**) showed some vibration patterns, which are in 3490 cm^{-1} from NH, 3310 cm^{-1} from-OH, 1650 cm^{-1} from stretching vibration of C=O, and 1560 cm^{-1} from stretching vibration of NH. The stretching vibration of NH is characteristic for chitin due to an NH group in -NH-CO- (acetylated amine group).

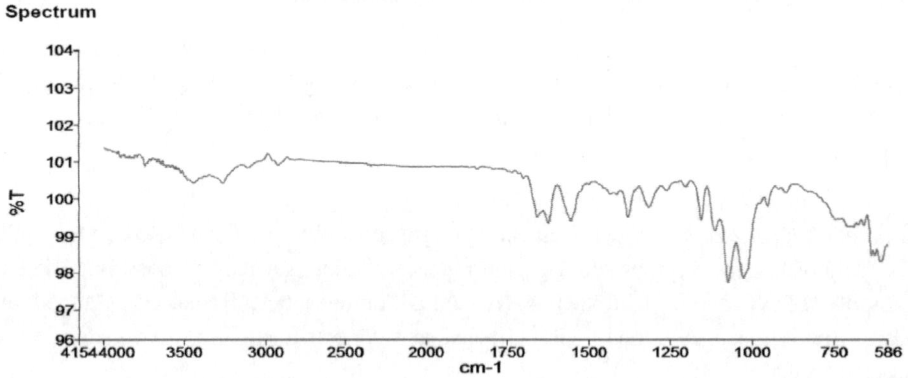

Fig. 1. FTIR Spectra of Chitin

Based on the Fig. 1 showed that vibration of CH_3 from chitin at 1310 cm^{-1} coincides with C-N amide vibration at 1400 cm^{-1}. The vibration of the amine group of chitin in the 3490 cm^{-} coincided with OH vibration because hydrogen bond in amine is weak and less polar. It makes N-H bond vibration becomes less intense than OH. Other vibration at 2970 cm^{-1} is a stretching from aliphatic C-H. This absorption is weak because chitin structure is dominated by R_3C-H (methine) which has a weak absorption (Sastrohamidjojo 1992). The presence of an absorption band at 1010-1150 cm^{-1} showed a vibration of C-O. Absorption pattern appears to indicate that the residue is a chitin as expected.

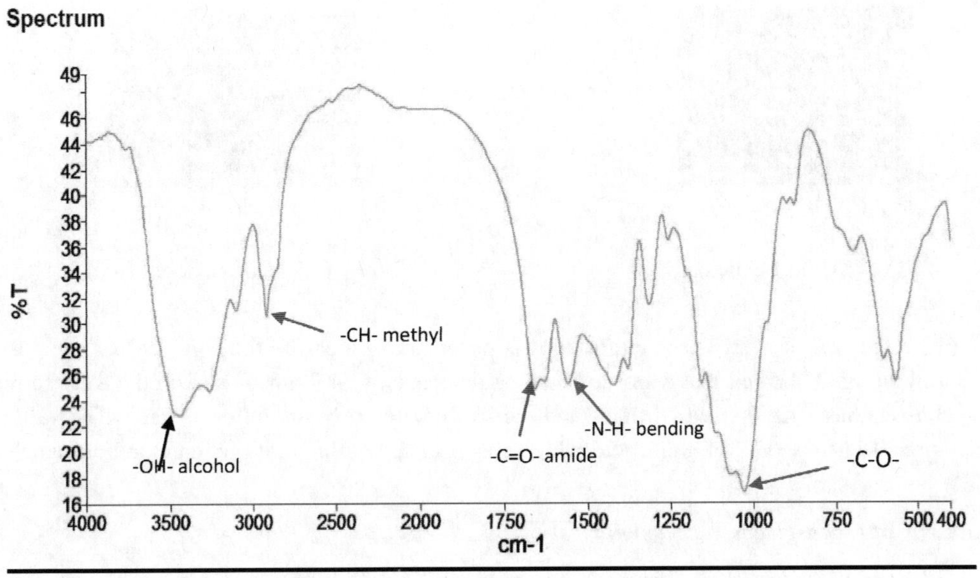

Fig. 2. FTIR Spectra of Chitosan

Based on Fig.2 that the changing of FTIR spectra of chitin and chitosan can be confirmed that deacetylation was successfully transformed chitin into chitosan. However, the carbonyl group absorption band at 1690 cm^{-1} showed that chitosan is not perfectly deacetylated. Due to that reason, it is a need to calculate the degree of deacetylation (DD) based on the baseline method, according to the following equation:

$$\% DD = 1 - \left[\frac{A_{1655}}{A_{3450}} x \frac{1}{1.33}\right] x\ 100\%$$

A_{1655} = absorbance at 1655 cm^{-1}wavelength number (amide absorption)
A_{3450} = absorbance at 3450 cm^{-1}wavelength number (hydroxyl absorption)
1.33 = A_{1655} for chitin which is fully deacetylated (100%).

Based on chitosan spectra (**Fig.2**), it is resulted in a degree of deacetylation of 87.78%. It means only about 89.98% residual chitin which is deacetylated into chitosan. The degree of acetylation of commercial chitosan is between 20-25%.In other words, the degree of deacetylation is between 70-90%. It can be concluded that deacetylation done in this research is in a good range.

The next stage is characterisation by *Scanning Electron Microscope* (SEM): the SEM data showed the morphological differences between chitosan and nanochitosan.

Fig 3. SEM of chitosan

Fig 4 SEM of Nanochitosan

To determine the size of synthesized chitosan, it is necessary to measure the chitosan particles using SEM. Based on fig. 3 showed that chitosan has a particle size of 64.9 nm. Based on the fig.4 showed that the chitosan has a nano-sized particles and the nano chitosan has a uniform shape. The value of encapsulation efficiency (EE) of gallic acid which encapsulated within chitosan nanoparticles is about 50.76%.

3.2 Inhibition of Alpha-glucosidase activity test

Inhibition test of the alpha-glucosidase enzyme was conducted to determine the antihyperglycemic activity of each extract. On this test, the alpha-glucosidase enzyme will hydrolyze *p*-nitrophenyl-α-D-glukopyranoside substrate into the yellow *p*-nitrophenol and glucose with the following the reaction:

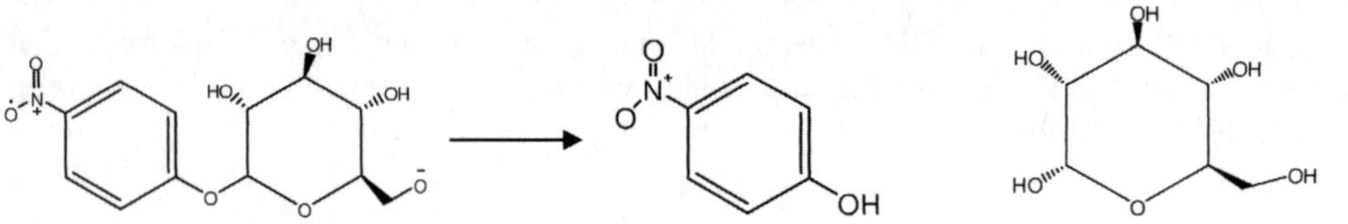

p-nitrophenyl- α-D-glucopyranose + α-glucosidase → *p*-nitrophenol + α-D-glucose

Fig. 5. An enzymatic reaction of α-glucosidase and *p*-nitrophenyl-α-D-glucopiranoside

Enzyme activity was measured based on the absorbance of the p-nitrophenol which has yellow color. In the presence of Gallic acid-chitosan nanoparticles which acts as an alpha-glucosidase inhibitor so the p-nitrophenol produced will reduce which was characterized by a lowering in the intensity of yellow color. The p-nitrophenol is a product of the enzymatic reaction of alpha-glucosidase with p-nitrophenyl-α-D-glucopyranose substrate. Gallic acid-chitosan nanoparticles can inhibit the action of the alpha-glucosidase enzyme which was characterized by a reduced concentration of p-nitrophenol.

It clearly is seen showed on **fig. 5** that 50 ppm of cinnamon oil could inhibit the enzyme alpha-glucosidase by 28.87 % and IC_{50} values were 54.94 ppm. IC_{50} values are numbers that indicate the concentration of the sample (in ppm) which are able to inhibit 50% of α-glucosidase enzyme activity. The results indicate that cinnamon oil could potentially inhibit α-glucosidase enzyme.

The alpha-glucosidase enzyme is an enzyme that plays a role in the termination of 1-4 glycoside bonds of carbohydrates to produce blood sugar (glucose). Inhibiting the alpha-glucosidase enzyme work will lead to prevent increasing concentration of glucose levels in blood after eating.

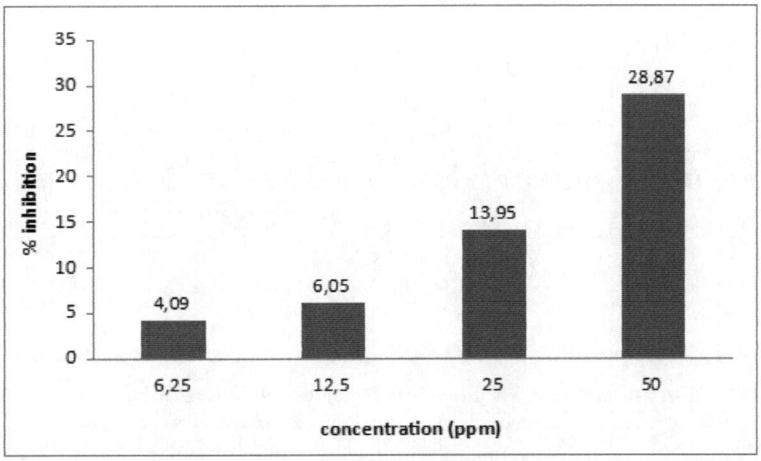

Fig.6 Percent inhibition of alpha-glucosidase enzyme from gallic acid-chitosan nanoparticles in varying concentration

4. Conclusions

Gallic acid can be encapsulated in nanoparticles of chitosan and proved that it has inhibited α-glucosidase. It showed that gallic acid encapsulated nanochitosan can inhibit 28.87 % at 50 ppm with IC50 = 54.94.

Refferences

[1] Mishra N 2013 Haematological and hypoglycemic potential *Anethum graveolens* seeds extract in normal and diabetic Swiss albino mice. *Vet World.* 6 502-7

[2] Wild S, Roglic G, Green A, Sicree R and King H 2004 Global prevalence of diabetes estimates for the year 2000 and projections for 2030. *Diabetes Care*. 27 1047-53

[3] Bosenberg LH 2008 The mechanism of action of oral antidiabetic drugs: a review of recent literature. *The Journal of Endocrinology Metabolism and Diabetes of South Africa*. 13 80-88

[4] Liu G S Xu and L Chen 2012 Chemical composition and bioactivities of a water-soluble polysaccharide from the endodermis of shaddock. *Int. J. Biol. Macromol* 51 763–766

[5] Ngo Dai-Hung, Vo, Thanh-Sang, Ngo, Dai-Nghiep, Kang, Kyong-Hwa, Je, Jae-Young, Pham, Hoang Nguyen-Duc, Byun, Hee-Guk, and Kim Se-Kwon 2015 Biological Effect of Chitosan and its Derivatives Review *Food Hydrocolloids*. 51 200-216

[6] Jacques AC, Pertuzatti PB, Barcia MT, Zambiazi RC, and Chim JF 2010 Estabilidade de compostos bioativos em polpa congelada de amora-preta (rubus fruticosus) cv. Tupy *Quim.* 1720–1725

[7] Pillai DS, Prabhasankar P, Jena BS, and Anandharamakrishnan C 2012 Microencapsulation of *Garcinia cowa* fruit extract and effect of its use on pasta process and quality *International Journal of Food Properties*.15 590–604

Pyrolysis result of polyethylene waste as fuel for solid oxide fuel cell with samarium doped-ceria (SDC)-carbonate as electrolyte

R J E Syahputra[1], F Rahmawati[2,], A P Prameswari[1] and R Saktian[1]

[1]Chemistry Department, Faculty of Mathematics and Natural Science, Sebelas Maret University, Jl. Ir Sutami No. 36 A, Jebres, Surakarta, Indonesia. 57126

[2]Research Group of Solid State Chemistry & Catalysis, Chemistry Department, Sebelas Maret University, Jl.Ir. Sutami 36 A Kentingan, Surakarta, Indonesia, 57126

Email :[1]rahmat_jes@student.uns.ac.id, [2]fitria@mipa.uns.ac.id

Abstract. In this research, the result of pyrolysis on polyethylene was used as fuel for a solid oxide fuel cell (SOFC). The pyrolysis result is a liquid which consists of hydrocarbon chains. According to GC-MS analysis, the hydrocarbons mainly consist of C7 to C20 hydrocarbon chain. Then, the liquid was applied to a single cell of NSDC-L | NSDC | NSDC-L. NSDC is a composite SDC (samarium doped-ceria) with sodium carbonate. Meanwhile, NSDC-L is a composite of NSDC with LiNiCuO (LNC). NSDC and LNC were analyzed by X-ray diffraction to understand their crystal structure. The result shows that presence of carbonate did not change the crystal structure of SDC. SEM EDX analysis for fuel cell before and after being loaded with polyethylene oil to get information of element diffusion to the electrolyte. Meanwhile, the conductivity properties were investigated through impedance measurement. The presence of carbonate even increases the electrical conductivity. The single cell test with the pyrolysis result of polyethylene at 300 – 600 °C, found that the highest power density is at 600 °C with the maximum power density of 0.14 mW/cm^2 and open circuit voltage of 0.4 Volt. Elemental analysis at three point spots of single cell NDSC-L |NSDC|NSDC-L found that a migration of ions was occurred during fuel operation at 300 – 600 °C.

1. Introduction

Increasing on energy consumption in this world became a problem that should be solved. It sparks many efforts to find alternatives solution to the problems. Such as converting some plants to fuel, animal wastes to become biogas, and also converting plastic waste to fuel. Research on polyethylene conversion through pyrolysis found that the polyethylene could be converted into gas for fuel, and continued to be condensed into oil [1]. Pyrolysis or thermal cracking offers suitable methods for long chain decomposition [2], to produce a chemical utilization of energy. The advantage of this method over combustion is a reduction in the volume of gas products by a factor of 5 to 20 which leads to considerable savings in the gas conditioning [3]. Many investigators have studied the pyrolysis of polyethylene. The observed similarity in the distribution and yields of volatile products from standard polymers and plastic wastes based on polyethylene suggests that the pyrolysis over the temperature range 500-800 °C can be used to convert these plastic wastes into corresponding monomers and hydrocarbons [4]. High-density Polyethylene (HDPE), the highest liquid yield is at 450 °C, highly volatile products are obtained at low temperature, and the products obtained at 500 °C and 550 °C are viscous liquid and wax [5]. Decomposition of low-density polyethylene by pyrolysis method found that the oil consists of 60% of alkane and 32% of alkene compounds, which are the typical compounds found in fuel [6].

It is known that plastic waste can not be decomposed naturally, and it becomes a huge problem for the environment [7]. Polyethylene is a kind of polymer that usually used for industrial and everyday life purposes. Therefore, the polyethylene waste is one of a major component of household waste. Meanwhile, it is known that pyrolysis on polyethylene could produce oil that consists of the hydrocarbon chain. The hydrocarbon chain is actually a potential liquid fuel. However, the technology that uses those liquid fuel has still not established yet. Therefore, a pessimistic statement might raise as the pyrolysis of polyethylene into fuel does not impact the increasing of energy production or become an alternative way for energy production, because the technology that uses the waste plastic oil is not ready [8]. The fuel cell can become a potential technology for energy conversion that may use those waste plastic oil. Due to as an electrochemical cell, the fuel cell has wide options of fuel,i.e.hydrogen fuel, methane, butane and also hydrocarbons. Those fuels could be converted into electrical energy in the fuel cell system [9].

In this research, an effort to convert the polyethylene waste into a liquid fuel was conducted. Then, the research continues to apply that liquid fuel as a fuel for a solid oxide fuel cell that consists of NSDC as electrolyte and NSDC-L as an electrode.

2. Experimental

2.1. Procedure of research

2.1.1. Pyrolysis of polyethylene the oil as the result

The polyethylene waste from a plastic bottle was cut into pieces and then was placed into the pyrolysis reactor. The pyrolysis process was conducted under 400-500 °C. The reactor setup was described in the previous paper [1]. The vapor formed was condensed by cooling water system and then collected in a container, as the polyethylene oil. The Oil was analyzed by Gas Chromatography-Mass Spectroscopy (GCMS): GC17A MSQP 5000 Shimadzu to identify the type of the produced hydrocarbon molecules and also the main component of the oil.

2.1.2. Preparation of single cell and fuel cell test

Samarium Doped-Ceria, $Sm_{0.2}Ce_{0.8}O_{1.9}$ (SDC), was synthesized by sol-gel method [10]. The precursors were cerium nitrate hexahydrate ($Ce(NO_3)_3.6H_2O$, 99 %, Sigma-Aldrich), $Sm(NO_3)_3.6H_2O$ (99.9 %, Sigma-Aldrich). The precursors were mixed stoichiometrically and dissolved with deionized water. Citric acid and polyethylene glycol as chelating agents were added under continuous stirring at 50°C until a transparent gel was formed, then the temperature was increased until 400 °C until a yellowish powder formed. The powder then was heated at 800 °C for 10 hours.

NSDC was prepared through the sol-gel method [10], in which the Cerium Nitrate Hexahydrate ($Ce(NO_3)_3.6H_2O$) and Samarium Nitrate Hexahydrate ($Sm(NO_3)_3.6H_2O$) blend with the molar ratio of Ce^{3+}: Sm^{3+} = 4:1 in 100 mL of solution. The Na_2CO_3 powder was firstly heated at 800 °C before it was dissolved in water to form 1.0 M of the solution. The mixed solution was heated at 100 °C and stirred for 2 h. The heating was continued to 800 °C for 2 hours to produce pure powder [11].

Synthesis of LiNiCuO catalyst by Sol-Gel method where ratio molar of Li^+: Ni^{+2}:Cu^{2+}= 2:7:1 from material Li_2CO_3: $CuCO_3$: $NiCO_3.2Ni(OH)_2.4H_2O$ in 1,0 Mol deionized water. Citric acid is added and stirred until homogeneous. The solution was heated at 100 °C to form sol then sintered at 800 °C for 2 hours.

The pellet of SDC and NSDC were prepared by pressing with a hydraulic press. Then it was sintered at 800 °C for 2 hours. The 1:1 weight ratio of NSDC and LiNiCuO was mixed with α - terpineol to form a homogeneous slurry. The slurry then applied to the both surface of NSDC and SDC by screen printing method and then heated at 600 °C for 2 hours to produce a single cell of NSDC - L | NSDC | NSDC – L. Ag mesh, as the current collector, was attached on both side with Ag paste as a sealer. Ag wire was attached to that Ag mesh and connect the cell with I-V meter (microvoltmeter, 89 Constant), a potentiometer up to 20kΩ. Scheme of the single cell with the electrical network is described in Figure 1(a). Meanwhile, the scheme of single cell performance measurement is described in Figure 1(b). The fuel cell test was conducted at 300 °C, 400 °C, 500 °C and 600 °C with polyethylene liquid as fuel and oxygen in the air as oxidator. The polyethylene liquid was converted into gas by heating at 150 °C. In this research, the gas reforming was conducted by evaporating polyethylene liquid and the polyethylene gas would be reformed inside the fuel cell reactor.

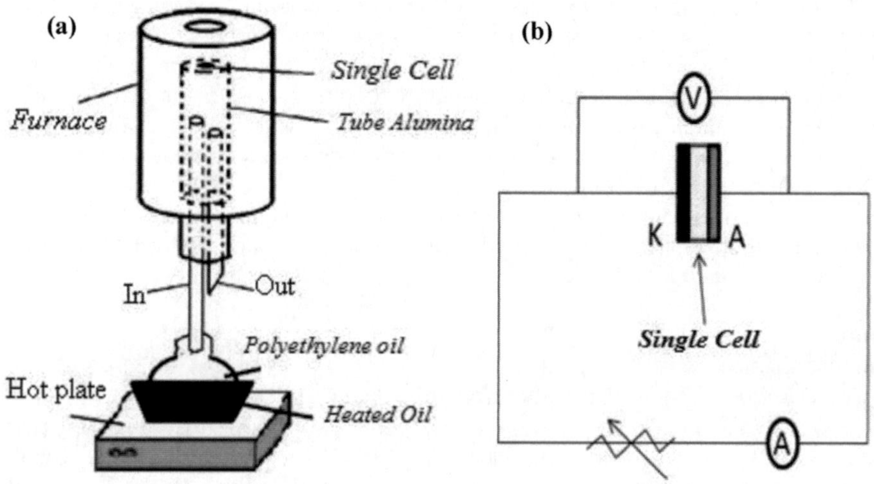

Figure 1. Scheme of single cell and the single cell test measurement

3. Result

The polyethylene liquid as produced from pyrolysis shows a yellow color liquid with physical properties as listed in Table 1.

Table 1. Physical Properties of the liquid polyethylene

No.	Properties	Identification
1	Color	Yellow
2	Density (g.cm^{-3})	0.760
3	Viscosity (Poise)	0.58
4	Ph	4

Meanwhile, molecular identification by GC-MS analysis results in a spectrum as depicted in Figure 2.

IOP Conf. Series: Materials Science and Engineering **172** (2017) 012043 doi:10.1088/1757-899X/172/1/012043

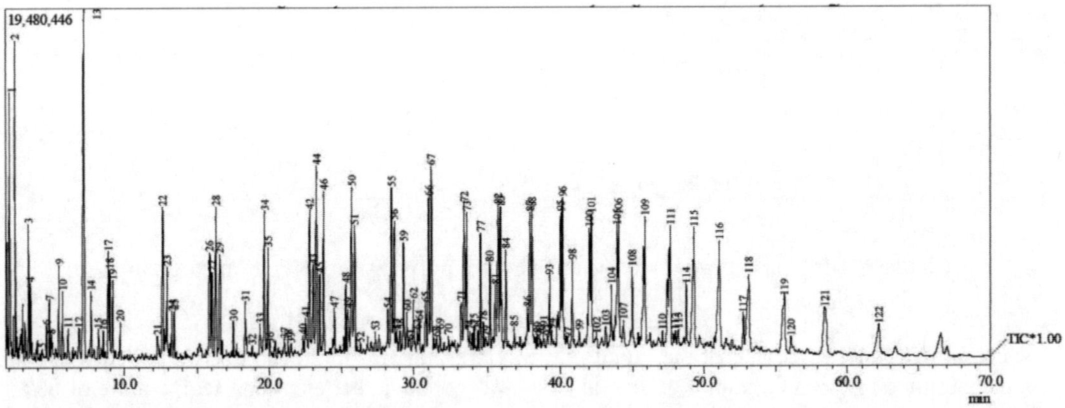

Figure 2. Chromatogram of the liquid polyethylene as produced by pyrolysis process

GC spectrum shows many peaks present in the liquid. It indicates that the liquid consist of some hydrocarbon with a various number of carbon chains. The Mass spectroscopy might identify some major peaks that present on the GC spectra. Some MS spectrums are depicted in Figure 3-6.

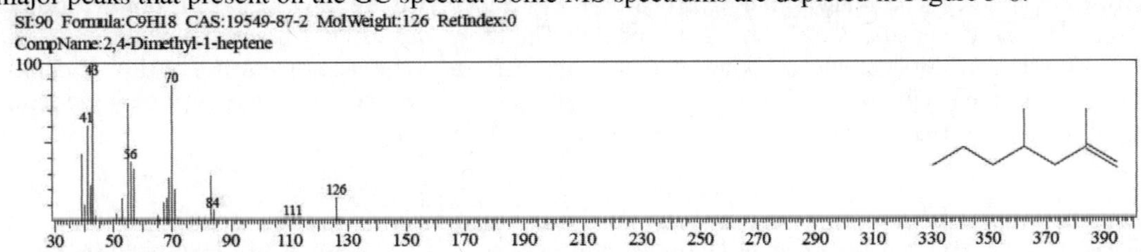

Figure 3. Mass-Spectroscopy of peak 13 from polyethylene oil chromatogram

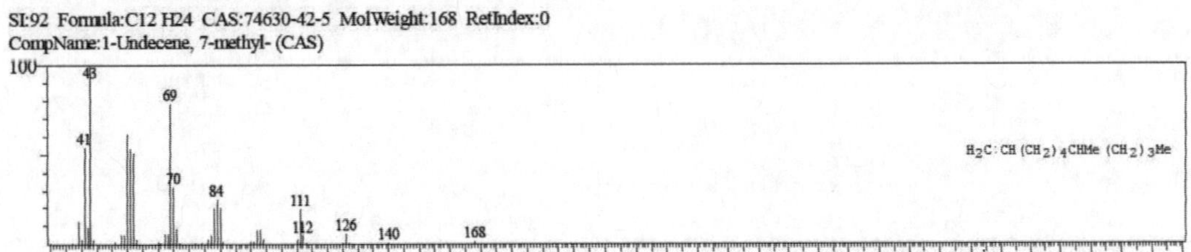

Figure 4. Mass-Spectroscopy of peak 44 from polyethylene oil chromatogram

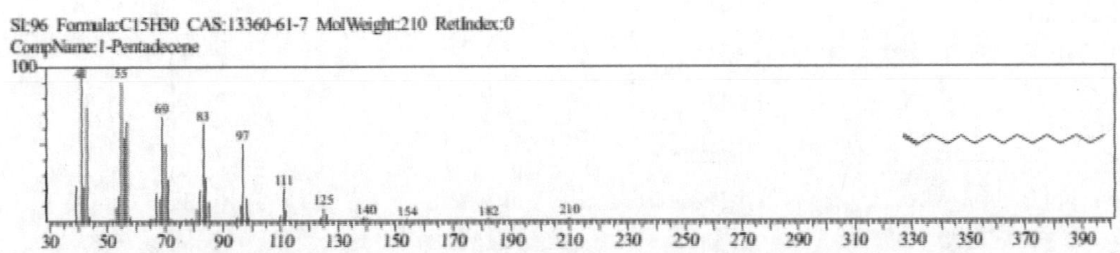

Figure 5. Mass-Spectroscopy of peak 55 from polyethylene oil chromatogram

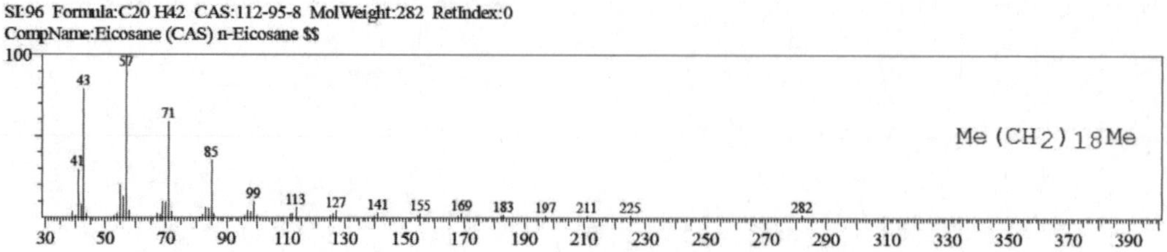

Figure 6. Mass-Spectroscopy of peak 101 from polyethylene oil chromatogram

The GC-MS analysis shows that the liquid polyethylene contains hydrocarbon compound with the carbon chain ranging from C_7 until C_{20}. The analysis was conducted at a column temperature of 60°C with split injection mode and electron ionization system, and also helium gas as the gas carrier. The analysis shows that the main compounds of the liquid polyethylene as detected by chromatogram are C_9H_{18} with a molecular weight of 126, $C_{12}H_{24}$ with a molecular weight of 168, $C_{15}H_{30}$ with a molecular weight of 210, and $C_{20}H_{42}$ with a molecular weight of 282. The result shows that some long-chain hydrocarbon molecules dominate the composition of the liquid polyethylene as the pyrolysis product. The long chain hydrocarbon molecules usually provide high boiling point temperature to convert the liquid into the gas phase. Therefore, that kind of liquid fuel actually needs to a posttreatment by catalytic cracking to reduce the carbon chain at least into C_7 until C_{15}. Short chain hydrocarbon will convert into gas easier than the long chain hydrocarbon, and it may increase its performance as fuel for the solid oxide fuel cell.

Table 2. Hydrocarbon compound of polyethylene pyrolysis oil

No	Compound of Hydrocarbon	Amount of Hydrocarbon (%)
1	C_5H_{10}	1.49
2	C_6H_{12}	1.69
3	C_7H_{14}	2.04
4	C_8H_{16}	2.18
5	C_9H_{18}	10.44
6	$C_{10}H_{20}$	3.8
7	$C_{11}H_{22}$	2.71
8	$C_{12}H_{24}$	14.11
9	$C_{13}H_{26}$	2.72
10	$C_{14}H_{28}$	9.06
11	$C_{15}H_{30}$	2.4
12	$C_{16}H_{32}$	9.61
13	$C_{17}H_{34}$	3.35
14	$\geq C_{18}H_{38}$	34.4

From this research, SDC and NSDC have been synthesized by the sol-gel method. Analysis of SDC powder was done by XRD to determine the crystal structure. The diffraction pattern of SDC compared to SDC diffraction standard from ICSD #28791 is shown in Figure 7. It can be seen that the peaks are in good agreement with the standard peaks, confirming that the SDC was successfully synthesized.

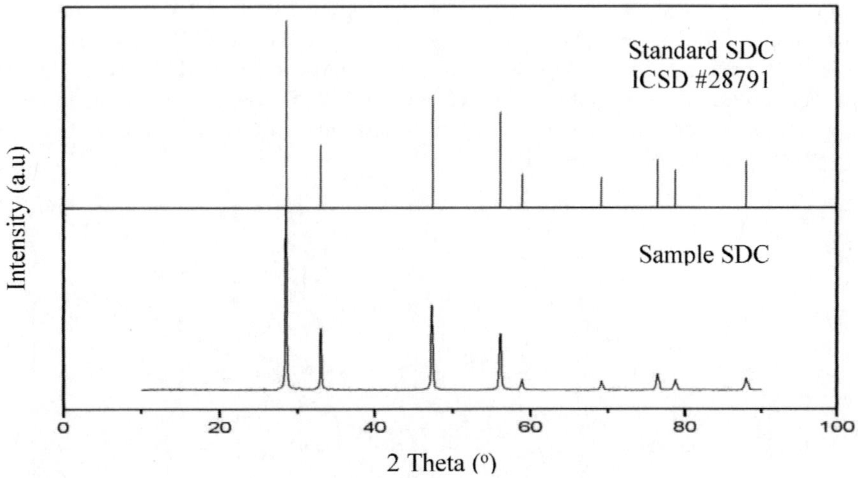

Figure 7. Diffractogram of Composite SDC and SDC standard

Meanwhile, NSDC is a mixture of SDC with Na_2CO_3 at a molar ratio of 1: 1. Before being mixed, Na_2CO_3 was calcinated at 800°C to release the H_2O molecules. The diffractogram (Figure 7) does not show any characteristic peaks of Na_2CO_3. It indicates that the Na_2CO_3 is in an amorphous structure [10, 12]. The amorphous Na_2CO_3 might coat the SDC particles [14] (Zhu et al., 2006). Le Bail refinement on NSDC diffraction data found that the NSDC has two phases, i.e SDC and Na_2O. The SDC is cubic with a space group of *FM 3 M*. Meanwhile, Na_2O also in a cubic structure with space group *FM 3 M*. The cell parameters are listed in Table 2. The cell parameter of SDC is 5.437909(2) Å, and it becomes 5.433043(2) Å after combined with Na_2CO_3. It indicates that the sodium ions, Na^+, or did not enter into the SDC structure. Both are combined as a composite which still carries they own properties.

Table 3. The cell parameters of SDC and NSDC as the result of Le Bail refinement with cubic structure and space group of *FM3M*

Parameter cell	SDC	NSDC	
	SDC cubic F M 3 M	SDC cubic F M 3 M	Na_2O F M 3 M
a(Å)	5.437909(2)	5.433043(2)	5.550000(0)
b(Å)			
c(Å)			
Volume (Å^3)	160.803619(1)	160.36852(9)	170.953888
Rp %	7.828	3.87	
Rwp %	7.926	3.80	

The single cell test of NSDC-L |NSDC| NSDC-L was conducted at 300, 400, 500, and 600 °C. The I-V curve is described in Figure 9. The result shows that the power density increases as the temperature increases. Meanwhile, the open circuit voltage, OCV reached the highest value on 500°C of temperature and was measured at 0.475 V. The values are listed in **Table 4**. The highest power density was found at 600 °C of operational temperature on 0.14 mW.cm^{-2}. However, the power density which is 0.14 mW.cm^{-2} is still too low for commercial purpose. GC-MS analysis shows that the polyethylene liquid mainly consists of C7-C20. Heating at 150 °C apparently only vaporize C7 and C8 which are the only small amount of the liquid fuel. Therefore, only small amount of fuel that would reform into hydrogen and carbon monoxide.

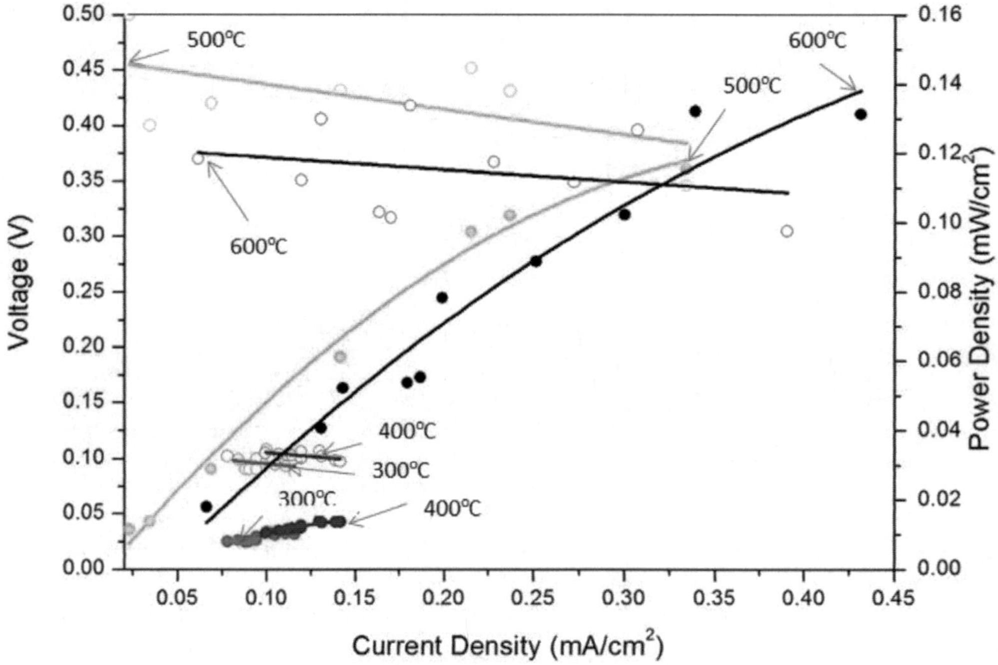

Figure 9. Effect of operation temperature on NSDC application loaded by polyethylene oil.

Table 4. The Open circuit voltage (volt) and power density (mW.cm^{-2}) at various temperature

Temperature (°C)	Open circuit voltage (volt)	Capacity (mW.cm^{-2})
300	0.10	0.0105
400	0.11	0.0140
500	0.475	0.1200
600	0.400	0.1400

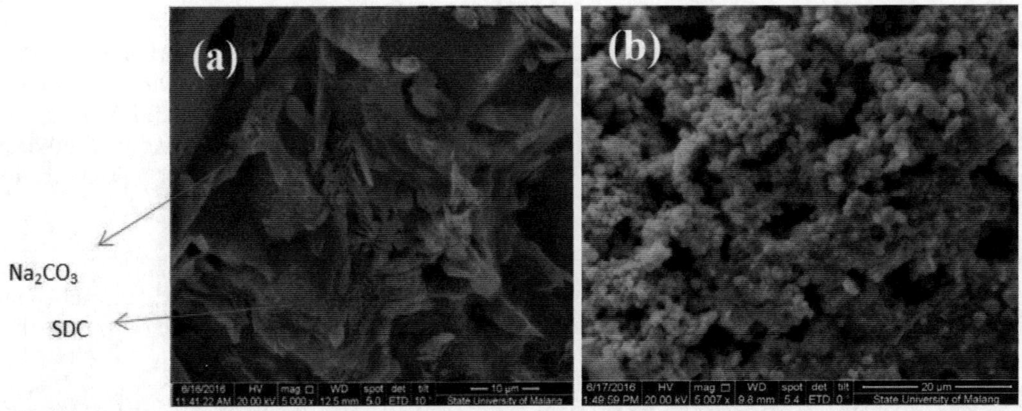

Figure 8. NSDC Image by SEM (a) before and after (b) being loaded with polyethylene oil

SEM analysis shows that there is a different morphology of NSDC electrolyte at before and after being loaded with polyethylene oil. Na_2CO_3 showed as an irregular chunk (**Figure 8a**) due to the preheating treatment changed sodium carbonate to sodium oxide. This is in accordance with the Le Bail refinement which confirms the presence of Na_2O instead of Na_2CO_3. The morphology after being loaded with polyethylene oil has a different form (**Figure 8b**). Heat treatment during fuel cell operation seemed to melt Na_2O chunks and the melted form layered the SDC particles.

Elemental analysis by EDX on three positions of the cathode-NSDC interface, in the middle of NSDC, and at the anode-NSDC interface shows that during fuel cell operation at 300 – 600 °C there was a migration of ions whether from the electrolyte, NSDC, to electrode, NSDC-L or vice versa. It can be seen in Table 5. That the Cerium content in electrolyte part (in the middle) increased from 08.03 % to 25.82 %, indicating the Cerium ions transport from the electrode to electrolyte. However, cerium ions content also increased on cathode and anode part. It might because of the decreasing of carbon content that caused the cerium part to be increased and also the migration from electrolyte part. Table 5 shows that the increasing of Ce content is highest among others. It confirms that the increasing was not only coming from rational effect but also because of ionic migration. Meanwhile, Cu ions are detected at the cathode and anode part, but those are not detected in the electrolyte part. It indicates that there was no migration of Cu ions from the electrode to the electrolyte.

Table 5. Elemental content of single cell NSDC-L(cathode)|NSDC|NSDC-L(anode) at before and after fuel cell test. The EDX analysis was conducted at three points.

Element	% Weight at Point					
	Cathode-NSDC interface		Middle (NSDC)		NSDC-anode interface	
	Before	After	Before	After	Before	After
C	18.50	07.80	13.56	06.39	13.32	05.64
O	42.00	35.92	33.33	31.91	35.83	24.00
Na	33.25	32.86	41.45	28.88	34.74	16.83
Ce	02.54	19.05	08.03	25.81	10.40	37.75
Sm	01.21	03.79	02.66	06.84	02.95	12.06
Ni	01.90	00.42	00.96	00.17	01.86	02.32
Cu	00.60	00.17	-	-	00.90	00.41

4. Conclusions

Polyethylene liquid that was produced through pyrolysis method contains hydrocarbon chain ranging C_5 - C_{20}. A single fuel cell of NSDC-L |NSDC|NSDC-L can convert polyethylene liquid into electricity and produce a maximum power density of 0.14 mW/cm^2 and open circuit voltage of 0.4 Volt at 600 °C. Gas reforming needs to be increased by heating treatment method and fuel cell design. SEM/EDX analysis of the three spots confirm that there was an ionic migration between electrode-electrolyte part.

References

[1] Syahputra RJE, Tri U, Arikasuci RSF, Widy A, and Khoirina DW 2015 Pemanfaatan limbah pecahan genteng sebagai katalis dalam reaksi pirolisis plastik polipropilena menjadi bahan bakar alternatif *Jurnal Rekayasa Kimia Lingkungan* Universitas Syiah Kuala

[2] Jahirul M, Rasul M, Chowdhury A, and Ashwath N 2012 *Biofuels Production*

[3] Kaminsky W 1993 Recycling of polymers by pyrolysis *Le Journal de Physique* IV 03 C7–1543–C7–1552 doi:10.1051/jp4:19937241

[4] Ying Ma 2009 *Ceria-based Nanocomposite Electrolyte for Low-Temperature Solid Oxide Fuel Cells* Thesis Royal Institute of Technology Sweden

[5] Wibawa ASA 2011 *Studi Sifat Minyak Pirolisis Campuran Sampah Biomassa dan Sampah Plastik Polypropylene (PP)* Jurusan Teknik Mesin UNS

[6] Wong SL, Ngadi N, Amin NAS, Abdullah TAT, and Inuwa IM 2015 Pyrolysis of low-density polyethylene waste in subcritical water optimized by response surface methodology *Environmental Technology* 1–10 doi:10.1080/09593330.2015.1068376

[7] Suhada H 2001 Fuel Cell Sebagai Penghasil Energi Abad 21 *Jurnal Teknik Mesin* 3 92-100.

[8] Achilias DS, Roupakias C, Megalokonomos P, Lappas AA, and Antonakou EV 2007 Chemical recycling of plastic wastes made from polyethylene (LDPE and HDPE) and Polyethylene (PP) *Journal of Hazardous Materials* 149 536–542.

[9] Steele BCH and Heinzel A 2001 Material for Fuel-Cell Technologies *Nature* 414 345.

[10] Rahmawati F, Winahyu F, Eddy H and Dani GS 2015 Oxygen Ion-Conductivity and Chemical Stability of Ceria-Based Electrolyte in Composite with Sodium Carbonate as electrolyte for Direct Biodiesel-Fuel Cells *Procedia Chemistry* 14. 164-170

[11] Carmine J and Dicks A 2003 *Fuel Cell System Explained Second Edition* John Wiley & Sons. England.

[12] Wang X, May, Raza R, Mamoun M, and Zhu B 2008 Novel core-shell SDC/amorphous Na_2CO_3 nanocomposite electrolyte for low-temperature SOFC *Electrochemistry Community* 10 1617–20

Acknowledgment

This research is a part of Hibah Kompetensi Project 2016 funded by Kemenristekdikti. Some of the data also project that was funded by Pertamina Corporation through OSN project. Authors acknowledge all of the financial supports.

IOP Conf. Series: Materials Science and Engineering **172** (2017) 012044 doi:10.1088/1757-899X/172/1/012044

Synthesis of *t*-Butyl (2*R*)-Hydroxyisovalerate, A Precursor of Aureobasidin B

R Maharani, D Puspitasari, Taufiqqurahman, D H P Huspa, A T Hidayat, D Sumiarsa, and I W Hidayat

Department of Chemistry, Universitas Padjadjaran, Jalan Raya Bandung-Sumedang Km.21 Jatinnagor 45363 Kabupaten Sumedang West Java Indonesia

Email: r.maharani@unpad.ac.id

Abstract. Aureobasidins are a family of cyclodepsipeptides have antifungal properties. They were isolated from the black yeast *Aureobasidium pullulans* R106 and over 30 derivatives have been successfully characterized. There are few publications reporting the total synthesis of aureobasidins. The limited reports of the synthesis of the aureobasidin derivatives are due to the difficult access to the preparations of precursors. The aim of this research is to synthesise a precursor of aureobasidin B, *t*-butyl (2*R*)-hydroxyisovalerate (*t*-Bu-Hiv), that is prepared for the total synthesis of aureobasidin B. The synthesis of AbB is planned to be undertaken by using a solid phase method, so the ester formation between *t*-Bu-Hiv and the Fmoc-β-hydroxymethylvaline will be carried out in solution phase to form depsidipeptide. The *t*-butyl group was used as protecting agent that is due to the straightforward elimination of the protecting group from the Fmoc-depsidipeptide. The *t*-Bu-Hiv acid was prepared from D-valine through diazotisation to form (2*R*)-acetyloxyisovaleric acid in 62.7% yield. Product of the first step was then protected by *t*-butyl group by using Boc-anhydride in *t*-butanol to give *t*-butil (2*R*)-acetyloxyisovalerate in 44% yield. In the last step, the acetyloxy group was eliminated by using potassium carbonate in methanol/water to give the desired product, *t*-Bu-Hiv in 33.5% yield. The *t*-Bu-Hiv is ready to be combined with Fmoc-β-hydroxymethylvaline to result in depsidipeptide that will be attached to the resin in the total synthesis of AbB. Each stage of this synthesis was controlled by thin layer chromatography and all products were purified by open column chromatography. All the synthesized products were characterized by various spectroscopic techniques, including infrared spectrophotometer, mass spectroscopy (ESI-MS), ^{1}H-NMR and ^{13}C-NMR.

1. Introduction

Aureobasidins, isolated from the black yeast *Aureobasidium pullulans* R106, are highly *N*-methylated cyclodepsipeptides, which have antifungal activity. Over 30 aureobasidin analogues have been successfully characterized including the major compound, aureobasidin A (AbA) (Figure 1). All of the aureobasidins are cyclic depsipeptides consisting of eight amino acids in which three or four of them are *N*-methylated and one alpha hydroxy acid is always present. All of the amino acids have the L-configuration.(1-3)

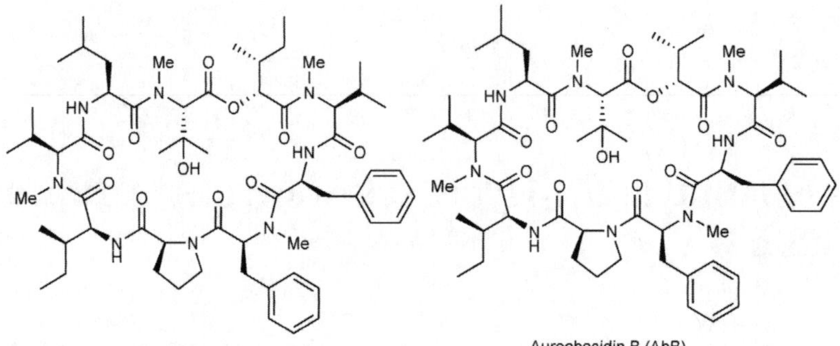

Figure 1. Structures of aureobasidin A (AbA) and aureobasidin B (AbB) .

There were few papers reporting total synthesis of aureobasidin compounds. In the 1990s, three articles described total synthesis of AbA in solution phase and another article was reported in 2014 for total synthesis of analogue of AbL, [2S,3S-Hmp]-AbL, carried out in a combination of solution- and solid-phase synthesis.(4-7) These few reported synthesis showed that the access into the aureobasidins was limited. One of the issues following the synthesis are the presence of some unusual residues that could be either commercially available with expensive prices or commercially unavailable and would need to be synthesized. Some of these residues are N-methylated amino acids and hydroxy acids. The later one would be our focused topic, particularly in preparation of the hydroxy acid.

Hydroxy acids, such as 2-hydroxy-2-methylpentanoic acid (D-Hmp) and 2-hydroxyisovaleric acid (Hiv), are moieties often found in aureobasidins and access to these types of compounds has been reported.(8-11) In this present work, Hiv as a residue of AbB (Figure 1) was being a target of synthesis, providing residue for the total synthesis of AbB that has not been synthesized priorly. A *t*-Bu-protected Hiv, was explored as this protecting group can be easily removed in acidic conditions. This strategy had been employed by Sleebs *et al.* (2011), when *t*-butyl L-phenyllactate was prepared for coupling to Fmoc-L-pipecolic acid under Mitsunobu conditions and by Maharani *et al.* (2014), when *t*-butyl (2*S*)-hydroxy-(3*S*)-methylpentanoate was prepared for coupling to Fmoc-MeVal under Steglich esterification.(7, 12)

2. Methodology

Methodology for the synthesis of *t*-butyl (2*R*)-hydroxyisovalerate was carried out through a methodology applied for the synthesis of *t*-butyl (2*S*)-hydroxy-(3*S*)-methylpentanoate.(7)

2.1. Formation of (2R)-acetyloxyisovaleric acid

D-valine (2 g;17.1 mmol) was dissolved in glacial acetic acid (25.6 mL). The stirred solution was occasionally cooled to keep the solution at room temperature. To this solution was added sodium nitrite (2.35 g;34 mmol) in several portions over 1 h. Once the addition was complete, the solution was stirred for 24 h. The reaction mixture was then evaporated and the crude was dissolved in ether. The ether solution was washed with water several times and then extracted with saturated sodium bicarbonate solution. The combined aqueous solution was acidified with 2 N hydrochloric acid and then extracted with ether. The organic layer was dried and evaporated to give (2*R*)-acetyloxy isovaleric acid (62.7%) as a colorless oil.

2.2. Formation of t-butyl (2R)-acetyloxyisovalerate

To a solution of (2*R*)-acetyloxy isovaleric acid (1.5 g;9.38 mmol) in *t*-butanol (5.95 mL) was added di-*tert*-butyldicarbonate (2.57 g;11.17 mmol) and 4-dimethylaminopyridine (0.14g;1.17 mmol). The

IOP Conf. Series: Materials Science and Engineering **172** (2017) 012044 doi:10.1088/1757-899X/172/1/012044

solution was stirred under nitrogen at room temperature for 4 h with the progress of reaction was monitored by TLC. The reaction mixture was evaporated and the resulting crude was dissolved in ethyl acetate. The ethyl acetate solution was washed with saturated sodium bicarbonate solution, dried with $MgSO_4$ and evaporated. The crude was purified by silica gel column chromatography (hexane:ethyl acetate = gradient eluent) to give pure *t*-butyl (2*R*)-acetyloxyisovalerate (43.9%) as a colorless oil.

2.3. *Formation of t-butyl (2R)-hydroxyisovalerate*
Compound *t*-butyl (2*R*)-acetyloxyisovalerate (0.2 g;67.5 mmol) was dissolved in water (3.39 mL) and methanol (2.54 mL). To this solution was added potassium carbonate (0.5 g;3.64 mmol) and the solution was stirred for 18 h with monitoring by TLC. The reaction mixture was evaporated and the resulting crude was evaporated and partitioned between water and ether. The organic layer was dried and evaporated to give *t*-butyl (2*R*)-hydroxyisovalerate (33.5%) as a colorless oil.

2.4. *Characterizations of synthesized products*
All of the synthesized products either intermediates or final products were characterized by spectroscopic methods, ESI-MS, IR, ^{1}H- and ^{13}C-NMR.

3. Results and Discussion
Preparation of *t*-butyl (2*R*)-hydroxyisovalerate started from a conversion of D-Val into its α-acetyloxy acid. According to a protocol explained by Kolasa and Miller, the deaminative hydroxylation of D-Val resulted in acetate.(13) The diazotisation reaction was carried out in the presence of glacial acetic acid/$NaNO_2$ generating 62.7% yield of the desired product (Figure 2). The product was characterized by ESI-MS, IR and NMR. ESI-MS analysis showed correct molecular mass of the acetate with $[M+H]^+$ 161,0609, representing the desired product. This data was supported by IR analysis showing the presence of broad O-H stretch with υ_{max} 3310 cm^{-1} that is characteristic for COOH group. Two C=O stretches at υ_{max} 1750 and 1730 cm^{-1} showed the existence of two carbonyl groups of ester and carboxylic groups, respectively. The presence of a signal at δ_H 2.15 ppm in the ^{1}H-NMR represented methyl protons in the acetate moiety. In addition, the presence of one additional carbonyl signal (δ_C 171.4 ppm) showed that the acetate was present in the structure.

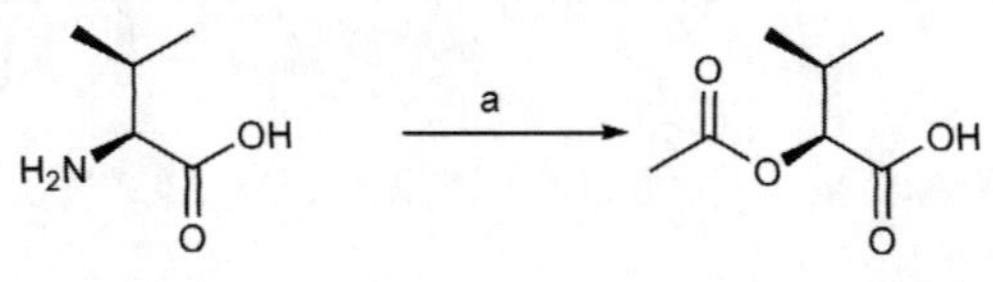

Figure 2. Synthesis of (2*R*)-acetyloxyisovaleric acid, (a) acetic acid, $NaNO_2$.

A mechanistic study by Sleebs describes the active intermediate, a diazonium ion, as being attacked by acetic acid instead of water (Figure 3).

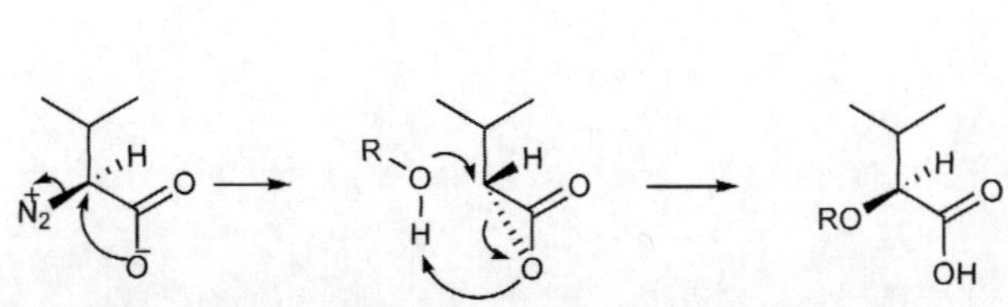

Figure 3. Mechanism for α-hydroxy carboxylic acid formation (R = H or R = Ac).

The *t*-Bu protective group was introduced to the (2*R*)-acetyloxyisovaleric acid by the addition of *t*-butanol, $(Boc)_2O$ and a catalyst, DMAP, in 43.9% yield (Figure 4). The structural characterisation of the resulting compound was based on ESI-MS, IR and NMR. ESI-MS analysis provided the molecular mass ion of *m/z* 217.0182, indicating the $[M+H]^+$ ion of the desired product, *t*-butyl (2*R*)-acetyloxyisovalerate. The absence of wavenumber of –COOH group has also supported that the carboxylic group has been successfully esterified. Nine upfield protons at δ_H 1.46 ppm in the ^{1}H-NMR with a singlet multiplicity were assigned as the three methyls corresponding to the *t*-butyl protective group. A downfield quaternary carbon at δ_C 82.0 ppm also indicated the presence of this group.

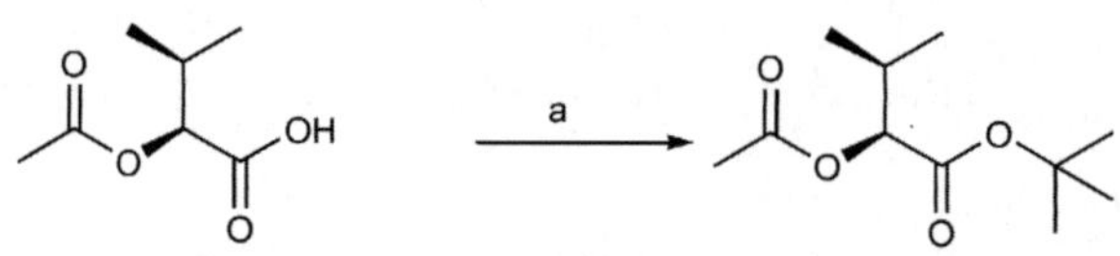

Figure 4. Synthesis of *t*-butyl (2*R*)-acetyloxyisovalerate, (a) *t*-butanol, $(Boc)_2O$, DMAP.

Once the *t*-butyl group was attached to the Hiv-acetate, the next task was to remove the acetyl group affording the corresponding *t*-butyl (2*R*)-hydroxyisovalerate. Through a simple deprotection reaction under basic conditions (K_2CO_3 in 50% MeOH in water), *t*-butyl (2*R*)-hydroxyisovalerate was afforded in 33.5% yield (Figure 5). The ESI-MS spectrum showed molecular mass ion of *m/z* 175,1189, indicating $[M+H]^+$ of the final product, *t*-butyl (2*R*)-hydroxyisovalerate (Figure 6). This data was supported by the IR spectrum, showing the presence of O-H stretch with wave number of 3517 cm^{-1} (Figure 7). This indicated that the acetyl group had been removed leaving the hydroxyl group. The successful reaction was also characterized by the loss of acetate signals in the ^{1}H-NMR and ^{13}C-NMR spectra (Figure 8 and 9).

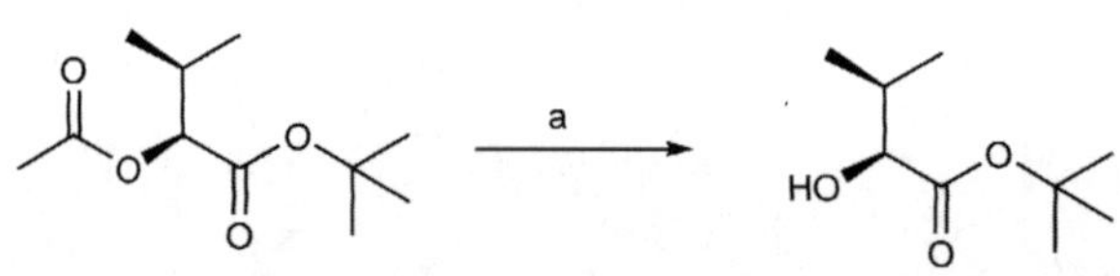

Figure 5. Acetate removal, (a) K_2CO_3 in 50% MeOH in water.

IOP Conf. Series: Materials Science and Engineering **172** (2017) 012044 doi:10.1088/1757-899X/172/1/012044

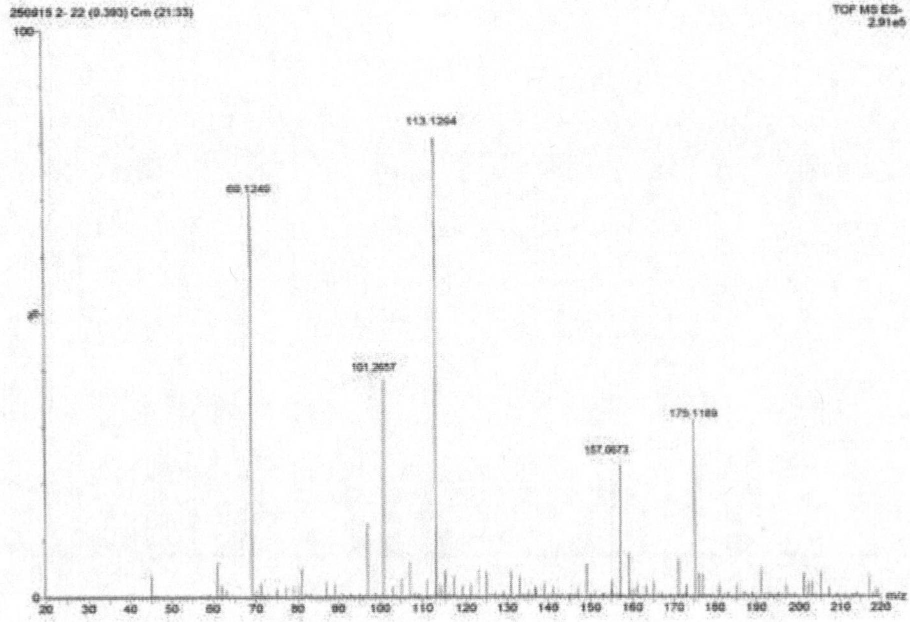

Figure 6. Mass spectrum of *t*-butyl (2*R*)-hydroxyisovalerate.

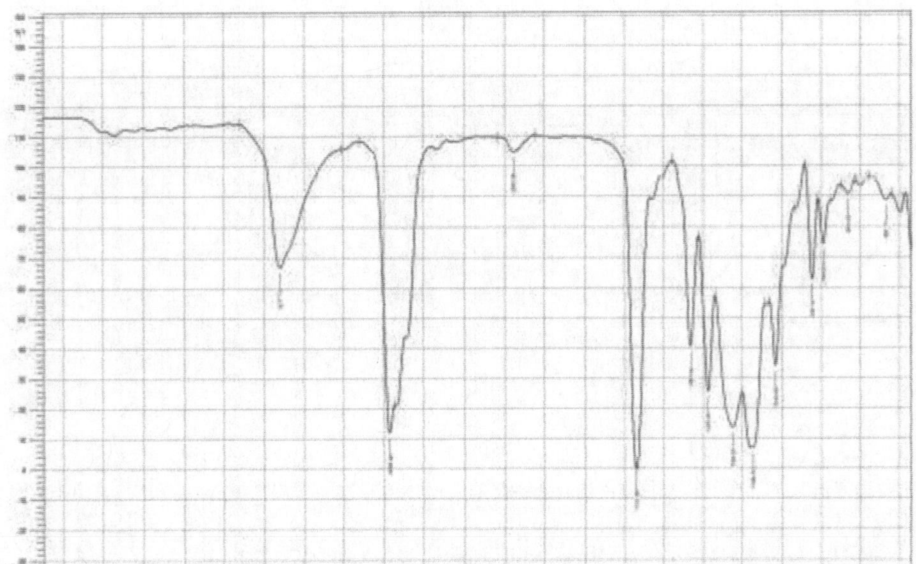

Figure 7. The infrared spectrum of *t*-butyl (2*R*)-hydroxyisovalerate.

IOP Conf. Series: Materials Science and Engineering **172** (2017) 012044 doi:10.1088/1757-899X/172/1/012044

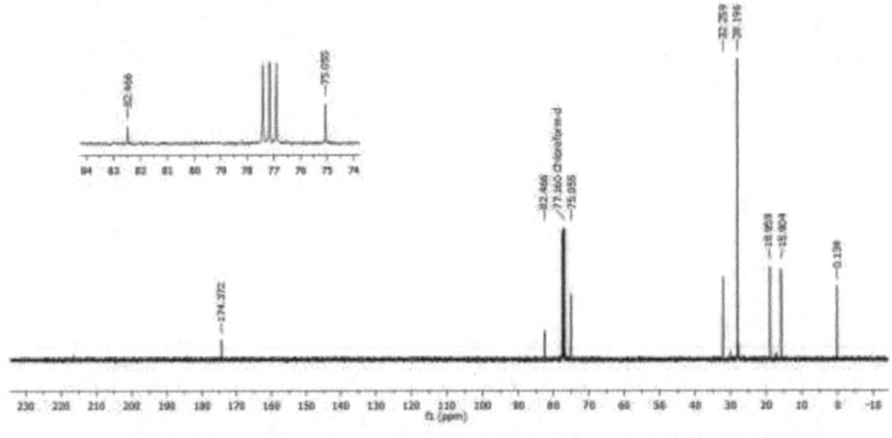

Figure 8. ^{13}C-NMR spectrum of *t*-butyl (2*R*)-hydroxyisovalerate.

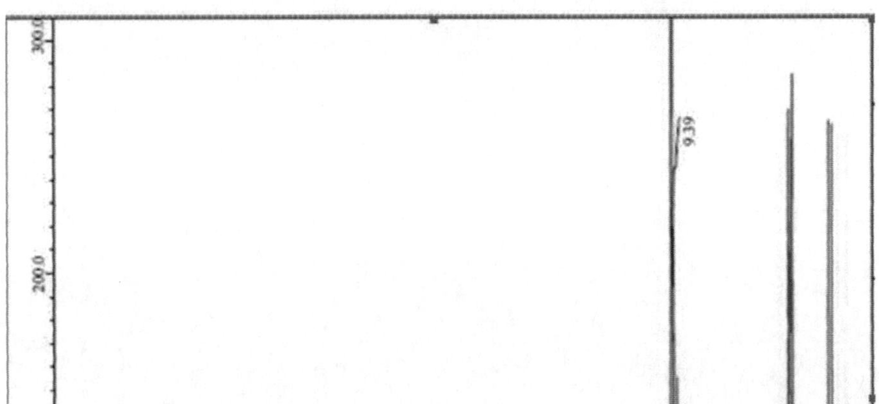

Figure 9. ^{1}H-NMR spectrum of *t*-butyl (2*R*)-hydroxyisovalerate.

4. Conclusion

Compound *t*-butyl (2*R*)-hydroxyisovalerate has been successfully synthesized through three consecutive steps, diazotisation, t-Bu attachment and acetate deprotection with 62.7%, 43.9%, and 33.5% yields, respectively.

All of the synthesized products, either intermediates or final product, were characterized by spectroscopic methods, IR, ESI-MS and NMR.

Acknowledgment

Author would like to thank KLN-DIKTI research grant for financial support, Ms. Elvira Hermawati, M.Si and Ms. Ida Nur Farida, S.Si, for the NMR and ESI-MS measurements.

References

1. Turner W and Rodriguez M 1996 *Current Pharmaceutical Design* **2(2)** 209-24.
2. Awazu N, Ikai K, Yamamoto J, Nishimura K, Mizutani S, and Takesako K 1995 *The Journal of antibiotics* **48(6)** 525-7.
3. Takesako K, Mizutani S, Sakakibara H, Endo M, Yoshikawa Y, and Masuda T 1996 *The Journal of antibiotics* **49(7)** 676-81.
4. Kurome T, Inami K, Inoue T, Ikai K, Takesako K, and Kato I 1996 *Tetrahedron* **52(12)** 4327-46.
5. Jao E, Cooper AB, Rane DF, Saksena AK, Desai J, and Wang J 1996 *Tetrahedron letters* **37(32)** 5661-4.
6. Schmidt U, Schumacher A, Mittendorf J, and Riedl B 1998 *The Journal of peptide research* **52(2)** 143-54.
7. Maharani R, Brownlee RT, Hughes AB, Abbott BM 2014 *Tetrahedron* **70(14)** 2351-8.
8. Snowden RL, Grenno R, and Vial C 2005 *Flavour and fragrance journal* **20(4)** 372-80.
9. Elgersma RC, Mulder GE, Posthuma G, Rijkers DT, and Liskamp RM 2008 *Tetrahedron Letters* **49(6)** 987-91.
10. Zubia A, Mendoza L, Vivanco S, Aldaba E, Carrascal T, and Lecea B 2005 *Angewandte Chemie* **117(19)** 2963-7.
11. Koh JT, Cornish VW, and Schultz PG 1997 *Biochemistry* **36(38)** 11314-22.
12. Sleebs MM, Scanlon D, Karas J, Maharani R, and Hughes AB 2011 *The Journal of organic chemistry* **76(16)** 6686-93.
13. Kolasa T and Miller MJ 1987 *The Journal of Organic Chemistry* **52(22)** 4978-84.

IOP Conf. Series: Materials Science and Engineering **172** (2017) 012045 doi:10.1088/1757-899X/172/1/012045

The Effect of Temperature on the Production of Nitrobenzene

R Agustriyanto, L Sapei, G Rosaline and R Setiawan

University of Surabaya, Surabaya, Indonesia

Email: rudy.agustriyanto@staff.ubaya.ac.id

Abstract. Nitrobenzene is a compound derived from benzene and quite fundamental in the petrochemical industry. One of which is in the manufacture of aniline. Aniline has an important role as dyes, synthetic rubber materials, rocket fuels and pharmaceuticals. Nitrobenzene can be synthesized by the nitration reaction of benzene. The purpose of the research is to design synthesis process of nitrobenzene by using Aspen Hysys simulator. The effect of changes in operating condition on nitrobenzene will be investigated. The benefits of the research are to obtain the optimum operating temperature for nitrobenzene production so that the use of utility can be minimized.

1. Introduction

Nitration reaction is important because it is the most general process for the preparation of aromatic nitro compounds. This reaction plays a role in the development of theoretical organic chemistry. It is of interest because of its characteristics as an electrophilic substitution [1]. Nitration of aromatics is a large scale industrial process with global production of about 4-5 million tons of commercially valuable chemicals (nitrobenzene, nitrotoluene, 2,4-dinitrotoluene) per year [2].

Nitrobenzene has been produced commercially since 1856 when Simpson, Maule and Nicholson started operation in England [3]. It was first synthesized in 1834 from benzene and fuming nitric acid. Since then the nitration of aromatic hydrocarbons has been the subject of numerous investigations, however, only a few investigators have determined data which are useful for engineering applications [4].

Nitrobenzene is produced by reacting benzene with a mixture of nitric acid and sulphuric acid at a temperature not exceeding 50°C, to avoid getting more than one nitro group [5]. The mixture is held at this temperature for about half an hour. The sulphuric acid is acting as a catalyst. Yellow oilly nitrobenzene is formed. The reaction is as follows:

$$C_6H_6 + HNO3 \rightarrow C_6H_5NO_2 + H_2O \tag{1}$$

or

(2)

The electrophile is the nitronium ion or the nitryl cation, which is formed by reaction between nitric acid and sulphuric acid:

IOP Conf. Series: Materials Science and Engineering **172** (2017) 012045 doi:10.1088/1757-899X/172/1/012045

$$HNO_3 + 2H_2SO_4 \rightarrow NO_2^+ + 2HSO_4^- + H_3O^+$$

The mechanism of electrophilic substitution are as follows:

Stage 1:

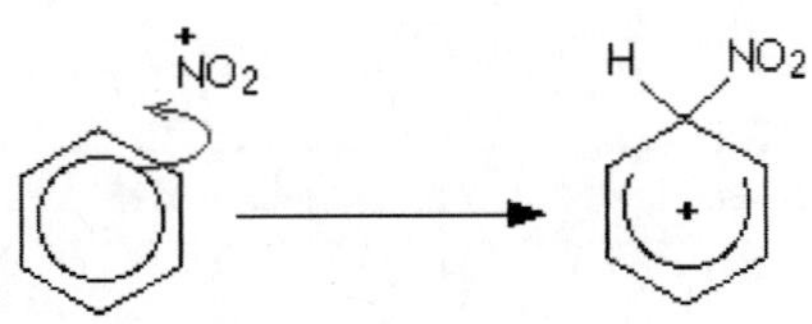

(3)

Stage 2:

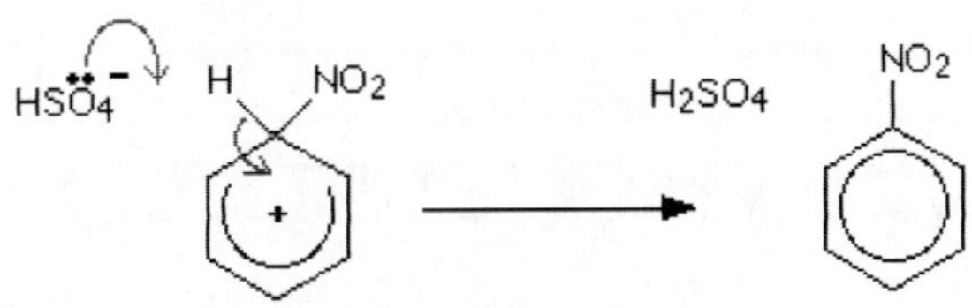

(4)

Chemical engineering research on nitration process of benzene has achieved many valuable improvements including better energy management of modern nitration plants and various clean nitration approaches focused on dispensing the use of sulphuric acid [6]. The catalyst (sulphuric acid) were costly to recycle and generated environmental problems making nitration of benzene as one of the most hazardous industrial processes. Therefore many researches focuss on the use of other material as catalyst [1,6-8].

Quadros et al [9] constructed a pilot plat for the continuous nitration. The following equations correlate E_a and k_o to sulphuric acid weight fraction S:

$$E_a = (-283.88S + 263.37) \times 1000 \quad (5)$$

$$k_o = \exp\left[166.64S^2 - 254.36S + 113.79\right] \quad (6)$$

By the availability of process simulator like Aspen Hysis, process design can be done more easily, since the derivation of the mathematical model for the process were difficult. The aim of this research is to simulate the nitration reaction of benzene at various temperature. The scope is mainly for the nitration reactor. The percentage of conversion will be shown over temperature for a specific loading duty in the reactor.

2. Method

In this study, Aspen Hysis were used to simulate process flow diagram for nitrobenzene production. Figure 1 shows the methodology framework.

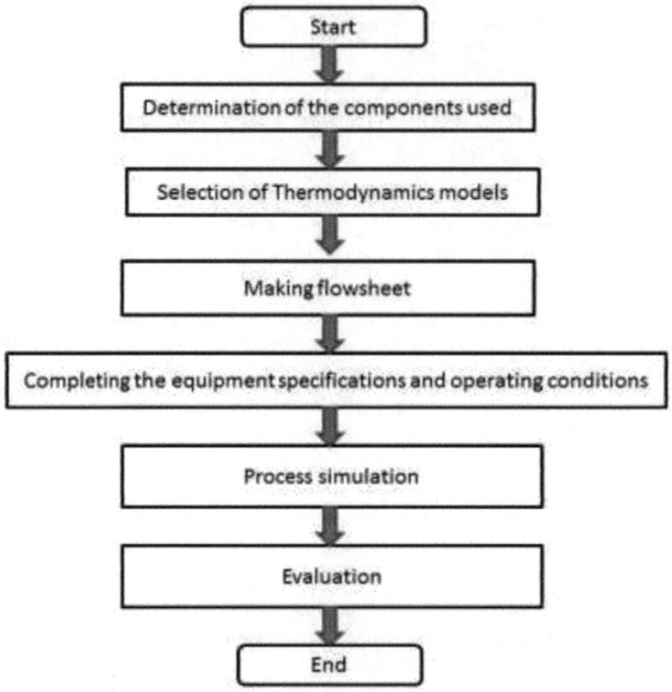

Figure 1. Methodology framework

Isothermal CSTR is used, and Table 1 shows the operating condition of the simulated system. Process Flow Diagram of the production system of nitrobenzene is shown in Figure 2.

Table 1. Operating conditions

Operating conditions	**Values**	
Mass flow benzene	1000	kg/h
Concentration of benzene	89.99	%
Mass flow of nitric acid	1173	kg/h
Concentration of nitric acid	65	%
Mass flow of sulphuric acid	2561	kg/h
Concentration of sulphuric acid	89	%
Reactor diameter	1.221	m
Reactor height	1.832	m
Pressure	101.3	kPa
Liquid volume in the reactor	80	%

IOP Conf. Series: Materials Science and Engineering **172** (2017) 012045 doi:10.1088/1757-899X/172/1/012045

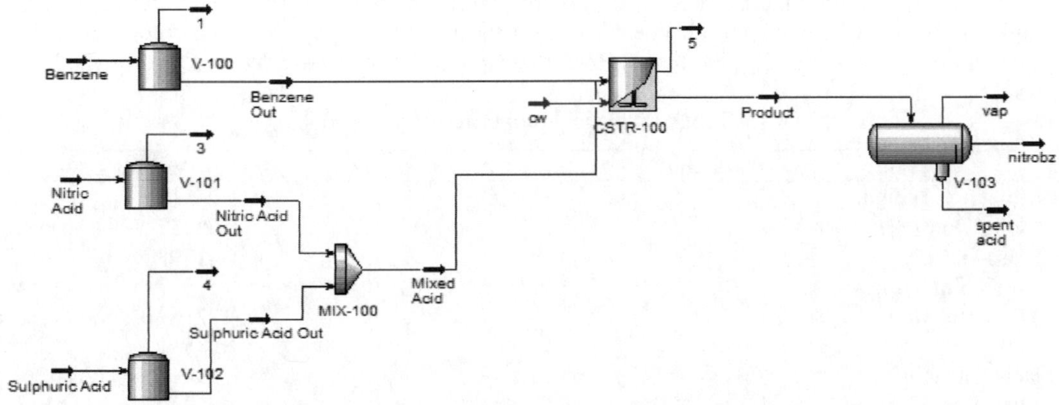

Figure 2. Simplified Nitration Process Flow Diagram

3. Results and Discussion

Figure 3 shows plot of benzene conversion as a function of temperature.

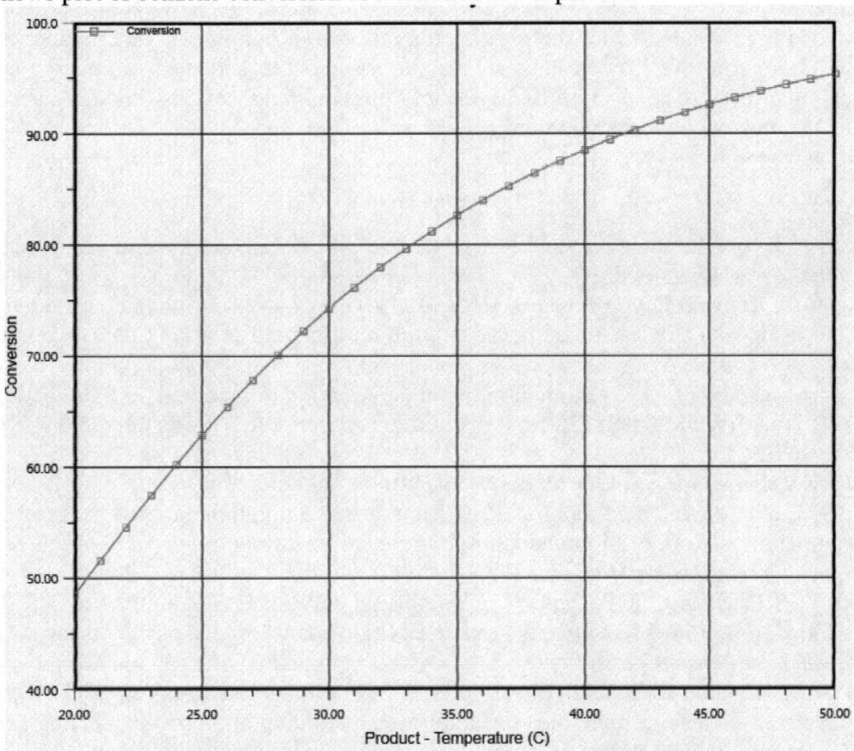

Figure 3. Plot of Conversion vs. Operating Temperature

These results were obtained while the physical properties of products were calculated automatically in Hysis as shown in Table 2. Properties of other streams can also be found.

It can be seen from Figure 3 that the conversion can reach 95.32% at 50°C and as low as 48.54% at 20°C. These conversions were obtained based on the loading given. The most crucial point in the simulation is the value of E_a and k_o used. Here, the parameter values provided by Quadros (2005) were used.

Table 2. Physical properties of the product

Name	Values	
Molecular Weight	52.11	
Molar Density	26.82	Kmole/m^3
Mass Density	1398	kg/m^3
Act. Volume Flow	3.387	M^3/h
Mass Enthalpy	-7169	kJ/kg
Mass Entropyc	0.8385	kJ/kg°C
Heat Capacity	100.5	kJ/kmole.°C
Mass Heat Capacity	1.929	kJ/kg.°C
Viscosity	2.309	cP
Thermal Conductivity	0.3902	W/mK
Viscosity Index	9.394	

4. Conclusions

Simulations have been conducted for the nitrification reaction of benzene at various temperatures by using Aspen Hysis software. The results are useful for process and energy optimization in the petrochemical industries. The possibility to explore the influence of residence time (hence the dimension of the reactor), ratio of sulphuric acid to nitric acid, and ratio of benzene to mix acid are open for further research.

5. References

[1] Hoggett J G, Moodie R B, Penton J R and Schofield K 1971 Nitration and aromatic reactivity, Cambridge University Press

[2] Koskin A P, Kenzhin R V, Vedyagin A A and Mishakov I V 2014 Sulfated perfluoro polymer–CNF composite as a gas-phase benzene nitration catalyst *Catalysis Communications* **53** 83-86

[3] Kobe K A and Mills J J 1953 Mononitration of benzene *Ind. Eng. Chem* **45**(2) 287-91.

[4] Biggs, R D and White R R 1956 Rate of nitration of benzene with mixed Acid *AIChE Jornal* **2**(1) 26-33

[5] http://www.chemguide.co.uk/mechanisms/elsub/nitration.html

[6] Gong S, Liu L, Cui Q and Ding J 2010 Liquid phase nitration of benzene over supported ammonium salt of 12-molybdophosphoric acid catalysts prepared by sol–gel method *Journal of Hazardous Materials* **178**(1-3) 404-8

[7] Ganjala V S P, Neeli C K P, Pramod C V, Khagga M, Rao K S R and Burri D R 2014 Eco-friendly nitration of benzenes over zeolite-β-SBA-15 composite catalyst *Catalysis Communications* **49** 82-86

[8] Kulal A B, Dongare M K and Umbarkar S B 2016 Sol–gel synthesized WO3 nanoparticles supported on mesoporous silica for liquid phase nitration of aromatics *Applied Catalysis B: Environmental* **182** 142-52

[9] Quadros P A, Oliveira N M C and Baptista C M S G 2005 Continuous adiabatic industrial benzene nitration with mixed acid at a pilot plant scale *Chemical Engineering Journal* **108**(1-2) 1-11

IOP Conf. Series: Materials Science and Engineering **172** (2017) 012046 doi:10.1088/1757-899X/172/1/012046

Anti-atherogenic properties of Deglet Noor Date seeds (*Phoenix dactylifera*) Methanol extract on Diet-Induced Hypercholesterolemic Rats.

S Saryono[1], J Eliyan[1], D Herdiati[1], AA Khikmatullah[1], CP Silvana[1], HPAdi[1]
[1]School of Nursing, Health Sciences Faculty, University of Jenderal Soedirman, Purwokerto

Email: sarbiokim@gmail.com

Abstract. This is the first study to investigate the completely anti-atherogenic effect of Deglet Noor Date seeds methanol extract administration on diet-induced hypercholesterolemic rats. About 24 male Wistar rats were divided into 6 groups. The normal control (NC) group, Hypercholesterolemic Control (HC) group was given high cholesterol diet, and Simvastatin Control (SC) group was given 0.18 mg/200g simvastatin after high cholesterol diet induction. The treatment groups of T0.25, T0.5 and T1 were given supplementation of 0.25, 0.5 and 1 g/kg of dates seed extract after high cholesterol diet induction, respectively for 21 days. Blood was collected from orbitals plexus vein for plasma lipid profile analysis. The levels of Total Cholesterol (TC), Low-Density Lipoprotein (LDL) and Atherogenic Index (AI) values were significantly decreased ($p<0.05$) on diet-induced hypercholesterolemic rats after supplemented with date seeds extract (T0.25, T0.5 and T1) but not in Triglycerides (TG). Along with that, High Density Lipoprotein (HDL) level was significantly increased ($p<0.05$). However, the T1 group was the best anti-atherogenic activity in compared to other groups. Results showed that plasma lipid profile was significant to get better after supplemented with date seeds extract.

1. Introduction

Death rate as a result of coronary heart disease is considered high in the world. In the previous epidemiology research, it was found that hypercholesterolemia is the main factor of this disease. High level of cholesterol will result in atherosclerosis that atherogenic index will increase leading to sluggish blood circulation and it a death of heart muscle [1]. The rise of total cholesterol level and low density of lipoprotein (LDL) will increase the risk of coronary heart disease. It is due to high consumption of a saturated fatty meal, low intensity of physical activities and unhealthy lifestyle. Scientific evidence show that oxidative stress is a cause of fat oxidation, functional change of endothelial that leading to atherosclerosis[2].

High consumption of fruits and vegetables is beneficial for health by preventing chronical diseases such as atherosclerosis, cancer, cardiovascular, diabetes, and neurodegenerative diseases. The antioxidant may decrease the risk of chronical diseases and protect health[3]. Most antioxidant capacities are from fruits which contain flavonols and polyphenol. Date seeds contain a lot of phytochemical compounds such as flavonols, isoflavones, flavones, anthocyanins, catechins and isocatechins which equal vitamin C and E or β-carotene, yet they have not been widely harnessed[4,5].

Many researches about the benefits of date extract have been conducted; however, the use of date seeds as the agent of anti-atherogenic by observing the whole indicators has not been proved. This research aims to investigate the effects of Deglet Noor date seeds as an anti-atherogenic for diet-induced hypercholesterolemic rats.

2. Material and methods

2.1. The preparation of sample and date seed extract

Deglet Noor date seeds were separated from the fruit. They were then washed with flowing water, laid out to dry for one day under the sun. About 2.5 kg of dried date seeds were then pounded and blended resulting in powder. This powder will be macerated for 72 hours with methanol, concentrated under the pressure of rota evaporator to reach the desired volume. Date seed powder is collected in glass objects, added with methanol as high as 2.5 cm above the surface of the date powder, stirring until homogeneous and covered with aluminum foil for 24 hours. The filtrate is transferred to erlemeyer and methanol is replaced back as the procedure until 3x24 hours. Filtrate the first, second and third-mingled, then evaporated with a rotary evaporator. The results of the date seed extract weighed and dissolved with 200 ml aquabidest. Extraction result forms a concentrated solution.

2.2. Experimental design

This research applied true experimental with a complete random design. The total of 24 Wistar male rats weighted around 150-200 g with 2-3 months of age were housed in a plastic cage covered with wire plaited cap. Rats were divided into six groups and acclimated for a week. Standard diet and aqua distillation were given through ad libitum. Air exchange and lighting were provided based on natural condition. The normal control group was given standard diet and the other groups were induced hypercholesterolemic with high fat diet of duck egg yolk for three weeks. After induction, group three, four and five were given date seeds extract of 0.25, 0.5, and 1 g/kg consecutively. A control group of simvastatin (SC) was given 0.18mg/200mg treatment for three weeks consecutively. Later, their blood was taken a day after the treatment.

2.3. Measurement of lipid profile

Blood samples were taken from rats after being starved for 8 hours on the last treatment day. The blood was taken as much as 3 cc from orbitalisplexus vein and then contained in EDTA vacutainer tube to check their Total cholesterol (TC), Triglycerides (TG), Low-Density Lipoprotein (LDL) and High-Density Lipoprotein (HDL). All profile lipid parameters applied Enzymatic Colorimetric Assay method. Atherogenic Index (AI) was measured with log 10(TG/HDLc).

2.4. Statistical Analysis

Analysis of Variance and LSD Post Hoc were carried out to compare group mean. The significance is accepted if $p<0.05$. Ethical approval for this research was obtained from the ethical committee of health research, medical faculty of Padjajaran University Bandung.

3. Research Result

Concentrations of TC, LDL,TG and AI after treatment for treatment and SC groups were lower compared with NC and HC groups significantly($p<0,05$) (Table 1). Hypercholesterolemic diet could increase TG, TC, AI and LDL levels before treatment. Date seeds extract supplementation could decrease, TC, AI and LDL significantly than TG level.

IOP Conf. Series: Materials Science and Engineering **172** (2017) 012046 doi:10.1088/1757-899X/172/1/012046

Table 1. Lipid profile asatherogenic indicator after 21 days experimental

Group	TG	HDL	TC	LDL	AI
Negative control (NC)	98.3	38.0	83	21.9	0.49
High fat diet (HC)	117.3	30.8	86.3	28.1	0.35
Date seed extract 0.25 g/Kg (T0.25)	103.5	45.3	72	9.8	0.31
Date seed extract 0.5 g/Kg (T0.5)	76.8	60.5	72.8	5.3	0.12
Date seed extract 1 g/Kg (T1)	73.8	57.3	70	4.2	0.09
Simvastatin0.18 mg/200g (SC)	69.0	60.3	70.3	16.3	0.17
Anova test	p=0.07	p=0.006	p=0.004	p=0.002	p=0.002

The analysis result of one-way ANOVA shows a significant difference of HDL level (p <0,05) in HDL level, total cholesterol, LDL, and atherogenic index; yet the difference was not significant for triglyceride level. Post hoc analysis of LSD reveals that 1 g/kg date seed extract was more significantly effective than other doses (p<0.05) in restoring rats' lipid profile after diet-induced hypercholesterolemia.

4. Discussion

It is the first research to use complete indicators to figure out the use of date seeds to restore lipid profile after diet-induced hypercholesterolemia rat. High saturated fat and cholesterol diets may increase triglyceride and total cholesterol blood levels, resulting in the increase of LDL synthesis by lipoprotein lipase (LPL) enzyme and decrease of HDL synthesis. High level of cholesterol may precipitate invascular and result in cholesterol oxidation in the arterial wall leading to atherogenic level increase (atherogenic index increase). Atherogenic index is a predictor of cardiovascular diseases [6]. Cholesterol will be fixated to triglyceride and apoprotein to generate lipoprotein. Therefore, high cholesterol diet can increase the level of LDL in specimens.

Triglyceride increase will result in the enhanced activities of *Cholesteryl Ester Transfer Protein* (CETP) which participates as a distribution mediator of triglyceride to HDL and ester cholesterol to LDL. Distribution of triglyceride to HDL generates HDL mature leading to the increase of LDL and decrease of HDL level. Antioxidant may also inhibit CETP activities. CETP mediates CE (cholesterol ester) transfer from HDL to pro-atherogenic apoB-lipoprotein and transfers ester cholesterol from triglyceride. CETP inhibition is necessary to develop to decrease the risk of coronary heart disease. The reduced CETP activities will be beneficial for the existence of HDL; thus triglyceride will not mostly be exchanged with HDL particle[7].

Date seeds contain various beneficial components for health[8,9]. The level of flavonol, polyphenol, unsaturated fatty acid, vitamin C and vitamin E, and fibers of date seeds are proved to be beneficial for anti-atherogenic[10]. This compounds act as antioxidants. The evident from previous studies shows that the consumption of date seeds can increase the body's antioxidant(3). Antioxidants prevent the free radicals oxidation thus decreasing LDL-ox. The research result reveals that date seed extract can decrease LDL, TC and AI level as well as increase HDL level.

Quercetin in date seeds as a flavonol has equal activities as α-tokoferol; thus protecting LDL from oxidation reaction. Quercetin reacts with free radicalby donating its proton and transforms into radical compounds; yet, unpaired electron resulted from delocalized by resonance thus the energy content will be low enough to become reactive radical. Moreover, Quercetin may also inhibit secretion of Apo-B100 to intestine, hence Apo- B number will decrease. It will cause the decrease of LDL generation in the human body.

Vitamin C content (ascorbic acid) in date seeds can also be used in hydroxylation reaction in the process of generating bile acids. The rise of bile acids will increase cholesterol excretion of LDL leading to the decrease of LDL cholesterol level in blood. The antioxidant of ascorbic acid also possesses the effect of slowing down autoxidation process through various mechanism such as metal ion fixation, oxidation, hyperoxide degradation to non-radical products, UV radiation absorption, and oxygen singlet deactivation[11].

Phenolic content in date seeds will inhibit the generation of HMG Ko-A reductase enzyme which functions in converting Asetil Ko-A to mevalonate in lipid metabolism in the human body. With no production of mevalonate in the body, cholesterol synthesis will be inhibited leading to the decrease of cholesterol concentration, triglyceride, total cholesterol and plasma LDL level and the rise of LDL receptor number. This LDL receptor works in Cholesterol LDL clearance; thus the increase of its level may cause the rise of LDL plasma cholesterol clearance[12]. Phenolic also works as chain reaction breaker, reacting with lipid radical thus transforming it to more stabilized products and as a preventive antioxidant to reduce initiation reaction speed, prevent lipid auto oxidation through fast hydrogen atom input in lipid radical[13]. Inhibition of HMG-CoA reductase activities may increase LCAT *(Lecithin Cholesterol Acyltransferase)* activities. LCAT is believed to be able to maintain cholesterol gradient which is not stratified between peripheral cells and HDL.

Phenolic is a type of polyphenol. According to recent research[14], polyphenol compounds may decrease cholesterol plasma level by inhibiting cholesterol absorption in the intestine and increase reactions in generating bile acid from cholesterol which is excreted through feces later.

Date seeds also contain Mono Unsaturated Fatty Acid (MUFA) and fibers of pectin[15,16]. Food fibers may fixate LDL cholesterol directly, fixate bile acid and inhibit enterohepatic circulation of bile acid. This mechanism will support the excretion of cholesterol LDL through feces. Methanol extract of date seeds will also decrease LDL serum level through phagocytosis process by preventing oxidized LDL stack on the wall of vascular with contained antioxidant[17].

Fatty acid contents (MUFA and PUFA) of date seeds participate in increasing HDL level. Main fatty acid components in date seeds consist of oleic acid (50.10%), linoleic acid (19.23%), lauric acid (10.24%), palmitic acid (9.83%), and stearic acid (7.51%)[5]. High MUFA diet affects the rise of HDL[18]. MUFA consumption of oleic acid in a high concentration will decrease CETP activities significantly[19].

Phenolic and gallic acid in date seed extract can increase activities of paraoxonase enzyme 1 which possesses protective effects towards LDL oxidation. A research conducted by Takaeidi *et al.*, (2014) revealed a significant difference on the increase of PON1 activities in the 1000mg/kg/weight date seed diet given to hypercholesterolemia rats in the control group[20]. PON1 enzyme prevents lipoprotein oxidation resulting in the prevention of oxidized HDL. The rise of PON1 activities has protective effects towards atherosclerosis process. Paraoxonase is an enzyme found in a human body participating in lipid metabolism. HPON1 enzyme may protect vena from arteriosclerosis by hydrolyzing derivate cholesterol or oxidized phospholipid in atherosclerotic lesions and in oxidized LDL. PON1 has two active points involving hydrolysis reaction and protecting LDL from oxidation.

5. Conclusions

All combinations of active compounds found in date seeds are assumed to be able to affect lipid profile by increasing HDL level and decreasing total cholesterol, triglyceride and LDL level, thus atherogenic index will decrease as well. This research reveals that effective dose of date seed extract to decrease LDL, triglyceride, and AI level is T1g/kg. This dose equals that of the groups which was given simvastatin (SC).

6. Acknowledgement

We gratefully acknowledge the expert technical assistance of Dr. Hernayanti.

References

[1] Ghuge G D and Zine R 2012 Atherogenic index of plasma in myocardial infarction in a rural population of Marathwada region *J. Evol. Med. Dent. Sci.* **1** 237
[2] Madamanchi N R, Vendrov A and Runge M S 2005 Oxidative stress and vascular disease

Arterioscler. Thromb. Vasc. Biol. **25** 29–38

[3] Saryono, Rahmawati E, Heryanto, Hapsari E D and Hidayat A I 2016 Antioxidant enzyme status on rat after date seeds (Phoenix dactylifera) steeping treatment *Int J Res Med Sci* **4** 1893–6

[4] Al-Tamim E 2014 Comparative study on the chemical composition of Saudi Sukkari and Egyptian Swei date palm fruits *J. Am. Sci.* **10** 149–53

[5] M. A, Razavizadeh R, Mohebbi G H and Barmak A 2012 Oil characteristic and fatty acid profile of seeds from three varieties of date palm (Phoenix dactylifera) cultivars in Bushehr-Iran *African J. Biotechnol.* **11** 12088–12093

[6] Kanthe P S, Patil B S, Bagali S, Deshpande A, Shaikh B and Manjunatha A 2012 Atherogenic Index as a Predictor of Cardiovascular Risk among Women with Different Grades of Obesity *Int. J. Collab. Res. Intern. Med. Public Heal.* **10**

[7] Tjandrawinata R R 2013 Future therapy on Down regulation expression through inhibition of cholesteryl ester transfer protein (CETP) in reducing the risk of cardiovascular disease *Med. Sci. J. Pharm. Dev. Med. Appl.* **26** 5–20

[8] Bouaziz M A, Basbes C, Blecker C and Attia H 2013 Chemical composition and some funtional properties of soluble fibro-protein extracts from Tunisian date palm seeds *African J. Biotechnol.* **12** 1121–31

[9] Vyawahare N, Pujari R, Khsirsagar A, Ingawale D, Patil M and Kagathara V 2009 Phoenix dactylifera: An update of its indegenous uses, phytochemistry and pharmacology *Internet J. Pharmacol.* **7**

[10] Platat C, Habib H M, Maqbali F D, Jaber N N and Ibrahim W H 2014 Identification of Date Seeds Varieties Patterns to Optimize Nutritional Benefits of Date Seeds *J. Nutr. Food Sci.* **1**

[11] Ratnawati H and Widowati W 2011 Anticholesterol Activity of Velvet Bean (Mucuna pruriens L.) towards Hypercholesterolemic Rats *Sains Malaysiana* **40** 317–21

[12] Charlton-Menys V and Durrington P N 2008 Human cholesterol metabolism and therapeutic molecules. *Exp. Physiol.* **93** 27–42

[13] Widowati D and Sarbini D 2010 The inhibitory effect of the Red Rosella Flower (Hibiscus sabdariffa linn) as an anti-atherosclerosis of the activation nf- κβ in cultured endothelial cells exposed low density lipoprotein oxidized *J. Penelit. Sains Teknol.* **11** 29–39

[14] Yokozawa T, Nakagawa T and Kitani K 2002 Antioxidative activity of green tea polyphenol in cholesterol-fed rats *J. Agric. Food Chem.* **50** 3549–52

[15] Efrat M and Aviram M 2010 Paraoxonase 1 interactions with HDL, antioxidants and macrophages regulate atherogenesis - A protective role for HDL phospholipids *Adv. Exp. Med. Biol.* **660** 153–66

[16] Najafi M H 2011 Date Seeds: A Novel and Inexpensive Source of Dietary Fiber *Icfeb 2011* **9** 323–6

[17] Ariantari N P, Chandra Y S and Swastini D A 2010 Cholesterol-lowering activity test of herbal honey products on the market in the diet-induced high fat rat *J Chem* **4**

[18] Kuipers R S, de Graaf D J, Luxwolda M F, Muskiet M H A, Dijck-Brouwer D A J and Muskiet F A J 2011 Saturated fat, carbohydrates and cardiovascular disease *Neth. J. Med.* **69** 372–8

[19] Jansen S, Lôpez-Miranda J, Castro P, López-Segura F, Marín C, Ordovás J M, Paz E, Jiménez-Perepérez J, Fuentes F and Pérez-Jiménez F 2000 Low-fat and high-monounsaturated fatty acid diets decrease plasma cholesterol ester transfer protein concentrations in young, healthy, normolipemic men *Am. J. Clin. Nutr.* **72** 36–41

[20] Takaeidi M R, Jahangiri A, Khodayar M J, Siahpoosh A, Yaghooti H and Rezaei, S. et al. 2014 The Effect of Date Seed (Phoenix dactylifera) Extract on Paraoxonase and Arylesterase Activities in Hypercolesterolemic Rats *Jundishapur J. Natl. Pharmacol.* **9** 30–4

IOP Conf. Series: Materials Science and Engineering **172** (2017) 012047 doi:10.1088/1757-899X/172/1/012047

Simple Colorimetric Sensor for Trinitrotoluene Testing

S Samanman[1]*, N Masoh[1], Y Salah[1], S Srisawat[1,2], R Wattanayon[1], P Wangsirikul[1] and K Phumivanichakit[1]

[1]Faculty of Science and Technology, Princess of Naradhiwas University, Muang, Narathiwat 96000, Thailand

[2]Pressident office, Princess of Naradhiwas University, Muang, Narathiwat 96000, Thailand

* Corresponding author's e-mail: saluma.s@pnu.ac.th or ssamanman@gmail.com

Abstract. A simple operating colorimetric sensor for trinitrotoluene (TNT) determination using a commercial scanner as a captured image was designed. The sensor is based on the chemical reaction between TNT and sodium hydroxide reagent to produce the color change within 96 well plates, which observed finally, recorded using a commercial scanner. The intensity of the color change increased with increase in TNT concentration and could easily quantify the concentration of TNT by digital image analysis using the Image J free software. Under optimum conditions, the sensor provided a linear dynamic range between 0.20 and 1.00 mg mL^{-1}($r = 0.9921$) with a limit of detection of 0.10± 0.01 mg mL^{-1}. The relative standard deviation for eight experiments for the sensitivity was 3.8%. When applied for the analysis of TNT in two soil extract samples, the concentrations were found to be non-detectable to 0.26±0.04 mg mL^{-1}. The obtained recovery values (93-95%) were acceptable for soil samples tested.

1. Introduction

2,4,6-trinitrotoluene (TNT) is one of the most widely used explosive compounds for the military, terrorist activities and mining. TNT is also highly toxic to many organisms because it could cause liver damages and many other health problems [1]. Therefore, the detection of TNT has attracted much attention due to these environmental and health related concerns. On the other hand, the identification of its post explosive residues in forensic chemistry is also needed for use as a guideline to help narrow any investigation associated with securities [2].

A wide range of methods including gas or liquid chromatographic analysis [3-4], electrochemical methods [5] spectrophotometry [6] and colorimetric sensor have been applied for the quantitative analysis of TNT. Among these, the colorimetric sensor is an interesting approach due to its fast response, simplicity, low cost, and easy preparation of standard solutions. Moreover, the color change can be easily detected by naked eyes. A number of commercial chemical kits such as DROPEX^{Plus} and EXPRAY have also been developed for TNT detection. However, these methods could be used only for qualitative analysis.

For a quantitative analysis and off-site detection, a novel quantitative analytical methodology based on digital image analysis of the photographs of the reaction test has been reported. The digital image was obtained by pocket camera [7] or mobile phone camera [8]. The reflected light from objects passes through and is detected by three RGB filters (Red Green and Blue) by using the Adobe Photoshop software [8].

The aim of this work is to demonstrate the application of a simple, rapid, cost-effective and quantitative analysis colorimetric sensor for TNT determination using sodium hydroxide (NaOH) as a tested reagent. The color change was recorded using commercial document scanner. The intensity of the red-violet color change will be increased after a reaction. This method can quantify the concentration of TNT by scanner image analysis using the Image J free software and can be applied for the detection of TNT contaminated soils.

2. Experimental

2.1. Materials and apparatus

2,4,6-Trinitrotoluene (TNT) and sodium hydroxide (NaOH) were purchased from Sigma-Aldrich (Steinheim, Germany). Acetonitrile was purchased from RCI Labscan (Bangkok, Thailand). Sodium phosphate diabasic AR and sodium dihydrogen orthophosphate AR were obtained from Lobachemie (Mumbai, India). The digital image was captured using the commercial scanner (Epson Perfection™ V39, Indonesia).

2.2 TNT detection

TNT in acetonitrile was used as the stock solution. The stock solution was further diluted with acetonitrile to appropriate concentrations in the range of 0.10 to 1.00 mg mL^{-1} to produce a range of TNT standards. A TNT color test was performed using NaOH as the reagent. One hundred microliters of each TNT standard solution was pipetted to a 96 well plate and two hundreds of 0.30 M NaOH in phosphate buffer pH 7.00 was added to obtain a red–violet product. The resultant colors were captured five minutes after color development using the commercial document scanner. Each concentration of prepared standards was repeated four times and each resultant color was recorded three times using the commercial scanner. The images were transferred to a computer (Microsoft Windows 7). The average color intensity of Red, Green, and Blue of each color product in each sensor was obtained using the "Crop" tool and the "Histogram" in Image J free software and transferred the data into a Microsoft Office Excel 2010 spreadsheet for subsequent data analysis. A calibration graph was prepared for each color. The limit of detection was calculated using standard methods [9] and precision was expressed as a percentage of the relative standard deviation of the RGB values for each color from the three analyzed images. The schematic presentation for TNT detection procedure is shown in Figure 1.

IOP Conf. Series: Materials Science and Engineering **172** (2017) 012047 doi:10.1088/1757-899X/172/1/012047

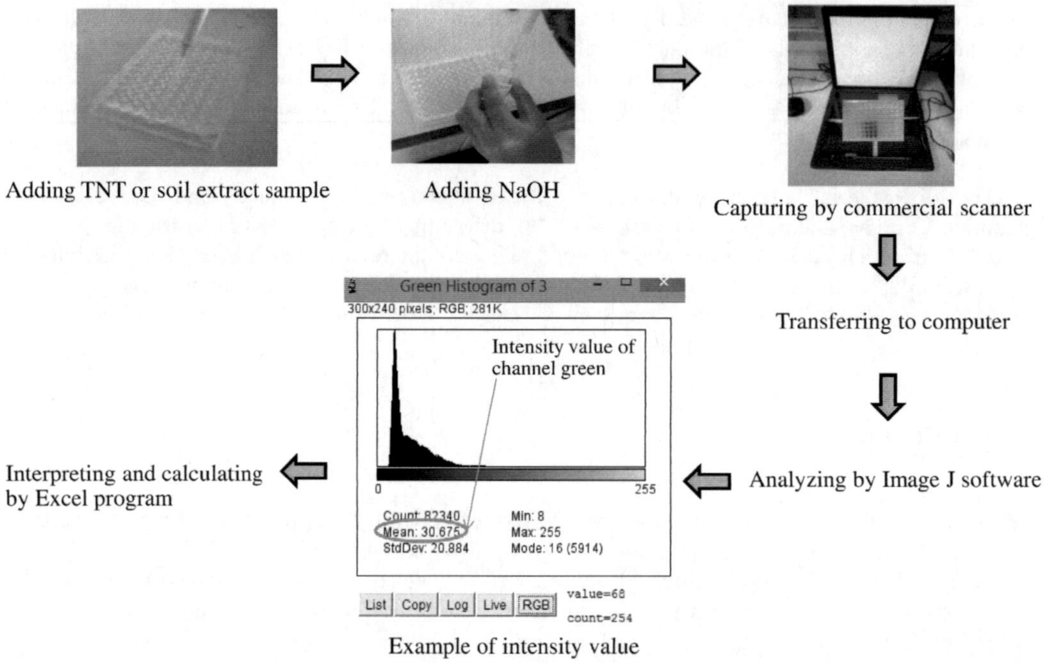

Figure 1. TNT colorimetric sensor detection

The initial detection conditions were 10 mins of reaction time, 150 μL of 0.6 M NaOH in phosphate buffer pH 8.00 and 150μL of TNT in acetonitrile. The detection response was the intensity of the color change of RGB values. All of these operating conditions were optimized to obtain the highest intensity color change of the colorimetric sensor as follows:

2.2.1 Reaction time
Optimization of standing time of the reaction between TNT and the tested reagent is necessary for the completion of the TNT reaction in basic medium. Hence, the effect of reaction time was studied in the range of 1-20 mins. At the optimum reaction time, the highest intensity color change was then selected for use in the next experiments.

2.2.2 Concentration of NaOH
The initial results showed that the maximum intensity color change of TNT detection was obtained in the alkaline medium due to the formation of Meisenheimer anion. Therefore, to succeed suitable basic media, the effect of different concentration of NaOH in the range of 0.15-1.20 M was investigated. Then at the optimum concentration of NaOH with the highest intensity color change was chosen.

2.2.3 The volume of test NaOH
The volume of test NaOH reagent was optimized to determine the volume required to produce the darkest color reaction for the least concentrated analyte solution. Therefore, to investigate this objective, the volume of the tested NaOH in the range of 100-250 μL was obtained.

2.3 Analytical performances

2.3.1Linear dynamic range and detection limit

Under optimum conditions, the linear dynamic range was investigated between 0.10 and 1.00 mg mL^{-1}. The limit of detection was evaluated from the equation $X_L = X_{bl}+3(S_{bl})$ where X_L was the signal at the limit of detection, X_{bl} and S_{bl} are the mean and the standard deviations of the blank response [9].

2.3.2 Repeatability

Repeatability was studied by comparing the sensitivity (slope of the calibration curves) of eight different experiments, and the relative standard deviation was calculated.

2.3.3 Soil samples analysis

A soil sample (1 g) was extracted with 2.5 mL of ethyl acetate and sonicated for 15 mins before filtering using Whatman No. 1 filter paper. The supernatant was spiked with a TNT standard solution to investigate the effect of the color of the soil filtrate on the TNT presumptive test color derived from the test. The extracted solution was analyzed according to the procedures described in the Section of 2.2.

3. Results

3.1 Optimization

No color change was observed for the reagent in the absence of TNT (blank). On the other hand, red-violet products were obtained from the NaOH test for the presence of TNT (Figure 2a). The color change is because, in the basic medium, the TNT reacts with NaOH to give the "Meisenheimer anion" which is a color product [10-11]. Therefore, it can be said that the color change was caused by the TNT complex from colorless to red-violet color (Figure 2b). The color change also can be easily detected by naked eyes. In summary, the optimum detection conditions were reported in Table 1.

Table 1. The optimum detection conditions for developing of TNT colorimetric sensor

Parameters	Studied ranges	Optimal values
Reaction time	1, 5, 10, 15 and 20 mins	5 mins
Concentration of test NaOH	0.15 , 0.30, 0.60 and 1.20 M	0.30 M
The volume of test NaOH	100, 150, 200 and 250μL	200 μL

3.2 Color test of TNT

Under the optimum conditions, the color change with the increasing TNT concentration was studied in the range of 0.10 to 2.00 mg mL^{-1}. The resulted series color change was then recorded using the commercial scanner, followed by analysis of their intensity of the color change using free software of Image J to get RGB profile (Figure1 example histogram of channel green). The relation between the intensity of color the change and TNT concentration was selected from the three colors of Red, Green or Blue with the highest sensitivity and best regression coefficient. The result showed the best-reducing color intensity was green, blue and red, respectively (Figure 2c). Then the channel green was used for the prepared soil extract analysis. The linear dynamic range was 0.20 to 1.00 mg mL^{-1} and the detection limit was determined and found to be 0.10±0.01 mg mL^{-1}.

IOP Conf. Series: Materials Science and Engineering **172** (2017) 012047 doi:10.1088/1757-899X/172/1/012047

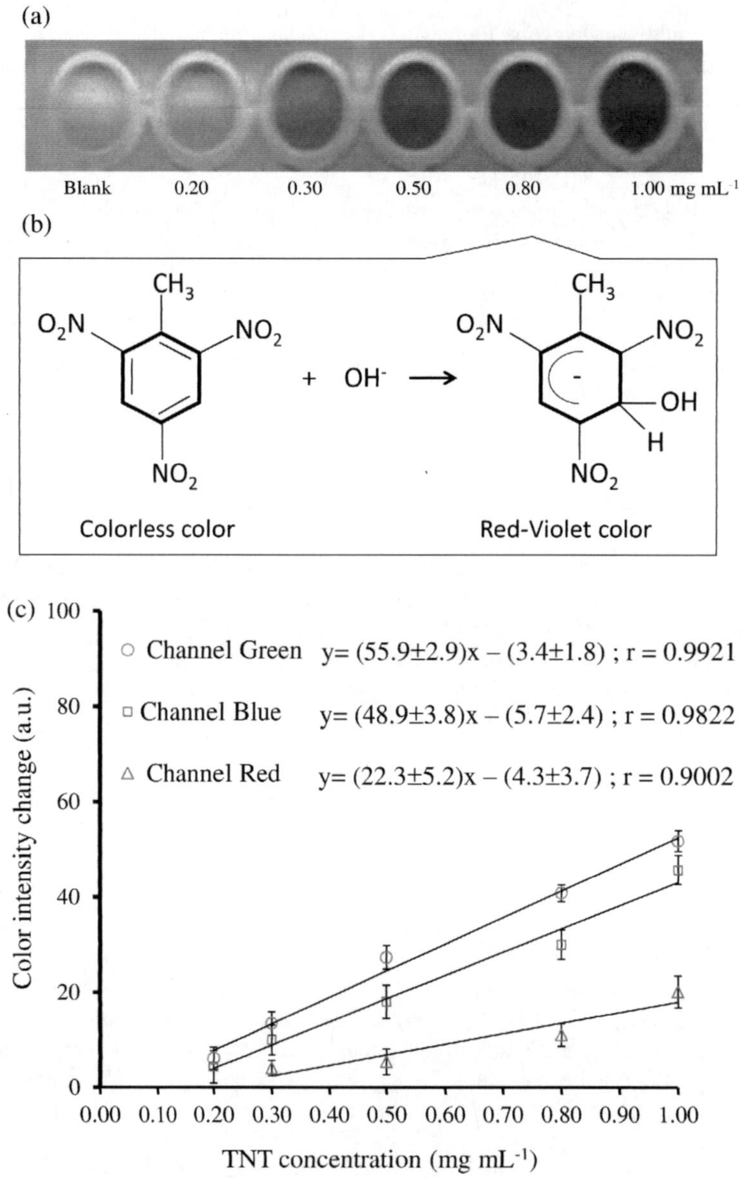

Figure 2. (a) The example color changes with the TNT concentration in a 96-well plate captured by the scanner. (b) The reaction of TNT in basic medium and (c) Calibration curves of TNT colorimetric sensor in a series TNT concentration (0.20 – 1.00 mg mL^{-1}) of (Δ) channel red, (o) green and (□) blue.

3.3 Repeatability

The repeatability was investigated through the sensitivity of eight experiments. Between the eight experiments, the relative standard deviations (RSDs) of 3.8% (Figure 3), indicated a good repeatability of the developed colorimetric sensor. This RSD was within the acceptance criteria (less than 8% RSD in the ppm level) [12].

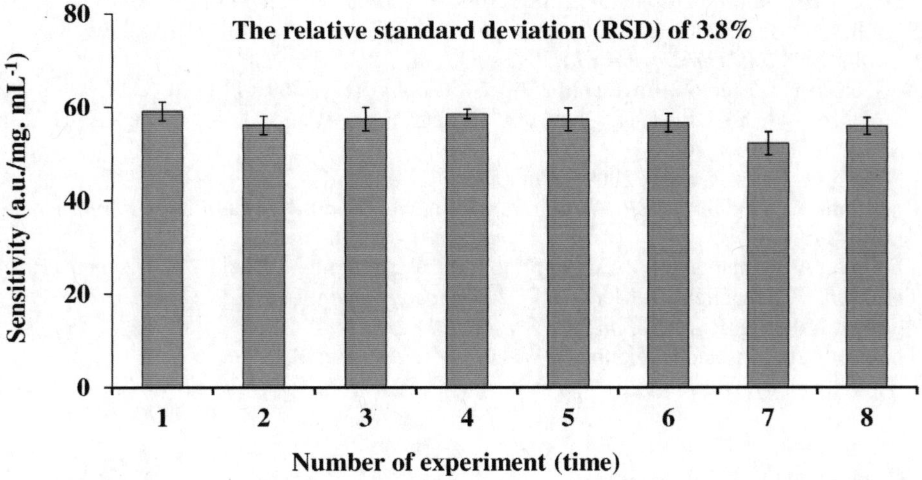

Figure 3. Repeatability of the sensitivities of the eight different experiments

3.4 Quantification of TNT in soil

The proposed colorimetric sensor was applied for the detection of TNT in prepared soil samples. Table 2 summarizes the results obtained for the two samples analyzed. TNT was detectable with the contents of non-detectable in sample No.1 and 0.26 ± 0.04 mg mL^{-1} in another sample. For validation of the proposed method, TNT standard solution was spiked into soil extract (0.50 mg mL^{-1}) as previously described. The obtained recovery values (93-95%) were acceptable for soil samples tested. This recovery value was within the acceptance criteria (between 90-107% in the ppm level) [12].

Table 2. Results for the determination of TNT in different soil samples.

Soil sample No.	Added (mg mL^{-1})	Found* (mg mL^{-1})	Recovery (%)
1	0.00	ND	-
	0.50	0.48±0.02	95±3
2	0.00	0.26±0.04	-
	0.50	0.72±0.02	93±4

* average ± standard deviation (n =3)
ND = non-detectable

4. Conclusion

In this work a simple and rapid (5 mins) was applied for the TNT testing based on the use of NaOH as a tested reagent and the used of the document scanner as a capturing digital image. This colorimetric sensor can be easily observed by naked-eyes. This work also provided a good repeatability and recoveries. Moreover, it was successfully applied for the analysis of TNT in soil extract samples. This can be used to indicate the quality of soil samples which is crucial for agricultural, environmental and health safety. Also this colorimetric sensor could be identified the TNT in explosive residues in forensic chemistry to be used as a guideline to help narrow any investigation associated with securities. This commercial scanner detector together with the image J software program could certainly be easily extended to different analytes using different designed colorimetric sensor reaction.

References

[1] L. Eyers, B. Stenuit, S.N. Agathos 2008 *Appl. Microbiol. Biotechnol.* **79** 489.
[2] S. Bell, Forensic Chemistry 2006 *Annu. Rev. Anal. Chem*
[3] P. Kolla 1994 *J.Chromatogr.A* **674** 309.
[4] A. Schreiber, J. Efer, W. Engewald 2000 *J. Chromatogr. A* **869** 411.
[5] T.-W. Chen, J.-Y. Xu, Z.-H. Sheng, K. Wang, F.-B. Wang, T.-M. Liang, X.-H. Xia 2012 *Electr. Commun.* **16** 30.
[6] E. Erçag, A. Üzer, R. Apak 2009 *Talanta* **78** 772.
[7] Choodum A., Kanatharana P., Wongniramaikul W., Niamh N. Daeid 2012 *Forensic Sci. Int.* **222** 340–345
[8] Choodum A., Kanatharana P., Wongniramaikul W., Niamh N. Daeid 2013 *Talanta* **115** 143
[9] Thomsen, V.,Schatzlein,D.,Mercuro,D. 2003 *Spectroscopy* **12** 112.
[10] A. Uzer, E. Ercag, R. 2008 *Forensic Sci. Int.* **174** 239.
[11] Thomas F. Jenkins and Marianne E. Walsh 1992 *Talanta* **39** 419
[12] I. Taverniers, M. De Loose, E. Van Bockstaele 2004 *Trends in Analytical Chemistry* **23** 480.

Acknowledgements

This research was supported by the Nation Research Council of Thailand. Partial support from and Faculty of Science and Technology, Princess of Naradhiwas University, Thailand are gratefully acknowledged. Thanks also to Miss Tshering Pelden for assistance with the English.

IOP Conf. Series: Materials Science and Engineering **172** (2017) 012048 doi:10.1088/1757-899X/172/1/012048

Synthesis and Characterization of Tetrakis(2-amino-3-methylpyridine)di(aqua)nickel(II)

S B Rahardjo and H P Aditya

Faculty of Mathematematic and Natural Sciences, Sebelas Maret University

Email: sentotbr@yahoo.com

Abstrak. Complex of tetrakis(2-amino-3-methylpyridine)-di(aqua)Nickel(II) nitrate had been synthesized by adding dropwise solution of $Ni(NO_3)_2.6H_2O$ (0,580 g ; 2 mmol) in 10 mL methanol into the solution of L (L=tetrakis(2-amino-3-methylpyridine)(0,864 g ; 8 mmol) in 30 mL methanol and stirring continuously for one hour at room temperature.The complex formula predicted is $[Ni(L)_4(H_2O)_2](NO_3)_2$. The infrared spectral data indicates that nitrogen atom of C=Npyr and oxygen atom of H_2O are coordinated to the central ion Ni^{2+}. The complex is paramagnetic and shows two maximum peaks which are $^3A_{2g}(F) \rightarrow {}^3T_{1g}(F)$ and $^3A_{2g}(F) \rightarrow {}^3T_{1g}(P)$ transition. Complex predicted has an octahedral geometry.

1. Introduction

Complex compounds have an important role in the application of science such as biology, analytical, medical, clinical, industrial and other fields. Some of the complex indicate that the N atom in the pyridine cluster(C=N) and N atom outside of pyridine (as NH primary) have the ability to be coordinated to the central atom of either one of the primary NH or pyridine cluster and is jointly coordinated to the central atom [1,2,3,4]. Pyridine derivatives, such as nicotinamide, nicotinalide, benzoylhydrazin and 2-amino-3-Methylpyridine (2-amino-3-picoline), have an N-donor atoms in primary NH groups and N atom in the pyridine cluster (C=N), both have the ability to be coordinated to the central atom. The most common structure is an octahedral and square-planar. The complex of Ni(II) with 2-amino-3-Methylpyridine (L)is expected to form an octahedral structure which can be used as an antibacterial material or other purposes.

2. Experimental

All chemicals were purchased from E.Merck and used without further purification.The Infrared spectra were recorded on a Spectrophotometers Prestige-21 Shimadzu.The nickel content was determined by Atomic Absorption Spectrometer (AAS) Shimadzu AA-6650. Spectra UV-VIS was performed with a Shimadzu UV-3601 spectrophotometer. Molar conductivity (Λ^*_m) of 1 mM solution in methanol was measured at 25° C with aJenway CE 4071 conductivity meter. The magnetic moment was measured with Auto Sherwood Scientific 10169 Magnetic Susceptibility Balance.

2.1.Complex Synthesis

A solution of nickel(II) nitrate hexahydrate (0,58 g; 2 mmol) in methanol (10 mL) was added dropwise to a solution of the ligand L(0,864 g; 8 mmol) in methanol (30 mL) and the mixture was stirred for 1 hour. The volume was reduced to third and left for about 72 hours. The green precipitate, which was separated, is washed with methanol and dried under vacuum.

3. Result and Discussion

3.1. Complex Synthesis

The mixture of solution $Ni(NO_3)_2.6H_2O$ (0,580 g ; 2 mmol) in methanol (10 mL) and solution L (2-amino-3-methylpyridine) (0,864 g ; 8 mmol) in methanol (30 mL) creates a green crystal. The electronic spectrum can be seen in Figure 1.

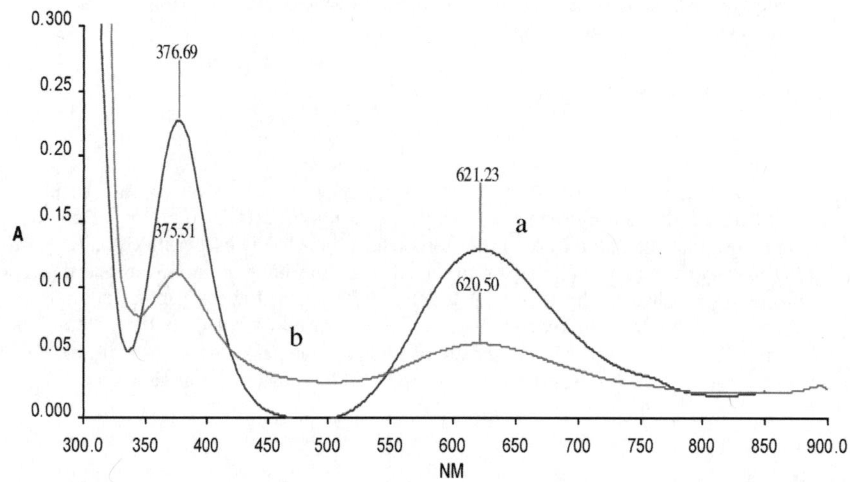

Figure 1. Electronic spectrum (a) solution of $Ni(NO_3)_2.6H_2O$ and (b) solutionof complex Ni(II)L within DMSO

Figure 1 shows a shift of maximum wavelength absorbance of$Ni(NO_3)_2.6H_2O$ (621 nm and 376 nm) to a smaller wavelength (620 nm and 375 nm) in the complex ofNi(II)L. The shift of λ_{max} reveals the formation of complexes between Ni(II) andL and also reveals that L is a stronger ligand than H_2O. The small shift of maximum wavelength indicates that the ligand Lis a weak crystal field as H_2O. Probably the ligand L and H_2O both are coordinated to Ni(II) which is also in the complex$[Ni(L^1)_2(H_2O)_2]$ (L_1= 3-methyl-4-nitro-1-phenyl-5-pyrazolone) [5] and $[Ni(L^2)(H_2O)]$ (L^2 = 1,1-diaminobutane-Schiff base) [6] as shown in Figure 2 and Figure 3.

IOP Conf. Series: Materials Science and Engineering **172** (2017) 012048 doi:10.1088/1757-899X/172/1/012048

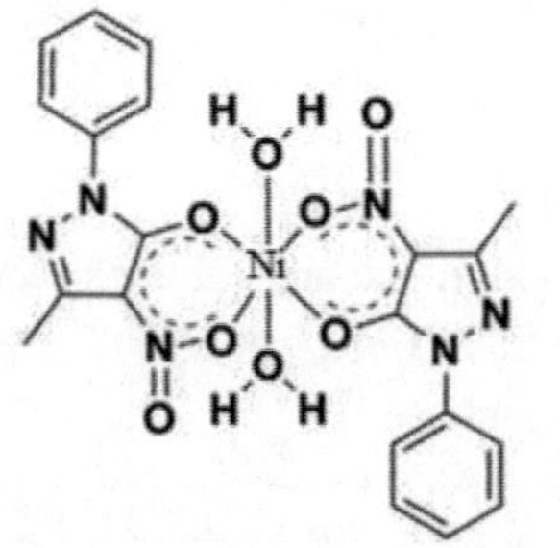

Figure 2. The complex structure of $[Ni(L^{1-})_2(H_2O)_2]$ (L^1= 3-methyl-4-nitro-1-phenyl-5 pyra-zolone) [5].

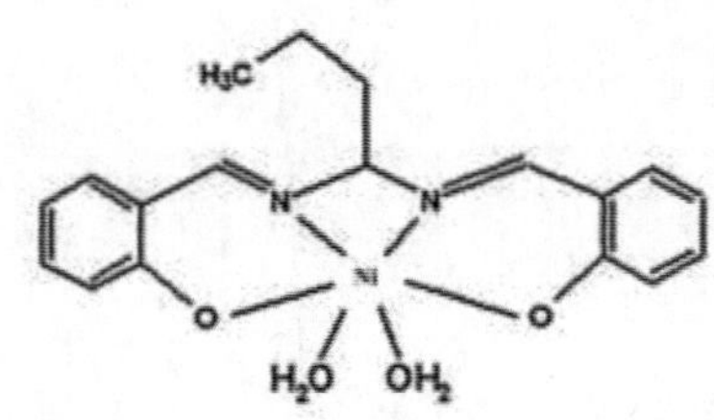

Figure 3. The complex structure of$[Ni(L^2)(H_2O)]$ (L^2 = 1,1-diaminobutane-Schiff base) [6]

3.2.Determination of Formula and Characteristics of The Complex

The measurements of the amount of Ni in the complex of Ni(II)Lis of 9,28±0,22%. If the result of the measurement is compared to the theoretical amount of Cu (9,27%), the estimated complex compound formula for Ni(II)L is possibly $Ni(L)_4.(NO_3)_2$.

3.3. Electrolytic Conductivity Measurement

The electrolytic conductivity measurement is shown in Table 1.

Table 1. The Electrolytic Conductivity of Standard Solution and complex $Ni(L)_4.(NO_3)_2.2H_2O$ $\pm 1.10^{-3}$ M in methanol

Solution	Λ^*_m ($S.cm^2.mol^{-1}$)	Number of ions
Metanol	0	-
$CuSO_4.5H_2O$	5.8± 0.2	1 : 1
$NiSO_4.6H_2O$	6.9 ± 0.1	1 : 1
$CuCl_2.6H_2O$	77.3 ± 0.7	1 : 2
$NiCl_2.6H_2O$	93.6 ± 0.6	1 : 2
$AlCl_3.6H_2O$	218.6 ± 0.4	1 : 3
$FeCl_3.6H_2O$	273.6 ± 1.4	1 : 3
Ni-L	87.3 ± 0.7	1 : 2

The Electrolytic Conductivity of the complex $Ni(L)_4.(NO_3)_2.2H_2O$ of copper(II) chloride and nickel(II) chloride indicates that the complex is electrolyte corresponds to 1 : 2 electrolyte and the nitrate is a counter ion and not coordinated to Ni(II) as in the $[Co(Hisn)_2(H_2O)_2](NO_3)_2$, Hisn =

acetone isonicotinoyl hydrazone [7]. Therefore, the complex may be formulated as $[NiL_4(H_2O)_2](NO_3)_2$.

3.4. Infra-Red Spectra

The infrared spectra of 2-amino-3-methylpyridine and complex $[NiL_4(H_2O)_2](NO_3)_2$ are shown in Figure4 and Figure 5 while infrared absorption bands are shown in Table 2.

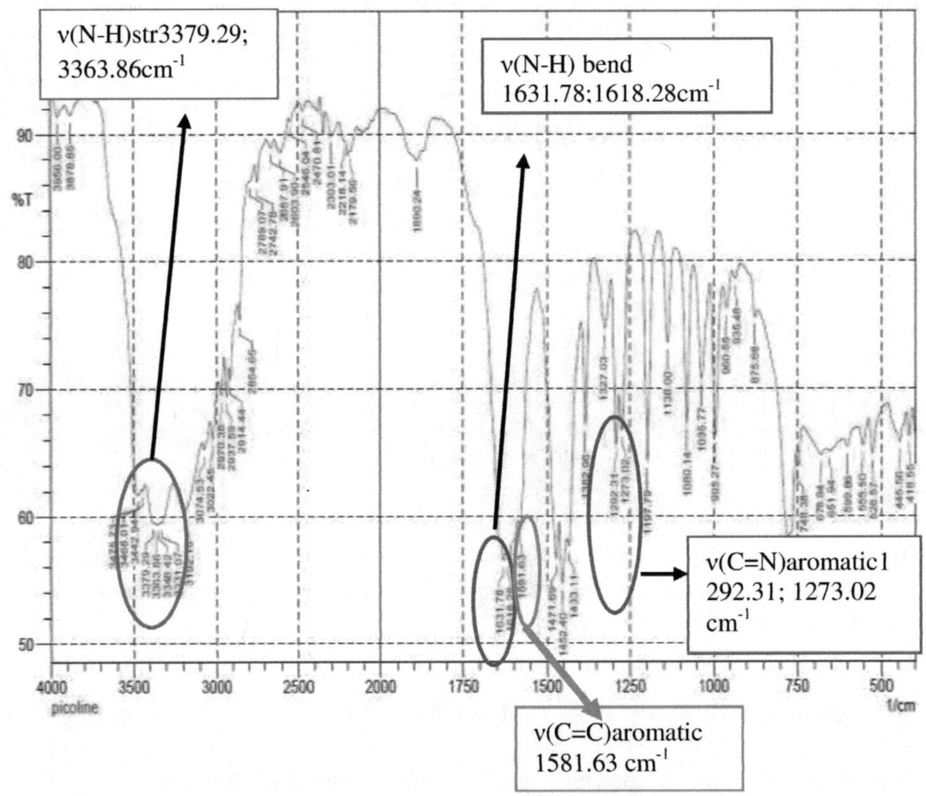

Figure 4. The infrared spectra of L

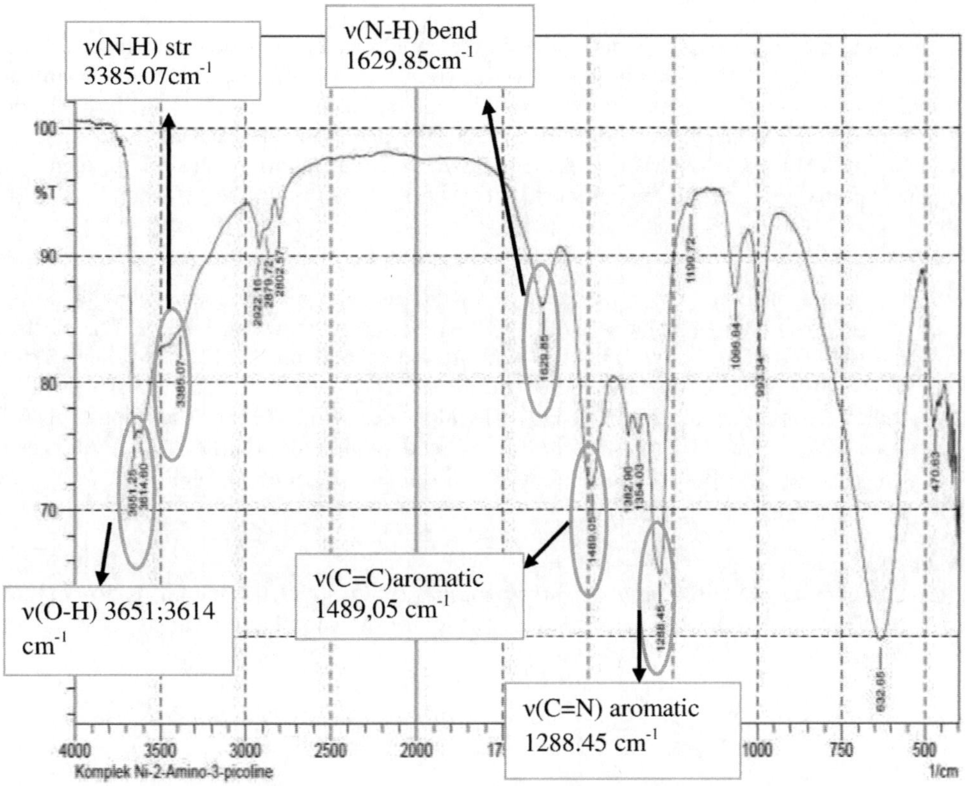

Figure 5. The infrared spectra of$[NiL_4(H_2O)_2](NO_3)_2$

Table 2. IR absorption bands (cm^{-1}) of 2-amino-3-methylpyridine and complex $[Ni(L)_4](NO_3)_2 n.H_2O$.

Compound	IR Band Spectra (cm^{-1})			
	(C=N)	(O-H)	(N-H)str	(N-H)bend
L	1292127	-	3379	1631
(2-amino-3-	3	-	3363	1618
Methylpyridine)				
$[Ni(L)_4]$ $(NO_3)_2$	1288	3651 3614	3385	1629

Figure 3 and 4 or Table 2 shows the shift of the IR spectra of NH stretching on the area 3379 cm^{-1} and 3363 cm^{-1} to 3385 cm^{-1}, NH bending is at 1631 cm^{-1} and 1618 $cm^{-1}$to 1629 cm^{-1}. Spectra in the area of 3651 cm^{-1} and 3614 cm^{-1}indicates O-H group. The existence of the O-H group absorption in the complex indicates H_2O molecules are in the complex of $[NiL_4(H_2O)2]$ $(NO_3)_2$. The shift of the IR absorption of C=N in this complex indicates that the nitrogen atom on the C=N group is coordinated to the central ion Ni $^{2+}$ as in the complex $[NiL^3]$, L^3= 4-amino-5- (pyridyl) -4H 1,2,4 triazole-3-thiol [8].

3.5. *Magnetic properties*

Effective magnetic moment measurements of the solid complex gave magnetic moments of 2.65 BM, which shows that the nickel(II) is in a high spin with two unpaired electrons. This value is similar to that of other complexes which have an octahedral geometry such as in [$Ni(L^4)_2Cl_2$] , L^4 = 3-aminocoumarin which has magnetic moments of 2.6 BM and with an octahedral geometry [9], $[Ni(OCOCH_3)_2(isn)_2(H_2O)]{\cdot}C_3H_7NO$(isn = isonicotinamide) (2.6 BM) an octahedral geometry [10]. Therefore the expected geometry of the complex $[NiL_4(H_2O)_2](NO_3)_2$ is also an octahedral.

3.6. *Electronic Spectra*

The electronic spectra of the complex recorded in DMSO and the maximum wavelength (λmaks), absorbance (A) and absorbtivitas molar (ε) with $Ni(NO_3)_2.6H_2O$ are shown in Table 2. The electronic spectra of $[NiL_4(H_2O)_2](NO_3)_2$ shows two wide absorption at 620 nm (16116 cm^{-1}) and 375 nm (26630 cm^{-1}). The absorption bands indicate*d-d* transition $^3A_{2g}(F) \rightarrow {}^3T_{1g}(F)$ and $^3A_{2g}(F) \rightarrow {}^3T_{1g}(P)$, as in the octahedral complex of $[Ni(HL^5)_2]$, L^5=2-Chloromethyl-3-(8-hydroxyquinolin-5-yl)-3(H)-quinozolin-4-one which shows absorption at 14962 cm^{-1} and 24548 cm^{-1}, $[Ni(HL^6)_2]$, L_4 = 6-bromo-2 [(4-phenyl) -1-yl) methyl]-3-[8-hydroxyquinoline-5-yl] quinazolin-4-one at 14896 cm^{-1} and 22800 cm^{-1} [11].

3.7. *Complex Structure*

The suggested structure of [Ni(*2-amino-3-methylpyridine*)$_4(H_2O)_2](NO_3)_2$ complex is an octahedral geometry with (C=N) and H_2O coordinated to nickel(II) ion as shown in Figure 6.

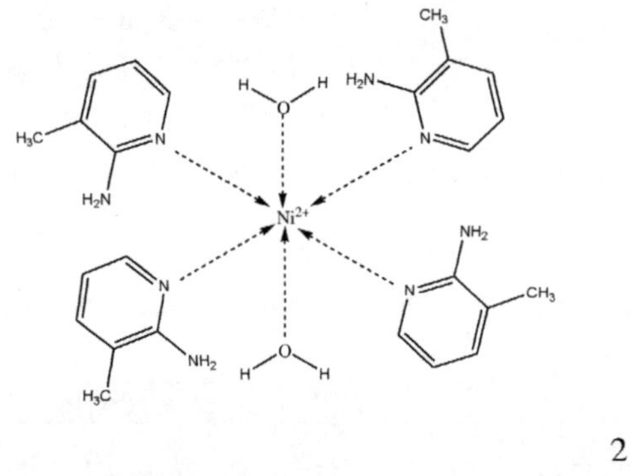

Figure 6. The suggested structure of $[Ni_4(H_2O)_2](NO_3)_2$ complex

According to IUPAC nomenclature, the suggested complex is named by *tetrakis*-(*2-amino-3-methylpyridine*)diaquanickel(II) nitrate.

References

[1] Back D F, de Oliveira G M, Fontana L A, Ram B F, Roman D, Iglesias B A 2015 One-pot synthesis, structural characterization, UV-Vis and electrochemical analyses of new Schiff base complexes of Fe(III), Ni(II) and Cu(II) *Journal of Molecular Structure* 264-71.

[2] Bharati P, Bharti A P A, Bhartya M K, Maiti B, Butcher R J, Singh N K 2013 Square planar Ni(II) complexes of pyridine-4-carbonyl-hydrazinecarbodithioate, 1-phenyl-3-pyridin-2-yl-isothiourea and 4-(2-methoxyphenyl)piperazine-1-carbodi-thioate involvingN–S bonding: An approach to DFT calculation and thermal studies *Polyhedron* **63** 156-66.

[3] Datta S, Seth D K, Gangopadhyay S, Karmakar P, Bhattacharya S 2012 Nickel complexes of some thiosemicarbazones: Synthesis, structure, catalytic properties and cytotoxicity studies *Inorganica Chimica Acta* **392** 118-30

[4] Kopel P, Starha R, Sindelar Z 1998 Mixed Ligand Nickel Complexes Involving Derivatives of Tetrahydroisoquinolinne and Phnenethylamine *Chemica Acta Universitatis Palackianae Olomucensis.*

[5] Xu R, Liang M, Yin F, Li S, Huang Y, Zhaoa Y, Liu J, Ma X 2014 In-situ nitration and syntheses of new Co(II) and Ni(II) complexes assembled from 3-methyl-1-phenyl-5-pyrazolone and cobalt or nickelnitrate under solvothermal conditions *Inorganic Chemistry Communications* **40** 120-23.

[6] Materazzi S, Risoluti R, Napoli A 2015 EGA–MS study to characterize the thermally induced decomposition of Co(II), Ni(II), Cu(II) and Zn(II) complexes with 1,1-diaminobutane-Schiff Base. *Thermochimica Acta* **606** 90-94.

[7] Roztocki K, Matoga D, Nitek W 2016 Cobalt(II) compounds with acetone isonicotinoyl hydrazone tautomer: Syntheses and crystal structure of complexes with free donor atoms *Inorganica Chimica Acta* **448** 86–92.

[8] Haddad R, Yousif Z E, Ahmed A 2013. Synthesis and Characterization of Transition Metal Complexes of 4-Amino-5-pyridyl-4H-1,2,4-triazole-3-thiol *SpringerPlus* **2** 510

[9] Kadhum A A H, Mohamad A B, Al-Amier A A, Takriff M S 2011 Antimicrobial and Antioxidant Activities of New Metal Complexes Derived from 3-Aminocoumarin *Molecules* **16** 6969-84.

[10] Dojer B, Pevec A, Jaglicˇic Z, Drofenik M, Kristl M 2016 Nickel(II) pyridinecarboxamide complexes: Solvothermal synthesis, crystal structures and magnetic properties *Inorganica Chimica Acta* **446** 124-31.

[11] Vashi, R T, Shelat C D, Patel H 2010 Synthesis And Antifungal Activity Of 6bromo2[(4phenyl)1yl)Methyl]3[8hydroxyquinoline5yl]3quinazolin4one Ligand And Its Metal Chelates *International Journal of Current Pharmaceutical Research* **2**(3) 2010.

Surface area and conductivity of polyaniline synthesized under UV irradiation

S Budi[1], E Fitri[1], M Paristiowati[1], U Cahyana[1], E Pusparini[1], H Nasbey[2], and A Imaddudin[3]

[1]Department of Chemistry, Faculty of Mathematics and Sciences, Universitas Negeri Jakarta, Jl. Rawamangun Muka, Jakarta 13220, Indonesia
[2]Department of Physics, Faculty of Mathematics and Sciences, Universitas Negeri Jakarta, Jl. Rawamangun Muka, Jakarta 13220, Indonesia
[3]Pusat Penelitian Metalurgi dan Material, Lembaga Ilmu Pengetahuan Indonesia, Puspiptek, Serpong, Indonesia

Email: setiabudi@unj.ac.id

Abstract. This paper reports our study on the synthesis of high electrical conductivity and surface area polyaniline using oxidative polymerization under UV light irradiation. The formation of emeraldine structures of polyaniline was revealed by major absorption bands of FTIR (Fourier transform infrared spectroscopy) spectra attributed to C-N stretching, C=C stretching in the benzenoid ring, C=C stretching in the quinoid ring and $QNH^{+}B$ stretching. XRD (X-ray diffractometer) measurements confirmed typical diffraction patterns with a crystallinity of 13% and 16% for polyaniline prepared under non-stirred and stirred reaction, respectively. SEM (Scanning electron microscope) studies showed more uniform morphology of polyaniline was obtained with stirring reaction process compare to those prepared without stirring. Surface analysis using SAA (surface area analyzer) showed that pure polyaniline with the relatively high surface area of ca.28 m^2/g was successfully prepared in this work. Based on four point probe measurement, the prepared polyaniline possesses high conductivity which is important in electrode application.

1. Introduction

A conductive polymer such as polyacetylene, polypyrrole, polythiophene and polyaniline are well-known materials that have been developed for much-advanced application in various devices. Amongst of these polymers, polyaniline in the form of emeraldine salt has attracted a remarkable attention due to its excellent properties, namely high conductivity, good electrochemical properties and redox reversibility, high electrical and environmental stability as well as simply in fabrication process [1, 2]. In addition, the commercial monomer for polyaniline is available in relatively low price. These advantageous characteristics make polyaniline has potential application in numerous devices such as chemicals sensor [1], toxic gas sensor [3], electrode [4], supercapacitor [5], solar cell technology [2] photovoltaics and thermoelectric energy conversion [6].

Polyaniline could be synthesized by means of oxidative polymerizatin and electrochemical polymerization techniques. Oxidative polymerization is a low cost and simple route for polyaniline preparation without any particular instrument made this approach used in many reported studies [7]. Physical and chemical properties of polyaniline are influenced by synthesis condition, e.g. solution concentration and composition, reaction temperature, stirring, dopant and pH of the solution system [8]. For instance, radical reaction rate controlling elongation processes in the polymer chain that affect molecular weight of polyaniline is known to rely on the reaction temperature. At which the molecular weight is recognized as one of the factors that determined polyaniline conductivity. Since electrical conductivity and surface area of polyaniline are important for its application, many attempts have been carried out to enhance the values of both properties, mostly by engineered polyaniline with other materials e.g. graphene [5, 9], titanium dioxide [10], and carbon nanotubes. However, based on some previously reported works [9, 11–13], conductivity and surface area of pure polyaniline are still need to be improved.

This study is intended to develop polyaniline with high conductivity and surface area through Ultra Violet (UV) light irradiation at initial reaction temperature of 0°C. UV exposure during polymerization can induce the formation of fine polyaniline structure [14] which may increase its surface area. Consequently, performance of the obtained polyaniline produced can be increased its performance in a suitable application.

2. Method

Polyaniline was synthesized by oxidative polymerization method using 0,2 M aniline solution prepared in 0,2 M HCl and 0,1 M Ammonium Peroxydisulfat (APS). These solutions were cooled down using ice bath until its temperature reach 0°C. Then, the ammonium peroxidisulfate solution was slowly poured into the aniline solution with continuous stirring on the magnetic stirrer. Further polymerization was performed under UV light irradiation using UV laminar air flow for 1 hour applied for both non-stirred and stirred reaction process. The obtained polyaniline powder was then separated from its solution by centrifugation, rinsed by aquadest and dried at room temperature subsequently. This powder was then used to make polyaniline film under vacuum mounting press at room temperature.

FTIR spectra of both powder and film form of the polyaniline were obtained using Fourier Transform Infrared Spectroscopy (Shimadzu IR Prestige-21) at wavenumber ranging from 400-4000 cm^{-1}. Crystallinity degree was examined by X-ray Diffractometer (PANalytical EMYREAN) using Cu Kα radiation with λ=1.54187. Morphological analysis was carried out with Scanning Electron Microscope (JEOL JSM-6510LA). Measurement electrical conductivity of the polyaniline film was performed using four point probe (Keithley 2700 Nanovoltmeter and Keithley 6200 Current Source). Surface area and porosity of the polyaniline were measured by Surface Area Analyzer (Micromeritic ASAP 2020).

3. Results and Discussion

FTIR spectra of the synthesized polyaniline through oxidative polymerization under UV light irradiation both from non-stirred and stirred reaction are presented in Figure 1. The figure indicates absorption characteristics of emeraldine salt form of polyaniline by the existence of the bands of N-H, C-H, C=C (Q), C=C (B), C-N, dan Q=NH^+-B that are assigned as follows. An absorption band observed at the wavenumber of 3400 was attributed to N-H stretching. The bands at 2900 cm^{-1} were attributed to stretching of C-H asymmetric and C-H symmetric. The bands at 1500 cm^{-1} and 1467 cm^{-1} for polyaniline obtained from non-stirred polymerization were associated to stretching of C=C quinoid and benzenoid, respectively. These bands were shifted to 1610 cm^{-1} and 1516 cm^{-1} for those obtained under stirring reaction. The bands at 1300 cm^{-1} and 1265 cm^{-1} were due to stretching of C-N. The bands at 1113 cm^{-1} and 1107 cm^{-1} were associated with the presence of Q=NH^+B which is the main characteristics of emeraldine salt form of polyaniline. These strong absorption bands related to good electrical conductivity of the obtained polyaniline [15].

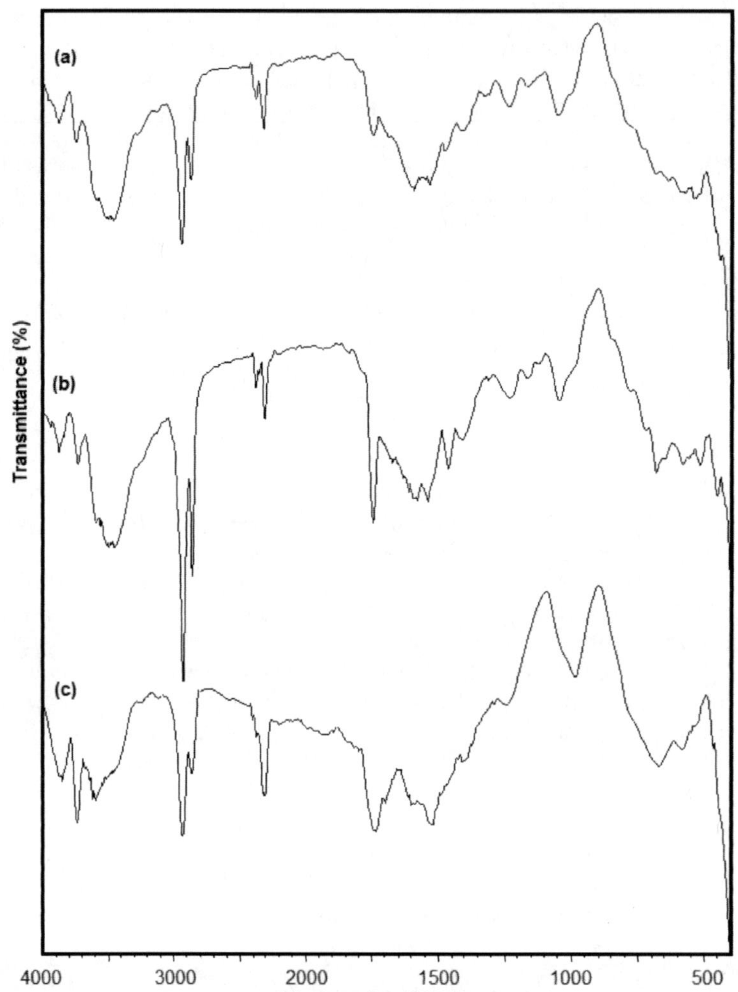

Figure 1. FTIR spectra of powder of polyaniline synthesized without stirring (a), with stirring (b), and film of the polyaniline (c)

Morphological analysis of the polyaniline was carried out using a scanning electron microscope (SEM). The micrograph shown in Figure 2 shows that polyaniline synthesized without stirring formed nanofiber with 100-200 nm diameter. Few granular polyaniline were also observed in the micrograph. In addition, the non-stirred reaction caused the difference of polymerization environment that resulted in not uniform morphology. Due to the direct exposure of UV irradiation, more rapid growth occurred at the surface compared to inner part of the system. Meanwhile, the stirred polymerization showed more uniform morphology with the fine diameter of 70–90 nm due to well distribution of UV light irradiation during polymerization process [5].

Figure 3 shows diffraction patterns of polyaniline obtained from the X-ray diffractometer (XRD) measurements. For polyaniline obtained from non-stirred reaction, the diffraction peaks appeared at 2θ 20.31° and 25.26°. The peaks were found to slightly shift to 2θ 20.55° and 25.59°, respectively, in the XRD pattern of polyaniline prepared under stirred reaction. After treatment with the mounting press, it is observed that there was no significant change in their diffraction patterns, although the first peak slightly shifted to 2θ 19.04° and 19.79° for the samples synthesized under non-stirred and stirred

IOP Conf. Series: Materials Science and Engineering **172** (2017) 012049 doi:10.1088/1757-899X/172/1/012049

reaction, respectively. Those are the main peaks indicating the formation of polyaniline, at which the first represents amorphous form and the second one corresponds to a crystalline structure of the polyaniline [16–20]. Benzenoid and quinoid rings in polyaniline chain have the major role to form a crystalline structure in the emeraldin structure of polyaniline [19]. The measurement results carried out using HighScore Plus software showed that crystallinity degree of non-stirred and stirred synthesized polyaniline were 13.21% and 14.56%, respectively.

Figure 2. SEM micrograph of polyaniline prepared without (a and b) and with stirring condition (c and d).

Surface area and porosity of the polyaniline were determined using surface area analyzer (SAA) Micromeritics ASAP 2020. Figure 4 shows type IV characteristics of nitrogen gas (N_2) adsorption-desorption of powder polyaniline. Hysteresis loops appeared at relatively high pressures of 0.8 P/P_0 for the polyaniline both prepared under non-stirred and stirred reaction. There was a remarkable rising of the adsorption volume when the condensation took place at high pressure. This hysteresis loop was caused by capillary condensation inside the mesopore [12]. From the curves, both polyanilines show similar hysteresis characteristics indicating typical porous which correspond to mesoporous material type. Distribution of pore size was then measured by Barrett-Joyner-Halenda (BJH) method. The distribution of polyaniline pore size showed average size at 19.8 nm and 17.1 nm respectively for polyaniline prepared under non-stirred and stirred reaction. These sizes were ranging from 2-50 nm and d>50 nm for both mesopore and macropore, respectively. BET surface area of the polyaniline synthesized without stirring was 27.07 m^2/g while with stirring was 28.68 m^2/g. This result shows that this approach produces high surface area of polyaniline compared to those reported previously i.e., 20.02 m^2/g dan 15.38 m^2/g [9, 12]. This condition may catalyze rapid electron transfer and high ion diffusion which could enhance electrochemistry performance of the polyaniline. In addition, it may also generate direct conduction path for electron [12].

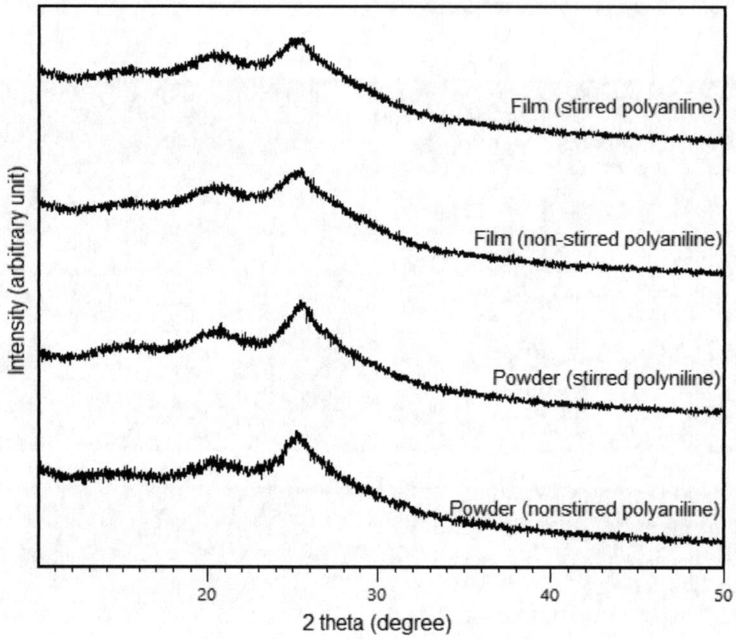

Figure 3. XRD pattern of synthesized polyaniline

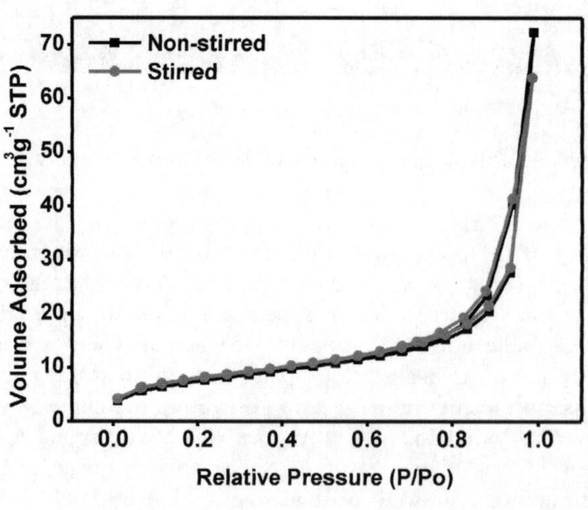

Figure 4. N_2 adsorption/desorption isotherm of the synthesized polyaniline

The conductivity of polyaniline measurement was carried out using four point probe. The obtained data was resistivity value (R). The conductivity was then calculated based on the following equation:

$$R = \frac{V}{I} = \rho \frac{x}{A} \quad (1)$$

$$\sigma = \frac{1}{\rho} \quad (2)$$

which R = resistance (Ω), ρ = resistivity (Ω.m), l = length (cm), A = cross-sectional area (cm^2) dan σ = conductivity (S/cm). The conductivity values of non-stirred and stirred polyaniline were 4.34 x 10^{-1} S/cm and 3.48 S/cm, respectively. The values indicated that polyaniline prepared under stirred reaction has higher conductivity than that of non-stirred. This condition could be related to morphology of the stirred reaction polyaniline which was fine and homogenous with a high surface area as revealed by the SEM and SAA results. These characteristics of polymer structures can improve electron transfer resulted in high conductivity of emeraldine salt of polyaniline. The result shows that electrical conductivity of pure polyaniline obtained from this work is higher than those reported in previous studies [11]. This suggests that UV irradiation can be employed to enhance the conductivity of emeraldine salt of polyaniline significantly.

4. Conclusion

Polyaniline in the conducting form emeraldine salt was prepared by oxidative polymerization under UV light irradiation with an initial polymerization temperature of 0°C. The obtained pure polyaniline shows high electrical conductivity as well as high surface area. It is found that stirring during polymerization process influences the growth characteristic and resulted in fine nanofiber and uniform morphology of the polyaniline. Though stirring do not affect crystallinity degree of polyaniline. The electrical conductivity of the emeraldine salt of polyaniline obtained from stirring reaction is eight times higher than those obtained from non-stirred polymerization. Owing to its high conductivity and surface area, the synthesized pure polyaniline might be a potentials material for electrode application.

References

[1] Jaymand M 2013 Recent progress in chemical modification of polyaniline. *Prog Polym Sci*; 38: 1287–1306.
[2] Rimbu G A, Stamatin I, Jackson C L, et al 2006 The morphology control of polyaniline as conducting polymer in fuel cell technology. *J Optoelectron Adv Mater*; 8: 670–674.
[3] Jakubik WP, Urbanczyk M, Maciak E, et al 2005 Polyaniline thin films as a toxic gas sensors in SAW system. *J Phys IV JP*; 129: 121–124.
[4] Dyachkova TP, Melezhyk a. V, Morozova ZG, et al 2012 Effects of the nature of oxidant and synthesis. *Вестник Тгту*; 718–730.
[5] Li ZF, Zhang H, Liu Q, et al 2013 Fabrication of high-surface-area graphene/polyaniline nanocomposites and their application in supercapacitors. *ACS Appl Mater Interfaces*; 5: 2685–2691.
[6] Du X, Xu Y, Xiong L, et al 2014 Polyaniline with high crystallinity degree: Synthesis, structure, and electrochemical properties. *J Appl Polym Sci*; 131: 6–13.
[7] Sapurina IY, and Shishov M A 2012 Oxidative Polymerization of Aniline: Molecular Synthesis of Polyaniline and the Formation of Supramolecular Structures. *New Polym Spec Appl*; 251–312.
[8] Shukla SK, Bharadvaja A, Tiwari A, et al 2010 Synthesis and characterization of highly crystalline polyaniline film promising for humid sensor. *Adv Mater Lett*; 1: 129–134.
[9] Sk MM, Yue CY, and Jena RK 2014 Synthesis of graphene/vitamin C template-controlled polyaniline nanotubes composite for high performance supercapacitor electrode. *Polym (United Kingdom)*; 55: 798–805.
[10] Parveen, Nazish, Ansari, M Omaish, and Cho MH 2015 Route to High Surface Area, Mesoporosity

of Polyaniline - Titanium Dioxide Nanocomposites via One Pot Synthesis for Energy Storage Applications. *Ind Eng Chem Res*; 55: 116–124.
[11] Ghani S a, Heah CY 2010 Conductive polymer based on polyaniline-eggshell powder (PANI-ESP) composites. *J Phys Sci*; 21: 81–97.
[12] Ayad M, and Zaghlol S 2012 Nanostructured crosslinked polyaniline with high surface area: Synthesis, characterization, and adsorption for organic dye. *Chem Eng J*; 204–205: 79–86.
[13] Bandgar DK, Khuspe GD, Pawar RC, et al 2012 Facile and novel route for preparation of nanostructured polyaniline (PANi) thin films. *Appl Nanosci*; 4: 27–36.
[14] De Barros RA, De Azevedo WM, and De Aguiar FM 2003 Photo-induced polymerization of polyaniline. *Mater Charact*; 50: 131–134.
[15] Sanches EA, Da Silva JMS, Ferreira JMDO, et al 2014 Structural and Morphological Characterization of Nanostructured Polyaniline Emeraldine-base form (EB-PANI). *J Mol Struct*; 1074: 732–737.
[16] Farrokhzad H, Gerven T Van, and Bruggen B Van Der 2013 Preparation and characterization of a conductive polyaniline / polysulfone film and evaluation of the effect of co-solvent. *Eur Polym J*; 49: 1–10.
[17] Lin Y W, and Wu T M 2009 Synthesis and characterization of externally doped sulfonated polyaniline/multi-walled carbon nanotube composites. *Compos Sci Technol*; 69: 2559–2565.
[18] Domingues S H, Salvatierra R V, Oliveira M M, et al 2011 Transparent and conductive thin films of graphene/polyaniline nanocomposites prepared through interfacial polymerization. *Chem Commun (Camb)*; 47: 2592–2594.
[19] Zareh E N, Moghadam P N, Azariyan E, and Sharifan I 2011 Conductive and biodegradable polyaniline/starch blends and their composites with polystyrene. *Polym J*; 20: 319–328.
[20] Farrokhzad H, Van Gerven T, and Van Der Bruggen B 2013 Preparation and characterization of a conductive polyaniline/polysulfone film and evaluation of the effect of co-solvent. *Eur Polym J*; 49: 3234–3243.

Acknowledgment
This work was supported by Universitas Negeri Jakarta and Directorat General of Higher Education (DGHE Ristekdikti) Republic of Indonesia through research fund no 34/SP2H//DRPM/LP-UNJ/HB/III/2016.

IOP Conf. Series: Materials Science and Engineering **172** (2017) 012050 doi:10.1088/1757-899X/172/1/012050

Synthesis and characterization of *p*-dimethylaminobenzaldehyde benzoylthiourea and study towards selective and sensitive fluorescent sensor for detection of iron (III) cation in aqueous solution.

S Hasan[1], N A Hamedan[1], A A A Razali[1], N H Uyup[1], and H M Zaki[1]

[1]Faculty of Applied Sciences, Universiti Teknologi MARA, 40450 Shah Alam, Selangor, Malaysia.

Email: sharizal187@perlis.uitm.edu.my

Abstract. Benzoyl thiourea derivatives ligand backbones contain Oxygen (O), Nitrogen (N), and Suphur (S) donor atoms can react with transition metal ions and form stable metal complexes. A new ligand was synthesized by the reaction of benzoyl isothiocyanate with diethylenetriamine and characterized by using Elemental Analysis (EA), Infrared (IR) spectroscopy and proton Nuclear Magnetic Resonance (^{1}H NMR) spectroscopy. The monofunctionalize ligand were synthesized using 1:1 condensation of *p*-dimethylaminobenzaldehyde with diethylenetriamine and confirmed by disappearance of carbonyl group, C=O at 1656 cm^{-1} with the appearance of peak C=N in the range 1633 - 1638 cm^{-1} in the IR spectrum. This monofunctionalize ligand was synthesis by using Schiff base technique. Benzoyl thiourea derivatives, **HN** are synthesis from the 1:1 condensation of benzoyl isothiocyanates with monofunctionalized ligand. **HN** ligand was verified by the presence of peaks *v*(N-H), *v*(C=O), *v*(C=N) *v*(C-N) and *v*(C=S) at 3317 - 3336 cm-1, 1612 - 1660 cm-1, 1550 -1589 cm-1, 1234-1366 cm-1 and 709-767 cm-1 respectively while ^{1}H NMR show peaks of alkane (CH_2), benzene (Ar-H), CONH, CSNH at 3.75, 6.73 – 7.33, 8.19, and 8.25 respectively. The total percent composition of C, H, N, O, and S using Elemental Analysis for **HN**, $C_{21}H_{27}N_5OS$ found was 85.23%. For application, further study on selectivity and sensitivity of **HN** ligand was conducted on several metal ions. The fluorescent emission spectroscopy shown that **HN** is a selective fluorescent sensor for Fe^{3+} and Cu^{2+} ions but not for other metal ions such as Co^{2+}, Ni^{2+}, Mn^{2+}, Cr^{3+} and Zn^{2+}. However, **HN** ligand shows more sensitive towards Fe^{3+} ions than Cu^{2+} ions which provide rapid detection of Fe^{3+} ions at concentrations as low as 4.5 x 10^{-7} M than 2.9 x 10^{-6} M of Cu^{2+} ions. As conclusion, **HN** ligand was expected to be useful as efficient chemical sensor for detection of Fe^{3+} ion.

1. Introduction

Nowadays, the development of chemosensors for selectivity and sensitivity quantification of environmentally and biologically for heavy and transition metal ions, has attracted a high attention [1]. They usually represent an environmental concern when present in uncontrolled amounts, but at the same time some of them such as iron, zinc, manganese, copper and cobalt are present as essential elements in biological systems. Among biologically important metal ions, iron is the most abundant transition metal ion present in the human body. As one of the most essential trace elements in

IOP Conf. Series: Materials Science and Engineering **172** (2017) 012050 doi:10.1088/1757-899X/172/1/012050

biological systems, Fe^{3+} performs a major role in many biochemical processes. Fe^{3+} ion is an important and essential for proper functioning of all living cells and acts as a cofactor in many enzymatic reactions to human as well as in specialized transport and storage of proteins. The deficiency of Fe^{3+} causes anaemia, liver damage, diabetes, hemochromatosis, Parkinson's disease and cancer [2]. However, iron overload is a clinical condition in which there is an excess of iron levels in the organism typically due to abnormally high iron absorption from the diet, periodical blood transfusions or wrong use of iron by the organism as result of the blocking action of determined cytokines [3]. If continued in time, high levels of Fe^{3+} within the body have been associated with increasing incidence of certain cancers and dysfunction of certain organs, such as heart, pancreas and liver. In recent years, several fluorescent sensors have been developed for the detection of transition and heavy metal ions, such as dimethylaminocinnamaldehyde-aminothiourea (DA) for detecting Ag^{+}, Hg^{2+} and Cu^{2+} [4], *p*-dimethylaminobenzaldehyde thiosemicarbazone for detecting Hg^{2+} [5], rhodamine derivatives (RBS) for detecting Cu^{2+} and F^{-} [6] and also salicylaldimine based receptor sensing of multi metal ions and anions [7]. Meanwhile, various sensors for Fe^{3+} have been reported [8][9][10]. However, to the best of our knowledge, the static quenching mechanism for aminothiourea-based chemosensor to monitor trace levels of iron ions is rarely reported. Fluorescence-based analytical methods are widely used for detection of metal ions in an early stage because they are sensitive, simple and less expensive. In this study, the fluorescent sensor for sensitive determination of iron is reported. In design of fluorescent sensors for these ions, aminothiourea is a preferred receptor because of its strong affinity [4]. Scheme 1 below shows the reaction of synthesis of ligand **HN** step by step.

Step 1: Preparation of benzoyl isothiocyanate

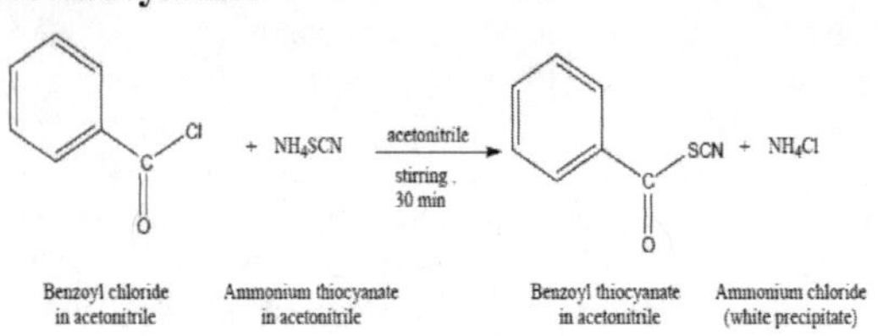

Step 2: Formation of monofunctional symmetrical triamine

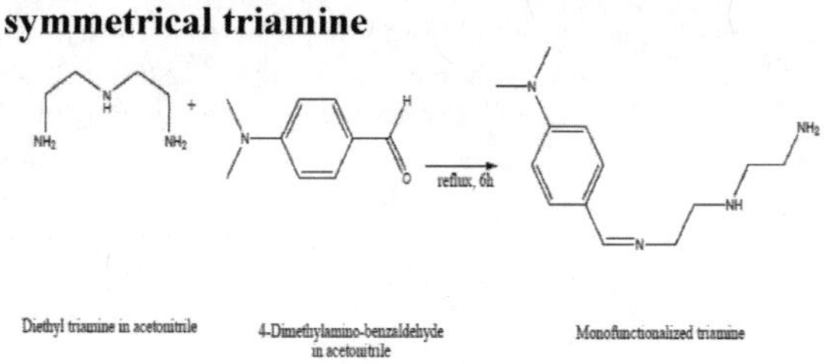

Step 3: Formation of difunctional symmetrical triamine, HN

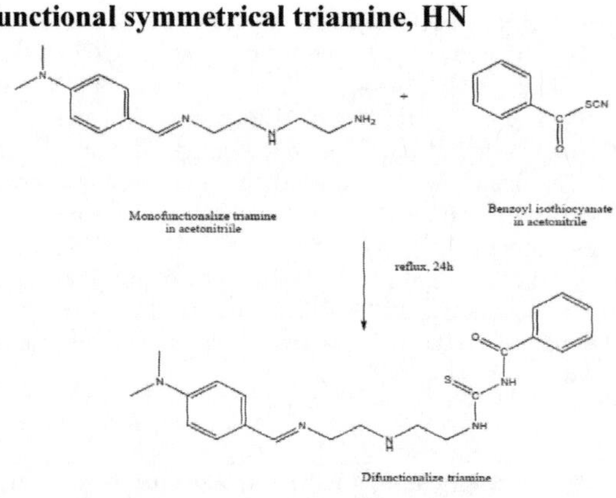

Scheme 1 Synthesis of ligand HN

1.1. Apparatus and reagents

The inorganic salts $Fe(Cl_2)_3.6H_2O$, $Mn(SO_4)_2$, $Cr(SO_4)_26H_2O$, $Co(Cl_2)_2.6H_2O$, $Ni(NO_3)_2.6H_2O$, $Cu(NO_3)_2.3H_2O$ and $Zn(Cl_2)_2$ were purchased from Sigma Aldrich. All solvents and other reagents were of analytical grade. ^1H NMR spectra are recorded on a Bruker Avance III 300 Spectrometer at room temperature. ^1H chemical shifts are reported in ppm downfield from tetramethylsilane (TMS), d scale with the solvent resonances as internal standards). Melting points were measured using BÜCHI Melting Point B-545 and the elemental analyses were conducted using CHNS Analyzer Flash EA 1112 series. The infrared spectra were performed using FTIR Perkin Elmer 100 Spectrophotometer in the spectral range of 4500-600 cm^{-1}.

1.2. General Procedure for Fluorescence spectrometer

The fluorescence spectra were recorded on Luminescence Spectrometer LS 55 equipped with quartz cuvettes of 1 cm path length. The excitation and emission slit widths were 10 nm. Stock solutions for analysis were prepared (0.1 M for Ligand **HN** (Acetonitrile/H_2O, 9:1 (v/v), pH = 7.0) immediately before the experiments. The fluorescence titration was performed with a series of 5×10^{-5} M solutions of **HN** containing various equivalents of Fe^{3+} ions. Binding studies were confirmed by job plot. The association constant (Ka) were calculated by the linear Benesi-Hildebrand equation from fluorescence (Eq. (1)) study by,

$$\frac{1}{F-F_0} = \frac{1}{(Fmin-F_0)} + \frac{1}{K(Fmin-F_0)} x \frac{1}{[M^{n+}]} \quad (1)$$

Where F, F_0 and F_{min} are the emission intensities, respectively in the absence and infinite concentration of the metal ion (Fe^{3+}).

*1.3. Synthesis of 1-Benzoyl-3-(2-{2-[(4-dimethylamino-benzylidene)-amino]-ethylamino}-ethyl)-thiourea, **HN**.*

A solution of benzoyl chloride (10 mmol; 1.2 ml) in 10 ml acetonitrile is added to a solution of ammonium thiocyanate (10 mmol; 0.80 g) in 40 ml acetonitrile under stirring for 30 minutes at room temperature. The white precipitate NH_4Cl is will bestow and filtered off. Diethylene triamine (50 mmol; 5.4 ml) is dissolved in 20 ml of acetonitrile and added of 4-dimethylaminobezaldehyde (50 mmol; 7.46 g) in 25 ml of acetonitrile. The reaction mixture is refluxed for 6 h at 100°C. Then, the acetonitrile is vaporized from the mixture by using rotary evaporator. The oily solution is dried to surrounding for an hour until it becomes solid. The percentage yield and melting point of the monofunctional symmetrical triamine are calculated and recorded. The benzoyl isothiocyanate (10 mmol) in 50 ml of acetonitrile is added dropwise to a round-bottomed flask containing the monofunctional ligand. Then, the mixture is refluxed for 24 hours at 100°C. On cooling, water is added to furnish as solid. The product is recrystallized twice.

HN: Yield 66%; dark orange solid, m.p 119 °C. IR (cm^{-1}): υ(C=O) 1660, υ(N-H) 3336, υ(C=N) 1550, υ(C=S) 1064. ^1H NMR (600 MHz, Acetone-d_6) δ 8.19 (s, 1H), 8.19 (s, 1H), 6.73-7.33 (dd, 6H), 2.75 (s, 6H), 2.06 (s, 1H). Anal. Calc. for $C_{21}H_{27}N_5OS$: C, 63.45; H, 6.85; N, 17.62; S, 8.06. Found: C, 60.01; H, 5.31; N, 17.54; S, 8.06.

2. Results and Discussions

2.1. Sensing in the presence of competing ions.

Ligand **HN** was applied as chemosensor by using Fluorescence Emission Spectroscopy (FES).to observe several parameters such as selectivity, sensitivity, stoichiometry and binding mode studies and also the competition.

2.1.1. Selectivity

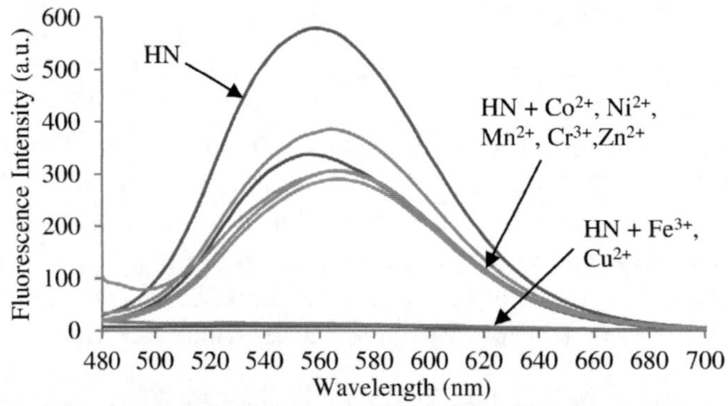

Figure 1. Fluorescence emission spectra of **HN** in the presence of of various metal ions in ACN/H_2O (10:1, v:v, pH 7.0) solution at room temperature.

For an excellent chemosensor, high selectivity is a matter of necessity. Among the various metal ions, the fluorescence emission spectra showed a noteworthy high selectivity to Fe^{3+} and Cu^{2+} with respect to the fluorescence quenching. Other metal ions developed no distinct significant fluorescence intensity changes. Related metal ions, including Fe^{3+}, Cu^{2+}, Co^{2+}, Ni^{2+}, Mn^{2+}, Cr^{3+} and Zn^{2+}, were used to evaluate the metal ion binding properties of **HN** in ACN/ H_2O (10:1/ v:v, pH 7.0) solution by fluorescence spectroscopy (Fig. 1).

2.1.2. Sensitivity

(a)

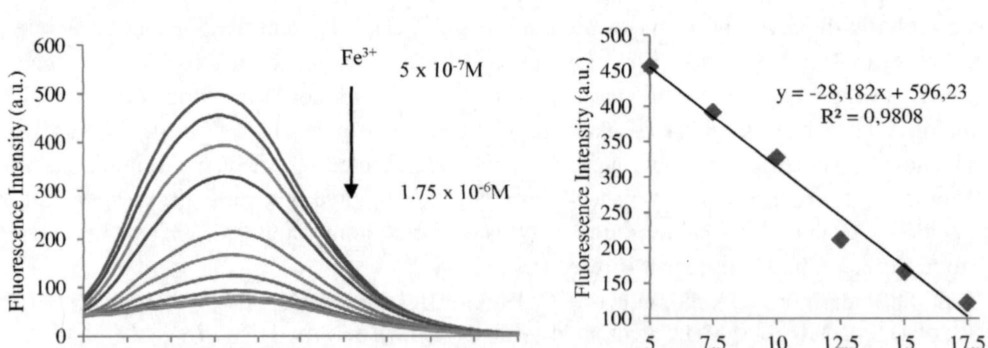

(b)

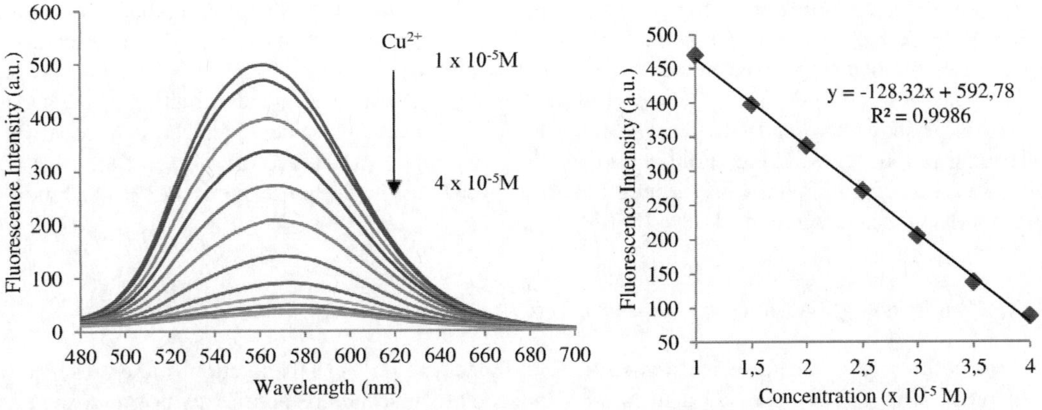

Figure 2. (a) Fluorescence emission spectra (λ_{ex} = 561 nm) of sensor **HN** upon addition of Fe^{3+} in aqueous solution ACN/ H_2O (10:1/ v:v, pH 7.0). (b) Fluorescence emission spectra (λ_{ex} = 561 nm) of sensor **HN** upon addition of Cu^{2+} in aqueous solution ACN/ H_2O (10:1/ v:v, pH 7.0)

HN can detect both Fe^{3+} ion and Cu^{2+} ion but **HN** more sensitive sensor towards Fe^{3+} ion than Cu^{2+} ion. Figure 2 shows the results of fluorescence titration of **HN** upon the gradual addition of Fe^{3+} (Fig. 2.(a)) and Cu^{2+} (Fig 2.(b)) in ACN/H_2O solution under excitation at λ = 561 nm. A linear relationship was observed to exist between the relative fluorescent intensity of **HN** and the concentration of Fe^{3+} in the range of 5 x 10^{-7}M-17.5 x 10^{-7}M with a detection limit of 4.5 x 10^{-7}M while the concentration of Cu^{2+} in the range of 1 x 10^{-5}M – 4 x 10^{-5}M with a detection limit of 2.9 x 10^{-6}M. According to the U.S. Environmental Protection Agency (EPA), the maximum content of iron in drinking water is ~5.357 μM. Therefore, detection of trace irons becomes very important in the chemical research[11].

2.1.3. Stoichiometry and binding mode studies

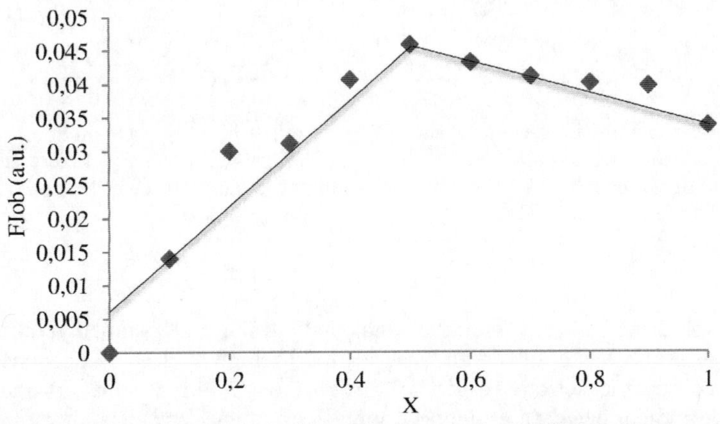

Figure 3. Job's plot for determining the binding stoichiometry in ACN/H_2O (10:1/ v:v, pH 7.0) solution, showing a 1:1 stoichiometry between **HN** and Fe^{3+}. The variation of which indicated that the stoichiometry of **HN**-Fe^{3+} complex was 1:1. The variation of Job's function (F_{Job}) at 561 nm was measured as a function of the molar ratio X ([Fe^{3+}]/([Fe^{3+}] + [chemosensor])).

Fe^{3+} ion was selected for further study in binding mode to the sensor **HN**. The stoichiometry formation of **HN**-Fe^{3+} complex was 1:1. To gain an insight into the stoichiometry of the **HN**-Fe^{3+} complex, the binding stoichiometry was determined by the Job plot. The Job's function F_{Job} is calculated according to the equation $F_{Job} = (1-X)\ F_o - F$. The plot of the luminescence versus the mole fraction of the added Fe^{3+} (Fig. 3) shows two parts that can be fitted by straight lines. The lines intersect at X equal 0.5. According to the Benesi-Hilderbrand equation $1/(F_o - F) = 1/\{K_a^*(F_o - F_{min})^*[Fe^{3+}]\} + 1/(F_o - F_{min})$, (where F_{min} is the minimum fluorescence intensity at 561 nm in the presence of Fe^{3+}), the binding constant (K_a) is calculated to be 4.58 x 10^5 M^{-1}.

2.1.4. Competition of Fe^{3+} toward other metal ions

The competition experiment, which was carried out by adding Fe^{3+} to **HN** solution in the presence of other metal ions (Fig. 4), showed that the Fe^{3+}-induced fluorescence response was not interfered by commonly coexistence ions. The result suggested that probe **HN** showed a remarkable selectivity toward Fe^{3+} over other competitive ions. However, Cu^{2+} also showed a remarkable selectivity but not as sensitive as Fe^{3+} after we used all the metal ions in 5 x 10^{-6}M.

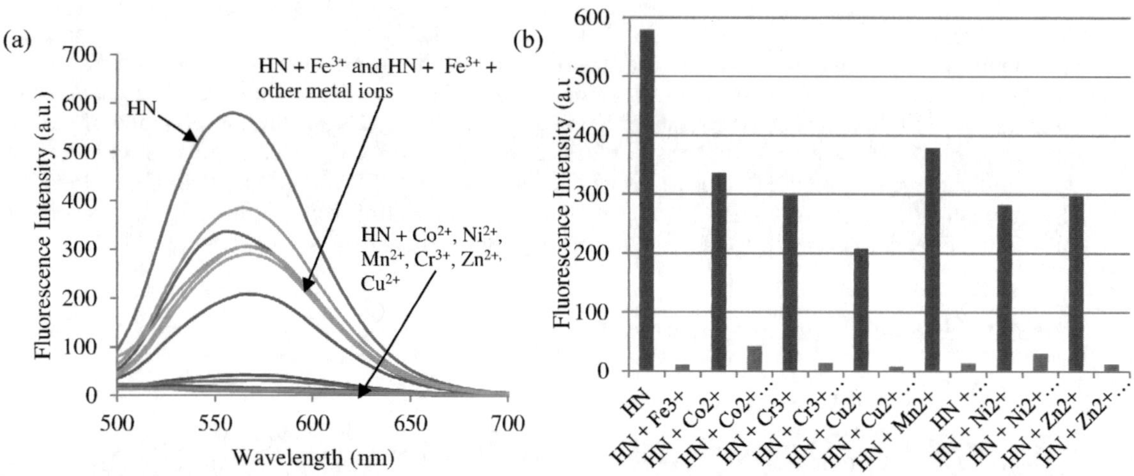

Figure 4. (a) Fluorescence emission spectra sensor **HN** (5 x 10^{-5}M) in ACN/H_2O solutions in the presence of Fe^{2+} and the miscellaneous cations Cu^{2+}, Cr^{3+}, Ni^{2+}, Zn^{2+}, $Mn^{2+}$and Co^{2+}. (b) Fluorescence intensity at wavelength of 561nm of sensor **HN** (5 x 10^{-5}M) in ACN solutions in the presence of Fe^{3+} and the miscellaneous cations Cu^{2+}, Cr^{3+}, Ni^{2+}, Zn^{2+}, $Mn^{2+}$and Co^{2+}.

3. Conclusion

In conclusion, **HN** bearing thiourea moiety as the binding site and *p*-dimethylaminobenzaldehyde moiety as the signal group was designed and synthesized to apply as fluorescent sensor for Fe^{3+} ion. This sensor showed specific selectivity for Fe^{3+} in ACN/H_2O binary solutions indicated that the aminobenzaldehyde moiety acted as a signal group (fluorophore) and played a crucial role in the process of fluorescent recognition. Investigation of the binding constant mechanism indicated that the sensor HN recognized Fe^{3+} by forming a stable 1:1 **HN**- Fe^{3+}complex. The coexistence of other cations did not interfere with the Fe^{3+} recognition process unless for Cu^{2+}. **HN** also sense the Cu^{2+} but not as sensitive as Fe^{3+}. Moreover, the detection limit of the sensor **HN** toward Fe^{3+} was 4.5 x 10^{-7}M and 2.9 x 10^{-6}M toward Cu^{2+}. This indicated that the sensor **HN** may be useful for preliminary detection of Fe^{3+} ion by simple solvent extraction in chemical and environmental applications.

References

[1] M She, Z Yang, B Yin, J Zhang, J Gu, W Yin, J Li, G Zhao, and Z Shi, A novel rhodamine-based fluorescent and colorimetric 'off-on' chemosensor and investigation of the recognizing behavior towards Fe^{3+} *Dye. Pigment.*, vol. 92, no. 3, pp. 1337–1343, 2012.

[2] N Roy, A Dutta, P Mondal, P C Paul, and S S Takhellambam, A new coumarin based dual functional chemosensor for colorimetric detection of Fe^{3+} and fluorescence turn-on response of Zn^{2+} *Sensors Actuators B Chem.*, vol. 236, pp. 719–731, 2016.

[3] F J Lázaro, L Gutiérrez, A R Abadía, M S Romero, and A López, Biological tissue magnetism in the frame of iron overload diseases *J. Magn. Magn. Mater.*, vol. 316, no. 2 SPEC. ISS., pp. 126–131, 2007.

[4] N Khoa, N Chi, N Thi, A Nhung, and N Tien, A highly sensitive fl uorescent chemosensor for simultaneous determination of Ag (I), Hg (II), and Cu (II) ions : Design , synthesis , characterization and application *Dye. Pigment.*, vol. 116, pp. 89–96, 2015.

[5] Y Yu, L Lin, K Yang, X Zhong, R Huang, and L Zheng, p -Dimethylaminobenzaldehyde thiosemicarbazone : A simple novel selective and sensitive fluorescent sensor for mercury (II) in aqueous solution vol. 69, pp. 103–106, 2006.

[6] H Yu, J Lee, S Angupillai, S Wang, S Feng, S Matsumoto, and Y Son, Spectrochimica Acta Part A : Molecular and Biomolecular Spectroscopy A new dual fluorogenic and chromogenic " turn-on " chemosensor for Cu^{2+}/F^{-} ions vol. 151, pp. 48–55, 2015.

[7] D Udhayakumari, S Saravanamoorthy, and S Velmathi, Colorimetric and fluorescent sensing of transition metal ions in aqueous medium by salicylaldimine based chemosensor *Mater. Sci. Eng. C*, vol. 32, no. 7, pp. 1878–1882, 2012.

[8] W Lu, H Jiang, F Hu, L Jiang, and Z Shen, A novel chemosensor based on Fe(III)-complexation for selective recognition and rapid detection of fluoride anions in aqueous media *Tetrahedron*, vol. 67, no. 41, pp. 7909–7912, 2011.

[9] H J Jung, N Singh, D Y Lee, and D O Jang, Single sensor for multiple analytes: chromogenic detection of I- and fluorescent detection of Fe^{3+}," *Tetrahedron Lett.*, vol. 51, no. 30, pp. 3962–3965, 2010.

[10] Z Zhang, S Lu, C Sha, and D Xu, A single thiourea-appended 1,8-naphthalimide chemosensor for three heavy metal ions: Fe^{3+}, Pb^{2+}, and Hg^{2+} *Sensors Actuators B Chem.*, vol. 208, pp. 258–266, 2015.

[11] C Yi, B Song, W Tian, X Cui, Q Qi, W Jiang, Z Qi, and Y Sun, Fluorescent sensor of fluorene derivatives having phosphonic acid as a fluorogenic ionophore: Synthesis and static quenched properties for Fe(III) *Tetrahedron Lett.*, vol. 55, no. 37, pp. 5119–5123, 2014.

Acknowledgements

We would like to thanks the Malaysian Ministry of Education (KPM) for providing generous financial support under RAGS grants: 600-RMI/RAGS 5/3 (2/2013) in conducting this study and Universiti Teknologi Mara (UiTM) for providing all the needed facilities.

IOP Conf. Series: Materials Science and Engineering **172** (2017) 012051 doi:10.1088/1757-899X/172/1/012051

Dimethylaminodiethylenetriamine Derivatives of Fluorescence Chemosenso for Detection of Zn^{2+} In Aqueous Solution

S N A M Adnan[1], S Hasan[1], S Zakaria[1] and Y M Yusof[1]

[1]Faculty of Applied Sciences Universiti Teknologi MARA 40450 Shah Alam, Selangor, Malaysia

Email: sharizal187@perlis.uitm.edu.my

Abstract. Fluorescent chemosensors for the detection and measurement of metal ions, especially for cations environmental interest such as Fe^{3+}, Co^{2+}, Mn^{2+}, Cu^{2+}, and Zn^{2+} are actively investigated because it shows simplicity, high sensitivity and fast response. New benzenyl derivative bearing pyridine group has been synthesized and studied as fluorescent chemosensor for Zn^{2+} ion. Chemosensor N-{2-[(4-Dimethylamino-benzylidene)-amino]-ethyl}-N'-pyridin-2-ylmethylene-ethane-1,2-diamine was synthesized by condensation of *p*-dimethylaminobenzaldehyde, diethylenetriamine and *o*-pyridinecarbaldehyde and characterized by FT-IR, ^{1}H-NMR and elemental analysis (CHN). FT-IR showed the appearance of peak azomethine (C=NH) at 1639.46 cm^{-1}, pyridine (C-N) at 1591 cm^{-1}and disappearance of NH_2 peak at 3278.78 cm^{-1} after the condensation reaction in between aldehyde and amine. ^{1}H-NMR signal at 8.19 ppm, 3.12 ppm and 8.08 ppm was assigned to C=NH, $N(CH_3)_2$ and C-N respectively, confirmed the formation of N-{2-[(4-Dimethylamino-benzylidene)-amino]-ethyl}-N'-pyridin-2-ylmethylene-ethane-1,2-diamine. The elemental analysis was found closed to the theoretical value and the percent composition of A is 91.82%. Sensor A exhibits high selectivity and sensitivity towards Zn^{2+}. Other metal ions such Cu^{2+}, Fe^{3+}, Co^{2+} and Ni^{2+} had no such significant effect on the fluorescence. The detection limit of N-{2-[(4-Dimethylamino-benzylidene)-amino]-ethyl}-N'-pyridin-2-ylmethylene-ethane-1,2-diamine for Zn^{2+} was 3.5 x 10^{-5} M. M. This sensor exhibits a very good fluorescence sensing ability to Zn^{2+} over a wide range of pH. Therefore it is capable of being a practical system for the monitoring of Zn^{2+} concentrations in real water sample.

1. Introduction

Zinc is the second most abundant transition metal ion in the human body, it plays an indispensable role in various biological activities such as gene expression, neural signal transmission, regulation of metalloenzymes, and DNA binding or recognition [1]. However, excessive amounts of zinc in human cause many severe diseases such as Alzheimer's disease [2], ischemic stroke [3] and epilepsy [4]. Hence it is very important to develop highly selective and sensitive chemosensors for the detection of Zn^{2+} in the environment as well as biological samples.

Owing to its essential roles in biological nutrition systems, the measurement of Zn^{2+} is an active field in analytical chemistry. Compared with numerous traditional analytical techniques, fluorescence molecule sensor display high sensitivity, selectivity, simplicity, and tunability thus have been developed as a powerful tool for detecting the trace amount of analytes. Chemosensors are commonly composed of two parts, a receptor (recognition element) responsible for the molecular recognition of the analyte and a fluorophore (fluorescence reporter) responsible of signaling the recognition event and both are connected through a spacer or also known as a linker [5]. Up to now, a number of fluorescence chemosensors for Zn^{2+} have been reported, which include the fluorophores such as rhodamine, coumarin, BODIPY, and the receptors such as dipicolylamine, and quinolone [6]. These chemosensors are capable of binding with Zn^{2+}. However, a problem with fluorophore chemosensor is their ability to selectively recognize metal ions especially involving selectivity towards one metal cation in mixing with other metal cations. The selectivity of chemosensor towards metal cation is depending on the energy produced by the compound in complexing with the metal cation.

Here we designed and synthesized N-{2-[(4-Dimethylamino-benzylidene)-amino]-ethyl}-N'-pyridin-2-ylmethylene-ethane-1,2-diamine as a chemosensor for Zn^{2+} which bears a diethylenetriamine and pyridyl moiety acting as the selective receptor of Zn^{2+} and 4-dimethylaminobenzenyl as fluorescence signal. The effects of selectivity, pH and competition ions were reported. Dimethylaminobenzyl moiety was selected in this study due to its suitability and performance as a fluorophore with high energy [7]. As for receptor, diethylenetriamine and 2-pyridinecarbaldehyde, were combined to be able to coordinate with metal ions. The coordination can likely influence the fluorescence of the molecule because of the presence of strong donor N atom.

2. Experimental

2.1. Materials and physical methods

All reagents and solvents were commercially available of analytical grade and were used without further purification. 4-dimethylaminobenzaldehyde, diethylenetriamine and 2-pyridinecarbaldehyde were obtained from Sigma. ^{1}H NMR spectra were recorded on a Bruker Varian 600 Hz using chloroform-d and acetone-d_6 solutions with TMS as an internal standard. The fluorescence spectra were recorded with a Luminescence Spectrometer LS 55 equipped with quartz cuvettes of 1 cm path length. The elemental analysis was carried out on Thermo Finnigan Flash EA 110 Elemental Analyzer. Infrared spectra were obtained on Perkin Elmer 100 spectrophotometer in the range 4000–600 cm^{-1}.

2.2. Synthesis of N-{2-[(4-Dimethylamino-benzylidene)-amino]-ethyl}-N'-pyridin-2-ylmethylene-ethane-1,2-diamine (A)

To prepare unsymmetrical Schiff base involves two steps; monocondensed and dicondensed.

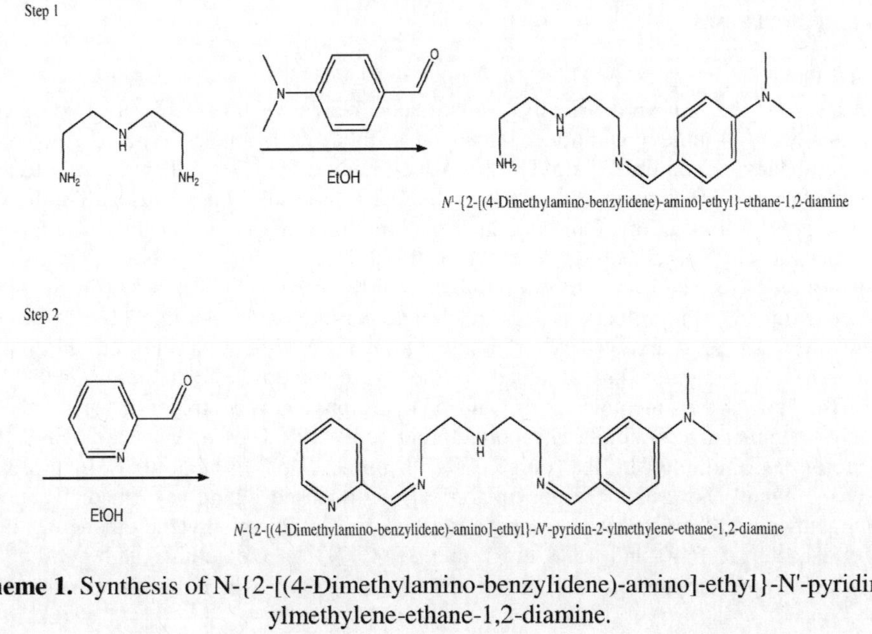

Scheme 1. Synthesis of N-{2-[(4-Dimethylamino-benzylidene)-amino]-ethyl}-N'-pyridin-2-ylmethylene-ethane-1,2-diamine.

Step 1
To the stirred solution of the diethylenetriamine (10 mmol) in ethanol (30 ml), was added dropwisely a solution of the 4-dimethylaminobenzaldehyde (10 mmol) in ethanol (30 ml). After addition was completed, the mixture was stirred and refluxed for 6 h. During this time the mixture turn to yellow solution. Then the solution was evaporated under reduced pressure, the oily product was dried in desiccator overnight to remove excess solvent to afford a yellow solid product. Yield: 2.2 g (84%). IR (cm^{-1}): 1633 (C=N), 3278.78 (NH_2), 1602.60 (Ar, C=C). ^1H NMR (600 MHz, Acetone-d_6) δ 9.74 (s, 1H), 8.96 – 8.81 (m, 2H), 8.44 (d, J = 7.8 Hz, 1H), 8.18 – 8.11 (m, 1H), 7.91 – 7.83 (m, 1H), 7.56 – 7.48 (m, 2H), 7.48 – 7.39 (m, 1H), 7.35 – 7.24 (m, 1H), 6.78 – 6.67 (m, 2H), 3.65 – 3.48 (s, 2H), 3.12 (s, 7H), 2.59 – 2.49 (m, 1H), 1.94 (s, 2H) ppm. Anal. Calc. for $C_{13}H_{22}N_4$: C, 66.63; H, 9.36; N, 23.91. Found: C, 60.65; H, 8.99; N, 22.27.
Step 2
To the stirred solution of the precursor (10 mmol) in ethanol (30 ml), a solution of 10 mmol in 30 ml ethanol of 2-pyridinecarbaldehyde was added and the solution was refluxed for 48 h. Then the solution was evaporated under reduced pressure, the oily product was dried in desiccator overnight to remove excess solvent to afford a yellow solid product. Yield: 3.3 g (76%). IR (cm^{-1}): 1633 (C=N), 3278.78 (NH_2), 1602.60 (Ar, C=C), 1521.13 (Ar, C=C). ^1H NMR (600 MHz, Chloroform-d) δ 8.08 (m, 4H), 3.59 (d, J = 6.2 Hz, 2H), 2.87 (d, J = 12.9, 4.5 Hz, 6H), 2.62 (s, 1H), 1.94 (d, J= 2H).

2.3. General procedure for fluorescence experiments
All fluorescence spectra were recorded on a Luminescence Spectrometer LS 55 after the addition of perchlorate metal salts in (CH_3OH/H_2O, 9:1 (v/v), Tris-HNO_3 = 10 mM, pH = 7.0), while keeping the ligand concentration constant (5×10^{-5} M). Solutions of metal ions were prepared from the perchlorate salts of Fe^{3+}, Co^{2+}, Ni^{2+}, Zn^{2+} and Cu^{2+}. The selectivity of N-{2-[(4-Dimethylamino-benzylidene)-amino]-ethyl}-N'-pyridin-2-ylmethylene-ethane-1,2-diamine (A) to analytes was measured with the concentration of N-{2-[(4-Dimethylamino-benzylidene)-amino]-ethyl}-N'-pyridin-2-ylmethylene-ethane-1,2-diamine (A) 5×10^{-5} M and 1×10^{-3} M analytes. The titration experiments were carried out by taking 5×10^{-5} M of N-{2-[(4-Dimethylamino-benzylidene)-amino]-ethyl}-N'-pyridin-2-ylmethylene-ethane-1,2-diamine (A) with the increasing concentration of Fe^{3+}, Cu^{2+}, Co^{2+}, Ni^{2+} and Zn^{2+} respectively.

3. Results and discussions

3.1. Synthesis
The synthesis of the unsymmetrically substituted benzenyl proceed in two steps, i.e., monocondensation of 4-dimethylaminobenzaldehyde with a diethylenetriamine to give N1-{2-[(4-Dimethylamino-benzylidene)-amino]-ethyl}-ethane-1,2-diamine, which are then allowed to react with 2-pyridinecarbaldehyde. The monocondensed product of amine also known as half unit where the primary amine group forms an imine or enamine bond and the other is unchanged, have a potential to react with other carbonyl moieties to form unsymmetrical ligands capable of binding transition metal ions. Condensation of the 'half unit' shows two bands in IR reign (3298.01-2809.63 cm^{-1}) [8]. These two bands are assigned to the primary amine stretchings. An intense band around 1638.80 cm-1 in the spectrum of the half unit and N-{2-[(4-Dimethylamino-benzylidene)-amino]-ethyl}-N'-pyridin-2-ylmethylene-ethane-1,2-diamine are assigned to the azomethine (C=N) vibrations [9]. ^1H NMR spectrum of the half unit azomethine (C=N) and NH_2 protons show a singlet at 8.19 and 2.17 ppm which closely resembles that of similarly reported compound [10]. The half unit, the ligand shows the aromatic proton as multiplet in the range 6.9-7.8 ppm. The absence of NH_2 proton in the monocondensed ligand shows that the unsymmetrical dicondensed ligand is formed. IR spectrum of N-{2-[(4-Dimethylamino-benzylidene)-amino]-ethyl}-N'-pyridin-2-ylmethylene-ethane-1,2-diamine displayed two bands at 1640 and 1588 cm-1 due to (C=N) of azomethine and pyridyl moieties, respectively. The pyridine proton has the signal at 9.23 ppm which confirmed the formation dicondensed ligand.

IOP Conf. Series: Materials Science and Engineering **172** (2017) 012051 doi:10.1088/1757-899X/172/1/012051

3.2. Selectivity

The fluorescence selectivity of N-{2-[(4-Dimethylamino-benzylidene)-amino]-ethyl}-N'-pyridin-2-ylmethylene-ethane-1,2-diamine (A) toward various metal ions (Fe^{3+}, Co^{2+}, Ni^{2+}, Cu^{2+}, and Zn^{2+}) were first studied in Tris-HNO_3 buffer solution (pH 7.0). N-{2-[(4-Dimethylamino-benzylidene)-amino]-ethyl}-N'-pyridin-2-ylmethylene-ethane-1,2-diamine (A) has strong fluorescence emission (λ_{max} = 450 nm). The fluorescence enhancement observed for Zn^{2+} as shown in Figure 1.

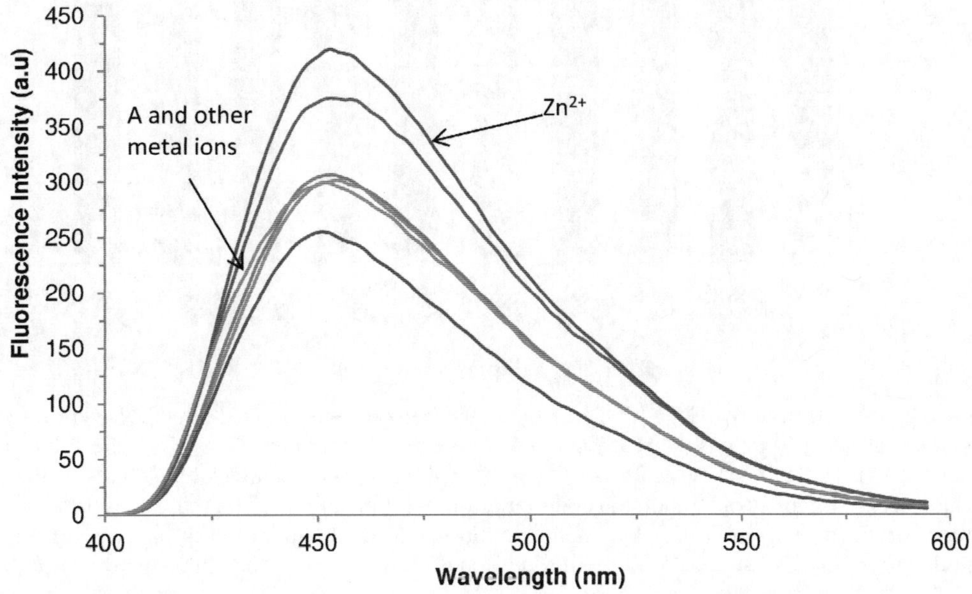

Figure 1. Fluorescence spectra of N-{2-[(4-Dimethylamino-benzylidene)-amino]-ethyl}-N'-pyridin-2-ylmethylene-ethane-1,2-diamine (A) upon addition of Zn^{2+} and other metal ions.

The increase in fluorescence intensity can be attributed due to the formation of the complex. The incorporation of Zn^{2+} effectively increases the conformational rigidity of the ligand and enhanced fluorescence intensity complex [11]. When same equivalent metal ions were added to N-{2-[(4-Dimethylamino-benzylidene)-amino]-ethyl}-N'-pyridin-2-ylmethylene-ethane-1,2-diamine (A), it was found that Fe^{3+}, Co^{2+}, Ni^{2+} and Cu^{2+} caused fluorescence quenched respectively. The results show significant separation in between Zn^{2+} and other metal. This method was used to determine the selectivity for metal ions reported in a previous study [12]. Therefore N-{2-[(4-Dimethylamino-benzylidene)-amino]-ethyl}-N'-pyridin-2-ylmethylene-ethane-1,2-diamine (A) has a better selectivity towards Zn^{2+} over other metal ions tested under the same conditions. The selectivity of N-{2-[(4-Dimethylamino-benzylidene)-amino]-ethyl}-N'-pyridin-2-ylmethylene-ethane-1,2-diamine (A) for Zn^{2+} over other metal was further investigated by competition experiments. High fluorescence intensity of Zn^{2+} and N-{2-[(4-Dimethylamino-benzylidene)-amino]-ethyl}-N'-pyridin-2-ylmethylene-ethane-1,2-diamine (A) affected by the presence of other metal ions. As shown in Figure 2 the individual fluorescence intensity of other metal ions were quenched. When comparing with the individually intensity of competition metals show the presence of Zn^{2+} ion in the mixture is higher than the intensity of metal itself with N-{2-[(4-Dimethylamino-benzylidene)-amino]-ethyl}-N'-pyridin-2-ylmethylene-ethane-1,2-diamine (A). The fluorescence intensity of N-{2-[(4-Dimethylamino-benzylidene)-amino]-ethyl}-N'-pyridin-2-ylmethylene-ethane-1,2-diamine (A) with Zn^{2+} decreasing in the presence of competition metal ion [13]. This shows that other metal ions are also capable of binding with N-{2-[(4-Dimethylamino-benzylidene)-amino]-ethyl}-N'-pyridin-2-ylmethylene-ethane-1,2-diamine (A). These results suggest that although the presence Fe^{3+}, Co^{2+}, Ni^{2+} and Cu^{2+} can quench the fluorescence, the total intensity is still good enough for detection of Zn^{2+}.

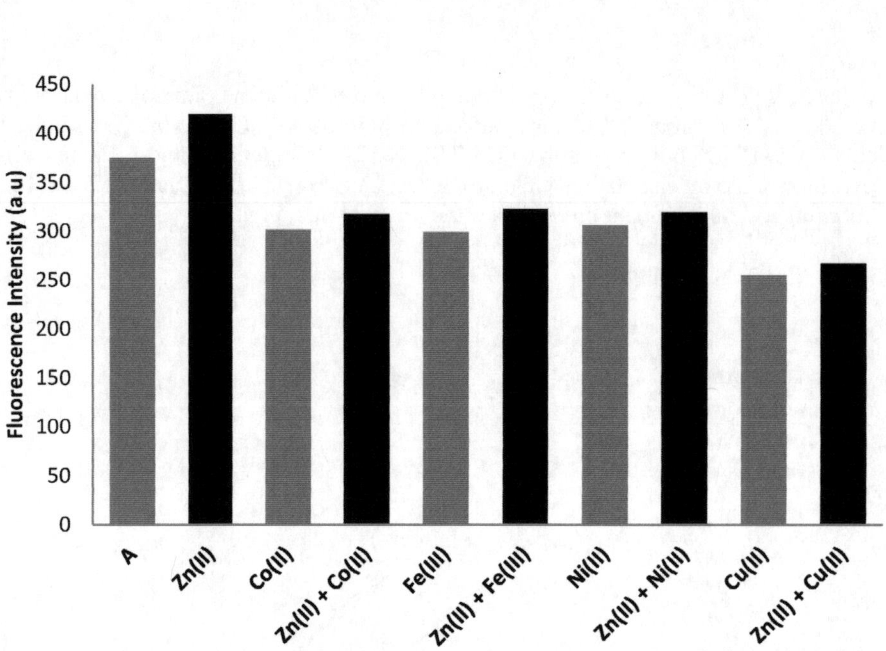

Figure 2. The selectivity of N-{2-[(4-Dimethylamino-benzylidene)-amino]-ethyl}-N'-pyridin-2-ylmethylene-ethane-1,2-diamine (A) for Zn^{2+} in the presence of other metal ions (Co^{2+}, Fe^{3+}, Ni^{2+} and Cu^{2+}) in CH_3OH/H_2O, 9:1 (v/v), Tris-HNO_3 = 10 mM, pH = 7.0. The black bars represent the fluorescence intensity of A upon the addition of different metal ions. The red bars represent fluorescence intensity of N-{2-[(4-Dimethylamino-benzylidene)-amino]-ethyl}-N'-pyridin-2-ylmethylene-ethane-1,2-diamine (A) upon the addition of different metal ions in the presence of Zn^{2+}.

3.3. Effects of pH

Any chemosensor to be considered for application in bio-imaging applications need to have high intensity of the spectroscopic signal at physiological pH. Since many chemosensors were sensitive to pH we studied the sensitivity of fluorescence emission in presence and absence of Zn^{2+} examined in a pH range from 3.0 to 12.0. The results of both N-{2-[(4-Dimethylamino-benzylidene)-amino]-ethyl}-N'-pyridin-2-ylmethylene-ethane-1,2-diamine (A) and N-{2-[(4-Dimethylamino-benzylidene)-amino]-ethyl}-N'-pyridin-2-ylmethylene-ethane-1,2-diamine (A)/Zn^{2+} shows high fluorescence intensities at pH range between 5.0-7.0 (Figure 3). These results indicate that Zn^{2+} could be clearly detected by the fluorescence spectra measurement at environmentally and physiologically relevant pH range (6.0–7.6) [16]. Chemosensor containing nitrogen donors are highly sensitive to environmental pH as the protonation degree of nitrogen are dependent on the pH. Any small change in pH environment of N-{2-[(4-Dimethylamino-benzylidene)-amino]-ethyl}-N'-pyridin-2-ylmethylene-ethane-1,2-diamine (A) near the physiological pH area were well tolerated.While in the basic conditions (pH > 7.0), the fluorescence intensity of N-{2-[(4-Dimethylamino-benzylidene)-amino]-ethyl}-N'-pyridin-2-ylmethylene-ethane-1,2-diamine (A)/Zn^{2+} was also gradually decreased may be due to the formation of Zn(II) hydroxide complex in this alkaline mixed solutions [14]. Hydrogen bonding form water cluster caused deprotonating, hindered the excitation of the energy which is responsible for the quenching of fluorescence intensity [15]. There is no specific explanation at pH range 3.0–5.0. Thus further study needed to investigate the effect of protonation towards the intensity of fluorophore. To ensure strong emission, the following experiments were carried out in solutions at pH 7.0.

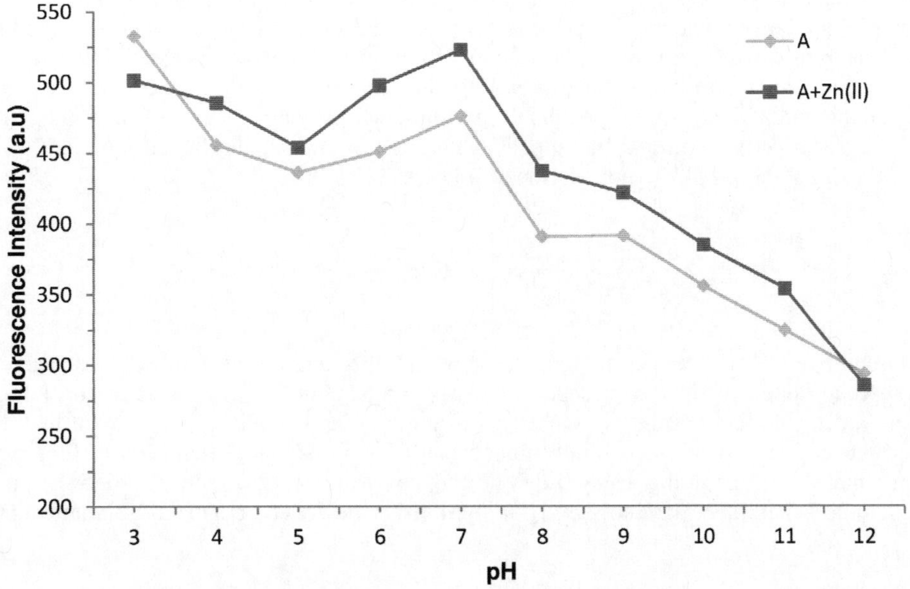

Figure 3. Influences of pH on the fluorescence spectra of N-{2-[(4-Dimethylamino-benzylidene)-amino]-ethyl}-N'-pyridin-2-ylmethylene-ethane-1,2-diamine (A) with Zn^{2+}.

3.4. Stoichiometry of complexation

For determination of stoichiometry between N-{2-[(4-Dimethylamino-benzylidene)-amino]-ethyl}-N'-pyridin-2-ylmethylene-ethane-1,2-diamine (A) and Zn^{2+}, Job's plot analyses were used.

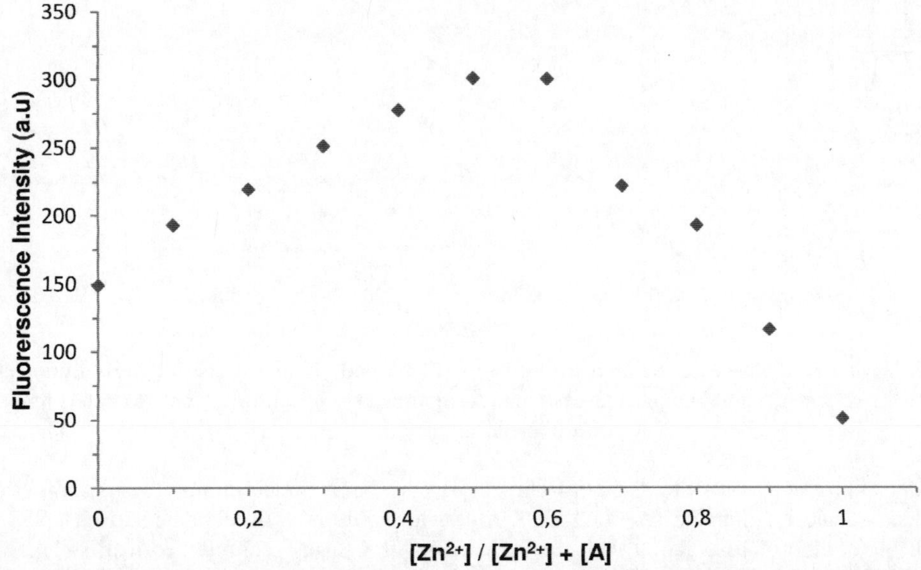

Figure 4. Job's plot of N-{2-[(4-Dimethylamino-benzylidene)-amino]-ethyl}-N'-pyridin-2-ylmethylene-ethane-1,2-diamine (A) with Zn^{2+} in CH_3OH/H_2O, 9:1 (v/v), Tris-HNO_3 = 10 mM, pH = 7.0.

The method is that keeping total concentration of N-{2-[(4-Dimethylamino-benzylidene)-amino]-ethyl}-N'-pyridin-2-ylmethylene-ethane-1,2-diamine (A) and Zn^{2+} at 10.0 mM and changing the molar ratio of Zn^{2+} from 0.1 to 0.9. From Figure 4 when molar fraction of Zn^{2+} was 0.5, the fluorescence maxima at 300 nm got to maximum, indicating that forming a 1:1 complex between N-{2-[(4-Dimethylamino-benzylidene)-amino]-ethyl}-N'-pyridin-2-ylmethylene-ethane-1,2-diamine (A) and Zn^{2+}. The stoichiometric ratio and the apparent binding constant of the ligand with Zn^{2+} were also determined using Benesi–Hildebrand relation as follows [2]:

$$\frac{1}{(F-F_0)}=\frac{1}{(F_\alpha-F_0)}+\frac{1}{K(F_\alpha-F_0)}\times\frac{1}{[Zn^{2+}]} \qquad (1)$$

where, F_0 and F are the fluorescence intensities in the absence and presence of Zn^{2+} respectively. F_α is the fluorescence intensity in the presence of excess amount of Zn^{2+}. Therefore, for 1:1 complex formation, the double reciprocal plot of $1/(F-F_0)$ against $1/[Zn^{2+}]$ should give a straight line; from the slope and intercept of which, the equilibrium constant (K) can be calculated. Inset of Figure 5 shows the representative linear fitting using Eq. (1) and confirm 1:1 stoichiometry between N-{2-[(4-Dimethylamino-benzylidene)-amino]-ethyl}-N'-pyridin-2-ylmethylene-ethane-1,2-diamine (A) and Zn^{2+}.

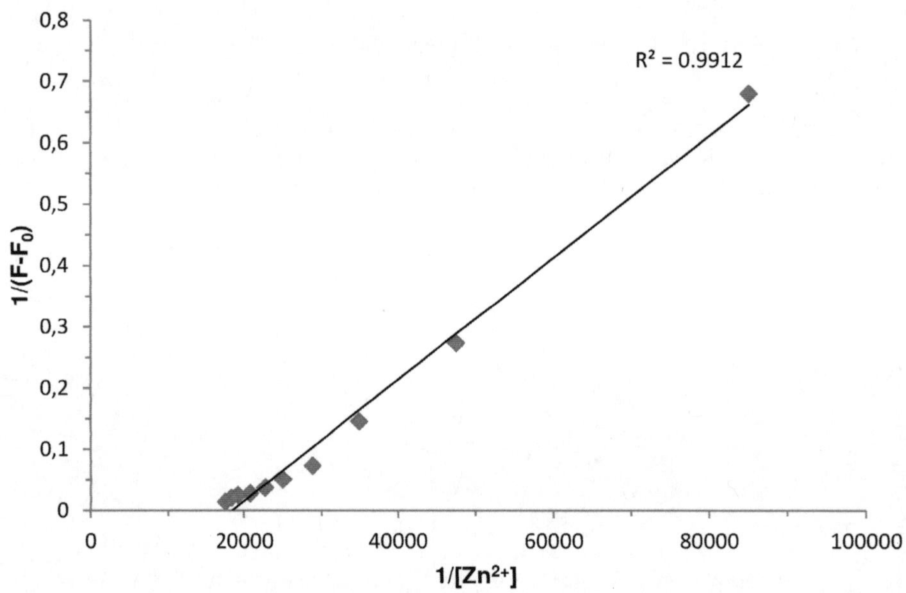

Figure 5. Variation of fluorescence intensity of N-{2-[(4-Dimethylamino-benzylidene)-amino]-ethyl}-N'-pyridin-2-ylmethylene-ethane-1,2-diamine (A) against concentration of Zn^{2+} to determine the association constant.

The association constant of N-{2-[(4-Dimethylamino-benzylidene)-amino]-ethyl}-N'-pyridin-2-ylmethylene-ethane-1,2-diamine (A) with Zn^{2+} in ethanol solution was calculated to be 1.837 x 10^4 M^{-1}. The high value of association constant K indicates that strong complex formation takes place between Zn^{2+} ion and N-{2-[(4-Dimethylamino-benzylidene)-amino]-ethyl}-N'-pyridin-2-ylmethylene-ethane-1,2-diamine (A). This is due to the presence of pyridyl group, which offers extra binding site. Moreover, this further corroborated 1:1 complex formation based on Job's plot analyses. The proposed mechanism of Zn^{2+}- N-{2-[(4-Dimethylamino-benzylidene)-amino]-ethyl}-N'-pyridin-2-ylmethylene-ethane-1,2-diamine (A) shown in Scheme 2.

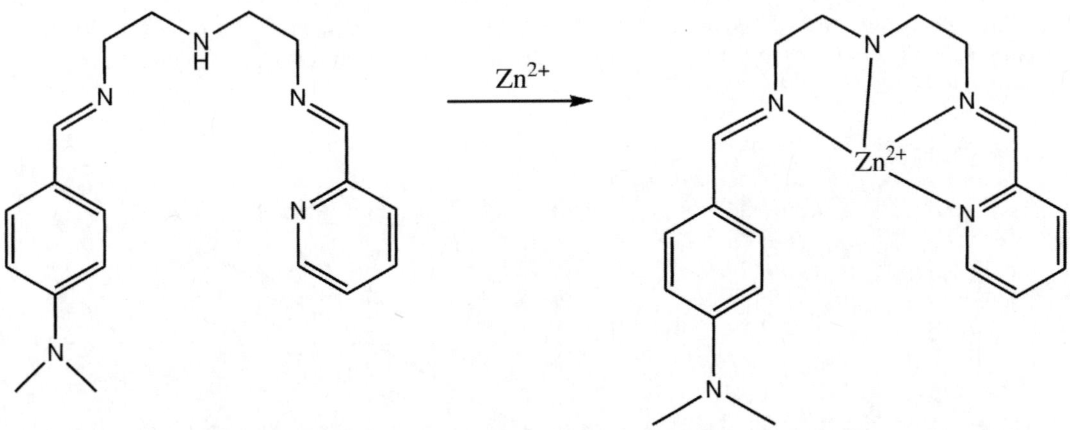

Scheme 2. The proposed mechanism for sensing of Zn^{2+} by N-{2-[(4-Dimethylamino-benzylidene)-amino]-ethyl}-N'-pyridin-2-ylmethylene-ethane-1,2-diamine (A).

3.5. Response of the chemosensor N-{2-[(4-Dimethylamino-benzylidene)-amino]-ethyl}-N'-pyridin-2-ylmethylene-ethane-1,2-diamine (A) toward different concentrations of Zn^{2+}

For many practical purposes, it is very important to detect the analytes at low concentrations. Even though N-{2-[(4-Dimethylamino-benzylidene)-amino]-ethyl}-N'-pyridin-2-ylmethylene-ethane-1,2-diamine (A) shows high fluorescence intensity with Zn^{2+} ion which was sensitive enough for practical uses.

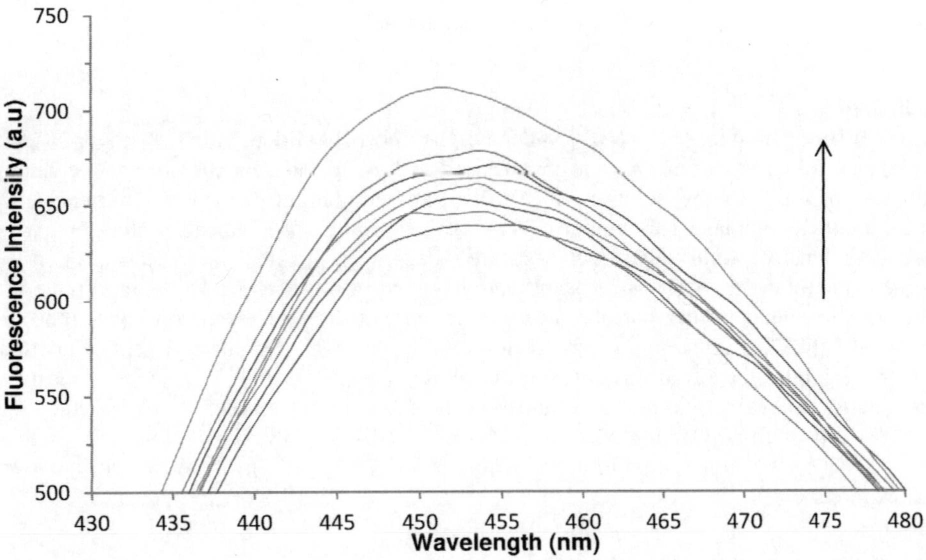

Figure 6. Fluorescence changes of N-{2-[(4-Dimethylamino-benzylidene)-amino]-ethyl}-N'-pyridin-2-ylmethylene-ethane-1,2-diamine (A) (5×10^{-5} M) in the presence of different concentrations of the Zn^{2+} ion.

The sensitivity of chemosensor was examined via the titration between chemosensor and different concentration of Zn^{2+} and achieved a highly selective linear response ($R^2 = 0.99$) to the targeted analyte (Figure 6). From the titration data, the limit of detection (LOD) of N-{2-[(4-Dimethylamino-benzylidene)-amino]-ethyl}-N'-pyridin-2-ylmethylene-ethane-1,2-diamine (A) was calculated by using the equations: LOD = (3 × standard deviation) / slope. Using the calibration curve (Figure 7), the LOD was found to be 3.5 x 10^{-5} M.

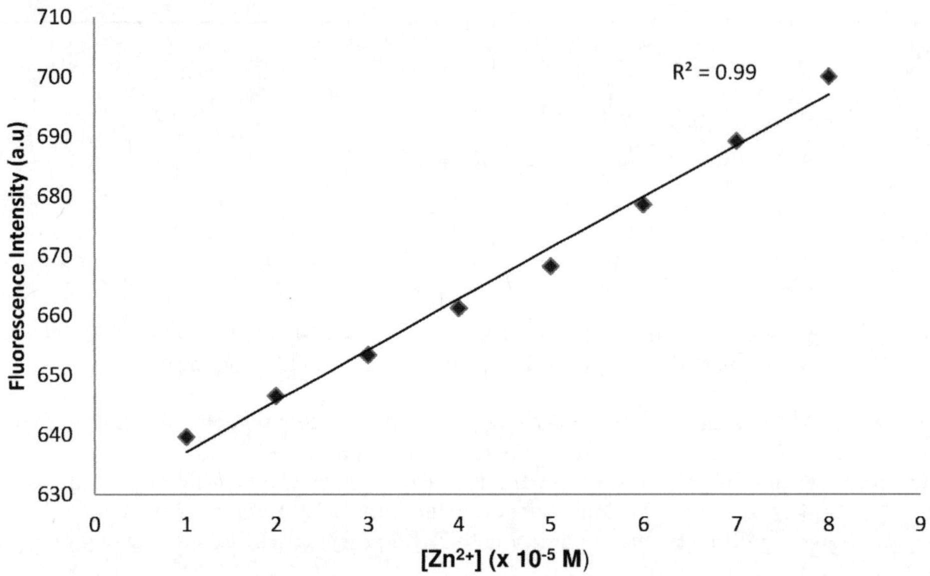

Figure 7. Calibration curve was obtained from the plot of fluorescence intensity with the addition of Zn^{2+} concentration.

4. Conclusion

A Schiff-base fluorescent N-{2-[(4-Dimethylamino-benzylidene)-amino]-ethyl}-N'-pyridin-2-ylmethylene-ethane-1,2-diamine (A) was synthesized and evaluated as a chemoselective Zn^{2+} sensor. The addition of Zn^{2+} to the solution of N-{2-[(4-Dimethylamino-benzylidene)-amino]-ethyl}-N'-pyridin-2-ylmethylene-ethane-1,2-diamine (A) resulted in a pronounced enhancement in the fluorescence intensity while other metal showed decreasing in the fluorescence intensity. The stoichiometric ratio and association constant were evaluated using Benesi–Hildebrand relation giving 1:1 stoichiometry. This further corroborated 1:1 complex formation based on Job's plot analyses. Further, N-{2-[(4-Dimethylamino-benzylidene)-amino]-ethyl}-N'-pyridin-2-ylmethylene-ethane-1,2-diamine (A) exhibits good fluorescence sensing ability to Zn^{2+} at pH=7.0, giving almost constant emission intensity in the physiological conditions. The detection limit is 3.5 x 10^{-5} M which indicated that N-{2-[(4-Dimethylamino-benzylidene)-amino]-ethyl}-N'-pyridin-2-ylmethylene-ethane-1,2-diamine (A) may be useful for preliminary detection of Zn^{2+} ions in chemical and environmental applications.

References

[1] K Li, X Wang, and A Tong, A 'turn-on' fluorescent chemosensor for zinc ion with facile synthesis and application in live cell imaging., *Anal. Chim. Acta*, vol. 776, pp. 69–73, May 2013.

[2] T S Singh, P C Paul, and H A R Pramanik, Fluorescent chemosensor based on sensitive Schiff base for selective detection of Zn^{2+}., *Spectrochim. Acta. A. Mol. Biomol. Spectrosc.*, vol. 121, pp. 520–6, Jan. 2014.

[3] Z Xu, J Yoon, and D R Spring, Fluorescent chemosensors for Zn^{2+}., *Chem. Soc. Rev.*, vol. 39, no. 6, pp. 1996–2006, 2010.

[4] Y Jin, S Wang, Y Zhang, and B Song, Highly selective fluorescent chemosensor based on benzothiazole for detection of Zn^{2+}, *Sensors Actuators B Chem.*, vol. 225, pp. 167–173, Mar. 2016.

[5] L Basabe-Desmonts, D N Reinhoudt, and M Crego-Calama, Design of fluorescent materials for chemical sensing, *Chem. Soc. Rev.*, vol. 36, no. 6, p. 993, 2007.

[6] H G Lee, J H Lee, S P Jang, I H Hwang, S J Kim, Y Kim, C Kim, and R G Harrison, Zinc selective chemosensors based on the flexible dipicolylamine and quinoline, *Inorganica Chim. Acta*, vol. 394, pp. 542–551, 2013.

[7] G Y Nagesh, K Mahendra Raj, and B H M Mruthyunjayaswamy, Synthesis, characterization, thermal study and biological evaluation of Cu(II), Co(II), Ni(II) and Zn(II) complexes of Schiff base ligand containing thiazole moiety, *J. Mol. Struct.*, vol. 1079, pp. 423–432, Jan. 2015.

[8] S Salehzadeh, R Golbedaghi, I S Tidmarsh, N K Al-Rasbi, H Adams, and M D Ward, Cd(II) and Mn(II) complexes of a new hexadentate Schiff base ligand derived from an asymmetric tripodal tetraamine and 2-pyridinecarboxaldehyde, *Polyhedron*, vol. 27, no. 18, pp. 3549–3556, Dec. 2008.

[9] A Biswas, M G B Drew, and A Ghosh, Nickel(II) and copper(II) complexes of unsymmetrical tetradentate reduced Schiff base ligands, *Polyhedron*, vol. 29, no. 3, pp. 1029–1034, Feb. 2010.

[10] O A M Ali, S M El-medani, D A Ahmed, and D A Nassar, Spectrochimica Acta Part A : Molecular and Biomolecular Spectroscopy Synthesis , characterization , fluorescence and catalytic activity of some new complexes of unsymmetrical Schiff base of 2-pyridinecarboxaldehyde with 2 , 6-diaminopyridine, *Spectrochim. ACTA PART A Mol. Biomol. Spectrosc.*, vol. 144, pp. 99–106, 2015.

[11] N Roy, H A R Pramanik, P C Paul, and T S Singh, A highly sensitive and selective fluorescent chemosensor for detection of Zn^{2+} based on a Schiff base., *Spectrochim. Acta. A. Mol. Biomol. Spectrosc.*, vol. 140, pp. 150–5, Apr. 2015.

[12] Z Li, Y Xiang, and A Tong, Ratiometric chemosensor for fluorescent determination of Zn^{2+} in aqueous ethanol., *Anal. Chim. Acta*, vol. 619, no. 1, pp. 75–80, Jun. 2008.

[13] Y J Na, I H Hwang, H Y Jo, S A Lee, G J Park, and C Kim, Fluorescent chemosensor based-on the combination of julolidine and furan for selective detection of zinc ion, 2013.

[14] H Liu, Y Dong, B Zhang, F Liu, C Tan, Y Tan, and Y Jiang, An efficient quinoline-based fluorescence sensor for zinc(II) and its application in live-cell imaging, *Sensors Actuators, B Chem.*, vol. 234, pp. 616–624, 2016.

[15] D Sarkar, A Pramanik, S Jana, P Karmakar, and T K Mondal, Quinoline based reversible fluorescent 'turn-on' chemosensor for the selective detection of Zn^{2+}: Application in living cell imaging and as INHIBIT logic gate, *Sensors Actuators B Chem.*, vol. 209, pp. 138–146, 2015.

[16] Y W Choi, G R You, J J Lee, and C Kim, Turn-on fluorescent chemosensor for selective detection of Zn^{2+} in an aqueous solution: Experimental and theoretical studies, *Inorg. Chem. Commun.*, vol. 63, pp. 35–38, Jan. 2016.

Acknowledgment
We express our sincere appreciation to the Ministry of Higher Education of Malaysia for providing generous financial support under the research grant RAGS UiTM 5/3/RAGS/(2/2013) and UiTM for providing all needed facilities.

The use of artificial neural network for modeling the decolourization of acid orange 7 solution of industrial by ozonation process

S Fatimah[1] **and W Wiharto**[2]

[1] Department of Chemical Engineering, Muhammadiyah University of Surakarta, Indonesia
[2] Department of Informatic, Sebelas Maret University, Indonesia

E-mail: sf120@ums.ac.id

Abstract. Acid Orange 7 (AO7) is one of the synthetic dye in the dyeing process in the textile industry. The use of this dye can produce wastewater which will be endangered if not treated well. Ozonation method is one technique to solve this problem. Ozonation is a waste processing techniques using ozone as an oxidizing agent. Variables used in this research is the ozone concentration, the initial concentration of AO7, temperature, and pH. Based on the experimental result that the optimum value decolourization percentage is 80% when the ozone concentration is 560 mg/L, the initial concentration AO7 is 14 mg/L, the temperature is 390 °C, and pH is 7,6. Decolourization efficiency of experimental results and predictions successfully modelled by the neural network architecture. The data used to construct a neural network architecture quasi newton one step secant as many as 31 data. A comparison between the predicted results of the designed ANN models and experiment was conducted. From the modeling results obtained MAPE value of 0.7763%. From the results of this artificial neural network architecture obtained the optimum value decolourization percentage in 80,64% when the concentration of ozone is 550 mg/L, the initial concentration AO7 is 11 mg/L, the temperature is 41 °C, and the pH is 7.9.

1. Introduction

Many industries use dyes and pigments to colourize their products. Some dyes with concentrations that they are discharged to the environment, are often non-poisonous, however, they give undesirable perspective to the water streams [1], whereas, some other dyes are poisonous and carcinogen and cause severe toxicity for aqueous microorganisms [2]. Dyes also can prevent from photosynthesis in aqueous ecosystems by absorbing sunlight. It is important to know that weaving industries use large amounts of water and in these industries more than 30–60 L of water per kilogram of dyed fabric is used and large amount of wastewater (16% of used water), is produced [3]. Also, total dye consumption in the weaving world industries is more than 107 kg/year and approximately 10% of this amount, 106 kg, is discharged to water streams [2]. Generally most of these compounds cause mutagenic and carcinogenic effects which subsequently lead to the generation of health disorders such as decreasing function of the kidney, reproductive system, liver, brain, and central nervous system. Dyes according to their charge and nature classified to categories including anionic (direct, acid, and reactive dyes), cationic (basic dyes), and non-ionic (disperse dyes). Among the anionic dyes are the brightest class of soluble dyes and generate serious environmental and health problems. These hazards and difficulties encourage the researchers to design novel approaches or material for their safe removal and subsequent achievement of clean aquatic environment.

Acid orange 7 (AO7) is one kind of azo dyes, which is widely used in textile industries. However, AO7 is difficult to be degraded by conventional methods due to its non-negligible toxicity, low biodegradability and dark colours. It is a mono azo dye and is widely used in dyeing of synthetic fibre and wool and cotton and in paper industries. Active azo dyes have double bound of nitrogen to nitrogen (–N=N–), which is bounded to an aromatic group so, they can produce harmful health effects and it is essential to have a proper method to remove this dye from wastewaters. The characteristics and structure of acid orange 7 were shown in Table 1.

Table 1. Characteristics and Structure Of Acid Orange 7

Name	Chemical structure	Molecular Formula	Colour index number	λ_{max} (nm)	M_w (g/mol)
Acid Orange 7	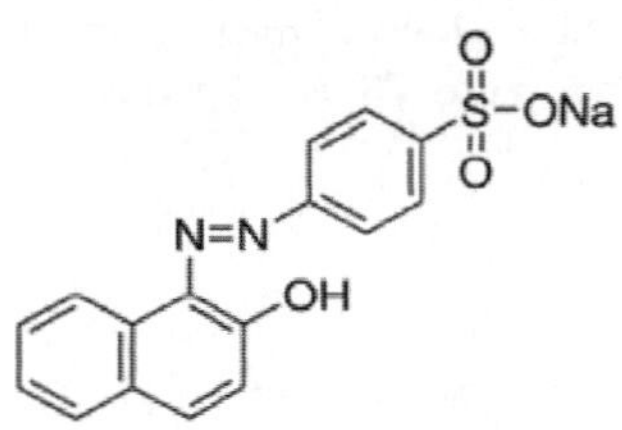	$C_{16}H_{11}N_2NaO_4S$	15510	484	350.32

A broad range of physicochemical methods has been proposed for treatment of dyeing effluents including biological [4], adsorption [5], membrane [6], coagulation-flocculation [7], oxidation-ozonation [8], and advance oxidation process (AOPs) [9][10]. Ozonation, one of the most used AOPs, is an oxidation process in which the oxidizing agent used is ozone (O_3). Interest in the use of ozone in wastewater treatment has increased considerably in recent years due to the numerous advantage so this process. Among them is the high oxidation potential of ozone even at low concentrations, with its high efficiency in the decomposition of organic matter and in the addition of oxygen to water [11]. Degradation process with ozone proceeds in two different ways, they are direct and radical reaction. At acidic pH, molecular ozone selectively attacks specific parts of organics molecules like double bonds or aromatic systems of reactive dyes in direct reaction. At alkaline pH, hydroxide radicals coming from ozone decomposition react unspesifically with organic compounds[12]. The alkaline to neutral pH of dyeing wastewater favours the unspecific radical reaction. The pH solution may decrease around the dye molecules due to the formation of organic acid such as the ozonation products. Therefore, the reaction mechanism shifts towards the selective direct oxidation [13].

The mechanism of O_3 reaction was studied by Hoigne [12]. Ozone react directly with pollutans via ozonolysis and through radical chain reaction. The reaction of ozone with hydroxide anions to intermediate radicals and to hydroxyl radicals is important for the oxidation of saturated organic compounds where no molecular ozonolysis is possible[14]. The efficiency of colour removal by ozone is dependent on various parameters such as the ozone dosage, initial dye concentration, initial pH, reaction time, and temperature of the solution[15]. In conventional methods used determined the influence of each one of these parameters, experiments were carried out by varying systematically the studied parameters and keeping constant the others. These should be repeated to all the influencing parameters, resulting in an unreliable number of experiments. In order to optimize the value number of experiments with the minimum number of experiments, was employed to find improved or optimal process settings in an efficient use of the experimental data. Using artificial neural network, it is possible to estimate linear, interaction, quadratic effects of the factors and to provide a prediction model for the responses.

In this work, the experimental system is modelled by using artificial neural networks. The artificial neural network (ANN) is a system of data processing based on the structure of a biological neural system. The prediction with ANN is made by learning of the experimentally generated data or using validated models [16]. The ANN is an artificial intelligence technique that mimics the human brain's biological neural network in the problem solving processes. Numerous applications of ANN have been successfully conducted to solve environmental problems since it is reliable and robust in capturing the nonlinear relationship between variables in complex system [17] [18] [19]. This novel modeling tool is newly grown and has not been used yet to model adsorption of AO7 using ozonations. It is applicable in most fields of science [20].

2. Material and Method

2.1. Data

In this study, the data used for experiments using the data in a study conducted Kasiri et. al [15]. The data attribute consist of 4 inputs and one output attribute, namely ozone concentrations (mg/L), AO7 concentration (mg/L), the degree of acidity of pH, temperature (^{o}C) and percentage docolorization.
Iran

2.1.1. Reagent

Ozone was continuously produced by an ozone generator (model 1000 mg/L, Nano Pakk Sayyal, Iran). The dyes of AO7 were obtained from Azar Paints Company, Tabriz,. Their structure and characteristics are given in Table. Sulphuric acid and sodium hydroxide were of laboratory reagent grade and were obtained from Merck Inc, Darmstadt, Germany.

2.1.2. Ozonation experimental setup

For decolourization process, ozonation was carried out in a batch reactor, 500 mL in volume, equipped with magnetic stirrer and a thermometer. Experimental setup for ozonation batch experiments is shown in figure 1.

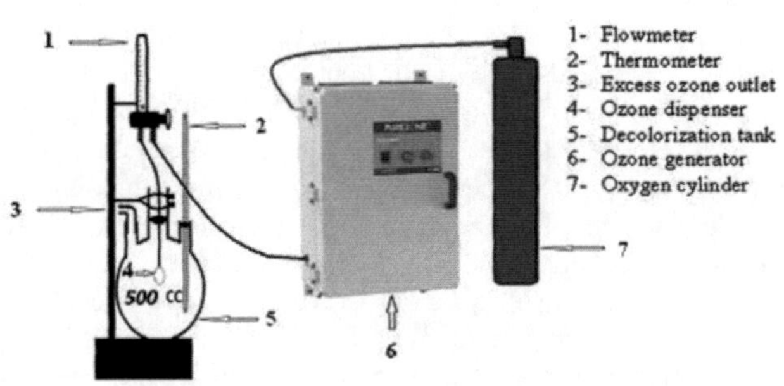

Figure 1. Experimental setup for ozonation batch experiments [15]

2.2. Experimental Design of Artificial Neural Network

Artificial Neural network (ANN) used in this study is backpropagation. Backpropagation neural network algorithms have a lot of training, in this study using a quasi newton one step secant [21]. The algorithm has the ability quickly to achieve convergent in the training process. Neural network back-propgation using supervised training models, so that the model requires that the data used for training in pairs, namely the input and the target. In this study, there are four inputs, namely ozone concentrations (mg/L), AO7 concentration (mg/L), the degree of acidity of pH and temperature (^{o}C), while the output of the system is the percentage decolourization. The training process is done by making changes to the weights ANN order to obtain output corresponding to the target used in training. Changes in weight training ANN with quasi Newton algorithm can be written in equation (1) below

$$W_{k+1} = W_k - A_k^{-1} * gW_k \quad (1)$$

Where A_k is the Hessian matrix, W_k is the k-th weighting. The algorithm does not perform a complete storage for hessian matrix, assuming that at each iteration, the previous hessian matrix is the identity matrix. This makes the training process becomes faster and does not require a large memory. Architecture of artificial neural network backpropagation with two hidden layers can be shown in figure 2.

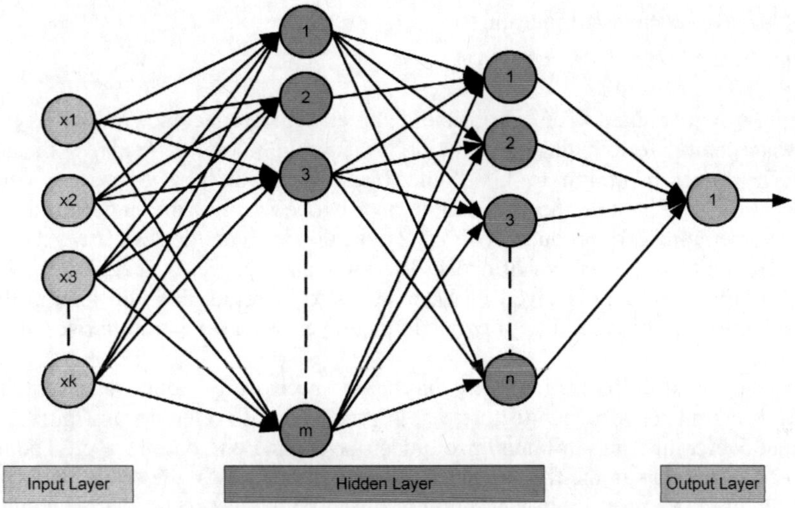

Figure 2. Neural Network Architecture System

In this study, to model percentage decolorisation using artificial neural network is divided into three processes, namely normalization and denormalize the data (preprocessing and post procesising), training and testing. The process of normalization of data is done to bring the value of the input attributes to within a certain range, and normalize the output side of the neural network to return values into their original range. In this study, using methods preprocesing and postprocessing mean and standard deviation. The next stage is done training the neural network. Training is done by using a neural network optimization algorithms backprogation with one step secant quasi newton. The algorithm is an alternative conjugate gradient to get the optimum value faster. The training process is done repeatedly, in order to obtain an optimum architecture. Furthermore, the last stage of testing of the system.

2.3. Evaluation Parameters

The performance evaluation were performed using SPSS version 23 for test of significance with 95% confidence level. In addition, it is also used to test the Mean Absolute Percentage Error (MAPE) is the absolute value of the data about the mean percentage error. The test to determine the ability of a neural network in doing prediction. The test can be formulated in equation (2) below

$$MAPE = \frac{\frac{\Sigma(\%decolorization\ predicted - \%decolorization\ experimental)}{\%decolorization\ experimental}}{n} \ x\ 100\% \quad (2)$$

After the evaluation of performance of the neural network, the neural network architecture that is produced is used to obtain the optimum value AO7 dye decolourization process.

3. Result and Discussion

One effective technique in the process of decolourization of waste containing dye AO7 is by Advance Oxidation Process (AOPs) or ozonation. The reaction mechanism of ozonation process has been studied by Hoigne [12]. In his research explained that the process of ozonation, ozone reacts with pollutants directly through ozonolysis process. This process involves a radical chain reaction of pollutants. The reaction between ozone with the hydroxide radical anion intermediates. These hydroxyl radicals that play an important role in degraded the pollutants. Illustration of hydroxyl radical formation can be described by equation (3-5) follows:

$$O_3 + UV(or\ hv < 310\ nm) \rightarrow O_2 + O(D^1) \quad (3)$$

$$O(D^1) + H_2O \rightarrow HO\bullet\ \text{(wet condition)} \quad (4)$$

$$O(D^1) + H_2) \rightarrow HO\bullet + \text{HO}\bullet \rightarrow \text{H}_2\text{O}_2\text{(in water)} \quad (5)$$

Decomposition of ozone due to sunlight in the damp air will produce active hydroxyl radicals. In the process of water tend to form hydrogen peroxide, so it would be more effective if the compound to be degraded have good absorption to UV light. Factors that affect the process ozonolysis on decolourization AO7 these include the concentration of ozone is used, the concentration of the dye AO7, pH and temperature. To produce hydroxyl radicals that can decolourise AO7 waste it is necessary also sufficient for the amount of ozone decolourization process. The greater the percentage of ozone concentration of ozone dissolved in water will also increase, this will lead to the ability to oxidize AO7 even greater. The solubility of ozone in the lower water temperature rises, this reaction is endothermic [22].

Modeling percentage decolourization AO7 has been successfully done using Artificial Neural Network. Artificial neural network architecture that is generated with reference to figure 2, comprising a number of input overcome four attributes, two hidden layers and one output layer. The hidden layer consists overcome 30 neurons in the first hidden layer and 25 neurons in the second hidden layer. The activation function used is tansig, logsig and purelin. Based on the test results showed that percentage decolourization resulting from ANN is not much different from the experimental results, as shown in table 2. In the Table shows the test results p-value> 0.05 which showed there was a significant difference between the prediction by the neural network with experiment.

Table 2. Test Significance Model ANN at Percentage Decolourization

		Paired Differences							
					95% Confidence Interval of the Difference				
		Mean	Std. Deviation	Std. Error Mean	Lower	Upper	t	df	Sig. (2-tailed)
Pair 1	Experimental – Predicted	0.00000	1.06458	0.19120	-0.39049	0.39049	0.000	30	**1.000**

ANN system testing, in addition to using t-test significance test, also using MAPE. The test results demonstrate the value with MAPE 0.7763%. MAPE value of less than 10% in the case of prediction can be tolerated. It can also be shown graphically percentage decolourization value of prediction and experiment result, as shown in figure 3. Artificial neural network model generated with reference to the results of performance testing with two parameters, namely the significance test t-test and the value of MAPE, artificial neural network model meets the criteria to be used in the prediction percentage decolourization the case of AO7 dye.

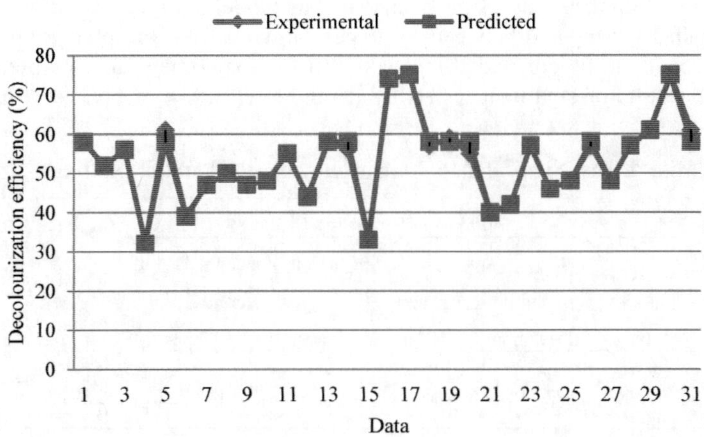

Figure 3. Comparation Experimental and Prediction.

By using architectural ANN 2 hidden layer testing is done to obtain optimum percentage decolourization. Testing by many as 14625 of data (results generated) values obtained percentage decolourization as shown in table 3. Results of optimum percentage decolourization not much different from that produced by the experiment. Based on the experiments showed that the percentage of decolourization AO7 by 80% with the condition of the ozone concentration of 500 mg/L, AO7 concentration of 14 mg/L, pH 7.6, and the temperature 39 °C. While predictions using ANN produce AO7 decolourization percentage of 80.64% with the condition of the ozone concentration of 550 mg/L, AO7 concentration of 11 mg/L, pH 7.9, and a temperature of 41 °C. This difference is caused by several things. First, the variable ozone concentration, pH and temperature. If the concentration of ozone increases then the temperature will increase the interaction between molecules of ozone increases. As an impact is the kinetic energy of the compound in solution will also increase. The increase of activation energy has resulted in the number of hydroxyl ions produced from the oxidation process increases. The existence of the number of hydroxyl ions will cause the pH to rise. Same with the research that has been conducted by Lu et.al [23] that the degradation of azo dyes effectively carried out under alkaline pH for hydroxyl radicals oxidizing properties will have an optimum at alkaline conditions. The second difference is the initial concentration AO7 variables. If the initial concentration of the lower AO7 it will produce increasingly higher percentage decolourization due to the presence of molecules of AO7 slightly. This affects the number of molecules AO7 oxidized by hydroxyl radicals. The ability of hydroxyl radicals as the oxidizing molecules will increase due to the availability AO7 slightly.

Table 3. Decolourization Efficiency at Optimum Values of The Process Parameters.

	Ozone concentration (mg/L)	Initial dye concentration (mg/L)	Initial pH	Temperature (°C)	% decolourization
Experimental	500	14	7.6	39	80.00
Predicted	550	11	7.9	41	80.64

Similar to the research that has been done by Kasiri et.al [15] whereas in the study the percentage of decolourization AO7 modeling using surface method area produces AO7 decolourization percentage of 77.9%. In this study, the use of ANN as modeling AO7 decolourization AO7 able to produce a percentage of 80.64%. This shows that the use of ANN is better than the surface method area to model the percentage of decolourization AO7.

To determine the level of confidence or significance of this modeling, the test statistic by using t-test. In the t-test which will be compared between the experimental results and predictions. If the p-value was greater than 0.05, it can be concluded that between the experiment and the predictions are not much different. The results of significance test by using t-test shown in table 4, where the p-value 0.376, or p-value > 0.05.

Table 4. Test Significance Prediction with ANN and Experiment

		Paired Differences							
		Mean	Std. Deviation	Std. Error Mean	95% Confidence Interval of the Difference: Lower	Upper	t	df	Sig. (2-tailed)
Pair 1	Experimental – Predicted	-9.9880	22.4427	10.0366	-37.8542	17.8782	-0.995	4	**0.376**

4. Conclusion
Model predictions using neural networks to predict the percentage of decolourization AO7 able to deliver results that are not much differ significantly with experimental results. This is indicated by the p-value between experiment and prediction generated is equal to 0.376. Under these conditions, the percentage of decolourization AO7 modeling can be done using ANN.

5. References
[1] Annadurai G, Juang J, and Lee D 2004 Use of cellulose-based wastes for adsorption of dyes from aqueous solutions *J. Hazard. Mater.* **92** 263–74.

[2] Wong Y, Szeto Y, Cheung W, and McKay G 2004 Adsorption of acid dyes on chitosan-equilibrium isotherm analyses *Process Biochem.* **39** 695–704.

[3] Sivaraj R, Namasivayam C, and Kadirvelu K 2000 Orange peel as an adsorbent in the removal of Acid violet 17 (acid dye) from aqueous solutions *Waste Manag.* **21** 105–110.

[4] Khataee A and Kasiri M 2011 Modeling of Biological Water and Wastewater Treatment Processes Using Artificial Neural Networks *CLEAN-Soil Air Water* **39** 742–749.

[5] Clercq J 2012 Removal of mercury from aqueous solutions by adsorption on a new ultra stable mesoporous adsorbent and on a commercial ion exchange resin *Int. J. Ind. Chem.* **3** 1–6.

[6] Harrelkas F, Azizi A, Benhammou A Y, and Pons M 2009 Treatment of textile dye effluents using coagulation–flocculation coupled with membrane processes or adsorption on powdered activated carbon *Desalination* **235** 330–339.

[7] Salari D, Niaei A, Khataee A, and Zarei M 2009 Electrochemical treatment of dye solution containing C.I. Basic Yellow 2 by the peroxi-coagulation method and modeling of experimental results by artificial neural networks *J. Electroanal. Chem.* **629** 117–125.

[8] Saruyu K, Swaminathan K, and Sandhya S 2007 Assesment of Degradation of Eight Commercial Reactive Azo Dyes Individually and in Aquous Solution by Ozonation *Dyes Pigments* **75** 362–368.

[9] Kasiri M and Khataee A 2012 Removal of organic dyes by UV/H2O2 process: modeling and optimization *J. Environ. Technol.* **33** 1417–1425.

[10] Rasoulifard M, Majidzadeh H, Demneh F, Babaei E, and Rasoulifard M 2012 Photocatalytic degradation of tylosin via ultraviolet-activated persulfate in aqueous solution *Int. J. Ind. Chem.* **16** 1–6.

[11] Souza S M A G, Bonilla K A S, and Souza A A 2010 Removal of COD and color from

hydrolyzed textile azo dye by combined ozonation and biological treatment *J. Hazard. Mater.* **179** 35–42.

[12] Hoigne J 1988 *Chemistry of Aqueous Ozone and Transformation of Pollutants by Ozonation and Advanced Oxidation Process* (Berlin: Springer).

[13] Kornmuller A, Karcher S, and Jekel M 2001 Cucurbituril for water treatment. Part II:: Ozonation and oxidative regeneration of cucurbituril *Water Res.* **35** 3317–3324.

[14] Chu L B, Xing X H, Yu A F, Sun X L, and Jurcik B 2008 Enhanced treatment of practical textile wastewater by microbubble ozonation *Process Saf. Environ. Prot.* **86** 389–393.

[15] Kasiri M B, Modirshahla N, and Mansouri H 2013 Decolourization of organic dye solution by ozonation; Optimization with response surface methodology *Int. J. Ind. Chem.* **4** 1–10.

[16] Fagundes-Klena M R, Ferria P, Martins T D, Tavaresb C R G, and Silvaa E A 2007 Equilibrium study of the binary mixture of cadmium–zinc ions biosorption by the Sargassum filipendula species using adsorption isotherms models and neural network *Biochem. Eng. J.* **34** 136–146.

[17] Turan N G, Mesci B, and Ozgonenel O 2011The use of artificial neural networks (ANN) for modeling of adsorption of Cu(II) from industrial leachate by pumice *Chem. Eng. J.* **171**1091–1097.

[18] Aber S, Amani-Ghadim A R, and Mirzajani V 2009 Removal of Cr(VI) from polluted solutions by electrocoagulation: Modeling of experimental results using artificial neural network *J. Hazard. Mater.* **171** 484–490.

[19] Salari D, Daneshvar N, Aghazadeh F, and Khataee A 2005 Application of artificial neural networks for modeling of the treatment of wastewater contaminated with methyl tert-butyl ether (MTBE) by UV/H2O2 process *J. Hazard. Mater.* **125** 205–210.

[20] Aber S, Daneshvar N, Soroureddin S, Chabok A , and Asadpour-Zeynali K 2007 Study of acid orange 7 removal from aqueous solutions by powdered activated carbon and modeling of experimental results by artificial neural network *Desalination* **211** 87–95.

[21] Wiharto, Aziz A, and Permana U 2015 Improvement of Performance Intrusion Detection System (IDS) Using Artificial Neural Network Ensemble *J. Theor. Appl. Inf. Technol.* **80** 191-201.

[22] Nugroho R and Ikbal 2005 Pengolahan air limbah berwarna industri tekstil dengan menggunakan teknik AOPs *JAI* **1**163–172.

[23] Lu X, Yang B, Chen J, and Sun R 2009 Treatment of wastewater containing azo dye reactive brilliant red X-3B using sequential ozonation and upflow biological aerated filter process *J. Hazard. Mater.* **161** 241–245.

IOP Conf. Series: Materials Science and Engineering **172** (2017) 012053 doi:10.1088/1757-899X/172/1/012053

Influence of Solvent on Liquid Phase Hydrodeoxygenation of Furfural-Acetone Condensation Adduct using Ni/Al_2O_3-ZrO_2 Catalysts

S M Ulfa, A Mahfud, S Nabilah, and M F Rahman

Chemistry Department, Faculty of Science, Brawijaya University, Jl. Veteran Malang, East Java, 65145

E-mail: ulfa.ms@ub.ac.id

Abstract. Influence of water and acidic protic solvent on hydrodeoxygenation (HDO) of the furfural-acetone adduct (FAA) over Ni/Al_2O_3-ZrO_2 (NiAZ) catalysts were investigated. The HDO of FAA was carried out in a batch reactor at 150°C for 8 hours. The NiAZ catalysts were home-made catalysts which were prepared by wet impregnation method with 10 and 20% nickel loading. The HDO reaction of FAA using 10NiAZ in water at 150°C gave alkane and oxygenated hydrocarbons at 31.41% with selectivity over tridecane (C13) in 6.67%. On the other hand, a reaction using acetic acid:water (1:19 v/v) in similar reaction condition gave only oxygenated compounds and hydrocracking product (C8-C10). The formation of tridecane (C13) was proposed by hydrogenation of C=O and C=C followed by decarboxylation without hydrocracking process. The presence of water facilitated decarboxylation mechanism by stabilized dehydrogenated derivatives of FAA.

1. Introduction

The limitation used in fossil fuels have led to the research of renewable material for supplying the energy demand. Different with other energy sources such as, solar, wind and geothermal; biomass is the most promising alternative resources since it can provide the liquid fuels directly by a sequence of chemical reactions [1, 2]. Upgrading of lignocellulosic biomass into diesel fraction have been intensively studied by catalytic hydrogenation-deoxygenation of furfural-acetone condensation adduct as chemical precursor [3].

Furfural is the bio-based-chemical which can be obtained from acid hydrolysis and dehydration of pentose. The condensation of furfural with ketone/aldehyde gives monomer/dimer of furanyl derivatives with carbon chain varied from C8-C15. From this reaction, the complex molecules with ketonic C=O, aliphatic C=C, and furanic C=C bond should be converted into valuable liquid hydrocarbons. Thus, the catalytic hydrogenation followed by deoxygenation is proposed in one-step reaction using Ni/Al_2O_3-ZrO_2 catalysts with the variation of solvent and nickel loading.

The selection of catalyst plays an important role. Among different metal loading, the noble metal such as platinum (Pt) and palladium (Pd) have been revealed as suitable metals for hydrogenation reaction [4, 5]. However, the high prices of these noble metal failed to impress the industrial scope. As a consequence, the development of non-noble metal catalyst seems more rational. The used of support also important to increase the activity of catalyst. The □-Al_2O_3 was previously used as a hydrogenating agent of oxygenated compounds, but severe coke formation was observed [6]. Moreover, it is not suitable for the reaction in the presence of water because of partially transform into boehmite [7]. To overcome these problems, the addition of ZrO_2 was reported increasing the activity of the catalyst by activated oxy-compounds and decreasing carbon deposition on the surface [8]. The crystal structure of ZrO_2 is relatively stable in water and organic solvent. Therefore, the reaction under liquid phase is favorable.

Herein, we reported liquid phase hydrodeoxygenation (HDO) of furfural-acetone adduct (FAA) using Ni/Al_2O_3-ZrO_2 in the presence of water and acidic solvent in one-pot reaction. The linear alkane as n-tridecane was identified as the selective product under water solvent.

2. Experimental

2.1. Materials

Furfural as starting material for condensation reaction was obtained from Sigma-Aldrich with >98% purity and used as received. Analytical grade of $Ni(NO_3)_2.6H_2O$, $ZrOCl_2.8H_2O$, γ–Al_2O_3, methanol, and dichloromethane were bought from Merck Germany. Acetone was purchased from Smart Lab Tbk. Technical grade of MgO and CaO was obtained from the local vendor and calcined at 650°C for 2 hours under N_2 flowing before used. N_2 gas in >99.99% purity for calcination and condensation reaction, Ultra-High-Purity (UHP) H_2 gas for hydrodeoxygenation (HDO) reaction was supplied by PT. Samator, Tangerang, Indonesia.

2.2. Methods

2.2.1. Preparation of Ni/Al₂O₃-ZrO₂ (10NiAZ and 20NiAZ)

Ni/Al_2O_3-ZrO_2 was prepared according to Zhang et al methods [9] with modification on the synthesis of the Al_2O_3-ZrO_2 support. The ratio of Al_2O_3/ZrO_2 in the support was 50:50 (wt/wt). Based on the stoichiometric calculation, to prepared 7 g of Al_2O_3-ZrO_2, 3.5 g of γ-Al_2O_3 was mixed with $Zr(OH)_4$ slurry which obtained from precipitation of 9.15 g $ZrOCl.8H_2O$ dissolved in deionized water with the dropwise addition of ammonium solution until pH equal to 8. Additional stirring the slurry for 12 h at 75°C, then filtered, dried for a night and calcined at 550°C for 5 h under N_2, by which the complex oxide of Al_2O_3-ZrO_2 was prepared. The sample was then designated as AZ support oxide.

Impregnation of nickel in the AZ was calculated for 10 and 20%, then each catalyst abbreviated as, 10NiAZ and 20NiAZ. The 10NiAZ was prepared by dissolved 2.96 g of $Ni(NO_3)_2.6H_2O$ in water and added with 5.96 g of AZ composite. On the other hand, the 20NiAZ was made from 7.14 g of nickel nitrate stirred with 7.20 g of AZ for 24 h at room temperature. After filtration dried the precipitate at 120°C for a night then calcined and reduced at 500°C for each 5 h under N_2 and H_2 gas.

2.2.2. Characterization

The catalysts obtained from previous methods were analyzed by X-ray diffraction (XRD) Pan Analytical Expert MPD. The measurements using Cu-Kα radiation at λ=1.5406 Å with the voltage and current held at 40 kV and 30 mA (2θ = 20–80°) at a scan rate of 1°/min. SEM data was collected from portable Hitachi TM3000 connected with EDX Swift ED3000 analyzer. The actual nickel content was analyzed using XRF PANalytical Type Minipal 4. FTIR spectrum was recorded by Shimadzu 8400S using KBr pellet in the range of 400–4000 cm^{-1}. The surface area and pore volume of the 10NiAZ and 20NiAZ are determined from BET adsorption isotherms using nitrogen at 77 K were measured by Quantachrome NOVA1200 analyzer after degassed at 300°C. Conversion and selectivity of hydrodeoxygenation (HDO) products were analyzed using Gas Chromatography Mass Spectrometer (GCMS) injected on GCMS QP5050A Shimadzu with FID detector and ZB-1MS column (30 m with ID = 0.25 mm).

IOP Conf. Series: Materials Science and Engineering **172** (2017) 012053 doi:10.1088/1757-899X/172/1/012053

2.2.3. Condensation reaction

Furfural-acetone adduct (FAA) as starting material for HDO reaction was synthesized based on Claisen–Schmidt condensation reaction according to previous methods [10]. Activated MgO and CaO were used as base-catalyst and deionized water as a solvent. The used of water is the most effective solvent according to Faba et al [11]. The reaction performed in sealed tube connected with a gas lever to removed oxygen and purged with an inert gas (N_2) to prevent oxidation. The reaction using reflux under N_2 atmosphere were also performed for comparison.

2.2.4. Hydrodeoxygenation (HDO) reaction

The HDO reaction was carried out in liquid phase reaction using sealed tube connected with a gas lever for inserting H_2 gas prior to reaction. The procedure of HDO reaction followed the previous report of our group [12]. For each run, 0.02 g of catalyst, 0.5 g FAA, and 20 mL of deionized water were put into the reactor, closed the lid and purged the air. Before the reaction, the reactor was filled with H_2 to 40 psi, then heated at an ambient temperature (150–200°C) for 8 h. The effect solvent is evaluated by modifying the acidity of solvent. An acidic protic solvent, that is, AcOH:H_2O (1:19 v/v) were used for HDO using 10NiAZ and 20NiAZ catalysts in a similar method. Conversion and selectivity of the products were calculated based on GCMS analysis using the equation below.

3. Results and discussion

3.1. Characterization of 10NiAZ and 20NiAZ

Figure 1 showed the morphology of the catalysts measured by Scanning Electron Microscopic (SEM). The particle size of AZ is larger than 10NiAZ or 20NiAZ. It means that impregnation of nickel broke the support and resulted in an irregular shape with smaller particle size. Agglomeration with irregular shapes is easily formed when impregnation method is used for the preparation of catalysts [13].

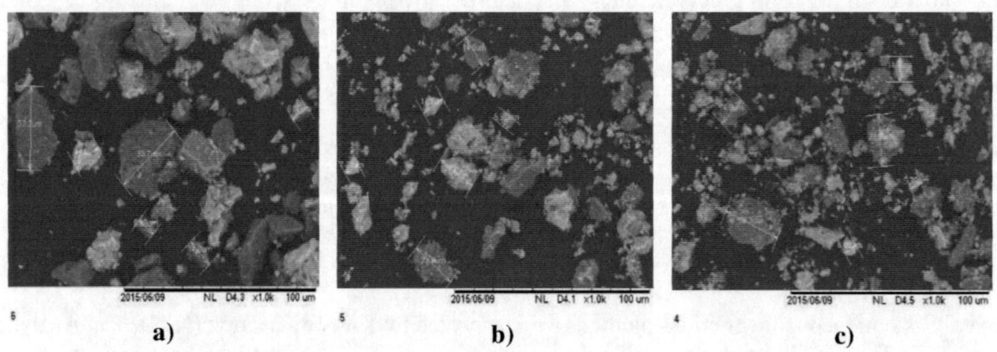

Figure 1. SEM micrograph of (a) AZ; (b) 10NiAZ; (c) 20NiAZ

The crystallinity of 10NiAZ and 20NiAZ were recorded by X-Ray Diffraction (XRD) Pan Analytical Expert MPD using Cu-Kα radiation at λ=1.5406 Å. The XRD profile of the catalyst showed in Figure 2. Compared with AZ, all catalysts gave the diffraction of metallic nickel appeared at 2θ = 44.44, 51.80, 76.28 (JCPDS 96-210-2279); no diffraction peaks of NiO at 2θ = 37°, 43°, and 64° were observed. We believed that NiO was completely reduced to Ni during the reduction at 500°C for 5 h. Furthermore, the diffraction of Al_2O_3 cannot be detected, except the sharp peak of ZrO_2 in cubic phase (JCPDS 96-900-9052). It can be inferred that nickel species strongly interact with AZ and increasing the crystallinity of the support as reported by Han, et al (2013) [14]. Different intensity of Ni diffraction correlated with nickel loading in both catalysts.

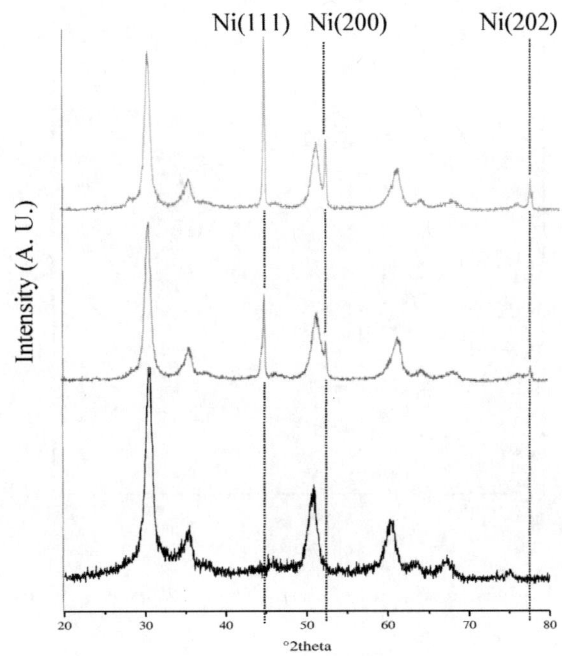

Figure 2. XRD profile of AZ (black), 10NiAZ (blue), and 20NiAZ (red)

According to N_2 isotherm (Figure 3), all the catalysts showed the type IV with hysteresis loop at P/Po = 0.6-1.0, typical of mesoporous catalyst. The BET surface (S_{BET}), pore volume and pore diameter of the catalysts showed in Table 1. The S_{BET} of AZ was found to be the larger (122.9 m^2/g) compared with 10NiAZ (48.6 m^2/g) and 20NiAZ (65.3 m^2/g). The introduction of nickel on AZ via impregnation induced a decreasing of S_{BET} caused by the sintering effect. On the other hand, 20NiAZ showed the higher S_{BET} and pore volume compared with 10NiAZ. This might be explained by unstable aggregation between nickel [13]. A higher surface area facilitated the reactant to access the active sites, decrease coke formation and improve catalytic activity [14].

Table 1. Surface area of the catalysts

Catalyst	S_{BET} (m^2/g)	Pore Volume (cc/g)	Pore Diameter (nm)	Metal Loading (%)[a]
AZ	122.945	0.330	5.639	-
10NiAZ	48.616	0.199	7.763	10.2
20NiAZ	65.314	0.250	6.580	17.58

[a]According to X-Ray Fluorescence analysis

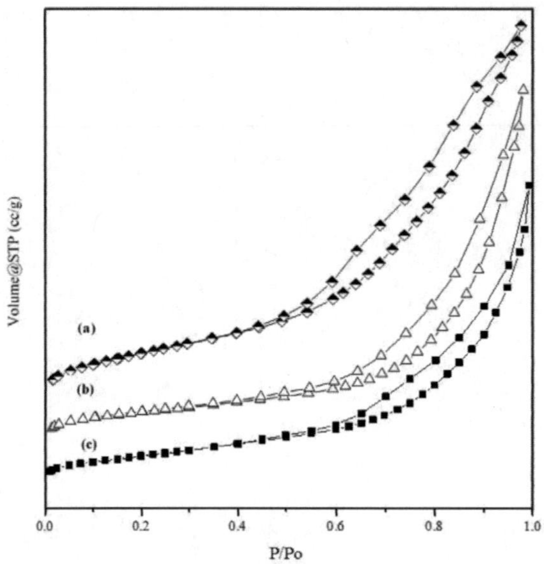

Figure 3. Isotherm adsorption of (a) AZ; (b) 10NiAZ; (c) 20NiAZ

3.2. Condensation reaction

The condensation reaction of furfural (**1**) and acetone was carried out using MgO and CaO solid bases catalyst. The yield of condensation products FAA after reaction for 6 h is depicted in Table 2. From the analysis of the products by GCMS, 2-(4-furyl)-3-buten-2-on (**2**, C8) and 1,5-bis-(furan-2-yl)-pentan-3-one (**3**, C13) was identified as main product with the most stable conformation, *trans*-C8 and *trans-trans*-C13 (Figure 4). The formation of *cis-cis*-C13 (**3a**) was observed as a minor product, similar to the finding of Faba *et al* (2012), the longer the condensation reaction, the most stable isomer were formed [11].

Among both methods, condensation with reflux using CaO exhibit a high yield product (33.98%) with a ratio of C8 and C13 is almost 1:1. In contrast, the reaction in a batch reactor gave 26.8% with selective formation of C8. This phenomenon was previously reported in our group and explained that vapor-phase condensation suppressed the formation of C13 by steric hindrance [15].

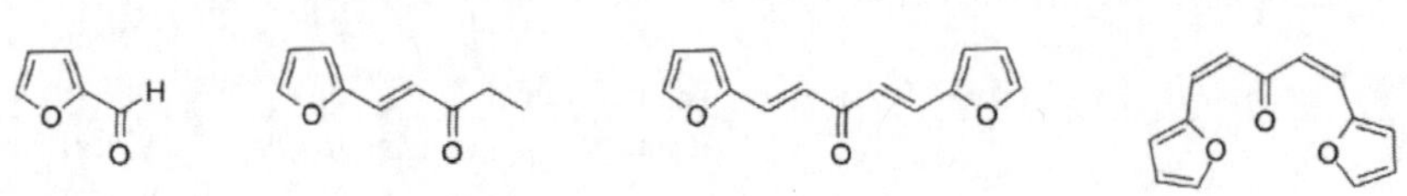

Figure 4. Isomer identification of condensation products

Table 2. The yield of FAA prepared by batch and reflux method

Method	Furfural (**1**) (mol)	Acetone (mol)	Catalyst	Yield of FAA (mol %)*	
				2	**3**
Batch reactor	1	2	MgO	17.39	8.69
Reflux	1	2	CaO	18.06	15.87

* quantitative analysis by GCMS

3.3. Catalytic Performance for Hydrodeoxygenation (HDO) reaction

The mixture of **2** and **3** (FAA) was subjected for direct aqueous HDO using the batch reactor, stirred and heated for 8 h at 150 and 200°C. Compound **2** and **3** have conjugated system, a furan ring, and carbonyl, then activity and selectivity of 10NiAZ and 20NiAZ can be evaluated for parallel reaction. The solvent for HDO was varied from neutral and acid to explore the effect of acidic solvent on the selectivity of the product.

Analysis of the product by GCMS showed the formation of alkane derivatives (**4**, **5**, and **6**) and oxygenated hydrocarbons (**7**, **8**, and **9**), depicted in Scheme 1. Parallel reaction suggested by the length of reaction time. The hydrocracking followed by hydrogenation, ring opening, and deoxygenation reaction was occurred giving the varied hydrocarbon derivatives [12].

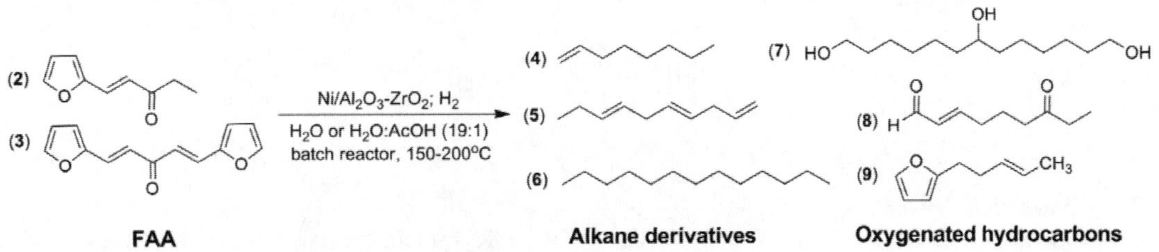

Scheme 1. HDO reaction of FAA

3.3.1. Effect of solvent

Total carbon yield and selectivity of HDO product are summarized in Table 3. Entry 1 showed that reaction in neutral water at 150°C using 10NiAZ gave the total carbon in moderate yield with selectivity over oxygenated hydrocarbons (**7**–**9**). Interestingly, the formation of n-tridecane (**6**) as linear alkane is detected only in this reaction. The similar reaction condition (entry 2) using AcOH/H_2O as a solvent didn't give compound **6**. The detected products were partial HDO compounds, that is linear alkene **4**–**5**. The formation of **6** proposed by hydrogenation of C=C, followed by ring opening of furan and deoxygenation of C=O bond of *trans-trans*-C13 (**2**) (Scheme 2) [16]. Increasing reaction temperature until 200°C (entry 4) only gave hydrocracking product **8 9**. Based on this result, HDO reaction is influenced by the kind of solvent rather than the reaction temperature. The higher the temperature promote the cracking of C–C bond afforded a molecule with lower weight or gas phase. Whereas, the used of neutral solvent endorsed decarbonylation/decarboxylation lead to higher molecular weight products or liquid phase [17]. However, the hydrogenation of simple molecules, such as phenol only gave hydrogenation product when using neutral solvent [18].

Table 3. Aqueous-phase hydrodeoxygenation of FAA[a]

Entry	Catalyst	T (°C)	Solvent	Total carbon yield (mol %)	Selectivity (mol %)					
					4	**5**	**6**	**7**	**8**	**9**
1	10NiAZ	150	H_2O	37.49	-	2.67	**6.67**	2.59	17.98	7.22
2	10NiAZ	150	AcOH:H_2O	56.94	22.4	10.06	-	2.68	8.08	13.72
3	20NiAZ	150	AcOH:H_2O	37.46	-	6.33	-	2.84	12.14	16.15
4	10NiAZ	200	AcOH:H_2O	22.14	-	-	-	-	9.96	12.18

[a] Reaction condition: FAA (0.5 g), H_2O (20 mL), catalyst (0.02 g), AcOH:H_2O (1:19 mL), 8 h reaction time, 40 psi H_2.

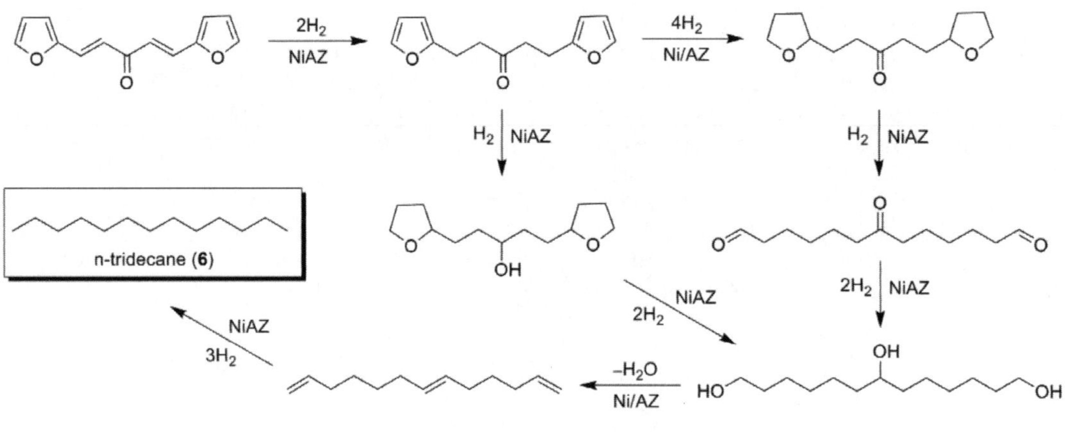

Scheme 2. Proposed reaction pathways for one-step HDO reaction of FAA

3.3.2. Effect of nickel loading

The HDO reaction using 10NiAZ and 20NiAZ in an acidic solvent (entry 3–4, Table 3) gave with various alkane and oxy-compounds as products with various selectivity. The used of 10NiAZ lead higher yield of products compared with 20NiAZ. This is out of hypothesis that the higher Ni loading should be given the better performance. Based on the physical characterization of 10NiAZ and 20NiAZ, it is showed that the latter catalyst has smaller pore diameter (6.580 nm) compared with previous (7.763 nm) (Table 1). By the used of bulky FAA-C13 (**3**) as a precursor, the reaction limited only on the surface of catalyst and reaction in the pore of catalyst didn't occur [19].

4. Conclusion

One pot hydrodeoxygenation reaction of furfural-acetone adduct using Ni/Al_2O_3-ZrO_2 to liquid alkane was investigated in a liquid phase. The solvent has influenced the formation of alkane by suppressing decarbonylation/decarboxylation and hydrocracking process. The reaction in the presence of water using 10% nickel loading at 150°C gave n-tridecane (6.67%) selectively. Addition of acidic solvent promote the hydrocracking process, thus the oxygenated compound and alkene derivatives were observed.

References

[1] Climent MJ, Corma A, Iborra S. Conversion of biomass platform molecules into fuel additives and liquid hydrocarbon fuels. Green Chem. 2014;16(2):516–47.

[2] Corma A, Iborra S, Velty A. Chemical routes for the transformation of biomass into chemicals. Chem Rev. 2007;107(6):2411–502.

[3] Corma A, de la Torre O, Renz M, Villandier N. Production of High-Quality Diesel from Biomass Waste Products. Angew Chem Int Ed. 2011 Mar 1;50(10):2375–8.

[4] Faba L, Díaz E, Vega A, Ordóñez S. Hydrodeoxygenation of furfural-acetone condensation adducts to tridecane over platinum catalysts. Catal Today. 2016 Jul 1;269:132–9.

[5] Shafaghat H, Sirous Rezaei P, Daud WMAW. Catalytic hydrogenation of phenol, cresol and guaiacol over physically mixed catalysts of Pd/C and zeolite solid acids. RSC Adv. 2015;5(43):33990–8.

[6] Sánchez-Sánchez MC, Navarro RM, Fierro JLG. Ethanol steam reforming over – (, La, Zr and Mg) catalysts: Influence of support on the hydrogen production. EHEC2005. 2007 Jul;32(10–11):1462–71.

[7] Elliot DC, Hart TR, Neuenschwander GG, Rotness LJ. Catalytic hydroprocessing of biomass fasts pyrolysis bio-oil to produce hydrocarbon products. 2009. 2009 Aug 5;28:441–9.

[8] Bui VN, Laurenti D, Afanasiev P, Geantet C. Hydrodeoxygenation of guaiacol with CoMo catalysts. Part I: Promoting effect of cobalt on HDO selectivity and activity. Appl Catal B

Environ. 2011 Jan 14;101(3–4):239–45.

[9] Zhang X, Zhang Q, Wang T, Ma L, Yu Y, Chen L. Hydrodeoxygenation of lignin-derived phenolic compounds to hydrocarbons over $Ni/SiO_2–ZrO_2$ catalysts. Bioresour Technol. 2013 Apr;134:73–80.

[10] Ulfa SM, Sari I, Kusumaningsih CP, Rahman MF. Structural Properties of $Ni/\gamma–Al_2O_3$ and $Cu/\gamma–Al_2O_3$ Catalyst and its Application for Hydrogenation of Furfurylidene Acetone. Procedia Chem. 2015;16:616–22.

[11] Faba L, Díaz E, Ordóñez S. Aqueous-phase furfural-acetone aldol condensation over basic mixed oxides. Appl Catal B Environ. 2012 Feb;113–114:201–11.

[12] Mahfud A, Utomo EP, Ulfa SM. Hydogenation/deoxygenation (H/D) reaction of furfural aceton condensation product using Ni/Al_2O_3-ZrO_2 catalyst. 2016. 2016 Aug 4;5(2):108–15.

[13] Rahemi N, Haghighi M, Babaluo AA, Jafari MF, Estifaee P. Synthesis and physicochemical characterizations of $Ni/Al_2O_3–ZrO_2$ nanocatalyst prepared via impregnation method and treated with non-thermal plasma for CO_2 reforming of CH_4. J Ind Eng Chem. 2013 Sep;19(5):1566–76.

[14] Han SJ, Bang Y, Yoo J, Kang KH, Song JH, Seo JG, et al. Hydrogen production by steam reforming of ethanol over mesoporous $Ni–Al_2O_3–ZrO_2$ aerogel catalyst. Int J Hydrog Energy. 2013 Nov;38(35):15119–27.

[15] Ulfa SM, Anjani RM, Amaniyah, Okamoto H. Selective Formation of Monomer and Dimer Product from Aldol Condensation of Furfural and Acetone by Reflux and Autoclave Batch Reactor using MgO Catalysts in Water. Proc 5th Annu Basic Sci Int Conf. 2015;31–4.

[16] Li Y, Huang X, Zhang Q, Chen L, Zhang X, Wang T, et al. Hydrogenation and hydrodeoxygenation of difurfurylidene acetone to liquid alkanes over Raney Ni and the supported Pt catalysts. Appl Energy. 2015 Dec 15;160:990–8.

[17] Behtash S, Lu J, Mamun O, Williams CT, Monnier JR, Heyden A. Solvation Effects in the Hydrodeoxygenation of Propanoic Acid over a Model Pd(211) Catalyst. J Phys Chem C. 2016 Feb 11;120(5):2724–36.

[18] Zhao C, He J, Lemonidou AA, Li X, Lercher JA. Aqueous-phase hydrodeoxygenation of bio-derived phenols to cycloalkanes. J Catal. 2011 May 16;280(1):8–16.

[19] He Z, Wang X. Hydrodeoxygenation of model compounds and catalytic systems for pyrolysis bio-oils upgrading. Catal Sustain Energy. 2012 Jan 19;1:28–52.

Acknowledgments
This work was supported financially by the PUPT research grant from Directorate General of Research, Technology and Higher Education number: 007/Add/SP2H/PL/DIT.LITABMAS/V/2015, We thanks to Prof. Didik Prasetyoko of ITS Surabaya for his guidance using SEM, XRD, and isotherms analysis and Ms. Indah Nur Pramesti for GCMS analysis.

Unsaturated 15 and 16 Membered Appended Naphthalene Macrocyclic Molecules for The Development of Fluorometric Chemosensors

S Hasan[1] ,S Salleh [1], S Hamdan[1] , and B Yamin[1]
[1]Faculty of Applied Sciences, Universiti Teknologi MARA, 40450 Shah Alam, Selangor, Malaysia.

e-mail:sharizal187@perlis.uitm.edu.my

Abstract. Unsaturated macrocyclic molecules have got an interest due to their potential in catalysis, ion exchange and electron transfer. Salicaldehyde derivatives macrocyclic molecules have a broad range in synthesis. In this study, two unsaturated macrocyclic molecules (L1 and L2) have been synthesized. The preparation of unsaturated macrocyclic ligands involve two steps; the reaction of salicylaldehyde with 1,2-dibromoethane or 1,4-dibromobutane to produce precursor, then cyclisation were completed using schiff base technique by adding diamines (naphthalene diamine). The ligands were characterized spectroscopically. In FT-IR spectrums, the form of the ligands mainly can be observed on the disappearance of the carbonyl group of aldehyde at approximately 1650cm^{-1} that was readily assigned to C=O group of salicylaldehyde with the replacement of C=N peak at 1684.13cm^{-1}. The success in producing macrocyclic ligands have been further characterized using fluorescence emission spectroscopy (FES) and revealed the typical emission of naphthalene at ~430nm. Fluorescence changes of L1 and L2 showed high selectivity for $Fe3^{+}$ and $Cu2^{+}$ respectively in the presence of other common metal ions, such as $Zn2^{+}$, $Ni2^{+}$ and $Co2^{+}$. Besides, the ligand was sensitive enough to detect the concentration of ferric ion with the detection limit down to 1.08×10^{-6} M and fluorescence change that was unaffected by the presence of other common coexisting metal ions. Complexation with Co(II) was also attempted.

Keywords: Macrocyclic Schiff base, Macrocyclic complexes. 1,2-bis(2-carboxyaldehyde phenoxy)ethane, Cobalt(II)complexes,1,8- Naphthalenediamine, Fluorophore.

1. Introduction

The macrocycles are important as a strong ligand in coordination with transition metal in coordination of chemistry. Macrocyclic ligand consist of different donor atoms that attached to the central ions and appended with an aromatic ring. Nitrogen, phosphorus, sulphur and oxygen is a donor atoms that is important in the coordination of the ligand with the metal ion [1]. The differences of donor atoms will give different force of bonding and different strength. Meanwhile, appended aromatic ring will delocalize the electron and result more efficient interactions between the molecules and the complex because of the molecule contain the π bond along the ring of the molecules.

Fluorophore is a chemical compound that able to re-emit light upon light excitation. Generally, it consists of aromatic groups or cyclic molecules. It can bind with the ligand to form selective and sensitive sensors in biological, clinical, and environmental interest. The investigation using fluorophore intrigued many researchers due to its low cost, easily performed, high sensitivity and versatility in typical of photoluminescence spectroscopy [2]. The fluorophore is able to send a signal in the responses of the selectivity of the metal. It combines with the ligand to form macrocyclic complexes. Among various fluorophore, 1,8-naphthalenendiamine is one of the useful tools due high stoke shift [3]&[4], which makes it appropriate for constructing macrocyclic complexes. Metal complexes have advantageous photophysical properties, such as high stability and relatively long lifetimes compared with those of organic fluorophores. Therefore, 1,8-naphthalenendiamine was chosen as a fluorophore due to their excellent photophysical properties, high fluorescence quantum yield and good coordination ability. Herein, we report a convenient and efficient route for the preparation of macrocyclic ligand by using Schiff base technique. It exhibits the stronger of fluorescence, so the molecule is more rigid and stable.

The purpose of this research is to construct 15 and 16 membered appended naphthalene macrocyclic complexes denoted as a macrocyclic ligand by using schiff base technique. In this research, schiff base is formed by reaction of salicylaldehyde and diamine derivatives because this reaction mostly form stable complexes with the transition metals. The interactions between the Co(II) complexes with N-donor ligands that containing binding units suitable for coordination of single metal ion. In addition, schiff base also capable of acting to as multidentate ligands for coordination with transition metal ions [5]. This technique has been known to form C=N in which will increase the stability of complex form by coordination of N with the metal and the electron will be delocalized among the three atoms (C, N and metal).

2. Experimental

2.1. Instruments

Elemental analysis was carried out on a Thermo Finnigan Flash EA 110 Elemental Analyzer. 1H Infrared spectra were obtained on Perkin Elmer Spectrum 100 FT-IR 1750X FTIR spectrophotometer on KBr discs in the range 4000–600 cm^{-1}. Fluorescence excitation and emission spectra were recorded on a Luminescence Spectrometer LS 55 using 1 cm pathlength cuvettes at room temperature. The excitation and emission slit widths were 15 nm. 5 x 10^{-5} M solution of the sensor L1 and L2 in CH3OH:H2O (9:1), Tris-HNO3= 10 mM, pH=7 and 1 x 10^{-3} M solutions of the cations were prepared respectively. Fluorescence titrations were carried out by the incremental addition of 40 μL- 400 μL guest solutions to 3ml of sensor L1 and L2 in the cuvette respectively. The receptors and cation solutions were prepared as 5 x 10^{-5} M in CH3OH:H2O (9:1), Tris-HNO3= 10 mM, pH=7.0 for Job's plot.

2.2. Materials

The salicylaldehyde derivatives used in the synthesis were prepared from salicylaldehyde, 1,2-dibromoethane or 1,3-dibromopropane and K_2CO_3 as shown in Scheme 1 and according to the literature method [6] and [7]. All the other chemicals and solvents were of analytical grade and used as received.

Scheme 1. Synthesis of 1,3-bis(2-carboxyaldehyde phenoxy)propane

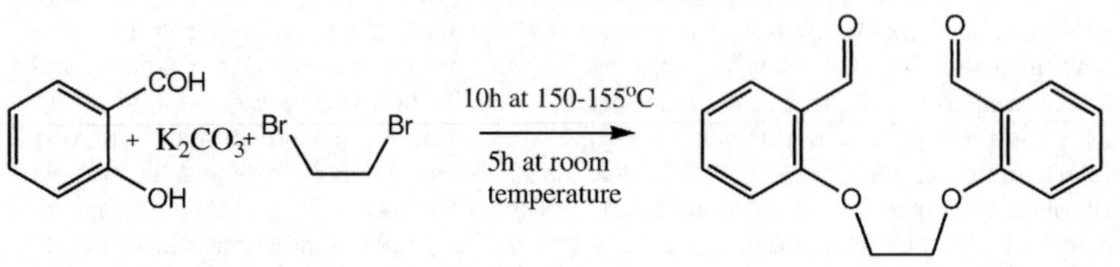

2.3. Synthesis of Schiff Base

Cautiously, diamine (5mmol) in methanol (40mL) was added drop wise to a stir solution of salicylaldehyde derivatives (1.5g, 5mmol) in methanol (60mL). Then, the reaction was reflux for 48 hours at 100°C. After the addition had completed, the reaction was left at room temperature for several hours and continued filter and wash with cool methanol and dried in the air. Figure 1 shows the step in synthesis of macrocyclic ligands, L1a and L2b

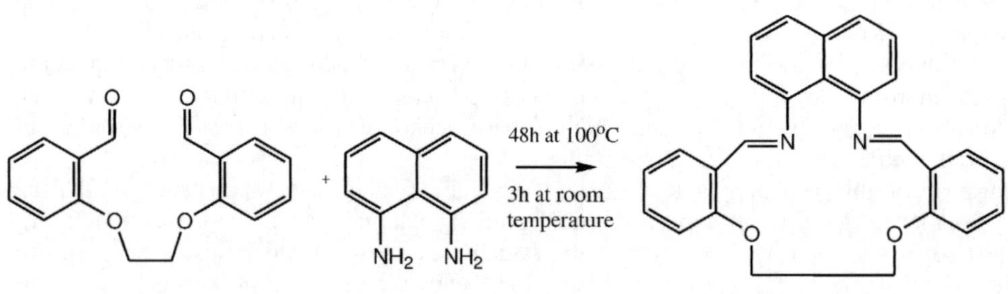

L1a

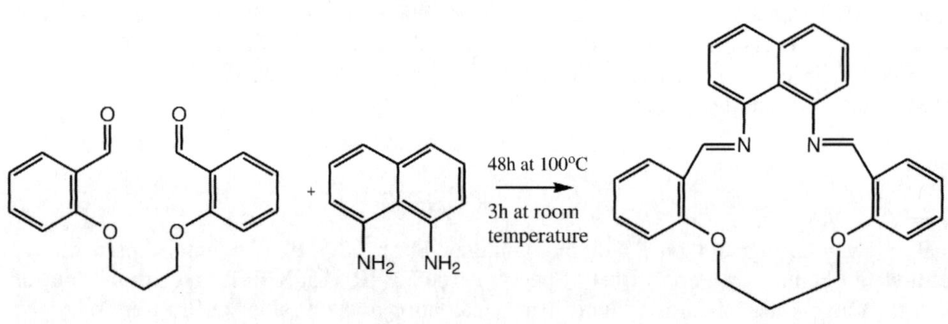

L2b

Figure 1. Step in synthesis of macrocyclic ligands, L1a and L2b

2.4. Synthesis of complexes

To a homogeneous ligand, Co(II)chloride-6-hydrate (1.85g, 5mmol) in methanol(60mL) was added dropwise to the macrocyclic of 1,2-bis(2-carboxylaldehydephenoxy)ethane (1.5g, 5mmol) in 40mL methanol. While stirring the mixture was reflux for 3-4 hours at 60°C. Then, the precipitate was filtered and washed with cool methanol dried to the air. The step in synthesis of macrocyclic complexes, L1a1 and L2b2 are shown in Figure 2.

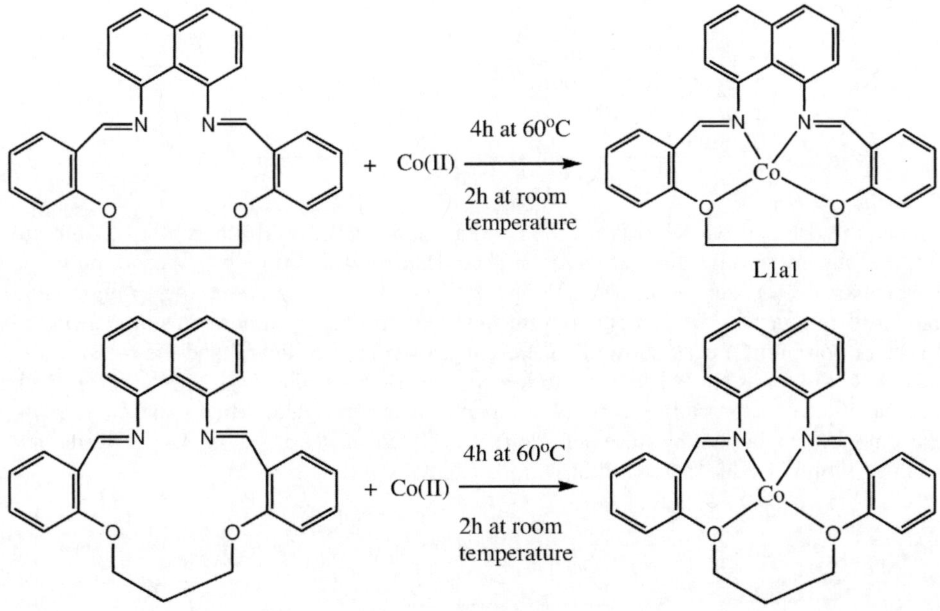

L2b2

Figure 2. Step in synthesis of macrocyclic complexes, L1a1 and L2b2

2.5. Association constant calculation

Generally, for the formation of 1:1 complexation species formed by the receptor and the guest cation, the following Benesi–Hildebrand equation was used to determining the association constants (Ka).

$$\frac{1}{(F-F_0)}=\frac{1}{(F_\alpha-F_0)}+\frac{1}{K(F_\alpha-F_0)}\times\frac{1}{[M^{n+}]}$$

Where F, F0 and Fα are the emission intensities, respectively in the absence and infinite concentration of the metal ion (Mn+) [8]. Fα is the fluorescence intensity in the presence of excess amount of M^{n+}. Therefore, for 1:1 complex formation, the double reciprocal plot of 1/(F F0) against 1/[M^{n+}] should give a straight line; from the slope and intercept of which, the equilibrium constant (K) can be calculated.

3. Result and discussion

3.1. Infrared Spectroscopy (FT-IR) of Macrocyclic ligands, L1a and L2b

The macrocyclic ligand L1a and L2b formations were obtained from the condensation of the carbonyl compound in methanol as a solvent with 1:1 mole ratio as stated in step in synthesis of macrocyclic ligands, L1a and L2b reaction in methodology. The macrocyclic result on FT-IR of L1a and L2b are shown in the figure 1.1a. There is one main feature in the infrared spectrum of the macrocyclic compound. The first one is disappearance of the characteristic (C=O) group at that has been replaced with (C=N) peak. This band range is shown in Table 1.

IOP Conf. Series: Materials Science and Engineering **172** (2017) 012054 doi:10.1088/1757-899X/172/1/012054

Table 1: FT-IR data for ligand L1a and L2b

Interpretation	Peak frequency (cm^{-1})	
	Macrocyclic compound	
	L1a	L2b
C=N	1633.25	1638.52

As shown in Table , the range of (C=N) peak for the L1a and L2b macrocyclic is between 1633cm^{-1} to 1638cm^{-1} which was has lower wave number from the disappearance peak of (C=O). According to the research done by (Ilhan, 2007) found that the (C=N) peak shifted towards to lower energy region due to the coordination of nitrogen in the compound. Besides, due to the appearance of nitrogen in the compound IR spectra on (C-O-C) peak showed a different between a free ligand and the macrocyclic which the bands of (C-O-C) has shifted towards to lower wave number with high energy level. Peak for the L1a macrocyclic that exist at 1633.25cm^{-1} compared to the free ligand that exists at 1290.44 cm^{-1}. This phenomenon explained by research done by (Ilhan, 2007), the (C-O-C) bands has undergone stretching vibrations due to the conformational changes in the compound.

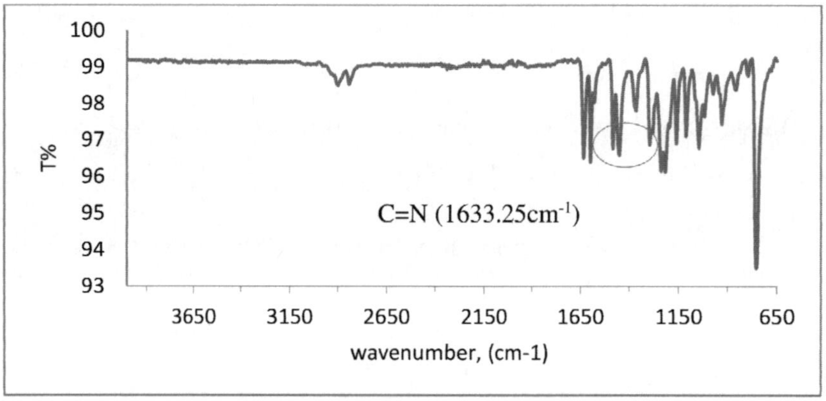

Figure 3a. FT-IR result on L1a macrocyclic compound

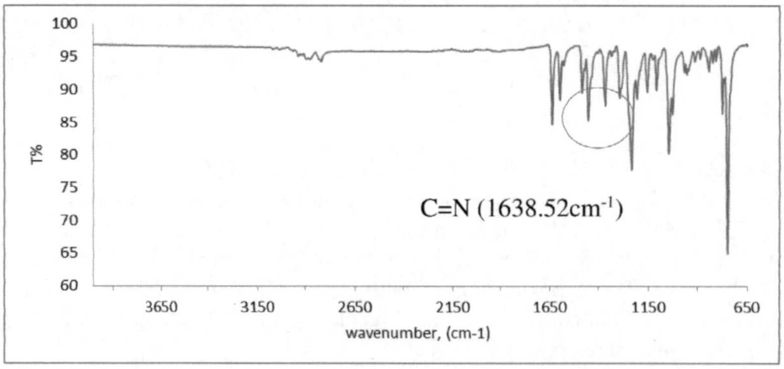

Figure 3.b. FT-IR result on L2b macrocyclic compound

3.2. Infrared Spectroscopy (FT-IR) of Macrocylic complexes, L1a1 and L2b2

For the L1a1 complexes the cobalt peak has appeared at 3224.15cm^{-1}. The peak has shifted to a lower wavenumber region which means lower energy. According to (Ebrahimi, 2014) the peak for complexes has shifted towards lower wavenumber side due to the coordination of the complexes to the metal ions through the lone pair of electrons on the nitrogen. In addition the peak of (C-O-C) band explained by research done by (Ilhan, 2007) has appeared at 1286.37cm^{-1} also has shifted to the lower wave number region because of the involvement of metal to the macrocyclic compound.The result on IR spectrum of L1a1 is shown in figure 3.c. The IR spectrum on the L2b2 macrocyclic complexes also has shown a medium broad peak as L1a1 macrocyclic complexes at 3225.25cm^{-1}. This peak also has shifted to the left to the lower wave number side. This phenomenon has been explained by (Ebrahimi, 2014) on his research that this phenomenon happens due to the lone pair of nitrogen as mentioned above.

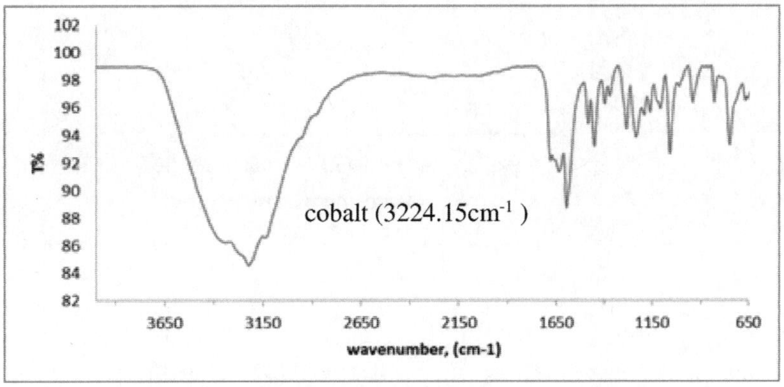

Figure 3c. FT-IR of macrocyclic complex, L1a1.

3.3 Elemental analyzer of Macrocyclic

Elemental analyzer test has been done on the macrocyclic compound since the macrocyclic compound exists in the form of solid. From the experimental composition and the values on C, H and N of the macrocyclic compound it can be said that Schiff base technique to produce macrocyclic compound are acceptable. From the table 4 the experiment data on the total percent error can be said that for the ligand L1a and L2b the percent composition is between 86 to 87% that is 86.164% and 87.261% respectively. The result is acceptable for the formation of the macrocylic. The experimental ratio that acquired for C:H and C:N also close to the theoretical ratio that confirmed the formation of the macrocyclic.

Table 2: Elemental composition on theoretical ratio, experimental ratio, total (%) error and total (%) composition

	Theoretical ratio (%)		Experimental ratio (%)			
Compound	C:N	C:H	C:N	C:H	Total (%) error	Total (%) composition
$C_{26}H_{20}N_2O_2$, L1a	26:2.0	26:20	26.1:2.0	17.83:29.0	9.379	86.184
$C_{27}H_{22}N_2O_2$, L2b	27:2.0	27:22	21.8:2.5	25.7:21.4	6.911	87.261

3.3. Fluorescence spectroscopic studies

Selectivity is an important characteristic of an ion-selective chemosensor. To investigate the selectivity of L1 and L2 the fluorescence selectivity experiments of various metal Fe^{3+}, Co^{2+}, Cu^{2+}, Ni^{2+} and Zn^{2+} were studied. The emission spectra of L1 and L2 after the addition of different metal cations in CH_3OH:H_2O are shown in Fig.4(a and b).

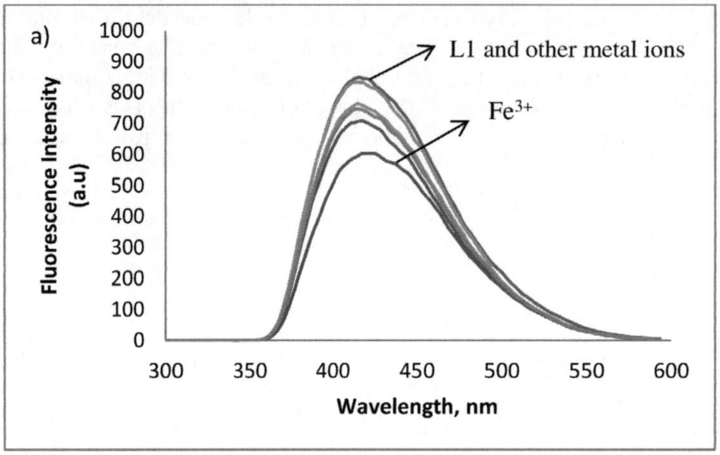

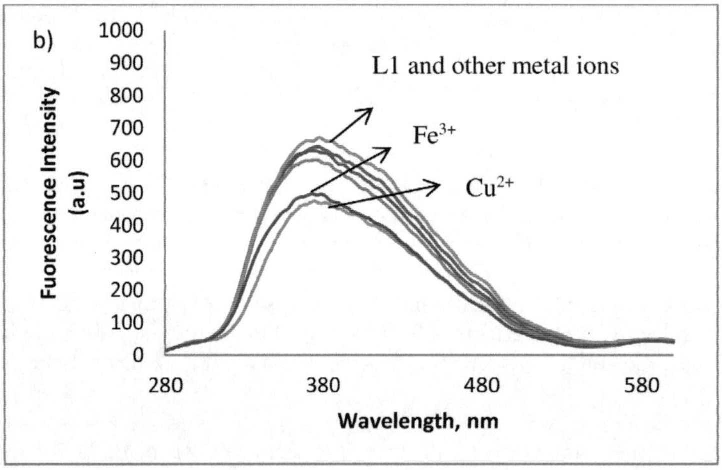

Figure 4(a and b). Fluorescence spectra of L1 (a) an L2 (b), in CH_3OH:H_2O

Based on the result of fluorescence spectra, upon addition of 5.0 x 10^{-3} M each of metal ions in CH_3OH:H_2O there are significant change on the fluorescence spectra of L1 and L2 were observed. The emission intensity of L1 increased significantly after addition of 5.0 x 10^{-3} M $Fe3^{+}$ ion. Meanwhile for the L2, there is a change in emission intensity after addition of 5.0 x 10^{-3} M $Fe3^{+}$ and $Cu2^{+}$. Change in emission intensity of sensors L1 and L2 were considered after addition of $Fe3^{+}$ ion and $Cu2^{+}$ ion is due to the formation of stable complex due high stoke shift, which makes it appropriate for constructing macrocyclic complexes. As it is revealed from selectivity studies above the response of sensors toward $Co2^{+}$, $Ni2^{+}$ and $Zn2^{+}$ were the same, but $Fe3^{+}$ and $Cu2^{+}$ showed different response. The emission intensity of sensors of $Fe3^{+}$ and $Cu2^{+}$ was lower than the other metal ions. Thus $Fe3^{+}$ and $Cu2^{+}$ were selected to study the fluorescence response profiles and detection limit. It exhibit the stronger of fluoroscence, so the molecule is more rigid and stable.

Further studies on sensitivity of compounds L1 and L2 were conducted on selected metal. As shown in Fig. 5a, 5b and Fig 6a, 6b the sensitivity studies of fluorescence intensities were performed using 5.0 x 10^{-5} M solution of L1 and L2 in CH_3OH:H_2O with different concentration of Fe^{3+} for L1 and Cu^{2+} for L2.

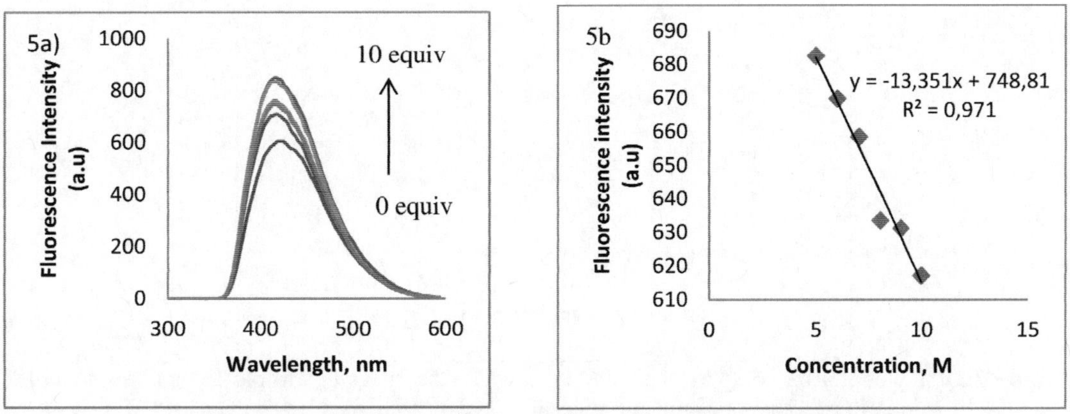

Figure 5(a and b). Sensitivity studies of fluorescence intensities

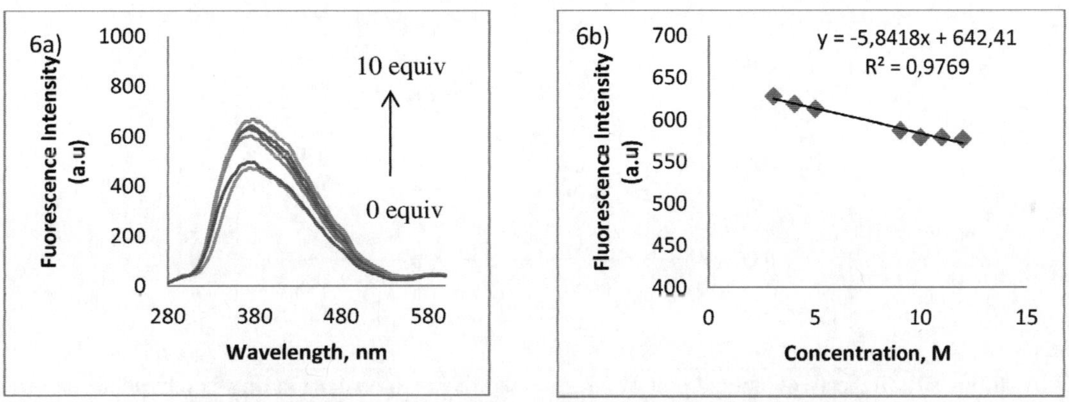

Figure 6(a and b). Sensitivity studies of fluorescence intensities

The results showed that the emission intensity was linearly changed with increasing ferric ion concentration for L1 and copper ion concentration for L2. To determine the detection limit, the relative fluorescence intensity was plotted as a function of $Fe3^{+}$ and Cu^{2+} concentration, the detection limit was calculated based on the fluorescence titration. The numerical value of limit of detection is calculated using 3 Sb/m [9], where Sb is standard deviations of the blank and m is slope the calibration curve. The limit of detection L1and L2 are $1.08\text{x}10^{-6}$ M and $1.2025\text{x}10^{-7}$ M respectively. According to the state regulations, the concentration of the iron ions in the sewage discharge is not more than 1.79x 10^{-6}M [10].

3.4. Job plot study

To determine the binding stoichiometry of L1 and Fe^{3+}, Job's method for absorbance measurement was carried out. Keeping the sum of the initial concentration of L1 and Fe^{3+}, the molar ratio of Fe^{3+} was varied from 0 to 1. As shown in Fig.7a, the fluorescence reached a maximum when the ratio of $[Fe^{3+}]/\{[Fe^{3+}]+[L1]\}$ was 0.5, indicating 1:1 stoichiometry of Fe^{3+} to L1 in the complex. From the fluorescence titration profile (Fig.7b), the binding constant (Ka) of L1 with Fe^{3+} ion was calculated according to the 1:1 model (2.9 x 10^5 M). According to the Fig. 8a and 8b, the result for the L2 and

IOP Conf. Series: Materials Science and Engineering **172** (2017) 012054 doi:10.1088/1757-899X/172/1/012054

Cu^{2+} also shows the same ratio and stoichiometry to the L1 and the binding constant (Ka) of L2 with Cu^{2+} ion was calculated according to the 1:1 model ($1.37x10^6$M). These values were accepted because in the range of those (10^3 – 10^5 M) which previously reported for binding chemosensor The high value of association constant Ka indicates that strong complex formation takes place between L1 and Fe^{3+} ion.

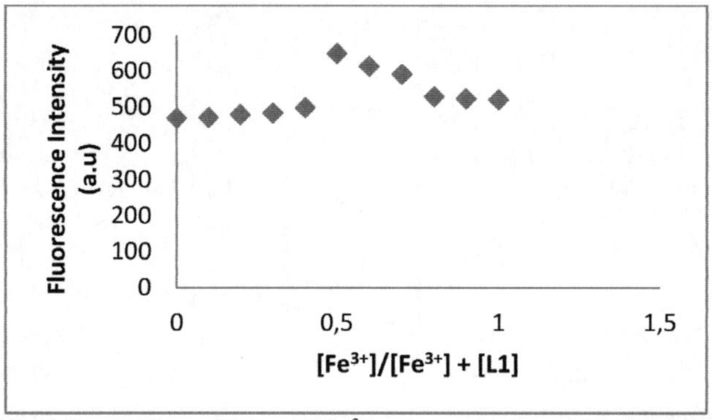

Figure 7a. Job plot of a 1:1 complex of L1 and Fe^{3+}, where the intensity was plotted against the mole fraction of Fe^{3+}. The total concentration of Fe^{3+} with receptor L1 was 10×10^{-5} M.

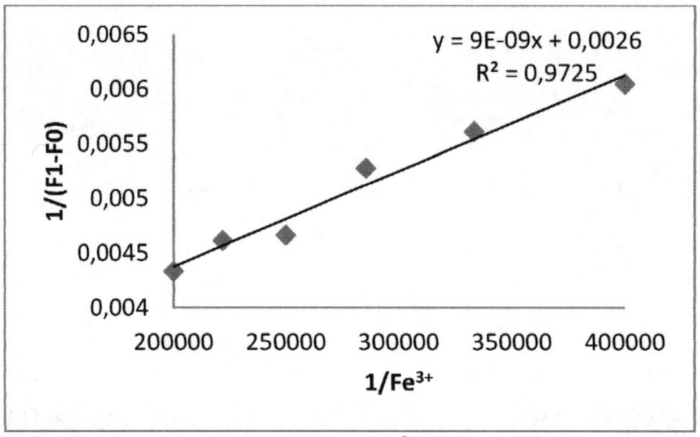

Figure 7b. Benesi-Hildebrand plot of L1 (5×10^{-5} M) based on 1:1 binding stoichiometry.

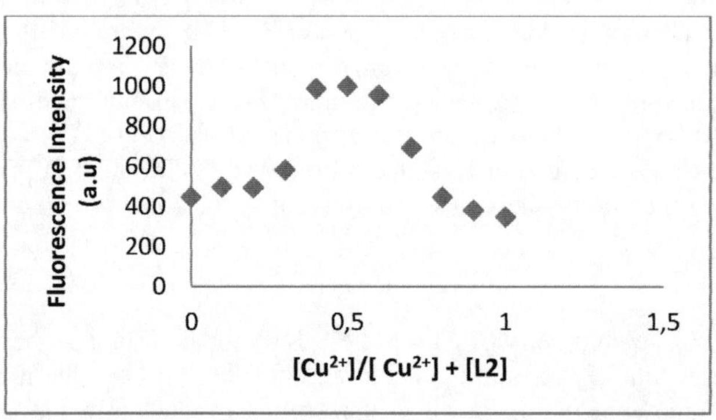

Figure 8a. Job plot of a 1:1 complex of L2 and Cu^{2+}, where the intensity was plotted against the mole fraction of Cu^{2+}. The total concentration of Cu^{2+} with receptor L2 was 10×10^{-5} M.

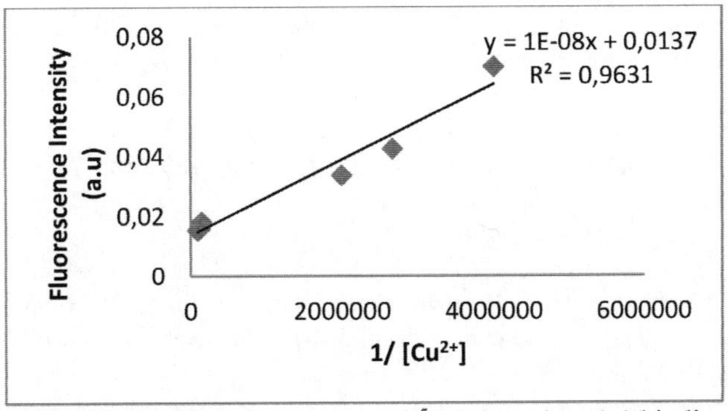

Figure 8b. Benesi-Hildebrand plot of L2 (5 x 10^{-5} M) based on 1:1 binding stoichiometry.

4. Conclusion

In this study, the macrocyclic ligands have been synthesis successfully by Schiff base technique. The formation of macrocyclic ligands have been proved by the FT-IR spectrum with the appearance of C=N in the range 1633cm^{-1} to 1638cm^{-1}. Elemental analyzer confirmed the theoretical and experimental value suggesting the attachment of diamine derivatives into the aldehyde precursor to form the macrocyclic. Fluorescence changes of L1 and L2 showed high selectivity for Fe^{3+} and Cu^{2+} respectively in the presence of other common metal ions, such as Zn^{2+}, Ni^{2+} and Co^{2+}. Besides, the ligand was sensitive enough to detect the concentration of ferric ion and copper ion with the detection limit down to 1.08 x 10^{-6}M and 1.2025x10^{-7}M respectively. However, the presence of other common coexisting metal ions especially cobalt ion which is used for the analysis of the complex were unaffected to the fluorescence change due to the low stoke shift and low of stability to form macrocylic complex.

5. References

[1] S Ilhan, H Temel, I Yilmaz, and M Sekerci, Synthesis, structural characterization and electrochemical studies of new macrocyclic Schiff base containing pyridine head and its metal complexes. *Jurnal of Organomtallic Chemistry.*, vol. 692, pp. 3855-3865, 2007.

[2] L Prodi, F Bolletta, M montalti, N Zaccheroni, Luminescent Chemosensor for transition metal ions. *Coordination Chemistry*., vol. 205, pp. 59–83, 2000.

[3] M Saleem, R Abdullah, A Ali, B Park, E Choi, I Hong, and K Lee, Facile synthesis, cytotoxicity and bioimaging of Fe(3+) selective fluorescent chemosensor. *Bioorg. Med. Chem.*, vol. 22, no. 7, pp. 2045–51, 2014.

[4] S Çetindere, S Tümay, A Kılıç, M Durmuş, and S Yeşilot, Hexa-BODIPY Linked-Triazole Based on a Cyclotriphosphazene Core as a Highly Selective and Sensitive Fluorescent Sensor for Fe2+ Ions. *J. Fluoresc.*, 2016.

[5] M Ashraf, K Mahmood, A Wajid, M Maah, and I Yusof, Synthesis, characterization and biological aactivity of Schiff bases. *Paper presented at the International Conference on Chemistry and Chemical Process.*

[6] S Ilhan, H Temel, Synthesis and spectral studies of macrocyclic Cu(ii), Ni(II) and Co(II) complexes by template reaction of 1,4-bis(3-aminopropoxy)butane with metal(II) nitrate and salicyldehyde derivatives. *Journal of Molecular Structure.*, vol. 891, pp. 157–166, 2008.

[7] S Ilhan, H Temel, and A KIlIc, Synthesis and spectral studies of macrocyclic Cu(II) complexes by reaction of various diamines, copper(II) perchlorate and 1,4-bis(2-carboyxaldehyde phenoxy) butane. *Journal of Coordination Chemistry.*, vol. 61(2), pp. 277–284, 2008.

[8] T Singh, P Paul, and H Pramanik, Fluorescent chemosensor based on sensitive Schiff base for selective detection of Zn2+. *Spectrochim. Acta. A. Mol. Biomol. Spectrosc.*, vol. 121, pp. 520–6, Jan. 2014.

[9] G You, G Park, S Lee, K Ryu, and C Kim, Chelate-type Schiff base acting as a colorimetric sensor for iron in aqueous solution. *Sensors Actuators, B Chem.*, vol. 215, pp. 188–195, 2015.

[10] L Zhao, X Xin, P Ding, A Song, Z Xie, J Shen, and G Xu, Fluorescent oligomer as a chemosensor for the label-free detection of Fe3+ and dopamine with selectivity and sensitivity. *Anal. Chim. Act.a,* 2016.

Acknowledgments

We express our sincere appreciation to the Ministry of Higher Education of Malaysia for providing generous financial support under the research grant FRGS UiTM 5/3/FRGS/(2/2014) and UiTM for providing all needed facilities.

Isolation and Antimicrobial Activity of Flavonoid Compounds from Mahagony Seeds (*Swietenia macrophylla*, King)

S Mursiti and Supartono

Semarang State University, Semarang, Indonesia

Email: kumalasari_berliana@yahoo.com

Abstract. Flavonoid is one of the secondary metabolites compounds in mahogany seeds. Mahogany seeds can be used as an antimicrobial. This study aims to determine the antimicrobial activity of flavonoid compounds from mahogany seeds against *Escherichia coli* (*E.coli*) and *Bacillus cereus* (*B.cereus*). Isolation of flavonoid compounds done step by step. First, the maceration using n-hexane, then with methanol. The methanol extract was dissolved in ethyl acetate and aquadest, then separated. Ethyl acetate extract evaporated Flavonoid compounds were. The testing of antimicrobial activity of flavonoid compounds using the absorption method. The results showed that the antimicrobial activity of flavonoid compounds from mahogany seeds shows the inhibitory activity and provide clear zone against bacteria *E.coli* with value Inhibitory Regional Diameter 18.50 mm respectively, and 14.50 mm to the bacteria. Based on the results of the study, it can be concluded that flavonoid compounds from mahogany seeds have antimicrobial activity against *E.coli* and *B.cereus*.

1. Introduction

Existing research shows that mahogany seeds contain compounds that have antidiabetic activity, usually used as well as insecticides, larvicides, nematicides, antipyretic, fungicides, antimicrobials and antioxidants [1]. Chemical constituents in mahogany seeds are alkaloids, saponins, flavonoids [2][3]. Research on antimicrobial activity test phase basil leaf extract n-hexane, ethyl acetate phase and the phase of 70 % ethanol and 70 % ethanol phase that shows antimicrobial activity *S. aureus* and *C. albicans* with agar diffusion method and dilution liquid [4]. However, the study did not report the active compounds that have antimicrobial activity. Results of other studies say that basil chloroform extract can inhibit the bacteria *Shigella dysenteriae* and methanol extract can inhibit microbe *Klebsiella pneumoniae, Salmonella paratyphy* and *S. aureus* with Inhibitory Regional Diameter (DDH) respectively 10 mm, 9 mm, and 7 mm, but does not mention the concentration used to test the antimicrobial activity [5].

These studies concluded that the part of the basil plant that often tested the activity of antimicrobial is part of basil leaves, whereas according to [6], not only the basil leaves just that contain secondary metabolites such as essential oils, but part of the basil plant such as trunks also contain secondary metabolites that may also have antimicrobial activity. So far have not found a research report stating about power test the antimicrobial activity on the stem of flavonoid from mahogany seeds against *B. cereus* and *E. coli* bacteria.

2. Research Methods

Flavonoid compounds isolation done step by step. First, the maceration using n-hexane, then with methanol. A total of 50 g of the methanol extract put in a separating funnel, then added each 50 mL of distilled water and ethyl acetate, then shaken, allowed to stand, and separated. The ethyl acetate phase was taken, evaporated to obtain a dry extract flavonoids. The extract was detected by TLC and KCV fractionated by using silica gel GF254 G60 with enhanced polarity eluent (n-hexane-EtOAc, EtOAc, EtOAc-MeOH). The fraction which still contains impurities further fractionated in a row using radial chromatography to obtain pure compounds. The testing of antimicrobial activity of flavonoid compounds using the absorption method.

Antimicrobial activity test using the absorption method is a modification of the order and that has been done [7] [8]. Nutrient medium so as to be used as a medium for bacterial growth is provided by heating NA back, then poured into a sterile petri dish aseptically. The bacteria are grown on medium NA by entering 1 mL of bacterial culture medium in NA then averaged over the surface of the agar medium. Paper discs with a diameter of 6 mm are dipped into isolated compounds were each concentration 10 mg/mL, 25 mg/mL, 50 μg/mL, 100 μg/mL, then allowed to stand for 1 hour, then placed on a saucer solder containing bacteria sterile. Negative controls using paper discs were dipped in distilled water, whereas the positive control using a paper disc dipped in wipol. The entire cup solder containing seeding bacteria were incubated for 12 hours at a temperature of 37 ^{o}C in reverse, then observed and measured the inhibition of bacterial growth in the area around the paper disc, followed by calculating the area.

3. Result and Discussion

Extraction of mahogany seed includes several stages of sample preparation and extraction process. The methanol extract of mahogany seeds obtained 10.7 % in the form of a brown powder with a distinctive aroma and taste very bitter. Qualitative test results showed that the methanol extract of mahogany seeds contain secondary metabolites are alkaloids, flavonoids, and saponins, and is consistent with the results of Mursiti's study (2009). There were containing flavonoid compounds.

Testing the antimicrobial activity of flavonoid compounds of mahogany seeds using two bacterias, *E. coli* and *B. cereus* with absorption method. Controls used in this method is a negative control (distilled water) and positive control (wipol). The test results showed that the methanol extract at a concentration of 10 mg/mL, 25 mg/mL, 50 μg/mL, 100 μg/mL showed activity against both bacteria *B. cereus* and *E. coli*. It is known by the diameter of the visible area of inhibition around the paper disc. The diameter of inhibitory regions experienced an increase means that the higher the concentration, the greater the concentration of active ingredient that serves as an antibacterial, so the ability to inhibit bacterial growth E. coli also getting bigger. According Dzidic [9] states that one of the mechanisms of bacterial resistance is inaktifikasi antibiotics by producing enzymes. One enzyme that can menginaktifikasi antibiotics are β-glukoronidase. E. coli is a bacteria that is capable of producing β-glukoronidase that allegedly active compound in flavonoid can be described by β-glukoronidase into other compounds that toxic for the bacteria. Higher levels of bioactive compounds that are generally bactericidal (lethal microbes) and a lower level usually is bacteriostatic (inhibits growth, not lethal microbes) [10]. This is consistent with research Khumaisah [11] which states that basil is less susceptible to E. coli and *Shigella sonnei*, but effectively inhibit *Salmonella* bacteria sonnei. This is caused by the bacterium *Salmonella sonnei* thought to have lower metabolic activity so slow to the synthesis of the ribosomal protein that antibacterial agents can freely enter and activity may be hampered. Data inhibitory activity by measuring the diameter of inhibitory regions (DDH) is presented in Table 1.

Table 1. Diameter inhibitory region (mm) mahogany seeds methanol extract against *E. coli* and *B. cereus*

Sample	Diameter (mm)					
	Escherichia coli			*Bacillus cereus*		
	100%	50%	25%	100%	50%	25%
D1	18.50	16.50	13.50	14.00	13.00	10.00
D2	18.00	16.50	13.50	15.00	13.00	10.50
D3	19.00	16.50	13.50	14.50	13.00	11.00
Average	18.50	16.50	13.50	14.50	13.00	10.50
Controle(+)						
D1	35.10	35.16	35.16	51.33	51.33	51.33
D2	35.50	35.50	35.50	51.00	51.00	51.00
D3	35.30	35.33	35.33	52.00	52.00	52.00
Average	35.30	35.33	35.33	51.44	51.44	51.44
Controle (-)	0	0	0	0	0	0

The results showed that the inhibition of the ethanol extract was higher against *E.coli* bacteria (gram-negative) as compared with *B. cereus* bacteria (gram-positive bacteria) value indicated by inhibition area diameter. This is due to differences in the sensitivity of bacteria to antibacterial influenced by the structure wall cell bacteria. Research of Pramuningtyas [12] concluded that Gram-positive bacteria tend to be more sensitive to the antibacterial because of the structure of the cell wall of gram-positive bacteria is simpler than the structure of the cell wall of gram-negative bacteria, making it easier for antibacterial compounds to enter the cell structure of the cell wall of gram-positive bacteria.

4. Conclusion

Based on the research that has been done, it can be concluded that the antimicrobial activity of flavonoid compounds from mahogany seed gives the largest clear zone where inhibition of *E.coli* bacteria (18.50 mm) higher than *B.cereus* (14.50 mm) with 100% concentration.

5. References

[1] Nurcahyanti, ADR and Timotius, KH 2011 Aktivitas Antioksidan dan Antibakteri Ekstrak Polar dan Non Polar Biji Selasih (Ocimum sanctum L.) *Jurnal Teknologi dan Industri Pangan.* **22** 1

[2] Mursiti, S 2009 Isolasi, Identifikasi, dan Elusidasi Struktur Senyawa Metabolit Sekunder dari Biji Mahoni serta Uji Aktivitas Antidiabetesnya, Research Report, Semarang: Unnes

[3] Dhulgande G, Birari AR and Dhale DA 2010 Preliminary Screening of Antibacterial and Phytochemical Studies of Ocimum americanum Linn. *Journal of Ecobiotechnology*, **2** 11-3

[4] Atikah N 2013 Uji Aktivitas Antimikroba Ekstrak Herba Kemangi (*Ocimum americanum* L) terhadap Staphylococcus aureus dan Candida albicans. *Thesis* UIN Syarif Hidayatullah Jakarta.

[5] Devi K, GK Devi, G Thirumaran, R Arumungan, and Anantharaman 2010 Antibacterial Activity of Selected Medicinal Plants from Parangipettai Coastal Regions Southeast Coast of Selected Medicinal Plants from Parangipettai Coastal Regions Southeast Coast of India. *Academic Journal of plant Sciences*. **3** 122-5

[6] Gupta N and Prakash P 2005 Therapeutic Uses of *Ocimum sanctum* Linn (Tulsi) with A Note On Eugenol and its Pharmacological Actions : Short Review. *Indian Journal Physiol Pharmacol.* **49** 125-131

[7] Chaturvedi P, D Singh, TR Kumar and VK Gupta 2012 Antimicrobial Activity of Some promising plants Oil, Molecules and Formulations. *Indian Journal of Experimental Biology.* **50** 714-7

[8] Radhakrishnan TM, V Raman, Samuel, P Saradhi, N Rao, VV Krishna and M Sudhakar 2012 Antibacterial, Antioxidant Activity and GC-MS Analysis of *Eupatorium odoratum*. *Asian Journal of pharmaceutical and Clinical Research*. **5** 2405-11

[9] Dzidic S, J.Suskovic and B Kos 2008 Antibiotic Resistence Mechanisms in Bacteria: Biochemical and Genetic Aspects. Food Technol. *Biotechnol*, **46** 211-21

[10] Binadja A, A Kamal and Sudarmin 2012 Aktivitas Antimikroba Senyawa Hasil Reaksi Hidrasi Kariofilena pada E. coli dan *S. aureus*. *Indonesian Journal of Chemical Science*. **1** 2

[11] Khumaisah LL, Asep K, Gebi D and Yuni A 2011 Komposisi Kimia dan Uji Aktivitas Antibakteri Minyak Kemangi (*Ocimum americanum* L.) terhadap bakteri *Eschericia coli*, *Shigella sonnei* dan *Salmonella enteridis*. *Berk penel Hayati*. **16** 101-10

[12] Pramuningtyas R and Rahadiyan WB 2009 Uji Aktivitas Antimikroba Ekstrak Etanol Cocor Bebek (*Kalanche pinnata*) terhadap Bakteri *Staphylococcus aureus* dan *Escherichia coli* secara In vitro. *Biomedika*. **2** 43-50

IOP Conf. Series: Materials Science and Engineering **172** (2017) 012056 doi:10.1088/1757-899X/172/1/012056

Enhanced the hydrophobic surface and the photo-activity of TiO_2-SiO_2 composites

S Wahyuni and A T Prasetya

Department of Chemistry, Faculty of Mathematics and Natural Sciences, Universitas Negeri Semarang, Indonesia.

Email: zifaluqi@gmail.com; sriwahyunikimia@mail.unnes.ac.id

Abstract. The aim of this research is to develop nanomaterials for coating applications. This research studied the effect of various TiO_2-SiO_2 composites in acrylic paint to enhance the hydrophobic properties of the substrate. Titanium dioxide containing silica in the range 20–35 mol% has been synthesized using sol–gel route. The XRD's spectra show that increasing SiO_2 content in the composite, decreasing its crystalline properties but increasing the surface area. TiO_2-SiO_2 composite was dispersed in acrylic paint in 2% composition by weight. The largest contact angle was 70, which produced by the substrate coated with TS-35-modified acrylic paint. This study also investigated the enhanced photo-activity of TiO_2-SiO_2 modified with poly-aniline. The XRD spectra show that the treatment does not change the crystal structure of TiO_2. The photo-activity of the composite was evaluated by degradation of Rhodamine-B with visible light. The best performance of the degradation process was handled by the composite treated with 0.1mL anilines per gram of TiO_2-SiO_2 composite (TSP-A). On the other side, the contact angle 70 has not shown an excellent hydrophobic activity. However, the AFM spectra showed that nanoroughness has started to form on the surface of acrylic paint modified with TiO_2-SiO_2 than acrylic alone.

1. Introduction

The photo-catalytic property of nano-crystalline TiO_2 has been explored by many researchers as high-technology applications [1–2]. Titanium dioxide (TiO_2), also known as titania, is well known for physical and chemical stability [3], non-toxicity, low cost and less photo-corrosion [4].Thus, titania has become a perfect choice for self-cleaning surfaces that leads to a large scale commercialisation of such products [5]. A common and simple method for preparing TiO_2 is the sol-gel method. The Addition of another oxide as dopants such as SiO_2 [6-7], to titania matrix has been found to further modify the photo-catalytic property of this material. The Addition of silica also enhances the thermal stability of the anatase phase by delaying the transformation to rutile by inhibiting the growth of titania particles. Most of the earlier work reported on titania-silica are about the ones synthesized from alkoxide precursors.

The wet ability of solid surfaces is a very important property governed by both the chemical composition and the geometric features of the surfaces. A surface with water contact angle (CA) greater than 90° is usually referred to as hydrophobic, and one with water CA higher than 140° is qualified as ultra-hydrophobic. The surfaces with very high water contact angles, particularly greater than 150°, are usually called super-hydrophobic surfaces,which have been extended to more applications. Current applications for self-cleaning paints involve two types of approaches: super-hydrophobicity and super-hydrophilicity [8]. In the former, liquid water forms spherical droplets that roll down the surface carrying dirt deposited on the surface. Both, hydrophilic and hydrophobic surfaces exhibit self-cleaning property through the nature of the interaction of water on the surface,the former by rolling droplets and the latter by sheeting water that carries away dirt. Hydrophobic self-cleaning has very interesting aspects. A self-cleaning surface requires high static contact angle,$\theta_s > 90°$ and a minimum inclination angle for the drop to roll off the surface [9].

Here we report, a simple sol–gel method to synthesize high temperature stable photoactive TiO_2-SiO_2 system,useful in the preparation of coatings on various substrates. Titania sol and silica sol were prepared separately through sol–gel route which was then mixed for getting different TiO_2-SiO_2 compositions [10]. The Hydrophobic property of TiO_2-SiO_2 nanoparticles in the water-based acrylic coating was evaluated. TiO_2-SiO_2 nanoparticles were then modified with aniline to form TiO_2-SiO_2-Pani. The photo-catalytic activity of TiO_2-SiO_2-Pani was evaluated by decomposing Rhodamine-B dye stuff as a model. The main application of photo-catalytic activity in organic coatings is in the decomposition of organic contaminants of water and air [11]. Illuminated anatase TiO_2 also exhibited super-hydrophilic properties, which were exploited for various applications such as self-cleaning, anti-fogging and antimicrobial effects in coatings [12-13].

2. Experiment

2.1. Material and Methods:

The chemicals used in this study were titanium tetra-isopropoxide (TTIP, 97%, Sigma-Aldrich) as a titanium precursor, tetra-ethyl ortho-silicate (TEOS, 98%,E-Merck) as silica source, NH_3(25 wt%, E-Merck), and anhydrous ethanol (C_2H_5OH, E-Merck), Aniline (E-Merck), acrylic paint.

2.2. General Procedure

2.2.1 Synthesis of TiO_2-SiO_2 (TS) composite. The TiO_2-SiO_2 (TS) composite was prepared by the following method. In the synthesis of Sol A, TTiP was used as a precursor and was mixed with ethanol, aqueous of 0.1 M HCl solution and deionized water in the volume ratio 1:2:2:2, respectively. The mixture was stirred for an hour and maintained in the pH range from 1 to 2. In Sol B, TEOS was mixed with 40mL ethanol under a nitrogen atmosphere and stirred for an hour. Sol A was then added into Sol B, and the mixture was stirred for 3 h at 60 °C [14]. The aqueous solution of PVA (1 %) was added to the above mixture at the rate of 0.5mL/min and was stirred for 2 h. The gel was dried at room temperature. Finally, the mixture was heated at 120 °C for an hour. The powder was calcined at 500°C for 3 hours. The composite produced was denoted as TS.

2.2.2 Preparation of TiO_2-SiO_2 (TS) mixed with acrylic. 0.5 g of TiO_2-SiO_2 nanoparticles (1mol %) were directly added to the 25 ml distilled water and dispersed using ultrasonic irradiating for 20 min. The dispersions were then added to the 25 g emulsion resins and 0.5 g coalescing agent, mixed for further 60 min at a constant rate [15]. Coating samples with a wet thickness of 200 ± 5 μm were applied on the acrylic substrate using a film applicator. The applied films were then allowed to dry at 23 ± 2 °C for about a week. These films were further used to investigate the structural, morphological, and wetting properties by contact angle measurement [16].

2.2.3 Preparation of TiO_2-SiO_2-Polianiline. TiO_2-SiO_2 nanoparticles (1 g) were dispersed in 190 ml of HCl aqueous solution (1·21 M) containing 0·3 ml aniline under ultrasonic vibration to reduce the aggregation of TiO_2-SiO_2 nanoparticles. APS (0·863 g) was dissolved in 15 ml of HCl aqueous solution (1·21 M) and added drop-wise to the solution containing aniline monomer under constant stirring [17]. The mixture was allowed to polymerize under stirring for 5 h at room temperature. The Reaction mixture was filtered under vacuum and washed with ethanol and water and then dried at 60 °C for 24 h to obtain PANI emeraldine salt (ES) as a green powder.

2.2.4 Evaluation of Photo-catalysis. The photo-catalytic activity of TiO_2 and TiO_2-SiO_2-Pani nanocomposite was evaluated by determining the Rhodamine-B photo-degradation. The aqueous solutions of Rh-B were used as a model pollutant for investigation of the photo-catalytic activity of samples. 50 mg of TiO_2 or TiO_2-SiO_2-Pani photo-catalysts was added to 50 ml of Rh-B aqueous solution (10 $mg{\cdot}L^{-1}$) under magnetic stirring. The suspension was stirred in dark condition at room temperature (25 °C) for 30 min to achieve adsorption equilibrium for Rh-B[18]. Afterward, the suspension was illuminated by a 100 W linear Xenon lamp for 80 min. Samples are collected once every 20 min, and theconcentration of the Rh-B was determined using UV–Vis spectrometer. To quantify the concentration of rhodamine B, the UV–vis spectrometer was calibrated to a concentration range between 2 mg/L–10 mg/L at a wavelength of 554 nm, which corresponds to the absorption maximum of RB.

2.3. Characterization

The roughness of the films (coating) was characterized using atomic force microscope (Park System XE-70 Non-contact). The tapping mode was selected during analysis at an analytical range of 5mm. Transmission electron microscopy (TEM) study was carried out using a JEOL JEM-1400. The samples of TEM were prepared by dispersing the final nanoparticles in ethanol; the suspension was then dropped on carbon-copper grids.Fourier-transform infrared spectra (FT-IR) of the sampleswere recorded on a spectrometer (SHIMADZU) in the range of 400–4000 cm^{-1}. Measurements were performed in the transmission mode in spectroscopic grade KBr pellets for all the powders. The composites X-ray diffraction (XRD) patterns were performed in the range of 2θ = 10–90° on a SHIMADZU diffractometer, using Cu Ka radiation (l = 0.15406 nm) as an X-ray source, operated at 40 kVand 30mA. Crystallite size of anatase TiO_2 can be calculated from the line broadening by Scherer's formula.

3. Results and Discussion

3.1. X-Ray Diffraction (XRD)

Figure 1 shows the XRD patterns of SiO_2, TiO_2, nanosized TiO_2-SiO_2(with varying weight percentage of SiO_2), and TiO_2-SiO_2-Pani (TSP) samples. All were calcined under the same condition in air at 500°C for 3 h. TiO_2 and nanosized TiO_2-SiO_2 showed a clear anatase-type crystal structure. The sharp peak and strong intensities indicated that crystallization was present, and the higher the percentage of TiO_2 gave the higher intensity of peaks. The peak associated with SiO_2 show that it is an amorphous form of silica. There is no difference between the XRD pattern of TiO_2-SiO_2-Pani and XRD pattern of pristine TiO_2 nanoparticles, which reveals deposition of PANI on the surface of TiO_2-SiO_2 nanocomposites which has no effect on the crystallinity of TiO_2 nanoparticles. Therefore, the polymorph of TiO_2 in the TiO_2-SiO_2-Pani nanocomposite is still anatase [18]. The crystallite sizes of Titania, TS, and TSP are listed in table 1.

Table 1. The crystallite size and BET surface area of TiO_2, TS, and TSP

Composite/nanoparticle	Crystallite size (nm)	Surface area (BET) m^2/g
TiO_2	19.57	98.0
TS-20	16.17	113.2
TS-25	15.53	184.4
TS-30	13.90	194.4
TS-35	11.90	214.8
TSP	21.13	-

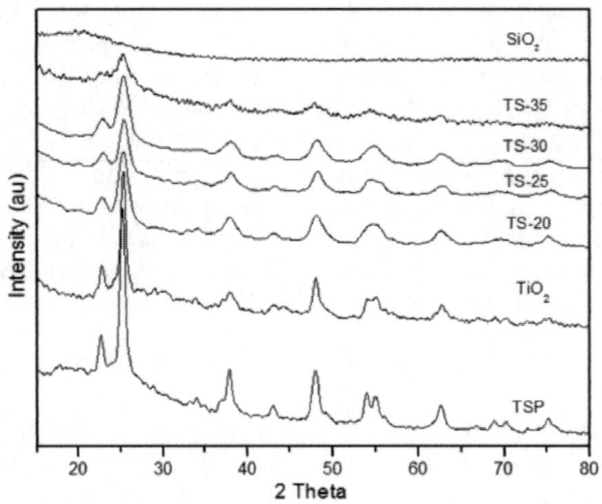

Figure 1. The XRD patterns of TiO_2, TS, and TSP

3.2. *Infrared Spectra*

The FT-IR spectra of pure titania and all the titania-silica samples calcined at 500 °C in the wavenumber range 4000–400 cm^{-1} is presented in figure 2. IR of pure silica was also taken for comparison. The absorption bands at around 3,440 and 1,635 cm^{-1} observed in all spectra are attributed to the stretching mode of water and hydroxyl group [6]. The peak around 1,080 cm^{-1} and 455 cm^{-1} in the case of silica added sample correspond to the asymmetric stretching vibration of Si–O–Si [19]. The peak at 950 cm^{-1} corresponding to the vibration of Si–O–Ti confirmed the formation of Si–O–Ti inorganic network between SiO_2 and TiO_2 [10, 19] in TS-30 sample. After TS composites mixed with acrylic paint, the general structure of the paint did not change sharply. This can be seen in figure 3.

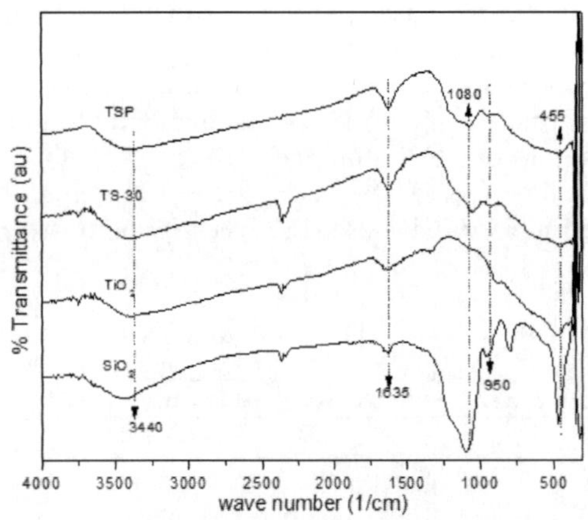

Figure 2. Spectra of TiO_2, SiO_2, TS, and TSP.

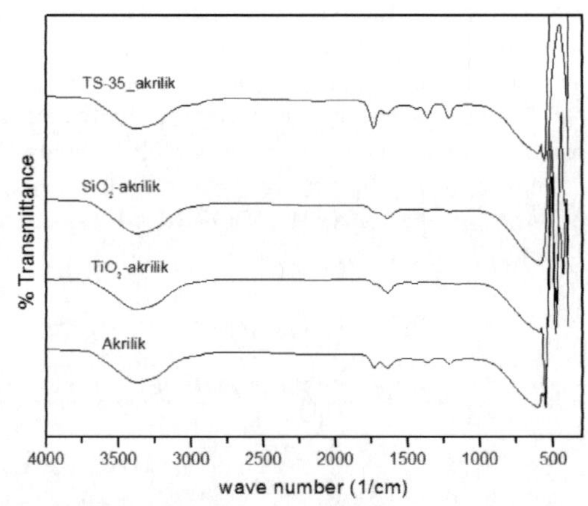

Figure 3. Spectra of acrylic, TiO_2-acrylic, SiO_2-acrylic, and TS-35-acrylic

3.3. *Contact Angle Measurement*

Static Contact Angle measurements performed on a solid surface without irradiation. Results of static contact angle measurements are shown in table 2 below. For the measurement of static water contact angle, acrylic paint-TiO_2-SiO_2 particles dispersed in the acrylic solvent and distributed on an acrylic substrate with a lateral size of 2 cm × 2 cm. A droplet of distilled water with a volume of 3 μL wasdeposited on the sample surface. For each sample, the water contact angle was measured twice after dropping and the averagevalue was recorded [20].

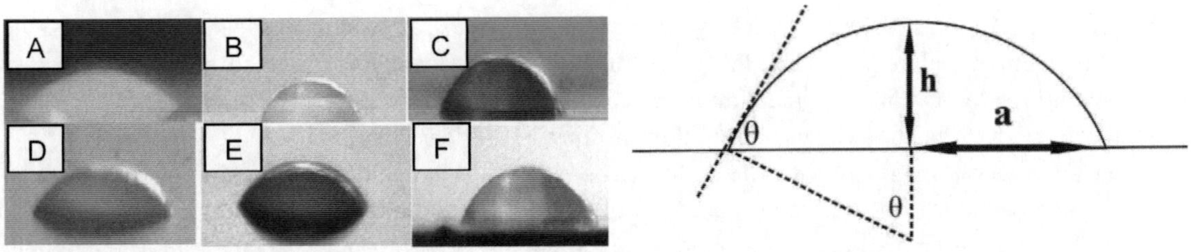

Figure 4. Water droplets photo

The amount of nanoparticle or nanocomposite (TiO_2, SiO_2, and TiO_2-SiO_2) in the acrylic paint is 2% by weight. The influence of the amount of composites in the acrylic paint has not been presented in this paper. The results of water contact angle measurements showed that the surface paint had not shown the hydrophobic nature because the contact angle is smaller than 90. The variation of mixed TiO_2- SiO_2 also influence to the contact angle magnification.Coating the surface with a TiO_2-SiO_2 modified acrylic paint has not given satisfactory results as the coating by PDMS reported by Kim, *et al.*, [20]. Kim, *et al.*, [20] report that PDMS-coating was resistant toward photo-catalytic decomposition: the very high water contact angle of the PDMS-coated TiO_2 was sustained with time, and no CO_2 evolution occurred in the presence of UV light.

Table 2. Static Water Contact Angle of the composite

sample	Contact angle
Acrylic (A)	46.65
Acrylic-SiO_2 (B)	67.65
Acrylic-TiO_2 (C)	72.94
Acrylic-TS-35 (F)	**70.02**
Acrylic-TS-25 (E)	64.43
Acrylic-TS-15 (D)	60.68

3.4. Atomic Force Microscope

The roughness of the coating surface was characterized usingatomic force microscope (Park system XE-70 Non-contact). Hydrophobic surfaces with static contact angle θs> 150° are usually developed by the combination of surface roughing and lowering of surface energy [21]. Figure 5 shows the coating surface with the widestnanoroughness is figure 5b, and followed by figure 5c and 5a.

a. Acrylic b. Acrylic+TiO_2 c. Acrylic+ TS-35

Figure 5. AFM spectra of acrylic, acrylic+TiO_2, acrylic+TS-35

3.5. Photo-degradation of Rhodamine-B

In addition to developing nanocomposite polymers to obtain material that has good hydrophobic properties, this study also developed a nanocomposite polymer having qualified photo-catalytic properties. The material that absorbed light of a broad wavelength becomes more effective and efficient. A mixed oxide TiO_2/SiO_2 is a more efficient photo-catalyst for the degradation of methylene blue than TiO_2 alone [18]. The PANI/TiO_2 core-shell nanocomposite showed better photo-catalytic activity for Me-O photo-degradation in aqueous solution compared to the pristine TiO_2 nanoparticles under both UV and visible light illuminations [17]. This research studies the possibility of enhancement in the photo-catalytic performance of TiO_2-SiO_2 composite if modified with PANI. One of the possible reasons for the improvement of PANI/TiO_2 photo-catalytic activity under both UV and visible light illuminations is the reduced aggregation state of TiO_2 nanoparticles in PANI/TiO_2 nanocomposite [17]. Therefore, adsorption of dye molecules on the nanocomposite is higher than pristine TiO_2 nanoparticles. Both PANI and TiO_2 nanoparticles absorb photons at their interface under irradiation. Since the CB (conduction band) of TiO_2 and lowest unoccupied molecular orbital (LUMO) of PANI are well matched for the charge transfer, the electrons generated by PANI $\pi \rightarrow \pi*$transition under visible light illumination can be injected into the CB of TiO_2 and the electrons in the VB of TiO_2 are delivered to PANI layer [22]. Under UV light illumination, photo-generated holes in the VB of TiO_2 can be transferred directly to highest unoccupied molecular orbital (HOMO) of PANI, as the Valence Band of TiO_2 matches well with the HOMO of PANI [17].

Figure 6 shows the photo-catalytic performance of TiO_2-SiO_2-Pani (TSP) for decomposing Rhodamine-B in aqueous solution.

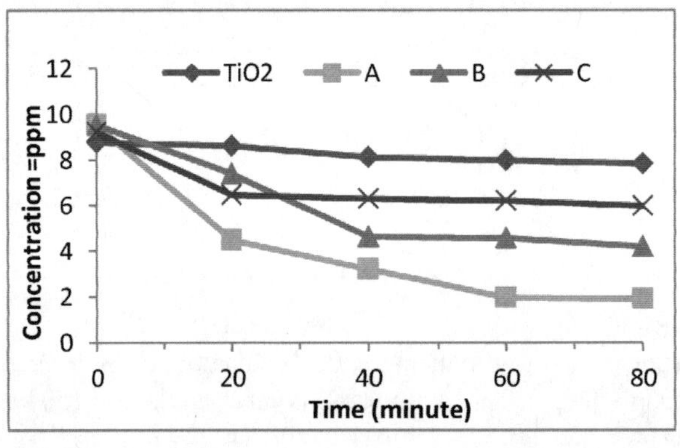

Figure 6. The photocatalytic performance of TiO_2 and TSP composites

Table 3. Percent degradation of Rh-B vs. Time

Time (minute)	TiO_2(%)	TSP-A(%)	TSP-B(%)	TSP-C(%)
0	9.23	2.25	2.19	5.39
20	10.89	54.25	24.26	33.60
40	15.96	67.05	52.51	35.42
60	17.52	79.89	53.33	36.16
80	18.86	80.41	56.97	38.76

IOP Conf. Series: Materials Science and Engineering **172** (2017) 012056 doi:10.1088/1757-899X/172/1/012056

4. Conclusions

SiO_2-TiO_2 mixed sols of different compositions was prepared through sonication technique route which was further used for coating with acrylic paint as a matrix binder. The crystallite size obtained showed that addition of silica had prevented the grain growth of TiO_2 considerably. The hydrophobic activity studies showed that the optimum composition for getting the largest contact angle is around 30 mol% SiO_2 added TiO_2. This optimum composition with acrylic paint as a matrix binder only produces a contact angle of about 70. The contact angle of this magnitude has not been enough for the hydrophobic criteria. In this study revealed that the modified acrylic paint with SiO_2-TiO_2 nanocomposite 2% by weight is not enough to produce a hydrophobic surface with a contact angle greater than 90. The poor results may be caused by the poor nanoscale roughness at the surface paint and by the improper composition of TiO_2 -SiO_2 nanocomposite with acrylic paint. In the application as a photo-catalyst, TiO_2-SiO_2 nanocomposite modified by PANI showed photo-catalytic performance in the visible light region. The optimum photo-catalytic performance obtained at the addition of aniline 0.1mL for each gram of TiO_2-SiO_2 (TSP-A) which decompose Rhodamine-B by 80.41%.

References

[1] Mao L Q, Li Q L, Dang H X and Zhang Z J 2005 Synthesis of nano-crystalline TiO_2 with high photo-activity and large specific surface area by sol–gel method *Mater Res Bull* **40** 201

[2] Sunada K, Kikuchi Y, Hashimoto K and Fujishima A 1998 Bactericidal and detoxification effects of TiO_2 thin film photocatalysts *Environ. Sci. Technol.* **32** 726

[3] Fujishima A and Honda K 1972 Electrochemical photolysis of water at a semiconductor electrode *Nature* **238** 37

[4] Jung K Y and Park S B 2004 Photoactivity of SiO_2/TiO_2 and $ZrO_2/$ TiO_2 mixed oxides prepared by sol–gel method *Mater. Lett.* **58** 2897

[5] Parkin I P and Palgrave R G 2005 Self-cleaning coatings *J. Mater. Chem.* **15** 1689

[6] Zhang X, Zhang F and Chan K Y 2005 Synthesis of titania-silica mixed oxide mesoporous materials, characterization and photocatalytic properties *Appl. Catal. A. Gen* **284** 193

[7] Aguado J, van Grieken R, Lopez-Munoz M J, Marugan J 2006 A comprehensive study of the synthesis, characterization and activity of TiO_2 and mixed TiO_2/SiO_2photocatalysts*Appl. Catal. A Gen.* **312**202

[8] Parkin, I and Palgrave R 2005 Self-cleaning coatings *J. Mater. Chem.* **15** 1689.

[9] Marmur A 2004 The lotus effect: Superhydrophobicity and metastability *Langmuir* **20** 3517

[10] Smitha V S, Manjumol KA, Baiju K V, Ghosh S, Perumal P and Warrier K G K 2010 Sol-gel route to synthesize titania-silica nano-precursors for photoactive particulates and coatings *J. Sol-Gel Sci.Technol.***54**, 203

[11] Yuranova T, Sarria, Jardim W, Rengifo J, Pulgarin C 2007 Photocatalytic discoloration of organic compounds on outdoor building cement panels modified by photoactive coatings *J. Photochem. Photobiol. A: Chem.* **188** 334

[12] Nakajima A, Hashimoto K, Watanabe T, Takai K, Yamauchi G, Fujishima A. 2000. Transparent superhydrophobic thin film with self-cleaning properties *Langmuir***16** 7044.

[13] Zhou M, Yu J and Cheng B 2006 Effects of Fe-doping on the photocatalytic activity of mesoporous TiO_2 powders prepared by an ultrasonic method *J. Hazard. Mater.* **137** 1838

[14] Venckatesh R, Balachandaran K and Sivaraj R 2012 Synthesis and characterization of nano TiO_2-SiO_2: PVA composite-a novel route *Int. Nano Lett.***2** 15

[15] Mirabedini A, Mirabedini S M, Babalou A A and Pazokifard S 2011 Synthesis, characterization and enhanced photocatalytic activity of TiO_2/SiO_2 nanocomposite in an aqueous solution and acrylic-based coatings *Prog. Org. Coat.***72** 453

[16] Zhang H, Lamb R and Lewis J 2005 Engineering nanoscale roughness on hydrophobic surface preliminary assessment of fouling behavior*Sci. Tech. Adv. Mater.***6**236

[17] Olad A, Behboudi S and Entezami, A A 2012 Preparation, characterization and photocatalytic activity of TiO_2/polyaniline core-shell nanocomposite *Bull. Mater. Sci.* **35**801

[18] Sirimahachai U, Ndiege N, Chandrasekharan R, Wongnawa S and Shannon M A 2010 Nanosized TiO_2 particles decorated on SiO_2 spheres (TiO_2/SiO_2): Synthesis and photocatalytic activities *J. Sol-Gel Sci.Technol.***56** 53
[19] Xu GQ, Zheng ZX, Wu Y C, Feng N 2009 Effect of silica on the microstructure and photocatalytic properties of titania*Ceram. Int.***35** 1
[20] Kim K D, Seo H O, Sim C W, Jeong M G, Kim Y D and Lim D C 2013 Preparation of highly stable superhydrophobicTiO_2 surfaces with completely suppressed photocatalytic activity *Prog. Org. Coat.***76** 596
[21] Genzer J and Efimenko K 2000 Creating long-lived superhydrophobic polymer surfaces through mechanically assembled monolayers *Science***290** 2130
[22] Min S, Wang F and Han Y 2007 An investigation on synthesis and photocatalytic activity of polyaniline sensitized nanocrystalline TiO_2 composites*J. Mater. Sci.***42** 9966

Acknowledgments
We are thankful to Laboratory of Instrumentation, Department of Chemistry, UNNES and Laboratory of Physical Chemistry, Department of Chemistry, UNNES, for providing all facilities to direct this work. The authors are grateful to the Directorate General of higher Education, for providing financial support.

IOP Conf. Series: Materials Science and Engineering **172** (2017) 012057 doi:10.1088/1757-899X/172/1/012057

Neutron radiography and tomography investigations on the porosity of the as-cast titanium femoral stem

Sutiyoko[1,2*], Suyitno[1], M Mahardika[1], F Akbar[3], Juliani[3], Setiawan[3] and Baroto[3]

[1]Department of Mechanical and Industrial Engineering, Gadjah Mada University, Yogyakarta, Indonesia
[2]Foundry Department, Manufacturing Polytechnic of Ceper, Klaten, Indonesia
[3]National Nuclear Energy Agency (BATAN), South Tangerang, Indonesia

Email : yoko_styk@yahoo.com; suyitno@ugm.ac.id; muslim_mahardika@ugm.ac.id; ozie@batan.go.id

Abstract. Gating system design in the centrifugal casting is one of the factors that influence the porosity of the femoral stem. The objective of this research is to analysis the porosity in the as-cast titanium femoral stem by neutron radiography and tomography. Three gating system designs which in three-ingates, four-ingates, and four-ingates by inversed position of the femoral stem were casted by a vertical centrifugal casting in investment mold. The porosity distribution in the titanium femoral stem was investigated by the neutron radiography film and followed by neutron tomography. The results indicate that there are large internal porosity in the subsurface region on both of the four-ingates designs but only small internal porosity on the three-ingates design. The large porosity also takes place in largest part of the femoral stem at all of the gating system designs. The product may be rejected due to the sub-surface porosity. The three-ingates design has the smallest risk on the reject product.

1. Introduction

Femoral stem is a part of hip prostheses, inserted in femur. It can be made from metallic biomaterials such as stainless steel 316L, Ti and Ti alloy, and CoCr alloys [1]. Titanium and Ti alloys have some excellent properties such low weight, high resistant to corrosion, good biocompatibility, as well as mechanical properties that are similar to gold alloys, but at a lower cost [2]. Several methods can be used to manufacture the femoral stem such as rapid prototyping [3] and casting [4]. Each method has advantages and limitations in process or in the final product.

The investment casting process is widely used to manufacture near net-shape component [5]. Therefore, the machining process can be minimized by this method and a complex shape can be manufactured in relatively low cost [6]. One of the serious problems in the casting is the porosity. Although the titanium femoral stem was casted in vacuum setting, the porosity may still occurs in this product. The porosity may be caused by shrinkage volume [7] and entrapped gas [8]. This defect reduces the mechanical properties and makes imperfect shape of the femoral stem.

There are several methods to determine the internal porosity in the materials. The Archimedes method usually used in the quantity of the porosity based on the density of material in air and liquid [9–11] This method is non-destructive test, quick and relatively inexpensive. It can only be used to determine a global density value relative to the reference fluid while the position of the porosity cannot be detected. A gas pycnometry method determines the porosity also in the density-based testing. The part density in an absolute sense is computed by measuring part of mass and part of volume separately. The gas displacement determines the part volume[10]. A thermography method detects the porosity of an electrode layer. This method based on heating and sensing the specimen by the irradiation power in the infra-red spectrum range [12]. An innovative non-destruction test measuring technique has been recently utilized for internal porosity detections. Computer tomography thanks to its capability for providing a completely analysis of shape, size, volume and distribution of pores within entire analyzed volume [13].

Neutron tomography detects the porosity of material by non-destruction test (NDT) method. Comparison between the X-ray and neutron tomography, the neutron is much better to penetrate the thick metal walls than the X-ray [14]. Neutron tomography is beneficial for deep investigation of explosive materials and narcotic hidden inside metals or materials with high atomic numbers [15]. Neutron film radiography results in two-dimensional visualization, while neutron tomography (NT) is a powerful method for three-dimensional visualization of materials. The different of neutron source as nuclear reactors, spallation sources, standard particles accelerators makes the facilities of the experimental and safety condition very different from each other [16]. A particular facility for tomography consists of scintillator screen, rotating table, charge coupled devices (CCD) digital video camera, and plane mirror [17]. The neutron tomography method had proved useful in many different areas such as geology, biology, cultural heritages, sciences and industrials application. A neutron image for radiography test, two-dimensional projections of the internal surface of the objects is acquired by films and track detectors. Processing of neutron image occurs by irradiation the object in a polychromatic neutron beam while a converter screen changes the transmitted neutron intensity into ionizing radiation. The final result is a cross-section image of the sample [18].

The neutron tomography is useful to obtain the detail inside the material such as the volume distribution of the biotic-muscovite grains inside the rock [19], and the porosity of titanium aerospace part [20]. The porosity usually becomes a common defect in the casting that cannot be detected by naked eye. The neutron radiography and tomography can be promoted as one of the methods to investigate the internal porosity of the casting result in more detail. The objective of this research is to obtain a detail of the internal porosity of the as-cast titanium femoral stem by the neutron radiography and tomography.

2. Experimental

2.1. Material and casting design

The commercially pure titanium femoral stems of this investigation were casted by the vertical centrifugal casting in the three designs of gating system as shown in figure 1. The first design has three ingates at cross section area: 78.5; 157.0; 128.5 mm^2, that the biggest cross section area was designed in the middle of the femoral stem. The second design was equal to the first design but the largest cross section area was divided into two similar ingates. In other word, it has four ingates. The third design was similar with second design but the femoral stem position was inversed. The as-cast femoral stems were removed from the gating system as shown in figure 2.

2.2. Experimental procedure

The neutron radiography and tomography facilities were installed at a 15 MW nuclear research reactor in Indonesia. The neutron radiography in this research used radiography film firstly to investigate the porosity quickly in 2D visualization. The main apparatus in the neutron tomography consists of scintillator screen Li6ZnS 200 μm, CCD camera 1024 × 1024 pixel resolution, TiO_2 high quality mirror, rotating table a 35-kg maximum load, rotary table control device and computer set. The apparatus schematic of the neutron tomography is shown in figure 3.

IOP Conf. Series: Materials Science and Engineering **172** (2017) 012057 doi:10.1088/1757-899X/172/1/012057

(a) (b) (c)

Figure 1. Three gating system designs of the femoral stem in the vertical centrifugal casting (a) first design (b) second design (c) third design

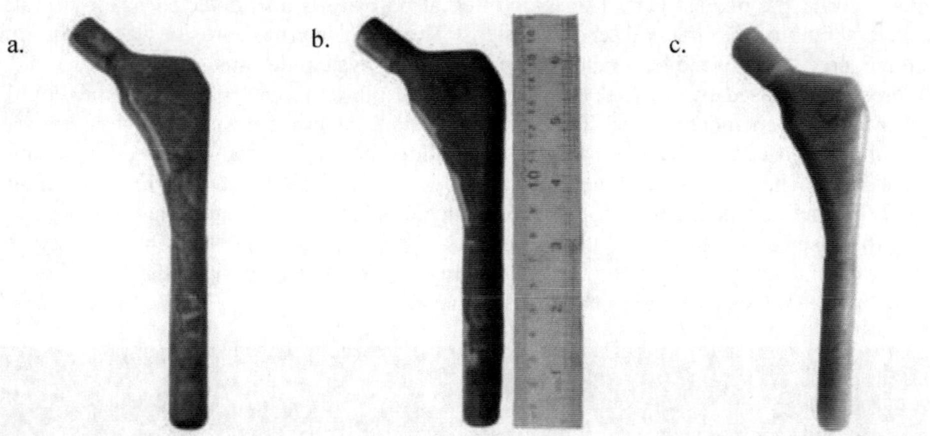

Figure 2. As-cast femoral stems without the gating system (a) first design (b) second design (c) third design

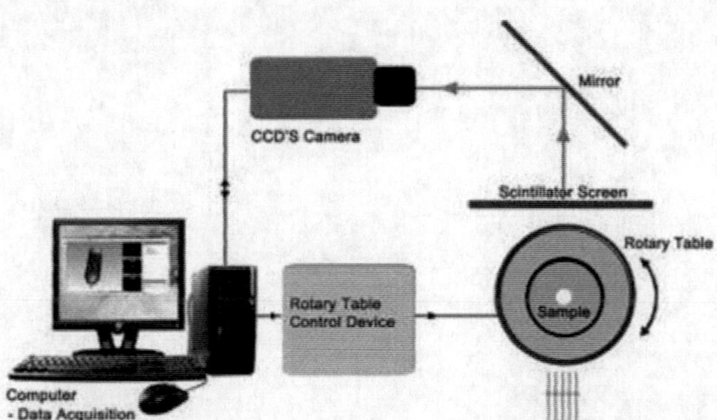

Figure 3. Apparatus schematic of neutron tomography [21]

In the neutron tomography, the femoral stem was rotated in 360 ° and bombarded by neutron beam to obtain 400 images. The images were taken using a CCD camera during 200 s/ image. These images were reconstructed by Octopus 8.8 to obtain the cross section images and 3D visualization was carried out using VG Studio Max 2.2. The porosity was analyzed based on the results of neutron radiography film and neutron tomography in the 3D visualization.

3. Result and discussion

The neutron radiography film of the three gating system designs are shown in figure 4. The first design has small porosity in near-surface region of the femoral stem but in the second and third designs have several large porosities also in near-surface region. These porosities take place at the femoral stem on the ingate location. The internal porosity in the second design is detected obscurely at the middle of the femoral stem as referred by the arrow. However, it cannot be detected in the first and third designs. The neutron tomography results of the femoral are shown in figure 5, while the middle slice longitudinally is shown in Fig.6. The colour bar beside of the femoral stem represents volumes of the internal porosity as 0-50 mm^3. According to figure 5 and figure 6, the large internal porosity in all of designs occur in the biggest part. The large internal porosity is also detected in the middle of the femoral stem at both of the first and second designs. The small internal porosity was spread out in the middle of the largest part of the femoral stem at all of gating system designs.

The porosity is caused by internal shrinkage [7] and the entrapped gas [8] in the liquid metal. Although the casting chamber was set in vacuum condition, the cover gas such as the inert argon gas which is commonly used in Ti casting process, may follow the liquid metal from melting chamber into the mold cavity. The gas also can be generated by metallurgical reactions [7]. According to the neutron radiography results, the porosity where the ingate is located, indicates that the ingate solidified faster than the casting. The internal shrinkage occurs in the casting and results the porosity. The first design has the smallest porosity near the ingate. In the first ingate design, the solidification time between the ingate and the casting is smaller than other designs.

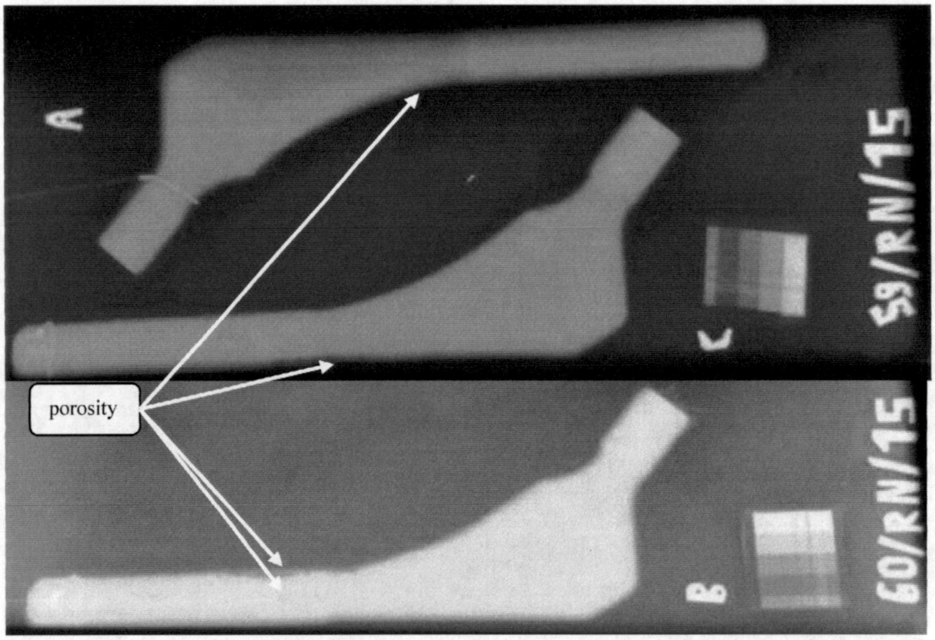

Figure 4. Neutron radiography film from the femoral stems; A is first design, B is second design, and C is third design

Based on the results of neutron tomography, the porosity volume about 50 mm^3 takes place in the largest part of the femoral stem at all of the ingate designs. It also occurs in the middle of the femoral stem at both the first and second designs. It shows that the largest part of the femoral stem solidifies slower than its surrounding, results the shrinkage porosity. It can be controlled using an engineering approach based on modulus. Modulus is ratio of the volume to surface area or crudely the heat to be removed by the removal rate [22].

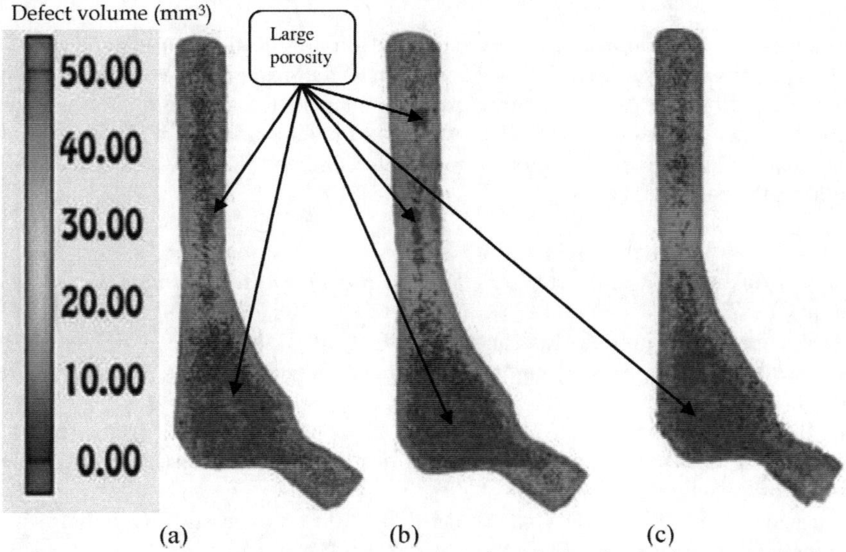

Figure 5. Neutron tomography transparently of the femoral stems (a) first design (b) second design (c) third design

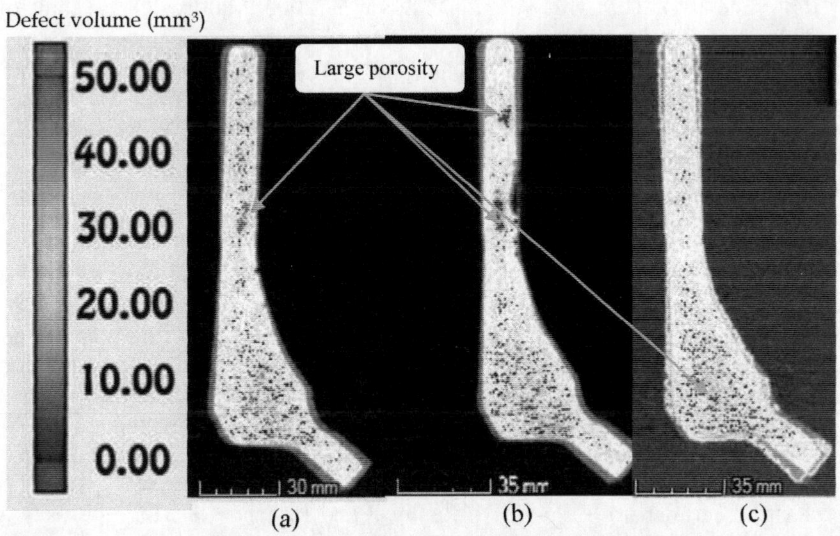

Figure 6. Middle slice longitudinally on the neutron tomography of the femoral stems (a) first design (b) second design (c) third design

Uniformity of liquid metal solidification is the key issue in controlling the position of defect [23]. The first design has two large porosities in middle of the femoral stem and a small porosity in subsurface region. The second design has three large porosities at the middle of the femoral stem and several large porosities in subsurface region. Both the several large porosities in sub-surface region and large porosities in the largest region occur at the third design. The large porosity in sub-surface region may cause a reject product at the casting. Based on the results of the three designs of gating system, the second design is the worse design while the first design may be promote as the better design.

4. Conclusion

The neutron radiography and tomography are the methods in non-destruction test (NDT) of materials. The neutron tomography represents the internal condition of the materials. The porosity of the femoral stem which was casted in three gating system designs is investigated to obtain the best gating system design. The large porosity distributed in the largest part of the femoral stem at all of the gating system designs while it occurs in sub-surface region at both the second and the third designs. The first design may be promote as the best design.

References

[1] Wong J Y, Bronzino D R and Peterson J D 2013 *Biomaterials principles and practices* (Boca Raton: CRC Press)
[2] Nogueira F, Fais L M G, Fonseca R G and Adabo G L 2010 The influence of short-heating-cycle investments on the quality of commercially pure titanium castings *J. Prosthet. Dent.* **104** 265–72
[3] Rahmati S, Farahmand F and Abbaszadeh F 2010 Application of rapid prototyping for development of custom – made orthopedics prostheses : An investigative study *Int. J. Adv. Des. Manuf. Technol.* **3** 11–6
[4] Ramos M and Simões J A 2009 CAD-CAM-RTV – lost-wax casting technology for medical implants *Rapid Prototyp. J.* **15** 211–5
[5] Vekariya R and Ravani R 2013 Investment Casting Process uzing fuzzy logic modelling *Int. J. Mech. Eng. Robot. Res.* **2** 232–41
[6] Atwood R C, Lee P D, Curtis R V and Di Silvio L 2005 Multiscale modeling of titanium *investment* cast dental prostheses *Mater. Sci. Eng. C* **25** 255–62
[7] Wu M, Sahm P R, Augthun M, Spiekermann H and Schädlich-Stubenrauch J 1999 Numerical *study* of porosity in titanium dental castings *J. Mater. Sci. Mater. Med.* **10** 519–25
[8] Magnitskiy A 1970 Casting properties of titanium alloys *Cleating-House Fed. Sci. Tech. Inf. Springf.* 5 and 92
[9] Jiang W, Fan Z, Chen X, Wang B and Wu H 2014 Combined effects of mechanical vibration and wall thickness on microstructure and mechanical properties of A356 aluminum alloy produced by expendable pattern shell casting *Mater. Sci. Eng. A* **619** 228–37
[10] Wits W W, Carmignato S, Zanini F and Vaneker T H J 2016 Porosity testing methods for the quality assessment of selective laser melted parts *CIRP Ann. - Manuf. Technol.* **65** 201–4
[11] Jiang W, Fan Z, Liu D and Wu H 2013 Influence of gas flowrate on filling ability and internal quality of A356 aluminum alloy castings fabricated using the expendable pattern shell casting with vacuum and low pressure *Int. J. Adv. Manuf. Technol.* **67** 2459–68
[12] Just P, Rost J, Echelmeyer T L, Ebert and Roscher M A 2016 A method to quantify coating thickness and porosity of electrodes for lithium-ion-batteries *Meas. J. Int. Meas. Confed.* **89** 312–5
[13] Khademzadeh S, Carmignato S, Parvin N, Zanini F and Bariani P F Micro porosity analysis in additive manufactured NiTi parts using micro computed tomography and electron microscopy *J.Mater Design* **90** 745–52
[14] Tudisco E, Hall S A, Charalampidou E M, Kardjilov N, Hilger A and Sone H 2015 Full-field Measurements of Strain Localisation in Sandstone by Neutron Tomography and 3D-Volumetric Digital Image Correlation *Phys. Procedia* **69** 509–15

[15] Abdullah M Z, Ashaari N A, and Aziz M A 2009 Development of a neutron tomography system for industrial applications *Meas. J. Int. Meas. Confed.* **42** 1017–26
[16] Abd el Bar W, Mahmoud I I, Konber H A and Mongy T 2015 Image reconstruction technique using projection data from neutron tomography system *Alexandria Eng. J.* **54** 1057–66
[17] Pugliesi R, Andrade M L G, Dias M S, Siqueira P T D and Pereira M A S 2015 Study of pixel damages in CCD cameras irradiated at the neutron tomography facility of IPEN-CNEN/SP *Nucl. Instruments Methods Phys. Res. Sect. A.* **804**, 59–63
[18] Pereira M A S, Schoueri R, Domienikan C, de Toledo F, Andrade M L G and Pugliesi R 2013 The neutron tomography facility of IPEN-CNEN/SP and its potential to investigate ceramic objects from the Brazilian cultural heritage *Appl. Radiat. Isot.* **75** 6–10
[19] Kichanov S E, Kozlenko D P, Ivankina T I, Rutkauskas A V, Lukin E V and Savenko B N 2015 The Neutron Tomography Studies of the Rocks from the Kola Superdeep Borehole," *Phys. Procedia* **69** 537–41
[20] du Plessis A and Rossouw P 2015 X-ray computed tomography of a titanium aerospace investment casting *Case Stud. Nondestruct. Test. Eval.* **3** 21–26
[21] Sutiarso, Akbar F, Gunawan and Indradjaja A 2012 Neutron tomography application in Batan Serpong for internal structure investigation of archeological object *Conference of science and materials technology* 181–4
[22] Monroe R 2005 Porosity in castings *AFS Trans.* **5** 1–28
[23] Lu S, Xiao F, Zhang S, Mao Y and Liao B 2014 Simulation study on the centrifugal casting wet-type cylinder liner based on ProCAST *Appl. Therm. Eng.***73** 512–21

Acknowledgment
This work was supported by Applied Research Grant in Research, Technology and High Education Ministry of Indonesia.

Synthesis of β-Calcium Pyrophosphate by sol-gel method

T Windarti[1], Taslimah[1], A Haris[1], Y Astuti[1] and A Darmawan[1]

[1]Research group of Material and Process, Chemistry Department, Diponegoro University, Indonesia

Email: tri.windarti@live.undip.ac.id

Abstract. Beta calcium pyrophosphate [β-CPP, β-$Ca_2P_2O_7$] can be used as bone graft extender in posterolateral lumbar fusion. In this research, β-CPP was synthesized by sol-gel method using phosphorus pentaoxide [P_2O_5] and calcium nitrate tetrahydrate [$Ca(NO_3)_2.4H_2O$] as phosphorus and calcium precursors. The reaction was carried out in ethanol medium with Ca/P ratio of 1.67. After 21 hours of reaction and 20 hours of drying at 80°C, white powder of amorphous calcium phosphate (ACP) was produced. Transformation of ACP to β-CPP was undertaken by firring at 400-800°C for 8 hours. Transformations of amorphous to microcrystalline, semicrystalline and crystalline structures occur at 400, 600 and 800°C, respectively. The β-CPP with the crystallite size of 61.71 nm, Ca/P ratio of 0.89 and Ca/O ratio of 0.21 was achieved by firing at 800°C. Morphology changes due to firing in which irregular shape of β-CPP at 400° changed to regular cuboid at 600 °C and above.

1. Introduction

Calcium phosphate compounds such as hydroxyapatite [HA, $Ca_{10}(PO_4)_6(OH)_2$] and β-tricalcium phosphate [β-TCP, $Ca_3(PO_4)_2$] are known as the biomaterial for bone filling due to their biocompatibility and osteoinductivity properties [1]. To achieve successful implantation in a difficult environment such as posterolateral lumbar fusion, osteoconductivity property must be combined with the porous structure of the biomaterial [2]. HA has a low resorption rate so that HA will remain in the body for a long time and hinders the constructive and reformative process as in a living bone tissue. While β-TCP even has good resorbability property, it has high biodegradability and might not persistent until the new bone formed [3]. β-calcium pyrophosphate [β-CPP, β-$Ca_2P_2O_7$] has biodegradability in between HA and β-TCP. Moreover β-CPP has all properties needed as biomaterial such as the same osteoinductivity degree with HA, bioactive and nontoxic. β-CPP was reported has an excellent fusion rate compared to other calcium phosphate ceramics. Many researchers suggested using β-CPP as an alternative material for bone filling [4,5,6]. On the other hands, only a few report of β-CPP synthesis method available.

The sol-gel method can be used in the synthesis of calcium phosphate compounds. The sol-gel method is known as a simple method to produce high homogeneity, nanosized, and semi-crystalline-phase product, also can be done at a relatively low temperature [7]. A study about the synthesis of HA with Ca/P molar ratio 1.5 – 1.67 showed that for Ca/P = 1.5 a secondary product as β-TCP was found. The amounts of β-TCP were decreased as the increase of Ca/P and HA was found as the only product for synthesis with Ca/P= 1.67 [8]. Synthesis of anhydrous dicalcium phosphate [DCPA, $CaHPO_4$] with

Ca/P= 1, ageing time of 24 hours and firing process at 600°C also produced β-TCP as a secondary product [9]. Liu et al [10] studied structure evolution of HA on sol-gel method and found that amorphous gel transformed into crystalline by firing at 400ºC and above. So it can be concluded that production of calcium phosphate by the sol-gel method is determined by several factors such as Ca/P molar ratio, ageing time, and firing temperature.

In this paper, β-CCP was synthesized by sol-gel method. As calcium and phosphorus precursors were used calcium nitrate tetrahydrate [$Ca(NO_3)_2$. $4H_2O$] and phosphorus pentoxide [P_2O_5], respectively.The reactions were carried out in ethanol medium with Ca/P ratio of 1.67. The reactions were conducted for 21 hours and continued by drying at 80°C for 20 hours to produce a white powder of amorphous calcium phosphate (ACP). Transformation of ACP to β-CPP was undertaken by firring at 400-800°C for 8 hours. The products were characterized by FTIR, XRD, and SEM-EDS to analyze calcium phosphate compound type, crystallite size and surface morphology.

2. Experimental procedures

Calcium nitrate tetrahydrate [$Ca(NO_3)_2$. $4H_2O$, Merck] and phosphorus pentaoxide [P_2O_5, Merck] were dissolved in ethanol [C_2H_5OH, Merck] to form 1.67 M of $Ca(NO_3)_2$. $4H_2O$ and 0.5 M of P_2O_5 solution.

One hundred ml of 0.5 M P_2O_5 solution was added slowly to 100 ml of 1.67 M $Ca(NO_3)_2$. $4H_2O$ solution. A gently mixing was conducted for 21 hours to form a stable gel phase. The Gel was washed with aquabidest and continued by drying process to produce amorphous calcium phosphate (ACP). The drying process of gel phase was proceed at 80°C for 20 hours. The dry powder of ACP was transformed to crystalline phase by firing process for 8 hours at various temperature of 400, 600 and 800ºC.

Products characterization were done by FTIR (Shimadzu) and XRD (Shimadzu XRD-7000). The mean crystallite size (d) of the particle base on XRD data was calculated by Scherer equation [11].

$$d_{scherrer} = \frac{0.95\lambda}{\beta cos\theta} \quad (1)$$

Where λ is the wavelength, β is the full wide on the half-maximum (FWHM) of β-CPP line and θ is the diffraction angle. To confirm the surface and calcium to phosphorus molar ratio (Ca/P) of products, SEM-EDS (JSM-6510LA) was used.

3. Results and discussion

There are several stages in the synthesis of β-CPP by sol-gel method. At first, calcium and phosphorus precursors were dissolved in ethanol separately. Dissolving precursors in ethanol would produce sol, a nano-sized colloidal suspensions of solid in liquid. Calcium nitrate tetrahydrate [$Ca(NO_3)_2$. $4H_2O$] would be ionized in ethanol and formed Ca^{2+}, NO_3^- and H_2O. When colloidal suspension of phosphorus pentaoxide [P_2O_5] was slowly mixed with calcium precursor solution, P_2O_5 would react with H_2O and produce H_3PO_4. Further ionization of H_3PO_4 would produce H^+, $H_2PO_4^-$, HPO_4^{2-} and PO_4^{3-} ions that depend on pH of solution. The slow reaction of those ions and Ca^{2+} in ethanol media would produce calcium phosphate particles. Then the particles would agglomerate each other and produce particulate in the form of a gel, a colloidal suspensions of liquid in solid. This stage was continued with drying process to slowly remove solvent from the gel and produced amorphous calcium phosphate (ACP). The last stage was firing the ACP to transform amorphous phase of calcium phosphate into crystalline structure.

FTIR spectra of ACP can be seen in figure 1. The spectra show specific peaks of calcium phosphate compounds. A peak at 956.69 cm^{-1} is the vibration adsorption v_1 of PO_4. Vibration v_3 and v_4 appear at wave number of 1041.56 and 570.93 cm^{-1}, respectively. A peak at 1041.56 cm^{-1} corresponds to asymmetric stretching of P-O. Meanwhile, the band at 956.69 cm^{-1} corresponds to symmetric P-O stretching of PO_4^{3-} ion. A broadband at 3448.72 cm^{-1} and a band at 1635.64 cm^{-1} indicate the presence of water. The spectra show the appearance of CO_3 at wave number of 1519.91 cm^{-1}.

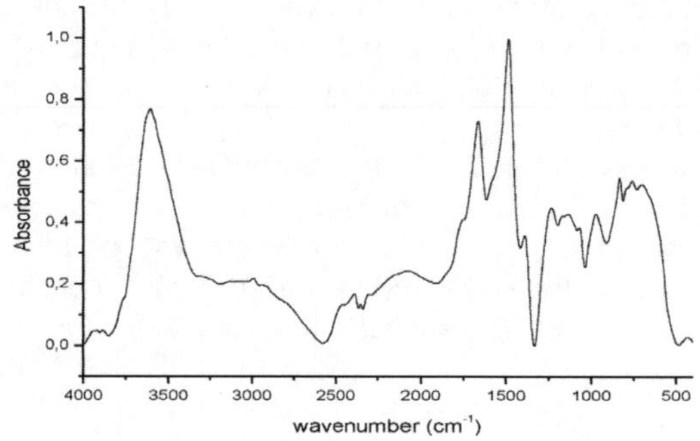

Figure 1. FTIR spectra of ACP

The reactions involved in the formation of β-CPP during the sol-gel procedures can be estimated as follows:

I. Solvation

$$Ca(NO_3)_2 . 4H_2O \longrightarrow Ca^{2+} + 2\ NO_3^- + 4\ H_2O$$
$$P_2O_5 + 3\ H_2O \longrightarrow 2\ H_3PO_4$$
$$H_3PO_4 \rightleftharpoons H^+ + H_2PO_4^-$$
$$H_2PO_4^- \rightleftharpoons H^+ + HPO_4^{2-}$$

II. Gelation

$$Ca^{2+} + 2\ H_2PO_4^- \longrightarrow Ca(H_2PO_4)_2$$
$$Ca^{2+} + HPO_4^{2-} \longrightarrow CaHPO_4$$

III. Drying

$$Gel \longrightarrow ACP$$

IV. Firing (600 and 800°)

$$2\ CaHPO_4 \longrightarrow Ca_2P_2O_7 + H_2O$$

The X-rays diffraction analysis was performed using the CuKα source operated at 30 kV and 30 mA. The diffractograms were recorded at 2θ = 5 – 90° with a scan speed of 2°/min and a step angle of 0.02°. XRD diffractogram of ACP with and without washing with aquabidest indicated that washing process was razed ACP crystal structure (figure 2).

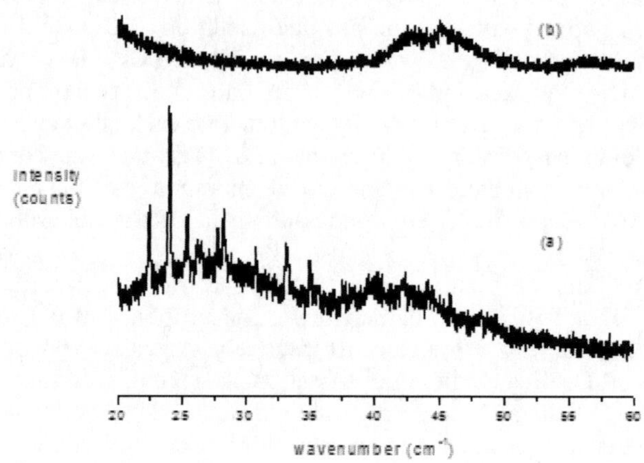

Figure 2. Diffractogram ACP with (a) and without (b) washing with aquabidest.

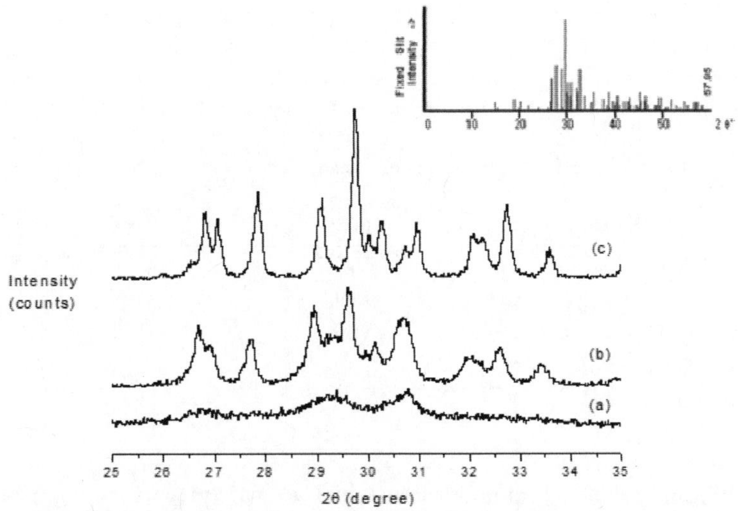

Figure 3. Diffractogram of products after firing process at 400°C (a), 600°C (b) and 800°C (c) Inset JCPDS card no 09-0346

Figure 3 showed the diffractogram of firing product at 400, 600 and 800 °C for 8 hours. The high peaks were occurred at 2θ = 25 - 35°. Transformations of amorphous to microcrystalline, semicrystalline and crystalline structures occur at 400, 600 and 800°C, respectively. The pattern of firing at 600 and 800 °C was similar. This signifies that the β-CPP crystal structure starts to form at 600°C but only diffractogram of firing at 800°C can be compared to JCPDS data. The strong peaks at 2θ = 27.84, 29.07, 29.75 and 33.59° indicated that product was β-calcium pyrophosphate [β-CPP, β-$Ca_2P_2O_7$] base on JCPDS card no. 09-0346. Those angles derived from X-ray diffraction of (202), (203), (008) and (205) β-CPP crystal planes. The β-CPP crystallite size was 61.71 nm.

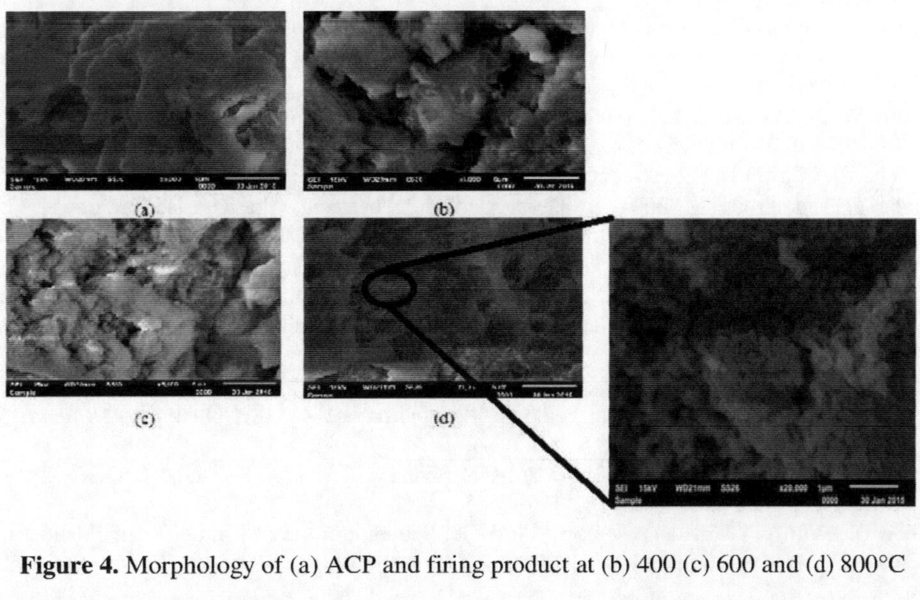

Figure 4. Morphology of (a) ACP and firing product at (b) 400 (c) 600 and (d) 800°C

IOP Conf. Series: Materials Science and Engineering **172** (2017) 012058 doi:10.1088/1757-899X/172/1/012058

Morphology changed due to firing, in which irregular shape of β-CPP at 400° changed to regular cuboid at 600 °C and above (figure 4). The Ca/P and Ca/O ratio of products decreased as firing temperature increases, from 1.06 to 0.88 for Ca/P ratio and 0.23 to 1.21 for Ca/O ratio (table 1). Those numbers are closed to Ca/P ratio and Ca/O ratio of β-CPP that are Ca/P= 1 and Ca/O=0.29.

Table 1. The Ca/P and Ca/O Ratio of ACP and Firing Products

Sample	Ca/P	Ca/O
ACP	1.06	0.23
After Firing 400°C	0.99	0.22
After Firing 600°C	0.89	0.21
After Firing 800°C	0.89	0.21

4. Conclusions

Synthesis of β-CPP can be done by sol-gel method with Ca/P ratio precursors of 1.67, reaction time of 21 hours, drying process at 80°C for 20 hours and firing temperature at 800°C for 8 hours. The β-CPP has crystallite size of 61.71 nm, Ca/P ratio of 0.89 and Ca/O ratio of 0.21. Morphology changes due to firing in which irregular shape of β-CPP at 400° changed to regular cuboid at 600 °C and above.

Acknowledgement

This work was supported by The Indonesia Ministry of Education and Culture (DIPA: 023.04.2.673453/2015).

References

[1] Zhang J, Liu W, Schnitzler V, Tancret F and Bouler J-M 2014 Calcium Phosphate Cement For Bone Substitution: Chemistry, Handling and Mechanical Properties *Acta Biomaterialia* **10** 1035-49

[2] Lee J H, Chang B-S, Jeung U-O, Park K-W, Kim M-S and Lee C-K 2011 The First Clinical Trial of Beta-Calcium Pyrophosphate as A Novel Bone Graft Extender In Instrumented Posterolateral Lumbar Fusion *Clinics in Othopedic Surgery* **3** 238-44

[3] Arcos D, Boccaccini A R, Bohner M, Diez-Perez A, Epple M, Gomez-Barrena E, Harrera A, Planell J A, Rodriguez-Manaz L and Vallet-Regi M 2014 The Relevance of Biomaterials to The Prevention and Treatment of Osteoporosis *Acta Biomateriallia* **10** 1793-05

[4] Tadic D and Epple M 2004 A Though Physicochemical Characterisation of 14 Calcium Phosphate-Based Bone Substitution Materials In Comparison to Natural Bone *Biomaterials* **25** 987-94

[5] Ryu H-S, Youn H-J, Hong K S, Chang B-S, Lee C-K and Chung S-S 2002 An Improvement in Sintering Property of β-Tricalcium Phosphate by Addition of Calcium Pyrophosphate *Biomaterials* **23** 909-14

[6] Lin F-H, Lin C-C, Lu C-M, Liu H-W, Sun J-S and Wang C-Y 1995 Mechanical Properties and Histological Evaluation of Sintered β-$Ca_2P_2O_7$ with $Na_4P_2O_7.10H_2O$ Addition *Biomaterials* **16** 793-02

[7] Brinker C J and Scherer G W 1990 Sol-Gel Science: The Physics and Chemistry of Sol-Gel Processing (Academic Press Inc. San Diego)

[8] Chen J, Wang Y 2011 A Simple Sol-Gel Technique for Synthesis of Nanostructured Hydroxyapatite, Tricalcium Phosphate and Biphasic Powders *Materials Letters* **65** 1923-26

[9] Eshtiagh-Hosseini H, Houssaindokht M R 2008 Preparation of Anhydrous Dicalcium Phosphate, DCPA, Through Sol–Gel Process, Identification and Phase Transformation Evaluation *Journal of Non-Crystalline Solids* **354** 3854-57

[10] Liu D, Yang M Q 2002 Structural Evolution of Sol–Gel-Derived Hydroxyapatite *Biomaterials,* **23** 1679-87

[11] Volkmer T M, Lengler F, Barreiro O, Sausa V C and Dos Santos L A 2013 Novel Method for The Obtainment of Nanostructured Calcium Phosphate Cements: Synthesis, Mechanical Strength and Cytotoxicity *Powder Technology* **235** 599-05

Preparation and Antibacterial Activity of Inclusion Complex Sulfonatocalix[4]arene-Xanthone against *Escherichia Coli*

T Kusumaningsih, M Firdaus, H F Suparjo

Department of Chemistry, Faculty of Mathematics and Natural Sciences Sebelas Maret University, Ir. Sutami street no. 36A, Kentingan, Jebres, Surakarta

Email: triana_kusumaningsih@staff.uns.ac.id

Abstract. The preparation of inclusion complex sulfonatocalix[4]arene (SC[4]A) with xanthone and measurement of antibacterial activity against *Escherichia coli* have been performed. Analysis of the structure of inclusion complex between SC[4]A and xanthone was investigated by using Fourier Transform Infrared Spectroscopy (FT-IR) and H Nuclear Magnetic Resonance (H-NMR). The solubility of the product in water was examined. Surface area and porous determination were analyzed by Surface Area Analyzer (SAA). Antibacterial activity test was performed for 0, 3, 6, 9, 12, and 24 hours. The results showed that antibacterial activity of the complex of xanthone SC[4]An against *Escherichia coli* was higher than that of xanthone it self.

1. Introduction

Infectious diseases have been known since ancient times. Infection is a major cause of disease in the world, especially in tropical regions like Indonesia because tropical regions have a high humidity that the microbes can flourish. Infectious diseases of the respiratory tract and digestive tract disease is an infection that most often occurs. Infections of the digestive tract caused by bacteria such as *E. coli*. Excess *E.coli* in the human gut are pathogenic that can cause diseases in the digestive tract, can also cause diarrhea, leading to severe intestinal bleeding [1]. Therefore, required an antibacterial compound that can inhibit the proliferation of the bacteria.

Indonesian people already know and use plants to treat a wide range of infections caused by microbes. Quite many types of plants are used as medicinal plants, one plant as a medicine is the skin of the mangosteen fruit (Garcinia mangostana L.). The skin of the mangosteen fruit contains flavonoids, xanthone and derivatives, as well as tannin [2]. According to research [3] and [4], xanthone is the most important chemical content contained in the skin of the mangosteen fruit. Xanthone (figure 1) has antioxidant activity [5], anti-inflammatory and hypo-allergenic [6], antifungal [7], anticancer [8] and antibacterial [9]. Xanthone has a low solubility in water and therefore its solubility in water should be improved to enhance its antibacterial activities.

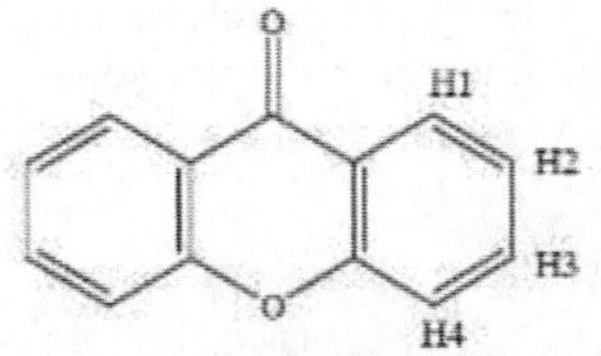

Figure 1. Chemical structures of xanthone

Sulfonatocalix[4]arene/SC[4]A were widely used in the pharmaceutical field because it is harmless and has good water solubility (up to 0.1 M) in aqueous media [10]. It was interesting for the inclusion complex of xanthone with SC[4]A to increase solubility xanthone. The Chemical structures of SC[4]A can be seen in figure 2. Synthesis of *p*-sulfonatocalix[n]arene (n = 4,6,8) can only be done by two methods, e.i sulfonation of *p*-Hidroxycalix[n]arene [11] and ipso-sulfonation method of *p*-tert-butilcalix[n]arene. Ipso-sulfonation method simpler than the method proposed by Shinkai but the reaction is often disrupted by the substitution incomplete and this method can only be used to get an instance of calix[8]arene because it has a low solubility [12].

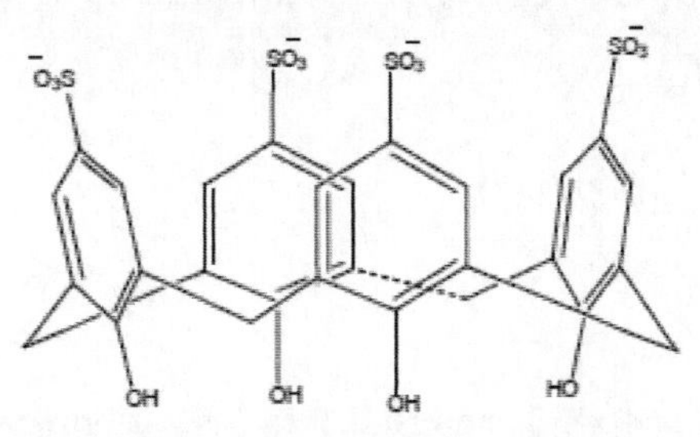

Figure 2. Chemical structures of SC[4]A

In this paper, the preparation of inclusion complex of xanthone with SC[4]A was discussed and the antibacterial activity of the inclusion complex SC[4]A-xanthone and xanthone against bacteria *E. coli* were examined.

2.Material and methods

2.1. Tools and chemicals
UV-Vis Spectrophotometer Double Beam (Perkin Elmer Lambda 25), infrared spectrometry (FTIR, Shimadzu, IR Prestige-21), nuclear magnetic resonance (^{1}H-NMR, Agilent NMR 400 MHz), Surface Area Analyzer (SAA, Nova 1200e). Bacterial cultures used was *E. coli* ATCC 25922. Bacterial media used was Luria Bertani Agar, Miller (Merck) and Luria Bertani Broth, Miller (Merck).

2.2. Methods

2.2.1. Preparation of inclusion complex SC[4]A-xanthone
SC[4]A was synthesized according to the literature procedures [13]. Inclusion complex SC[4]A-xanthone at 1:1 M ratio were obtained by mixing SC[4]A and xanthone in a solution mixture of methanol and water (v:v = 1:9) and stirred for 48 hours at room temperature [10]. The mixture was evaporated to remove methanol and water, then dried in vacuum to obtain SC[4]A-xanthone complex. Complex SC[4]A-xanthone was analyzed using FTIR, ^{1}H NMR and SAA.

2.2.2. Solubility Test

Inclusion complex SC[4]A-xanthone (50 mg) was dissolved into 5 mL of water and then was stirred for 1 hour at room temperature. The insoluble complex was separated by filtration and the filtrate was evaporated to dryness and the residue was weighed. The same treatment was also carried out on the xanthone, C[4]A, and SC[4]A.

2.2.3.Antibacterial test
LB Broth (1.25 g) was dissolved into 50 mL of distilled water, then sterilized by autoclaving at 121 °C for 15 minutes. A total of 1 ose bacteria were grown in LB Broth by incubating at 37°C for 24 hours at 150 rpm [14]. Xanthone (0.1 g) was dissolved in distilled water and added of LB Broth 2.5% (w/v), and then sterilized by autoclaving at 121 °C for 15 minutes. Formed mixture was cooled and added 0.5 mL of the bacterial inoculum 1%. Samples were incubated at 37°C with agitation of 150 rpm (Lahmer, *et. al*., 2012). Antibacterial activity was investigated by measuring absorbance sample solution for 0, 3, 6, 9, 12, and 24 hours using UV-Vis spectrophotometer at λmax 600 nm. The control is a solution of bacteria in the media. The inclusion complex SC[4]A-xanthone was also examined for antibacterial activity test using the above procedures. The % Inhibition of antibacterial activity was calculated by the Equation (1):

$$\% \text{ Inhibition} = [(At-Ao)-(Bt-Bo) / (At-Ao)] \times 100\% \quad \text{.................................(1)}$$

Ao = OD control at 0 hours At
= OD control at t hours
Bo = OD sample at 0 hours Bt
= OD sample at t hours

3. Results and discussion
Evaluation of the complex
The reaction between SC[4]A and xanthone resulted in a solid brownish white, has a melting point of > 350 °C and has a yield of 84.075%. Table 1 shows a summary of wave number (cm^{-1}) and the corresponding functional groups of C[4]A, SC[4]A, and SC[4]A-xanthone obtained from FT-IR spectra.

Table 1. The wave number (cm^{-1}) of functional groups for C[4]A, SC[4]A, and SC[4]A-xanthone

Functional groups	Wave number (cm^{-1})		
	C[4]A	SC[4]A	SC[4]A-xanthone
OH	3149	3417	3443
C-H *Stretch*	2868, 2933	2871, 2959	3016
C=C *Aromatic*	1465, 1593	1487, 1618	1459, 1607
Sulfonate (SO_3^-)	-	1049, 1122, 1202	1047, 1121, 1208
C-O-C *stretch*	-	-	1331, 1346
C=O	-	-	1656

The FTIR spectra of SC[4]A showed a new peak which appears at 1049, 1122, and 1202 cm^{-1} from sulfonate (SO3) [15]. In the spectra of SC[4]A-xanthone appears the new groups are carbonyl groups indicated that the vibration C=O stretch at 1656 cm^{-1} which is the characteristic absorption for carbonyl on xanthone [16] and 1331 and 1346 cm^{-1} (COC stretch) where both is the group of xanthone.

The ^{1}H NMR measurement of the xanthone, SC[4]A, and complex SC[4]A-xanthone were employed to obtain the interactions between xanthone with SC[4]A. ^{1}H NMR spectrum SC[4]A in figure 3 showed the presence of four peaks of protons that peak A is δH 7.67 ppm, peak B at δH 4.16 ppm, peak C at δH 3.71 ppm and peak D at δH 3.43 ppm. ^{1}H NMR spectra were obtained suspected that peak A is a proton from the group on the benzene ring (α-H), and peak B, C, D is a proton from the group CH_2 (β-H). Integration of peak A is 1.00 and peak B, C and D when the integration is summed were obtained 1.02 that the ratio is 1:1. The conformation of SC[4]A affects the results of ^{1}H NMR spectra were obtained. It is suspected that the most likely conformation of SC[4]A is cone and cone partial. The ^{1}H NMR spectrum of SC[4]A-xanthone also showed the four protons that peak A is δH 7.65 ppm, peak B at δH 4.10 ppm, peak C at δH 3.63 ppm and peak D at δH 3.40 ppm. The fourth peak is one proton signal of a benzene ring (peak A) and three proton signal of CH_2 (peak B, C and D). The ^{1}H NMR spectrum of both compounds SC[4]A and SC[4]A-xanthone seems like there is no difference peaks are visible, but when done magnification ^{1}H NMR spectrum SC[4]A-xanthone at the chemical shift δH 7-8 ppm it will show two small peaks that peak H2 appeared at 7.45 ppm and peak H3 at chemical shift of 7.85 ppm. Peak H2 and H3 in ^{1}H NMR spectrum of SC[4]A-xanthone is proton of CH aromatic in xanthone.

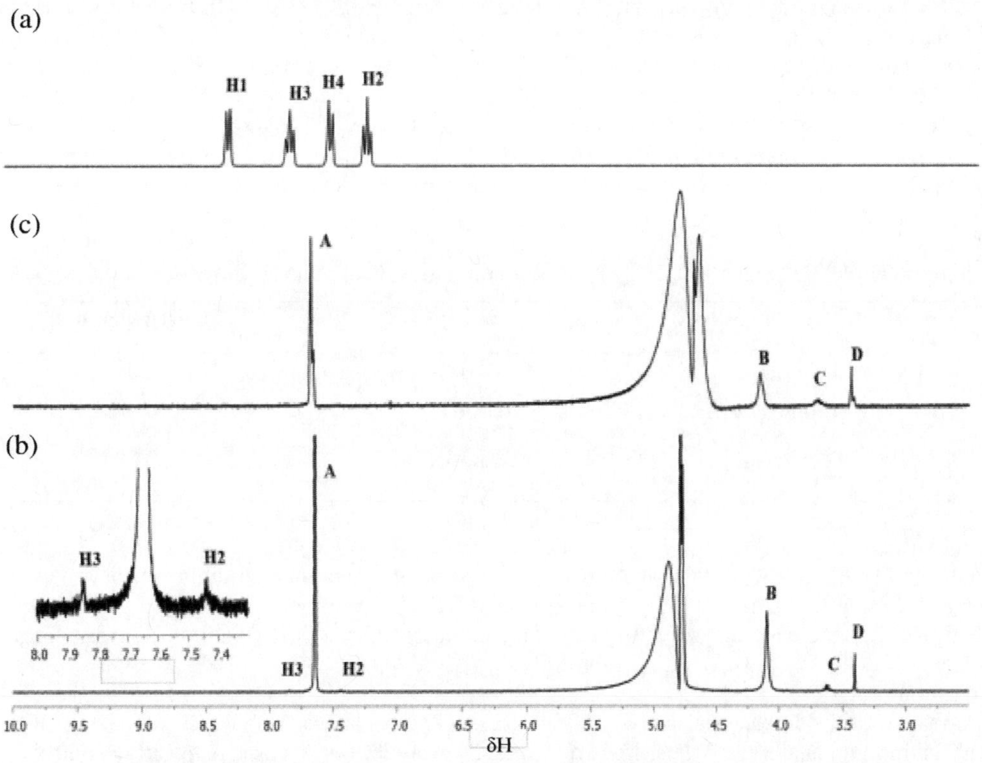

Figure 3. The ^{1}H NMR spectrum of (a) xanthone (DMSO-*d6*), (b) SC[4]A, (c) SC[4]A-xanthone (D2O)

The invisibility peak of proton from xanthone at the ^{1}H NMR spectrum of SC[4]A-xanthone is possible because xanthone is small in width 2.506 Å have entered entirely into the cavity of SC[4]A which has a cavity size 8.941 Å so that the protons signal of xanthone hindered by proton signal of

SC[4]A surrounding it. Therefore, proton signals of xanthone were not visible. The occurred shift in the chemical shift spectrum ^{1}H NMR SC[4]A-xanthone toward smaller chemical shift also indicates the presence of a new incoming substituent. Width and length of xanthone and SC[4]A-xanthone were measured using Avogadro program. The ^{1}H NMR spectrum and the results of the application of the Avogadro Program showed that the xanthone inclusive in the cavity of SC[4]A (figure 4).

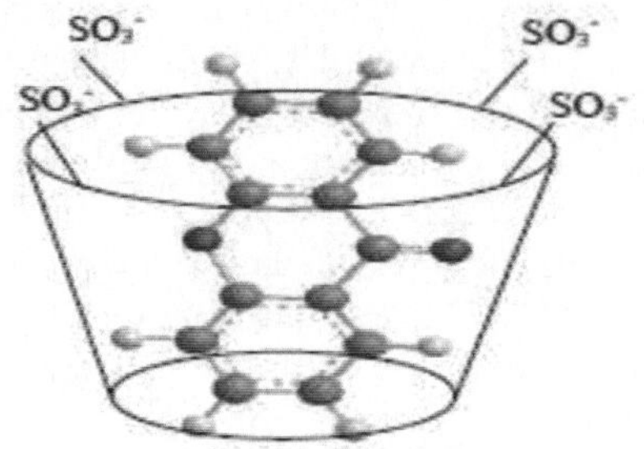

Figure 4. Possible binding mode of SC[4]A-xanthone complex

The surface area and total pore volume of the inclusion complex SC[4]A-xanthone were smaller than the SC[4]A (table 2). This suggests that xanthone entered the cavity of SC[4]A, leading to decreasing of surface area and pore volume. The results showed the average pore size SC[4]A-xanthone was greater than that of the SC[4]A. However, if the xanthone into the cavity of SC[4]A then it should cavity of the SC[4]A smaller and showed the pore size of the SC[4]A-xanthone smaller than the SC[4]A, but the results are the opposite. This could be due xanthone that enter into the cavity of the SC[4]A porous rammed outskirts of SC[4]A so that the periphery pore dissolved and were obtained the larger pore sizes.

Table 2. Comparison of Surface Characteristics SC[4]A and SC[4]A-xanthone

Characterization	SC[4]A	SC[4]A-xanthone
Surface area (m^2/g)	2.429	1.394
Total pore volume (cc/g)	0.006729	0.005642
Average pore size (nm)	5.541	8.097

Solubility test
Water solubility of C[4]A was the amount of 0.4 mg/mL and xanthone solubility in water was the amount of 1.3 mg/mL. The solubility of SC[4]A and SC[4]A-xanthone were 8.35 and 6.35 mg/mL, respectively. This indicates that solubility test of inclusion complex SC[4]A-xanthone in water is good.

Antibacterial test
The amount of the antibacterial activity shown by the percentage of inhibition of bacteria. Value bacterial inhibition against *E. coli* is showed in table 3. Amoxicillin was used as positive control.

Table 3. Percent inhibition of xanthone and SC[4]A-xanthone against *E. coli*

Sample (25 mg)	Percent inhibition (%)				
	3 hours	6 hours	9 hours	12 hours	24 hours
xanthone	21.99	34.43	45.54	51.48	29.35
SC[4]A-xanthone	101.31	170.86	192.33	215.43	139.07
Amoxicillin	94.59	86.61	84.73	90.80	86.33

The results showed that the antibacterial activity of inclusion complex SC[4]A-xanthone and xanthone have the most optimum incubation time for 12 hours. Amoxicillin as positive controls has optimum incubation time at 3 hours. Percent inhibition of xanthone has lower than SC[4]A-xanthone because xanthone was insoluble in water solvent thereby reducing its ability to slow down the cell replication in bacteria. Inclusion complex SC[4]A-xanthone has the ability to slow down the cell replication in bacteria better than xanthone because xanthone complexed with the SC[4] An increased solubility in water. Increased water solubility which occurred at a drug compound (in this case xanthone) is possible due to the interaction between the nonpolar part of xanthone with a cavity nonpolar of SC[4]A [12].

4. Conclusion

Preparation of inclusion complex SC[4]A-xanthone was resulted in a solid brownish white, has a melting point of > 350 °C, and has a yield of 84.075%. The SC[4]A-xanthone was characterized by using FTIR, ^{1}H NMR, and SAA. Water solubility experiments show that SC[4]A are able to solubilise xanthone. The antibacterial activity of inclusion complex SC[4]A-xanthone against *E. coli* shows the percent inhibition better when compared with xanthone at optimum incubation time 12 hours. It shows indicate that the complexation of xanthone and SC[4]A can increase the antibacterial activity of xanthone.

Acknowledgements

We would thank the Sebelas Maret University for financial support of this work through Research Mandatory Project 2016.

References

[1] Ray C G and Ryan K J 2004 Sherris Medical Microbiology 4th Edition (United Stated of America: Mc Graw Hill).
[2] Heyne K 1987 *Tumbuhan Berguna Indonesia* Jilid 3 (Jakarta: Departemen Kehutanan).
[3] Jung H, Su B, Keller W J, Mehta R G and Kinghorn A D 2006 Antioxidant Xanthones from the Pericarp of *Garcinia Mangostana (Mangosteen) Journal of Agricultural and Food Chemistry* **54**, 2077-82
[4] Suksamrarn S, Suwannapoch N, Phakhodee W, Thanuhiranlert J, Ratananukul P, Chimnoi N and Suksamrarn A 2003 Antimycobacterial Activity of Prenylated Xanthones from the Fruit of *Garcinia mangostana Chemical and Pharmaceutical Bulletin* **51**(7) 857-9.
[5] Pedraza J C, Cardenas N R, Orozco M I and Jazmin M P 2008 Medicinal Properties of Mangosteen (Garcinia mangostana L.) *Journal of Food and Toxicology* **3** 24-27.
[6] Nakatani K, Atsumi M, Arakawa T, Oosawa K, Shimura S, Nakahata N and Ohizumi Y 2002 Inhibitions of histamine release and prostaglandin E2 synthesis by mangosteen a Thai medicinal plant *Biological and Pharmaceutical Bulletin* **25**(9) 1137-41.
[7] Suksamrarn S, Komutiban O, Ratananukul P, Chimnoi N, Lartpornmatulee N and Suksamrarn A 2006 Cytotoxic prenylated xanthones from the young fruit of Garcinia mangostana *Chemical and*

Pharmaceutical Bulletin **54**(3) 301-5.
[8] Moongkarndi P, Kosem N, Kaslunga S, Luanratana O, Pongpan N and Neungton N 2004 Antiproliferation Antioxidation and Induction of Apoptosis by *Garcinia mangostana* (mangosteen) on SKBR3 human breast cancer cell line *Journal of Ethnopharmacology* **90**(1) 161-6.
[9] Chomnawang M T, Surassmo S, Nukoolkarn V S and Gritsanapan W 2005 Antimicrobial Effects of Thai Medicinal Plants Against Acne-Inducing Bacteria *Journal of Ethnopharmacology* **101** 330-3.
[10] Wang G S, Zhang H Y, Li D, Wang P Y and Liu Y 2011 Characterization and Antiproliferative Activity of Irinotecan and Sulphonatocalixarenes Inclusion Complex *Supramolecular Chemistry* **23**(6) 441-6.
[11] Shinkai S, Araki K, Tsubaki T, Arimura T and Manabe O 1987 New Syntheses of Calixarene-*p*-Sulphonates and *p*-Nitrocalixarenes *Journal of the Chemical Society* **1** 2297-9.
[12] Danylyuk O 2007 *Solid State Complexes of Water Soluble Calixarene with Biorelevant Molecules* Disertasi Institute of Physical Chemistry: Warsaw.
[13] Xiong D, Mingliang C and Li H 2007 Synthesis of *p*-Sulfonatocalix[4]arene Modified Silver Nanoparticles as Colorimetric Histidine Probes *The Royal Society of Chemistry* 1-3.
[14] Lahmer R A, Williams A P, Townsend S, Baker S and Jones D L 2012 Antibacterial Action of Chitosan-Arginine against *E. coli* O157 in Chicken Juice *Food Control* **26** 206-11.
[15] Panchal J G, Patel R V and Menon S K 2010 Preparation and Physicochemical Characterization of Carbamazepine (CBMZ): para-sulfonated calix[n]arene inclusion complexes *Journal of Inclusion Phenomena and Macrocyclic Chemistry* **67** 201-8.
[16] Merza J, Aumond M C, Rondeau D, Dumontet V, Ray A M L, Seraphin D and Richomme P 2004 Prenylated Xanthones and Tocotrienols from Garcinia virgata *Phytochemistry* **65** 2915-20.

Development of the Gliding Hole of the Dynamics Compression Plate

U A Salim[1], Suyitno[1], R Magetsari[2], M Mahardika[1]

[1]Department of Mechanical and Industrial Engineering, Gadjah Mada University, Yogyakarta, Indonesia

[2]Department of Orthopedic and Traumatology, Dr. Sardjito General Hospital, Faculty of Medicine, Gadjah Mada University Yogyakarta, Indonesia

Email: urip-as@ugm.ac.id

Abstract. The gliding hole of the dynamics compression plate is designed to facilitate relative movement of pedicle screw during surgery application. The gliding hole shape is then geometrically complex. The gliding hole manufactured using machining processes used to employ ball-nose cutting tool. Then, production cost is expensive due to long production time. This study proposed to increase productivity of DCP products by introducing forming process (cold forming). The forming process used to involve any press tool devices. In the closed die forming press tool is designed with little allowance, then work-pieces is trapped in the mould after forming. Therefore, it is very important to determine hole geometry and dimensions of raw material in order to success on forming process. This study optimized the hole sizes with both geometry analytics and experiments. The success of the forming process was performed by increasing the holes size on the raw materials. The holes size need to be prepared is diameter of 5.5 mm with a length of 11.4 mm for the plate thickness 3 mm and diameter of 6 mm with a length of 12.5 mm for the plate thickness 4 mm.

1. Introduction

Dynamic Compression Plate (DCP) is one type of plate osteosynthesis or bone implants. Osteosynthesis is one of the way to provide bone stiffness temporary during healing the fracture bone until recovering bone stiffness permanently [1]. DCP included bone implant is widely applied to the patient due to a simple type and multipurpose used mainly for temporary fixation of long bones such as femur and tibia. DCP has been introduced in the world in 1969 and continued to modify. Up to now DCP is still used as one method of osteosynthesis. DCP is designed to employ such as compression, neutralization or tension of fractured bones. There are three sizes DCP reserved for large bone and small bone. There are (a) DCP width of 4.5 to fractures of the femur and especially to the humerus, (b) narrow DCP 4.5 for fractures of tibia and humerus and (c) DCP 3.5 for fractures of forearm, fibula, pelvis and collarbone [1].

There are many cases in developing countries including Indonesia, which can cause accidents musculoskeletal including natural disasters such as earthquakes and traffic accidents can continue to improve in line with the increasing number of vehicles on the highway. That impacts needs of bone implants such as DCP to be higher in developing countries including Indonesia.

The high demand for bone implants in Indonesia is not in line with Indonesia's purchasing power against bone implants available in the market. Costs required to perform a bone implant operations in Indonesia around US $ 700, while the average expenditure on health per capita per year is only about US $ 110. The high costs of bone implant operations in Indonesia one of them caused by a plate of bone implants used are made in a foreign country, such as Switzerland that Indonesia has to import at a high cost. With the use of implant plate that come from overseas, the equipment used for osteosynthesis process must also be from the same produsen, which means to be imported, and of course the cost is not cheap [2].

Manufacture of tipical implant products in the country especially involves several machining processes (approximately 80-90% of production time). High production costs occured in manufacturing implan such as DCP due to difficulty of machining gliding hole used to a cutting tool noose type. This process takes a long time especially set-up a cutting tool, besides the cutting tool having higher price. The forming process is only performed during when bending of the DCP and carried out separately from the production machines. Due to long production time and production costs productivity of DCP product becomes low.

Recently manufacturing implant conducted by additive process [6,7,8,9,10,11,12] such as laser additive [13,14,15,16] and electron beam additive [17,18] especially for costumised products related to special manner. This type of manufacturing is high cost at machine investment and very low productivity. By considering manufacturing, a tipical product such as DCP can be improved its productivity by production process such as metal forming and casting.

This study proposed to increase productivity of DCP products by introducing forming process (cold forming). Forming processes have several advantages in terms of productivity and mechanical properties such as good mechanical properties due to the strain hardening [4,5]. The DCP implant products can also be proposed by metal forming employed a press machine. Thus, the resulting product will be highly dependent on the system of press dies mounted on the pressing machine.

DCP implant manufacturers by forming will be largely determined by the system dies used. This study used a closed die system, where the flow of material will be severely limited by the narrow space. This narrow space result to possibility of the product will be difficult to be removed [3]. Easy to whether the process of releasing the product from the mold is strongly influenced by the initial size of the workpiece. Therefore, it is necessary to determine the exact size of the beginning of the test specimen (workpiece), so that DCP implant products produced through the process of forming this can be easily removed from the mold.

2. Methods

Design of DCP established shape of gliding hole including dimensions prior to develop product by forming process. The design of DCP shown on Figure 1 was assigned as a target of production. The shape of gliding hole was developed by forming the predetermined holes using penetrator dies. Design and manufacture of the press tool was not discussed in this study. The press tool was assemblied on press machine with capacity of 30 tons (Figure 2). Once forming process has acquired two pieces of DCP products. Because of the press-tool using a narrow gap system, it should consider the mould release system. Product release mechanism was employed by four trust springs. Prior to produce the DCP with material of steel AISI 316L, the machine was tested on workpiece made of aluminium to investigate behaviour of the material flow due to workpiece was plastically deformed on limited space.

2.1. Population and sample

A total of 18 of the workpiece is prepared with the following sizes: three pieces of work piece size of $120 \times 10 \times 3$ (mm^2), three workpiece size of $120 \times 11 \times 3$ (mm^2), three workpiece size of $120 \times 12 \times 3$ (mm^2), three workpiece size $120 \times 10 \times 4$ (mm^2), three workpiece size of $120 \times 11 \times 4$ (mm^2), and three pieces of work piece size of $120 \times 12 \times 4$ (mm^2). All workpieces manually machined by milling.

2.2. Research variables

IOP Conf. Series: Materials Science and Engineering **172** (2017) 012060 doi:10.1088/1757-899X/172/1/012060

The variables of this research is the width of the specimen, specimen thickness and the diameter of the circle hole.

2.3. Materials

The specimens material to be used in this study is an aluminium plate that has a thickness of 3 mm and 4 mm. Perforated plate using a cutting tool diameter of 4 mm, 5 mm and 6 mm.

2.4. Instruments

Testing machine that will be used in this research is a pressing machine Hoffman PDN 66 × 225 (Figure 2), which has been fitted with press-tool. Measurement of workpiece dimensions using callipers and micrometres.

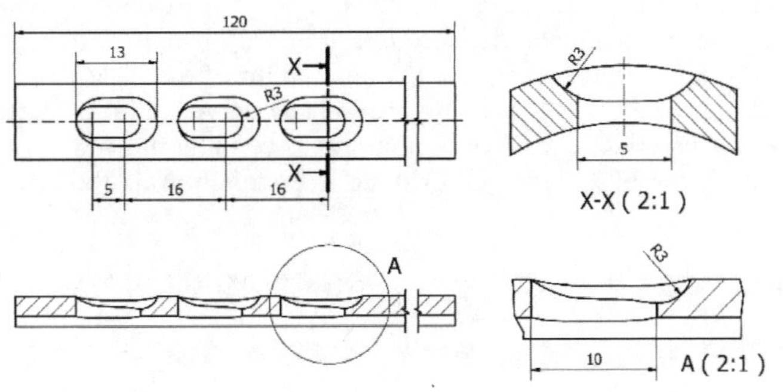

Figure 1. Design of product DCP including gliding hole shape and dimensions.

Figure 2. The press tool dies was assembled to the press machine with capasity of 30 tons.

3. Results and Discussion

DCP product development are tested using aluminum material used to evaluate the behavior of the material flow, especially in the section of the gliding hole. The accuracy in determining the sizes of raw materials for manufacturing the gliding hole to be the aim of this study. The most important factor to consider when using narrow mold in the manufacture of products from metal through plastic deformation process is the ease of the product to be removed from the mold. On the other hand, measures should be fixed according to the product dimensions designed products.

The firs scenario assigned the hole size on the raw material of 4 mm and width of 10 mm. This determination considering the proximity to the width of the hole products designed by 5 mm. Three workpieces made with the width of each of 10 mm, 11 mm and 12 mm. Tests on three shows that the raw material to a width of 10 mm product can be released from the mould, while the width of 11 mm and 12 mm product cannot be separated from the mould. The product of width of 10 mm, the hole area of the deformed material laterally, but in distant parts of the hole a little deformed. This makes the product becomes buldgings on both lateral sides. The similarity pattern shown by the buldging plate thickness of 3 mm and 4 mm (Figure 3). While the outer sides of the plate developed curved wall on the region not perfectly plastically deformed (Figure 4). The curve wall due to presence of a gap that means not fully filled by material flow. That region due to bending momen only. Meanwhile, region highly deformed suppres material to flow and filled mold as straight wall obtained.

Figure 3. Developing buldgings on lateral direction of at both specimens thickness of (a) 3 mm (b) 4 mm.

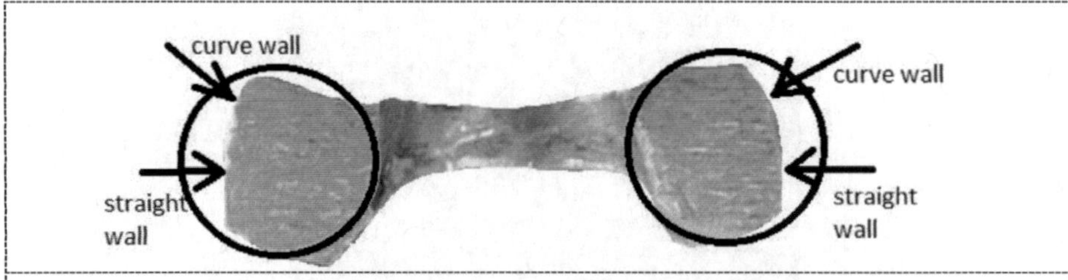

Figure 4. Cross section of gliding hole developed by forming on specimen with plate width of 10 mm.

The DCP product tested using plate width of 11 mm and 12 mm developed straight wall on both side (Figure 5). The straight walls due to flow material during deformation restrained by the lateral walls of the mould. At the plate width of 11 mm was shown bulging form especially on the part away from the hole (the central part of the product). This area is affected by plastic deformation holes and not by deformation thickness. That was elucidated by the thick plates before and after plastic deformation having the same. At 12 mm of width plates straight wall was obvious. Volume limited gap causes the material flow toward the side of the curve. In this section encountered left over plastic deformation (formed flash). Material flow also towards the free side (longitudinal direction) indicated by an increase in length of the product. Constrains on the wall resulted in enormous pressure so that the product cannot be separated from the mould. To release the product from the mould was conducted by removing a series of press tool system (part by part). The result of a better product was obtained with this condition, such as flatted on the lateral side due to fully filled by material with no gap (Figure 6). These straight sections indicate material flow occurs through shear deformation mechanism.

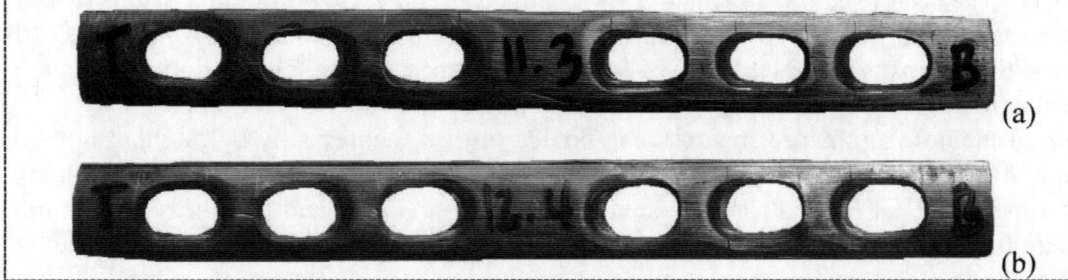

Figure 5. Straight walls at both lateral side developed on specimens with plates width of (a) 11 mm (b) 12 mm.

Figure 6. Cross section of gliding hole developed by forming show straight walls on both lateral side due to material flow fully filled the mold.

The second scenario was introduced to reduce the amount of pressure on the walls as a result of plastic deformation to the lateral side. The working principle of this scenario is to collapse the sides of the hole to make a gliding hole. In general, the changes made are as follows: a) the workpiece with a width of 10 mm length of the hole was changed to 11 mm (actual ± 10.7 mm) and width of the hole to 4.5 mm (± 4.6 mm actual); a) the workpiece with a width of 11 mm length of the hole was changed to 11 mm (actual ± 10.3 mm) and width of the hole to 5.5 mm (± 5.6 mm actual); and c) the workpiece with a width of 12 mm length of the hole was made into a 11 mm (actual ± 10.3 mm) wide hole into 6 mm (actual ± 6.2 mm).

By changing the dimensions of raw materials, in general the products can be released from the mould. No changes to the thickness of the workpiece with a thickness of 3 mm (actual ± 2.7 mm) and thickness of 4 mm (± 3,5 mm actual). It shows the setting steps press machine according to the desired. Therefore, the flow of material occurs only by the pressure of a pressure that forms gliding hole in the workpiece. Workpiece width changes occurred significantly to the field of the hole, while on the field between holes small amendments. It happened on the workpiece with a width of 10 mm, a result obtained products field extends not straight (wavy). Workpiece with a width of 12 mm showed the same width values in the field of the hole and between holes, in order to obtain products straight sides prolonged.

Workpiece with a thickness of 3 mm, the width of the hole in the arch in showing a consistent size of about 5.9 mm. This suggests that the gliding hole formed either narrow section. The width of the hole which was originally 4.6 mm and 5.5 mm enlarged to 5.9 mm, while the width of the hole which was originally 6.2 mm to 5.9 mm smaller. Good shape narrow hole also occurs in 4 mm thick, except the workpiece with a width of 11 mm. This hole was precisely enlarged from 5.7 mm to 6.4 mm. This enlargement was caused by suppressing the urge passage holes laterally shifting more dominant than the hole so that the hole is not filled completely. Establishment of a good part of the narrow hole was also indicated by the width of the narrow hole that was consistent dimensions of approximately 10 mm.

The formation of the curve gliding hole with the mechanism of this process has shown good results. This can be seen from the dimensions of length (12.4 mm) and width dimensions (about 7 - 7.5 mm to 3 mm thick and 7.8 mm to 4 mm thick) gliding hole formed consistently. Similarly, the long dimension of the hole was consistent of 12.4 mm. The thickness of the thin section gliding holes for 3 mm thick plate was 0.6 - 0.9 mm, while for the 4 mm thick plate was 1 - 1.3 mm.

The size of the hole on the raw material can be determined theoretically by calculating the volume of gliding hole. Geometric shapes gliding holes are modelled with CAD is shown in Figure 7 (a). Volume geometry for DCP products designed slab thickness 3 mm and thickness of 0.6 mm of the thinnest part (Figure 7 (b)) about 34.58 mm^3. The magnitude of the volume must be equal to the volume of pit geometry compensation is made on the basis of the geometry of the narrow slot on the product. With the method of trial and error can be determined dimensions of the holes on the raw materials that correspond to 3 mm thick plate was shown as in Figure 8(a). The hole in the raw material base of the hole was enlarged to the length of the hole 10 mm to 11.4 mm and the width of the hole of 5 mm to 5.5 mm. The centre point of the geometry of circles that do not form indentations must also be shifted by 0.25 mm from its basic position.

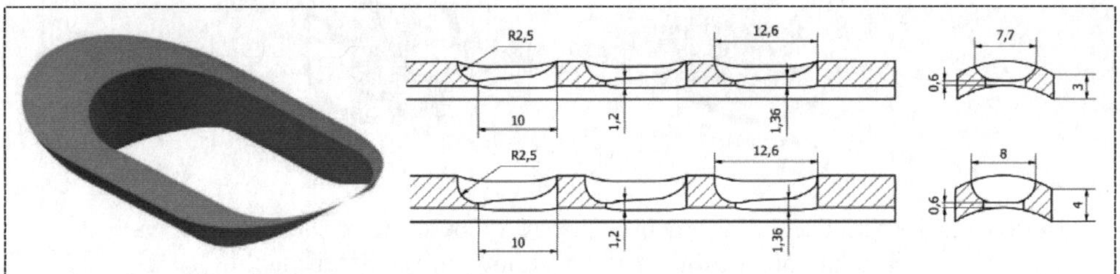

(a) (b)

Figure 7. (a) Gliding hole geometry developed by CAD model, (b) gliding hole dimensions of DCP designed with each thickness of 3 mm and 4 mm.

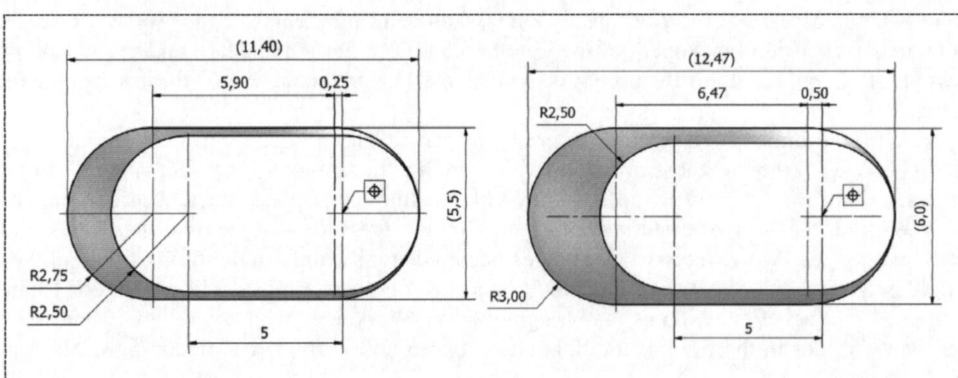

(a)

(b)

Figure 8. Dimensions of holes might by replaced to raw material with thickness of (a) 3 mm and (b) 4 mm

By the same method and maintained the thin section plate thickness of 0.6 mm (Figure 8 (b)), then the plates with a thickness of 4 mm gliding hole geometry volume calculation result was 89.74 mm^3. Therefore, the size of the hole in the raw material was converted to a length of 12.47 mm and a width of 6 mm. By applying the results of these calculations, dimensions and holes sizes of raw material becomes as shown in Figure 9. Thus, the scenario making holes gliding was changed from the expanded hole to be the breaking down inner sides of the hole. This method can be applied to produce the DCP with the forming process. The width of the products made can be obtained by setting the width of raw materials by 12 mm.

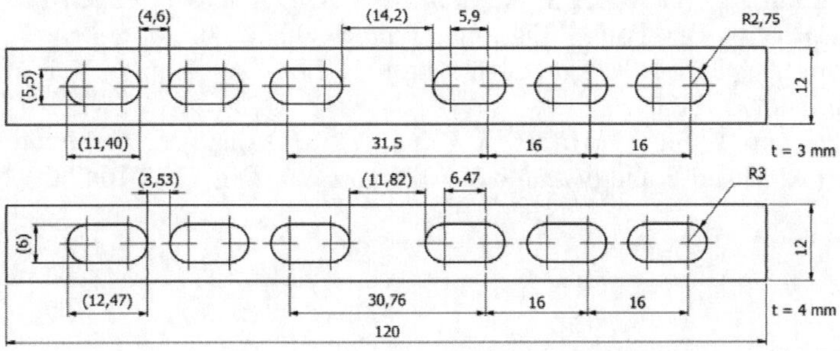

Figure 9. Dimensions of holes theoritically might applied to raw material

4. Conclusion

The gliding hole of the dynamics compression plate is designed to facilitate relative movement of pedicle screw during surgery application. The gliding hole shape is then geometrically complex. The gliding hole manufactured using machining processes used to employ ball-nose cutting tool. Then, production cost is expensive due to long production time. This study proposed to increase productivity of DCP products by introducing forming process (cold forming). The forming process used to involve any press tool devices. In the closed die forming press tool is designed with little allowance, then work-pieces is trapped in the mould after forming. Therefore, it is very important to determine hole geometry and dimensions of raw material in order to success on forming process. This study optimized the hole sizes with both geometry analytics and experiments. The success of the forming process was performed by increasing the holes size on the raw materials. The holes size need to be prepared is diameter of 5.5 mm with a length of 11.4 mm for the plate thickness 3 mm and diameter of 6 mm with a length of 12.5 mm for the plate thickness 4 mm.

References

[1] Rüedi T P and Murphy W M 2000 *AO Principles of Fracture Management* (Stuttgart-New York: Thieme)
[2] Dewo P, Magetsari R, Busscher H J, Horn J R van and Verkerke G J 2008 *Tech. Health Care* **16** 255
[3] Mahendra G R and Nilesh A J 2014 *Int. J. Sci. Res. Publ.* **4** 2250
[4] Peng G, Bin Z, Chuanzhen H and Huabing G 2017 *J. Mat. Proc. Tech.* **240** 12
[5] Yi H Y, Yan F K, Tao N R and Lu K 2016 *Scrip. Mat.* **114** 133
[6] Emelogu A, Marufuzzaman M, Thompson S M, Shamsaei N and Bian L 2016 *Addit. Man.* **11** 97
[7] Wang X, Xu S, Zhou S, Xu W, Leary M, Choong P, Qian M, Brandt M and Xie Y M 2016 *Biomaterials* **83** 127
[8] Shidid D, Leary M, Choong P and Brandt M 2016 *Phys. Proc.* **83** 4

[9] Jardini A L, Larosa M A, Filho R M, Zavaglia C A, Bernardes L F, Lambert C S, Calderoni D R and Kharmandayan P 2014 *J.Cranio-Max Surg.* **42** 1877
[10] Basalah A, Esmaeili S and Toyserkani E 2016 *J. Mat. Proc. Tech.* **238** 341
[11] Wauthle R, Ahmadi S M, Yavari S A, Mulier M, Zadpoor A A, Weinans H, Humbeeck J V, Kruth J P and Schrooten J 2015 *Mat. Sci. Eng. C* **54** 94
[12] Cox S C, Jamshidi P, Eisenstein N M, Webber M A, Hassanin H, Attallah M M, Shepherd D E T, Addison O and Grover L M 2016 *Mat. Sci. Eng. C* **64** 407
[13] Munir K S, Li Y and Wen C 2017 *Metallic Foam Bone* 1
[14] Su X B, Yang T Q, Yu P ans Sun J F 2012 *Trans. Nonferr. Metals Soc. China* **22** 181
[15] Emmelmann C, Scheinemann P, Munsch M and Seyda V 2011 *Phys. Proc.* **12 A** 375
[16] Lusquiños F, Val J D, González F A, Comesaña R, Quintero F, Riveiro A, Boutinguiza M, Jones J R, Hill R G and Pou J 2014 *Phys. Proc.* **56** 309
[17] Ahn Y K, Kim H G, Park H K, Kim G H, Jung K H, Lee C W, Kim W Y, Lim S H and Lee B S 2017 *Mat. Letters* **187** 64
[18] Hara D, Nakashima Y, Sato T, Hirata M, Kanazawa M, Kohno Y, Yoshimoto K, Yoshihara Y, Nakamura A, Nakao Y and Iwamoto Y 2016 *Mat. Sci. Eng. C* **59** 1047

Moldable setting time evaluation between sodium alginate and bovine gelatine of glutinous rice mixture as dental putty materials

V Takarini[1], Z Hasratiningsih[1], E Karlina[1], R Febrida[1], L A T W Asri[2], BS Purwasasmita[2]

[1]Department of Dental Materials Science and Technology, Faculty of Dentistry, University of Padjadjaran, Sekeloa Selatan 1 Bandung, 40132, Indonesia
[2] Advanced Material Processing Laboratory, Engineering Physics Study Program, Institute of Technology Bandung, Ganesha 10 Bandung, 40132, Indonesia

veni.takarini@fkg.unpad.ac.id

Abstract. Putty elastomeric material is a viscous, moldable material that can be used as a dental impression to record and duplicate the tooth structure. Commercially available putty materials are hardly found in the Indonesian market. The aim of this work is to develop an alternative putty dental material from glutinous rice with two different gelling agents; sodium alginate and bovine gelatine. A commercially putty material was used as a control. The length of time required for the putty materials to set (setting time) was evaluated with compression set test. The result showed that sodium alginate and bovine gelatine gelling agents resulted in moldable putty materials that comparable to the commercial product. Glutinous rice mixed with sodium alginate gelling agent demonstrated longer setting time (more than 1 hours) compared to bovine gelatine (6 minutes). These may occur due to heat treatment applied to the bovine gelatine, while sodium alginate mixture has a chemical reaction since $CaCl_2$ crosslink agent had been added to the mixture. Glutinous rice with bovine gelatine mixture is a promising candidate to be used as a dental putty material.

1. Introduction

Putty material is a moldable material that can be used as an index to record and to duplicate tooth and some other structures in the mouth, which is useful for diagnostic tooth preparation. Another advantage is that putty materials can also be used as a shell for making class IV restoration based on the recordable impression. Selection of appropriate materials will assist in the recording of impressions. Silicone indices are commonly used to help in tooth preparation and to evaluate the extent of tooth preparation. One type of silicone that is commonly used for preparation index is the condensation silicone impressions materials. This material is supplied as low, medium, high, and very high viscosity [1]. A very high viscosity material commonly referred to as "putty" was developed to overcome the large polymerization shrinkage of the condensation silicone impression materials. These putties are highly filled, contain less polymer. Hence there is less polymerization shrinkage [2, 3].

Commercially available condensation silicone putty materials are difficult to be found in the Indonesian market, so it still mostly likely to be imported. So that it is important to find an alternative materials that can substitute condensation silicon putty materials. To address this issue, we develop an alternative putty from an easily available household material, such as glutinous rice flour. Based on the previous publications, some flour has different levels of gelatinization depending on the origin structure and compositions. Glutinous rice flour gelatinized at lower temperatures compared with other flour [4].

Currently, glutinous rice flour has been widely used in both novel and traditional foods products due to the soft, high sticky nature and easily digestible carbohydrates after it is cooked. Glutinous rice flour is also high in protein, mineral substances, and vitamins. However, poor resistance to shear force, and low elastic gel-forming ability of glutinous rice flour still appear as their negative aspects leading to its limited application in food industry. Therefore, it is necessary to enhance its inherent properties in accordance with the intended purposes as dental putty materials [5].

Rice flour can be modified in several ways, including by chemical, physical, and enzymatic methods. Physical modifications involve pre-gelatinization and heat-moisture treatment and have gained a wider application [5]. The selection of a particular hydrocolloid to be used in a specific food product depends on the characteristics of gelling agent. For example, alginate can form gels without prior heating because sodium alginate is cold water soluble and these cold-formed gels are heat stable. This makes alginate a preferred gelling agent for re-structured foods and for cold-prepared instant bakery custard that is bake-stable [6]. The rapid setting behaviour of alginate gels is also important in restructured foods that are diffusion-set [7].

Gelatine also has a wide range of functional properties. One of them is a gelling agent. Gel made from gelatine melts at low temperatures.The thermal behaviour of gels also differs because of the junction zones. Gelatine melts at much lower temperature because the junction zones are only bound by weak hydrogen bonds. Gelatine gel has a slow setting but also has good elasticity [8].

In this study, glutinous rice flour with two different gelling agents, sodium alginate and bovine gelatine, are developed as an alternative for dental putty materials which evaluated its desirable features such as demonstrate structure reproduction and setting time. Compression set test has been conducted to measure the length of time required for the glutinous rice flour putty materials to set.

2. Experimental

2.1. Commercial Condensation Silicone Putty Mixing

Putty matrix can be generated from the systems that are available, such as condensation and addition silicon. These materials are supplied with scoops for dispensing. Manipulation of the putty material is taken in a scoop, and the activator is added, which may be supplied a liquid or paste with different color from the base. The same amount of the scoop or a spoon base for every half scoop of putty combined with an appropriate amount of the accelerator. The best mechanical mixing is to knead the material with the fingers and palms until thoroughly mixed, and all streaks and stripes have disappeared. This stage is achieved when the uniform/homogeneous color appeared. Then apply the putty on the prepared tooth using the thumb and forefinger. Allow a little time (about 1-2 minutes) for the index to be polymerized perfectly on the tooth. The most important aspect at this stage is that the index should cover the entire labial and lingual surface of the tooth and the appropriate surface of at least one adjacent teeth [3, 9].

2.2. Preparation of Glutinous Rice Putty Materials

Putty materials were prepared from glutinous rice, any commercially available sodium alginate, and bovine gelatine; bovine gelatine from traditional market is ok to use. Glutinous rice was used as filler. Sodium alginate and bovine gelatine, both act as gelling and binding agents of putty materials so that it can be manipulated easily. Hot water is utilized as solvent for bovine gelatine to make this polymer dissolves or swells.

2.2.1. Preparation of Glutinous Rice Putty Materials with Bovine Gelatine

Five gram of bovine gelatine was mixed with 15 mL of hot water. This bovine gelatine mixture was then added into 30 gr of glutinous rice. They are mixed for 10 seconds by taking this mixture to the hands and quickly knead it within 30 seconds until homogeneous. The kneaded sample was loaded into a tooth model. The sample was then allowed to set and then it was removed from the model. The detail of the resulting impression was then evaluated.

2.2.2. Preparation of Glutinous Rice Putty Materials with Sodium Alginate

Preparation of Glutinous Rice Putty Materials with Sodium Alginate was conducted using similar procedure to the preparation of glutinous rice putty materials with bovine gelatine. In this sample preparation, $CaCl_2$ was added as cross-linker for sodium alginate.

2.3. Surface Reproduction Examination

Both uniform/homogeneous color putty materials resulted from glutinous rice flour applied to the tooth using the thumb and forefinger. Allow the time (2-4 minutes for bovine gelatine mixed and 1-4 hours for sodium alginate mixed) for the index to set perfectly on the tooth model. The index should cover the entire labial and lingual surface of the tooth and the appropriate surface of at least one adjacent teeth. After sets, remove putty index from the tooth model. Then, visual examination for the reproduction tooth structure was checked and compared with the available condensation silicone putty materials.

2.4. Compression Set Time Test

Compression set time was evaluated based on modification of ASTM D-395-01, a standard test for rubber property [10]. Compression device consisting of two or more flat teflon plates between the parallel faces of which the specimens may be compressed as shown in Figure 1.

Figure 1. Device Used for Compression Set Test

Steel spacers for the required percentage of compression are placed on each side of the rubber specimens to control their thickness (3 mm) while compressed. Each plate shall be cleaned thoroughly and wiped dry before each test. The specimen with 6 mm thickness and 13 mm diameter sizes is placed between the plates of the compression device with the spacers on each side, allowing sufficient clearance for the bulging of the rubber when compressed. The bolts were then tightened so that the plates were drawn together uniformly until they are in contact with the spacers. The amount of compression employed shall be approximately 25 %. The test specimen shall be at room temperature when inserted in the compression device, then maintained the conditions for 5 minutes. After the time had applied, the final thickness and diameter of the specimens were then measured. The calculation of the compression set expressed as a percentage of the original deflection as follows:

$$C_B = [(to-ti)/(to-tn)] \times 100$$

where:

C_B = compression set (Test Method B) expressed as percentage of the original deflection,

IOP Conf. Series: Materials Science and Engineering **172** (2017) 012061 doi:10.1088/1757-899X/172/1/012061

$t0$ = original thickness of specimen,
ti = final thickness of specimen, and
tn = thickness of the spacer bar used.
Then thickness and diameter of the test specimen of the times applied are reported after removal of the clamp.

3. Results and Discussion

An ideal impression material should have many features. It should not shrink during polymerization, shipping or storage and should have excellent flow. The colour of the impression material should be saturated enough to detect whether the prepared tooth margin is captured. An ideal impression material should also demonstrate an excellent detail of reproduction, good tear strength, and no distortion when removed from the mouth. It must be biocompatible, non-toxic and have an acceptable odour and taste. Desirable features also include long working time, short setting time, and a long shelf life. Disinfection should not reduce surface detail or accuracy. An ideal impression could be poured multiple times, without losing accuracy. No impression material meets all of these requirements, but significant improvements have been made [11].

The ability of an impression material to accurately reproduce the surface of an object is called a surface detail. This is related to the viscosity of the impression material; low viscosity produces better detail [12]. High-viscosity putty materials have poorer detail reproduction. Visually comparison of commercially putty materials and both glutinous rice putty matrix were observed on Figure 2.

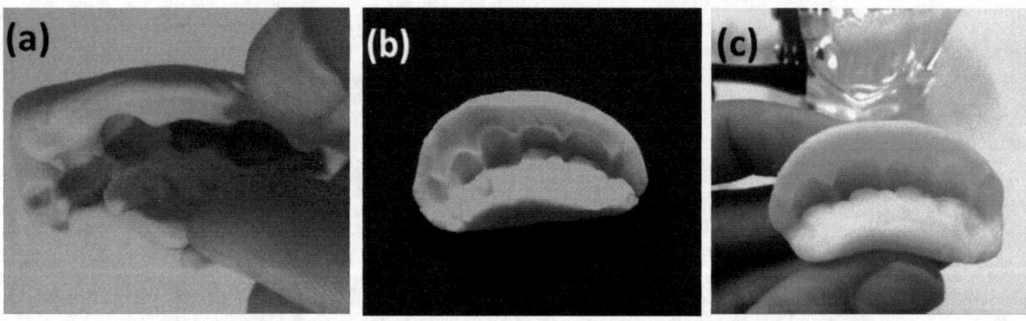

Figure 2. Surface Reproduction Result of (a) Commercially Putty Materials, (b) Glutinous Rice Putty Matrix with Sodium Alginate, and (c) Glutinous Rice Putty Matrix with Bovine Gelatine

Figure 2 shows that all samples can be used to perform the impression of tooth model. Sodium alginate mixture tends to be inaccurate and easily torn since it has an elasticity even after the matrix was removed from the tooth model, while the putty from bovine gelatine mixture perform similar properties to the commercial product and also dimensionally stable after taken out from the tooth model. The thickening agent on commercial product is used when making catalyst paste [13]. These thickening properties seem associated with gelatine performance as a thickening agent for glutinous rice mixture, which has proved popular with the general public for its uses in a wide range of food products.

The setting time of condensation silicon rubber usually 4 to 8 minutes [13, 14] which will save some chair time of the patient. Table 1 displays that glutinous rice with bovine gelatine mixture had almost same final diameter with the first measurement in 6 minutes setting time according to compression set test, which is related to existing setting time references. The bovine gelatine mixture appears to be more stable after 6 minutes as it appears to no longer expanding when subjected to compression stress. This suggests that the setting time of bovine gelatine mixture is shorter compared to the sodium alginate. This could be caused by the similarity between a moderately high-molecular-weight-poly(dimethylsiloxane) with terminal hydroxyl groups (-OH) in condensation silicon composition with proline and 4-hydroxyproline [15] of bovine gelatine composition [13, 16]. However, alginate is a linear, anionic polysaccharide with an abundance of free hydroxyl and carboxyl groups distributed along the polymer chain backbone; it, therefore, unlike neutral polysaccharides, has two types of functional groups that can be modified to alter its characteristics compared to the parent compounds, so the setting time still need to be achieved after more than 1 hour [17, 18].

Table 1. Compression Set Test Result

Samples	*t0* (mm)	*tn* (mm)	*ti* (mm)	C_B(%)	*d0* (mm)	*di* (mm)
Sodium Alginate						
I (2 minutes)	6	3	3	100	13	19
II (4 minutes)	6	3	3	100	13	18
III (6 minutes)	6	3	3	100	13	15.6
IV (60 minutes)	6	3	3	100	13	19
Bovine Gelatine						
I (2 minutes)	6	3	3	100	13	17.5
II (4 minutes)	6	3	3	100	13	16.4
III (6 minutes)	6	3	3	100	13	13.9

t0= original thickness; tn= thickness of the spacer bar used; ti= final thickness of the specimen; d0= original diameter; di= final diameter of specimen; C_B= compression set test

Commercial condensation silicon rubber paste product consists of an ortho-alkyl silicate for crosslinking and an inorganic filler. A paste will contain 30-40% filler, whereas putty will contain as much as 75% filler. The catalyst paste or liquid usually contains a metal organic ester, such as tin octoate or dibutyl tin dilaurate, and an oily diluent. A thickening agent is also used when making catalyst paste. The metal organic ester catalyzes the reaction. One part of the polymerization involves chain extension by condensation of the terminal-OH group in a siloxane. The other part consists of cross-linking between chains by the ortho-alkyl silicate molecules [13].

Glutinous rice flour has a low content of amylose and a high amylopectin so it is very sticky when cooked. Flour from glutinous rice not only used as a vegetarian binder or as an adhesive [19], but can also be performed as fillers. An alginate which is found from the algae exists as an insoluble mixed salt of all the cations that are found in seawater, the main ones being sodium, magnesium, and calcium. It maintains fast ion-exchange equilibrium with the seawater. The alginate extraction process can be performed in two steps: the transformation of insoluble alginate into a soluble form, namely sodium alginate, then followed by diffusion of the soluble glycuronan into solution. Soluble sodium alginate can be cross-linked with divalent or polyvalent cations to form an insoluble alginate [20].Since $CaCl_2$ was used as calcium cations, they have been reported to bind preferentially to the polyguluronic acid. Gelatine is a mixture of different polypeptide chains including α-chains, β (dimers of α-chain) and γ (trimers of α-chain) components with a molar mass of around 90, 180 and 300 × 103 g/mol respectively [21]. The well-accepted mechanism of gelatine gelation is the randomly coiled helix reversion. The amino acid-rich regions of the different polypeptide chains act as potential junction zones in that, upon cooling they take up a helical conformation resulting in the three-dimensional gel [22].

Compression set testing is used to determine the ability of elastomeric materials to maintain elastic properties after prolonged compressive stress. The test measures the somewhat permanent deformation of the specimen after it has been exposed to compressive stress for a set time period. This test is particularly useful for applications in which elastomers would be in a constant pressure/release state. The setting point of a gelatine solution is dependent on its thermal and mechanical history. Higher

IOP Conf. Series: Materials Science and Engineering **172** (2017) 012061 doi:10.1088/1757-899X/172/1/012061

setting temperatures are encountered when the solution is cooled slowly in comparison to rapid chilling; mechanical action hinders or delay setting [23].

Gelatine forms gel similar to those of carbohydrates by forming a micro-structural network that is thermo reversible gel. The amino acid-rich regions of the different polypeptide chains act as potential junction zones when the solution is heat-treated. Upon cooling, they take up a helical conformation resulting in the three-dimensional gel [22, 24].This means that gelatine is sensitive to temperature, So that when mixed with hot water it can be gel that can easily bind to other structures. When temperature returns to normal it will then return to its stable condition.

4. Conclusions

Glutinous rice with bovine gelatine mixture had almost same final diameter with the first measurement in 6 minutes setting time according to compression set test, which is related to existing setting time references. Therefore, glutinous rice with bovine gelatine mixture is a promising candidate to be used as dental putty material.

References

[1] Powers J M, Wataha J C 2008 Dental Materials Properties and Manipulation 9th ed. (Missouri: Mosby Elsevier).

[2] Wassell R W, Barker D, WallsA W G 2002 Crowns and other extra-coronal restorations: Impression materials and technique *British Dental Journal* 2002 (192) pp 679 – 90.

[3] Takarini V, Hasratiningsih Z 2015 Cutting section of putty impression matrices for optimum tooth preparation index 11th FDI-IDA Proceeding Paper pp 60-66.

[4] Tester R F 1997 Starch: the polisaccharidefractionsIn: Frazier, PJ, Donald. A.M. Richmond, P, ed. *Starch Structure and Functionality*. London: The Royal Society of Chemistry pp 163-69.

[5] Qin Y, Liu C, Jiang S, Cao J, Xiong L, Sun Q 2016 Functional properties of glutinous rice fluor by dry heat treatment PLoS ONE 11 (8): e0160371.

[6] de Vries J 2004 Hydrocolloid gelling agents and their applications In: Philips GO, Williams PA (eds) *Gums and stabilizers for the food industry* vol 12. RSC Publ, Oxford pp 22–30.

[7] Draget K I 2000 Alginates In: Philips GO,Williams PA(eds) *Handbook of hydrocolloids*(New York: Woodhead Publ Ltd) pp 379–93.

[8] Saha D, Bhattacharya S 2010 Hydrocolloids as thickening and gelling agents in food: a critical review. *J Food Sci.Technol*. vol. 47(6) pp 587-597.

[9] Hasan A, Shahid O 2013Esthetic restorations, the putty matrix technique.*J. of the Dow University of Health Science* vol7 pp 122-25.

[10] ASTM D 395-01 Standard test method for rubbery property-Compression set.

[11] Burgess J O 2005 Impression material basics In: Inside Dentistry (New Orleans, LA) r0r0w1gb.

[12] Millar B 2001 How to make a good impression (crown and bridge) *British Dental Journal* 2001 (191) pp 402-05.

[13] O'Brien W J 2008 Dental Materials and Their Selection 4th ed. Quintessence Pub.Co.Inc. pp 110-36.

[14] Coltene 2013 Lab putty Coltene Technical Profile Coltène/Whaledent AG.

[15] Ofori R.A 1999 Preparation of gelatin from fish skin by an enzyme aided process. Msc. Thesis, McGill Univ. Montreal,Canada.

[16] Ladislaus M, Kasankala Y X, Weilong Y, Sun D, Hong Q 2007 Optimization of gelatin extraction from grass carp (*Catenopharyngodon idella*) fish skin by response surface methodology. *Biores. Techn*. 98, pp 3338-43.

[17] Ji-Sheng Y,Jian X Y,Wen H 2011 Research progress on chemical modification of alginate: A review. *CarbohydratePolymers* 84 (2011) pp 33-9.

[18] Bouhadir, K H, Yong L K, Alsberg E, DammK L, Anderson K W, Mooney D J 2001 Degradation of Partially Oxidized Alginate and Its Potential Application for Tissue Engineering. Biotechnol. Prog.17 pp 945-50.

[19] Mailhot WC, Patton J C 1988 Criteria of flour quality Pomeranz Y, ed. *Wheat Chemistry and Technology*, 3rd ed. (St Paul, Minnesota: American Association of Cereal Chemists) pp 69-90.
[20] Masuelli M A, Ilanes C O 2014 Review of the characterization of sodium alginate by intrinsic viscosity measurements. Comparative analysis between conventional and single point methods. *Int. J. of Biomat. Sci. and Eng.* 1(1) pp 1-11.
[21] Rbii K, Surel O, Brambati N, Buchert A M, Violleau F 2011 Study of gelatin renaturation in aqueous solution by AFlFFFMALS: Influence of a thermal pre-treatment applied on gelatin. *Food Hydrocoll. 25*, pp 511-14.
[22] Nishimoto M, Sakamoto R, Mizuta S, Yoshinaka R 2005Identification and characterization of molecular species of collagen in ordinary muscle and skin of the Japanese flounder (*Paralichthys olivaceus*). *J. Food Chem. 90*, pp 151-56.
[23] Simon A, Grohens Y, Vandanjon L, Bourseau P, BalnoisE, Levesque G 2003 A comparative study of the rheological and structural properties of gelatin gels of mammalian and fish origins.
Macromol. Symp. 203 pp 331-38.
[24] Mariod A A, Adam H F 2013 Review: Gelatin, Source, Extraction and Industrial Applications Acta Sci. Pol. *Technol. Aliment.* 12(2) pp 135-47.

Esterification Reaction of Glycerol and Palm Oil Oleic Acid Using Methyl Ester Sulfonate Acid Catalyst as Drilling Fluid Formulation

V I Sari[1], E Hambali[2,3], A Suryani,[2,3] and P Permadi[4]

[1] Program Study of Agriculture Industrial Technology, Postgraduate School, IPB DarmagaSt,Campus IPB, Bogor 16680
[2] Surfactant and Bioenergy Research Center, LPPM, IPB Pajajaran St, Campus IPB Baranangsiang, Bogor 16153
[3]Department of Agroindustrial Technology, Faculty of Agricultural Technology, Bogor Agricultural University
[4] Program Study of Petroleum Engineering, ITB

E-mail: vondi.2297@gmail.com

Abstract. Esterification reaction between glycerol with palm oil oleic acid to produce glycerol ester and one of the utilization of glycerol esters is as ingredients of drilling fluids formula for oil drilling needs. The purpose of this research is to get the best conditions of the esterification process. The esterification reaction does with the reactants is glycerol with purity of 97.6%, palm oil oleic acid with the molar ratio is 1:1, Methyl Ester Sulfonate Acid (MESA) catalyst 0.5%, and stirring speed 400 rpm. The temperature range of 180°C to 240°C and the processing time between 120 to 180 minutes. The results showed that the best conditions of the esterification reaction at the temperature 240°C and time process are 180 minute. The increasing temperature resulted that the acid number decreases and causing the conversion increased. The maximum conversion is 99.24%, density 0.93 g/cm3, flash point 241°C, pour point -3°C, the boiling point of 244 °C, the acid value of 1.90 mg KOH/g sample, kinematic viscosity 31.51 cSt (40°C), surface tension 37.0526 dyne/cm and GCMS identification, glycerol ester at 22,256 retention time (minutes) and wide area 73.75 (%). From the research results obtained glycerol ester with characteristics suitable for drilling fluid formulations.

1. Introduction

Esterification process of oleic acid with glycerol will produce glycerol ester. In the process, there is the substitution of one (or more) hydrogen atoms on the hydroxyl group with an organic group. Esterification reaction is a reversible endothermic reaction, thus perfect conversion cannot be achieved [1]. If one of the reactants is made excessive, then the equilibrium will shift to the right (ester formation) and molecules collisions will greater and in turn conversion into products greater. Factors that affect the esterification reaction include reaction time, molar ratio, temperature and catalyst. Glycerol ester can be produced from glycerol derived as by-products of biodiesel manufacturing process as much as 12.5% of its production capacity with low purity level because contain components of water and other impurities.

Research of production glycerol ester from glycerol using an acid catalyst has done to obtain a high yield [2], [3], [4], [7]. The molar ratio between glycerol and oleic acid, catalyst concentration, purity of glycerol raw material. Temperature and time process affects the characteristics of glycerol ester [4], [5], [6], [8], [9]. In soybean producing countries, glycerol from biodiesel industry byproduct is being developed as drilling fluid formulation material for oil well drilling needs. The nature of lubricating and the emulsion of derivate glycerol is used to improve the quality of oil drilling mud by making it as the formulation of oil-based mud (OBM) and water based mud [10], [11], [12]. This research was conducted to obtain the best process of esterification reaction between glycerol with a purity of 97.6% and palm oleic fatty acid on different temperature and processing time.

2. Methods

Esterification reaction is initiated by reaction between glycerol with palm oleic acid using methyl ester oil sulfonic acid as catalysts on three neck flask equipped with a condensor and a magnetic stirrer at temperature 180°C, 210°C and 240°C and reaction time of 120, 150 and 180 minutes [3], [5], [6], [7]. The ratio of glycerol with palm oleic acid is fixed variable, 1:1 [5], with MESA catalyst concentration 0.5% w/w [5]. MESA is the catalyst that is highly acidic with pH 2.2 which derived from palm oil derivatives.The flow rate of nitrogen gas is 100 cc/minute and stirring speed at 400 rpm [5]. Characteristics of glycerol ester as reaction products are known by analyzing the acid number, density with a density meter DMA 4500M Anton Paar, viscosity kinematic (ASTM D 445), the boiling point (ASTM D 86), flash point (ASTM D 92), pour point (ASTM D 97), and chemical composition of glycerol ester with Gas Chromatography Mass Spectrometry. The characteristics are compared with standard base oil that used as drilling fluid formulation.

3. Results and Discussion

3.1 Effect of acid numbers on reaction

Acid number analysis was conducted to determine the residual acid contained in the reaction mixture in produced glycerol ester. Such fatty acids residue was correlated with the content of glycerol esters formed during the reaction. Effect of temperature increases and reaction time process on the acid number can be seen in Figure 1. In general, there is a decline in acid number in certain range of time and temperature. The longer the esterification process is the lower acid number. This is because of the longer reaction time of esterification process, the higher fatty acids converted into glycerol ester. In other hands, longer reaction causing more frequent collisions among reactant molecules and in turn the conversion into products increases. Likewise with different temperature treatment. The higher the temperature is the decline in acid number. The lowest acid numbers found at reaction time 180 minutes and temperature at 240°C, namely 1.90 mg KOH/g sample. A decrease in acid number indicates the conversion of fatty acids into the ester. For comparison, oleic acid had acid number 194-204 mg KOH/g sample. After esterification, the acid number of product mix is 1.90 to 1.96 mg KOH/g sample at temperature treatment 240°C. This is in accordance with the yield of the reaction.

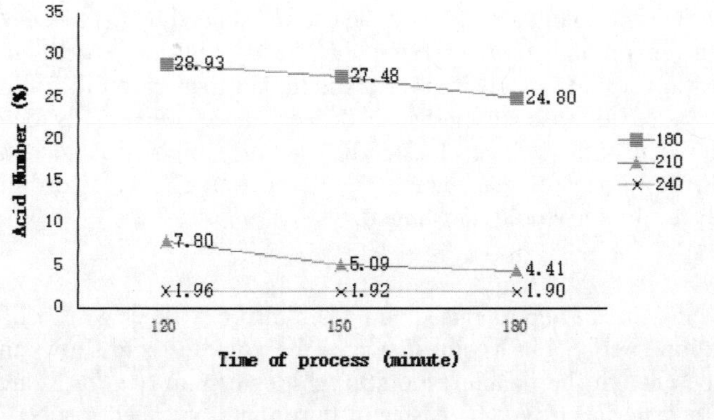

Figure 1. Acid Numbers glycerol ester by temperature and time of esterification process

IOP Conf. Series: Materials Science and Engineering **172** (2017) 012062 doi:10.1088/1757-899X/172/1/012062

The yield of the reaction product is the amount of yield in a chemical reaction or a percentage of yield compared to the raw materials processed. In the esterification process, the yield does not reach 100% because the reaction is reversible and in turn perfect conversion cannot be achieved. Water vapor is one factor that reduces the yield of glycerol ester product if not separated from the reactor because it can hydrolyze glycerol esters into glycerol and carboxylic acids. Based on calculations, the highest yield obtained at reaction time 180 minutes and temperature 240°C with nearly maximum yield 99.24%. The high reaction yield showed that the conversion of fatty acids into ester is high. The higher the purity of glycerol, the more ester bonds are formed so that less fatty acids are formed which shows the declining acid numbers and increased reaction yield.

3.2 Kinematic viscosity of glycerol ester

Kinematic viscosity is a number that indicates the amount of resistance in a liquid material to flow or size of shearing resistance of a liquid material. Viscosity test was conducted to determine the resistance of glycerol ester to flow. Figure 2 shows the effect of temperature and reaction time on the kinematic viscosity. The finding show treatment of temperature increase and reaction time produce viscosity values ranging from 31.5 to 75.6 cSt. The higher the temperature and the longer processing time produce the lower viscosity value. The viscosity values found in this research was lower than the viscosity values generated from previous studies ranged between 81-105 cSt [17].

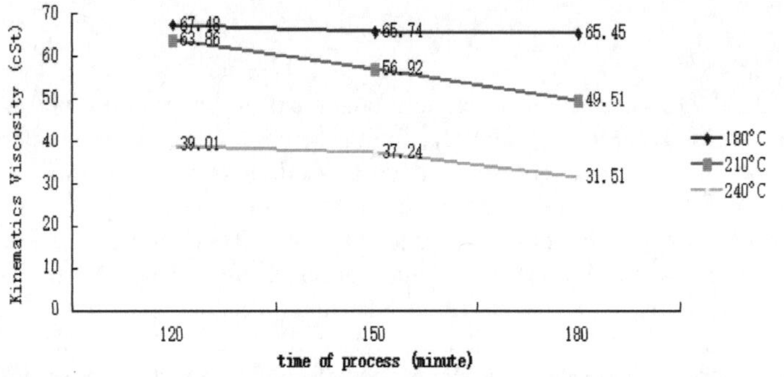

Figure 2. Kinematic Viscosity glycerol ester by temperature and esterification process

The high value of the kinematic viscosity is influenced by carbon chain of oleic fatty acid used in the esterification reaction. Kinematic viscosity values are influenced by the length of the carbon chain, the position and number of double bonds of the fatty acids or alcohols is used in synthesizing ester [13]. The longer the carbon chain is the higher kinematic viscosity value, while the presence of double bonds will reduce the kinematic viscosity value. Kinematic viscosity is also influenced by intermediate compounds such as monoglycerides and diglycerides which having sufficiently high of polarity and molecular weight [14], [17]. The kinematic viscosity of glycerol esters is higher than base oil drilling fluid in marketplace that is 4.5 cSt. High kinematic viscosity indicates the resistance of materials to the changes in temperature. It shows that the resulting glycerol ester has a fairly high resistance on temperature changes, so that when applied to processes that require high temperatures such as drilling fluid OBM, the physical properties of glycerol ester would unchanged.

3.3 Flash point of glycerol ester

Flash point is the temperature at which a number of vapors raised and when mixed with air will form a combustible mixture when given activation energy. Flashpoint indicates the volatility and shows the lowest temperature at which there will be enough flammable vapors to ignite when an *ignition* source is applied. The lower the flashpoint of a material, it will be easier to burn therefore need a specific handling in storage. Results of testing flash point at different temperature and time can be seen in Figure 4.

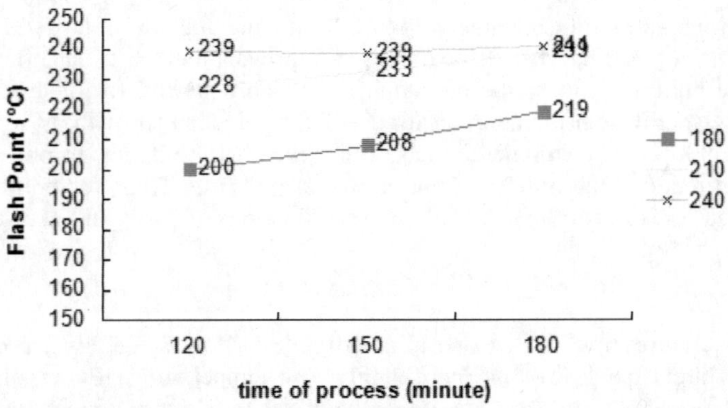

Figure 3. Flash point glycerol ester by temperature and time of esterification process

Generally, glycerol ester produced in this research has high flash point value namely 202°C - 240°C. Figure 4 show that the higher the temperature and the longer the process, the resulting flash point is higher. The higher flashpoint after esterification indicates that there has been a decrease in light fractions content of glycerol ester. In the esterification process, more flammable glycerol components will release a proton from the hydroxyl group to produce activated complex compounds. The release of water and protonated then generate higher flash point of the ester. Flashpoint is influenced by the content of light fractions (residual alcohol) [19], [20], [21]. The lower content of light fraction is the higher the temperature required for ester can be lit. The higher the flash point of a material is the better and the easier for fire handling. The lowest flash point was found from glycerol ester by treatment at 150°C and 120 minutes while the highest was at temperature treatment of 240°C and processing time for 180 minutes. However, the overall flash point finding in this research is still meeting the specification as standardized for oil based mud, namely minimum 85°C for saraline.

3.4 Boiling point of glycerol ester

The boiling point is physicochemical properties that show temperature of glycerol ester starts to boil. Boiling point shows the force of molecules attractions in a fluid, where in the higher attraction force will increase the value of the boiling point. The boiling point is influenced by molecular weight and hydrogen bonding [31], [32]. Figure 4 shows the boiling point of glycerol esters in different time and temperature.

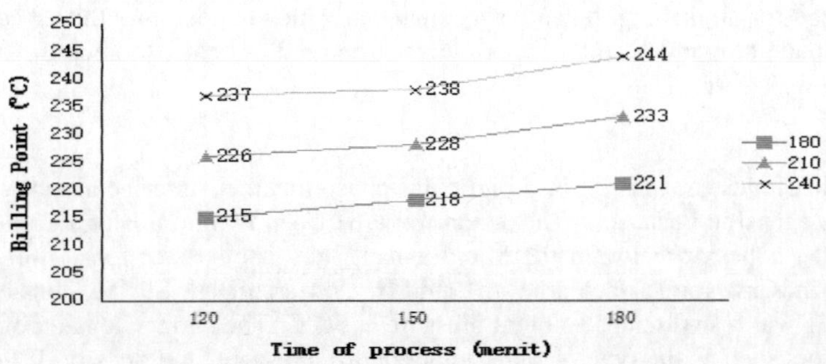

Figure 4. Bolling point of glycerol ester by temperature and time of esterification process

From Figure 4 shows that time and temperature of esterification process affect the boiling point of glycerol ester. The higher of the temperature is the higher value of the boiling point. The higher temperature and the longer reaction time, the boiling point is higher. The highest boiling point found at the temperature 240°C and process time 180 minutes, namely 244°C. In the manufacture of base fluid for drilling fluid, boiling point parameter was important to be observed to prevent the use of base fluid at temperature conditions above the boiling point of the base fluid itself. Particularly, for oil-based base fluid, the high boiling point of glycerol ether is important for use in deep drilling for with oil well temperature and pressure more than 300°C [27].Commercial base fluid like Saraline begins to boil at 200°C. Other commercial base fluid requires the minimum boiling point at 200°C. Thus, for boiling point parameters, glycerol ester produced is in compliance with the specifications of base fluid used as drilling fluid in oil drilling.

3.5 Pour point of glycerol ester
Pour point indicates the lowest temperature in which a material is still able to flow. The lower pour point of a material, such materials are capable of low increasingly low temperatures. Pour point indicates that at the lowest temperature, oil can still flow and when the oil is at its lowest temperature, the oil will be frozen and cannot flow. The analysis results of glycerol ester pour point with different temperature and time is shown in Figure 5.

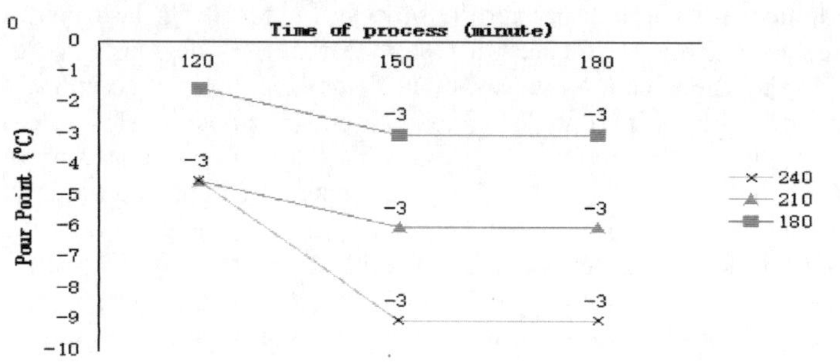

Figure 5. Pour point of glycerol ester by temperature and time of esterification process

Figure 5 shows that pour point of glycerol ester with different temperature and process time is fluctuated, ranging from 0°C, -1.5°C to-3°C. The best average of pour point is process time of 150°C and 180°C namely -3°C. The commercial base fluid has pour point of -20°C and the maximum of 18°C. Thus, for the pour point parameters, glycerol ester produced in compliance with the specifications of drilling fluid in oil drilling. Especially for oil-based mud (OBM), the lower pour point of glycerol esters is important for use in drilling, particularly offshore drilling. By knowing the pour point of a base oil that is used for oil drilling, it can be calculated at which temperature the oil can still be pumped or not or to calculate how much water vapor (steam) required as heater to keep the oil still can be pumped out from the well-bore

3.6 Interfacial tension
Interfacial tension is the force that holds the surface of a particular phase together. Interfacial tension is always smaller than the surface tension because the adhesion force between two liquids that do not mix each other is greater than the adhesion between the liquid and the air. The surface tension of a liquid is the energy required to increase the surface area of liquid by unit area [19]. Surface tension between water and glycerol esters was tested under ambient pressure at 30°C. The surface tension of a liquid is depends on several factors such as the type of fluid, temperature, pressure, and density. If the

liquid molecules are large, such as water, the surface tension is usually high. Likewise, the density, the higher the density of the liquid, the higher the energy required to break the wall surface of the liquid due to the increased attractiveness among molecules. Interfacial tension between water and glycerol esters was tested using tensiometer. Droplet radius of Glycerol ester is 1.475 mm, and the value of interfacial tension is 37.0526 dyne/cm. The test of interfacial tension shows that glycerol esters are not water soluble.

3.7 Chemical composition analysis of glycerol ester by GCMS method

GC-MS testing methods were performed to determine the composition of fatty acid constituent of glycerol ester derived from esterification reaction. Detection of fatty acids types and triglycerides is using gas chromatography (GC), followed by mass spectrometry (MS) analysis. GC method was performed for separation, quantification, and analysis of fatty acids by first made its fatty acid derivatives, as well as MS analysis to determine the fragmentation of saturated and unsaturated fatty acids, and the location of double bonds of fatty acids. Analysis by GCMS method was performed on glycerol ester derived from the greatest conversion of esterification reaction. According to the analysis results, it appears that the number and types of esters from the perfect separation of esterification produce peaks with different retention times (Figure 6). Short chain esters are polar than the long chain. Laws of like dissolve like ester stated that the type of esters with longer chain will be retained in the column, while the short-chain ester will pass along eluted out of the column. Short chain polar will appear earlier than a nonpolar long chain. Fatty acid having a single carboxylic groups and nonpolar hydrocarbon chains, causing the fat does not dissolve in water [1], [31], [33].

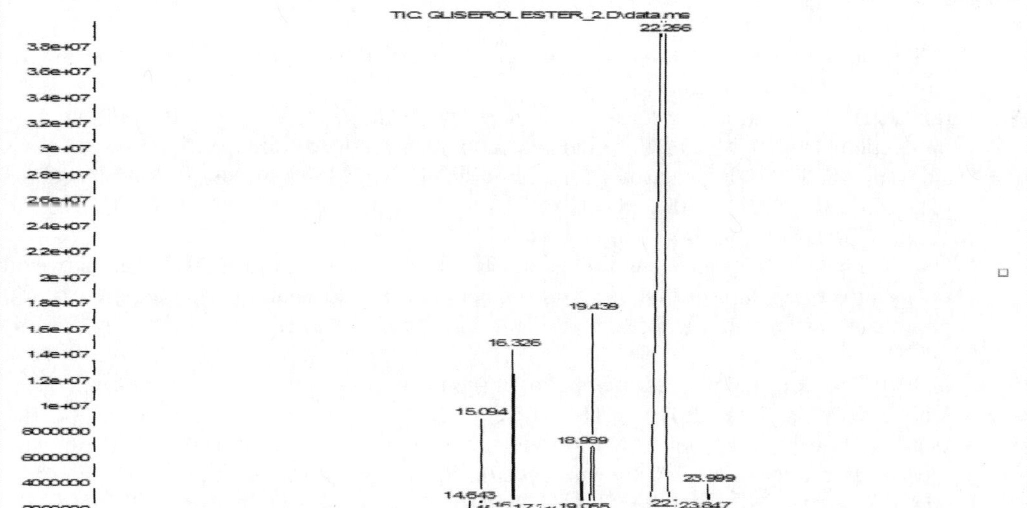

Figure 6.The results of GCMS chromatogram of glycerol ester

Chromatogram analysis shows that glycerol ester consists of ester compounds with 3 highest peaks. First, linoleic acid as the first highest peak that is an unsaturated fatty acid with two double bonds had a retention time of 16.32 minutes and the area 14.32%. Second, palmitic acid, that is long-chain saturated fatty acid with a retention time of 19.42 minutes and area 6.08%. The last highest peak is glycerol monooleate with a retention time of 22.256 (minutes) and area 73.75 (%).

4. Conclusion

Esterification reaction between glycerol as the byproduct of biodiesel industry with palm oleic acid produce higher value-added glycerol ester. One potential use of glycerol esters is as one component in the formulation of drilling fluids for oil drilling needs. To be used in petroleum drilling industry, the glycerol must have characteristics in accordance with the base fluid derived from petroleum derivatives. Raw materials used in the esterification reaction are glycerol with a purity of 97.68%, which is reacted with palm oleic acid and methyl ester sulfonic acid (MESA) as the catalyst from palm oil derivatives. The results showed that the success of the esterification reaction is influenced by the increasing temperature and processing time. The best characteristic of glycerol ester is obtained from the reaction at a temperature of 240ºC and processing time of 180 minutes, with the largest rate of decline in acid number and generates maximum conversion of 99.24%, density of 0.93 g/cm^3, flash point 241ºC, pour point -3ºC, boiling point 244ºC, acid value 1.90 mg KOH/g sample, kinematic viscosity 31.51 cSt (40°C), and surface tension 37.0526. Glycerol esters are used as the base oil in the drilling fluid is a product that is more environmentally friendly and renewable because it comes from palm oil derivatives.

Acknowledgement

The author would like to thank to the Educational Fund Management Institution for research funding support, tools and laboratory analysts in SBRC LPPM-IPB for its assistance in the research.

References

[1] Kirk RE and Othmer DF 1994 *Encyclopedia of Chemical Technology*. Fourth Edition. New York (US): The Interscience Pub.

[2] Pardi 2005 Optimation glycerol monooleat production from byproduct of biodiesel production [Thesis]. Medan: Postgraduate School University of Sumatera Utara.

[3] Handayani AS, Marsudi S, Nasikin M, and Sudibandriyo, M 2006 Esterification reaction of oleic acid and glycerol with acid catalyst. *Jurnal Sains Materi Indonesia:* 102-105.

[4] Kreutzer, U R 1984 *J. Am Oil Chem Soc* **61** 343

[5] Erliza H, Giovanni NP, Ani S, Pudji P, Bonar THM, Ari I S, and Mira R 2014 Prelimary study on the effects of palm methyl ester sulfoic acid catalyst concentration on the esterification process of glycerol 80% with palm oil oleic acid. *American J. of Oil and Chem Techn* **5** 45-47

[6] Susi 2010. The potential utilization of palm oil as monoasilglycerol emulsifier. *Agroscientae*, Vol 17 No 3 December 2010.

[7] Mostafa, Ashraf M and Wael A 2013 Production of mono-, di-, and triglycerides from waste fatty acids through esterification with glycerol. Bioscience and Biotechnology **4** 900-907

[8] Milda G, Violeta M and Eglė S 2011 Esterification of By-products of Biodiesel Fuel Production with Methanol and Technical Glycerol Using Biocatalysts. Environmental Research, Engineering and Management **56** 28 – 34

[9] Dheerendra S, Prafull P, Anuradda G and Sanjay M 2013 Esterification of Oleic Acid with Glycerol in the Presence of Supported Zinc Oxide as Catalyst.Industrial & Engineering Chemistry Research. DOI: 10.1021/ie401636v

[10] Soares VLP, Elizabeth RL, Jorge de ARJr, Luciano NB, and Regina SVN 2011 *New Applications for Soybean Biodiesel Glycerol.* Universidade Federal do Rio de Janeiro Brazil. Applications and Technology, Prof. Tzi-Bun Ng (Ed.). InTech

[11] Westfechtel A, Diana M, and Heinz M 2012 Oligoglycerol fatty acid ester additives for water-based drilling fluids. *U.S. Patent No. 8,148,305.*

[12] Muller H, Maker D, and Herzog N 2012 Oligoglycerol Fatty Acid Ester Thickeners for Oil-Based Drilling Fluids. *US Patent No. 8153562 B2.*
[13] Bajpai D and Tyagi VK 2006 Biodiesel: Source, Production, Composition, Properties and its Benefits. *Joul of Oleo Sci* **10** : 487-502.
[14] Kinney CR 1938 A system corelating molecular structure of organic compounds with their boiling points. *Chemical Laboratory Utah University.* No 51.
[15] Mittelbach M and Remschmidt C 2004 *Biodiesel the Comprehensive Hand Book.* 2nd edition. Boersedruck Ges. MBH, Vienna. ISBN: 3-200-00249-2.
[16] Fessenden R J and Fessenden. 1982 *Organic chemical.* 3th edition. Erlangga.
[17] Gerhard K and Steidley KR 2011 Kinematic viscosity of fatty acid methyl esters: Prediction, calculated viscosity contribution of esters with unavailable data, and carbon–oxygen equivalentss. *Fuel.* **84** 3217–3224
[18] Ryen C 2011 Composition and Properties of Drilling and Completion Fluids. DOI: 10.1016/B978-0-12-383858-2.00011-1 © 2011 Elsevier.Inc.
[19] Adams NJ and Charrier T 1985 *Drilling Engineering: A Complete Well Planning Approach.* PennWell Books: Tulsa, Oklahoma.
[20] Amoco Production Company 1994 *Drilling Fluids Manual.* Texas: Amoco Production Company.
[21] [AOAC] Official Method of Analysis of the Association of Official Analytical Chemist 1999 Washington: AOAC.
[22] [API] American Petroleum Institute 2003 Specification for drilling-fluid materials – Specification 13A. Washington DC: American Petroleum Institute.
[23] [API] American Petroleum Institute 2004 API Recommended Practice Standard Procedure for Field Testing Oil-Based Drilling Fluids 13B-1 3rd Edition: Exploration and Production Departement. Washington DC: American Petroleum Institute.
[24] [ASTM] American Society for Testing and Material 2002 Standard Test Methods for Flash Point by Pensky-Martens Closed Cup Tester. Washington DC: American Society for Testing and Material.
[25] [ASTM] American Society for Testing and Material. 2002 Standard Test Method for Pour Point of Petroleum Products. Washington DC: American Society for Testing and Material
[26] Marbun BTH, Aristya R, Corina AN and Hutapea PA 2013 Innovation of purified glycerol palm oil in drilling fluid system. *American J. of Oil and Chem. Techn.* volume 1 Issue 10, December 2013.
[27] Teow 2010 The Comparison of Saraline, Sarapar, and Diesel Performance as Base Oil at High Temperature and High Pressure. *Proceedings of Malaysian Science and Technology Congress.* University Teknologi Malaysia.
[28] Dalmazzone C, Audibert-Hayet A, Langlois B and Touzet S 2007 *Oil based drilling fluid comprising a temperature-stable and non-polluting emulsifying system.* Patent US No. 7247604
[29] Lee L, Patel AD, and Stamatakis E 2001 *Glycol Based Drilling Fluid.* US Patent No 6291405.
[30] Gunstone, Frank D, John LH, and FB Padley. 1994. *The Lipid Handbook.* London: Chapman and Hall.
[31] Gelosa D, Ramaioli M, Valente G and Morbidelli M 2003 Chromatographic reactors: esterification of glycerol with acetic acid using acidic polymeric Resins. *Ind. Eng. J Chem. Res* **42** 6536-6544
[32] Darnoko D and M Cheryan, 2000, Kinetics of Palm Oil Transesterification in a Batch reactor, *JAOCS 77*, 1263-1267.
[33] Othmer K 1990 *Encyclopedia of chemical technology*, 4th edition. Vol 1: A to Alkaloids. John Wiley & Sons Inc.Qwb

Effect of silanized-chitosan on flammability, mechanical, water absorption and biodegradability properties of pseudo-stem banana fiber and montmorillonite filled waste polypropylene biocomposite

W E Prasetyo[1], A Prihandoko, S Pujiasih, A Widianto, N Rahmawati, O A Saputra, and D S Handayani
Chemistry Department, Faculty of Mathematic and Natural Sciences, Sebelas Maret University Surakarta, Indonesia.

E-mail: wahyuekoprasetyo95@student.uns.ac.id

Abstract. Growing consciousness for an eco-friendly environment has revived the interest to develop composite fibers from biobased products. In this study, flammability, mechanical, water absorption and biodegradability properties of chitosan filled biocomposite waste polypropylene (wPP) reinforced with pseudo-stem banana fiber (PBF) and montmorillonite (MMt) biocomposites has been conducted investigate. It was successfully processed in solution method. Chitosan was chemically treated with glycidyloxypropyltrimethoxysilane (GPTMS) to improve interfacial adhesion between chitosan and wPP. The chitosan treated with GPTMS content in the biocomposites were varied from 0 to 7% (dry wt. basis). Flammability, tensile strength and water absorption index of biocomposites were measured according to ASTM D635, ASTM D638, and ASTM D570 respectively. To study the nature of its biodegradability, the biocomposites were technically buried in garbage dump land. The results show that the addition of treated chitosan 3-GPTMS has improved thermal properties such as Time to Ignition (TTi), Burning Rate (BR), and Heat release (HR) of treated biocomposites compared with neat PP and untreated biocomposite with treated chitosan. The treated biocomposites exhibit higher tensile strength and Young's modulus, but lower elongation at break compared with neat PP and untreated biocomposites with treated chitosan. The biocomposites show a reduction in the rate of water uptake with higher loading of CH.

1. Introduction

Nowadays, plastics especially polypropylene (PP) has been widely used in many fields such as mineral water packaging, electronic and electric industry, automobile and housing due to its good mechanical properties, excellent electrical resistance, low toxicity, low density and good chemical resistance[1-2]. However, PP is not a biodegradable polymer and makes potential hazard to the environment [3]. The degradation of waste PP (wPP) requires a long time, and most of them end up overburdening on landfill. This can make industrial waste creating disposal problems over the world [4]. Together with the other industrial wastes, the effect on the ecological equilibrium of these wastes increases every day and may reach levels to the point of threatening natural life. Therefore, the recycling of the industrial wastes is not only an attempt to decrease environmental pollution but also an effort to increase economical effectiveness. Waste polypropylene (wPP) can be transformed into another useful product such as biocomposite by adding natural fiber, which has higher economic value and saving the environment from pollution [5]. Because of these environmental issues, a variety of biomaterial such as biocomposites has been developed. But the main drawbacks of PP are its inherent flammability and its poor melt strength limits which restrict its application in many fields for safety consideration. The PP also lacks functional groups necessary for good adhesion to synthetic or natural fillers. Consequently, the improvement of flame retardancy of PP is not only necessary but also urgent [6-9].

In the recent years, significant attention has been given to the use of natural fibers and polymer as reinforcing composite materials [10]. Both of natural fiber and chitosan can be obtained from agricultural residues that represented a disposal problem because they can lead to environmental pollution and biomass spoilage. Compared with synthetic fibers, natural fibers, and polymer are low in cost, low density, have highly specific properties and low abrasive nature. Other advantages they are environment-friendly, biodegradable, easily available [11-17]. There are many natural fibers have been used as filler or to reinforce polymer composites, such as cotton, coir, sisal, jute, hemp, sisal, bamboo, kenaf banana and chitosan [18-19]. Pseudo stem banana fiber (PBF) is one of potential candidate natural fiber for biocomposites. Al-Qureshi [20] using banana fiber mixed with epoxy resin and hardener to develop a truck model called Manaca. Corbiere-Nicollier et al. [21] have investigated physically and mechanically of banana fibercement biocomposites. It was reported that kraft pulped banana fiber biocomposite has good flexural strength. Maleque et al (2007) [22] also used pseudo-stem banana woven fabric to reinforce epoxy composite. The results show that the mechanical properties of biocomposites were increases compared to the virgin epoxy. Also, Geethama et al [23] reported that the tenacity and elastic modulus of banana fiber, and are in the range of 529-759 MPa, and 820 GPa, respectively. Due to low density, high tensile strength, high tensile modulus, and low elongation at break of banana fibers, biocomposites based on these fibers have very good potential use in the various sectors, such as automotive industries, construction materials, machinery, household applications, and various plastic products for future industrial applications [24-25]. In the other hand, natural fibers and natural polymer also have some drawbacks, such as large variations in mechanical properties, moisture absorption, and sensitivity to UV radiation, and low resistance to impact between the hydrophobic polymeric matrix and the hydrophilic fibres. However, the main loss of these natural fibre/polymer biocomposites seems to be the incompatibility between the hydrophobic PP matrix and hydrophilic natural fibres [26]. Consequently, the use of compatibilizers or coupling agents is necessary to improve the adhesion between fibre and matrix which yield biocomposites with poor properties [26-29].

Chitosan has very kind properties as a biomaterial. It is the naturally abundant polysaccharide and renewable polymers which have excellent properties such as, biodegradability, bio-compatibility, non-toxicity, and adsorption [30-32]. Due to its superior properties, chitosan has been used as reinforcement in composite materials. It also can be modified by several chemicals to make biocomposites with enhanced properties [33]. The addition of coupling agents or compatibilizers have been used to improve dispersion, adhesion, and compatibility between hydrophilic filler and hydrophobic polymer [34]. The coupling agent provides a stable bond between two otherwise poorly bonding surfaces. A coupling agent is a chemical that functions at the interface to create a chemical bridge between the reinforcement and matrix. Silanization is a convenient method that can be used to cover the biomaterials surface through the covalent binding of organofunctional alkoxysilane molecules and surface hydroxyl groups [35-36]. Glycidoxypropyltrimethoxysilane (GPTMS) and aminopropyltriethoxysilane (APTES) are among of the coupling agents and currently they are most commonly used as coupling agents. Salmah et al [36] have reported that the use of coupling agent like APTMS was able to improve the mechanical properties of polypropylene-chitosan composite. Liu et al 2005 reported that the hydrophilicity organic-inorganic chitosan membrane was able reduced by adding GPTMS [37].

Due to the lack of information about the thermal, physical and mechanical properties of some biocomposites materials, in the present study, the feasibility of using GPTMS to silanize chitosan was evaluated. The effect of the chemical modification of chitosan with GPTMS on the flammability, mechanical properties, water absorption and biodegradability properties of PP/SPP/MMt/chitosan biocomposites were measured. Characterization was analyzed by Fourier transform infrared (FTIR). Examinations with Universal testing machine (UTM), and flammability test were conducted to investigate the physical and thermal properties of PBS/CF biocomposites. Biodegradability was conducted by burial test method.

2. Experimental

2.1. Materials
Raw material waste polypropylene (wPP) was obtained from waste drinking water cup. It was prepared by cleaning the waste drinking cups with ethanol and chopped in 2x2 cm^2. The PBF used was obtained from Banana agricultural field in Salatiga, Central of Java, Indonesia. It was pre-treated by delignification with NaOH 4% (w/v). Then, it was cleaned with ethanol, dried in oven at 40 °C, and grounded into 100 mesh particle size. Chitosan was obtained from PT. Biotech, Indonesia with an average size of 80 mm with the degree of deacetylation (DD) of 88%. The silane coupling agent, γ-glycidyloxypropyltrimethoxysilane (GPTMS), Maleic Anhydride and montmorillonite were obtained from Sigma-Aldrich, Singapore. Acetic acid, Natrium Hydroxide, and ethanol were purchased from Merck. The distilled water was supplied from Chemistry Laboratory of Sebelas Maret University. The formulation of untreated and treated PP/PBF/MMt/chitosan biocomposites with GPTMS in this study is shown in Table 1.

Table 1. The formulation of untreated and treated PP/PBF/MMt/chitosan biocomposites

Ingredients (php)	Biocomposites					
	W1	**W2**	**W3**	**W4**	**W5**	**W6**
wPP	60	60	60	60	60	60
PP-g-MAH	-	10	10	10	10	10
PBF	-	20	17	17	15	13
CH	-	-	3	-	-	-
CH-GPTMS	-	-	-	3	5	7
MMt	-	10	10	10	10	10

2.2. Chitosan modification
The coupling agent GPTMS was diluted in ethanol to make 20% solution. The amount of GPTMS used was 3% by weight of filler. The filler was charged into a mixer and solution was added slowly to ensure uniform distribution of GPTMS. After completion of the GPTMS addition, the filler was continuously mixed for another 30 min. The treated filler was then dried at 100 °C for about 5 h to remove residual acetic acid.

2.3. Preparation of pp/chitosan biocomposites
The synthesis of wPP/wPP-g-MAH/PBF/MMt/CH and wPP/wPP-g-MAH/PBF/MMt/CH-GPTMS biocomposites were reactively processed using benzoyl peroxide (BPO) in solution method using xylene as solvent. The formulation of biocomposites synthesis can be seen in Table 1.

2.4. Characterization and testing of biocomposites

2.4.1. FTIR analyze. Fourier transform infrared (FTIR) spectroscopy analysis of untreated and treated chitosan were carried out with a FTIR (Fourier Transform Infra Red) IRPrestige-21 SHIMADZU Spectroscopy to determine functional group. The biocomposites sample was dispersed in dry KBr powder and was ground to obtain fine particles. The KBr technique was applied. The scanned range was 400-4000 cm^{-1}.

2.4.2. Flammability properties. The flammability of the wPP and biocomposites were tested according to ASTM D635 to determine their time for ignition (TTi), burning rate (BR), and heat release (HR).

2.4.3. Mechanical properties. Tensile strength (TS), tensile modulus (TM), and elongation at break (Eb) of the films were measured by using Universal Testing Machine (model H5KT, with a 1 kN load cell, Tinius-Olsen, Horsham, USA). The samples were cut using ASTM procedure D 638-99. Tensile tests were carried out according to ASTM D-638 type IV.

2.4.4. Water absorption and biodegradability properties. Water absorption studies were performed following the ASTM D 570-95 standard.The biocomposites films were soaked in aquadest for 24 h. The samples were taken out periodically, wiped out the water on the surface of the sample, and weighed immediately to find out the content of water absorbed. Biodegradability was determined by burial test method with cellulolytic bacteria enriched with garbage soil for two months. Soil used in the test was 1:1 mixture of black soil and leaf mold for gardening. Then it was measured the weight loss (LW) of the biocomposites after biodegradation process.

3. Result and discussion

3.1. Material characterization

Chemical analysis of biocomposites and its modification were conducted using FTIR. This technique is very important to analyze the functional group of each material. Moreover, it was also used by the researcher for proving the chemical interaction in compounding.

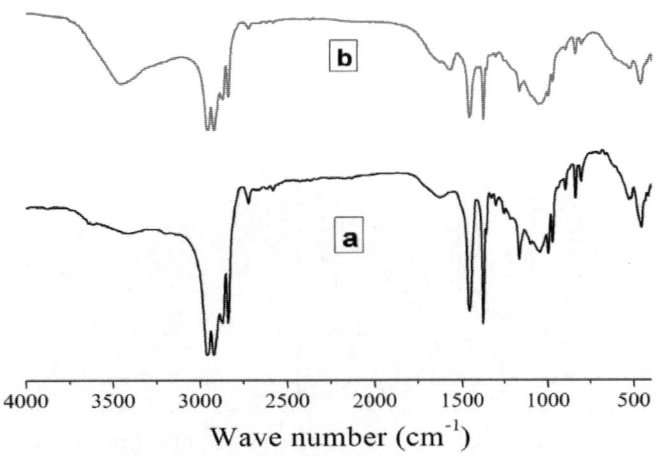

Figure 1. Infra-red spectrum of (a) wPP/wPP-g-MAH/PBF/MMt, and (b) wPP/wPP-g-MAH/PBF/MMt/CH-GPTMS

Figures 1. (a) and (b) exhibits the FTIR spectrum for wPP/wPP-g-MAH/PBF/MMt and wPP/wPP-g-MAH/PBF/MMt/CH-GPTMS biocomposites respectively. It can be seen that wPP/wPP-g-MAH/PBF/MMt (figure 1a) biocomposites exhibit the main characteristic peaks at 3405 cm^{-1} that is corresponded to stretching vibration of hydroxyl group, 2864 cm assigns to C–H stretch vibration, 1633 cm^{-1} associated to C=O stretching vibration of carbonyl groups, 2723 cm^{-1} typicals of PP; 2843, 1446, 1369 and 996 cm^{-1} from methine, methylene and methyl groups [38]. In IR spectrum of wPP/wPP-g-MAH/PBF/MMt/CH-GPTMS biocomposites (figure 10 b) shows the FTIR spectra of treated chitosan with GPTMS. The appearance of a new absorption peak at lower frequency 1627 cm^{-1} corresponds to the $-NH_2$ bending vibration. The peak intensity at 3418 cm^{-1} increases to 3450 cm^{-1}, due to the hydrogen bonding interaction between hydrolyzed amino silane and the –OH group in chitosan. This is further supported by the decreased absorbance for $-CH_2$ bending vibration 1466 to 1459 cm^{-1} which is due to interaction between the amine group of chitosan and OH group of GPTMS. The increasing at peak 1098 to 1163 cm^{-1} indicates the GPTMS has bonded to the chitosan.

3.2. Flammability properties

Figures 2, 3, and 4 exhibit the flammability properties of wPP, wPP/wPP-g-MAH/PBF/MMt composites without chitosan, wPP/wPP-g-MAH/PBF/MMt/CH without the coupling agent, and the wPP/wPP-g-MAH/PBF/MMt/CH composites with coupling agent. Results in figure 2 indicate that of the fillers studied, the addition of 3 php treated CH with GPTMS was the most successful in inhibiting combustion delaying the time to ignition from 1.57 to 6.1 sec and also reducing the HR (figure. 4) from 70 to 59 % for W1 and W4 respectively. The highest TTI, the lowest BR, and HR indicate that the material composites have poor flammability [39]. Both of biocomposites without (W2) and with untreated chitosan (W3) have a higher flammability than biocomposite with treated chitosan (W4, W5, W6). It indicates that the addition of 3 php CH to the wPP increased the flammability about 188%. But, the addition of the 3 php treated CH with coupling agent, GPTMS, affected the flammability when compared to the biocomposites with and without the coupling agents. It could imply that the coupling agents affected the dispersion of the fillers in the polymer matrix. The optimum formula reached at loading of 3% chitosan.

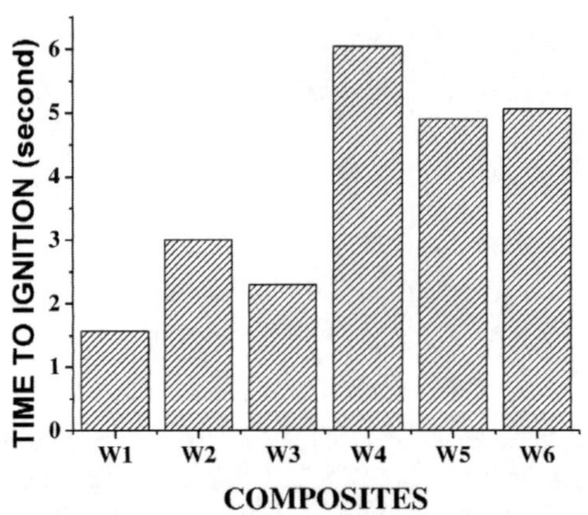

Figure 2. Time to ignition of wPP, untreated, and treated biocomposites.

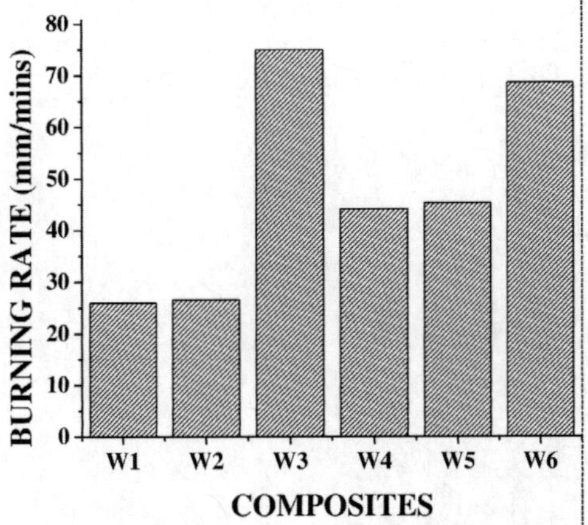

Figure 3. Burning rate of wPP, untreated, and treated biocomposites.

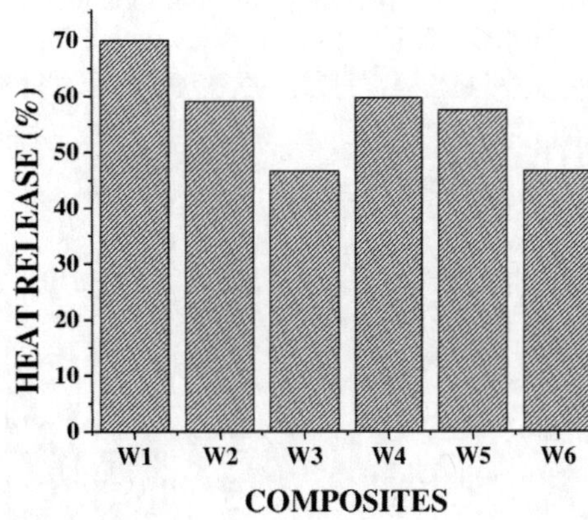

Figure 4. Heat release rate of wPP, untreated, and treated biocomposites.

3.3. Mechanical properties

The main aspect that affects the mechanical properties of the fiber reinforced material is the fiber-polymer matrix interfacial adhesion. Figure 5 exhibits the effect of filler loading on the tensile strength of untreated and treated wPP/wPP-g-MAH/PBF/MMt/CH composites. Compared with wPP/wPP-g-MAH/PBF/MMt composite (W2), the formula W3 has lower TS. The reduced value of tensile strength is due to poor interfacial bonding between hydrophilic filler and hydrophobic polymer matrix [36]. But, in the same composition at 3 php CH (treated with GPTMS), the tensile strength of composite W4 was increased until 75 %. These results exhibit that with the increasing of filler content the tensile strength of composites decreased. This might be due to poor interfacial bonding between filler and matrix polymer at higher filler content. Chemical modification of chitosan by silanization with GPTMS was carried out to reduce the weak bonding between filler and polymer matrix. It also can improve tensile strength of the composites. This is allowed, because that the addition of GPTMS can form a good siloxane network, so it has a high tensile strength due to strong chemical interaction between the matrix of chitosan by the presence of a cross-linked network of the silane groups [37]. The chemical treatment of chitosan with GPTMS was proven effective in enhancing the dispersion, adhesion, and compatibility of a system consisting of hydrophilic chitosan and hydrophobic matrix. It can be seen that the tensile strength of treated composite is higher compared to untreated composites. However, the increasing filler loading of chitosan decreased the TS of W5 and W6 biocomposites. It is because of the poor interfacial bonding between the hydrophilic filler and the hydrophobic matrix polymer.

The results of elongation at break of wPP and biocomposites at different filler loading are depicted in Figure 6. It can be seen that chitosan addition (W3) decreased of % E until 238 %. This is because the addition of chitosan reduced the mobility and increased the brittleness of the composites. But in the same composition, the addition of 3 php treated chitosan (W4) with GPTMS can lead into increased of % E until 58 % when compared with biocomposite without coupling agent (W3). The modification with GPTMS has increased the tensile strength of biocomposites with the enhancement in rigidity and decrement of the ductility of composites, which consequently lowered the elongation at break of the composites [36].

Figure 7 exhibits the effect of filler content on Young's modulus of wPP, wPP, wPP/wPP-g-MAH/PBF/MMt composites without chitosan, wPP/wPP-g-MAH/PBF/MMt/CH without the coupling agent, and the wPP/wPP-g-MAH/PBF/MMt/CH composites with coupling agent. It can be seen that biocomposite without chitosan (W2) has lower MY than wPP. But with the presence of chitosan treated

IOP Conf. Series: Materials Science and Engineering **172** (2017) 012063 doi:10.1088/1757-899X/172/1/012063

with GPTMS can lead into increasing of MY. This is because increasing the filler content significantly increases the stiffness of composites [36]. The Young's modulus of treated wPP/wPP-g-MAH/PBF/MMt/CH composites with GPTMS is higher than untreated composites. This indicates that the presence of GPTMS has enhanced the compatibility between chitosan-polypropylene.

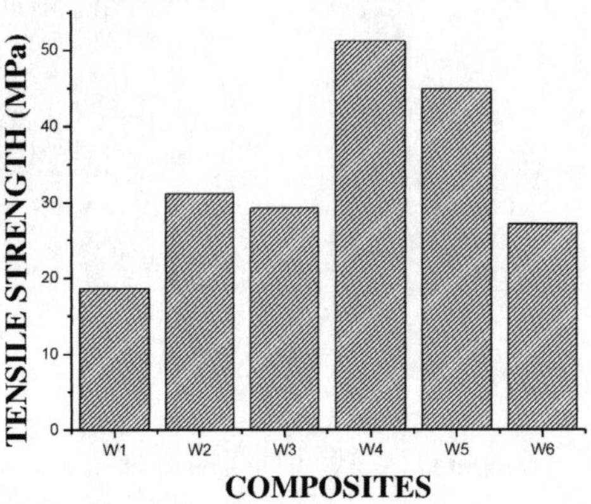

Figure 5. The tensile strength of wPP, untreated, and treated biocomposites.

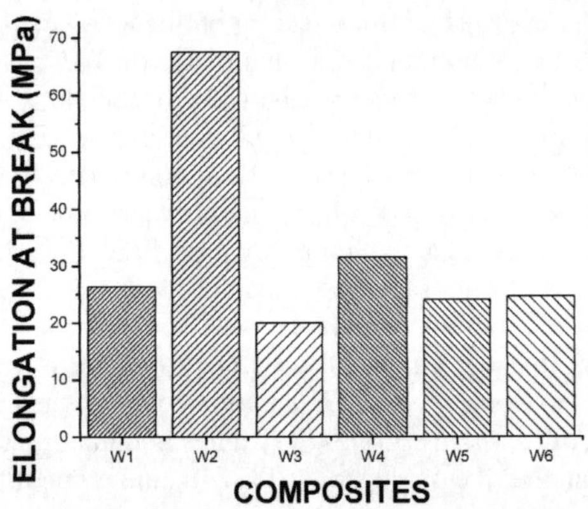

Figure 6. Elongation at Break of wPP, untreated, and treated biocomposites.

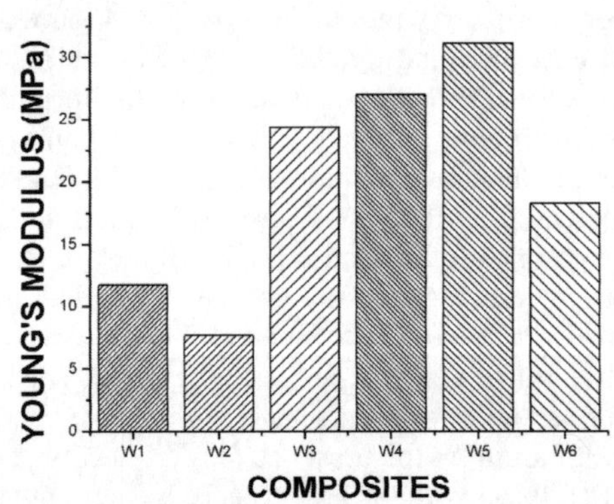

Figure 7. Young's Modulus of wPP, untreated, and treated biocomposites.

3.4. *Water absorption and biodegradability*

Poor resistance of the fibres to water absorption can have undesirable effects on the mechanical properties and the dimensional stability of the biocomposites. Water not only interacts with the matrix physically, chemically, or regarding plasticization, but also attacks the fiber/matrix interface, where it can cause de-bonding [41]. It is very important to study in detail about the water absorption behaviour to estimate not only the consequences that the water absorbed may have but also how this water uptake can be minimized in some ways. The amount of water absorbed in the biocomposites was calculated by the weight difference between the samples exposed to water and the dried samples. Figure 8 shows the percentage of water absorbed for all the samples. The matrix had little effect on the amount of water absorbed as wPP demonstrated around the same water content, still 0.0 % after 24 h of exposure.

However, both biocomposites with (W3) and without (W2) chitosan have higher water absorbed, 16 and 13 % respectively. It indicates that the chitosan addition increased the hydrophilic character on biocomposites. The hydrophilic character of natural fibers is responsible for the water absorption in the biocomposites. Therefore, a higher content on fibers leads to a higher amount of water absorbed. But, the presence of treated chitosan with GPTMS decreased higher amount of water absorbed. It has better resistant to water both in long-term immersion in water. This is because of the coupling agent like GPTMS can build chemical bonds (silanol bonds) and hydrogen bonds, which reduce the moisture-caused fiber-matrix de-bonding [42].

In general, degradation is defined as a detrimental change in its appearance, mechanical, physical properties, and chemical structure of a molecule [43]. The results of biodegradability test can be seen in Figure 9. The presence of PBF and CH can improve the degradation properties. It is indicated that PBF and CH improved the degradation of the biocomposites. As compared to the biocomposites with (W4) and without (W3) silane treatments, the untreated biocomposite showed a slightly higher degradation rate than the treated biocomposite. The improved interfacial adhesion between PBS matrix and CF caused by the silane coupling agent reduces the possibility of water and microorganisms or enzymes (i.e., cellulase and lipase) to penetrate the samples [44], thus resulted in the lower degradation rate.

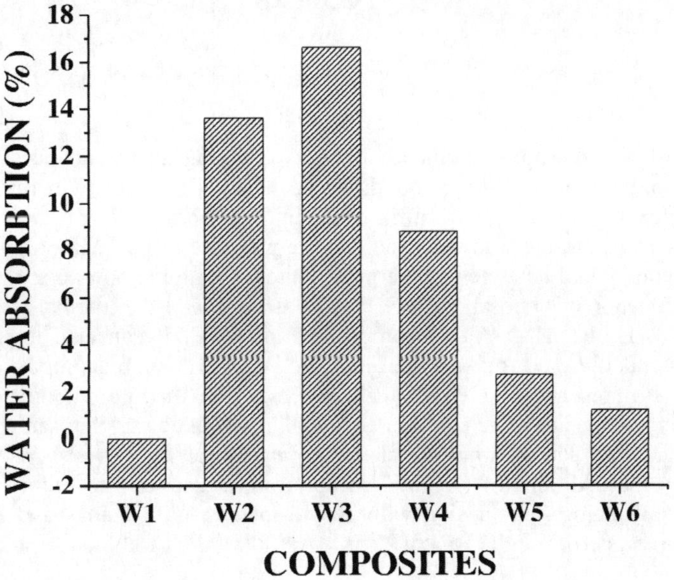

Figure 8. % Water absorption of wPP, untreated, and treated biocomposites.

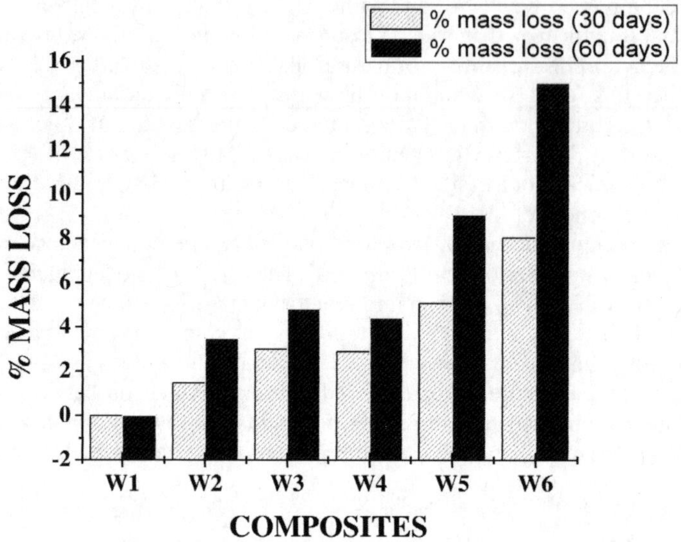

Figure 9. % Mass loss of wPP, untreated, and treated biocomposites.

4. Conclusion

The application of y-glycidoxypropyltrimethoxy-silane as coupling agent enhances the fiber-matrix adhesion because silanol bonds and hydrogen bonds are provided. This leads to increase the interfacial interaction between filler particles and resin matrix which in turn improves the physical and mechanical properties. The results of the present study showed that the presence of GPTMS on biocomposites can reduce the interfacial energy and homogenize the polar chitosan with the non-polar synthetic polymer phase. Chemical modification of chitosan with GPTMS has improved the thermal stability properties both of wPP/wPP-g-MAH/PBF/MMt/CH biocomposites and wPP. It can also improve the tensile strength and Young's modulus, but reduce the elongation at break of biocomposites. The thermal properties show that the presence of GPTMS has increased the thermal stability of wPP/wPP-g-MAH/PBF/MMt/CH biocomposites. The effect of chemical modification of chitosan with GPTMS was obvious from FTIR analysis. The hydrophilic character of natural fiber is also reduced by adding GPTMS. Additionally, the additions of GPTMS on chitosan is also able to reduce the hydrophilic character of chitosan enhancing its adhesion with a hydrophobic wPP matrix and led to improved mechanical and thermal properties of the wPP/wPP-g-MAH/PBF/MMt/CH biocomposites.

Acknowledgement

The authors are very grateful for financial support through PKM-DIKTI grants 2016 from Indonesian General of Higher Education.

References

[1] Sain M M, Balatinecz J, and Law S 2000 *Journal of Applied Polymer Science.* **77** 260-8
[2] Gong J, Niu R, Wen X, Yang H, Liu J, Chen X, Sun ZY, Mijowska E, and Tang T 2015 *RSC Advances* **5** 5484-93
[3] Yildirim M Ş, Bíçer Y, and Yildiz C 1996 *Journal of Porous Materials* **3** 189-91
[4] Xu Y X, Kim K M, Hanna M A, and Nag D 2005 Industrial crops and Products **21** 185-92
[5] Suharty N S, Almanar I P, Dihardjo K, and Astasari N 2005 *Procedia Chemistry* **4** 282-7

[6] Gilman JW, Jackson CL, Morgan AB, Harris R, Manias E, Giannelis EP, Wuthenow M, Hilton D, and Phillips SH 2000 *Chemistry of Materials* **12** 1866-73
[7] Kashiwagi T, Harris R H, Zhang X., Briber R M, Cipriano B H, and Raghavan, S R, Awad W H and Shields, J R, 2004 *Polymer* **45** 881-891.
[8] Song P A, Xu L, Guo Z, Zhang Y, and Fang Z 2008 *Journal of Materials Chemistry* **18** 5083-91
[9] He Q, Yuan T, Yan X, Ding D, Wang Q, Luo Z, Shen TD, Wei S, Cao D, and Guo Z. 2014 *Macromolecular Chemistry and Physics* **215** 327-40
[10] Taj S, Munawar M A, and Khan S 2007 *Proceedings-Pakistan Academy of Sciences* **44** 129
[11] Ayrilmis N, Jarusombuti S, Fueangvivat V, Bauchongkol P, White RH 2011 *Fibers and Polymers* **12** 919-26
[12] Bledzki A K, and Gassan J 1999 *Progress in polymer science* **24** 221-74
[13] Srinivasa P C, Ramesh M N, Kumar K R, and Tharanathan R N 2003 *Carbohydr. Polym.* **53** 431–438
[14] Salmah H, Azieyanti A N 2011 *J. Reinfor. Plast. Compos.* **30** 195–202
[15] Srinivasa P C, Ramesh M N, Kumar K R, and Tharanathan R N *Carbohydr. Polym.* **53** 431–438
[16] Husseinsyah S, Amri F, Husin K, and Ismail H 2011 *Journal of Vinyl and Additive Technology* **17** 125-31
[17] Tanjung F A, Husseinsyah S, and Hussin K 2014 *Fibers and Polymers* **15** 800-8
[18] Satyanarayana K G, Sukumaran K, Mukherjee P S, Pavithran C. and Pillai S G K 1990 *Journal of Cement and Concrete Composites* **12** 117-136
[19] Monteiro SN, Calado V, Rodriguez RJ, Margem FM. *Journal of Materials Research and Technology* **1** 117-26.
[20] Al-Qureshi H A 1999 *InSecond International Wood and Natural Fibre Composites Symposium* 1-8
[21] Corbière-Nicollier T, Laban B G, Lundquist L, Leterrier Y, Månson JA, and Jolliet O 2001 *Resources, Conservation and Recycling* **33** 267-87.
[21] Espert A, Camacho W, and Karlson S 2003 *Journal of Applied Polymer Science* **89** 2353-60
[22] Maleque M, Belal F Y, Sapuan S M 2007 *The Arabian journal for science and engineering* **32** 359-64
[23] Geethama V G, Mathew K T, Lakshminarayana R, and Thomas S 1998 *Journal of Polymer Science* **65** 1483-1491
[24] Bhattacharyya D, and Jayaraman K 2003 *Polymers & polymer composites* **11** 433-40
[25] Wallenberger F T, and Weston N O 2004 *Materials Source Book*
[26] Espert A, Vilaplana F, Karlsson S 2004 *Applied science and manufacturing* **35** 1267-76
[27] Gassan J, and Bledzki A K 2000 *Appl Compos Mater* **7** 373–85
[28] Gassan *J. A Compos* 2002 *Appl Sci Manufact* **33A** 369–74
[29] Gauthier R, Joly C, Coupas AC, Gauthier H, Escoubes M 1998 *Polymer Composites* 19 287-300
[30] Xu Y X, Kim K M, Hanna M A, and Nag D 2005 *Industrial crops and Products* **31**;21 185-92
[31] Dutta P K, and Dutta J 2004 *Journal of scientific and industrial research* **63** 20-3
[32] Hudson S M, and Smith C 1998 *InBiopolymers from renewable resources* 1998-118
[33] Klotzboch T, Wart M, Ansari Y, and Minteer S D 2006 *J. Membr. Sci.* **282** 276-283
[34] Arbelaiz A, Fernandez B, Ramos J A, Retegi A, Llano-Ponte R, Mondragon I 2005 *Composites Science and Technology* **65** 1582-92
[35] Li G, Zhang L, Wang C, Zhao X, Zhu C, Zheng Y, Wang Y, Zhao Y, Yang Y 2014 *Carbohydrate polymers* **101** 718-26
[36] Salmah H, Faisal A, Kamarudin H 2011 *International Journal of Polymeric Materials* **60** 429-40
[37] Saputra O A, Prasetyo W E, Pramono E, Handayani D S, Kurnia, Pujiasih S, Arifagama I 2015 *Prosiding Seminar Nasional Kimia* 181
[38] Xie Y, Hill CA, Xiao Z, Militz H, Mai C 2010 *Applied Science and Manufacturing* **41** 806-19
[39] Suharty N S, Wirjosentono B, Firdaus M, Handayani D S, Sholikhah J, Maharani YA 2008 *Journal of Physical Science* **19** 105-15

[40] Suharty NS, Almanar IP, Dihardjo K, Astasari N 2012 *Procedia Chemistry* **31** 282-7
[41] Nunna S, Chandra P R, Shrivastava S, and Jalan A K 2012 *Journal of Reinforced Plastics and Composites* **31** 759-69
[42] Pérez-Pacheco E, Cauich-Cupul J I, Valadez-González A, Herrera-Franco P J 2013 *Journal of materials science* **48** 1873-82
[43] Kyrikou I, and Briassoulis D 2007 *Journal of Polymers and the Environment* **15** 125-50
[44] Calabia B P, Ninomiya F, Yagi H, Oishi A, Taguchi K, Kunioka M, Funabashi M 2003 *Polymers* **5** 128-41

IOP Conf. Series: Materials Science and Engineering **172** (2017) 012064 doi:10.1088/1757-899X/172/1/012064

Electrosynthesis of Metal-Organic Frameworks (MOFs)Based on Nickel(II) and Benzene 1,3,5-Tri Carboxylic Acid (H_3BTC): An Optimization Reaction Condition

W W Lestari[1], I D Winarni[1] and F Rahmawati[1,2]

[1]Research Group of Porous Materials for Sustainability, Chemistry Department, Sebelas Maret University, Jl. Ir. Sutami 36 A Kentingan, Jebres, Surakarta, Indonesia 57126
[2]Research Group of Solid State Chemistry & Catalysis, Chemistry Department, Sebelas Maret University, Jl.Ir. Sutami 36 A Kentingan Surakarta, Indonesia 57126

Email: witri@mipa.uns.ac.id

Abstract. Electrosynthesis of metal-organic frameworks based on nickel(II) and benzen 1,3,5-tricarboxylic acid (H_3BTC) to form [$Ni_3(BTC)_2$] has been conducted. This study aims to determine the optimum electro-synthetic conditions of [$Ni_3(BTC)_2$] by varying the solvents, electrolytes, as well as the voltages. The optimum condition was determined based on the percent yield of the product which upon washing and drying at room temperature showed pale green precipitate. The materialhas high crystallinityaccording to XRD analysis with the main peak observed at 2θ 19° and 28° and appropriate with [$Ni_3(BTC)_2$] pattern (CCDC No. 1274034). The refinement results using Le Bail methods revealed the Rp and Rwp values are 3.29% and 3.47%, respectively. The coordination between nickel(II) and carboxylate moeties of the linker has been characterized using FTIR and showed significant shift from 1723 $cm^{-1}$to 1608 cm^{-1}. The compound has thermal stability up to 400 °C according to TG/DTA analysis. The SEM analysis confirmed that compound has morphology nanoplates shape with a thickness of 75 ± 0.023 nm. Another interesting feature of the obtained material is the occupancy of the reversible frameworks, which proved after methanol absorption. The optimum condition of the electro-synthesis of [$Ni_3(BTC)_2$] achieved in the methanol with TBATFB (0.1 M) as electrolyte, and the voltage of 15 V at room temperature with a yield of 99.99%.

1. Introduction

Metal-organic frameworks (MOFs) or called as porous coordination polymers are crystalline hybrid porous materials constructed from metal ions or metal oxide clusters and bi-or poly dentate organic linkers based on carboxylate, pyridil, phophonat, sulfonat end group through strong covalent coordinative bonding to form infinite two or three-dimensional structures [1,2]. The tunability in designing MOFs structure by varying the metal cations and functionalized organic linkers resulted very rich variety in MOFs topology and function [1]. The occupancy of unsaturated metal center, large surface area and high porosity found in MOFs make them very potential as functionalized host materials and can be applied in gas storage and separation [3], catalysis [4], drug release [5], sensors and photoluminescence [6], nanoreactors [7], and molecular recognition [8].

Several methods have been developed in MOFs synthesis including microwave-assisted synthesis, mechano and sonochemical, slow evaporation, solvo-hydrothermal, and electrochemistry [9]. The innovative electrochemical synthetic method of MOFs has been pioneered by Mueller *et al.*since 2006 who synthesized several MOFs based on Zn^{II}, Cu^{II}, Fe^{III}, Co^{II} and Mg^{II} with a series of organic linkers under batch and continuous flow operation [10, 11]. In addition, due to the easier handling, efficient reaction time and safety issue, the electro-synthetic method of MOFs is potential to be applied in industrial scale [10]. Several factors affect in the electro-synthesis of MOF among other types of solvents, electrolytes, and applied voltage [12]. Some type of MOFs for instance HKUST-1 [13], MIL-100(Fe), MIL-53(Al), NH_2-MIL-53(Al) and ZIF-8have been synthesized and optimized through electrochemical method [14]. However, MOFs based on benzene 1,3,5-tri carboxylic Acid (H_3BTC) linker with nickel(II) to form [$Ni_3(BTC)_3$] (figure 1) which is isostructural with [$Zn_3(BTC)_3$] and [$Co_3(BTC)_3$] reported by Yaghi *et al.* [15] has not been deeply investigated for electrochemical synthesis. Kumar *et al.* reported the overview the electrosynthesis of [$Ni_3(BTC)_3$]. However, the XRD and other data are not sufficient and not well explained [16]. Thus, in this work we report the optimization reaction condition of electrosynthesis of [$Ni_3(BTC)_3$] by varying the solvents, electrolytes, voltage and study the chemical and physical properties of the obtained material. The existence of unsaturated metal center based on nickel(II) after removal of the coordinated water molecules in [$Ni_3(BTC)_3$] cause the material promising to be used in catalysis under mild reaction condition.

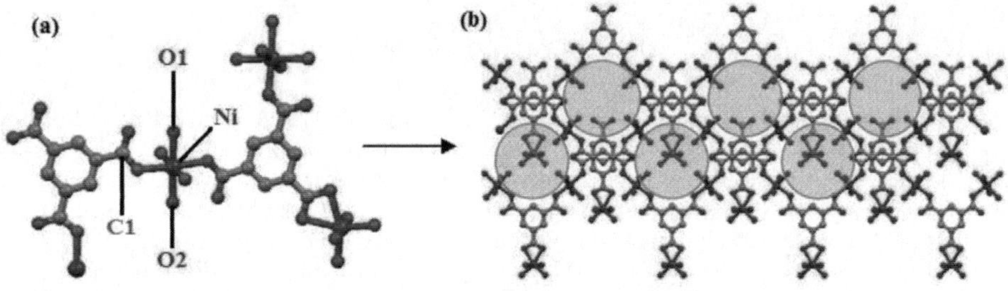

Figure 1. Monomeric unit of [$Ni_3(BTC)_2$] (a) and frameworks depiction of [$Ni_3(BTC)_2$] along c axis (b)[15].

2. Experimental

2.1. Materials and methods

High purity (99.99%) of nickel plate with dimension (16x1x0.01 cm^3) was used as working electrodes. Benzene 1,3,5-tricarboxylic acid (H_3BTC, 95%), tetrabutylammonium tetrafluoroborate (TBATFB, 99%), tributylmethylammonium methyl sulfate (MTBS, 99%) were purchased from sigma Aldrich and used without any further purification. Solvents such as ethanol, methanol and dimethyl formamide (DMF) was obtained from Merck with high grade and used as purchased. The nickel plates were initially activated by immersing in HNO_3 (1 M) then scrubbed with sand paper till shiny before used in electro-synthesis.

2.2. Electrosynthesis of [$Ni_3(BTC)_2$]

H_3BTC (2.10 g, 0.1 M) and the supporting electrolyte TBATFB (0.1 M) was dissolved in 100 mL of solvent (solvent variation were: methanol, ethanol:water (1:1), or DMF) and stirred for 15 minutes till formed an homogenous solution then placed in an electrochemical cell with two nickel plates used as electrode with distance ca. 3 cm a part as shown in figure 2. Electrolysis was run under constant voltage for 150 minutes at room temperature to form a pale green precipitate of [$Ni_3(BTC)_2$] at the cathode. The precipitate was dried first at room temperature for 24 hours and then further dried in an oven at 100 °C for 12 hours and activated at 200 °C for 2 hours. Pale green precipitate turned into yellow precipitate after activation till 200 °C but the color remain unchanged under activation up to 150 °C. To determine the optimum conditions of the synthesis, the applied voltage were varied from 13 to 17 V, while TBATFB concentration varied from 0.02 to 0.1 M. Other than TBATFB, MTBS (0.1 M) was also applied as supporting electrolyte.

Figure 2. Experimental set up of electrosynthesis of [$Ni_3(BTC)_2$].

To test the reversibility of the frameworks, the activated [$Ni_3(BTC)_3$] were immersed in methanol for 24 hours then filtered and dried at room temperature.

2.3. Characterization

The electrosynthesized products were characterized using powder X-Ray diffraction (JEOL JDX-3530 diffractometer) using Cu–Kα radiation (λ = 1.5406 Å), with the voltage and current were held at 40 kV and 30 mA (2θ = 5–50°) at a step size of 0.02°/s. Thermal stability of compound was analyzed using STA Linseis PT-1600 under nitrogen atmosphere with heating rate of 10 °C/minute. Scanning Electron Microscopy FEI type Inspect S50 were used to determine the morphology of the material. FTIR spectra were recorded on Shimadzu IR Prestige-21 PC using KBr pellet in the range of 400–4000 cm^{-1}. BET surface area determined based on BET adsorption isotherms of nitrogen at 77 K using a static volumetric apparatus Micrometrics, NOVA 1200e after the sample activated at 150 °C.

3. Results and discussion

3.1. Optimization reaction condition in electrosynthesis of [$Ni_3(BTC)_2$]

The formation of [$Ni_3(BTC)_2$] through electrochemical method occurred based on redox reaction. The nickel(II) cations formed from the oxidation of Ni metal plate as anode, while the H_3BTC ligand deprotonated and released protons as H^+ and BTC^{3-} indicated by the decrease of pH value from 4.09 to 2.89. The presence of H^+in the system will be further reduced to be hydrogen and released as a side product indicated by the bubble formation in the electrochemical cell. The nickel(II) coordinated further with BTC^{3-} to form main product appeared as pale green precipitate found in cathode. Explanation about the redox process is clearly shown in scheme 1.

Anode	: $Ni_{(s)}$	$\rightarrow Ni^{2+}{}_{(aq)} + 2\,e^-$
Deprotonation	: $2\,H_3BTC_{(l)}$	$\rightarrow 2\,BTC^{3-}{}_{(aq)} + 6\,H^+{}_{(aq)}$
Cathode	: $3\,H^+(aq) + 3\,e^-$	$\rightarrow {}^3/_2\,H_{2(g)}$

Total reaction : $3\,Ni_{(s)} + 2\,H_3BTC_{(l)} + 6\,H^+{}_{(aq)} \rightarrow \underbrace{3\,Ni^{2+} + 2\,BTC^{3-}}_{[Ni_3(BTC)_2]_{(s)}} + 6\,H^+{}_{(aq)} + 3\,H_{2(g)}$

Scheme 1. The electrochemical synthetic process of [$Ni_3(BTC)_2$].

To obtain the maximum product of [$Ni_3(BTC)_2$] several reaction parameters have been varied, such as solvents, voltages, and electrolytes in electrosynthesis for 1.5 h. Data regarding the optimization condition and the corresponding yield are listed in table 1 to table 3.

Table 1. The influence of solvents variation and TBATFB concentration toward reaction yield under contast voltage of 15 V.

Electrolyte concentration (M)	Yield (%)		
	methanol	ethanol : water (1:1)	DMF
TBATFB 0.02 M	- (no precipitate)	-	-
TBATFB 0.1 M	99.99	8.19	-

Table 2. The influence of voltage variation toward reaction yield of electrosynthesis in methanol with TBATFB (0.1 M) as electrolyte.

Voltage (V)	13	14	15	17
Yield(%)	51. 39	43.80	99.99	69.70

Table 3. The influence of electrolytes variation toward reaction yield of electrosynthesis in methanol under constant voltage of 15V.

Electrolyte (0.1 M)	TBATFB	MTBS
Yield (%)	99.99	60.99

According to table 1 to table 3 the optimum condition of electrosynthesis of [$Ni_3(BTC)_2$] achieved at 15 V in methanol with TBATFB (0.1 M) as the electrolyte at ambient temperature. The addition of electrolyte is an important factor not only acting as a media for ions migration but also increasing the charge mobility required in a solution [17]. The increase in electrolyte concentration is directly proportional to the speed of ion transport in the reaction system. The use of methanol as solvent produced the highest yield in the reaction (reaching 99.99%). In this case, methanol and ethanol are the types protic polar solvent that has the ability in dissolving metal cations and H_3BTC ligand better than DMF which is an aprotic polar solvent [18]. In addition, the solvent also affects the deprotonated ligand, in which the use of suitable solvents will make it easier to deprotonate the ligand [12]. The higher voltage applied in the reaction system gave the higher energy supply, thus increase the reaction yield. However, electrosynthesis of [$Ni_3(BTC)_2$ higher than 15 V tend to decrease the reaction product. The reason could be that if the voltage was too high lead to the deformation of the frameworks [13].

3.2. Material Characterization

The obtained [$Ni_3(BTC)_2$] in several voltage and electrolyte variation were characterized by XRD analysis to confirm phase purity and crsytallinity of the product compared to the standart pattern (CCDC No. 1274034). Two main peaks are observed at 2θ 19 dan 28° according to figure 3 and indicates that [$Ni_3(BTC)_2$] have been synthesized successfully.

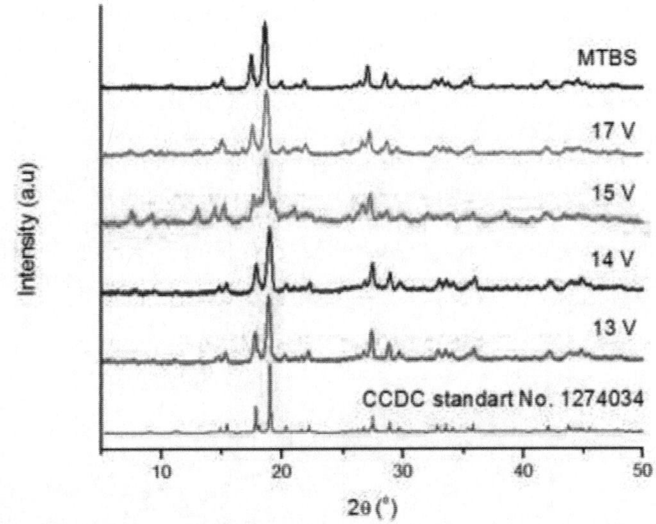

Figure 3. Diffractogram of $[Ni_3(BTC)_2]$ under voltage and electrolyte variations.

Refinement using Le Bail method been carried out to confirm the phase purity of the product and showed that the Rp and Rwp values are 3.29% and 3.47%, respectively. Two phase standard have been applied as comparison, the standard pattern of $[Ni_3(BTC)_2]$ and the data of the ligand, H_3BTC, JCPDS No. 45-1880 and it shown the major peaks correspond to $[Ni_3(BTC)_2]$ and the peaks that not in accordance with $[Ni_3(BTC)_2$ standard could be identified as unreacted ligand with 2θ observed at 2θ13°. Purification through washing or by refluxing the product could be applied in this regard. FTIR analysis have also been conducted and confirmed the formation of $[Ni_3(BTC)_2]$ can be identified from the significant shift of the absorption peak of CO group from 1723 to 1608 cm^{-1} indicate that deprotonation of the carboxylate moieties of the ligand and coordination with nickel(II) have occurred (figure 4). In addition, sharp absorption band at 726 $cm^{-1}$is characteristic for *stretching* vibration of Ni-O in the structure. O-H stretching appeared as broadband observed at 2800-3500 indicating the coordinated water molecule in $[Ni_3(BTC)_2]$ before an activation process.

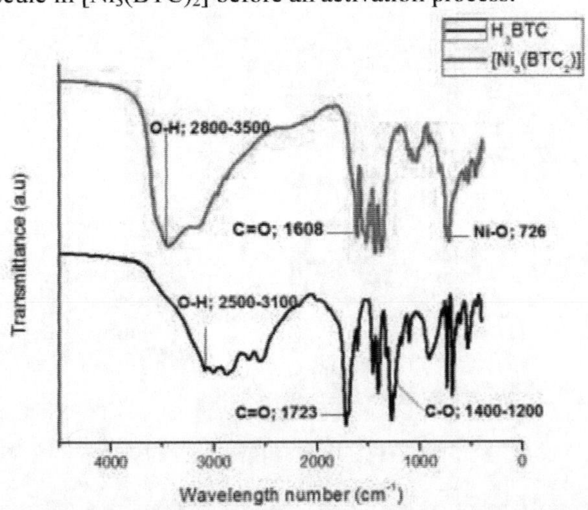

Figure 4. FTIR spectra of H_3BTC ligand and synthesized $[Ni_3(BTC)_2]$ at optimum condition.

To know the water molecules content either as coordinated water or as guest molecule in the surface or pore of H_3BTC thermogravimetric analysis had been conducted. The TG/DTA profile (figure 5) showed 3 step of mass loss with the first step (22.34 %) occurred from RT to 175 °C indicating the removal of ten water molecules as guest molecules occupied in the pore and surface of [$Ni_3(BTC)_2$]. Secondly, (5.02%) in region 175 to 260 °C corresponds to the loss of two coordinated water molecules from the nickel(II) based MOFs affect the change of coordination number from 6 to 4 indicating by the change of the color from pale green to yellow. The third weight loss (36.98 %) observed at a temperature range from 260 to 405 °C assigned as decomposition process of the organic linker. This result is in line with the publication reported by Reimer *et al* [19]. In general, [$Ni_3(BTC)_2$] is thermally stable to 400 °C and afterward, the compound totally decomposed (73.62%) into NiO[20].

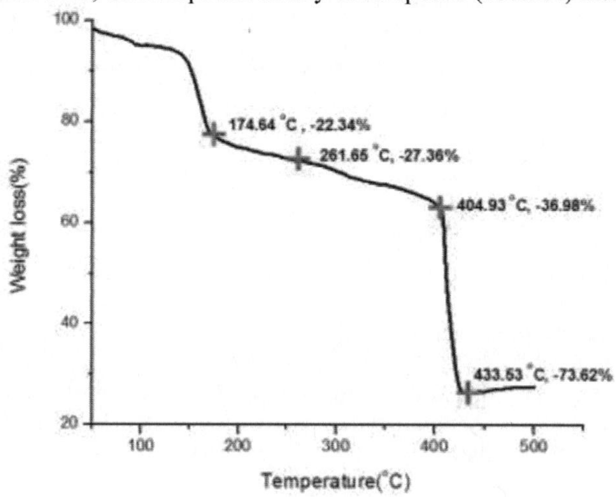

Figure 5. Thermogravimetry analysis of [$Ni_3(BTC)_2$]

The scanning electron microscopy confirmed that the synthesized [$Ni_3(BTC)_2$] has good crystallinity and nanoplate shape with particle size distribution of 40 – 130 nm and average particle thickness of 75±0.023 nm (figure 6).

Figure 6. SEM image of [$Ni_3(BTC)_2$] and thickness histogram of [$Ni_3(BTC)_2$].

3.3. Reversibility test of [$Ni_3(BTC)_2$] frameworks
Color change of [$Ni_3(BTC)_2$] from pale green into yellow after activation until 200 °C for 2 hours indicates the change of coordination number in [$Ni_3(BTC)_2$] from six to four, consequently the

geometry of the compound changed from octahedral into *square* planar [21] (figure 7 b). Moreover, the frameworks structure tend to collapse from 3dimensional structure into sheet structure (2-dimesion) as shown from X-Ray diffractogram. The change of sharp peak intensity into broad peak confirmed that the crystallinity of the compound totally changed into amorf. However, after immersed in methanol for 24 h, the color turned back into the pale green which stipulates the reversibility of the frameworks after contacting with the solvent as observed by XRD analysis (figure 7 c). Thus, the stability of the frameworks is highly dependent on the availability of solvents in the structure. In addition, BET surface area of [$Ni_3(BTC)_2$] obtained from nitrogen sorption isotherm measurement was 4.365 m^2/g.

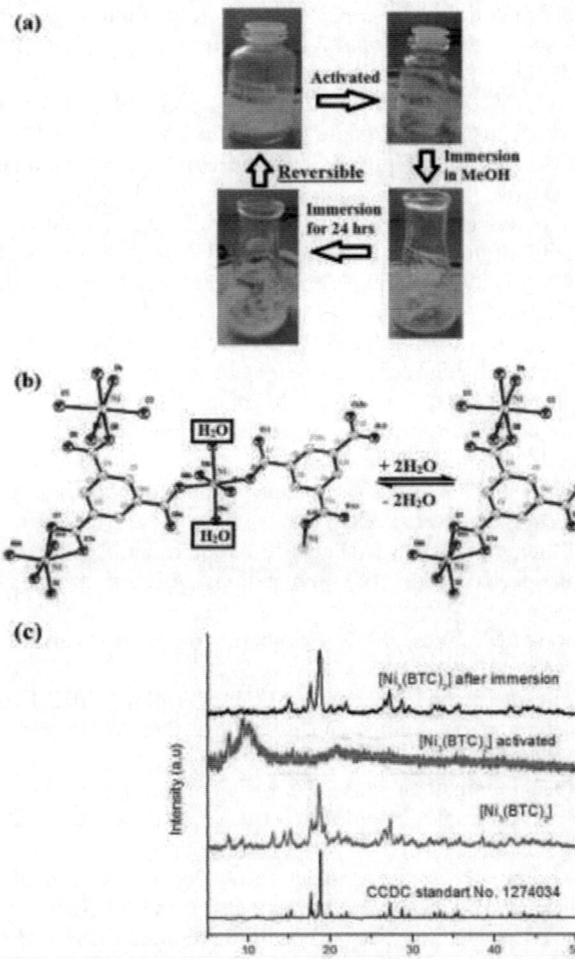

Figure 7. (a) Reversible color change of [$Ni_3(BTC)_2$] after activation and immersion, (b) the change of the geometry and(c)X-Ray diffractogram of [$Ni_3(BTC)_2$] before and after activation and immersion.

4. Conclusions and Outlook

Electrosynthesis of [$Ni_3(BTC)_2$] has been succesfully optimized. The optimum synthetic condition for 1.5 h according to the high percent of yield occurred at 15 V in methanol with TBATFB (0.1 M) as supporting electrolyte. The compound is thermally stable up to 400 °C and potentially to be used as surface catalyst under mild reaction condition due to the occupancy of unsaturated metal center.

Acknowledgements
The authors gratefully acknowledge to MSc. Wahyu Prasetyo Utomo from ITS Surabaya for XRD measurements and ISINAS RISTEK Grant 2015 in the scheme of fundamental research and MRG Grant PNBP UNS 2016 for financial support.

References

[1] Gangu K K, Maddila S, Mukkamala S B,Jonnalagadda S B 2016 A review on contemporary Metal–Organic Framework materials *Inorganica Chimica Acta* **446** 61–74
[2] Öhrström L 2015 Let's Talk about MOFs—Topology and Terminology of Metal-Organic Frameworks and Why We Need Them *Crystals* **5** 154-62
[3] Rosi N L, Eckert J, Eddaoudi, M, Vodak D T, Kim J, O'Keeffe M and Yaghi O.M 2003 Hydrogen storage in microporous metal–organic framework *Science* **300** 1127–29
[4] Lee J, Farha O K, Roberts J, Scheidt K A, Nguyen S T and Hupp J T 2009 Metal organic framework materials as catalysts. *Chemical Society Reviews* **38** 1450–59
[5] Horcajada P, Gref R, Baati T, Allan P K, Maurin G, Couvreur P, Ferey G, Morris R E and Serre C 2012 Metal–Organic Frameworks in Biomedicine *Chemical Reviews* **112** 1232–68
[6] Cui Y, Yue Y, Qian G and Chen B 2011 Luminescent functional metal–organic frameworks *Chemical Reviews* **112** 1126–62
[7] LlabrésiXamena F X and Gacson J 2013 *Metal-Organic Frameworks as Heterogeneous Catalysts* (Cambridge: RSC publishing) page 310-36
[8] Qiu L -G, Li Z -Q, Wu Y, Wang W, Xu T and Jiang X 2008 Facile synthesis of nanocrystals of a microporous metal-organic framework by an ultrasonic method and selective sensing of organoamines *Chemical Communications* **35** 3642–4
[9] Lee Y -R, Kim J and Ahn W -S 2013 Synthesis of metal-organic frameworks: A mini review. *Korean Journal of Chemical Engineering* **30**(9) 1667–80
[10]Mueller U, Schubert M, Teich F, Puetter H, Schierle-Arndt K and Pastré J 2006 Metal–organic frameworks—prospective industrial applications *Journal Materials Chemistry* **16**(7) 626–36
[11] Mueller U, Puetter H, Hesse M, Wessel H, Schubert, M, Huff J and Guzmann M, WO2005049892-A1; EP1687462-A1, 2005
[12] Joaristi A M, Juan-alcan J, Serra-crespo, P, Kapteijn, F and Gascon J 2012 Electrochemical Synthesis of Some Archetypical Zn^{2+}, Cu^{2+}, and Al^{3+} Metal Organic Frameworks.*Crystall Growth and Design* **12** 3489-98
[13] Kumar R S, Kumar S S and Kulandainathan M A 2013 Efficient electrosynthesis of highly acttive $Cu_3(BTC)_2$-MOF and its catalytic application to chemical reduction. *Microporous and Mesoporous Materials* **168** 57–64
[14] Gascon J, Aguado S and Kapteijn F 2008 Manufacture of Dense Coatings of $[Cu_3(BTC)_2]$ (HKUST-1) on α-Alumina. *MicroporousMesoporous Material* **113**(1-3) 132-8
[15] Yaghi O M, Li H and Groy T L 1996 Construction of Porous Solids from Hydrogen Bonded Metal Complexes of 1,3,5-Benzenetricarboxylic Acid**.** *Journal of The American Chemical Society* **118**(38) 9096–101
[16] Kumar R S Studies on Electrochemical Synthesis of Metal-Organic Frameworks and It Multivarious Applications. *Doctoral Thesis* Chapter 7: 123-43
[17] Izutsu K 2009 *Electrochemistry in Nonaqueous Solutions*; (New York, Wiley-VCH Verlag GmbH & Co. KGaA) 171
[18] Loudon G. M 2002 Organic Chemistry 4th ed. (New York: Oxford University Press) 317

[19] Reimer N, Gil B, Marszalek B and Stock, N 2012 Thermal post-synthetic modification of Al-MIL-53–COOH: systematic investigation of the decarboxylation and condensation reaction *Crystall EngineeringCommunications* **14** 4119
[20] Jin L N, Liu Q and Sun W Y 2013 Room temperature solution-phase synthesis of flower-like nanostructures of $[Ni_3(BTC)_2 \cdot 12H_2O]$ and their conversion to porous NiO *Chinese Chemical Letters* **24**(8) 663–7
[21] Alfred R P 1990 Anhydrous Metal Chlorides *Inorganic Syntheses* **28** 22

Characterization of Nanoencapsulated *Centella asiatica* and *Zingiber officinale* Extract Using Combination of Malto Dextrin and Gum Arabic as Matrix

Y Meliana, S B Harmami, W K Restu

Polymer Chemistry Group, Research Center for Chemistry, Indonesian Institute of Sciences Kawasan Puspiptek Serpong, Tangerang Selatan, Banten 15314 Indonesia

Email: yenn001@lipi.go.id

Abstract. This research investigated nanoencapsulation of *Centella asiatica* and *Zingiber officinale* extract. The encapsulated extract was used as a complex matrix of multi-layered interfacial membranes between malto dextrin and gum Arabic. Characterization of nanoencapsulation using Transmission Electron Microscope (TEM), Fourier Transform Infrared Spectroscopy (FTIR) and BET surface area (SA) showed the morphology, functional group and cumulative adsorption in the surface area of pores. The TEM image of the nanoencapsulated powders of *Centella asiatica* and *Zingiber officinale* extract showed a nearly spherical shape with the particle size of 664 nm from its average radius.

1. Introduction

Nanoencapsulation allows protection of sensitive and bioactive pharmaceuticals component, as well as cosmetics and food ingredients from unfavorable environmental conditions, eradication of incompatibilities, solubilization or masking of unpleasant taste or odor. It can be produced as capsules in the range of submicrometer (from 10 to 1000 nm) in the colloidal systems [1-8]. The delivery system is defined as one in which a bioactive material is entrapped in a carrier to control the rate of bioactive release. Carriers can protect a bioactive component from unfavorable environmental conditions e.g. oxidation, pH and enzymes degradation [8-10]. Nanocarriers provide more surface area and have the potential to enhance solubility and improve bioavailability. Furthermore, it can ameliorate controlled release and targeting of the encapsulated food and cosmetics ingredients [11,12].

Centella asiatica can stimulate collagen biosynthesis and increase the incorporation of proline, thus modulating the metabolism of connective tissue. Asiaticoside, this pure triterpene from *Centella asiatica* is known to stimulate vascular and perivascular, with specific activity on collagen type I, whose reduction significant in aging and increased the tensile strength of newly formed skin [13-16]. This active agent is soluble in water and easy to defect at the high temperature. Ginger (*Zingiber officinale*) is known to have antioxidant properties which help to neutralize free radicals in human body. An antioxidant compound in a sufficient amount, substantially maintains the activity of collagenase and elastase in the skin [17]. This both active agent can be protected by encapsulation of malto dextrin and gum arabic with the proper ratio.

Gum arabic, a natural composite polysaccharide derived from exudates of A. Senegal and A. seyal trees, is one of the most common wall material used in encapsulation due to its low viscosity, good emulsifying, high stabilizing and film-forming properties [18,19]. It has the ability of better retention of volatile substances and effective protection against oxidation[20,21]. Gum arabic has a highly branched structure which results in compact spheroidal conformation. Other matrix materials used was maltodextrin which has numerous properties that allow them to be used for diverse purposes in both the food and pharmaceutical industries. Their ability to retain water and form gels explains their choice as efficient food stabilizers in the microencapsulation of vitamins, minerals, colourants as well as fat and oils. However, the poor film-forming ability, hygroscopicity and turbidity of maltodextrins account for their inability to protect volatile compounds during spray drying [22-25]. Therefore, this work studied the nanoencapsulation of *Centella asiatica* extract and *Zingiber officinalle* extract by spray drying with malto dextrin and gum arabic as wall materials. The nanoencapsulation was evaluated for the particles size, BET surface and transmition electron microscope to observe the nanostructure of encapsulated powders. Meanwhile, FTIR finds out the chemical structure of the nanoencapsulation.

2. Material and method

2.1. Material

A sample of *Centella asiatica* (L.) was obtained from plantation area in Solo (Indonesia) with *Centella asiatica* extract using standardized Gotukula VDE by product number: 1614X623N. *Zingiber officinale* herbs extract procured from market herbs (Indonesia), furthermore *Zingiber officinale* extract using standardized Red Ginger VDE by product number: 740X226N with extract grade was nutraceutical. Gum arabic was purchased from Merck KGaA (Germany) with spray dried powder. Ethanol food grade purchased from PT. Brataco (Indonesia) with the purity of 70%. Maltodextrin was purchased from PT. SetiaGuna (Indonesia) and distilled water was obtained from a Milli-Q water purification system (Research center for chemistry, Indonesia). All of the materials mentioned above employed directly except the gum arabic and malto dextrin which was milled by a ball mill.

2.2. Method

2.2.1. Milling of raw materials. Malto dextrin and gum arabic were prepared by milling in a high-energy ball mill machine type Hem-E3D for 4 hours with the ball to powder ratio 10:1, speed of 1450 rpm and cycle time of 10 minutes in order to obtain more fine raw materials and the nano-sized scale.

2.2.2. Preparation of carrier agent solution. The carrier agent, i.e. gum arabic and malto dextrin which would be used as wall solutions were dissolved in distilled water and also were stirred with Thermo scientific cimarec for ±20 minutes until homogen. The solution of carrier agents was then mixed with the core materials i.e*Centella asiatica* and *Zingiber officinale* extract (30% solid contents). The carrier agent and active agent were in a weight ratio of 4: 1.

2.2.3. Preparation of active agent. The core materials (active agent) consisted of asiaticoside extract 2.5% w/w and gingerol extract 1% w/w of solid contents. Furthermore, active agent solution was prepared with ethanol and then stirred with Ultra-Turrax homogenizer Yellow line DI 25 digital for ± 15 minutes with a speed of 10,000 rpm until it became homogeneous.

2.2.4. Preparation of nanoencapsulation. Carrier agent solutions were combined with the active agent containing asiaticoside extract (2.5% w/w solid content) and gingerol extract (1% w/w solid content). Active agent slowly poured drop by drop into 50% of carrier agent solutions and stirred to reach homogeneity with Ultra-Turrax homogenizer Yellow line DI 25 digital for ±30 minutes with a speed of 15,000 rpm. Then, the remaining carrier agent solution was slowly added to 30% solid contents and stirred with Ultra-Turrax homogenizer Yellow line DI 25 for ±20 minutes with a speed of 15,000 rpm until homogen. The last step was spray drying process where 1000-2000 ml of feed solutions were prepared.

2.2.5. Spray drying process. The feed solutions of nanoencapsulation were spray dried to get the product in powder form. The dried process operated at inlet air temperature of 100 – 180°C and outlet air temperature of 60 – 80°C. The temperature of feed solutions was 25°C. The principle of this technique was based on dissolving or dispersing the active agent in solution.

2.3. Characterization of nanoencapsulation

2.3.1. Particle size analyzer (PSA). 5 mg nanoencapsulation was dispersed in 3 ml distilled water. The particle size was determined by Beckman Coulter DelsaNano C (USA). The measurement of particles size distribution and a polydispersity index of samples was done at 25°C and sample was read three times.

2.3.2. Fourier transform infra-red (FT-IR) spectroscopy. FT-IR spectroscopy was a technique based on the vibration of the atoms of a molecule. An infrared spectrum was commonly obtained by passing infrared radiation through a sample and determination what fraction of the incident radiation was absorbed in a particular energy. Chemical structure of the obtained nanoencapsulation by FT-IR spectroscopy using IR Prestige-21 Shimadzu.

2.3.3. Transmission electron microscope (TEM). Transmission electron microscope (TEM) was applied to observe the morphology of nanoencapsulate product, carrier agent and solution of nanoencapsulation before spray drying process. A number of samples were placed on one surface of a double-faced adhesive tape stuck to the sample stage, then coated with copper under a vacuum condition to enhance their electric ability and observed on a JEOL (JEM-1400) electron microscope.

2.3.4. BET surface area (SA) analysis. Surface area affects for example dissolution rates of pharmaceuticals, the adsorption capacity of air, water purifiers, the activity of an industrial catalyst and the processing of most powders and porous materials. For determining the ability adsorption of nanoencapsule, it can be obtained with surface area measurment by TriStar II 3020 Version 2.00.

3. Resultsand discussion

3.1. Particle size analyzer

In the analysis, the particle size has a multi modal particle size distribution. Table 1 shows the results of particle size and polydispersity index. Raw materials of maltodextrin and gum arabic that were milled previously have a radius of 555.6 nm and 708.95 nm, respectively. Meanwhile, the polydispersity indexes are 0.411 and 0.511, respectively. The smallest particle was produced by nanoencapsulation of *Centella asiatica* and *Zingiber officinale* extract has the mean radius of 664.1 nm and has a low polydispersity index of 0.485. It indicates a fairly uniform size that represents homogenous particles. This uniform size also was influenced by the composition of the carrier agent as the wall in encapsulated material. The appropriate composition can obtain a better stability and accuracy release.

Tabel 1. Particle size and polydispersity index

Sample	Particle size in radius (nm)	Polydispersity index (PDI)
Maltodextrin	555.6	0.411
Gum arabic	708.95	0.511
Nanoencapsulation of of *Centella asiatica* and *Zingiber officinale* extract	664.1	0.485

3.2. Fourier transform infra-red (FTIR) spectroscopy

Figure 1 shows the FT-IR spectra of nanoencapsuled materials containing active agent (*Centella asiatica* and *Zingiber officinale* extract) and the carrier agent. The FT-IR spectra of *Zingiberofficinale* extract (Fig. 1a) shows the peak at a wave number of 1605 cm^{-1} which is identified as C=C vibration of in an aromatic ring. Meanwhile, the peaks at wave numbers of 2930 cm^{-1} and 3475 cm^{-1} are stretching vibration of C-H and –OH. The results are confirming that the gingerolwas contained in *Zingiber officinale* extract. The IR spectra of *Centella asiatica* extract (Fig. 1b) shows some principal peaks at wave numbers 1050 cm^{-1}, 1450 cm^{-1}, 1750 cm^{-1}, 2950 cm^{-1} and 3450 cm^{-1} refer to the stretching of C=C, C-O-C, C-C, C-H and –OH that confirming the purity of *Centella asiatica* extract was in the form of compounds asiaticoside. From the spectra, we can see that (Fig.1c-e) the intensity of peaks in 1637 cm^{-1} came from the aromatic ring (C=C) vibration sources from gingerols. Stretching vibration of –OH at peaks in 3437 cm^{-1} and stretching vibration of C=O at 1627 cm^{-1}, stretching vibration of –OH at wave number 3437 cm^{-1} from gum arabic and malto dextrin. Others peak in 2927 cm^{-1} was stretching vibration of C-H on the benzene ring from asiaticoside. It can be said that the encapsulated material contained the extract and also the matrix from the certain peaks that it showed.

3.3. Transmition electron microscope (TEM)

Morphological analysis of the nanoencapsule product, carrier agent and solution of nanoencapsulation before spray drying process were conducted using transmission electron microscope (TEM). Figure 2(a) shows some irregularly shaped particles and also some spherical shape.The morphology of the outer surface of the capsule (Fig. 2(b)) show the present of carrier agents i.e combination of gum arabic and malto dextrin and also the morphology of the encapsulated-material. The morphology of carrier agent without an active agent was clearly observed as a spherical shape, respectively.The nanoencapsule product was successfully formed with an active agent from *Centella asiatica* extract and *Zingiber officinale* extract as it is shown in Figure 2(c) that describes spherical shapes with a smooth surface. The TEM images do not show any pores or crack in the nanoencapsulated-form of *Centella asiatica* and *Zingiber officinale* extract with the carrier agent as wall material. The figure also showed the morphology of outer surface of the nanoencapsule which resulted from the interaction of inhomogeous reaction kinetics, fluid-induced shear force and shell-determined elastic forces [26]. However, the size of nanoencapsulated material was slightly bigger than the carrier agent itself. This is due to the ability of encapsulated active materials to expanse the size of nanoparticles [27].

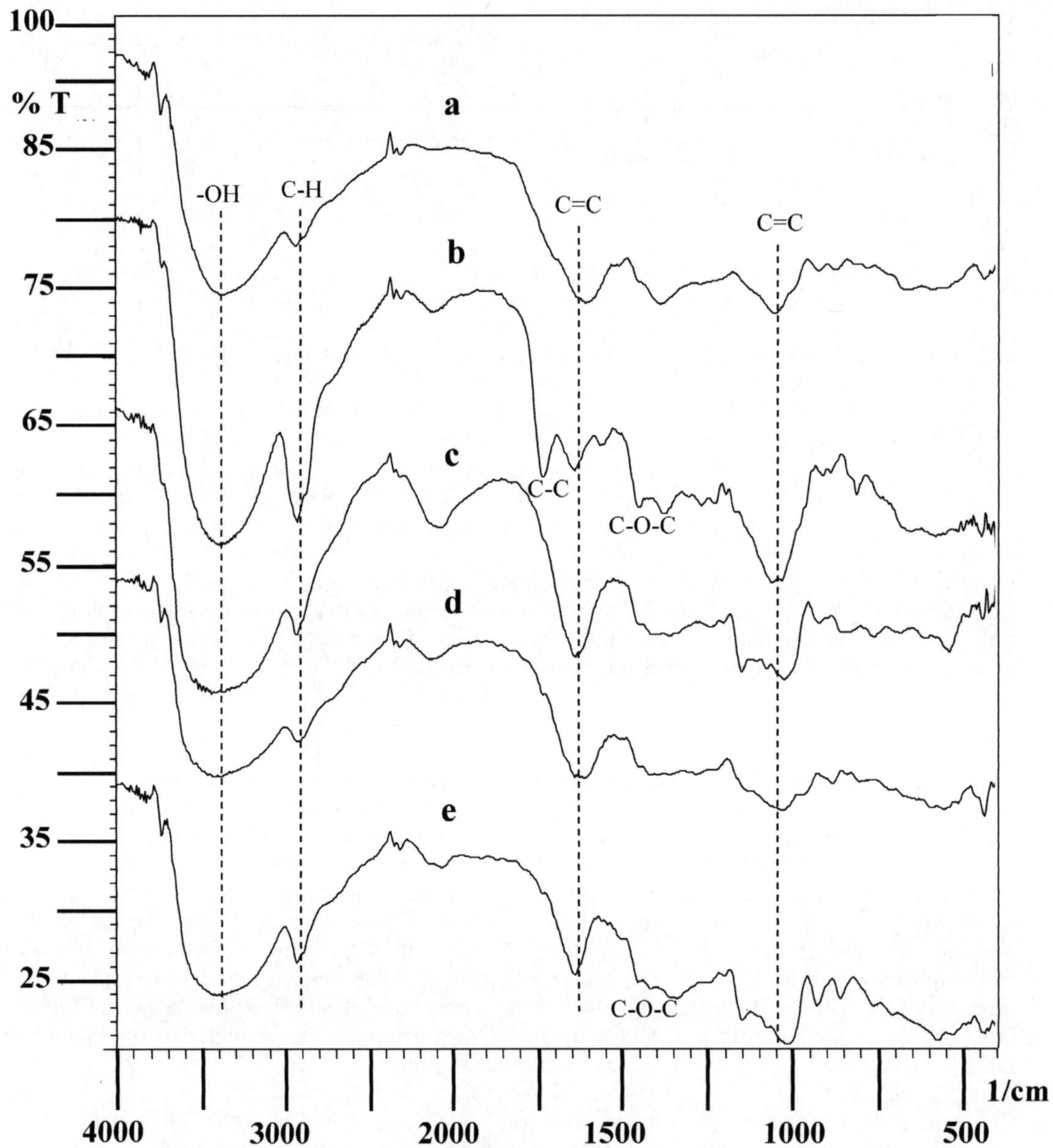

Figure 1. FT-IR spectrum: (a) *Zingiber officinale* extract; (b) *Centella asiatica* extract; (c) Gum arabic; (d) Dextrin and (e) nanoencapsulation of *Centella asiatica* extract and *Zingiber officinale* extract

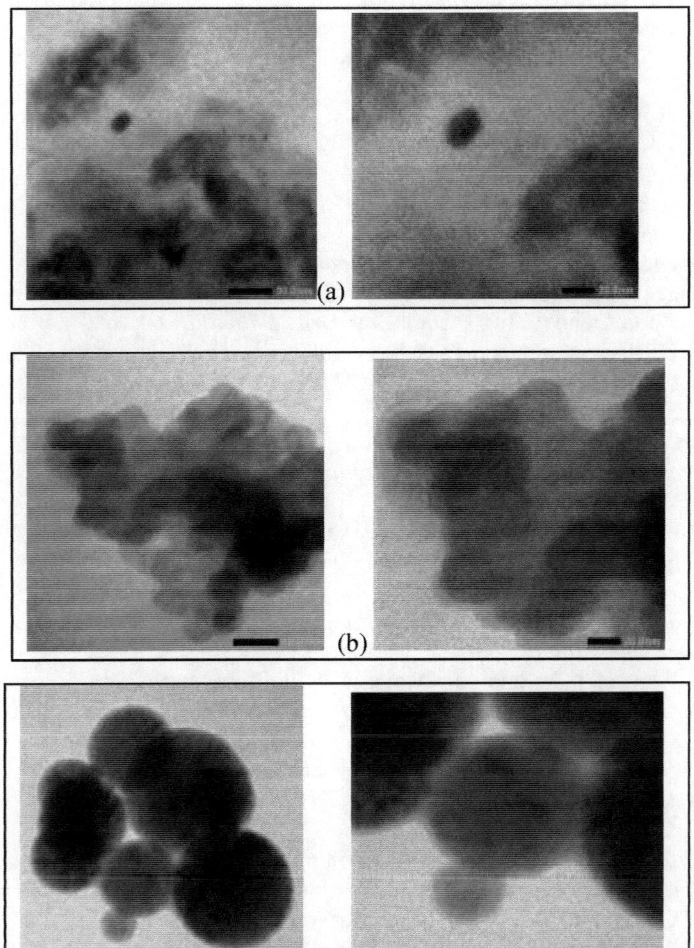

Figure 2. Transmission electron microscope (TEM) image: (a) solution of nanoencapsulation before spray drying process, (b) The wall (carrier agent), (c) nanoencapsule powder of extract *Centella asiatica* and *Zingiber officinale*

3.4. BET Surface area (SA) of nanoencapsule

A surface area analysis of nanoencapsulated-powder was conducted to describe the changes in multilayer adsoption of inert molecules onto a surface of nanoencapsule. The results obtained a single point surface area of nanoencapsule that was 0.3876 m^2/g, cumulative adsorption volume of pores was 1.6x10-4 cm^3/g. The analysis assumes that adsorption occurs by multilayer formation and that the number of adsorbed layers was infinite [28].

4. Conclusion

Nanaoencapsule of *Centella asiatica* and *Zingiber officinale* extract using malto dextrin and gum arabic as a matrix has an average radius of 664.1 nm. Morphology of this product has a spherical shape and smooth surface with the surface area at 0.3876 m^2/g. This work would promote the application of this nanoencapsulation system product for cosmetic and pharmaceutical industries.

Acknowledgements
The financial support to the work by Indonesian Institute of Sciences, Republic of Indonesia, is gratefully acknowledged.

References

[1] Fathi M, Mozafari M R and Mohebbi M 2012 *Trends Food Sci. Technol.***23** 13
[2] Farokhzad O C and Langer R 2009 *ACS Nano***3** 16
[3] Liu Z, Jiao Y, Wang Y, Zhou C and Zhang Z 2008 *Adv. Drug Deliv. Rev.***60** 1650
[4] Müller R H, Petersen R D, Hommoss A and Pardeike J 2007 *Adv. Drug Deliv. Rev.***59** 522
[5] Lee B H 2014*Fundamental of Food Biotechnology: Micro- and Nanoencapsulation*Oxford: Wiley Blackwell,Chapter 1, Second Edition, 131
[6] Shah R B *et al.* 2007 *Int. J. Pharm.***341** 189
[7] Shimoni E 2009*Nanotechnology for foods: delivery systems. Global issues in food science and technology* San Diego: Academic Press 411
[8] Fang Z and Bhandari B 2010 *Trends Food Sci. Technol.***21** 510
[9] Ghosh A, Mandal A K, Sarkar S, Panda S and Das N 2009 *Life Sci.***84** 75
[10] Zimet P and Livney Y D 2009 *Food Hydrocoll.***23** 1120
[11] Mozafari M R2006 *Bioactive entrapment and targeting using nanocarrier technologies: an introduction. Nanocarrier technologies*Netherlands: Springer 1
[12] Weiss J, Gaysinsky S, Davidson M and McClements J 2009 *Nanostructured encapsulation systems: food antimicrobials.Global Issues in Food Science and Technology*San Diego: Academic Press425
[13] World Health Organization 1999 WHO monographs on selected medicinal plants.Essent. Med. Heal. Prod. Inf. Portal**1** 183
[14] Kartnig T 1988*Spices and Med. Plants***3**46
[15] Bradwejn JT et al. 2000 J. of Clin.Physopharmacol. **20** 680
[16] Jew, S. S. *et al.* 2000*Bioorg. Med. Chem. Lett.***10** 119
[17] Zancan K C, Marques M O M, Petenate A J and Meireles M A A 2001*J. Supercrit. Fluids***24** 57
[18] Ali B H, Ziada A and Blunden G 2009*Food Chem. Toxicol.***47** 1
[19] Sarkar S, Gupta S, Variyar P S, Sharma Aand Singhal R S 2013*Carbohydr. Polym.***95** 177
[20] Gabas A L, Telis V R N, Sobral P J A and Telis-Romero J 2007*J. Food Eng.***82** 246
[21] Righetto A M and Netto F M 2005*Int. J. Food Prop.***8** 337
[22] Tombs M P and Harding SE 1998 *An introduction to the polysaccharide. In Biotechnology*Gunpowder Square, London: Taylor and Francis Ltd. 88
[23] Toure A, Zhang X M, Jia C S and Dong Z J2007 *Intl. J. Dairy Sci.* **2** 387
[24] Chronakis I S1998*Crit. Rev. Food Sci. Nutr.***38**599
[25] Raja K C M, Sankarikutty B, Sreekumar M, Jayalekshmy A and Narayanan S 1989*Starch/Stärke***41** 298
[26] YangJ, KellerM W, MooreJ S, WhiteS R and SottosN R2008 *Macromolecules*, **41** 9650
[27] Okonogi S, Sirithunyalug Jand Chen Y 2008X*VI th International Conference on Bioencapsulation*, Dublin, Ireland. **P03**1
[28] Adamson AW and Gast A P1997*Physical Chemistry of Surfaces, 6th ed*Newyork: Wiley John Willey & Sons, Inc.

IOP Conf. Series: Materials Science and Engineering **172** (2017) 012066 doi:10.1088/1757-899X/172/1/012066

Enrichment of Nanodiamond Surfaces with Carboxyl Groups for Doxorubicin Loading and Release

Y Astuti[1], F D Saputra[1], S Wuning[1], Arnelli[1], and G Bhaduri[2]

[1]Chemistry Department, Science and Mathematics Faculty, Diponegoro University, Central Java, Indonesia 50275
[2]School of Chemical Engineering and Advanced Materials, Newcastle University, UK

Email: yayuk.astuti@live.undip.ac.id

Abstract. In their pristine state, nanodiamond crystals produced via detonation techniques containing several functional groups present on the surface including amine, amide, alcohol, carbonyl, and carboxyl. These functional groups facilitate nanodiamond to interact drugs so as to nanodiamond is potential for medical application such as drug delivery. Even though research on the use of nanodiamond for this application has been conducted widely, research on the effect of enrichment of nanodiamond surface with carboxyl functional groups for drug loading and release has not been explored extensively. Therefore, in this paper, the effect of carboxyl-terminated nanodiamond (ND-COOH) on drug loading and release will be presented. The enrichment of nanodiamond with carboxyl groups was undertaken by treating nanodiamond with sulphuric acid and nitric acid. The results show that the doxorubicin (DOX) loading and release efficiencies of ND pristine are higher than that of ND-COOH.

1. Introduction

Nanodiamond can be synthesized by several methods such as shock wave, chemical vapour deposition (CVD), high pressure and high temperature (HPHT) and detonation. Synthesis using detonation method generates diamond with the particle size of about 4-5 nm [1]. Due to their excellent properties of biocompatibility, non-toxicity, large surface area, nitrogen vacancy, high thermal conductivity, high hardness nanodiamonds have found applications in various fields such as drug delivery, photonic devices, cellular bio-imaging and composite [2]. One of the most interesting chemical properties of nanodiamond is that its surface can be easily modified with other molecules or functional groups[3, 4] making them suitable for medical application such as drug delivery [5, 6].

Huang et al. [7] were the first to use nanodiamonds for drug delivery. They successfully loaded doxorubicin hydrochloride (DOX), one of the drugs commonly used in cancer chemotherapy, onto the surface of functionalized nanodiamond and successfully delivered the drug into human cells of colorectal carcinoma with efficacy [7]. Recently, significant efforts have been undertaken to understand and improve the chemical properties of nanodiamond surface and its effect on the efficiency of drug loading using simulation tools such as density functional-based tight binding (DFTB) and molecular dynamics (MD) [8-10]. It has been suggested that replacement of the functional groups present on the surface of nanodiamonds such as amine, amide, alcohol, carbonyl, and carboxyl with the

IOP Conf. Series: Materials Science and Engineering **172** (2017) 012066 doi:10.1088/1757-899X/172/1/012066

carboxylic acid functional group provides as a good mediator for the interaction between nanodiamond surface and drugs [10-12].

Research on the development of nanodiamond as drug carrier has been widely conducted for 5 years [13-19]; however a detailed study of the effect of nanodiamond surface enrichment with specific functional groups to the drug loading and release efficiencies of this material has not been extensively explored. Therefore, this paper presents the study of the relationship between the nanodiamond surface enriched with carboxyl groups and its ability for the drug (DOX) loading and release.

2. Experimental section

2.1 Materials

Nanodiamond purchased from Changsha 3 Better Ultra-hard Materials Co.,Ltd, China with the size of about 5-50 nm, 98% H_2SO_4, HNO_3 and aquabidest. All reagents are of analytical grades purchased from Merck Index Indonesia.

2.2 Preparation of carboxyl-enriched nanodiamond

Synthesis of ND-COOH was carried out according to the procedures in Ref. [20] with slight modification. 500 mg of ND-pristine had been heated in a mixture of strong acid (98% H_2SO_4 and 65% HNO_3 at the ratio 3 : 1) at 90 °C in water bath for 10 hours, equipped with a stirrer. To eliminate the remaining acid content, the addition of distilled water into the sample. Nanodiamond with acid separation is done by centrifugation at 6000 rpm. Subsequently, the supernatant was dried in an oven at a temperature of 35 ° C for 1 hour and 80 ° C for 4 hours.

2.3 Characterisation

Several characterization techniques were applied including FTIR, TEM, TGA and PSA. FTIR was performed using FTIR Prestige 21 Shimadzu, meanwhile TEM images were obtained using JEOL JEM 1400. A thermogravimetric analyzer (TGA 4000 Perkin Elmer) was used to measure the thermal weight loss of functional groups at the temperature range of 530–900 °C with heating rate of 5 °C/min under nitrogen stream.

2.4 Doxorubicin loading and release

2.4.1 Preparation of standard solution

The standard solution was prepared by dissolving doxorubicin (DOX) into aquabidest into buffer acetate (pH 7.4) at different concentrations: 10 ppm, 15 ppm, 20 ppm, 25 ppm, 30 ppm and 35 ppm. The measurement of the absorbances of every concentration was conducted using UV-Vis spectrophotometer (UV-Vis Shimadzu UV-1201) at the wavelength of 480 nm.

2.4.2 Doxorubicin loading onto nanodiamond

Doxorubicin loading onto nanodiamond was carried out by dissolving DOX into buffer phosphate (pH 7.4) and then stirred for 30 minutes. After that, different concentrations of DOX was added 10, 20, 30, 40, 50, 60 ppm onto 0.003 g ND. The mixture was then stirred back for 1 hour. Moreover, the mixture was then centrifuged for 10 minutes and 6000 rpm in order to separate the nanodiamond loaded with DOX and unadsorbed DOX. Subsequently, DOX unadsorbed was analyzed using UV-Vis spectrophotometer at 480 nm to determine the loading DOX on ND surface.

2.4.3 Releasing of nanodiamond

Samples applied for drug *release* were ND-DOX and ND-COOH-DOX from DOX loading. The samples were re-suspended in 10 mL buffer phosphate(pH 7.4) and stirred for 30 seconds. After that, they had been incubated for 1 hour at 37^0C. The next step was the samples had been stirred back for 10 seconds and centrifuged 6000 rpm for 10 minutes to separate nanodiamond and filtrate. The filtrates containing DOX were then analyzed using UV-Vis spectrophotometer at 480 nm.

3. Results and Discussion

3.1 Characterisation of nanodiamond after treatment

3.1.1 Fourier Transform InfraRed (FTIR)

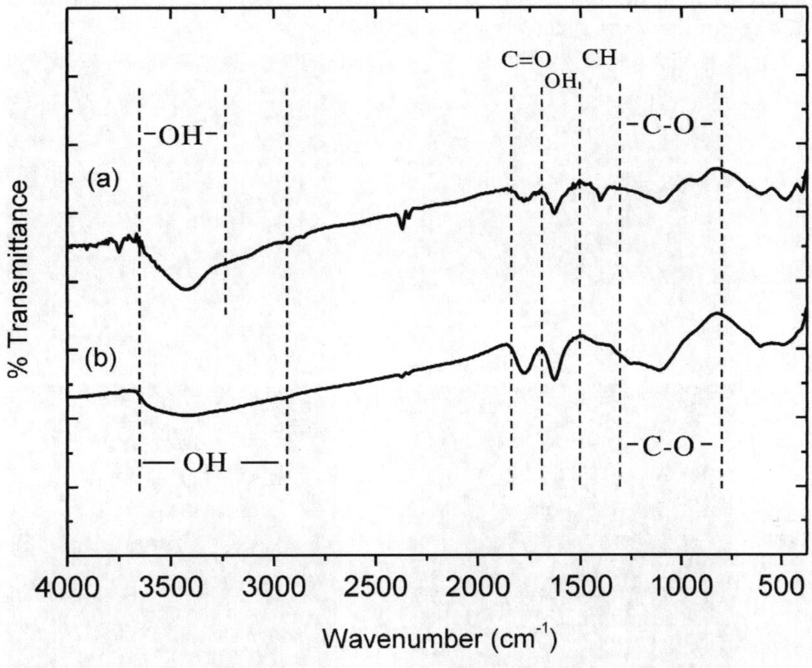

Figure 1. FTIR spectra of (a) nanodiamond untreated (top), (b) nanodiamond modify with sulphuric acid and nitric acid (bottom).

The FTIR spectra of ND-pristine in Figure 1(a) show several peaks with the intense peak observed at 1103.28 cm^{-1} indicating the presence of C-O-C vibration mode. In addition, the peaks observed at 1404.18 cm^{-1} and 2924.09 cm^{-1} are attributed to the bending vibration and overextend of C-H bond, respectively. The vibration mode at 1627.92 cm^{-1} and 3425.58 cm^{-1} are assigned to bending and stretching vibration of -OH, respectively. The presences of C=O and C-H bonds stretching vibrations were observed respectively at 1720.50 and 2924.09 cm^{-1} [21].

Figure 1(b) shows the FTIR spectra of the ND-COOH. The peaks observed at 1111.00 to 1242.16 cm^{-1}, 1627.92 cm^{-1}, 1774.51 cm^{-1}, and 3425.58 cm^{-1} indicate the presence of C-O-C bond vibration, -OH bending vibration, the vibration modes of the C=O and -OH stretching vibration, respectively [22]. These two later peaks proved the existence of the -COOH group attached onto nanodiamond surface [20]. Both samples contain –COOH functional groups; however, the treatment of pristine ND by sulphuric acid and nitric acid resulted in enrichment of ND with –COOH groups.

3.1.2 Transmission Electron Microscopy (TEM)

As seen in Figure 2(a), before treatment using H_2SO_4 and HNO_3, the diamond nanoparticles tend to agglomerate; however, after oxidation using the mixture of both chemical reagents, the agglomerate was then separated into small clusters consisting of particles with the size of 4-5 nm which can be seen in Figure 2(c). The TEM results suggest that the use of sulphuric acid and nitric acid not only oxidized the nanodiamond surface to produce carboxyl terminated nanodiamond but also deagglomerated the nanodiamond into small clusters, even individual particles. This is required since the agglomerated nanodiamond is inappropriate for medical application especially for drug carrier [2].

As seen in Figure 2(b) and (d), the diffraction pattern shows that both samples have the crystal structure of (111), (220), and (311). This indicates that there is no change in the crystal structure of nanodiamond due to enrichment with carboxyl functional groups on its surface which is good for drug carrier application.

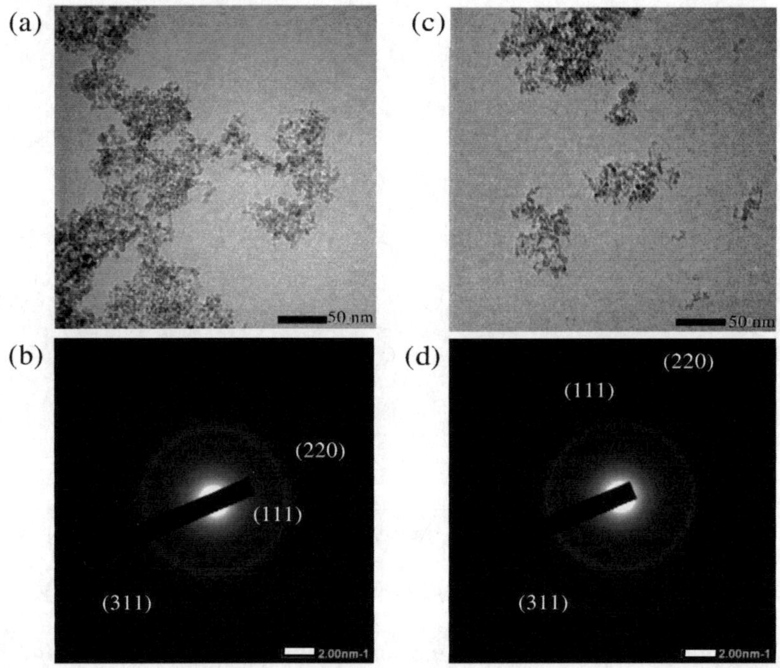

Figure 2. Bright field (BF) TEM images (upper) and diffraction pattern (lower) of as received-ND (a) (b) and of carboxyl-enriched nanodiamond (c) (d).

3.1.3 Thermo Gravimetric Analysis (TGA)

Thermograms from TGA analysis of nanodiamond before and after enrichment with carboxyl functional groups are presented in Figure 3. Thermogravimetry analysis was conducted at temperatures between 50–900 °C at a heating rate of 5 °C min^{-1}. As seen in Figure 3(a) and Table 1 for pristine nanodiamond, there is a slight weight loss (approximately 26.51%) between 70–110 °C, which can be associated with the evaporation of water due to hygroscopic hydroxyl groups. The second insignificant decline occurred at temperatures between 150-300 °C with the mass change of about 18.39%, attributed to heat-related degradation of functional groups. Meanwhile, Figure 3(b), it can be seen that there is a weight loss at the temperature between 95-220 °C (47.87%), which is much larger than that of pristine nanodiamond, indicating a much heavier mass of water associated with a carboxyl group. The second weight loss occurred between 295–450 °C, assigned to heat-related degradation of the molecule.

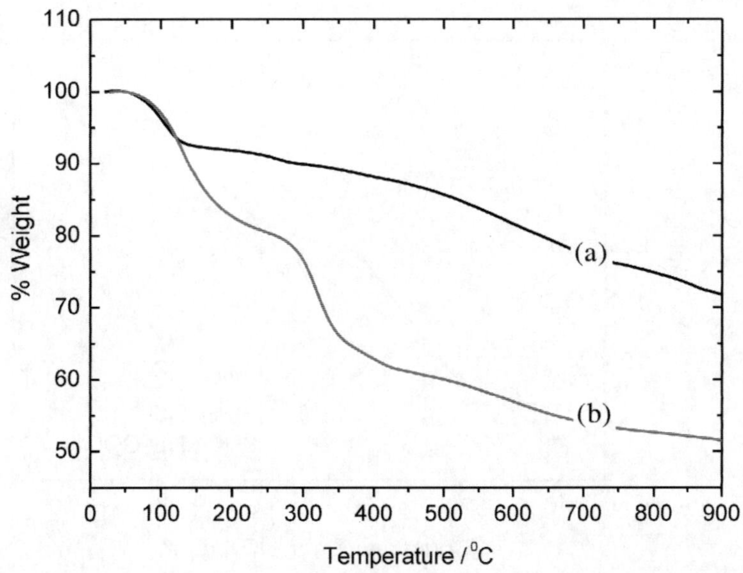

Figure 3. Thermogram of (a) as received nanodiamond and (b) carboxyl-enriched nanodiamond

Table 1. The mass change of pristine nanodiamond (ND) and carboxyl enriched nanodiamond (ND-COOH)

Samples	Onset (^{0}C)	Mass Change (%)
ND	72.5	26.51
	256.0	18.39
ND-COOH	95.9	47.87
	298.7	31.73

3.2 DOX Loading

In order to investigate the adsorption efficiency of ND-COOH and ND, both samples were applied to adsorb DOX. The adsorption of DOX was conducted in different concentration. The results of DOX adsorption is presented in Figure 5 and Table 3. The results show that after being modified with sulphuric acid and nitric acid, the adsorption efficiency of ND-COOH is higher than that of ND. Therefore, it can be said that enrichment of nanodiamond surfaces with carboxyl groups enables nanodiamond to adsorb DOX more efficiently.

DOX loading on both pristine and COOH-enriched nanodiamonds at different concentration is presented in Figure 4. It can be seen that the higher the concentration of DOX, the higher DOX loaded on both samples. However, DOX loaded on ND pristine is little higher than on N-COOH. This difference could be due to the distinction of the functional groups present on particle surfaces of both samples which influence the interaction between DOX and nanodiamond surfaces.

IOP Conf. Series: Materials Science and Engineering **172** (2017) 012066 doi:10.1088/1757-899X/172/1/012066

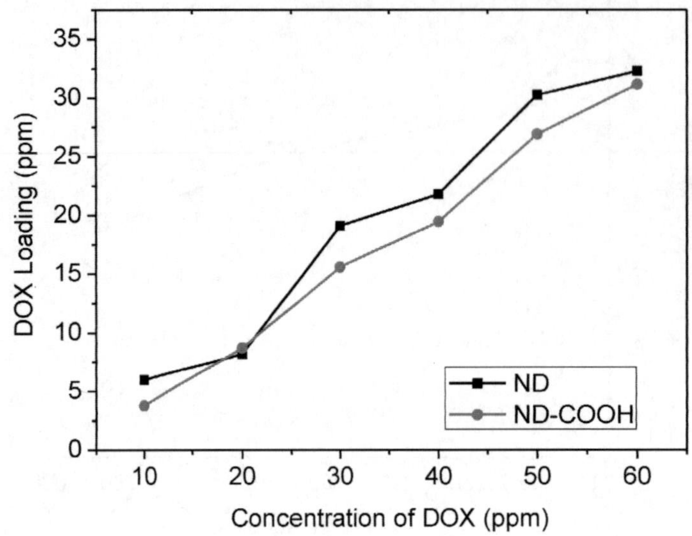

Figure 4. DOX loading on ND and ND-COOH in different concentrations

3.3 DOX Release

DOX release from both ND pristine and ND-COOH at two different time intervals is presented in Figure 6 for 60 ppm DOX loading. It can be observed from Figure 5 that after one hour the drug release from the pristine ND was ~14% more than the treated ND, whereas after two hours the pristine ND had ~55% more drug release as compared to treated ND. This result shows that DOX is more strongly bonded to the pristine ND than ND-COOH groups of the treated ND. As mentioned previously, the pristine nanodiamond surfaces contain several functional groups such as hydroxyl, carboxyl, ether, carbonyl, etc.[2] while in carboxyl-enriched nanodiamond, nanodiamond surfaces mainly contain carboxyl groups. Therefore, this difference may influence the interaction between DOX and nanodiamond surfaces.

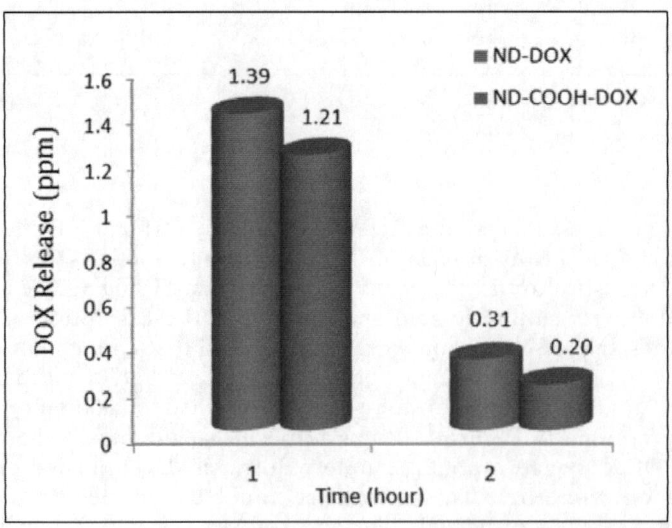

Figure 5. DOX release from pristine ND and ND-COOH

4. Conclusion

Nanodiamond without surface modification shows better loading and release efficiencies to the doxorubicin than that of nanodiamond surface enriched with a carboxyl group. The difference on the functional groups contained on nanodiamond surfaces in both samples is the main reason of the interaction between DOX and nanodiamond surface. Pristine nanodiamond contains several functional groups, namely, -OH, -COOH, -CO, etc. On the other hand, carboxyl-enriched nanodiamond, its surface contains mostly carboxyl group. The interaction energy between DOX and functional group contained on nanodiamond surface should be determined in order to convince this argument. The computational calculation of the interaction energy is on going.

References

[1] Shenderova O A and Gruen D M, *Ultrananocrystalline Diamond.* William Andrew Publishing, Norwich, New York. 2006.

[2] Mochalin V N, Shenderova O, Ho D and Gogotsi Y 2012 The properties and applications of nanodiamonds *Nat. Nanotechnol.* **7** 11-23.

[3] Astuti Y, Poolton N R J and Šiller L 2013 Raman and Photoluminescence Spectroscopic Study of 1-Undecene Functionalized Nanodiamonds *J K-Theory* **1** 1597.

[4] Astuti Y, Poolton N R J, Butenko Y V and Šiller L 2014 Evaporation and alignment of 1-undecene functionalised nanodiamonds *J Lumin* **156** 41-48.

[5] El-Say K M 2011 Nanodiamond as a drug delivery system: Applications and prospective *J Appl Pharm Sci* **1** 29-39.

[6] Zhu Y, Li J, Li W, Zhang Y, Yang X, Chen N, Sun Y, Zhao Y, Fan C and Huang Q 2012 The biocompatibility of nanodiamonds and their application in drug delivery systems *Theranostics* **2** 302-312.

[7] Huang H, Pierstorff E, Osawa E and Ho D 2007 Active nanodiamond hydrogels for chemotherapeutic delivery *Nano Lett.* **7** 3305-3314.

[8] Barnard A S 2008 Self-assembly in nanodiamond agglutinates *J. Mater. Chem.* **18** 4038-4041.

[9] Lai L and Barnard A S 2012 Interparticle Interactions and Self-Assembly of Functionalized Nanodiamonds *J Phys. Chem. Lett.* **3** 896-901.

[10] Kim H, Man H B, Saha B, Kopacz A M, Lee O-S, Schatz G C, Ho D, Liu W K 2012 Multiscale simulation as a framework for the enhanced design of nanodiamond-polyethylenimine-based gene delivery *J. Phys. Chem. Lett.* **3** 3791-3797.

[11] Chow E K, Zhang E Q, Chen M, Lam R, Robinson E, Huang H J, Schaffer D, Osawa E, Goga A and Ho D 2011 Nanodiamond therapeutic delivery agents mediate enhanced chemoresistant tumor treatment *Sci. Transl. Med.* **3** 73ra21.

[12] Schmidlin L, Pichot V, Comet M, Josset S, Rabu P and Spitzer D 2012 Identification, quantification and modification of detonation nanodiamond functional groups *Diam. Relat. Mater.* **22** 113-117.

[13] Chen M, Pierstorff E D, Lam R, Li S-Y, Huang H, Osawa E and Ho D 2009 Nanodiamond-mediated delivery of water-insoluble therapeutics *ACS Nano* **3** 2016-2022.

[14] Li Y, Zhou X, Wang D, Yang B and Yang P 2011 Nanodiamond mediated delivery of chemotherapeutic drugs *J. Mater. Chem.* **21** 16406-16412.

[15] Zhang X, Wang S, Fu C, Feng L, Ji Y, Tao L, Li S and Wei Y 2012 PolyPEGylated nanodiamond for intracellular delivery of a chemotherapeutic drug *Polym. Chem.* **3** 2716-2719.

[16] Xiao J, Duan X, Yin Q, Zhang Z, Yu H and Li Y 2013 Nanodiamonds-mediated doxorubicin nuclear delivery to inhibit lung metastasis of breast cancer *Biomaterials* **34** 9648-9656.

[17] Li Y, Tong Y, Cao R, Tian Z, Yang B and Yang P 2014 In vivo enhancement of anticancer therapy using bare or chemotherapeutic drug-bearing nanodiamond particles *Int. J. Nanomed.* **9** 1065-1082.

[18] Man H B, Kim H, Robinson E, Liu W K, Chow E K and Ho D 2014 Synthesis of nanodiamond-daunorubicin conjugates to overcome multidrug chemoresistance in leukemia *Nanomed.-Nanotechnol.* **10** 359-369.
[19] Mochalin V N, Pentecost A, Li X-M, Neitzel I, Nelson M, Wei C, He T, Guo F and Gogotsi Y 2013 Adsorption of drugs on nanodiamond: Toward development of a drug delivery platform *Mol. Pharm.* **10** 3728-3735.
[20] Jee A Y and Lee M 2009 Surface functionalization and physicochemical characterization of diamond nanoparticles *Curr. Appl. Phys.* **9** e144-e147.
[21] Larionova I, Kuznetsov V, Frolov A, Shenderova O, Moseenkov S and Mazov I 2006 Properties of individual fractions of detonation nanodiamond *Diam. Relat. Mater.* **11–12** 1804-1808.
[22] Lau X C, Desai C and Mitra S 2013 Functionalized nanodiamond as a charge transporter in organic solar cells *Sol. Energy* **91** 204-211.

Acknowledgement
We would like to thank the Ministry of Research, Technology, Higher Education of Republic Indonesia for the financial support with the grant no. DIPA-023.04.1.673453/2015.

IOP Conf. Series: Materials Science and Engineering **172** (2017) 012067 doi:10.1088/1757-899X/172/1/012067

Hardness evaluation of PMMA reinforced with two different calcinations temperatures of ZrO_2-Al_2O_3-SiO_2 filler system

Z Hasratiningsih[1], V Takarini[1], A Cahyanto[1], Y Faza[1], L A T W Asri[2], and B S Purwasasmita[2]

[1]Department of Dental Materials Science and Technology, Faculty of Dentistry, University of Padjadjaran, Sekeloa Selatan 1 Bandung, 40132, Indonesia.
[2] Advanced Material Processing Laboratory, Department of Engineering Physics, Institute of Technology Bandung, Ganesha 10 Bandung, 40132, Indonesia.

Email: zulia.hasratiningsih@fkg.unpad.ac.id

Abstract. Polymethyl methacrylate (PMMA) is one of the materials used for the temporary crown while making fixed partial dentures. Unfortunately, it has low mechanical properties. This study aim's to improve PMMA hardness by adding ZrO_2-Al_2O_3-SiO_2 filler system in two different calcination temperatures. Thirty-two disc form samples with 8mm diameter x 3mm thickness were made from two category-filler based that had been previously evaluated using Scanning Electron Microscope (SEM). Each category was divided into six groups of three respectively from a controlled and different concentrations of 7(A), 9(B), 11(C), 13(D), and 15(E) weight % of reinforced filler system, then tested with Vickers Hardness Tester. Filler particles that calcined at 550° can increase to 700° consist of fine crystalline and amorphous phases; however, the sample shows the highest hardness about 20,19 VHN even though the increase is only 7,5% compared to control. Meanwhile, the sample that calcined at 700°C exhibiting the highest hardness about 15,66 VHN corresponds to sample D, it has increased 25% compared to the control. This is correlated with microstructure result that has more growth crystalline particles. The results were analyzed by ANOVA which determined were not statistically significantly different ($p<0.05$). Therefore, it can be concluded that the 13% reinforced ZrO_2-Al_2O_3-SiO_2 filler system calcined at 700°C shows the highest hardness increase compare to calcination temperature 550-700°C. Although it is not significantly different.

1. Introduction

Dental aesthetic plays a very important role in life since most of the people want their teeth to look natural. Even Skinner, in 1959 has already noted that the aesthetic quality of the restoration is important to dental health as well as the mental health of the patient. Nowadays, aesthetic consideration are still primary factors for seeking various dental treatments, especially in indirect restoration as their steadily grows to restore the dentition naturally [1], [2]. Implant supported ceramic restoration is one of the treatments chosen for partially and fully edentulous patients since this fixed prosthesis type almost similar to the natural dentition. Besides, its success rate has been shown to be over 90% [2].

Implant fixed prosthesis could be made with two steps approach, which is firstly the surgery that takes several months to wait for the osseointegration process and healing phase of the soft tissues surroundings. Those conditions are very important in implant treatments as their successful outcomes, which are crucial due to the high costs of ceramic dental implants. Usually, the conventional protocols should take 3 to 6 months to allow the osseointegration process to take place [2], [3]. Secondly, the fabrication of crown or fixed partial denture (bridge) is a laboratory - based procedure, which needs time–consuming procedures, highly trained technician, and several weeks or some months may lapse between the preparation of the teeth and the cementation of the permanent restoration [4], [5], [6]. Since the treatment could take several months for final restoration, thus a temporary restoration be provided. This temporary restoration then needs to be strong, durable and aesthetic to last throughout the duration of the treatment [2], [7].

Acrylic resin or Polymethyl methacrylate (PMMA) is a polymer that is commonly used in dentistry, especially for indirect restorations such as denture base material and provisional crown jackets [8], [9], [10], [11]. This material has been chosen based on its aesthetics, fabrication method, and economic point of view. However, its low mechanical properties limited the clinical use of restorative materials [10], [11], [12].

Several attempts have been done to increase the mechanical properties, such as the addition of structural component including metal and ceramic, by adding filler particles with different elements such as zirconia, silica, alumina, glass fiber, tin and copper to PMMA [12], [13], [14], [15], [16], [17]. The incorporation of inorganic nanoparticles into PMMA alters the properties of the end product depending on the type, sizes, and shapes, concentration, and interaction within the polymer matrix itself [16].

In presuming a long term use of the provisional crown, in 2015, a study has been done to increase the mechanical properties of PMMA by the reinforcement with self-synthesized of ZrO_2-Al_2O_3-SiO_2 filler system using sol–gel technique and coating with two types of silane coupling agent. It was found that type and percentage of coupling agent affected the PMMA filled [18].

Based on the research data above, the effect of different temperature treatment to the fillers have not been studied yet. This study aimed to continue the work which is comparing the hardness of PMMA reinforced with ZrO_2-Al_2O_3-SiO_2 filler system that synthesized and calcined at two different calcination temperatures (at 550°C for 1 hour and increased to700°C while the other directly heated to 700°C) using different ratios of PMMA powder: The variations ratio of PMMA to filler system are 7, 9, 11, 13 and 15 wt%.

2. Materials and Methods

The research procedures were divided into 3 stages consists of the preparation of ceramic particles of SiO_2-Al_2O_3-ZrO_2, provided PMMA samples filled with the ceramic particles, and then tested with the hardness test.

2.1. Preparation of Ceramic Particles

a. Synthesis of ceramic nanoparticles. Ceramic nanoparticles composed of zirconia-alumina-silica were prepared using sol-gel techniques as mentioned by Hasratiningsih et al (2015). Aluminum nitrate nonahydrate, tetraethyl orthosilicate, and zirconium chloride were used as precursors for alumina, silica, and zirconia, respectively. Distilled water and 1%w/v chitosan were used as solvent and dispersant, respectively. Briefly, the precursor with the proportion of 10% alumina, 70% silica, and 20% zirconia was employed and mixed homogeneously followed by hydrolysis and condensation process. The prepared ceramic was ultra stirred and homogenized using ultrasonic homogenizer, and also dried at 100°C. Afterward, heat treatment was applied. The ceramic powder obtained were divided into two calcination temperatures. Firstly, the powder was calcined at 550°C for 1 hour then the temperature raised to 700°C. Secondly, the other powder was heated directly to 700°C for two hours. Both were heated in an electric furnace (Cress Electric Furnace C1228/935, Carson City, Nevada). The powder obtained then characterized to evaluate its structure and morphology [18].

b. Filler coating process. The ceramic filler system of Silica - Alumina - Zirconia was silanated with chitosan 1%. The coupling process was developed based on Hasratiningsih work (2015) with minor modification. Briefly, prepared the chitosan 1% solution using 1 gram of chitosan powder, 2 ml of 100% acetic acid and 98 ml aquabidest. Mix the chitosan powder and acetic acid in aquabidest use magnetic stirrer for ± 30 minutes until all the chitosan has dissolved. Then put both ceramic powders into two beaker glass and soaked in coupling agent followed by centrifugation. Next, the powder was dried in *vacuo*. The powder obtained was then crushed and stored at room temperature.

2.2. PMMA Sample Preparation

Two categories of samples according to two different calcination temperatures were prepared (each category consists of 18 samples including control). A total of thirty six samples were made using a conventional method as processing acrylic dentures using heat cured acrylic resin (De Trey®Stellon C Rapid Shade 3). Briefly, the wax patterns were made in molds (made of Teflon® PTFE) with a dimension of disc form with 8 mm in diameter and 3 mm in thickness. Gypsum molds were prepared by investing the wax pattern using dental stone in dental flasks then immersed in boiling water. Afterward, the softened wax was taken out and flushed with boiling water added to detergent to remove any impurities and to facilitate the application of separating medium (Cold Mold Seal). The mold cavities obtained were used for the preparation of test samples.

The ratio of PMMA heat cured resin polymer and monomer used was of 1:2 by weight was mixed and packed into the mold and used as a control. The flasks were put in the water at room temperature then raised until boiled within 1 hour. After that, the flasks were kept for 1 hour to continue the polymerization. The samples were deflasked after bench cooling. PMMA added with zirconia-alumina-silica coated by coupling agents were prepared with aforementioned procedure, except the coated zirconia-alumina-silica were mixed with resin then packed in the mold. The percentage of each filler (two different calcination temperatures) used were 7, 9, 11, 13 and 15% consecutively then cured. Each sample was prepared accordingly. Prior to hardness test, all samples were stored in water bath 37°C for 24 hours.

Table 1. Categories of grouping samples.

Category I Groups	CategoryII Groups	Description
A	A1	Unfilled acrylic resin (PMMA) without ZrO_2-Al_2O_3-SiO_2 filler system as control
B	B1	PMMA with 7% ZrO_2-Al_2O_3-SiO_2
C	C1	PMMA with 9% ZrO_2-Al_2O_3-SiO_2
D	D1	PMMA with 11% ZrO_2-Al_2O_3-SiO_2
E	E1	PMMA with 13% ZrO_2-Al_2O_3-SiO_2
F	F1	PMMA with 15% ZrO_2-Al_2O_3-SiO_2

2.3. Hardness Test

The novel composites produced were then subjected to the hardness test. The tests were performed onto the surface of each sample by means of Mickrovickers and determined using Vickers hardness tester machine, LECO - Japan M - 400 - H1/H2/H3 with a load of 200 grams for 15 seconds (ADA Specification No. 27). Five indentations were equally placed over the surface of each specimen and not closer than 1 mm to the adjacent indentation. The tests were conducted at the Laboratory of metallurgical physics and ceramics, Faculty of Mining and Petroleum Engineering, Institute of Technology Bandung.

3. Results and Discussion

Temporary crown and bridge materials are mostly made from uncross-linked acrylic polymer. The composition and manipulation are essentially the same as for denture bases. Mostly the acrylic polymers in dental prostheses are homopolymers (one type of monomer unit) that have no crosslinks. Compared to cross link resin this PMMA has lower mechanical properties [4]. This study aimed to develop PMMA reinforced with better mechanical properties. The samples prepared in this study mostly by manual while some of the procedures are technique sensitive so the samples made according to the categories were prepared in one series (i. e control &different groups).

Mokhtar *et al* 2013 observed the crystallization of tetragonal zirconia start showing up at 500°C and the pattern of this crystallization are more intense upon increasing the calcination temperature according to XRD analysis. The intensity will continue until 700°C, but at the temperature above 700°C, the monoclinic crystalline will start to appear. The filler prepared using calcination temperature at 700°C based on the theory that high mechanical properties associated with the high intensity of the crystallinity of filler particles. The higher the crystallinity of the filler the better the mechanical properties are. The crystal structure and phase formed are also influenced by calcination temperature and heat treatment process that the crystallinity increased with increasing calcination temperature [18], [19].

Zhang and Choy (2015) figured that 100% tetragonal zirconia phase was formed at temperature 700^{0}C in Yttria-Stabilized Zirconia. At temperature 800^{0}C tetragonal phase decreased to 97.1% followed by unexpected monoclinic phase was coming up. This condition was followed by an exothermic peak at temperature ~790^{0}C. The higher the temperature used, the higher amount of monoclinic crystalline was formed. Although in this study was using alumina instead of yttria as a stabilizer, it seems that phenomenon aforementioned was found as well. Besides zirconia, silica was added to compensate the opacity of zirconia and alumina. The additions of the other oxide to zirconia, like silica, not only to improve optical properties but also alumina can act as stabilizing oxides as well, which then given treatment temperatures [17], [18], [20], [21], [22].

Silica has 3 different crystal phases, such as quartz, tridymite, and cristobalite. The silica when is heated, a change in crystalline form occurs at a transition temperature characteristic of the particular form of silica. The best silica crystal phase, according to Ted Pella (2015) is beta - quartz which generates scale 7 on the Mohs hardness. Beta-quartz crystal phase will be formed at a temperature of 573°C - 870°C [9]. At high temperature (550 - 700°C) the peaks and intensity of the zirconium crystal phase suggest that tetragonal phase coexists with the monoclinic and cubical phase [18]. Both are main reasons for calcination temperature that were used in this research consist of two different temperature. A temperature of 550°C for 1 hour then raised to 700°C temperature and the other one that directly raised from room temperature to 700°C then maintained it for 2 hours.

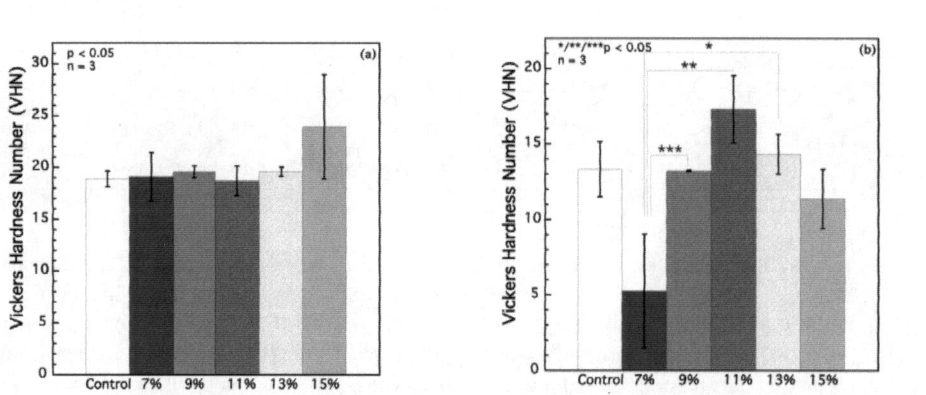

Figure 1. (a) Diagram of hardness test result that calcined at 550°C for 1 hour then raised to 700°C; (b) Diagram of hardness test result that calcined at 700°C

Figure 1 shows the diagram of Vickers hardness number of the samples which calcined at 550°C for 1 hour then raised to 700°C (a) and samples that calcined only at 700°C (b). There was no statistically significant for group 1. Meanwhile, group 2 shows statistically significant between 7%, 9% and 11%

samples. However, these significances were not revealed when compared to control in each group. These were supported by ANOVA analyzed. In other words, the temperature parameter that was employed in this study has not given any different effect of increasing the Vickers hardness number.

The reason based on this hardness test results are quite related with the prediction of transformation toughening mechanism. The additions of Al_2O_3 can discontinue transformation that shows the volume fraction of stress-induced transformation is definitely not uniform across the zone width. The other prediction is that system of ZrO_2-Al_2O_3 can transform to stacks of monoclinic twins. In both classes of material, the adjacent pairs of twins provided an extremely effective means of accommodating the shear component of the transformation shape strain [20].

Ayad, Badawi and Fatah (2008) found the addition of zirconia treated with zirconate (5% and 15%) increased the transverse strength of acrylic resin but no significant difference was detected for surface hardness. However, Ahmed and Ebrahim (2014) suggested that Zirconium oxide nano-fillers was increased the flexural strength and hardness of the acrylic resins significantly using 7%wt ZrO_2 concentration but did not explain about silanization.

Shirkavand and Moslehifrad (2014) in their study found the various ratio of TiO2 reinforced PMMA increasing the mechanical properties in certain concentration then decreased by additional more amount of TiO_2. While Asopa et al (2015) studied using zirconia filler found the same phenomena on the strength properties of the PMMA reinforced. On the other hand, the hardness was decreased.

Our result showed support Ahmed and Ibrahim work (2014), but in contrary with Asopa et al (2015). In this study filler used were ZrO_2-Al_2O_3-SiO_2 coated with chitosan 1% as a coupling agent. Coupling agents are used to linking the filler to the polymer. Silanization of the filler surface gives the ability of composite to distribute stress from the matrix to fillers. A silane/coupling agent acts as a sort of intermediary between organic materials to inorganic materials. The hydrogen bonding allows bonding between organic and inorganic material as well as promoting adhesion by forming chemical bonding with resin [9], [21].

Both factors filler system and coupling agent used increased the filler: PMMA ratio and hardness at a certain ratio. The reason as [10] because the additional of fillers (TiO2 in their study) act as impurities and the strength decreases as a result of the extra additive. Improper dispersion of the filler in PMMA matrix unfavorably affects the reaction of monomers, leading to increased levels of uncured monomer, which acts as a plasticizer. The increased in filler content cause these particles to agglomerate. The agglomerated compounds can act as stress concentrating center in the matrix and adversely affect the mechanical properties of the polymerized material [10].

That statement supported by Asopa et al (2015) that the increase in the filler fraction does not necessarily lead to increase in the strength. However, because excessive filler fraction creates more defects the material weakens. So that percentage of filler used (zirconia) for reinforcement should be such that the filler particles should disperse event into the resin matrix without interrupting the continuity of the resin matrix. The effect of higher filler content on reducing strength can be explained on the basis that after reaching saturation point the resin cannot incorporate furthermore filler particles. Any attempt to add filler particles after reaching saturation of matrix leads to an interruption in the resin matrix continuity and thus causing a decrease in the strength of the reinforced specimens. On the other hand, there finding the addition fillers in any ratio decreased the hardness value compared to control group [12]. In this study, the decrease of hardness was found when the filler added more than 11 wt% of the filler calcined at 550 and 13 wt%.

4. Conclusions

The 13% reinforced ZrO_2-Al_2O_3-SiO_2 filler system at 700°C shows the highest hardness increased about 25% compare to calcination temperature 550 - 700°C. Although it has not statistically significant difference, this reinforcement may improve PMMA mechanical properties.

References

[1] Roberson T M, Heymann H O, and Swift E J 2006 Sturdevant's Art and Science of Operative Dentistry, 5th ed, Mosby. St Louis, pp 498 & 625.

[2] Santosa R E 2007 Provisional Restoration Option in Implant Dentistry, *Aust Dent J* pp 52 : (3): 234 –242.

[3] ADI (Association of Dental Implantology), Dentist's Guide to Implantology, 2012, p 16.

[4] Power J M and Wataha J C 2008 Dental Materials Properties and Manipulation, 9th ed, Mosby Elsevier, St Louis, p 311.

[5] Guth J F, Zuch T, Zwinge S, Engels J, Timmelmayr M, Edelhoff D 2013 Optical properties of manually and CAD/CAM – fabricated polymers. *Dent. Mat J* pp 32(6):865 -871.

[6] Shen J Z, T Kosmac 2014 Advanced Ceramics for Dentistry. Trends Edition. The United States of America. Elsevier In. Pp 52-72.

[7] Regish K M, Sharmaa D, and Prithviraj 2011 Review Article Techniques of Fabricating of Provisional Restoration: An Overview; *Int. J Dent*, Article ID 134659, 5 pp: doi;10.1155/2011/134659.

[8] ADA Council on Scientific Affairs : Direct and Indirect Restorative Materials, JADA; 2003 pp 134 : 463– 472.

[9] Powers J M and Sakaguchi R L, J M Powers 2012 Craig's Restorative Dental Material. 13th Edition. The United States of America. Elsevier Mosby. Pp 54-61.

[10] Shirkavand S and Moslehifard E 2014 Effect of TiO2 nanoparticles on Tensile Strength of Dental Acrylic Resins, *JDent Res DentClin DentProspect*pp 8(4):197 - 203

[11] Duymus Z Y, Karaalioglu F O, Sulaiman F 2014 Flexural Strength of Provisional Crown and Fixed Partial Denture Resin Both With and Without Reinforced Fiber. J Mater Sci Nanotechnol 2(1):102doi:10.15744/2348 – 9812.302.

[12] Asopa V, Suresh S, Khandelwal M, Sharma V, Asopa S S, Kaira L S 2015 A Comparative Evaluation of Properties of Zirconia Reinforced High Impact Acrylic Resin with that of High Impactacrylic Resin: *Saudi J Dent Res* pp 6, 146 – 151.

[13] Atla J, Manne P, Gopinadh A, Samphat A, Muvva S B, Kishore K, Shandeep C, and Chittamsetty H 2013 The Effect of Al_2O_3 Addition on the Thermal Diffusivity of Heat Activated Acrylic Resin, *J Clin and Diagnostic Re*, pp (8):1797 -1798

[14] Alnamel H A and Mudhaffer M M 2014 The effect of Silicon dioxide Nano-Fillers Reinforcement on Some Properties of Heat Cure Polymethyl Methacrylate Denture Base Material, *J Bagh College Dent* pp 26(1)32 – 36.

[15] Ahmeed M A and Ebrahim M I 2014 Effect of Ziconium Oxide nano-Fillers Addition on the Flexural Strength, Faracture Toughness, and hardness of heat- Polymerized Acrylic Resin. World *J Nano Sci and Engineering*, pp 4, 50 – 57.

[16] Hameed H K and Rahman H A 2015 The Effect of Addition Nano Particle ZrO_2 on Some Properties of Autoclave Processed Heat Cure Acrylic Denture Base Material; *J Bagh College Dent* pp 27(1)32 – 39.

[17] Rasikh M K, Hasratiningsih Z, Karlina E, and Purwasasmita B S 2015 Hardness Test of Poly Methyl Methacrylate (PMMA) Filled With Self Synthesized Ceramic Fillers of SiO_2-Al_2O_3-ZrO_2 for Indirect Restorative Materials. International Dental Scientific Conference & Expo, *International Dental Student Oral Presentation Dies Forum 2015 Proceeding Paper*.

[18] Hasratiningsih Z, Cahyanto A, Takarini V, Karlina E, Djustiana N, Febrida R, Usri K, Faza Y, Hardiansyah A, and Purwasasmita B S 2016 Basic Properties of PMMA Reinforced Using Ceramics Particles of ZrO_2-Al_2O_3-SiO_2 Coated with Two Types of Coupling Agents. *Key Eng Material*, pp 696: 93-98.

[19] Mokhtar M, Basahel S, and Ali T T 2013 Effect of synthesis methods for mesoporous zirconia on its structural and textural properties, *J Mater Sci* pp 48;2705-2713.

[20] Hannink R H J, Kelly P M, and Muddle B C 2000 *J. Am.* Ceram. Soc., 83 [3] 461–87.

[21] Anusavice, Shen, Rawls 2013 Phillip's Science of Dental Materials 12th ed. Elsevier pp 202, 286.

[22] Zhang P and Choy K L 2015 The Synthesis of Single Tetragonal Phase Zirconia by Sol-Gel Route. *Int Journal of Engineering Research & Science (IJOER)* pp 20-21

Acknowledgment
The authors were grateful to Directorate of Higher Education Ministry of Research, Technology and Higher Education, the Republic of Indonesia through Padjadjaran University for funded the research.

Bacillus thuringiensis HCB6 Amylase Immobilization by Chitosan Beads

Zusfahair[1*], D R Ningsih[1], D Kartika[1], A Fatoni[1,2] and A L Zuliana[1]

[1]Department of Chemistry, Faculty of Mathematics and Natural Sciences, Universitas Jenderal Soedirman, Purwokerto, Indonesia.
[2]Center for Maritime Biosciences Studies, Universitas Jenderal Soedirman, Purwokerto, Indonesia

*email: zusfahair@gmail.com

Abstract. The purpose of this study was to optimize the amylase immobilization using a chitosan bead and to characterize immobilized amylase of *Bacillus thuringiensis* Bacteria HCB6. This study was started of amylase production, continued by immobilization optimization including ratio of chitosan:enzymes, enzyme-matrix contact time, substrate concentration, pH effect, incubation temperature effect, reaction time, and stability of immobilized enzyme. Amylase activity assay was dinitro salicylic (DNS) method. The results showed the optimum chitosan:enzyme ratio was 2.5: 1 (v/v), immobilization contact time of 18 hours and immobilization efficiency of 87.93%. Furthermore, immobilized amylase of *B. thuringiensis* HCB6 showed optimum substrate concentration of 1.5%, optimum pH of 6, optimum incubation temperature of 37 ° C, and the reaction time of 30 minutes. The Michaelis-Menten constant K_M value for free and immobilized amylase were 5.30% and 1.33% respectively. Immobilized amylase can be used up to five times with the remaining activity of 43.3%.

1. Introduction

Amylase is an enzyme catalyse the breakdown of glycoside bonds of starch into simple sugars. Amylase plays an important role in the industrial application. This enzymes can be produced from plants, animals and microorganisms [1]. Amylase was also reported can be isolated from a waste, for example the liquid waste of tapioca, which rich of complex carbohydrate. The amylase producer bacteria from tapioca liquid waste has been previously reported as *Bacillus thuringiensis* HCB6 [2].

Generally, enzyme production costs is relatively high, due to the complex procedure of their isolation from living organism which would increase the cost of enzymatic base process. In another side, the conventional enzymatic process of single-use enzymatic reaction was also contributed to increase the cost of enzymatic base application. One strategy to reduce the cost of enzymatic process is to use the enzyme in immobilized form, which allow to reuse the enzyme in several reactions. Many strategies have been reported in the enzyme immobilization, including the immobilization matrix, crosslinker and the form of immobilized enzyme-matrix. Several material for enzyme immobilization such as bentonite [3], sheparose, gelatin and chitosan were generally used [4]. Chitosan is one of the interesting material, which a natural polymer allowed high biocompatibility, low cost, renewable, and easy to obtain from chitin by deacetylation with alkali [5]. The lower level of N-acetyl groups (<40%) provides greater solubility in pH below 6.5, nontoxic, available in different forms (powder, gel, fibers, and membranes), and easy to derivate and demonstrating high affinity protein.

Enzyme immobilization on the chitosan matrix could be performed based on the formation of chemical bonds between free amine groups of the chitosan and carboxyl groups of the enzyme, tailored with crosslinkers such as tripolyphosphate (TPP), glutaraldehyde, isocyanate derivatives, bisdiazobenzidina, N, N-ethylene bismalemida and N, N-polymethyl bisoodoaseomida. The efficiency of enzyme immobilization using a chitosan supporting matrix with TPP as a crosslinking was influenced by various factors. This report studied several factors in the amylase immobilization using chitosan matrix, such as the ratio chitosan-enzyme and contact time of enzyme-chitosan beads. Immobilized amylase was also studied including an optimal substrate concentration, pH effect, temperature effect, reaction time and stability of the immobilized amylase.

2. Materials and Methods

2.1. Materials
Chitosan from crab shell and sodium tripolyphosphate were from Sigma-Aldrich (Steinheim, Germany), acetic acid, nutrient agar, sodium chloride, starch, DNS (dinitro-salicylic acid), sodium hydroxide, potassium sodium tartrate, disodium hydrogen phosphate, hydrogen disodium phosphate, sodium acetate, citric acid and sodium citrate were Merck (Germany). Bacteria isolate of *Bacillus thuringiensis* HCB6 was collection of Biochemistry Laboratory, Department of Chemistry, Jenderal Soedirman University, Indonesia.

2.2. Apparatus and Measurements
UV-Vis spectrophotometer (Shimadzu UV-1800, Japan) was used to measure the glucose concentration using DNS method. Bacteria as amylase sources was growth in the incubator (Memmert, Schawbach, Germany). Extracellular amylase separation was perform using centrifuge (Vulcan Quantum Q100).

2.3. Amylase production
Amylase used in this study was produced from *B. thuringiensis* HCB. In brief, bacterial isolate was inoculated in a 100 ml inoculum medium, incubated on a shaker incubator for 18 hours at room temperature. This inoculum medium was then poured into 400 mL of production medium and incubated on a shaker for 48 hours at room temperature. Crude amylase was separated by centrifuging for 15 minutes, with speed of 7,000 rpm, at 4 °C. Supernatant as a crude amylase was kept at 4 °C when it is not in use.

2.4. Enzyme Assay

Determination of amylase activity was performed by using the DNS method. Solution of 400 mL crude extract enzyme was added to the sample test tube. Starch solution (1%) as a substrate of 350 mL was added in the other test tube. Both tubes were then incubated at 37 °C for 10 minutes. Both solutions were mix and incubated for next 15 min at 37 °C. DNS reagent of 750 mL was added into the solution and incubated for 5 min at 100 °C. The mixture was then cooled for 20-60 minutes. Control assay was also prepare by the absent of starch solution as substrate. The solution both samples and control was diluted with 3 mL of distilled water and mix well, continued by measuring using spectrophotometer at 575 nm. Enzyme activity was then calculated by one unit amylase activity defined as 1 μmol formation of reducing sugar per mL (0.18 mg reducing sugar) per minute

$$Amylase\ activity = \frac{reducing\ sugar\ of\ sample - reducing\ sugar\ of\ control}{0.18\ x\ incubation\ time\ (min)}$$

2.5. Chitosan bead preparation
Chitosan solution of 2% (w/v) was prepare in dilute acetic acid of 1% (v/v), dissolve with heating at 60°C and filtered. This solution was then dropped in 50 mL of 1% TPP using micropipette and allowed for 30 minutes to complete the bead formation. Beads formed was then filtered, rinse with distilled water and then stored at -20 °C overnight. The chitosan beads was then storage at 4 °C when it was not in use. Amylase immobilization was performed by activation of chitosan bead in a solution of 1% TPP for 20 minutes. The beads were separated by filtration and then rinsed with distilled water, and immerse in the amylase solution for immobilization. This amylase immobilization was performed at 4 °C for overnight.

2.6. Amylase immobilization Optimization

2.6.1. Enzyme – chitosan ratio
The amylase on the chitosan bead preparation was used the same procedure of chitosan beads preparation. The chitosan beads were then activated using TPP solution for 20 min, and soaked in the enzyme solution to allow enzyme immobilization with various volume ratio of chitosan:enzyme of 0.5: 1; 1: 1; 1.5: 1; 2: 1; 2.5: 1 and 3: 1.Enzyme-chitosan with the highest amylase activity was then selected for further study. Immobilized amylase activity was also calculated its immobilization activity by comparing to the free amylase activity.

2.6.2. Immobilization contact time
Using the optimum enzyme-chitosan ratio, the immobilization contact time was then studied. Longer enzyme-chitosan beads contact time should better enzyme-chitosan bonding. Various contact times were studied, including 6, 12, 18, 24 and 30 hours. Optimum immobilization contact time was determined by the highest activity with the shortest contact time.

2.6.3. Immobilized Amylase Activity
The immobilized amylase activity assay was performed using a simple enzyme reactor composed of substrate container, pipe flow, tube reactors, flow control valves, and products container. The tube reactor was filled with 150 units of immobilized enzyme, pre-incubation at 37 °C for 10 min. Starch solution of 1% (w/v) was filled into the substrate container which was higher position than the enzyme reactor to allow substrate flow by the gravitation. Substrate of 5 mL was allowed to flow to the reactor containing chitosan beads by opening flow controller. The process of enzymatic reactions performed at 37 ° C for 15 minutes. The flow controller was reopened after 15 minutes to remove the product which was then kept in products container. Reducing sugar release was then determined using DNS method.

2.7. Amylase characterization
Substrate concentration variations was first studied in the amylase characterization. Starch solution as substrates concentration studied were 0.5 1.0, 1.5, 2.0, and 2.5% (w/v). The substrate concentration effect was studied similar to the enzyme activity assay. Furthermore, under the optimum substrate concentration, the pH effect was also studied. Various pH 3, 4, 5, 6 and 7 were used in this study. Besides substrate concentration and pH, temperature effect on the amylase activity was also studied. The study was performed under the optimal substrate and pH with various incubation temperature of 32 °C, 37 °C, 42 °C and 47 °C. The incubation time was then also studied in the various incubation of 5, 10, 15, 20 and 25 minutes.

2.7.1. Immobilized amylase reusability

This study was similar to the immobilized amylase activity assay, under the optimum condition, the immobilized amylase was reused to catalyse of 5 mL substrate uninterrupted catalysing, which each step of 15 min incubation time. The resulted reducing sugar was tested and compared each catalysing batch relatively compared to the first enzyme reaction.

3. RESULTS AND DISCUSSION

3.1. Immobilization of Amylase

Immobilization of enzymes was performed to enhance the enzyme efficient as biocatalyst, which can be reusable. Amylase was immobilized using TPP which had multivalent anions to allow reaction between the anion groups with positive charged of amine groups of chitosan. This polycations of chitosan solution formed when it was dissolved in acetic acid [6]. The chitosan beads was prepared using an ionic gelation method which was the process of beads formation of particles by electrostatic interactions between the positive charge and negative charge polyanion polymer that act as crosslinker [6]. The chitosan beads were activated using the cross-linker to prepare the amine group needed on the enzyme immobilization [7]. The amylase immobilization was the optimized to improve the enzyme activity.

3.2. Optimization of Amylase Immobilization

Enzyme immobilization optimization studied was the enzyme-chitosan ratio. Results showed that increasing chitosan-enzyme ration would also increase the immobilization efficient from 0.5:1 to 2.5:1. However, the immobilization coefficient was decrease in the higher ratio of chitosan of 3:1 (**Fig. 1**). The addition of chitosan volume would lower the efficiency of immobilization. This decline was because the increase in support may increase of the steric hindrance and diffusion resistance of the enzymes, leading to some inhibition in catalyzing the reaction, thus, the enzymatic activity recovery gradually decreased [8].

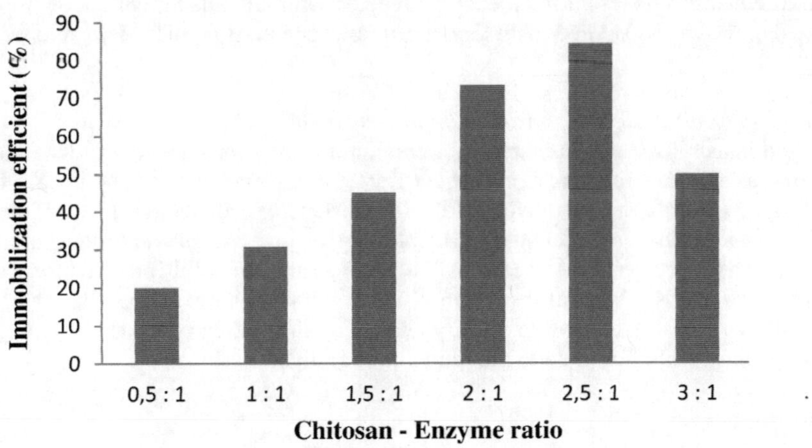

Figure 1. Enzyme-chitosan ratio effect on the immobilized *B. thuringiensis* HCB6 amylase immobilization coefficient.

3.3. Immobilization contact time optimization

Amylase immobilization efficiency increased with the increasing contact time between chitosan beads and the enzyme (**Fig. 2**). In the short time, the degree of crosslinking was not enough, the cross-linked enzyme-chitosan was low and lead to the low immobilized enzyme activity. The higher enzyme-chitosan contact time would increase the cross-linking degree, thus, enzyme activity was also increased. Extra contact time was decreased the efficiency of immobilization, which may due to the configuration of the immobilized enzyme changed by the excessive cross-linking, promoted the denaturation of enzyme protein [9].

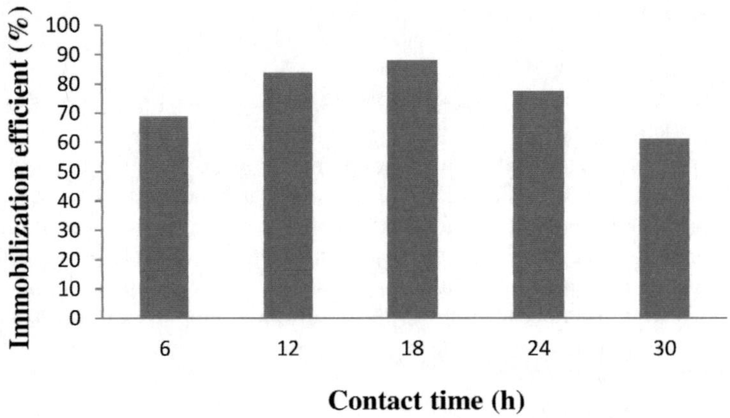

Figure 2. Immobilization contact time effect on the immobilized amylase activity.

3.4. Characterization of Immobilized Amylase

Immobilized amylase was tested in an enzyme reactor assembled using continuous flow system. This immobilized amylase was studied including substrate concentration, pH effect, temperature effect and stability.

3.4.1. The effect of substrate concentration on the immobilized amylase activity

Both free and immobilized amylase activity were increase with the substrate concentrations, up to rich their maximum enzymatic reaction rate of about 1.5% for free amylase and about 3.5% for immobilized amylase (**Fig. 3**). Furthermore, amylase kinetic constant of Michaelis-Menten constant (K_M) was calculated using Lineweaver-Burk equation, by plotting the invers of substrate concentration and inverse of enzyme activity. The result showed that K_M for free and immobilized amylase were respectively 5.30% and 1.32%. %. K_M value of the immobilized amylase was lower than that of free amylase. The lower K_M may be due to the amylase affinity change by the enzyme conformation change which was higher accessibility of the substrate to the immobilized amylase [10].

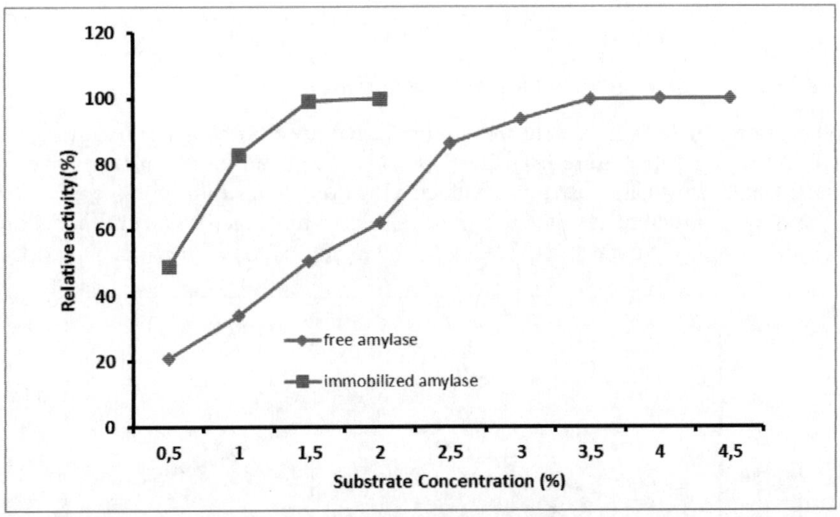

Figure 3. Substrate concentration effect on the *B. thuringiensis* HCB6 free and immobilized amylase.

3.5. *The Effect of pH on amylase activity*

The optimum pH on the amylase immobilization was shifted from 6.0 of free enzyme to 7.0 of immobilized enzyme (**Fig 4.**). This could be resulted from the change in acidic and basic amino acid side chain ionization in the microenvironment around the active site, which was the caused by the newly formed interactions between basic residues of enzyme and TPP during cross linking [11]. Shi-Lin Cao [12] suggests that the shift in the value of the optimum pH may be related to the strong interaction between the enzyme and matrices such as hydrogen bonding and electrostatic interactions. These shifts may be attributed to secondary interactions, such as ionic, polar interaction and hydrogen bonding, between the enzyme and the polymeric matrix [12].

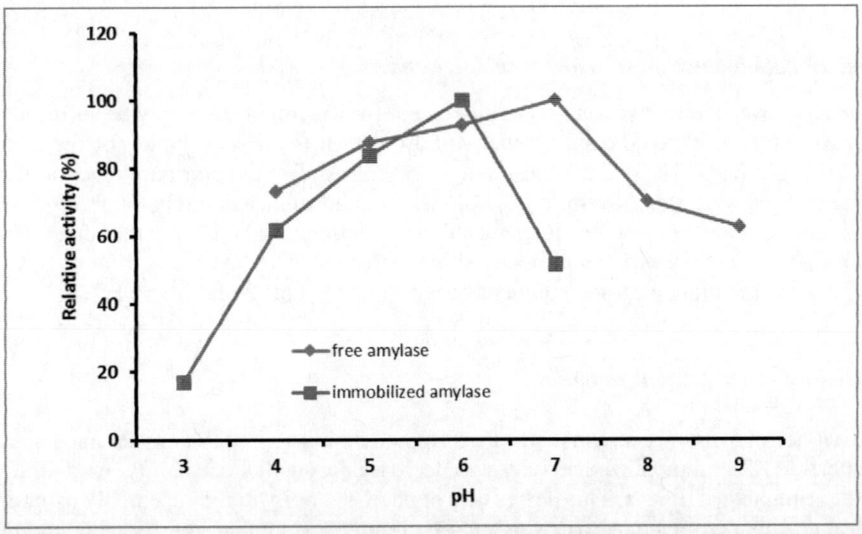

Figure 4. pH effect on the *B. thuringiensis* HCB6 amylase activity of free and immobilized enzyme.

3.6. *The effect of temperature on the amylase activity*

Both free and immobilized amylase showed optimum activity at 37 °C and subsequently enzyme activity decreased. Increasing temperature from 27 °C to 37 °C resulted increasing energy of each molecule, thus, resulted high enzymatic reaction product. However, temperature above 37 °C may started to degrade the enzyme structure, resulted lower amylase activity (**Fig. 5**). The optimum temperature of the free and immobilized enzyme was at 37 °C, which was similar to the previously reported [13-15].

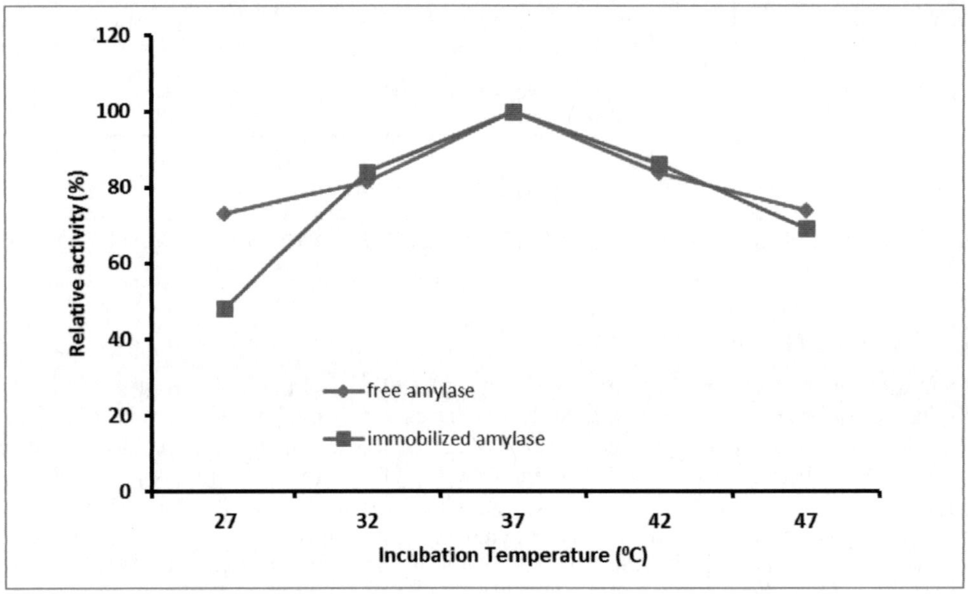

Figure 5. Temperature effect on the *B. thuringiensis* HCB6 amylase.

3.7. *The effect of reaction time on amylase activity*

The contact between the enzyme and the substrate to form a complex of enzyme substrate is affected by the reaction time, in which the short time would result in the least complex of the enzyme substrate formed and vice versa. The decreasing of amylase activity after the optimum reaction time may due to the damage of enzyme structure from environmental exposure, such temperature, light, and chemicals [16]. Immobilized amylase showed longer optimum reaction time of 30 min that that free amylase of 25 min. This may due to the diffusion effects which were caused by the supporting matrix in the form of chitosan, so the substrate-enzyme requires a longer time to contact and react [17].

3.7.1. Reuse of immobilized enzyme

The activity of immobilized amylase was then studied their stability by uninterrupted use to convert the given substrate. The immobilized activity was decrease during this reusability, with up to the fifth used of 43.3% remaining activities. This decreasing of catalyst ability in the repeatedly using may due to the release of enzymes from the matrix, which totally reduce the amount of enzyme bound on the chitosan beads.

4. CONCLUSION

Amylase from *B. thuringiensis HCB6* has been immobilized using a supporting matrix of chitosan bead, with the optimum ratio of chitosan: enzyme was 2.5: 1 and immobilization contact time of 18 hours, and resulted immobilization efficiency of 87.9%. Furthermore, immobilized amylase showed optimum substrate concentration of 1.5%, optimum pH of 6, optimum incubation temperature at 37 °C and optimum reaction time of 30 min, with amylase activity of 12.2 U / mL. Calculated K_M for free and immobilized amylase were 5.30% and 1.32% respectively. Immobilized amylase showed high stability can be used up to five times with the remaining activity of 43.3%.

References

[1] Kathiresan K and Manivannan S 2006 -Amylase production by Penicillium fellutanum isolated from mangrove rhizosphere soil *African journal of Biotechnology* **5**

[2] Zusfahair Z, Ningsih D R, Kartika D and Fatoni A 2016 Amylase from Bacillus thuringiensis isolated from Tapioca Waste: Isolation, Partial Purification and Characterization *Malaysian Journal of Fundamental and Applied Sciences* **12**

[3] Ghiaci M, Aghaei H, Soleimanian S and Sedaghat M 2009 Enzyme immobilization: Part 1. Modified bentonite as a new and efficient support for immobilization of Candida rugosa lipase *Applied Clay Science* **43** 289-95

[4] Datta S, Christena L R and Rajaram Y R S 2013 Enzyme immobilization: an overview on techniques and support materials *3 Biotech* **3** 1-9

[5] Lorenzoni A S, Aydos L F, Klein M P, Ayub M A, Rodrigues R C and Hertz P F 2015 Continuous production of fructooligosaccharides and invert sugar by chitosan immobilized enzymes: comparison between in fluidized and packed bed reactors *J. Mol. Catal. B: Enzym.* **111** 51-5

[6] Bhumkar D R and Pokharkar V B 2006 Studies on effect of pH on cross-linking of chitosan with sodium tripolyphosphate: a technical note *Aaps Pharmscitech* **7** E138-E43

[7] Fatoni A, Numnuam A, Kanatharana P, Limbut W, Thammakhet C and Thavarungkul P 2013 A highly stable oxygen-independent glucose biosensor based on a chitosan-albumin cryogel incorporated with carbon nanotubes and ferrocene *Sens. Actuators, B* **185** 725-34

[8] Cao S-L, Huang Y-M, Li X-H, Xu P, Wu H, Li N, Lou W-Y and Zong M-H 2016 Preparation and characterization of immobilized lipase from pseudomonas cepacia onto magnetic cellulose nanocrystals *Scientific reports* **6**

[9] Chen H, Zhang Q, Dang Y and Shu G 2013 The effect of glutaraldehyde cross-linking on the enzyme activity of immobilized-galactosidase on chitosan bead *Advance Journal of Food Science and Technology* **5** 932-5

[10] Zhou Y, Wang L, Wu T, Tang X and Pan S 2013 Optimal immobilization of β-glucosidase into chitosan beads using response surface methodology *Electronic Journal of Biotechnology* **16** 6-

[11] Aytar B S and Bakir U 2008 Preparation of cross-linked tyrosinase aggregates *Process Biochem.* **43** 125-31

[12] Kara F, Demirel G and Tümtürk H 2006 Immobilization of urease by using chitosan–alginate and poly (acrylamide-co-acrylic acid)/κ-carrageenan supports *Bioprocess and biosystems engineering* **29** 207-11

[13] Arya S and Srivastava S 2006 Kinetics of immobilized cyclodextrin gluconotransferase produced by Bacillus macerans ATCC 8244 *Enzyme Microb. Technol.* **39** 507-10

[14] Ibrahim A S, El-Tayeb M A and Al-Salamah A A 2010 Characterization of immobilized alkaline cyclodextringlycosyltransferase from a newly isolated Bacillus agaradhaerens KSU-A11 *African Journal of Biotechnology* **9** 7550-9

[15] Swarnalatha V, Esther R A and Dhamodharan R 2013 Immobilization of α-amylase on gum acacia stabilized magnetite nanoparticles, an easily recoverable and reusable support *J. Mol. Catal. B: Enzym.* **96** 6-13

IOP Conf. Series: Materials Science and Engineering **172** (2017) 012068 doi:10.1088/1757-899X/172/1/012068

[16] Łaba W and Rodziewicz A 2010 Keratinolytic potential of feather-degrading Bacillus polymyxa and Bacillus cereus *Pol. J. Environ. Stud* **19** 371-8
[17] Anwar A, Qader S A U, Raiz A, Iqbal S and Azhar A 2009 Calcium alginate: a support material for immobilization of proteases from newly isolated strain of Bacillus subtilis KIBGE-HAS *World Applied Sciences Journal* **7** 1281-6

Acknowledgments
We would like to thank the Directorate of Higher Education through the Fundamental Research Grants Contract Number DIPA-042.06-0/2016 on 7 December 2015.

Institute of Physics
Dirac House, Temple Back
Bristol BS1 6BE UK

ISSN: 1757-8981
ISBN 978-1-5108-3675-4

CURRAN ASSOCIATES INC.
proceedings.com